COMPREHENSIVE
MEDICINAL CHEMISTRY II

COMPREHENSIVE MEDICINAL CHEMISTRY II

Editors-in-Chief

Dr John B Taylor
Former Senior Vice-President for Drug Discovery, Rhône-Poulenc Rorer, Worldwide, UK

Professor David J Triggle
State University of New York, Buffalo, NY, USA

Volume 8

CASE HISTORIES AND CUMULATIVE SUBJECT INDEX

Volume Editors

Dr John B Taylor
Former Senior Vice-President for Drug Discovery, Rhône-Poulenc Rorer, Worldwide, UK

Professor David J Triggle
State University of New York, Buffalo, NY, USA

ELSEVIER

AMSTERDAM BOSTON HEIDELBERG LONDON NEW YORK OXFORD
PARIS SAN DIEGO SAN FRANCISCO SINGAPORE SYDNEY TOKYO

Elsevier Ltd.
The Boulevard, Langford Lane, Kidlington, Oxford OX5 1GB, UK

First edition 2007

1.05   PERSONALIZED MEDICINE © 2007, D Gurwitz
2.12   HOW AND WHY TO APPLY THE LATEST TECHNOLOGY © 2007, A W Czarnik
3.40   CHEMOGENOMICS © 2007, H Kubinyi
4.12   DOCKING AND SCORING © 2007, P F W Stouten

The following articles are US Government works in the public domain and not subject to copyright:
1.08   NATURAL PRODUCT SOURCES OF DRUGS: PLANTS, MICROBES, MARINE ORGANISMS,
       AND ANIMALS
6.07   ADDICTION

British Library Cataloguing in Publication Data
A catalogue record for this book is available from the British Library

Library of Congress Catalog Number: 2006936669

ISBN-13: 978-0-08-044513-7
ISBN-10: 0-08-044513-6

For information on all Elsevier publications
visit our website at books.elsevier.com

Printed and bound in Spain

06 07 08 09 10 10 9 8 7 6 5 4 3 2 1

Disclaimers

Both the Publisher and the Editors wish to make it clear that the views and opinions expressed in this book are strictly those of the Authors. To the extent permissible under applicable laws, neither the Publisher nor the Editors assume any responsibility for any loss or injury and/or damage to persons or property as a result of any actual or alleged libellous statements, infringement of intellectual property or privacy rights, whether resulting from negligence or otherwise.

Knowledge and best practice in this field are constantly changing. As new research and experience broaden our knowledge, changes in practice, treatment and drug therapy may become necessary or appropriate. Readers are advised to check the most current information provided (i) on procedures featured or (ii) by the manufacturer of each product to be administered, to verify the recommended dose or formula, the method and duration of administration, and contraindications. It is the responsibility of the practitioner, relying on their own experience and knowledge of the patient, to make diagnoses, to determine dosages and the best treatment for each individual patient, and to take all appropriate safety precautions. To the fullest extent of the law, neither the Publisher, nor Editors, nor Authors assume any liability for any injury and/or damage to persons or property arising out or related to any use of the material contained in this book.

Contents

Contents of all volumes ix

Preface xvii

Editors-in-Chief xix

Contributors to Volume 8 xxi

Contributors of all Volumes xxiii

Personal Essays

8.01 Introduction
W T COMER, *TorreyPines Therapeutics, Inc., La Jolla, CA, USA* 1

8.02 Reflections on Medicinal Chemistry Since the 1950s
H TIMMERMAN, *Vrije Universiteit, Amsterdam, The Netherlands* 7

8.03 Medicinal Chemistry as a Scientific Discipline in Industry and Academia: Some
Personal Reflections
C R GANELLIN, *University College London, London, UK* 17

8.04 Some Aspects of Medicinal Chemistry at the Schering-Plough Research Institute
A K GANGULY, *Stevens Institute of Technology, Hoboken, NJ, USA* 29

Case Histories

8.05 Viread
R OLIYAI and M FARDIS, *Gilead Sciences Inc., Foster City, CA, USA* 53

8.06 Hepsera
R OLIYAI and M FARDIS, *Gilead Sciences Inc., Foster City, CA, USA* 59

8.07 Ezetimibe
J W CLADER, *Schering-Plough Research Institute, Kenilworth, NJ, USA* 65

8.08 Tamoxifen
V C JORDAN, *Fox Chase Cancer Center, Philadelphia, PA, USA* 83

8.09 Raloxifene
J LEWIS-WAMBI and V C JORDAN, *Fox Chase Cancer Center, Philadelphia,
PA, USA* 103

8.10 Duloxetine   123
      K B THOR, *Dynogen Pharmaceuticals, Inc., Research Triangle Park, Durham, NC,*
      USA and I YALÇIN, *Eli Lilly and Co., Indianapolis, IN, USA*

8.11 Carvedilol   137
      R R RUFFOLO and G Z FEUERSTEIN, *Wyeth Research Laboratories,*
      Collegeville, PA, USA

8.12 Modafinil, A Unique Wake-Promoting Drug: A Serendipitous Discovery in Search of a
      Mechanism of Action   149
      F RAMBERT, J F HERMANT, and D SCHWEIZER, *Cephalon France,*
      Maisons-Alfort, France

8.13 Zyvox   157
      S J BRICKNER, *Pfizer Inc., Groton, CT, USA*

8.14 Copaxone   173
      D TEITELBAUM, R ARNON, and M SELA, *Weizmann Institute of Science,*
      Rehovot, Israel

8.15 Ritonavir and Lopinavir/Ritonavir   187
      D J KEMPF, *Abbott, Abbott Park, IL, USA*

8.16 Fosamax   199
      A A RESZKA, *Merck Research Laboratories, West Point, PA, USA* and
      G A RODAN[✠], *University of Pennsylvania School of Medicine, Philadelphia, PA, USA*

8.17 Omeprazole   213
      P LINDBERG, *AstraZeneca R&D, Mölndal, Sweden*

8.18 Calcium Channel α_2–δ Ligands: Gabapentin and Pregabalin   227
      A J THORPE and C P TAYLOR, *Pfizer Global Research and Development, Ann Arbor,*
      MI, USA

      Cumulative Subject Index   247

[✠]Deceased.

Contents of all Volumes

Volume 1   Global Perspective

Historical Perspective and Outlook
1.01   Reflections of a Medicinal Chemist: Formative Years through Thirty-Seven Years Service in the Pharmaceutical Industry
1.02   Drug Discovery: Historical Perspective, Current Status, and Outlook
1.03   Major Drug Introductions
The Impact of New Genomic Technologies
1.04   Epigenetics
1.05   Personalized Medicine
1.06   Gene Therapy
Sources of New Drugs
1.07   Overview of Sources of New Drugs
1.08   Natural Product Sources of Drugs: Plants, Microbes, Marine Organisms, and Animals
1.09   Traditional Medicines
1.10   Biological Macromolecules
1.11   Nutraceuticals
Animal Experimentation
1.12   Alternatives to Animal Testing
The Role of Industry
1.13   The Role of Small- or Medium-Sized Enterprises in Drug Discovery
1.14   The Role of the Pharmaceutical Industry
1.15   The Role of the Venture Capitalist
1.16   Industry–Academic Relationships
Drug Discovery: Revolution, Decline?
1.17   Is the Biotechnology Revolution a Myth?
1.18   The Apparent Declining Efficiency of Drug Discovery
Healthcare in the Social Context
1.19   How Much is Enough or Too Little: Assessing Healthcare Demand in Developed Countries
1.20   Health Demands in Developing Countries
1.21   Orphan Drugs and Generics
Ethical Issues
1.22   Bioethical Issues in Medicinal Chemistry and Drug Treatment
1.23   Ethical Issues and Challenges Facing the Pharmaceutical Industry
Funding and Regulation of Research
1.24   The Role of Government in Health Research
1.25   Postmarketing Surveillance

Intellectual Property
1.26   Intellectual Property Rights and Patents
Subject Index

Volume 2   Strategy and Drug Research

Introduction
2.01   The Intersection of Strategy and Drug Research
2.02   An Academic Perspective
2.03   An Industry Perspective
Organizational Aspects and Strategies for Drug Discovery and Development
2.04   Project Management
2.05   The Role of the Chemical Development, Quality, and Regulatory Affairs Teams in Turning a
        Potent Agent into a Registered Product
2.06   Drug Development
2.07   In-House or Out-Source
2.08   Pharma versus Biotech: Contracts, Collaborations, and Licensing
2.09   Managing Scientists, Leadership Strategies in Science
2.10   Innovation (Fighting against the Current)
2.11   Enabling Technologies in Drug Discovery: The Technical and Cultural Integration of the
        New with the Old
2.12   How and Why to Apply the Latest Technology
2.13   How and When to Apply Absorption, Distribution, Metabolism, Excretion, and Toxicity
2.14   Peptide and Protein Drugs: Issues and Solutions
2.15   Peptidomimetic and Nonpeptide Drug Discovery: Receptor, Protease, and Signal
        Transduction Therapeutic Targets
2.16   Bioisosterism
2.17   Chiral Drug Discovery and Development – From Concept Stage to Market Launch
2.18   Promiscuous Ligands
Targets
2.19   Diversity versus Focus in Choosing Targets and Therapeutic Areas
2.20   G Protein-Coupled Receptors
2.21   Ion Channels – Voltage Gated
2.22   Ion Channels – Ligand Gated
2.23   Phosphodiesterases
2.24   Protein Kinases and Protein Phosphatases in Signal Transduction Pathways
2.25   Nuclear Hormone Receptors
2.26   Nucleic Acids (Deoxyribonucleic Acid and Ribonucleic Acid)
2.27   Redox Enzymes
List of Abbreviations
List of Symbols
Subject Index

Volume 3   Drug Discovery Technologies

Target Search
3.01   Genomics
3.02   Proteomics
3.03   Pharmacogenomics
3.04   Biomarkers
3.05   Microarrays
3.06   Recombinant Deoxyribonucleic Acid and Protein Expression
3.07   Chemical Biology

Target Validation
3.08   Genetically Engineered Animals
3.09   Small Interfering Ribonucleic Acids
3.10   Signaling Chains
3.11   Orthogonal Ligand–Receptor Pairs
Informatics and Databases
3.12   Chemoinformatics
3.13   Chemical Information Systems and Databases
3.14   Bioactivity Databases
3.15   Bioinformatics
3.16   Gene and Protein Sequence Databases
3.17   The Research Collaboratory for Structural Bioinformatics Protein Data Bank
3.18   The Cambridge Crystallographic Database
Structural Biology
3.19   Protein Production for Three-Dimensional Structural Analysis
3.20   Protein Crystallization
3.21   Protein Crystallography
3.22   Bio-Nuclear Magnetic Resonance
3.23   Protein Three-Dimensional Structure Validation
3.24   Problems of Protein Three-Dimensional Structures
3.25   Structural Genomics
Screening
3.26   Compound Storage and Management
3.27   Optical Assays in Drug Discovery
3.28   Fluorescence Screening Assays
3.29   Cell-Based Screening Assays
3.30   Small Animal Test Systems for Screening
3.31   Imaging
3.32   High-Throughput and High-Content Screening
Chemical Technologies
3.33   Combinatorial Chemistry
3.34   Solution Phase Parallel Chemistry
3.35   Polymer-Supported Reagents and Scavengers in Synthesis
3.36   Microwave-Assisted Chemistry
3.37   High-Throughput Purification
Lead Search and Optimization
3.38   Protein Crystallography in Drug Discovery
3.39   Nuclear Magnetic Resonance in Drug Discovery
3.40   Chemogenomics
3.41   Fragment-Based Approaches
3.42   Dynamic Ligand Assembly
Subject Index

Volume 4   Computer-Assisted Drug Design

Introduction to Computer-Assisted Drug Design
4.01   Introduction to the Volume and Overview of Computer-Assisted Drug Design in the Drug
           Discovery Process
4.02   Introduction to Computer-Assisted Drug Design – Overview and Perspective for the Future
4.03   Quantitative Structure–Activity Relationship – A Historical Perspective and the Future
4.04   Structure-Based Drug Design – A Historical Perspective and the Future

Core Concepts and Methods – Ligand-Based
4.05   Ligand-Based Approaches: Core Molecular Modeling
4.06   Pharmacophore Modeling: 1 – Methods
4.07   Predictive Quantitative Structure–Activity Relationship Modeling
4.08   Compound Selection Using Measures of Similarity and Dissimilarity
Core Concepts and Methods – Target Structure-Based
4.09   Structural, Energetic, and Dynamic Aspects of Ligand–Receptor Interactions
4.10   Comparative Modeling of Drug Target Proteins
4.11   Characterization of Protein-Binding Sites and Ligands Using Molecular Interaction Fields
4.12   Docking and Scoring
4.13   De Novo Design
Core Methods and Applications – Ligand and Structure-Based
4.14   Library Design: Ligand and Structure-Based Principles for Parallel and Combinatorial Libraries
4.15   Library Design: Reactant and Product-Based Approaches
4.16   Quantum Mechanical Calculations in Medicinal Chemistry: Relevant Method or a Quantum Leap Too Far?
Applications to Drug Discovery – Lead Discovery
4.17   Chemogenomics in Drug Discovery – The Druggable Genome and Target Class Properties
4.18   Lead Discovery and the Concepts of Complexity and Lead-Likeness in the Evolution of Drug Candidates
4.19   Virtual Screening
4.20   Screening Library Selection and High-Throughput Screening Analysis/Triage
Applications to Drug Discovery – Ligand-Based Lead Optimization
4.21   Pharmacophore Modeling: 2 – Applications
4.22   Topological Quantitative Structure–Activity Relationship Applications: Structure Information Representation in Drug Discovery
4.23   Three-Dimensional Quantitative Structure–Activity Relationship: The State of the Art
Applications to Drug Discovery – Target Structure-Based
4.24   Structure-Based Drug Design – The Use of Protein Structure in Drug Discovery
4.25   Applications of Molecular Dynamics Simulations in Drug Design
4.26   Seven Transmembrane G Protein-Coupled Receptors: Insights for Drug Design from Structure and Modeling
4.27   Ion Channels: Insights for Drug Design from Structure and Modeling
4.28   Nuclear Hormone Receptors: Insights for Drug Design from Structure and Modeling
4.29   Enzymes: Insights for Drug Design from Structure
New Directions
4.30   Multiobjective/Multicriteria Optimization and Decision Support in Drug Discovery
4.31   New Applications for Structure-Based Drug Design
4.32   Biological Fingerprints
Subject Index

Volume 5   ADME-Tox Approaches

Introduction
5.01   The Why and How of Absorption, Distribution, Metabolism, Excretion, and Toxicity Research
Biological and In Vivo Aspects of Absorption, Distribution, Metabolism, Excretion, and Toxicity
5.02   Clinical Pharmacokinetic Criteria for Drug Research
5.03   In Vivo Absorption, Distribution, Metabolism, and Excretion Studies in Discovery and Development
5.04   The Biology and Function of Transporters
5.05   Principles of Drug Metabolism 1: Redox Reactions

5.06   Principles of Drug Metabolism 2: Hydrolysis and Conjugation Reactions
5.07   Principles of Drug Metabolism 3: Enzymes and Tissues
5.08   Mechanisms of Toxification and Detoxification which Challenge Drug Candidates and Drugs
5.09   Immunotoxicology
Biological In Vitro Tools in Absorption, Distribution, Metabolism, Excretion, and Toxicity
5.10   In Vitro Studies of Drug Metabolism
5.11   Passive Permeability and Active Transport Models for the Prediction of Oral Absorption
5.12   Biological In Vitro Models for Absorption by Nonoral Routes
5.13   In Vitro Models for Examining and Predicting Brain Uptake of Drugs
5.14   In Vitro Models for Plasma Binding and Tissue Storage
5.15   Progress in Bioanalytics and Automation Robotics for Absorption, Distribution, Metabolism, and Excretion Screening
Physicochemical tools in Absorption, Distribution, Metabolism, Excretion, and Toxicity
5.16   Ionization Constants and Ionization Profiles
5.17   Dissolution and Solubility
5.18   Lipophilicity, Polarity, and Hydrophobicity
5.19   Artificial Membrane Technologies to Assess Transfer and Permeation of Drugs in Drug Discovery
5.20   Chemical Stability
5.21   Solid-State Physicochemistry
In Silico Tools in Absorption, Distribution, Metabolism, Excretion, and Toxicity
5.22   Use of Molecular Descriptors for Absorption, Distribution, Metabolism, and Excretion Predictions
5.23   Electrotopological State Indices to Assess Molecular and Absorption, Distribution, Metabolism, Excretion, and Toxicity Properties
5.24   Molecular Fields to Assess Recognition Forces and Property Spaces
5.25   In Silico Prediction of Ionization
5.26   In Silico Predictions of Solubility
5.27   Rule-Based Systems to Predict Lipophilicity
5.28   In Silico Models to Predict Oral Absorption
5.29   In Silico Prediction of Oral Bioavailability
5.30   In Silico Models to Predict Passage through the Skin and Other Barriers
5.31   In Silico Models to Predict Brain Uptake
5.32   In Silico Models for Interactions with Transporters
5.33   Comprehensive Expert Systems to Predict Drug Metabolism
5.34   Molecular Modeling and Quantitative Structure–Activity Relationship of Substrates and Inhibitors of Drug Metabolism Enzymes
5.35   Modeling and Simulation of Pharmacokinetic Aspects of Cytochrome P450-Based Metabolic Drug–Drug Interactions
5.36   In Silico Prediction of Plasma and Tissue Protein Binding
5.37   Physiologically-Based Models to Predict Human Pharmacokinetic Parameters
5.38   Mechanism-Based Pharmacokinetic–Pharmacodynamic Modeling for the Prediction of In Vivo Drug Concentration–Effect Relationships – Application in Drug Candidate Selection and Lead Optimization
5.39   Computational Models to Predict Toxicity
5.40   In Silico Models to Predict QT Prolongation
5.41   The Adaptive In Combo Strategy
Enabling Absorption, Distribution, Metabolism, Excretion, and Toxicity Strategies and Technologies in Early Development
5.42   The Biopharmaceutics Classification System
5.43   Metabonomics

5.44   Prodrug Objectives and Design
5.45   Drug–Polymer Conjugates
List of Abbreviations
List of Symbols
Subject Index

Volume 6   Therapeutic Areas I: Central Nervous System, Pain, Metabolic Syndrome, Urology, Gastrointestinal and Cardiovascular

Central Nervous System
6.01   Central Nervous System Drugs Overview
6.02   Schizophrenia
6.03   Affective Disorders: Depression and Bipolar Disorders
6.04   Anxiety
6.05   Attention Deficit Hyperactivity Disorder
6.06   Sleep
6.07   Addiction
6.08   Neurodegeneration
6.09   Neuromuscular/Autoimmune Disorders
6.10   Stroke/Traumatic Brain and Spinal Cord Injuries
6.11   Epilepsy
6.12   Ophthalmic Agents
Pain
6.13   Pain Overview
6.14   Acute and Neuropathic Pain
6.15   Local and Adjunct Anesthesia
6.16   Migraine
Obesity/Metabolic Disorders/Syndrome X
6.17   Obesity/Metabolic Syndrome Overview
6.18   Obesity/Disorders of Energy
6.19   Diabetes/Syndrome X
6.20   Atherosclerosis/Lipoprotein/Cholesterol Metabolism
6.21   Bone, Mineral, Connective Tissue Metabolism
6.22   Hormone Replacement
Urogenital
6.23   Urogenital Diseases/Disorders, Sexual Dysfunction and Reproductive
        Medicine: Overview
6.24   Incontinence (Benign Prostatic Hyperplasia/Prostate Dysfunction)
6.25   Renal Dysfunction in Hypertension and Obesity
Gastrointestinal
6.26   Gastrointestinal Overview
6.27   Gastric and Mucosal Ulceration
6.28   Inflammatory Bowel Disease
6.29   Irritable Bowel Syndrome
6.30   Emesis/Prokinetic Agents
Cardiovascular
6.31   Cardiovascular Overview
6.32   Hypertension
6.33   Antiarrhythmics
6.34   Thrombolytics
Subject Index

Volume 7   Therapeutic Areas II: Cancer, Infectious Diseases, Inflammation & Immunology and Dermatology

Anti Cancer
7.01   Cancer Biology
7.02   Principles of Chemotherapy and Pharmacology
7.03   Antimetabolites
7.04   Microtubule Targeting Agents
7.05   Deoxyribonucleic Acid Topoisomerase Inhibitors
7.06   Alkylating and Platinum Antitumor Compounds
7.07   Endocrine Modulating Agents
7.08   Kinase Inhibitors for Cancer
7.09   Recent Development in Novel Anticancer Therapies
Anti Viral
7.10   Viruses and Viral Diseases
7.11   Deoxyribonucleic Acid Viruses: Antivirals for Herpesviruses and Hepatitis B Virus
7.12   Ribonucleic Acid Viruses: Antivirals for Human Immunodeficiency Virus
7.13   Ribonucleic Acid Viruses: Antivirals for Influenza A and B, Hepatitis C Virus, and Respiratory Syncytial Virus
Anti Fungal
7.14   Fungi and Fungal Disease
7.15   Major Antifungal Drugs
Anti Bacterials
7.16   Bacteriology, Major Pathogens, and Diseases
7.17   β-Lactam Antibiotics
7.18   Macrolide Antibiotics
7.19   Quinolone Antibacterial Agents
7.20   The Antibiotic and Nonantibiotic Tetracyclines
7.21   Aminoglycosides Antibiotics
7.22   Anti-Gram Positive Agents of Natural Product Origins
7.23   Oxazolidinone Antibiotics
7.24   Antimycobacterium Agents
7.25   Impact of Genomics-Emerging Targets for Antibacterial Therapy
Drugs for Parasitic Infections
7.26   Overview of Parasitic Infections
7.27   Advances in the Discovery of New Antimalarials
7.28   Antiprotozoal Agents (African Trypanosomiasis, Chagas Disease, and Leishmaniasis)
I and I Diseases
7.29   Recent Advances in Inflammatory and Immunological Diseases: Focus on Arthritis Therapy
7.30   Asthma and Chronic Obstructive Pulmonary Disease
7.31   Treatment of Transplantation Rejection and Multiple Sclerosis
Dermatology
7.32   Overview of Dermatological Diseases
7.33   Advances in the Discovery of Acne and Rosacea Treatments
7.34   New Treatments for Psoriasis and Atopic Dermatitis
Subject Index

Volume 8   Case Histories and Cumulative Subject Index

Personal Essays
8.01   Introduction
8.02   Reflections on Medicinal Chemistry Since the 1950s

Preface

The first edition of *Comprehensive Medicinal Chemistry* was published in 1990 and was intended to present an integrated and comprehensive overview of the then rapidly developing science of medicinal chemistry from its origins in organic chemistry. In the last two decades, the field has grown to embrace not only all the sophisticated synthetic and technological advances in organic chemistry but also major advances in the biological sciences. The mapping of the human genome has resulted in the provision of a multitude of new biological targets for the medicinal chemist with the prospect of more rational drug design (CADD). In addition, the development of sophisticated in silico technologies for structure–property relationships (ADMET) enables a much better understanding of the fate of potential new drugs in the body with the subsequent development of better new medicines.

It was our ambitious aim for this second edition, published 16 years after the first edition, to provide both scientists and research managers in all relevant fields with a comprehensive treatise covering all aspects of current medicinal chemistry, a science that has been transformed in the twenty-first century. The second edition is a complete reference source, published in eight volumes, encompassing all aspects of modern drug discovery from its mechanistic basis, through the underlying general principles and exemplified with comprehensive therapeutic applications. The broad scope and coverage of *Comprehensive Medicinal Chemistry II* would not have been possible without our panel of authoritative Volume Editors whose international recognition in their respective fields has been of paramount importance in the enlistment of the world-class scientists who have provided their individual 'state of the science' contributions. Their collective contributions have been invaluable.

Volume 1 (edited by Peter D Kennewell) overviews the general socioeconomic and political factors influencing modern R&D in both the developed and developing worlds. Volume 2 (edited by Walter H Moos) addresses the various strategic and organizational aspects of modern R&D. Volume 3 (edited by Hugo Kubinyi) critically reviews the multitude of modern technologies that underpin current discovery and development activities. Volume 4 (edited by Jonathan S Mason) highlights the historical progress, current status, and future potential in the field of computer-assisted drug design (CADD). Volume 5 (edited by Bernard Testa and Han van de Waterbeemd) reviews the fate of drugs in the body (ADMET), including the most recent progress in the application of 'in silico' tools. Volume 6 (edited by Michael Williams) and Volume 7 (edited by Jacob J Plattner and Manoj C Desai) cover the pivotal roles undertaken by the medicinal chemist and pharmacologist in integrating all the preceding scientific input into the design and synthesis of viable new medicines. Volume 8 (edited by John B Taylor and David J Triggle) illustrates the evolution of modern medicinal chemistry with a selection of personal accounts by eminent scientists describing their lifetime experiences in the field, together with some illustrative case histories of successful drug discovery and development.

We believe that this major work will serve as the single most authoritative reference source for all aspects of medicinal chemistry for the next decade and it is intended to maintain its ongoing value by systematic electronic upgrades. We hope that the material provided here will serve to fulfill the words of Antoine de Saint-Exupery (1900–44) and allow future generations of medicinal chemists to discover the future.

'As for the future, your task is not to foresee it but to enable it'
Citadelle (1948)

John B Taylor and David J Triggle

Editors-in-Chief

John B Taylor, DSc, was formerly Senior Vice President for Drug Discovery at Rhône-Poulenc Rorer. He obtained his BSc in chemistry from the University of Nottingham in 1956 and his PhD in organic chemistry at the Imperial College of Science and Technology with Nobel Laureate Professor Sir Derek Barton in 1962. He subsequently undertook postdoctoral research fellowships at the Research Institute for Medicine and Chemistry in Cambridge (US) with Sir Derek and at the University of Liverpool (UK), before entering the pharmaceutical industry.

During his career in the pharmaceutical industry Dr Taylor spent more than 30 years covering all aspects of research and development in an international environment. From 1970 to 1985 he held a number of positions in the Hoechst Roussel organization, ultimately as research director for Roussel Uclaf (France). In 1985 he joined Rhône-Poulenc Rorer holding various management positions in the research groups worldwide before becoming Senior Vice President for Drug Discovery in Rhône-Poulenc Rorer.

Dr Taylor is the co-author of two books on medicinal chemistry and has more than 50 publications and patents in medicinal chemistry. He was joint executive editor for the first edition of Comprehensive Medicinal Chemistry, a visiting professor for medicinal chemistry at the City University (London) from 1974 to 1984 and was awarded a DSc in medicinal chemistry from the University of London in 1991.

David J Triggle, PhD, is the University Professor and a Distinguished Professor in the School of Pharmacy and Pharmaceutical Sciences at the State University of New York at Buffalo. Professor Triggle received his education in the UK with a BSc degree in chemistry at the University of Southampton and a PhD degree in chemistry at the University of Hull working with Professor Norman Chapman. Following postdoctoral fellowships at the University of Ottawa (Canada) with Bernard Belleau and the University of London (UK) with Peter de la Mare he assumed a position in the School of Pharmacy at the University at Buffalo. He served as Chairman of the Department of Biochemical Pharmacology from 1971 to 1985 and as Dean of the School of Pharmacy from 1985 to 1995. From 1996 to 2001 he served as Dean of the Graduate School and from 1999 to 2001 was also the University Provost. He is currently the University Professor, in which capacity he teaches bioethics and science policy, and is President of the Center for Inquiry Institute, a secular think tank located in Amherst, New York.

Professor Triggle is the author of three books dealing with the autonomic nervous system and drug–receptor interactions, the editor of a further dozen books, some 280 papers, some 150 chapters and reviews, and has presented over 1000 invited lectures worldwide. The Institute for Scientific Information lists him as one of the 100 most highly cited scientists in the field of pharmacology. His principal research interests have been in the areas of drug–receptor interactions, the chemical pharmacology of drugs active at ion channels, and issues of graduate education and scientific research policy.

Contributors to Volume 8

R Arnon
Weizmann Institute of Science, Rehovot, Israel

S J Brickner
Pfizer Inc., Groton, CT, USA

J W Clader
Schering-Plough Research Institute, Kenilworth, NJ, USA

W T Comer
TorreyPines Therapeutics, Inc., La Jolla, CA, USA

M Fardis
Gilead Sciences Inc., Foster City, CA, USA

G Z Feuerstein
Wyeth Research Laboratories, Collegeville, PA, USA

C R Ganellin
University College London, London, UK

A K Ganguly
Stevens Institute of Technology, Hoboken, NJ, USA

J F Hermant
Cephalon France, Maisons-Alfort, France

V C Jordan
Fox Chase Cancer Center, Philadelphia, PA, USA

D J Kempf
Abbott, Abbott Park, IL, USA

J Lewis-Wambi
Fox Chase Cancer Center, Philadelphia, PA, USA

P Lindberg
AstraZeneca R&D, Mölndal, Sweden

R Oliyai
Gilead Sciences Inc., Foster City, CA, USA

F Rambert
Cephalon France, Maisons-Alfort, France

A A Reszka
Merck Research Laboratories, West Point, PA, USA

G A Rodan*
University of Pennsylvania School of Medicine, Philadelphia, PA, USA

R R Ruffolo
Wyeth Research Laboratories, Collegeville, PA, USA

D Schweizer
Cephalon France, Maisons-Alfort, France

M Sela
Weizmann Institute of Science, Rehovot, Israel

C P Taylor
Pfizer Global Research and Development, Ann Arbor, MI, USA

D Teitelbaum
Weizmann Institute of Science, Rehovot, Israel

K B Thor
Dynogen Pharmaceuticals, Inc., Research Triangle Park, Durham, NC, USA

A J Thorpe
Pfizer Global Research and Development, Ann Arbor, MI, USA

H Timmerman
Vrije Universiteit, Amsterdam, The Netherlands

I Yalçin
Eli Lilly and Co., Indianapolis, IN, USA

*Deceased.

Contributors of all Volumes

N J Abbott
King's College London, London, UK

S Abdelhadi
University of Medicine and Dentistry of New Jersey, Newark, NJ, USA

D J Abraham
Virginia Commonwealth University, Richmond, VA, USA

M H Abraham
University College London, London, UK

B Abrahamsson
AstraZeneca, Mölndal, Sweden

F C Acher
Université René Descartes – Paris V, Paris, France

M Afshar
Ariana Pharmaceuticals, Paris, France

S Agatonovic-Kustrin
The University of Western Australia, Perth, WA, Australia

T Ahrens
Discovery Partners International AG, Allschwil, Switzerland

I L Alberts
De Novo Pharmaceuticals, Cambridge, UK

A A Alex
Pfizer Global Research and Development, Sandwich, UK

F H Allen
Cambridge Crystallographic Data Centre, Cambridge, UK

N E Allen
Indiana University School of Medicine, Indianapolis, IN, USA

D Amaratunga
Johnson & Johnson Pharmaceutical Research & Development LLC, Raritan, NJ, USA

T B Andersson
Institute of Environmental Medicine, Karolinska Institutet, Stockholm, Sweden

R Apweiler
EMBL Outstation European Bioinformatics Institute, Hinxton, Cambridge, UK

R Arnon
Weizmann Institute of Science, Rehovot, Israel

A M Aronov
Vertex Pharmaceuticals Inc., Cambridge, MA, USA

T Arrhenius
Del Mar, CA, USA

P Artursson
Uppsala University, Uppsala, Sweden

M Ashton
Evotec (UK) Ltd, Abingdon, UK

R P Austin
AstraZeneca R&D Charnwood, Loughborough, UK

A Avdeef
pION Inc., Woburn, MA, USA

M A Avery
University of Mississippi, University, MS, USA

E R Bacon
Worldwide Discovery Research, Cephalon, Inc., West Chester, PA, USA

S J Baker
Anacor Pharmaceuticals, Palo Alto, CA, USA

M Baldus
Max Planck Institute for Biophysical Chemistry, Göttingen, Germany

A J Barker
AstraZeneca, Macclesfield, UK

R Barker
Association of the British Pharmaceutical Industry (ABPI), London, UK

M Baroni
Molecular Discovery, Pinner, UK

J F Barrett[*]
Merck Research Laboratories, Rahway, NJ, USA

N Barton
GlaxoSmithKline Pharmaceuticals plc, Harlow, UK

P Barton
AstraZeneca R&D Charnwood, Loughborough, UK

G M Battle
Cambridge Crystallographic Data Centre, Cambridge, UK

M M T Bauer
Boehringer Ingelheim Pharma, Biberach, Germany

I R Baxendale
University of Cambridge, Cambridge, UK

---

[*]Deceased.

P Bazzini
Prestwick Chemical, Illkirch, France

D P Behan
Arena Pharmaceuticals, Inc., San Diego CA, USA

A S Bell
Pfizer Global Research and Development, Sandwich, UK

C F Bennett
ISIS Pharmaceuticals, Carlsbad, CA, USA

H M Berman
Rutgers – The State University of New Jersey, Piscataway, NJ, USA

F Bernhard
University of Frankfurt/Main, Germany

M Bertrand
Technologie Servier, Orléans, France

T Blackburn
Helicon Therapeutics Inc., Farmingdale, New York, USA

J M Blair
Anzac Research Institute, University of Sydney, NSW, Australia

F E Blaney
GlaxoSmithKline Pharmaceuticals plc, Harlow, UK

R Blasius
Hôpital Kirchberg, Luxembourg City, Luxembourg

W F Bluhm
University of California, La Jolla, CA, USA

B Bolon
GEMpath Inc., Cedar City, UT, USA

M-J Bossant
Technologie Servier, Orléans, France

P E Bourne
University of California, La Jolla, CA, USA

T A Bowdle
University of Washington, Seattle, WA, USA

D J Bower
University of Dundee, Dundee, UK

D Bozyczko-Coyne
Worldwide Discovery Research, Cephalon, Inc., West Chester, PA, USA

S J Brickner
Pfizer Inc., Groton, CT, USA

C A Briggs
Abbott Laboratories, Abbott Park, IL, USA

K E Browman
Abbott Laboratories, Abbott Park, IL, USA

D Brown
Alchemy Biomedical Consulting, Cambridge, UK

D G Brown
Pfizer Global Research and Development, Sandwich, UK

J N Burrows
AstraZeneca R&D, Södertälje, Sweden

L A Cabanilla
Tufts University, Boston, MA, USA

S Cagnani
University of Parma, Parma, Italy

J S Caldwell
Genomics Institute of the Novartis Research Foundation, San Diego, CA, USA

D Callahan
The Hastings Center, Garrison, NY, USA

Q Cao
Pfizer Research Technology Center, Cambridge, MA, USA

T Carlomagno
Max Planck Institute for Biophysical Chemistry, Göttingen, Germany

G Caron
University of Torino, Torino, Italy

E Carosati
University of Perugia, Italy

D Cavalla
Arachnova Ltd, Cambridge, UK

C L Cavallaro
Bristol-Myers Squibb, Princeton, NJ, USA

M R Chance
Case Western Reserve University, Cleveland, OH, USA

S K Chanda
Genomics Institute of the Novartis Research Foundation, San Diego, CA, USA

S Chatterjee
Worldwide Discovery Research, Cephalon, Inc., West Chester, PA, USA

G Chinea
Center for Genetic Engineering and Biotechnology, Havana, Cuba

P Ciapetti
Prestwick Chemical, Illkirch, France

M Cik
Johnson & Johnson Pharmaceutical Research & Development, Beerse, Belgium

J W Clader
Schering-Plough Research Institute, Kenilworth, NJ, USA

J Cockbain
Frank B Dehn & Co, Oxford, UK

G Colmenarejo
GlaxoSmithKline, Madrid, Spain

G Colombo
University of Ferrara, Ferrara, Italy

P Colombo
University of Parma, Parma, Italy

R D Combes
FRAME, Nottingham, UK

J E A Comer
Sirius Analytical Instruments Ltd, Forest Row, UK

W T Comer
TorreyPines Therapeutics, Inc., La Jolla, CA, USA

R D Connell
Pfizer Global Research and Development, Groton, CT, USA

G M Cragg
NCI, Frederick, MD, USA

M T D Cronin
Liverpool John Moores University, Liverpool, UK

G Cruciani
University of Perugia, Italy

L Cucurull-Sanchez
Pfizer Global Research and Development, Sandwich, UK

A W Czarnik
University of Nevada, Reno, NV, USA

L Da Ros
GlaxoSmithKline, Verona, Italy

M Danhof
Leiden University, Leiden, The Netherlands

M Darnbrough
London, UK

Y T Das
Rutgers – The State University of New Jersey, New Brunswick, NJ, USA

R E Davis
3-D Pharmaceutical Consultants, San Diego, CA, USA

E De Clercq
Rega Institute for Medical Research, Leuven, Belgium

M J De Groot
Pfizer Global Research and Development, Sandwich, UK

S E DePrimo
Pfizer Global Research and Development, San Diego, CA, USA

P M Dean
De Novo Pharmaceuticals, Cambridge, UK

W A Denny
University of Auckland, Auckland, New Zealand

N Deshpande
University of California, La Jolla, CA, USA

L Di
Wyeth Research, Princeton, NJ, USA

M Dicato
Centre Hospitalier de Luxembourg, Luxembourg City, Luxembourg

M Dickins
Pfizer Global Research and Development, Sandwich, UK

M Diederich
Hôpital Kirchberg, Luxembourg City, Luxembourg

J A Dodge
Lilly Research Laboratories, Indianapolis, IN, USA

C K Donawho
Abbott Laboratories, Abbott Park, IL, USA

P Dorr
Pfizer Global Research and Development, Sandwich, UK

T J Dougherty
Pfizer Global Research and Development, Groton, CT, USA

A M Doweyko
Bristol-Myers Squibb, Princeton, NJ, USA

J B Dunbar Jr
Pfizer Inc., Michigan Laboratories, Ann Arbor, MI, USA

C R Dunstan
Anzac Research Institute, University of Sydney, NSW, Australia

J A Dykens
EyeCyte Therapeutics, Encinitas, CA, USA

M L Eaton
Stanford University, Graduate School of Business, Stanford, CA, USA

A El-Gengaihy
University of Medicine and Dentistry of New Jersey, Newark, NJ, USA

M J Elices
PharmaMar USA, Cambridge, MA, USA

A A Elmarakby
Medical College of Georgia, Augusta, GA, USA

T Engel
Chemical Computing Group AG, Köln, Germany

P W Erhardt
University of Toledo, OH, USA

G Ermondi
University of Torino, Torino, Italy

C Esser
Institut für Umweltmedizinische Forschung, Heinrich-Heine University, Düsseldorf, Germany

N Eswar
University of California at San Francisco, San Francisco, CA, USA

S Evangelista
Menarini Ricerche SpA, Firenze, Italy

D Fabbro
Novartis Institutes for BioMedical Research, Basel, Switzerland

B Faller
Novartis Institutes for Biomedical Research, Basel, Switzerland

M Fardis
Gilead Sciences Inc., Foster City, CA, USA

H-J Federsel
AstraZeneca, Södertälje, Sweden

R D Feldman
Robarts Research Institute, London, ON, Canada

R E Fessey
AstraZeneca R&D Charnwood, Loughborough, UK

G Z Feuerstein
Wyeth Research Laboratories, Collegeville, PA, USA

S Fidanze
Abbott Laboratories, Abbott Park, IL, USA

R Flaumenhaft
Harvard Medical School, Boston, MA, USA

J L Flippen-Anderson
Rutgers – The State University of New Jersey, Piscataway, NJ, USA

M M Flocco
Pfizer Global Research and Development, Sandwich, UK

A Foreman
pION Inc., Woburn, MA, USA

A C Foster
Neurocrine Biosciences Inc., San Diego, CA, USA

G B Fox
Abbott Laboratories, Abbott Park, IL, USA

R Fraczkiewicz
Simulations Plus, Inc., Lancaster, CA, USA

Q C Franco
StratEdge, Washington, DC, USA

S Freedman
Elan Pharmaceuticals, South San Francisco, CA, USA

C Funk
F. Hoffmann-La Roche Ltd, Basel, Switzerland

J D Gale
Pfizer Global Research and Development, Sandwich, UK

A Galetin
University of Manchester, Manchester, UK

C R Ganellin
University College London, London, UK

A K Ganguly
Stevens Institute of Technology, Hoboken, NJ, USA

C García-Echeverría
Novartis Institutes for BioMedical Research, Basel, Switzerland

S Garland
GlaxoSmithKline Pharmaceuticals plc, Harlow, UK

G F Gebhart
University of Pittsburgh, Pittsburgh, PA, USA

D R Gehlert
Eli Lilly and Company, Indianapolis, IN, USA

L Gianellini
Nerviano Medical Sciences S.r.l, Nerviano, Italy

D A Giegel
Celgene Corporation, San Diego, CA, USA

B Giethlen
Prestwick Chemical, Illkirch, France

V J Gillet
University of Sheffield, Sheffield, UK

A M Ginsberg
Global Alliance for TB Drug Development, New York, NY, USA

D Giron
Novartis Pharma AG, Basel, Switzerland

P J Goadsby
Institute of Neurology, The National Hospital for Neurology and Neurosurgery, London, UK

K R Gogas
Neurocrine Biosciences Inc., San Diego, CA, USA

H Göhlmann
Johnson & Johnson Pharmaceutical Research & Development, Beerse, Belgium

A Good
Bristol-Myers Squibb, Wallingford, CT, USA

M Gopalakrishnan
Abbott Laboratories, Abbott Park, IL, USA

D W Green
Amgen Inc., Cambridge, MA, USA

N Greene
Pfizer Global Research and Development, Groton, CT, USA

C Griesinger
Max Planck Institute for Biophysical Chemistry, Göttingen, Germany

E Griffen
AstraZeneca R&D, Macclesfield, UK

R H Griffey
ISIS Pharmaceuticals, Carlsbad, CA, USA

R Griffith
Genelabs Technologies, Inc., Redwood City, CA, USA

D E Grigoriadis
Neurocrine Biosciences Inc., San Diego, CA, USA

D Gurwitz
Tel-Aviv University, Tel-Aviv, Israel

S Haider
University of Oxford, Oxford, UK

E D Hall
Spinal Cord and Brain Injury Research Center, University of Kentucky Medical Center, Lexington, KY, USA

L H Hall
Eastern Nazarene College, Quincy, MA, USA and Hall Associates Consulting, Quincy, MA, USA

L M Hall
Hall Associates Consulting, Quincy, MA, USA

M M Hann
GlaxoSmithKline R&D, Stevenage, UK

C Hansch
Pomona College, Claremont, CA, USA

C Hartmann
Max Planck Institute for Informatics, Saarbrücken, Germany

M J Hartshorn
Astex Therapeutics, Cambridge, UK

D R Hawkins
Huntingdon Life Sciences, Alconbury, UK

B G Healey
Serono Research Institute, Rockland, MA, USA

B R Hearn
Kosan Biosciences, Hayward, CA, USA

R A Hegele
Robarts Research Institute, London, ON, Canada

J R Henry
Lilly Research Laboratories, Indianapolis, IN, USA

J F Hermant
Cephalon France, Maisons-Alfort, France

A Hersey
GSK Medicines Research Centre, Stevenage, UK

M Hewitt
Liverpool John Moores University, Liverpool, UK

G A Hicks
Novartis Pharmaceuticals Corporation, East Hanover, NJ, USA

R Hilgenfeld
University of Lübeck, Lübeck, Germany

R F Hirschmann
University of Pennsylvania, Philadelphia, PA, USA

T Hogg
University of Lübeck, Lübeck, Germany

P Honore
Abbott Laboratories, Abbott Park, IL, USA

A L Hopkins
Pfizer Global Research and Development, Sandwich, UK

M M Hopkins
University of Sussex, Brighton, UK

S Hoving
Novartis Institutes for BioMedical Research, Genome and Proteome Sciences, Basel, Switzerland

J R Howard
University of Pennsylvania, Philadelphia, PA, USA

R D Hubbard
Abbott Laboratories, Abbott Park, IL, USA

C Hubschwerlen
Actelion Pharmaceuticals Ltd, Allschwil, Switzerland

D M Huryn
University of Pennsylvania, Philadelphia, PA, USA

J D Imig
Medical College of Georgia, Augusta, GA, USA

J Irelan
Genomics Institute of the Novartis Research Foundation, San Diego, CA, USA

M Y Ismail
Paratek Pharmaceuticals, Inc., Boston, MA, USA

W Jahnke
Novartis Pharma AG, Basel, Switzerland

H R Jalian
David Geffen School of Medicine at UCLA, Los Angeles, CA, USA

M F Jarvis
Abbott Laboratories, Abbott Park, IL, USA

H Jones
F. Hoffmann-La Roche Ltd, Basel, Switzerland

D M Jonker
Novo Nordisk A/S, Bagsværd, Denmark

R P Joosten
Centre for Molecular and Biomolecular Informatics, Nijmegen, The Netherlands

V C Jordan
Fox Chase Cancer Center, Philadelphia, PA, USA

M R Jurzak
Merck KGaA, Darmstadt, Germany

T Kaneko
Pfizer Global Research and Development, Groton, CT, USA

C O Kappe
Karl-Franzens-University Graz, Graz, Austria

D B Kassel
Takeda San Diego, Inc., San Diego, CA, USA

J Keefer
NovaScreen Biosciences Corporation, Hanover, MD, USA

W W Keighley
Pfizer Global Research and Development, Sandwich, UK

T Kelly
Boehringer-Ingelheim Inc., Ridgefield, CT, USA

D J Kempf
Abbott, Abbott Park, IL, USA

P D Kennewell
Swindon, UK

E H Kerns
Wyeth Research, Princeton, NJ, USA

H Kessler
TU München, Garching, Germany

L B Kier
Virginia Commonwealth University, Richmond, VA, USA

J Kim
David Geffen School of Medicine at UCLA, Los Angeles, CA, USA

J F Kirmani
University of Medicine and Dentistry of New Jersey, Newark, NJ, USA

H A Kirst
Consultant, Indianapolis, IN, USA

J Klages
TU München, Garching, Germany

C Klammt
University of Frankfurt/Main, Germany

G J Kleywegt
University of Uppsala, Uppsala, Sweden

P Klimko
Alcon Research, Ltd., Fort Worth, TX, USA

L J S Knutsen
Worldwide Discovery Research, Cephalon Inc., West Chester, PA, USA

R Koenig
Genomics Institute of the Novartis Research Foundation, San Diego, CA, USA

A Kraft
University of Sussex, Brighton, UK

R T Kroemer
Sanofi-Aventis, Centre de Recherche de VA, Vitry-sur-Seine, France

H Kubinyi
University of Heidelberg, Heidelberg, Germany

T Kulikova
EMBL Outstation European Bioinformatics Institute, Hinxton, Cambridge, UK

P A Lachance
Rutgers – The State University of New Jersey, New Brunswick, NJ, USA

G Lange
Bayer CropScience, Frankfurt, Germany

A Lanoue
Ariana Pharmaceuticals, Paris, France

R A Laskowski
European Bioinformatics Institute, Wellcome Trust Genome Campus, Hinxton, Cambridge, UK

T Lavé
F. Hoffmann-La Roche Ltd, Basel, Switzerland

A R Leach
GlaxoSmithKline Research and Development, Stevenage, UK

S M Lechner
Neurocrine Biosciences Inc., San Diego, CA, USA

V J Lee
Adesis, Inc., New Castle, DE, USA

L Lehman
Gilead Sciences, Inc., Foster City, CA, USA

J-M Lehn
ISIS Université Louis Pasteur, Strasbourg, France

T Lengauer
Max Planck Institute for Informatics, Saarbrücken, Germany

H Lennernäs
Uppsala University, Uppsala, Sweden

A J Lewis
Novocell, Inc., Irvine, CA, USA

D F V Lewis
University of Surrey, Guildford, UK

J Lewis-Wambi
Fox Chase Cancer Center, Philadelphia, PA, USA

S V Ley
University of Cambridge, Cambridge, UK

A Li Wan Po
Centre for Evidence-Based Pharmacotherapy, Nottingham, UK

I Lieberburg
Elan Pharmaceuticals, South San Francisco, CA, USA

P Lindberg
AstraZeneca R&D, Mölndal, Sweden

K Lindpaintner
F. Hoffman-La Roche, Basel, Switzerland

E Littler
MEDIVIR UK Ltd, Little Chesterford, UK

D J Livingstone
ChemQuest, Sandown, Isle of Wight, UK

L Lou
Genelabs Technologies, Inc., Redwood City, CA, USA

T A Lyle
Merck Research Laboratories, West Point, PA, USA

Z Ma
Global Alliance for TB Drug Development, New York, NY, USA

M M Mader
Lilly Research Laboratories, Indianapolis, IN, USA

T V Magee
Pfizer Global Research and Development, Groton, CT, USA

R Magolda
Wyeth Research, Princeton, NJ, USA

S Mahdi,
University of Sussex, Brighton, UK

M-S Maira
Novartis Institutes for BioMedical Research, Basel, Switzerland

V G Manolopoulos
Democritus University of Thrace, Alexandroupolis, Greece

M J Marino
Worldwide Discovery Research, Cephalon Inc., West Chester, PA, USA

S Markison
Neurocrine Biosciences Inc., San Diego, CA, USA

G R Marshall
Washington University, St. Louis, MO, USA

M-J Martin
EMBL Outstation European Bioinformatics Institute, Hinxton, Cambridge, UK

P A Martin
Nottingham University, Nottingham, UK

Y C Martin
Abbott Laboratories, Abbott Park, IL, USA

J S Mason
Lundbeck Research, Valby, Copenhagen, Denmark

P Matsson
Uppsala University, Uppsala, Sweden

W McCarthy
Neurocrine Biosciences Inc., San Diego, CA, USA

J G McGivern
Amgen Inc., Thousand Oaks, CA, USA

S L McGovern
M.D. Anderson Cancer Center, Houston, TX, USA

S McKenna
Serono Research Institute, Rockland, MA, USA

H-Y Mei
Neural Intervention Technologies, Inc., Ann Arbor, MI, USA

A W Meikle
University of Utah School of Medicine, Salt Lake City, UT, USA

H Meltzer
Psychiatric Hospital Vanderbilt, Nashville, TN, USA

E Messersmith
Elan Pharmaceuticals, South San Francisco, CA, USA

B Metzger
Feinberg School of Medicine, Northwestern University, Chicago, IL, USA

A Miagkov
NovaScreen Biosciences Corporation, Hanover, MD, USA

A D Miller
Imperial College London, London, UK

C-P Milne
Tufts University, Boston, MA, USA

L A Mitscher
University of Kansas, Lawrence, KS, USA

S Modi
GlaxoSmithKline Research and Development, Stevenage, UK

J Moll
Nerviano Medical Sciences S.r.l, Nerviano, Italy

B Moloney
Evotec (UK) Ltd, Abingdon, UK

W H Moos
SRI International, Menlo Park, CA, USA and University of California–San Francisco, San Francisco, CA, USA

D Moras
Institut de Génétique et de Biologie Moléculaire et Cellulaire, Illkirch, France

F Morceau
Hôpital Kirchberg, Luxembourg City, Luxembourg

I Mori
Pfizer Global Research and Development, Nagoya, Japan

A A Mortlock
AstraZeneca, Macclesfield, UK

D Motiejunas
EML Research, Heidelberg, Germany

M Muda
Serono Research Institute, Rockland, MA, USA

S A Munk
Ash Stevens Inc., Detroit, MI, USA

K M Muraleedharan
University of Mississippi, University, MS, USA

B J Murphy
SRI International, Menlo Park, CA, USA

C W Murray
Astex Therapeutics, Cambridge, UK

R M Myers
University of Cambridge, Cambridge, UK

D C Myles
Kosan Biosciences, Hayward, CA, USA

Y Nakagawa
Novartis, Emeryville, CA, USA

D L Nelson
Eli Lilly and Company, Indianapolis, IN, USA

M L Nelson
Paratek Pharmaceuticals, Inc., Boston, MA, USA

J J Nestor
TheraPei Pharmaceuticals, Inc., San Diego, CA, USA

S Neuhoff
Uppsala University, Uppsala, Sweden

A H Newman
National Institutes of Health, Baltimore, MD, USA

D J Newman
NCI, Frederick, MD, USA

C G Newton
BioFocus DRI, Saffron Walden, UK

R Newton
Incyte Corporation, Wilmington, DE, USA

J K Nicholson
Imperial College, London, UK

P Nightingale
University of Sussex, Brighton, UK

U Norinder
AstraZeneca R&D, Södertälje, Sweden

B Oesch-Bartlomowicz
University of Mainz, Mainz, Germany

F Oesch
University of Mainz, Mainz, Germany

M Olah
University of New Mexico School of Medicine, Albuquerque, NM, USA

R Oliyai
Gilead Sciences Inc., Foster City, CA, USA

C Oostenbrink
Vrije Universiteit, Amsterdam, The Netherlands

T I Oprea
University of New Mexico School of Medicine, Albuquerque, NM, USA

J K Osbourn
Cambridge Antibody Technology, Cambridge, UK

E Ottow
Schering AG, Berlin, Germany

J K Ozawa
SRI International, Arlington, VA, USA

I K Pajeva
Bulgarian Academy of Sciences, Sofia, Bulgaria

J P Palma
Abbott Laboratories, Abbott Park, IL, USA

A Pannifer
Pfizer Global Research and Development, Sandwich, UK

G V Paolini
Pfizer Global Research and Development, Sandwich, UK

Y Parmentier
Technologie Servier, Orléans, France

N Parrott
F. Hoffmann-La Roche Ltd, Basel, Switzerland

G Pasut
University of Padua, Padua, Italy

M A Pearson
Novartis Institutes for BioMedical Research, Basel, Switzerland

A Pedretti
Institute of Medicinal Chemistry, University of Milan, Milan, Italy

P J Peeters
Johnson & Johnson Pharmaceutical Research & Development, Beerse, Belgium

S Perschke
NovaScreen Biosciences Corporation, Hanover, MD, USA

S Phillips
Urodoc Ltd, Herne Bay, UK

S D Pickett
GlaxoSmithKline, Stevenage, UK

D M Pollock
Medical College of Georgia, Augusta, GA, USA

P Preziosi
Catholic University of the Sacred Heart, Rome, Italy

T Prisinzano
University of Iowa, Iowa City, IA, USA

N Proctor
Simcyp, Sheffield, UK

J R Proudfoot
Boehringer Ingelheim Inc., Ridgefield, CT, USA

M Pruess
EMBL Outstation European Bioinformatics Institute, Hinxton, Cambridge, UK

N Pullen
Pfizer Global Research and Development, Sandwich, UK

A I Qureshi
University of Medicine and Dentistry of New Jersey, Newark, NJ, USA

F Rambert
Cephalon France, Maisons-Alfort, France

O Ramström
Royal Institute of Technology, Stockholm, Sweden

E Ratti
GlaxoSmithKline, Verona, Italy

U Regenass
Discovery Partners International AG, Allschwil, Switzerland

J-P Renaud
AliX, Illkirch, France

A A Reszka
Merck Research Laboratories, West Point, PA, USA

A E Rettie
University of Washington, Seattle, WA, USA

T I Richardson
Lilly Research Laboratories, Indianapolis, IN, USA

M Ridderström
AstraZeneca R&D Mölndal, Mölndal, Sweden

C Roberts
Genelabs Technologies, Inc., Redwood City, CA, USA

S Robertson
Cambridge Crystallographic Data Centre, Cambridge, UK

G A Rodan✱
University of Pennsylvania School of Medicine, Philadelphia, PA, USA

R Root-Bernstein
Michigan State University, East Lansing, MI, USA

P J Rosenthal
University of California, San Francisco, CA, USA

D P Rotella
Wyeth Research, Princeton, NJ, USA

G P Roth
Abbott Bioresearch Center, Worcester, MA, USA

R B Rothman
National Institutes of Health, Baltimore, MD, USA

R R Ruffolo
Wyeth Research Laboratories, Collegeville, PA, USA

P Russo
University of Salerno, Fisciano, Italy

H Rüterjans
University of Frankfurt/Main, Germany

R Rydel
Elan Pharmaceuticals, South San Francisco, CA, USA

G Sachs
University of California at Los Angeles, CA, USA, and Veterans' Administration Greater Los Angeles Healthcare System, Los Angeles, CA, USA

M Sahl
MJSahl Consulting, Philadelphia, PA, USA

A Sali
University of California at San Francisco, San Francisco, CA, USA

J Sallantin
CNRS, Montpellier, France

M S P Sansom
University of Oxford, Oxford, UK

P Santi
University of Parma, Parma, Italy

A M Sapienza
Simmons College, Boston, MA, USA

M Sathyamoorthy
NovaScreen Biosciences Corporation, Hanover, MD, USA

T K Sawyer
ARIAD Pharmaceuticals, Cambridge, MA, USA

✱Deceased.

R A Scherrer
BIOpK, White Bear Lake, MN, USA

J-M Scherrmann
University of Paris, Paris, France

L Schmeltz
Feinberg School of Medicine, Northwestern University, Chicago, IL, USA

U Schmitz
Genelabs Technologies, Inc., Redwood City, CA, USA

G Schnapp
Boehringer Ingelheim Pharma, Biberach, Germany

D M Schnur
Bristol-Myers Squibb, Princeton, NJ, USA

B Schnurr
Discovery Partners International AG, Allschwil, Switzerland

G E Schulz
Albert-Ludwigs-Universität, Freiburg im Breisgau, Germany

D Schweizer
Cephalon France, Maisons-Alfort, France

A Scriabine
Yale University School of Medicine, New Haven, CT, USA

M J Seibel
Anzac Research Institute, University of Sydney, and Concord Hospital, Concord, NSW, Australia

M Sela
Weizmann Institute of Science, Rehovot, Israel

C Selassie
Pomona College, Claremont, CA, USA

P Seneci
Desenzano del Garda, Italy

N A Sharif
Alcon Research, Ltd., Fort Worth, TX, USA

S J Shaw
Kosan Biosciences, Hayward, CA, USA

W Shi
Case Western Reserve University, Cleveland, OH, USA

J M Shin
University of California at Los Angeles, CA, USA, and Veterans' Administration Greater Los Angeles Healthcare System, Los Angeles, CA, USA

A R Shoemaker
Abbott Laboratories, Abbott Park, IL, USA

J A Sikorski
AtheroGenics, Alpharetta, GA, USA

M Skingle
GlaxoSmithKline, Stevenage, UK

J S Skotnicki
Wyeth Research, Pearl River, NY, USA

P M Smith-Jones
Memorial Sloan Kettering Cancer Center, New York, NY, USA

D A Smith
Pfizer Global Research and Development, Sandwich, UK

P P S So
University of Toronto, Toronto, ON, Canada

F Sonvico
University of Parma, Parma, Italy

M Spigelman
Global Alliance for TB Drug Development, New York, NY, USA

R V Stanton
Pfizer Research Technology Center, Cambridge, MA, USA

P F W Stouten
Nerviano Medical Sciences, Nerviano, Italy

M T Stubbs II
Martin Luther University, Halle, Germany

J L Sturchio
Merck & Co., Inc., White House Station, NJ, USA

K Sugano
Pfizer Inc., Aichi, Japan

G J Swaminathan
European Bioinformatics Institute, Wellcome Trust Genome Campus, Hinxton, Cambridge, UK

E E Swayze
ISIS Pharmaceuticals, Carlsbad, CA, USA

S Takahashi
David Geffen School of Medicine at UCLA, Los Angeles, CA, USA

J Taskinen
University of Helsinki, Helsinki, Finland

S Tavelin
Umeå University, Umeå, Sweden

C M Taylor
Washington University, St. Louis, MO, USA

D A Taylor
East Carolina University, Greenville, NC, USA

C P Taylor
Pfizer Global Research and Development, Ann Arbor, MI, USA

A J Tebben
Bristol-Myers Squibb, Princeton, NJ, USA

B Tehan
GlaxoSmithKline Pharmaceuticals plc, Harlow, UK

D Teitelbaum
Weizmann Institute of Science, Rehovot, Israel

M-H Teiten
Hôpital Kirchberg, Luxembourg City, Luxembourg

B Testa
University Hospital Centre, Lausanne, Switzerland

I V Tetko
Institute of Bioorganic and Petrochemistry, Kiev, Ukraine

M Thanou
Imperial College London, London, UK

W J Thomsen
Arena Pharmaceuticals, Inc., San Diego, CA, USA

K B Thor
Dynogen Pharmaceuticals, Inc., Research Triangle Park, Durham, NC, USA

A J Thorpe
Pfizer Global Research and Development, Ann Arbor, MI, USA

J-P Tillement
Faculty of Medicine of Paris XII, Paris, France

H Timmerman
Vrije Universiteit, Amsterdam, The Netherlands

N P Todorov
De Novo Pharmaceuticals, Cambridge, UK

R A Totah
University of Washington, Seattle, WA, USA

W F Trager
University of Washington, Seattle, WA, USA

D Tremblay
AFSSAPS Consultant (French Agency for Drugs), France

D J Triggle
State University of New York, Buffalo, NY, USA

D Trist
GlaxoSmithKline, Verona, Italy

A Tropsha
University of North Carolina at Chapel Hill, Chapel Hill, NC, USA

J Trzaskos
Bristol-Myers Squibb Corporation, Lawrenceville, NJ, USA

J V Turner
The University of Queensland, Brisbane, Qld, Australia

R Tuttle
El Cerrito, CA, USA

D Ullmann
Evotec AG, Hamburg, Germany

A-L Ungell
AstraZeneca R&D Mölndal, Mölndal, Sweden

P H Van der Graaf
Pfizer Global Research and Development, Sandwich, UK

W F van Gunsteren
Eidgenössische Technische Hochschule, Zürich, Switzerland

M M H van Lipzig
Vrije Universiteit, Amsterdam, The Netherlands

J van Oostrum
Novartis Institutes for BioMedical Research, Genome and Proteome Sciences, Basel, Switzerland

H van de Waterbeemd
AstraZeneca, Macclesfield, UK

F M Veronese
University of Padua, Padua, Italy

S A G Visser
AstraZeneca R&D, Södertälje, Sweden

G Vistoli
Institute of Medicinal Chemistry, University of Milan, Milan, Italy

D Voloboy
pION Inc., Woburn, MA, USA

H Voshol
Novartis Institutes for BioMedical Research, Genome and Proteome Sciences, Basel, Switzerland

G Vriend
Centre for Molecular and Biomolecular Informatics, Nijmegen, The Netherlands

S Waddington
University College London, London, UK

R C Wade
EML Research, Heidelberg, Germany

A S Wagman
Novartis Institutes for BioMedical Research, Inc., Emeryville, CA, USA

M J A Walker
University of British Columbia, Vancouver, BC, Canada

I Wall
GlaxoSmithKline Pharmaceuticals plc, Harlow, UK

O B Wallace
Lilly Research Laboratories, Indianapolis, IN, USA

B Walther
Technologie Servier, Orléans, France

F L Wang
NovaScreen Biosciences Corporation, Hanover, MD, USA

J Wang
Novartis Institutes for Biomedical Research, Cambridge, MA, USA

J Wasley
Simpharma LLC, Guilford, CT, USA

A A Wasunna
The Hastings Center, Garrison, NY, USA

W W Weber
University of Michigan, Ann Arbor, MI, USA

H Weinmann
Schering AG, Berlin, Germany

A Weissman
NovaScreen Biosciences Corporation, Hanover, MD, USA

M P Wentland
Rensselaer Polytechnic Institute, Troy, NY, USA

C G Wermuth
Prestwick Chemical, Illkirch, France

W Wierenga
Neurocrine Biosciences, Inc., San Diego, CA, USA

M Wiese
University of Bonn, Bonn, Germany

P Willett
University of Sheffield, Sheffield, UK

J P Williams
Neurocrine Biosciences Inc., San Diego, CA, USA

M Williams
Feinberg School of Medicine, Northwestern University, Chicago, IL, USA

M Williams
*Northwestern University, Chicago, IL, USA and Worldwide Discovery Research, Cephalon, Inc.,
West Chester, PA, USA*

I D Wilson
AstraZeneca, Macclesfield, UK

T J Winchester
Pfizer Global Research and Development, Sandwich, UK

S Winiwarter
AstraZeneca R&D Mölndal, Mölndal, Sweden

P R Wolpe
University of Pennsylvania, Philadelphia, PA, USA

T P Wood
Pfizer Global Research and Development, Sandwich, UK

P Worland
Celgene Corporation, San Diego, CA, USA

J F Worley III
Vertex Pharmaceuticals Inc., San Diego, CA, USA

P M Woster
Wayne State University, Detroit, MI, USA

J-M Wurtz
Institut de Génétique et de Biologie Moléculaire et Cellulaire, Illkirch, France

M G Wyllie
Urodoc Ltd, Herne Bay, UK

I Yalçin
Eli Lilly and Co., Indianapolis, IN, USA

T Yednock
Elan Pharmaceuticals, South San Francisco, CA, USA

L S Young
California Pacific Medical Center Research Institute, San Francisco, CA, USA

I Zamora
Universitat Pompeu Fabra, Barcelona, Spain

E Zass
Informationszentrum Chemie Biologie Pharmazie, Zürich, Switzerland

N T Zaveri
SRI International, Menlo Park, CA, USA

Q Zhang
University of California, La Jolla, CA, USA

H Zhou
Anzac Research Institute, University of Sydney, NSW, Australia

X-X Zhou
Medivir AB, Huddinge, Sweden

R J Zimmerman
Zimmerman Consulting, Orinda, CA, USA

K P Zuideveld
F. Hoffmann-La Roche Ltd, Basel, Switzerland

8.01 Introduction

W T Comer, TorreyPines Therapeutics, Inc., La Jolla, CA, USA

8.01.1	**Introduction**	1
8.01.2	**Case 1**	1
8.01.3	**Case 2**	1
8.01.4	**Case 3**	3
8.01.5	**Case 4**	3
	References	4

8.01.1 Introduction

The role of a medicinal chemist in drug discovery is the design, synthesis, and registration of the best compound for treating a particular disease condition. The job is not finished when the most potent compound for a given target (receptor, ion channel, enzyme) is identified. The best compound of the series for bioavailability to the target must also be found, the active compound with least metabolism or most predictable pharmacokinetics, and certainly the compound with fewest other effects or greatest selectivity for the desired effect must be identified. The convergence of these properties is the best compound for the disease until a more relevant mechanism or target is found that provides a more effective compound.

My perspectives on drug discovery have developed over nearly 50 years that I have worked in the field, but many of the views were clear early in this time before combinatorial chemistry, high-throughput screening, and cloned and expressed receptor subtypes. The case histories I describe here focus on whether to discontinue or redirect projects; only the time required to complete these projects is shortened by new technologies, but the strategic lessons are still valid today.

8.01.2 Case 1

My first project in industry was to inhibit the body's synthesis of cholesterol because my employer sold the resin Questran which absorbed bile salts as an effective way of lowering serum cholesterol, but compliance was poor because of constipation and the unpleasant task of 'eating sand' several times a day. Hydroxymethylglutaryl (HMG) CoA reductase was known to be the rate-limiting step of cholesterol synthesis so inhibition of this enzyme was the obvious target and assays were quickly established. A series of aryl pentadienoic acids was optimized within a year; they were effective by oral dosing to cholesterol-fed rabbits which were sacrificed and fatty deposits in the aorta measured for drug effect. But what was the disease and how would we measure patient benefit? This was 1962 before ultrasound was used to measure a decrease in fatty deposits in femoral arteries as an indicator of coronary artery disease; the link between serum cholesterol levels and coronary artery disease had not yet been made. The project was put on hold until the company developed a clinical strategy for disease intervention, but that was not a high priority and the clinicians waited for someone else to show the way. A few years later, compactin was identified by Sankyo and mevinolin by Merck as research clinicians quantitated fatty blockages with ultrasound; the statins were born. Two lessons emerged from this shelved project: (1) having clinical strategies and measurements of disease progression is a fundamental part of the drug discovery process and must be available for a seamless and timely development; (2) stepping back from the cutting edge of an emerging field concedes failure because you are no longer first and the competition may be better.

8.01.3 Case 2

Also in the early 1960s, I joined a company effort to explore beta-adrenergic agents because of clinical use of epinephrine for cardiac rescue and isoproterenol for bronchodilation in asthma. Pharmacologic identification of the first beta-blocker, sotalol, concurrent with ICI's pronethalol which soon yielded to the more potent propranolol, hit the same wall as inhibiting cholesterol synthesis – what is the disease condition and clinical measurement? The marketing director clearly asked, "What kind of patient needs their betas blocked?" In 1964, the British physician Brian Pritchard

Figure 1   A = HO, R = *i*Pr, isoproterenol; A = MeSO$_2$NH, R = *i*Pr soterenol; A = HOCH$_2$, R = *t*Bu, albuterol.

reported significant lowering of blood pressure in hypertensives by propranolol, opposite to what was predicted because of its predominant effect on cardiac function rather than peripheral vasculature. By following rather than leading with clinical trials, sotalol was registered first in Europe but its sales were dwarfed by propranolol. In later years, probing clinical research established a significant role for sotalol as an antiarrhythmic, a use not anticipated by early animal pharmacology.

More lessons emerged from the adrenergic studies. Our major thesis for structurally modifying the catecholamines was to replace phenolic –OH groups with the methanesulfonamide group in hopes of achieving some organ or functional selectivity. With sotalol (4-MeSO$_2$NH–) we achieved good beta-blocking potency, some beta$_1$/beta$_2$-selectivity, and good oral bioavailability. The more interesting experiment was replacing one phenolic –OH at a time with MeSO$_2$NH in the catecholamines to retain beta agonist potency (**Figure 1**). The 4-OH, 3-NHSO$_2$Me analog was equipotent to isoproterenol with much longer $t_{1/2}$ and more beta$_2$-selective, whereas the 3-OH, 4-NHSO$_2$Me was over 10 000 times less active. The more acidic proton of MeSO$_2$NH had to be *meta* to the phenethanolamine for beta-adrenergic potency, and the group is more ionized at blood pH than the catechol, whereas the ionized group *para* to the side chain does not function as an agonist. Furthermore, the *meta* MeSO$_2$NH group is not a good substrate for catechol-*o*-methyltransferase (COMT) methylation, which explains why these bioisosteres were not rapidly metabolized and have a longer $t_{1/2}$. Three compounds from this series (different amine substituents) were clinically evaluated and found to be very safe, effective, and selective agonists.[1,2] Further studies showed the 3-MeSO$_2$NH group to be a useful bioisostere in phenylalanines, but just as it was not a good substrate for COMT it did not permit decarboxylation with L-DOPA decarboxylase. This meant the methanesulfonamide bioisostere of L-DOPA would not provide an alternative Parkinson's drug, nor did this methanesulfonamide bioisostere provide an effective long-lasting alternative to estradiol, and its usefulness remained primarily in the catecholamines.

A major lesson came belatedly from the clinical trials and toxicological studies of the beta agonists. Soterenol, the bioisostere of isoproterenol, had almost completed phase III clinical trials as a bronchodilator for asthmatics, and mesuprine was in phase III for peripheral vascular disease and premature labor as a uterine relaxant when preliminary results from chronic toxicity and carcinogenicity studies became available. At necropsy 20 of 50 rats in the high dose group had benign tumors, mesovarial leiomyomas, which appeared like a third ovary. No tumors were found after 18 months, only after 24 months. With a heightened concern for tumors and carcinogenicity at the Food and Drug Administration (FDA) in the 1960s, all clinical trials were stopped and all patients who had received drug were to be followed for their lifetime. The FDA even assumed the MeSO$_2$NH group was responsible for the leiomyomas, although the rats were dosed with 10 000 times the amount of compound that would have killed them with epinephrine. The concern at that time was not benefit/risk ratio or separation between active and toxic dose, but an absolute concern for tumors. All development projects with sulfonamido beta-adrenergic agonists were stopped and no products were marketed.

About 7 years later, Glaxo developed a similar compound salbutamol/albuterol, which differed from isoproterenol and soterenol only by having the *meta* OH or MeSO$_2$NH isostere replaced by CH$_2$OH and *t*-butylamine in place of *i*-propylamine. The amine substituent was a trivial change because all three series of catecholamines had *t*-butyl > *i*-Pr > Me in potency with no appreciable difference in pharmacologic or kinetic/disposition properties. The pK_a of the catechol or 'mixed catechol' group was MeSO$_2$NH > OH > CH$_2$OH, but all somewhat acidic due to the *para* OH, and only the OH was readily metabolized by COMT to an inactive analog. At that time, the Mead Johnson (Bristol-Myers) management agreed to repeat the carcinogenicity studies to see if the leiomyomas were truly drug related, even though it cost nearly 3 years and > $2 million at that time. The new study included four arms with a small group necropsied at 18 months: (1) were the results reproducible? (2) were they reproducible with a different strain of rats? (3) were they reproducible if drug was administered orally by gavage or in feed? (4) were they reproducible if a similar dose of albuterol was used instead of soterenol? The answer to all questions was yes with nearly identical results. Albuterol had been licensed to Schering Plough for the US market, and the carcinogenicity study was conducted by Schering Plough, but no mesovarial leiomyomas were found. We called the head of R&D at Schering Plough who asked their pathologist to reexamine the preserved rat ovaries from their study. Although surprised, they found a similar incidence of leiomyomas when they knew what to look for. At that time, our tissue pharmacologists measured the effect of soterenol on isolated mesovarial tissue and found the density of beta-adrenergic receptors to be much greater than the density in

8.02 Reflections on Medicinal Chemistry Since the 1950s

H Timmerman, Vrije Universiteit, Amsterdam, The Netherlands

8.02.1	**The Years of Learning**	7
8.02.2	**Pharmacochemistry at the Vrije Universiteit Amsterdam**	9
8.02.3	**Being a Professor in Amsterdam**	10
8.02.4	**The 'New' Histamine Receptors**	12
8.02.5	**Histamine, Histamine Receptors, and Ligands as Research Tools**	12
8.02.6	**Looking Back**	13
8.02.7	**My Histamine**	14
References		14

8.02.1 The Years of Learning

When I started to study chemistry at the Vrije Universiteit in Amsterdam in 1956, it was just one hundred years since the history of modern synthetic medicines had started. Perkin had synthesized the first synthetic dye (1856), and in Germany pharmacology became more and more an experimental discipline. Series of chemically related compounds became available, and were subjected to a pharmacological screening. When in 1868 Crum Brown and Fraser published their famous paper "On the connection between chemical constitution and physiological action," some scientists became euphoric and predicted that soon new medicines would be designed and, moreover, for each disease a specific medicine would become available.

The optimism during the last part of the nineteenth century was not based on evidence. At that time the chemical structure of compounds was poorly understood and also the knowledge of physiological and pathological mechanisms was very limited. Developing new medicines remained for almost a complete century a matter of trial and much error.

When I arrived as a fresh student, the subdiscipline of medicinal chemistry was hardly recognized in the Netherlands. In the USA, the Division of Medicinal Chemistry (American Chemical Society) got its name in 1948; earlier it was known as the Division of Pharmaceutical Chemistry (1909–20), and from 1920 as the Division of Medicinal Products. At the Dutch universities there were no courses in medicinal chemistry, but at the Vrije Universiteit, my alma mater, the organic chemist Wijbe T. Nauta was a lecturer, and at the same time research director of the Dutch pharmaceutical company Brocades (now acquired by Yamanouchi-Fujisawa).

Nauta has been in the Netherlands – and Europe – instrumental in the process that eventually led to the birth of the new discipline of medicinal chemistry. Nauta, however, preferred the use of the word 'pharmacochemistry.' He had two reasons: first, a medicinal chemist (pharmacochemist) does not make medicines, but active ingredients of medicines, pharmaca. Second, Nauta liked to include the fields of crop protectants in this discipline. In the Netherlands and several other countries the term pharmacochemistry is in use as more or less the equivalent of medicinal chemistry.

The manner in which Nauta transferred from a 'classical' organic chemist into a medicinal chemist may be an example of what happened in the years after World War II when medicinal chemistry emerged as a new chemical subdiscipline. Immediately after World War II, the countries that had suffered got free access to patents owned by German companies. The Dutch pharmaceutical wholesaler Brocades asked Nauta to explore the possibilities of starting a research program on the basis of selected patents. This happened, and met with good success, and subsequently Nauta initiated the search for new active molecules. As an organic chemist he had worked on tetraphenylethanes and studied the influence of substitution patterns in the aromatic groups on radical formation from these derivatives. When Parke-Davis introduced the diphenylmethane derivative diphenhydramine as an antihistamine (H_1 blocker), Nauta investigated the influence of substitutions in the benzhydryl group on the histamine-blocking properties. He identified compounds with a strongly enhanced effect, and 'gaps' in the Parke-Davis' patents as well. Subsequently, alkyl-substituted diphenhydraminines were introduced (e.g., orphenadrine for Parkinsonism, based rather on antimuscarinic properties, however).

Figure 1   Professor E.J. (Eef) Ariëns at a Camerino–Noordwijkerhout symposium in Camerino.

Nauta brought his success at Brocades to his university group in the early 1950s. He was absolutely convinced that for a successful research program seeking new active agents, pharmacological studies should be included: not separated groups for chemistry and pharmacology, but a department to which both the disciplines contribute.

In the late 1950s, I joined Nauta's group as an undergraduate and later became a PhD student. I had bought at a booksale a book called *Chemie in Dienst der Mensheid*, which may be translated as "Chemistry at the service of mankind." I was extremely impressed by the stories about Ehrlich, Domagk, the penicillins and other antibiotics, DDT, and on many other scientists and compounds. At that time several of my fellow students considered the choice of this kind of organic chemistry quite negative, the easy way. Applied science was apparently not a popular choice.

I was very lucky to join Nauta's group because of the vision of its leader. But there was more. At the same time the Dutch pharmacologist Eef (E.J.) Ariëns (**Figure 1**) published the book *Molecular Pharmacology*. The way Ariëns treated pharmacology has influenced my way of thinking very much. The ligand–receptor interactions were described in simple mathematics, the pharmacological testing was carried out on isolated organs in a very reproducible manner, and, most importantly, structure–activity relationships (SARs) got much (at best it is qualitative) attention. It is the combination of synthetic chemistry, pharmacology at the molecular level, and the study of the relationship(s) between structure and biological activity, which defines the current discipline of medicinal chemistry.

During the period I worked on my PhD thesis, the pioneering work of Corwin Hansch changed the face of medicinal chemistry in a fundamental way. The SAR studies were transformed into quantitative SAR (QSAR) studies. Many scientists became as euphoric as those who predicted about 75 years earlier that new medicines would be designed. The enthusiasm reached a remarkable level: the IUPAC Committee on Medicinal Chemistry deemed it necessary to initiate a study on the question whether the Hansch approach would allow one to predict the activity of a given compound to such a level that there would not be any innovative aspect in proposing the given compound(s) as a potential medicinal agent. If such were the case, patents for new biological compounds might become complicated – or even impossible – such was the fear. The fear was not justified, concluded the IUPAC study.

Indeed, the Hansch approach has influenced the field of SARs enormously, together with other methods of study. Too often, however, it is forgotten, even nowadays, that the QSAR approaches have a very important limitation. Comparing activities of compounds only makes sense when one has made sure that the biological activities of all compounds do stem from one and the same interaction between the compounds and the site of interaction in the biological system. This limitation makes, e.g., a QSAR study of chemically unrelated compounds using LD_{50} values as the biological parameter an absolute nonsense. But often the limitation is neglected, though it is clear that in many cases the condition is not fulfilled at all. In studies applying mutants of given receptors it has, e.g., been shown that the influence of the mutation may differ even to an absolute extent from one compound to another compound of a chemically closely related series of derivatives. In such cases, a QSAR approach does not make any sense any more! The results have a statistical meaning only, just as in the case of QSARs of LD_{50} values. QSAR studies, with series of compounds without the guarantee that when all have a similar way of interfering with the target, may be considered as an example of what Ariëns has called for another case 'sophisticated nonsense' (see later discussion).

Medicinal chemistry does require the scientist's understanding of both chemical and pharmacological principles. In this sense the pharmacologist Ariëns was a medicinal chemist. This eminent scientist defended on the same day two PhD theses: one in chemistry and the other in medical sciences. One of the many contributions of Ariëns to medicinal chemistry concerns the role of chirality in drug action. The different levels of activity of enantiomers had long been

known, but the important practical consequences had been neglected. In the 1920s, the well-known pharmacologist Cushny considered it of no practical interest that in a given compound the activity resided in one isomer: it is not needed, he claimed, to use the pure enantiomer, you just double the dose when using the mixture. Ariëns explained in a very effective way that in case one isomer does not contribute to a given biological profile, it should be removed. He used the term isomeric ballast for racemic mixtures of both medicines and agricultural preparations. Scientists who used racemic mixtures in pharmacokinetic studies, not knowing whether they were measuring concentrations of one of the other isomer or of mixtures, were condemned as practitioners of 'sophisticated nonsense.'

8.02.2   Pharmacochemistry at the Vrije Universiteit Amsterdam

It was in this atmosphere of a continuous meeting of pharmacologists and (originally) synthetic chemists that I learned the metier of medicinal chemistry and I saw that a new discipline had been born. This new discipline became a star. Its products have contributed to a large extent to the increase of quality of life, even to solving some medical problems. Medicinal chemistry became a mature science, both independent and interdependent, as Alfred Burger once said.

It is currently relatively easy to identify compounds that interfere with a given target; the problem is of course that it is necessary to ensure that the target is meaningful and can be used as the point of attack for a certain disease. In other words, when a target is available the modern approaches in medicinal chemistry allow the identification of a potential medicinal agent. In a way it is permissible to say that the critical step in the process aiming at new medicine is no longer the identification of an active compound or a series of active compounds for a given target. The new issue is much more the question of whether an active compound can be converted into a medicine which is better than the already available therapies. Has medicinal chemistry then no further role to play? Can it not contribute in other ways than by the routinely making compounds which have an attractive level of activity at a certain target? No, I would say, medicinal chemistry will continue to play a major role in drug research and development; the way this role will be played is however changing all the time, especially because of the increased knowledge in the field of life sciences.

Both research and development in medicinal chemistry should become more and more transdisciplinary. As in the years of Ariëns, biological phenomena still require special attention. Let us focus for a while on receptors as targets for medicines; receptors – especially G protein-coupled receptors (GPCRs) – are by far the largest class of drug targets in any case. In my own research program I have always used the histamine receptors as examples for receptors – now GPCRs – in general, and I would like to describe how within a period of about 25 years the field changed dramatically, especially as a consequence of more precise information from the pharmacology of these histamine receptors. I would also like to stress that in this essay I use the histamine field as an example.

Nauta started to work on antihistamines in the 1950s in the traditional way: making compounds and testing compounds. He detected qualitative regularities in the relationship between the substitution pattern and the histamine-blocking properties of extended series of especially alkyl-substituted diphenhydramines. At that time nothing was known about the structure of 'receptor.' In the classic book *Molecular Pharmacology* (Ariëns and associates), a receptor was compared with a beautiful lady to whom you might write letters; you received occasionally an answer, but nobody could claim ever to have seen this remarkable lady. On another occasion – a meeting of the New York Academy of Sciences in 1967 – Ariëns, who had concluded that it was unlikely that agonists and antagonists of a given receptor would bind in the same way to the receptor, said: "when I am talking about receptors, I am talking about something I know nothing about."[1] Indeed, at that time the lock-and-key theory was the accepted concept, without any information about the nature of the lock, never mind the mechanism of the lock.

It was Nauta who proposed, possibly being the first to do so, that a receptor might be a protein in a helix shape. He and his associates published a model[2] in which histamine and the antagonist diphenhydramine were shown to bind in a reversible way to selected amino acid units of a hypothetical protein (**Figure 2**). This proposal unfortunately did not get much attention and it was not followed up by Nauta and his team either.

The Nauta's group – of which I had become a member – continued the search for 'better' antihistamines, including nonsedating ones. Several interesting compounds were identified, such as a quaternary compound (pirfonium) with high activity; poor oral activity made the compound unfit for development as a medicine, but due to its quaternary ammonium function it has been used as a research tool.

In fact, it was not difficult to identify highly potent antihistamines, but at the same time it was not easy to add an extra feature making the compound interesting for development as a therapeutic agent. In the diphenhydramine series, we arrived at pA_2 values up to around 11.0, but no compound except for the early ones went into clinical use.

When the Hansch approach was introduced, the group in Amsterdam started to follow it. It was Roelof Rekker who came up with the so-called 'fragmental constant methodology' to calculate log *P* values. His method was different from

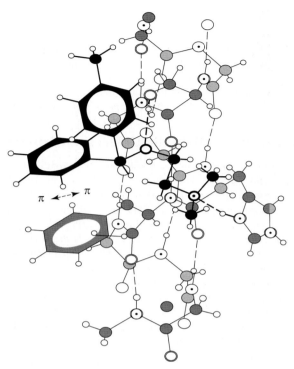

Figure 2   The receptor scheme of Nauta, showing the proposed histamine receptor to which diphenhydramine is bound.

the Hansch system, as Rekker was using a value (established from large series of compounds) for each, including a hydrogen atom, fragment of a molecule,[3] whereas Hansch (and Leo) 'neglected' a value for hydrogen atoms. Rekker and Hansch never settled the straightforward competition between them; however, Rekker became the most-cited scientist of the Faculty of Chemistry of the Vrije Universiteit Amsterdam. The QSAR studies of Rekker, especially using antihistamines as examples, continued into the 1980s, but in the histamine field they did not lead to a discovery of any better compound.

When the research team of Sir James Black at Smith, Kline & French in the UK identified a new class of antihistamines (the H_2 antagonists) during the 1970s, the Amsterdam group failed to pick up the new theme. Ariëns, who was a consultant of the Brocades company (already known then as Gist-brocades, for which company I worked at that time, and of which Nauta was still the director of research), asked that attention should be given to the new ideas, but Nauta was apparently not persuaded by Arien's pharmacological expertise and did not take up this suggestion. How unfortunate this attitude was; the group lost contacts in the (small) world of histaminologists, which focused more and more on what became known as the H_2 receptors than on H_1, the classic field.

8.02.3   Being a Professor in Amsterdam

Toward the end of the 1970s, Nauta had to retire (because of his age, in accordance with the strict Dutch laws) and I was appointed as his successor (**Figure 3**). I decided to revive the (anti) histamine research of the group. As I did not feel any need to develop a new medicine I selected to work on H_2 agonists rather than antagonists. It is my sincere conviction that academics should use their freedom for selecting research fields; they should never imitate what is done by their colleagues working in the pharmaceutical industry; they would be in a poor position anyhow when doing so. At that time it had been proposed that for activating the H_2 receptor, a proton transfer via the tautomeric species of the imidazole nucleus of the histamine molecule was essential. We could show by the relatively high H_2 agonist activity of properly subsisted thiazole analogs of histamine that this could not be true. In the meantime we identified a highly selective H_2 (versus H_1) agonist amthamine, a substituted thiazole analog of histamine, a compound that became a much-used research tool.

What had happened with the H_1 antagonist in the 1950s was seen in the H_2 blockers in the 1980s. The first compounds were moderately active (cimetidine), but soon extremely potent H_2 antagonists were identified. Again,

Figure 3 The author pays tribute to his professor W. Th. (Wijbe) Nauta during a lecture at his alma mater, the Vrije Universiteit Amsterdam.

it was shown that when an interesting target is available, medicinal chemistry will come up with ligands. Not long thereafter it seemed that the histamine book could be closed for the second time; medicinal chemistry of the field was finished.

However, one question related to the clinical profile of the classical antihistamines (H_1 antagonists) had so far not been solved. All compounds caused severe sedation, but the mechanism by which it was caused was unknown. Many investigators thought that the effect was not related to an interaction with the histaminergic system but was rather caused by blockade of the muscarinic receptor (at that time only one type). It was generally accepted that histamine had no function as a neurotransmitter.

But this opinion changed when Schwartz showed by elegant ex vivo studies that there was a clear relationship between the level of occupation of histamine receptors in the central nervous system (CNS) and the level of sedation caused by H_1 antagonists. The results of Schwartz led to two important conclusions: histamine is likely a neurotransmitter and H_1 antagonists cause sedation by blocking H_1 receptors in the brain. Both conclusions proved to be true soon thereafter.[4]

Subsequently, H_1 blockers attracted new attention of the pharmacochemical industry. The first nonsedation or second-generation H_1 blocker, terfenadine, was found by chance; it had been developed as a Ca-entry-blocking agent. The nonsedating properties of this moderately active H_1 antagonist were the result of a poor capacity to enter the brain; this approach had been tried earlier, but without success. But now the principle has been proved to be productive; new nonsedating compounds reached the market and became blockbusters; after terfenadine, for example, cetirizine and loratadine were introduced.

All new-generation derivatives caused only minimal blockade of H_1 receptors in the CNS, and textbooks stated that this was 'because of a high hydrophilicity.' However, the nonsedating compounds showed log P values which according to a rule of thumb in respect of lipophilic character would allow them to pass the blood–brain barrier readily. We tried to explain the finding by applying the Δlog P theory: a high Δlog P (log P octanol–water minus log P cyclehexane–water) would mean a high hydrogen-binding capacity and therefore a strong binding to plasma proteins and consequently a poor

CNS penetrating capacity. We could explain the findings only by using, besides the $\Delta\log P$, additional properties of the compounds. It seemed almost impossible to design non-CNS-penetrating compounds; it was largely a matter of chance. Later on the real mechanism became clear, when it was shown that all second-generation compounds are substrates of the P-gP transporter. Indeed, the new compounds had been the result of especially good luck.

8.02.4   The 'New' Histamine Receptors

In the early 1980s Schwarz showed that histamine, being a neurotransmitter, could regulate its own synthesis and release from specific neurons by an interaction with the newly defined H_3 receptor. This presynaptic receptor was subsequently shown to be both a histamine autoreceptor and a heteroreceptor present on other neurons, such as cholinergic and dopaminergic neurons. The receptor is especially, but not exclusively, located in the CNS. Shortly after the H_3 receptor had been postulated, histaminergic neurons were observed by applying immunocytochemical techniques. Selective highly potent ligands, both agonists and antagonists, were identified and potential uses of such ligands were predicted. The pharmaceutical industry showed minimal interest, however. It was the time of 'cloning the genes,' but efforts of several teams – including ours in Amsterdam – to clone the H_3 receptor gene failed. We have, when presenting our findings on new H_3 ligands, often been asked: "Are you sure there is indeed such a thing as an H_3 receptor?" We were sure, especially as the level of selectivity of agonists (H_3 versus H_1 or H_2) was very high.

This situation lasted until the late 1990s, when the gene was finally found, and indeed, found in a database. Now it became clear why high selectivity was often seen: the homology with both H_1 and H_2 is extremely low, much lower than for example that observed for the subtypes of the dopamine or muscarine receptors. Now, the industry became interested and several companies, including big pharma started research programs, especially for antagonists. Large series of patent applications appeared, but as of now (2006) no H_3 ligand has reached the marketplace as a medicine.[5]

But, again, not all aspects of the histamine-related physiology or pathology had been addressed by the medicinal chemist. Histamine has strong bronchoconstrictory properties but the efficacy of histamine H_1 blockers against in allergy-related bronchoconstriction is rather limited. The reason of this lack of efficacy is not only due to the fact that other factors besides histamine are involved in the process (e.g., acetylcholine), but is also as a consequence of the absence of an anti-inflammatory property in H_1 antagonists; in asthma inflammation plays a major role and histamine has proinflammatory properties. Just a few years ago, around the turn of the century, a newly detected histamine receptor, H_4, was found to be present especially on leukocytes. For histamine antagonist ligands for use in asthma it is most likely that compounds which block both H_1 and H_4 receptor properties will be needed; information on such compounds has so far not been published. As the homology between H_3 and H_4 receptors is significant, it is not easy to identify selective H_4 ligands.

8.02.5   Histamine, Histamine Receptors, and Ligands as Research Tools

The story of the histamine receptors so far identified is an intriguing one. The role of histamine in allergic reactions was discovered around 1930. The compounds developed in the 1940s and 1950s as antihistamines were not very active against the asthmatic condition, owing to the lack of efficacy and a high incidence of side effects, especially drowsiness. The latter problem has been solved; the first might find a solution in the coming years. The story can be used for explaining the need of interaction between chemical and biological disciplines when the aim is to find new medicines and also when 'better' medicines are desired. Recent developments in 'receptorology' may lead to compounds with an improved profile for use in medicinal preparations.

Since the introduction of the lock-and-key hypothesis for interactions between a ligand and an enzyme (Emil Fischer) or receptor (Ehrlich, Clark), agonists of a receptor have been described as a key able to open a lock (e.g., receptor), whereas an antagonist fits the lock, but cannot open it, thereby inhibiting the real key from reaching the lock. Ariëns used the following metaphor. An agonist is a piano player, who sits on a piano stool and plays the piano; an antagonist is only able to sit on the stool, but he blocks this stool for the real piano player. We may add that there are more people who do not play the piano than there are piano players; it is indeed easier to identify antagonists of a receptor than to find new agonists; in an agonist two properties have to be present: affinity and intrinsic activity; for an antagonist the former suffices.

The lock-and-key paradigm has however been seriously challenged during recent years. Our group has contributed to these new developments, again using the histamine receptors and their ligands as examples. I am referring to such principles as spontaneously active receptors, inverse agonism versus classical antagonism. In the early 1980s the term inverse agonism had appeared in the literature; it was used to explain the opposite effects of ligands that stimulated

the so-called diazepam receptor (ion channel linked) and ligands that inactivated this system, the two classes were called agonists and inverse agonists. At the same time compounds that blocked both the agonist and the inverse agonists became available. In the 1990s, a comparable principle was shown to be operative in GPCRs. Several GPCRs show a certain level of spontaneous activity (i.e., production of the second messenger, e.g., cyclic AMP in the absence of an agonist); an agonist increases this level. Several compounds, especially many of those known as antagonists, have the opposite effect and they behave as inverse agonists.

We were able to show that most of the known H_2 antagonists are in fact inverse agonists; most of them, but not all, e.g., burimamide, are antagonists of both histamine and the inverse agonists. For both effects, blocking an agonist or an antagonist, burimamide behaved as it should be, the same level of antagonism (i.e., affinity). Later on comparable situations were observed for H_1 (all established H_1 antagonists are in fact inverse agonists) and H_3 receptors; H_4 receptors also showed a spontaneous activity.

Spontaneous activity of receptors is especially observed in artificial systems, i.e., receptors expressed in isolated cells. It reaches sometimes high levels when certain mutants are used. This spontaneous activity has been shown to be the cause of certain diseases: due to a mutation, a receptor is always 'on'; inverse agonists can switch off such systems. In pharmacological experiments inverse agonism has been shown to be operative in vivo for, e.g., H_3 receptors, using thioperamide as the inverse agonist.

The crux of the new model, the difference from the old principle, is of course that many ligands that were considered to be silent at their receptor (they blocked the lock only) are in fact active as an inverse agonist, causing the opposite effect of an agonist. Is such a mechanism important? It might very well be. Just as agonists can downregulate receptor densities, the inverse agonists have been shown to upregulate the system. Such an upregulation may have serious consequences, as it has for the downregulation, and may even be dangerous when therapy with such a compound is stopped abruptly. Indeed, true antagonists may have an advantage, especially for long-lasting treatments.

The developments around agonism versus inverse agonism constitute a perfect example of the benefit of transdisciplinary research for both parties. The new principle would however have been difficult to prove without the availability of several ligands. At the same time the principle opens new vistas for drug development. And, again, the relatively easily accessible histamine receptors have been very useful for understanding the matter. It is sometimes too easily forgotten that for mapping out biological systems the availability of biologically active compounds has been essential: atropine and adrenaline were needed to differentiate the autonomous neuron system into the parasympathetic and the sympathetic components; it has not been the reverse.

8.02.6 Looking Back

During a period of about 40 years I have been very lucky in having excellent teachers, and also in the fact that new theories and models, and especially equipment, have become available. I refer to molecular pharmacology in the beginning and later on the molecular biology. I saw the arrival of QSAR, though I later became quite skeptical about the meaning of the approach; molecular modeling using individual compounds would seem to deliver more useful information.

I have, from the beginning, also been skeptical about combinatorial synthetic methods. For me, the technological achievements were impressive, but for a true medicinal chemist it seemed to be rather a step backward, not making the required compounds, but the compounds which were possible to be synthesized; back to screening large numbers of compounds. That was also done in the nineteenth and the first part of the twentieth century.

When medicinal chemistry is considered to be a science that can produce new and better medicines, the medicinal chemist should admit that there is a need to live close to the pharmacologists. The new ideas come especially from physiology and pharmacology. Medicinal chemistry when applying its skills properly will be able to come up with new attractive compounds to be used in therapy, as soon as targets have been defined.

One new development emerging from modern pharmacology causes concern for me. Through the enormous increase of the output of gene-profiling technologies more and more mutations in important genes encoding proteins can be identified as the real cause of a certain condition; such a condition, however, in the given patient may be seen only in a relatively small group of 'comparable' patients. Individualized medicine, 'tailor-made treatment,' is what then becomes possible. Such individualized medicines are something like the opposite of blockbuster drugs. The consequence is obviously that because the costs for development of an individualized medicine will be as high as for a drug which could become a blockbuster, such tailor-made treatment will become extraordinarily expensive. It is an odd situation: here we have something like an orphan disease for which nevertheless a medicine will be developed. Who can pay for this? This question seems seldom to be asked.

It is indeed not the need for a drug that determines whether a drug will be developed; it is the need for a drug to be used by patients who can pay for it. In a way, it is just because AIDS emerged in the Western world at a relatively early

period that anti-HIV agents have been developed. If such had not been the case, if the outbreak of AIDS had been limited to Africa, it is likely that no active medicine would have become available. Who can bear such costs as around €15 000 per year for an anti-HIV treatment, the costs for 1-year treatment with a fusion protein inhibitor developed in the early years of the current century? It is a very disturbing situation that enormous amounts of money are used to reduce the health problems of only a small part of the world's population, knowing that with the same money life-threatening conditions for much larger groups could be cured. As I said earlier in this paragraph, for emerging countries it is not the science, but the political and economical will that determines whether the major health problems in such countries will be improved or not.

Would it be correct to blame the pharmaceutical industry in particular for this situation? It is certain that nobody can justify a situation where medicines that are available are not made available, just because of cost. But who has to bear the costs? Let us compare this situation with another major problem of the emerging countries: famines and the enormous stocks of food in the Western countries. Who has to make the food available? The producers, the farmers? Or would it be an international organization like the Food and Agriculture Organization or the World Health Organization? And could such a distribution system be organized for medicines as well? It is my true conviction that too often the pharmaceutical industry is blamed for the medicines issue, while for food the responsibility is seen to be society-wide.

8.02.7   My Histamine

I have often considered histamine to be my amine. It has brought much to me. The results of histamine research in physiology, pharmacology, and medicinal chemistry have throughout the years contributed much to the progress in these fields. The effective drugs from this research are very useful in allergic diseases (H_1) and in treating gastric ulcers (H_2, though the proton-pump inhibitors took over the role of H_2 blocking agents). General principles, developed by searching the histamine field, are, however, probably even a more important result of about hundred years of histamine-related investigations than the new medicines developed in this field. I am pleased that I was able to contribute during the last decades a little to a continuation of a program that started a century ago.

References

1. Ariëns, E. J.; Simonis, A. M. *Ann. NY Acad. Sci.* **1967**, *144*, 824–865.
2. Nauta, W. Th.; Rekker, R. F. In *Handbook of Experimental Pharmacology: Histamine and Anti-Histamines*; Roche, X. Y., Silva, X. Y., Eds.; Springer, Verlag: Berlin, Germany, 1978, pp 215–245.
3. Rekker, R. F. *The Hydrophopic Fragmental Constant*; Elsevier: Amsterdam, The Netherlands, 1977.
4. Zhang, M.-Q.; Leurs, R.; Timmerman, H. In *Burger's Medicinal Chemistry and Drug Discovery*; Wolff, M. E., Ed.; Wiley: New York, 1997, pp 495–559.
5. Leurs, R.; Bakker, R. A.; Timmerman, H.; de Esch, I. J. P. *Nature Drug Disc.* **2005**, *4*, 107–120.

Biography

Henk Timmerman (b. 1937) obtained his PhD in 1967 at the Vrije Universiteit Amsterdam under the guidance of Wijbe T Nauta. He worked until 1979 for the Dutch pharmaceutical company Gist-brocades, until at the end of that year he succeeded his teacher as professor of pharmacochemistry at the Vrije Universiteit Amsterdam.

His research concentrated on histamine receptors and their ligands, studying receptor mechanisms, and SARs: in his research group the interplay between the disciplines contributing to medicinal chemistry received much attention. Several of the new ligands the group developed became research tools, such as the H_2 agonist amthamine and the H_3 antagonist clobenpropit.

Henk Timmerman is (co-)author of about 500 published papers or chapters; he supervised 35 PhD students. He has been on many national and international bodies, committees, etc., having been, e.g., president of the Royal Netherlands Chemical Society (KNCV) and the European Federation of Medicinal Chemistry (EFMC). He has been honored by, e.g., the honorary membership of the KNCV and honorary degrees of the Gadjah Mada University in Yogyakarta (Indonesia) and the Medical University of Lodz (Poland). In 2002 he retired from his chair in Amsterdam.

Comprehensive Medicinal Chemistry II
ISBN (set): 0-08-044513-6

ISBN (Volume 8) 0-08-044521-7; pp. 7–15

8.03 Medicinal Chemistry as a Scientific Discipline in Industry and Academia: Personal Reflections

C R Ganellin, University College London, London, UK

8.03.1	**Introduction**	**17**
8.03.2	**SmithKline & French Laboratories**	**18**
8.03.2.1	Early Years	18
8.03.2.2	Structure–Activity Analysis	19
8.03.2.3	Dynamic Structure–Activity Analysis	19
8.03.2.4	University Collaborations	20
8.03.2.5	Diamino-Nitroethene as a Bioisostere	20
8.03.2.6	Medicinal Chemistry Summer School	21
8.03.2.7	Research Post-Cimetidine	21
8.03.3	**University College London**	**23**
8.03.3.1	University College London Medicinal Chemistry	23
8.03.3.2	Potassium Ion Channels	23
8.03.3.3	Cholecystokinin	24
8.03.3.4	Histamine H_3 Receptors	25
8.03.4	**Conclusion**	**26**
References		**26**

8.03.1 Introduction

There was probably a strong genetic determinant for me to become a chemist. My father was a chemist working for a firm of public analysts (Dr Bernard Dyer & Partners) in the City of London and one of my maternal uncles (A. Cluer) was an oil chemist. I have a cousin (Dr Brian D Ross) who is a medically qualified biochemist and Director of Magnetic Resonance Spectroscopy at Huntingdon Medical Research Institutes, Pasadena, CA, and my daughter is a pharmacist. I was born in East London and studied Chemistry at Queen Mary College, London University, receiving a PhD in organic chemistry with Professor Michael JS Dewar for research on tropylium chemistry. I started my PhD research in 1955 and the tropylium cation (**1**) had been synthesized for the first time by von E Doering and Knox the year before.[1] I found this research to be very challenging and extraordinarily exciting. For example, I discovered that I could oxidize cyclooctatetraene (C_8H_8) directly to the tropylium cation ($C_7H_7^+$).[2]

1 Tropylium cation

I was amazed to see this apparently simple loss of CH! Part of the challenge for me at the time was that I was working as an organic chemist synthesizing molecules that had inorganic properties, since tropylium existed as an ionic salt. This apparent contradiction had a profound influence in making me very aware of the relationship between chemical structure and chemical properties. There was, of course, a theoretical basis for this phenomenon, which had been analyzed and predicted by Hückel[3] in 1938 and so I was also introduced to the concept of using molecular orbitals and computational

chemistry to make explanations and predictions. This was in any case the forte of Michael Dewar, who was using theoretical chemistry to predict and explain the course of reactions in organic chemistry.[4] Calculations had to be rather simple in those days because we did not have access to electronic computers. I mention this because I believe that it removed a barrier for me as an organic chemist so that, later on, I was not afraid to collaborate with computational chemists. I believe that education is a mixture of imparting information and techniques but also of removing barriers to further learning. Furthermore, I subsequently realized that working with Michael Dewar had strongly imprinted me with an interest in chemical properties. Also there was no spoon-feeding; we were left to get on with our work with only very occasional discussion and so one learnt to develop one's own resourcefulness to overcome the many problems which PhD research posed. I do not know whether these observations are causally related to the subsequent manner in which I conducted my own research or whether it is simply a convenient post hoc rationalization.

8.03.2   SmithKline & French Laboratories

8.03.2.1   Early Years

I joined SK&F in 1958 because it happened to be near where I was living at the time in London and its local organization, although small, appeared to me to be flexible and have potential for expansion; I was also impressed by the people I would report to, namely David Jack (later to become Research Director at Allen & Hanburys and then Glaxo Research Laboratories) and Geoffrey Spickett (who eventually became Research Director at Laboratorios Almirall in Barcelona, Spain). The research, however, turned out to be unexciting. It was based upon testing compounds in animal models using very-low-throughput screening. We sought a novel antidepressant, or an analgesic, or anticonvulsant, or anti-inflammatory drug. The chemical leads were either thrown up by the screens or they were reported compounds and we synthesized analogs. For example, we made partial structures of reserpine[5] and tetrabenazine.[6] We made some interesting indene isosteres of indoles[7] and some 4-piperidones.[8] These were patented and taken as far as clinical studies as potential antidepressants, but none was successful enough for full clinical trials, let alone to reach the market. I found this to be very disappointing because of the complete lack of any biological mechanistic understanding.

There was certainly no quantitative way of relating chemical properties to the biological properties. I began to wonder whether this pharmaceutical research could ever be as exciting again as I had found the research during my PhD years. Meanwhile SK&F moved out of London to Welwyn Garden City and then underwent an internal reorganization that brought in several scientists from ICI Pharmaceuticals. In particular James Black came in like a breath of fresh air to head Pharmacology and I instantly knew that here was somebody with whom I wanted to collaborate. Black initiated several lines of research but eventually all work was concentrated (because of limited resources) on to the search for compounds to block a new type of histamine receptor, the putative H_2 receptor. This was a novel approach to controlling the secretion of gastric acid: the therapeutic aim was to treat peptic ulcer disease. This was a real challenge for the medicinal chemists because there was no chemical lead other than the chemistry of the natural transmitter, histamine, and I was given the responsibility of directing the chemistry. Much has already been published[9–11] about the work that gave rise to the prototype drug burimamide[12] (**2**) that was used to characterize histamine H_2 receptors for the first time, and to validate the pharmacology and provide proof of principle in human volunteer studies. The further development of the failed drug metiamide[13] (**3**) and the first of the modern 'blockbuster' drugs, cimetidine[14,15] (**4**), has also been well publicised. So I will, instead, discuss aspects of my philosophy as a medicinal chemist at this time.

2 Burimamide	R=H, X=CH_2, Y=S
3 Metiamide	R=CH_3, X=S, Y=S
4 Cimetidine	R=CH_3, X=S, Y=N–C≡N

8.03.2.2   Structure–Activity Analysis

Having been trained as an organic chemist I had to learn medicinal chemistry 'on the job.' I realized that there was a major problem in communicating with other disciplines that arises from the language, and concepts, which we have been taught and then take for granted. The words we use with other scientists appear to be the same but our own appreciation of them depends upon the context in which they were presented to us. The nuances and shades of meaning can differ, sometimes markedly. In my discussions with Black, I soon learnt that I could not take the words for granted or at face value, but had to get behind them to understand what he meant, that is, to appreciate what lay underneath the words and what were the ideas that he was expressing. The discussions were very stimulating and from them I learnt about the chemical issues that were of importance in pharmacology. Of course we also discussed the results from testing our compounds and what were the implications to us as chemists for projecting to the next compounds to synthesize. To the chemist, the first appreciation of this is based on pattern recognition. However, sometimes the results appeared to defy rational explanation and then Black would throw down his final challenge: "my rats can see a big difference in these structures, why can't you chemists see it?"

For me, the essence of medicinal chemistry is structure–activity analysis. We have to combine two disciplines continuously: we have to select what to make and identify how to synthesize it. These two aims are not always achievable in an ideal form and sometimes we have to compromise or reconcile contradictions. This is why it is preferable to retain the two disciplines within each medicinal chemist. Furthermore we have to understand the language and limitations that express 'activity.' This will not be news to most medicinal chemists but in my experience the major problem for chemists is to 'read' the chemistry contained in the structures of their compounds. The problem lies in the way chemistry is taught at university. As organic chemists we mainly learn about synthesis and reactions, that is, making and breaking covalent bonds. In biology, most of the productive drug–receptor interactions are noncovalent. Of course enzymes eventually react covalently but, even so, the initial interaction is noncovalent.

In relating chemical structure to biological activity, I realized that we should use properties as a bridge; that is, to relate chemical properties to the biological properties. This may appear obvious but it poses the major issue of how to appreciate the chemical properties. Most chemists think that their problem is to understand biology but it is my contention that their real problem is to understand the chemistry. This analysis leads naturally into the dual exercise of determining the chemical properties of drug molecules and of trying to discern which properties are most critical for biological activity. This becomes part of the iterative process whereby one continuously analyzes for relationships between chemical properties and biological activity, then predicts the next compounds to be made and then determines how to synthesize them. It places a different emphasis in comparison with the approach that selects compounds on the basis of synthetic availability. This does not mean that one should ignore synthetic accessibility but it does mean that the research should be property driven, not synthetically driven.

Molecular interactions between molecules are determined fundamentally by molecular size, shape, and charge distribution. Inspecting molecules from this standpoint leads to an awareness that they rarely have a unique description and that there may be several different forms or species in equilibrium.

8.03.2.3   Dynamic Structure–Activity Analysis

We know that writing a chemical structure on paper is misleading since the molecule may not be planar and the formula does not adequately represent a three-dimensional stereochemistry; also various conformers may exist in equilibrium. Furthermore, many drug molecules have ionizable protons and these too can give rise to prototropic equilibria. An interesting question arises from these considerations: if one changes drug structure to alter the equilibria, can one relate the consequences to changes in biological activity? If the answer is yes, then perhaps this may provide some insight into the chemical mechanism of drug action and give a method for drug design. These thoughts led me to formulate[16] a concept of DSAA and to apply it to histamine. That histamine may interact with its receptor as a monocation was also used as an argument to relate this property to partial agonism at histamine receptors; this led to the search for a noncationic compound.[10,11] The result gave a pure antagonist which led on to burimamide (2), the first H_2 receptor histamine antagonist.[12] Although histamine, being an imidazole derivative, is tautomeric, I was surprised to

find that nobody had published on its tautomeric properties, and so we measured the tautomeric ratio for histamine.[17] Later on the same approach of DSAA was used to develop metiamide (**3**) from burimamide.[13]

One of the problems I had experienced early on at SK&F was the difficulty of having chemical properties measured. The analytical laboratory had a background of chemical analysis to service chemical and pharmaceutical production and the resources were arranged specifically to conduct these tasks. There was simply no conception of what might be needed to investigate the open-ended questions posed by a medicinal chemistry research laboratory. So I became involved in an unedifying internal struggle to set up a physical-organic chemistry laboratory that would establish evidence for structures and purity, but that also had some spare capacity to make measurements and answer some of the questions posed by medicinal chemistry. The key was to have it report to Research Management instead of Production. Eventually we were able to do this and to put in charge an excellent organic chemist, Dr PMG Bavin, who built up a department that in the 1960s could study conformation and pK_as by nuclear magnetic resonance, could determine hydrogen-bonding by infrared spectroscopy, and could measure solvent/water partition and pK_a values.

8.03.2.4   University Collaborations

As part of our studies of histamine's properties I started collaborations with university scientists. We needed more information about properties but I did not wish to divert our resources from synthesis into these studies. At that time the UK government had instituted research grants which combined university and industrial collaboration in PhD degrees (Cooperative Awards in Science and Engineering: CASE) and I started a project with Dr Graham Richards at Oxford University for molecular orbital computations on histamine,[18,19] for both its conformational and tautomeric preferences. I also started working with Dr Keith Prout,[20,21] a crystallographer at Oxford University. The advantage of such collaborations was not only the data that they produced, but also the perspective these colleagues gave about properties; this was gained through discussion with them as chemists who viewed chemistry in a different manner, especially when it came to nonbonded interactions between atoms in molecules.

Another consequence of considering properties is that it allows one to formulate questions where the answers can be found in the literature, especially when this concerns quantitative data. This became very fruitful when seeking pK_a data and applying the Hammett equation to predict the pK_a values of novel compounds. This approach led us to identify that cyanoguanidine could be a bioisostere for thiourea and led to the synthesis of cimetidine[15] (**4**).

Meanwhile we were also seeking properties that might provide a quantitative correlation between structure and H_2 receptor antagonist activity by using multiparameter correlation analysis. Since methyl on the thioureido group (the 'polar group') increased the potency to give burimamide or metiamide, and since thioureas were much more potent than ureas (by about 20-fold) we wondered whether desolvation of the drug molecule was playing a critical role prior to its binding to the receptor. Support for this idea was obtained from an excellent correlation between activity of the whole molecule and octanol/water partition of the polar group for a series where only the structure of the polar group was changed. There were, however, certain outliers, which are discussed below.

8.03.2.5   Diamino-Nitroethene as a Bioisostere

About this time we also became interested in replacing the N of cyanoguanidine by a CH on the grounds that it might be less readily hydrated and have a higher octanol/water partition ratio and therefore may be more potent. However this leads to another problem because the resulting diamino-cyanoethene tautomerizes to a cyanoacetamidine (**Figure 1**). One of my younger colleagues, HD Prain, drew my attention to a publication describing a synthesis of diamino-nitroethene (**5**). Encouraged to make it, he synthesized the corresponding analog of cimetidine but it was only similarly active, i.e., it was not more potent and, surprisingly, the group had a much lower octanol/water partition than did the corresponding cyanoguanidine. Based upon this latter finding, the compound should have been much less active. These considerations of DSAA and pK_a analysis led us to identify the diamino-nitroethene (**5a**, $R = NO_2$) bioisosteric group,[16] which we published in a patent.[22] It was subsequently taken up by Glaxo Laboratories to produce the second marketed H_2 receptor

Figure 1 $R = CN$, the diamino-cyanoethene (**5a**) in equilibrium with the cyanoacetamidine, **5c**, via the cyanoacetamidinium conjugate acid, **5b**. $R = NO_2$, the diamino-nitroethene (**5a**) in equilibrium with the nitroacetamidine, **5c**, via the nitroacetamidinium conjugate acid, **5b**.

antagonist 'blockbuster' drug, ranitidine[23] (**6**), since the Glaxo researchers found that replacing cyanoguanidine by diamino-nitroethene in their furan series unexpectedly increased potency by 10-fold.

6 Ranitidine

Outliers from a linear correlation can be extremely valuable because they can indicate that another chemical property is affecting the biological activity. In the above case, a cyclic analog was much less potent, possibly due to tautomerism into another structural form. On the other hand, the diamino-nitroethene was much more potent than predicted (see **Figure 1** in [16]). This led us to investigate quantitatively the polarity of the polar group and we collaborated with Professor Ted Grant of Queen Elizabeth College, London University, who measured the dipole moments[24] in water, but this did not provide a correlation. Now, dipoles are also directional, and so we used molecular orbital calculations to predict the dipole vector; this latter gave an excellent correlation which depended on an optimum orientation of the dipole together with the octanol/water partition value.[25] Intuitively this sounds very sensible, i.e., that an interaction of a very polar molecule should require an appropriate orientation, but it is rather unusual to be able to dissect it out and demonstrate a relationship. It would be very interesting to do this for other series of H_2 antagonists where the ring structure–activity relationship follows a different pattern from the imidazoles, as occurs for the furan series with ranitidine.[26]

8.03.2.6   Medicinal Chemistry Summer School

My interest in medicinal chemistry as a scientific discipline also led me to become involved with the Royal Society of Chemistry (RSC) summer school. My colleague, Dr AM Roe, was Chairman of the RSC Education Committee which advised the RSC on the various courses mounted for postgraduate chemists on specialist topics. These were usually residential and each lasted for approximately 1 week. One such course had started as a week's summer school in medicinal chemistry but it had not attracted enough chemists from the pharmaceutical industry, possibly because it appeared to be overly dependent on techniques of structure determination and chemical analysis. Anthony Roe invited me to think up a suitable syllabus.

Our aim was to provide a rapid and concentrated conversion course for recently hired postdoctoral research chemists in the pharmaceutical industry; in the main these were organic chemists who needed to know what was required to become a practicing medicinal chemist. My past experience had shown me that such courses usually dealt with diseases and their test models but I decided to avoid this approach. It seemed to me that the basic discipline for medicinal chemists was to understand structure–activity analysis and the interface with the other disciplines involved in drug discovery. Thus the core lectures would be on physicochemical properties (octanol/water partition, pK_a and hydrogen-bonding, conformational analysis), multiparameter correlation analysis and computation, bioassay, receptors and enzymes, drug disposition (DMPK, drug metabolism and pharmacokinetics) and the drug development process. We would also include several case histories of drug discovery (something which I had previously encountered in a Society for Drug Research symposium). Lecturers were mainly industrial, and the number of participants was limited to around 100 to foster a more intimate and informal atmosphere.

The first summer school of this type was mounted in 1981 and the result was a resounding success. It has since been repeated every other year, and is always oversubscribed, with a healthy participation of delegates from continental Europe. It has received further accolade by providing the model for the annual course put on in the USA since 1987 at Drew University, Madison, NJ. It has also been publicised within the International Union of Pure and Applied Chemistry (IUPAC).[27]

8.03.2.7   Research Post-Cimetidine

Our aims at SK&F after cimetidine had been to make improved compounds for inhibiting gastric acid secretion. I was appointed Director of Histamine Research and the aims were broadened to explore other potential therapeutic applications of histamine H_2 receptor ligands, as in inflammation, the central nervous system, and immunology.

The vasculature contains both H_1 and H_2 receptors and it is important to block both simultaneously if the aim is to suppress the histamine-induced inflammation (vasodilatation and edema) in the skin. To this end we set out to combine H_1 and H_2 antagonist properties in the same ligand; by using our discovery of pyridine analogs of imidazole and cyclized

versions of acylguanidines which gave 2-amino-4-pyrimidones (isocytosines), we were able to use appropriate substituents in the pyridine and isocytosine rings to provide a compound, icotidine (**7**), which was equipotent as an antagonist at both H_1 and H_2 receptors.[28] Unfortunately it ran into problems during safety studies in animals and was discontinued.

An unexpected finding with icotidine, however, was that it had no propensity to enter the brain. This gave us an opportunity to use molecular modification to engineer out the H_2 antagonist property and increase the H_1 receptor potency. Again, this depended on the appropriate use of substituents; temelastine (**8**) went into development as an H_1 antihistamine that did not penetrate into the brain and did not cause the usual drowsiness associated with conventional antihistamines.[29] Although the compound was very effective in rats and guinea-pigs it was, unfortunately, too short-lived for human studies and was abandoned.[30] Temelastine has a very unusual structure for an antihistamine since it is not a basic tertiary amine and the main part of its structure is very similar to those of some H_2 receptor antagonists. Although the H_1 and H_2 receptors have only 40% amino acid sequence homology[31] for the transmembrane domains, it is fascinating to find potent antagonists that are so similar in chemical structure.

| | | | pA_2 | |
			H_1	H_2
7	Icotidine	$R^3=OCH_3$ $R^5=H$	7.8	7.5
8	Temelastine	$R^3=CH_3$ $R^5=Br$	9.55	c. 5.9

Almost all of the potent H_2 receptor antagonists are very polar molecules which do not penetrate into the brain. Since there are H_2 receptors in the brain, we decided to seek an H_2 receptor antagonist that would penetrate. This led us to study a series of compounds by radiolabeling them and determining brain penetration The compounds were carefully selected and it became apparent that having a high octanol/water partition value (log P) was no guarantee of brain penetration. Even H_2 antagonist compounds with log $P > 4$ did not penetrate. This was surprising and ran counter to the wisdom which Corwin Hansch had propounded.[32,33] My colleagues Drs RC Mitchell and RC Young were able to show that hydrogen-bonding was critical in reducing brain penetration and they obtained a very interesting inverse correlation between the difference between octanol/water and cyclohexane/water partitions and brain penetration – the so-called Δ log P hypothesis. Using this hypothesis we were able to design zolantidine (**9**), where we had reduced the donor hydrogen-bonding capability (small Δ log P) so that the compound became a very effective brain-penetrating H_2 receptor antagonist.[34] Sadly, this interesting compound could not be pursued into human studies because of the management decision to abandon this project.

9 Zolantidine

The results from immunological studies of possible involvement of H_2 receptors were rather difficult to pin down. We now know that there are both H_2 and H_4 receptor components involved in the immunological actions of histamine, for example in interleukin-16 release from human T lymphocytes,[35] and it is probable that the effects of the H_4 receptor dominate.

About this time it was my feeling that management pressure was too intense and so I followed up an offer that I had received from Sir James Lighthill, the Provost of University College London (UCL), to take a Chair of Medicinal Chemistry in the Chemistry Department. There was not an adequate financial backing for this position and subsequent negotiations led to the establishment of the SK&F Chair of Medicinal Chemistry. I took this up in 1986. This turned out to be a remarkable year for me personally, since I had been awarded the DSc degree from London University for my studies on the medicinal chemistry of histamine, and had also been admitted as a Fellow of the Royal Society (the UK National Academy of Science).

8.03.3   University College London

8.03.3.1   University College London Medicinal Chemistry

It is not that usual to go from industry to academia and I joined UCL at a difficult time since there was considerable financial pressure on the universities in general. The pressures on the Chemistry Department had led to its contraction from 35 academic (faculty) staff to 21. The Department no longer funded research and the majority of grant applications to the UK government-funded Science Councils were refused for lack of funds, even though the Council may have deemed them worthy of support. If technical staff left the Department they were not replaced. The infrastructure of the building was also slowly deteriorating. This was the result of Prime Minister Margaret Thatcher's policy and the consequential decline in the support for physical sciences in British universities. Yet the UCL Chemistry Department increased the number of students and the number of publications, although at a cost to the academic staff: they had to spend more time teaching and as a result they tended to do 'safe' research. Furthermore their pay was continuously eroded as it did not keep up with increases in the cost of living.

The Department was unusual in that it provided an undergraduate education in medicinal chemistry, which had been initiated by Professor James Black (when Head of the UCL Pharmacology Department), together with Professor Charles Vernon, a biological chemist in the Chemistry Department. It sacrificed much of the usual inorganic chemistry teaching from the BSc Chemistry degree and replaced it by units (or half-units) in physiology, pharmacology, and biochemistry during the 3-year period. In the final year, the students attended a half-unit on the principles of drug design. I injected a healthy amount of the physical-organic basis for structure–activity analysis into this course and also invited industrial speakers to give case histories of drug discovery (leading to about 30–35 lectures). The students also did a 3-month practical research project for which they wrote a thesis and sat an oral examination.

My appointment to a Chair in Medicinal Chemistry was meant to give a bigger research presence to the subject. Yet I arrived with no research funding and no equipment or glassware. I also knew that the Science Research Council was not interested in funding medicinal chemistry, first because one did not develop novel organic synthesis methodology and second, because the Council took the view that this is what the pharmaceutical industry did. I also faced another problem. It is usual that an academic is appointed to a Chair after having produced a good volume of interesting research and having generated an accelerating research momentum, which will be expanded on taking up the professorship. In my case I had to leave all my research behind to SK&F and start again from scratch. I decided not to choose projects that might be in competition with work that was conducted in the pharmaceutical industry. My approach was to collaborate with biologists to help them solve a biological problem by providing appropriate compounds; my hope was that this might lead to new areas for drug discovery by producing useful chemical substances as tools which could ultimately give prototype drugs. In essence this was ligand design for new areas of biology. Critically important in an academic setting was whether the project could be funded.

The work at UCL encompassed a wide range of biological applications, from G protein-coupled receptors (for histamine and serotonin), enzyme inhibitors for cholecystokinin-inactivating peptidase and human immunodeficiency virus (HIV)-aspartyl peptidase, potassium ion channels, through to phosphatidyl inositol transfer protein, transport P, and persistent sunscreens. This involved collaborations with various biologists but two, in particular, stood out for their excellent science and manner of working with chemists.

8.03.3.2   Potassium Ion Channels

At UCL, Professor DH Jenkinson in the Pharmacology Department had strong interests in calcium-activated potassium ion channels. We were fortunate to obtain a 5-year grant from the Wellcome Trust to fund both medicinal chemistry and pharmacology (electrophysiology). In chemistry this provided the seed that we were able to grow by accommodating project students and academic visitors over many years.

The small-conductance Ca^{2+}-activated K^+ channel (SK_{Ca}) is found in many cell types and was originally defined electrophysiologically using a natural peptide toxin apamin. To find a simpler molecule, the drug dequalinium was taken as a μmol L^{-1} lead. Since dequalinium is a 4-aminoquinoline, the influence of the amino group was investigated in a small series of substituted analogs and an excellent correlation was obtained between blocking potency and the σ_R substituent constant.[36] This was extended to a much larger series in which activity was correlated with the energy of the Lowest Unoccupied Molecular Orbital (LUMO).[37] The effects of conformational restriction in the linking chain were also investigated[38] and then the dequalinium analogs were cyclized to give tetra-aza-cyclophanes that were particularly interesting. Thus, UCL 1530 (**10**) provided the first evidence[39] for pharmacological differentiation between the SK_{Ca} channels in liver and neuronal cells, while UCL 1684 (**11**) was the first[40] nonpeptidic nanomolar inhibitor ($IC_{50} = 3$ nmol L^{-1}), and this was followed[41] by UCL 1848 (**12**).

10	UCL 1530	$A= -(CH_2)_{10}$	$A= -(CH_2)_{10}$
11	UCL 1684	$A= -CH_2-\langle\text{benzene}\rangle-CH_2-$	$A= -(CH_2)_{10}$
12	UCL 1848	$A= -(CH_2)_{10}$	$A= -(CH_2)_{10}$

8.03.3.3   Cholecystokinin

My other highly productive collaboration was with Professor Jean-Charles Schwartz, Director of an Institut National de la Santé et de la Recherche Médicale (INSERM) Unit in Paris, France. I had previously had a strong collaboration with him on histamine research when I was at SK&F. We were very lucky to share a large grant from the Upjohn Company (Kalamazoo, MI, US) for a proposal to discover an inhibitor of the peptidase that inactivates the octapeptide neurotransmitter, cholecystokinin-8 (CCK-8). It was hypothesized that such a compound would prolong the natural lifetime of CCK-8 and promote a feeling of satiety, thereby reducing food intake in a natural way.

The enzyme had not been fully purified but its activity was isolated from rat brain in Schwartz's laboratory and characterized as a serine proteinase. Compounds were to be synthesized at UCL and tested at INSERM. Our approach was to avoid incorporating a serine-reactive group (for a transition state or irreversible inhibitor) but to seek a reversible inhibitor since this should be more likely to be selective and nontoxic. To do this we used molecular probes to seek noncovalent molecular interactions with the enzyme active site; the aim was to achieve closely matched stereospecific interactions between the enzyme and the putative inhibitor. The strategy was first to characterize the binding opportunities of the enzyme subsites using a series of systematically varied dipeptides and tripeptides by screening commercially available compounds supplemented by some which we synthesized.[42] Peptides were selected from those with alkyl or aryl side chains to determine the accessible volume for binding and to probe the potential for hydrophobic interactions. Dipeptides were also derivatized at the NH_2 or CO_2H termini.

From the above work there emerged a submicromolar dipeptide amide (P_1P_2NHR) as a lead. The side chains of the amino acids (in $P_1P_2NH_2$) were then optimized with respect to activity by synthesizing and testing analogs as primary amides in which the two amino acids were systematically varied to afford Abu.Pro.NHR (R = H), then R was optimized. Fusion of a benzene ring to Pro gave an indoline derivative, butabindide (**13**), a prototype drug which is a selective competitive reversible nanomolar inhibitor ($K_i = 7\,\text{nmol L}^{-1}$) that does not contain a serine-reactive group.[43] This compound was shown to be active in potentiating the action of CCK-8 and to reduce food intake (as a result of the satiating effect of CCK-8) in starved mice. Analogs of butabindide then yielded potent subnanomolar inhibitors.

13 Butabindide

As a result of the above work, the identity of the proteinase that inactivates CCK-8 was shown to be tripeptidylpeptidase II (TPP II), a known enzyme of previously unknown function. Thus, searching for the noncovalent interactions around the enzyme active site and exploiting hydrophobic effects led to a potent, reversible

competitive and selective peptidase inhibitor, the first known inhibitor of TPP II. This strategy has the potential to provide a general approach to the design of peptidase inhibitors provided that the enzyme possesses an accessible lipophilic pocket, even though the structure of the enzyme may be unknown.

8.03.3.4 Histamine H_3 Receptors

I also collaborated with Professor Schwartz and his colleagues on designing ligands for the histamine H_3 receptor. For many years all the known appropriately active ligands were imidazole derivatives. For several reasons a nonimidazole was preferred but all attempts to replace imidazole by other heterocycles led only to inactive or weakly active compounds. We therefore went back to first principles and applied thoughts that had been proposed by EJ Ariens[44,45] in the 1960s.

It is possible to convert an agonist into an antagonist by introducing additional groups into the molecule which can locate binding sites in the vicinity of the receptor. Whether the resulting molecule will be a partial agonist or a pure antagonist probably depends on whether the agonist moieties continue to engage the receptor in the critical manner required to elicit a receptor response. If they do not, then the molecule will be an antagonist and one may question whether the agonist moieties actually make any useful contribution to the affinity. If the additional groups are correctly positioned and interact appropriately with the receptor, the resultant molecule should achieve a considerable increase in affinity.

For histamine, the thought arose that it might be possible to convert histamine into an antagonist by the addition of appropriate groups, and then to remove the imidazole ring to yield a nonimidazole antagonist molecule. It therefore seemed to be worthwhile applying this analysis to the interaction of histamine at the H_3 receptor. The difficulty of the approach resides in finding out what may be appropriate groups to incorporate into the histamine molecule and in which positions they should be introduced to achieve a sufficient increase in affinity.

Of various attempts made, the one that appeared to hold promise was the finding that N^α-(4-phenylbutyl)histamine (14) was a pure antagonist of histamine at the H_3 receptor with a $K_i = 0.63 \,\mu\text{mol L}^{-1}$. Removal of the imidazole ring from this structure led to the synthesis and testing of N-ethyl-N-(4-phenylbutyl)amine (15) which, remarkably, was found to have a $K_i = 1.3 \,\mu\text{mol L}^{-1}$ as an H_3 receptor histamine antagonist. The removal of the imidazole ring had led merely to a twofold drop in affinity and had successfully produced the necessary lead to generate a nonimidazole H_3 receptor histamine antagonist. Inserting an O or S atom in the chain at the position α to the phenyl ring simplified the synthesis for a structure–activity exploration.[46] Investigating the effect of substituents in the phenyl ring, and altering the chain length and the type of amino group led to the very potent antagonist,[47,48] UCL 2173, N-(3-p-acetylphenoxy-propyl)-$trans$-3,5-dimethyl-piperidine (16), $K_i = 1.8 \,\text{nmol L}^{-1}$, $\text{ED}_{50} = 0.12 \,\text{mg kg}^{-1}$, which, in vivo, is considerably more potent than the reference drug, thioperamide. These discoveries were made before the availability of the human recombinant receptor. However, following the cloning[49] of the human H_3 receptor cDNA in 1999, many pharmaceutical companies set up high-throughput screens to seek other nonimidazole H_3 receptor antagonists and several such compounds have since entered the drug development process.[48]

14 N^α-(4-phenylbutyl)histamine

15 N-ethyl-N-(4-phenylbutyl)amine

16 UCL 2173 (N-3-p-acetylphenoxy-propyl)
-trans-3, 5-dimethyl-piperidine)

8.03.4   Conclusion

This has been a personal account focusing on the approaches I have taken in attempting to discover new potential drugs as medicines. The success rate has been extraordinarily low, but at the very least, one has tried to construct molecules having specific biological properties that may serve as tools to help unravel physiological mechanisms. The key to success has been to collaborate with outstanding biologists and outstanding chemists.

Along the way, one has also aimed at helping to develop the scientific discipline of medicinal chemistry and to inspire others also to enjoy research. I cannot help wondering, though, on how things would have turned out if James Black had not come to SK&F and if I had not been involved in the discovery of cimetidine. Would I have found other opportunities for new drug design, or would I have retreated from medicinal chemistry and gone back to researching problems in organic chemistry?

Research has been very stimulating and it has sometimes been very exciting; it has been taxing and occasionally it has generated strong emotions. Of one thing I am sure: I have been one of the fortunate few who has been paid to work on a hobby. The real bonus has been to be involved in a discovery (cimetidine) that helped millions of people fight their disease.

Would it have been like that now, or could it be like that in the future? Nowadays there seems to be so much pressure on chemists in the pharmaceutical industry. They appear to be ruled by technologies such as high-throughput screening for lead generation, parallel synthesis to develop their structure–activity relationship database, and ready-made computer programs to assist structure–activity analysis. Will they take time to stand back and think as scientists, or will they be regarded as technicians carrying out instructions? Medicinal chemists will have to be very careful in the future not to let the technologies dominate them.

References

1. Doering, W. von E.; Knox, L. H. *J. Am. Chem. Soc.* **1954**, *76*, 3203–3206.
2. Ganellin, C. R., Pettit, R. *J. Chem. Soc.* **1958**, 576–581.
3. Hückel, E. *Grunzuge der Theorie ungesattigter und aromatische Verbindung*; Verlag Chemie: Berlin, 1938, pp 77–85.
4. Dewar, M. J. S. *The Electronic Theory of Organic Chemistry*; OUP: London, UK, 1949.
5. Woodward, R. B.; Bader, F. E.; Bickel, H.; Frey, A. J.; Kierstead, R. W. *J. Am. Chem. Soc.* **1956**, *78*, 2023–2025.
6. Brossi, A.; Lindlar, H.; Walter, M.; Schnider, O. *Helv. Chim. Acta* **1958**, *41*, 119–139.
7. Ganellin, C. R.; Jack D.; Spickett, R. G. W. New Benzylindene Derivatives and Method of Preparing the Same. British Patent 953,194 (1964) and 959,704 (1964).
8. Ganellin C. R.; Spickett, R. G. W. New Piperidine Derivatives and Processes for Preparing the Same. British Patent 861,862 (1962) and 960,895 (1964).
9. Cooper, D. G.; Young, R. C.; Durant, G. J.; Ganellin, C. R. Histamine Receptors. In *Comprehensive Medicinal Chemistry*; Emmett, J. C., Ed.; Pergamon Press: Oxford, UK, 1990; Vol. 3, pp 323–421.
10. Ganellin, C. R.; Durant, G. J.; Emmett, J. C. *Fed. Proc. Fed. Am. Soc. Exp. Biol.* **1976**, *85*, 1924–1930.
11. Ganellin, C. R. *J. Appl. Chem. Biotechnol.* **1978**, *28*, 183–200.
12. Black, J. W.; Duncan, W. A. M.; Durant, G. J.; Ganellin, C. R.; Parsons, M. E. *Nature (Lond.)* **1972**, *236*, 385–390.
13. Black, J. W.; Durant, G. J.; Emmett, J. C.; Ganellin, C. R. *Nature (Lond.)* **1974**, *248*, 65–67.
14. Brimblecombe, R. W.; Duncan, W. A. M.; Durant, G. J.; Emmett, J. C.; Ganellin, C. R.; Parsons, M. E. *J. Int. Med. Res.* **1975**, *3*, 86–92.
15. Durant, G. J.; Emmett, J. C.; Ganellin, C. R.; Miles, P. D.; Prain, H. D.; Parsons, M. E.; White, G. R. *J. Med. Chem.* **1977**, *20*, 901–906.
16. Ganellin, R. *J. Med. Chem.* **1981**, *24*, 913–920.
17. Ganellin, C. R. *J. Pharm. Pharmacol.* **1973**, *25*, 787–792.
18. Ganellin, C. R.; Pepper, E. S.; Port, G. N. J.; Richards, W. G. *J. Med. Chem.* **1978**, *16*, 610–616.
19. Farnell, L.; Richards, W. G.; Ganellin, C. R. *J. Med. Chem.* **1975**, *18*, 662–666.
20. Prout, K.; Critchley, S. R.; Ganellin, C. R. *Acta Crystallogr. Sect. B* **1974**, *30*, 2884–2886.
21. Prout, K.; Critchley, S. R.; Ganellin, C. R.; Mitchell, R. C. *J. Chem. Soc., Perkin Trans.* **1977**, *2*, 68–75.
22. Durant, G. J.; Emmett, J. C.; Ganellin C. R.; Prain, H. D. Heterocyclic Substituted 1,1-Diamino-Ethylene Derivatives. British Patent 1,421,792 (1976).
23. Bradshaw, J.; Brittain, R. T.; Clitherow, J. W.; Daly, M. J.; Jack, D.; Price, B. J.; Stables, R. *Br. J. Pharmacol.* **1979**, *66*, 464P.
24. Young, R. C.; Ganellin, C. R.; Graham, M. J.; Grant, E. H. *Tetrahedron* **1982**, *38*, 1493–1497.
25. Young, R. C.; Durant, G. J.; Emmett, J. C.; Ganellin, C. R.; Graham, M. J.; Mitchell, R. C.; Prain, H. D.; Roantree, M. L. *J. Med. Chem.* **1986**, *29*, 44–49.
26. Daly, M. J.; Price, B. J. Ranitidine and Other H2-Receptor Antagonists: Recent Developments. In *Progress in Medicinal Chemistry*; Ellis, G. P., West, G. B., Eds.; Elsevier: Amsterdam, 1983; Vol. 20, pp 337–368.
27. Ganellin, C. R. *Chem. Int.* **1995**, *17*, 212–214.
28. Ganellin, C. R.; Blakemore, R. C.; Brown, T. H.; Cooper, D. G.; Durant, G. J.; Harvey, C. A.; Ife, R. J.; Owen, D. A. A.; Parsons, M. E.; Rasmussen, A. C.; Sach, G. S. *N. Engl. Soc. Allergy Proc.* **1986**, *7*, 126–133.
29. Cooper, D. G.; Durant, G. J.; Ganellin, C. R.; Harvey, C. A.; Meeson, M. L.; Owen, D. A. A.; Sach, G. S.; Wilczynska, M. A. Structure–Activity Studies in the Design of the Selective H1-Receptor Histamine Antagonist, SK&F 93944. In *Proceedings of the 8th International Symposium on Medicinal Chemistry*; Dahlbom, R., Nilsson, J. L. G., Eds.; Swedish Pharmaceutical Press: Stockholm, 1985; Vol. 2, pp 198–199.
30. McCall, M. K.; Nair, N.; Wong, S.; Townley, R. G.; Lang D.; Weiss, S. J. *J. Allergy Clin. Immunol.* **1985**, *75*, abstract 256.
31. Yamashita, M.; Fukui, H.; Sugama, K.; Horio, Y.; Ito, S.; Mizuguchi, H.; Wada, H. *Proc. Natl. Acad. Sci. USA* **1991**, *88*, 11515–11519.

32. Glave, W. R.; Hansch, C. *J. Pharm. Sci.* **1972**, *61*, 589.
33. Hansch, C.; Bjorkroth, J. P.; Leo, A. *J. Pharm. Sci.* **1987**, *76*, 663.
34. Young, R. C.; Mitchell, R. C.; Brown, T. H.; Ganellin, C. R.; Griffiths, R.; Jones, M.; Rana, K. K.; Saunders, D.; Smith, I. R.; Sore, N. E. et al. *J. Med. Chem.* **1988**, *31*, 656–671.
35. Gantner, F.; Sakai, K.; Tusche, M. W.; Cruikshank, W. W.; Center, D. M.; Bacon, K. B. *J. Pharmacol. Exp. Ther.* **2002**, *303*, 300–307.
36. Galanakis, D.; Davis, C. A.; Del Rey Herrero, B.; Ganellin, C. R.; Dunn, P. M.; Jenkinson, D. H. *BioMed. Chem. Lett.* **1995**, *5*, 559–562.
37. Galanakis, D.; Davis, C. A.; Ganellin, C. R.; Dunn, P. M. *J. Med. Chem.* **1996**, *39*, 359–370.
38. Campos Rosa, J.; Galanakis, D.; Ganellin, C. R.; Dunn, P. M. *J. Med. Chem.* **1996**, *39*, 4247–4254.
39. Campos Rosa, J.; Beckwith-Hall, B. M.; Galanakis, D.; Ganellin, C. R.; Dunn, P. M.; Jenkinson, D. H. *BioMed. Chem. Lett.* **1997**, *7*, 7–10.
40. Campos Rosa, J.; Galanakis, D.; Piergentili, A.; Bhandari, K.; Ganellin, C. R.; Dunn, P. M.; Jenkinson, D. H. *J. Med. Chem.* **2000**, *43*, 420–431.
41. Chen, J.-Q.; Galanakis, D.; Ganellin, C. R.; Dunn, P. M.; Jenkinson, D. H. *J. Med. Chem.* **2000**, *43*, 3478–3481.
42. Ganellin, C. R.; Bishop, P. B.; Bambal, R. B.; Chan, S. M. T.; Law, J. K.; Marabout, B.; Mehta Luthra, P.; Moore, A. N. J.; Peschard, O.; Bourgeat, P. et al. *J. Med. Chem.* **2000**, *43*, 664–674.
43. Rose, C.; Vargas, F.; Facchinetti, P.; Bourgeat, P.; Bambal, R. B.; Bishop, P. B.; Chan, S. M. T.; Moore, A. N. J.; Ganellin, C. R.; Schwartz, J. C. *Nature (Lond.)* **1996**, *380*, 403–409.
44. Ariens, E. J.; Simonis, A. M. *Arch. Int. Pharmacodyn. Ther.* **1960**, *127*, 479.
45. Ariens, E. J.; Simonis, A. M.; van Rossum, J. M. One Receptor System. In *Molecular Pharmacology*; Ariens, E. J., Ed.; Academic Press: New York, 1964; Vol. 1, pp 212–226.
46. Ganellin, C. R.; Leurquin, F.; Piripitsi, A.; Arrang, J.-M.; Garbarg, M.; Ligneau, X.; Schunack, W.; Schwartz, J.-C. *Arch. Pharm. Pharm. Med. Chem.* **1998**, *331*, 395–404.
47. Schwartz, J.-C.; Arrang, J.-M.; Garbarg, M.; Lecomte, J.-M.; Ligneau, X.; Schunack, W.; Stark, H.; Ganellin, C. R.; Leurquin, F.; Sigurd, E. Non-Imidazole Alkylamines as Histamine H$_3$ Receptor Ligands and their Therapeutic Applications. European Patent 1,100,503 (2004), example 67.
48. Cowart, M.; Altenbach, R.; Black, L.; Faghih, R.; Zhao, C.; Hancock, A. A. *Mini-Revs Med. Chem.* **2004**, *4*, 979–992.
49. Lovenberg, T. W.; Roland, B. L.; Wilson, S. J.; Jiang, X.; Pyati, J.; Huvar, A.; Jackson, M. R.; Erlander, M. G. *Mol. Pharmacol.* **1999**, *55*, 1101–1107.

Biography

C Robin Ganellin studied chemistry at Queen Mary College, London University (BSc and PhD) and in 1960 was a Research Associate with Prof A C Cope, at the Massachusetts Institute of Technology (MIT). Then he joined SmithKline & French Laboratories (SK&F) in the UK, and from 1966 collaborated with Sir James Black, to lead the chemical research for the discovery of the H$_2$-receptor histamine antagonists. He is coinventor of the drug cimetidine (Tagamet) which revolutionised the treatment of peptic ulcer disease. He became Director of Histamine Research and, subsequently, Vice-President for Research, at SK&F Welwyn, UK. In 1986 he was elected as a Fellow of the Royal Society and appointed to the SK&F Chair of Medicinal Chemistry at University College London, where he is now Emeritus Professor. He is an author of some 250 scientific publications and named coinventor on over 160 US patents. He has received various awards for medicinal chemistry. He is a past Chairman of the UK Society for Drug Research (1985–87). He was a member of the IUPHAR Committee for Receptor Nomenclature and Drug Classification (1990–98), was President of the Medicinal Chemistry Section of IUPAC (1999–2001), and is currently Chairman of the IUPAC Subcommittee on Medicinal Chemistry and Drug Development.

Comprehensive Medicinal Chemistry II
ISBN (set): 0-08-044513-6

ISBN (Volume 8) 0-08-044521-7; pp. 17–27

8.04 Some Aspects of Medicinal Chemistry at the Schering-Plough Research Institute

A K Ganguly, Stevens Institute of Technology, Hoboken, NJ, USA

8.04.1	**Introduction**	**29**
8.04.2	**The Discovery of Ezetimibe**	**30**
8.04.2.1	The Synthesis and Biological Activities of Initial Leads	31
8.04.2.2	The Discovery of SCH 48461	31
8.04.2.3	The In Vivo Activity of SCH 48461	32
8.04.2.4	The Design of Ezetimibe	33
8.04.2.5	The Synthesis of Ezetimibe	34
8.04.2.6	The Biological Activity of Ezetimibe	34
8.04.2.7	The Results of Clinical Trials of Ezetimibe	35
8.04.2.7.1	The mechanism of action of ezetimibe	35
8.04.3	**The Discovery of Posaconazole**	**36**
8.04.3.1	The Synthesis of Initial Leads	37
8.04.3.2	The Synthesis and Antifungal Activities of SCH 45012 and Its Isomers	38
8.04.3.3	The Improved Synthesis of SCH 51048	41
8.04.3.4	The Design of Posaconazole	41
8.04.3.5	The Synthesis of Posaconazole	41
8.04.3.6	The Antifungal Activity of Posaconazole	42
8.04.4	**The Discovery of Lonafarnib**	**42**
8.04.4.1	Overview	42
8.04.4.2	Screening of the SPRI Library of Compounds to Discover Initial Leads	43
8.04.4.3	Structure–Activity Relationships and the Discovery of SCH 44342	43
8.04.4.4	X-ray Crystallography Results and the Design of Future Analogs	45
8.04.4.5	Lonafarnib and Its Biological Activity	46
8.04.4.6	Synthesis of Lonafarnib	47
8.04.5	**Summary**	**49**
References		**49**

8.04.1 Introduction

Since the discovery of penicillins during the Second World War, pharmaceutical companies around the world have made spectacular contributions toward curing many diseases. In the area of infectious diseases, the introduction of newer antibiotics has saved many lives, although challenges lie ahead because of the emergence of bacterial resistance to these antibiotics. Similarly, the use of protease inhibitors has helped millions of AIDS patients around the globe, and in this area also there is ongoing research to discover better drugs. A few CCR5 antagonists are in clinical trials, and they are expected to inhibit viral entry into the cells. Hepatitis C infection is being treated with interferon and ribavarin. Interferon was the first biotechnology-derived product to be introduced in human medicine. The work in biotechnology started in earnest in the 1980s, with some in the industry not completely convinced of its importance. Today, biotechnology-based companies are making major contributions to medicine, in areas as diverse as growth factors, arthritis, diabetes, and vaccines.

The introduction of enzyme inhibitors such as angiotensin-converting enzyme (ACE) inhibitors and statins has revolutionized the management of cardiovascular diseases, and the general concept of drug discovery using enzyme inhibitions has been used for drugs in other disease conditions.

Receptor antagonists, for example H_1 antagonists, have been in use in the clinic for a long period of time, and, more recently, using the knowledge of G protein-coupled receptors, several new drugs have reached the clinic, and many more exciting drugs in this category are in development.

Besides infectious and cardiovascular diseases, cancer is one of the major causes of death in the world. Cytotoxic agents, including paclitaxel and temzolomide, have been in use for a number of years, and the longevity of cancer patients has greatly improved. Recently, understanding how kinases work intracellularly, several pharmaceutical companies have discovered novel anticancer agents. Imatinib is already in the clinic, and others are in various stages of development.

No attempt has been made here to present a catalog of drugs synthesized; instead, an attempt has been made to very briefly capture the trend in drug discovery.

Several new technologies have been introduced in the drug discovery process since the start of the 1990s. Amongst these, combinatorial synthesis, genomics, proteomics, and high-throughput assays must be highlighted. Many articles and reviews have been written on these subjects. It is our expectation that, with time, one would learn how to use these technologies to the fullest extent, which is already happening in the pharmaceutical industry, and, as a result, many new drugs involving these new technologies are expected to reach clinical trials soon. There has been a lot of unreal expectation that the use of these technologies would shorten the time required to discover drugs, and, therefore, there has been disappointment in some corners.

With the advent of biotechnology, the cloning and purification of receptors and enzymes has almost become a routine in the industry. Biologist are able today to establish an in vitro assay in a short period of time, and, with the help of high-throughput assays, an active lead is generally – but not always – found in the collection of compounds in the corporate libraries. It is not an uncommon experience in the industry to find a lead structure from their past collections, because the compounds in the file were analogs of biologically active compounds. It has therefore become very important to have good-quality compounds in the library, and, here, chemists have a challenge and responsibility to cleverly use combinatorial chemistry and synthetic organic chemistry to achieve this goal.

In recent years, the number of structures of proteins solved by x-ray analysis has vastly increased. In some instances this information has greatly helped in the drug discovery process. In an ideal world, the x-ray crystal structure of a protein of interest bound to a ligand would be available prior to chemists starting their work. This situation, however, is rarely the case, and certainly nonexistent in receptor-based drug design.

This perspective article is based on my Hershberg award lecture (American Chemical Society, 2003) and also on several reviews and articles written by my colleagues. I have attempted to capture different journeys taken toward the discovery of ezetimibe (Zetia), posaconazole (Noxafil), and lonafarnib (Sarasar). A common theme among these drugs is that they are all used in curing diseases where previously there was an unmet need for a pharmaceutical. The work on ezetimibe began with finding an acyl coenzyme A-cholesterol O-acyltransferase (ACAT) inhibitor, and ended with the discovery of a compound with potent activity in inhibiting the absorption of cholesterol. The mechanism of its action, which is different to ACAT inhibition, was discovered after the drug was approved by the US Food and Drugs Administration (FDA) for its use in the clinic. Vytorin, a combination of ezetimibe and simvastatin, jointly developed by Schering-Plough and Merck, has also been approved by FDA. Posaconazole is a novel azole antifungal that has demonstrated broad-spectrum activity in the clinic, particularly against *Aspergillus*, infection which is difficult to cure with existing drugs. *Aspergillus* infection is common amongst AIDS and cancer patients. It is hoped that posaconazole will soon be approved by the FDA for use in humans. In the case of lonafarnib, the discovery of an anticancer agent was based on the inhibition of farnesyl protein transferase (FPT), an important enzyme involved in the posttranslational modification of Ras that is required for the protein to attach to the cell membrane. Mutation of Ras has been found in a significant number of human tumors. During our work, the x-ray crystal structure of FPT bound to an initial lead compound was available, which allowed us to rationally design and synthesize lonafarnib. Lonafarnib is presently undergoing clinical trial against several cancer targets.

8.04.2   The Discovery of Ezetimibe

Extensive clinical trials and epidemiological studies have unequivocally established the importance of lowering the low-density lipoprotein (LDL) level in the treatment and prevention of coronary heart disease. Lowering of the LDL level has been achieved in humans using 3-hydroxy-3-methylglutaryl coenzyme A (HMG-CoA) reductase inhibitors, an

enzyme responsible for the biosynthesis of cholesterol in the liver. Several extremely potent HMG-CoA reductase inhibitors (statins), including lovastatin, simvastatin, and atorvastatin, have been extensively used in the clinic with much success. Bile acid sequestrate inhibitors, including resins such as colestipol and cholesteryl amine, have also been used, with limited success. There are two sources of the cholesterol in our body: approximately 70% is from biosynthesis in the liver, and the remaining 30% comes from the food that we consume. Intensive effort led to the discovery of ezetimibe[1] at the Schering-Plough Research Institute (SPRI) as a potent inhibitor of cholesterol absorption. It has been approved by the FDA as a monotherapy for lowering the serum cholesterol level. In addition, Vytorin, a combination of ezetimibe and simvastatin, has also been approved recently as a potent agent for lowering serum cholesterol. Vytorin was developed jointly by the Schering-Plough Corporation and the Merck Corporation.

Before our work began in this area it was known[2] that ACAT was involved in cholesterol trafficking in hamsters; however, its relevance in nonrodents was unclear. In the hamster model,[3] when the animals were fed with on a high-cholesterol diet they showed a significant increase in cholesterol ester in their liver without much change in their serum cholesterol level. Among the initial compounds synthesized, 1 and 2 showed in vitro ACAT inhibition, and also showed in vivo activity in the hamster model by lowering the cholesterol ester level in the liver without changes in the serum cholesterol level. As the enzyme inhibitory activity of 2 was considerably superior to 1, it was decided to prepare further conformationally constrained azetidine analogs, represented by structure 3.

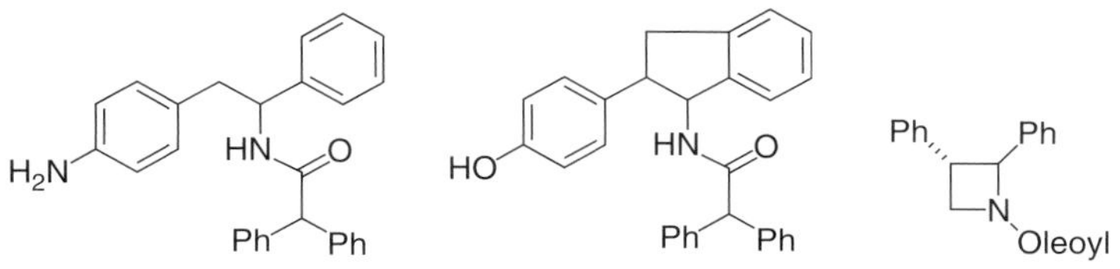

8.04.2.1 The Synthesis and Biological Activities of Initial Leads

The ester enolate derived from 4 was condensed with the Schiff base 5, to yield a mixture of 6 and 7 (Scheme 1). Compound 6 was treated with ceric ammonium nitrate, and the reaction product was reduced and then acylated, to yield 3.[4,5]

Several azetidinones were synthesized, and their activities determined. Compounds represented by structure 7, when administered orally in the hamster model, showed modest serum cholesterol lowering activity, even though the ACAT activities of these analogs were no better than, for example, 1 and 2. The activities of 1, 2, 3, 6, and 7 are presented in Table 1.

8.04.2.2 The Discovery of SCH 48461

After synthesizing several analogs of 7, it became clear that there was no correlation between ACAT activity and the ability of these analogs to lower the serum cholesterol level when they were administered orally in the hamster. At this

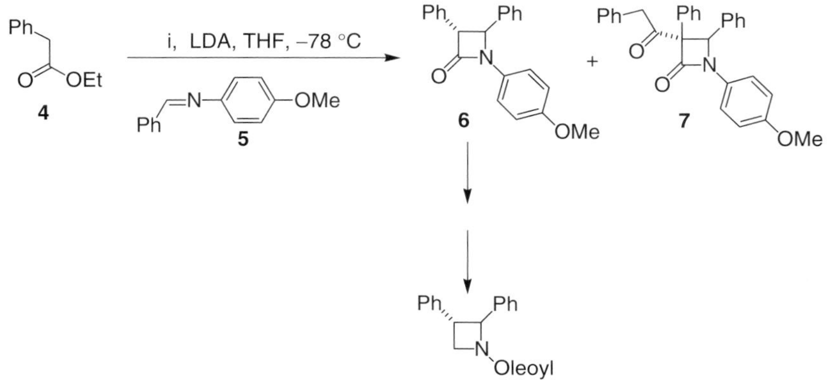

Scheme 1 Reprinted with permission from Clader, J. W. *J. Med. Chem.* **2004**, *47*, 2. Copyright (2004) American Chemical Society.

Table 1 Activities of azetidinones **1–3**, **6**, and **7**

Parameter	Compound				
	1	*2*	*3*	*6*	*7*
ACAT IC$_{50}$	900 nM	40 nM	4% inhibition at 25 μM	64% inhibition at 25 μM	4% inhibition at 25 μM
Hamster (100 mg kg^{-1}) serum cholesterol level	NE	NE	NE	NE	−10%
Cholesterol ester	−80%	−88%	NE	NE	−26%

NE, not effective.
Reprinted with permission from Clader, J. W. *J. Med. Chem.* **2004**, *47*, 2. Copyright (2004) American Chemical Society.

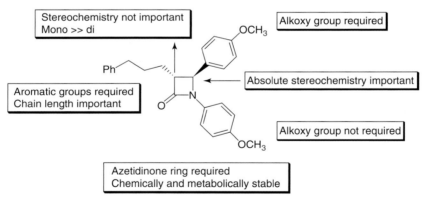

ACAT: IC$_{50}$: 26 μM

Cholestrol-fed animal models, ED$_{50}$:

Hamster:	2.2 mg kg^{-1}
Rat:	2.0 mg kg^{-1}
Monkey:	0.2 mg kg^{-1}
Dog:	0.1 mg kg^{-1}

SCH 48461 **8**

Figure 1 SCH 48461: structure, ACAT inhibition and ED$_{50}$ values. (Reprinted with permission from Clader, J. W. *J. Med. Chem.* **2004**, *47*, 5. Copyright (2004) American Chemical Society.)

Stereochemistry not important
Mono >> di

Alkoxy group required

Absolute stereochemistry important

Aromatic groups required
Chain length important

Alkoxy group not required

Azetidinone ring required
Chemically and metabolically stable

Figure 2 Structure–activity relationship of SCH 48461. (Reprinted with permission from Clader, J. W. *J. Med. Chem.* **2004**, *47*, 7. Copyright (2004) American Chemical Society.)

point it was decided that the structure–activity relationships in azetidinones needed to be established following in vivo results, which was challenging for both chemists and biologists. Extensive work in this area led to the discovery of SCH 48461 (**8**) as the most potent inhibitor of cholesterol absorption.[6] The activity of SCH 48461 is summarized in **Figure 1**, and a summary of the structure–activity relationship in this series is presented in **Figure 2**.

8.04.2.3   The In Vivo Activity of SCH 48461

The effect of SCH 48461 in cholesterol-fed rhesus monkeys is summarized, along with control animals, in **Figure 3**. The total serum cholesterol level in the control animals steadily increased over a period of 3 weeks compared with the monkeys dosed with SCH 48461 at 1 mg kg^{-1} over the course of the same period. The serum cholesterol levels did not show any significant change in the SCH 48461 group, and remained at the baseline. At the end of 3 weeks the control

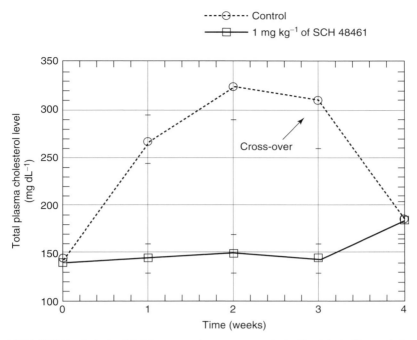

Figure 3   Effect of SCH 48461 in cholesterol-fed rhesus monkeys versus controls. (Reprinted with permission from Clader, J. W. *J. Med. Chem.* **2004**, *47*, 5. Copyright (2004) American Chemical Society.)

animals were administered SCH 48461 at $1\,mg\,kg^{-1}$, and it was observed that their cholesterol levels returned back to the baseline in a short period of time. The withdrawal of SCH 48461 from the second group of monkeys resulted in the rise of their serum cholesterol levels, again in a very short period of time. These results unequivocally established that SCH 48461 is a potent inhibitor of cholesterol absorption in various species of animals, and, based on its lack of ACAT inhibitory activity, it was obvious that SCH 48461 inhibited cholesterol absorption by an unknown mechanism. We shall return to this point later on in this chapter.

Although the above results were very promising, SCH 48461 however underwent extensive metabolism in vivo, thus it became important to identify the structures of the major metabolites, and then incorporate the information toward the design and synthesis of future analogs. Biologists at the SPRI established,[7] in a cleverly designed experiment using an intestinally cannulated bile duct diverted rat model, that one of the major active metabolites of SCH 48461 was the glucoronide of compound **9**.

9

8.04.2.4   The Design of Ezetimibe

Based on the activity of **9**, the phenols **10**, **11**, **12**, **13**, **14**, and **15** were synthesized, and their in vivo activities determined. The results are summarized in **Figure 4**.

Based on the structure–activity relationship as described in **Figure 2** and also on the biological activities of the possible metabolites as described in **Figure 4**, several new analogs were synthesized. Ezetimibe was found to be the most potent among all the analogs synthesized in inhibiting the absorption of cholesterol. The structure of ezetimibe[8] and a summary of its design is presented in **Figure 5**.

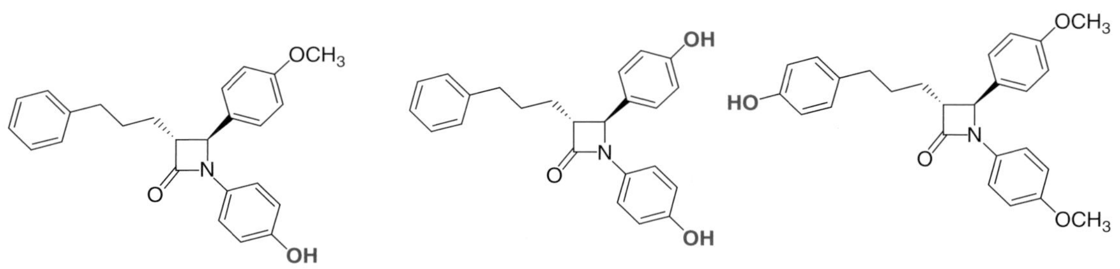

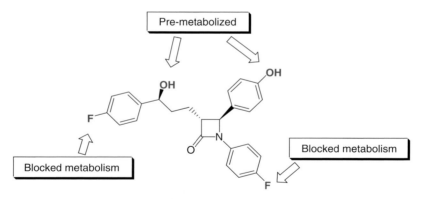

Hamster (50 mg kg^{-1})
Cholesterol ester   **10** −68%   **11** −78%   **12** −16%

Hamster
Cholesterol ester   **13** ED$_{50}$: 0.89 mg kg^{-1} **14** 5.00 mg kg^{-1} **15** 1.87 mg kg^{-1}

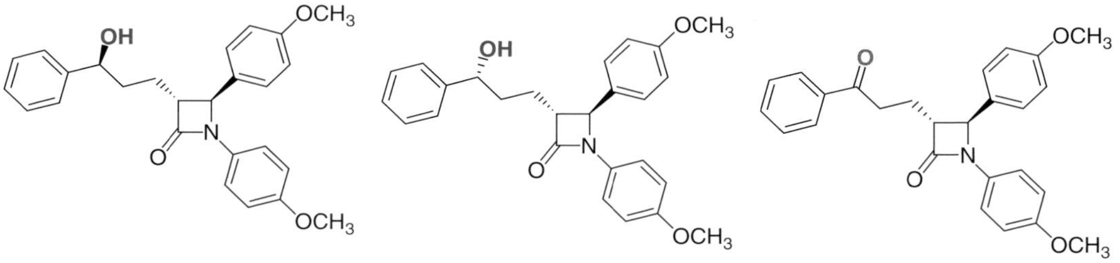

Figure 4   In vivo activities (in the hamster) of phenols **10–15**. (Reprinted with permission from Clader, J. W. *J. Med. Chem.* **2004**, *47*, 7. Copyright (2004) American Chemical Society.)

Pre-metabolized

Blocked metabolism

Blocked metabolism

Blocked metabolism

Figure 5   Structure–activity relationship of ezetimibe. (Reprinted with permission from Clader, J. W. *J. Med. Chem.* **2004**, *47*, 7. Copyright (2004) American Chemical Society.)

8.04.2.5   The Synthesis of Ezetimibe[9]

The keto acid **16** was converted to **17**, and then reduced to **18** using Corey's oxazaborolidine catalyst (**Scheme 2**). The titanium enolate of the silylated derivative of **18** was condensed with the silylated phenolic Schiff base **19**, to yield **20**. The basic nitrogen atom of **20** was silylated in situ, and then treated with tetrabutyl ammonium fluoride, to form the azetidinone ring. Final deprotection of the silyl groups with acid yielded ezetimibe.

8.04.2.6   The Biological Activity of Ezetimibe

The cholesterol absorption inhibitory property of ezetimibe was compared with SCH 48461, and the results are summarized in **Table 2**. In every species studied, ezetimibe showed dramatic improvement in activity when compared with SCH 48461, and, in monkeys, ezetimibe showed the greatest activity.

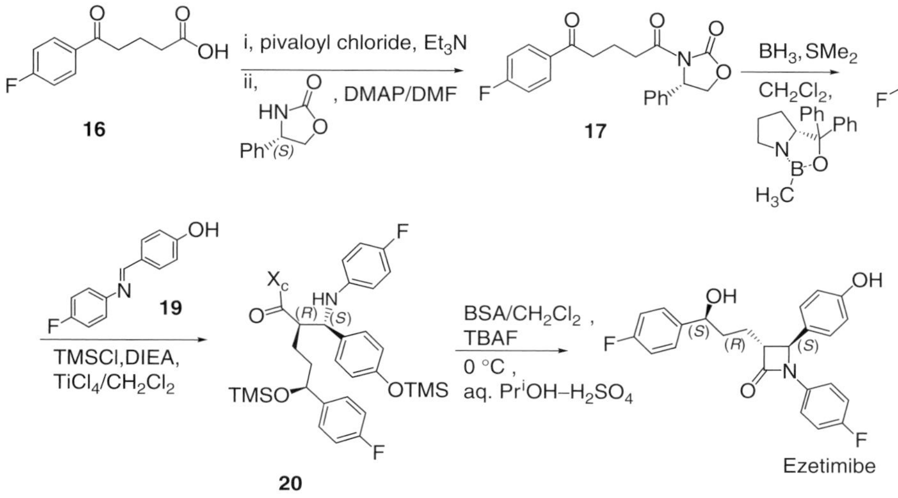

Scheme 2 Reprinted with permission from Clader, J. W. *J. Med. Chem.* **2004**, *47*, 8. Copyright (2004) American Chemical Society.

Table 2 Cholesterol absorption inhibition of ezetimibe and SCH 48461[a]

Species	ED_{50} $(mg\,kg^{-1})$	
	SCH 48461	Ezetimibe
Hamster	2.2	0.04
Rat	2.0	0.03
Monkey	0.2	0.0005
Dog	0.1	0.007

Reprinted with permission from Clader, J. W. *J. Med. Chem.* **2004**, *47*, 7. Copyright (2004) American Chemical Society.
[a] Blood levels significantly lower for ezetimibe.

In parallel, we studied the synergistic effect of ezetimibe, along with a statin for lowering cholesterol levels. Thus, ezetimibe ($0.007\,mg\,kg^{-1}$) and lovastatin ($5\,mg\,kg^{-1}$) were administered orally to two different sets of chow-fed dogs for 14 days. Neither ezetimibe nor lovastatin showed significant activity; however, the combination showed a dramatic reduction in serum cholesterol levels (**Figure 6**). Based on all these results, ezetimibe has progressed to the clinic as a monotherapy agent and also in combination with simvastatin. The combination drug, named Vytorin, was jointly developed by Schering-Plough and Merck.

8.04.2.7   The Results of Clinical Trials of Ezetimibe

Based on all these observations, ezetimibe was advanced alone and also in combination with simvastatin and atorvastatin in the clinic, and the results are presented in **Figure 7**. In the clinic,[10–13] ezetimibe (10 mg) when administered alone reduced serum cholesterol levels by 18.5%, on average, and in combination with simvastatin (10 mg), serum cholesterol levels were reduced by 51.9%.

Based on safety studies and clinical response, ezetimibe has been approved for human use as a monotherapy, and Vytorin (ezetimibe plus simvastatin) has also been approved by the FDA for reducing serum cholesterol levels.

8.04.2.7.1 The mechanism of action of ezetimibe
Scientists at the SPRI have recently discovered that ezetimibe blocks the activity of the cholesterol transporter NPC1L1 that is expressed at the apical surface of enterocytes.[14] It is believed to be the transporter for dietary cholesterol absorption. As a further proof, it was demonstrated that in NPC1L1 knockout animals, ezetimibe was ineffective in preventing the absorption of cholesterol.

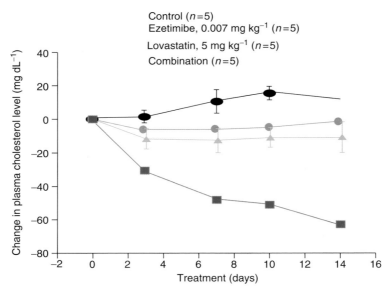

Figure 6 Effects of ezetimibe and lovastatin on cholesterol levels. (Reprinted with permission from Clader, J. W. *J. Med. Chem.* **2004**, *47*, 7. Copyright (2004) American Chemical Society.)

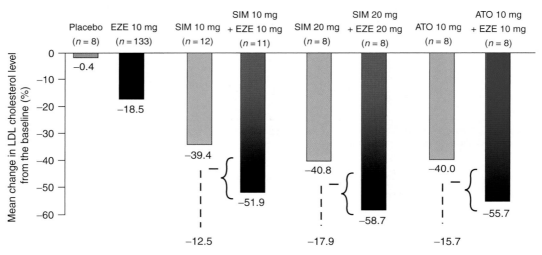

Figure 7 Results from a clinical trial of ezetimibe (EZE) alone and in combination with simvastatin (SIM) and atorvastatin (ATO) on cholesterol levels. (Reprinted with permission from Clader, J. W. *J. Med. Chem.* **2004**, *47*, 8. Copyright (2004) American Chemical Society.)

8.04.3   The Discovery of Posaconazole

Posaconazole[15] (**21**) is a novel triazole that has broad-spectrum antifungal activity against *Aspergillus* spp., *Cryptococcus* spp., *Histoplasma* spp., and a variety of other pathogens. In the clinic, posaconazole has been found to be well tolerated, with common side effects being gastrointestinal in origin. Life-threatening opportunistic fungal infections occur in AIDS patients, and also in patients undergoing chemotherapy for cancer, or those who have undergone organ transplants. The older antifungals do not work well with these patients, and the use of amphotericin B, a broad-spectrum antifungal, is limited in its use by its inherent toxicity. Posaconazole has been demonstrated in extensive clinical trials to be a potent orally active antifungal agent that works very well in the above-mentioned patient population.

21

Based on the above observations, Schering-Plough has submitted a New Drug Application to the FDA, and is waiting for its approval to use posaconazole in the clinic.

It has been established for a long time that 'azoles'[16] (i.e., fluconazole[17] (**22**), ketoconazole (**23**), itraconazole (**24**), and saperconazole (**25**)) work as antifungals by inhibiting the biosynthesis of ergosterol using fungal cytochrome P450 enzyme lanosterol 14α-demethylase. Ergosterol is an essential component of the fungal cell membrane, and therefore inhibiting its synthesis will prevent the fungus growing. It is imperative for the above process to be selective over mammalian cytochrome P450, to avoid toxic side effects.

22   **23**

Fluconazole (**22**) is orally active, and is very widely used against *Candida* and *Cryptococcus* infections; however, it lacks activity against the important pathogen *Aspergillus*. Ketoconazole (**23**) was the first example of an orally active antifungal used in the clinic, but exhibits hepatotoxicity and also interferes with testosterone biosynthesis.

Itraconazole[18] (**24**) is an orally active broad-spectrum antifungal that shows activity against *Aspergillus*.

24 X = Cl
25 X = F

8.04.3.1   The Synthesis of Initial Leads

Our aim at the SPRI was to discover an orally active antifungal, comparable to fluconazole (**22**) in its activity against *Candida* and also having activity against fluconazole-resistant strains of *Candida glabrata* and *Candida krusei*. In addition, our compound should have activity against *Aspergillus*, and should be superior to itraconazole (**24**), be safe, and not be an inducer of human cytochrome P450 enzymes.

As has been pointed out already, it will be important to have an azole moiety in our new antifungal; however, it should not have the 1:3-dioxolane ring system present in the structures of **23**, **24**, and **25**, because it is expected that such a ring system might induce instability of the drug under the acidic conditions in the stomach. Thus, we wished to explore whether other oxygen-containing heterocycles represented by **26**, **27**, **28**, **29**, and **30** will possess activities of interest. We have synthesized[19] all these novel structures with appropriate R groups in the racemic form, and determined their activities, and the results are as follows: **26**, **27**, and **28** were more active than **23** and **24**, and, among these compounds, **27** was the most active. In this chapter we focus on the synthesis of this class of compounds that led to the discovery of posaconazole (**21**).

26   **27**   **28**

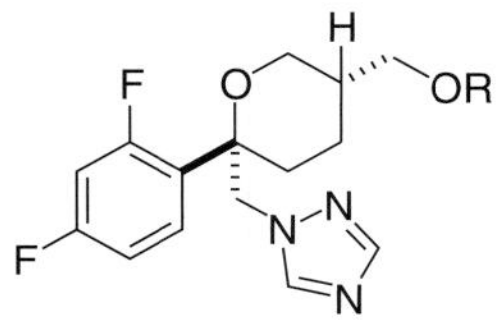

29 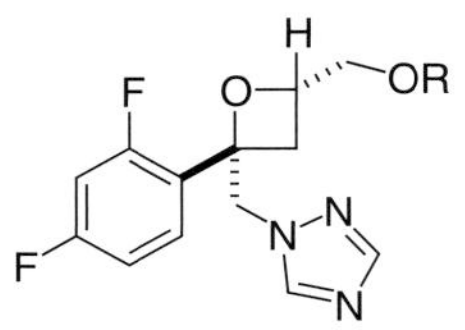   **30**

In our initial studies the importance of the aromatic spacer group –OR, and the presence of an azole moiety for its broad-spectrum activity, potency, and oral bioavailability, became apparent. Among the initial compounds synthesized, SCH 45012 (**31**)[20] was more active than itraconazole (**24**) and saperconazole (**25**) against systemic *Candida* and pulmonary *Aspergillus* infection models.

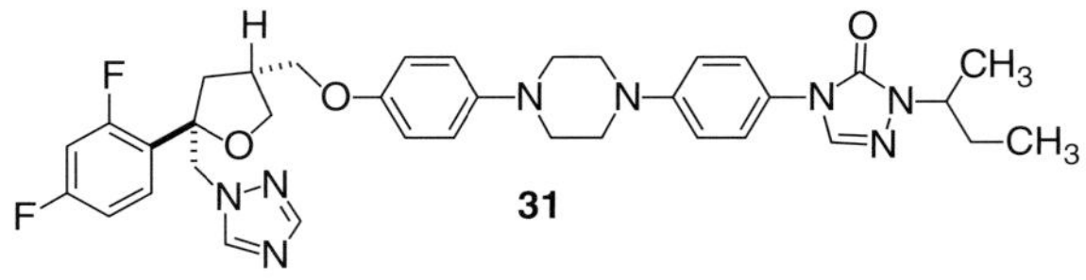

31

8.04.3.2   The Synthesis and Antifungal Activities of SCH 45012 and Its Isomers

SCH 45102 has three chiral centers, two of which have an established *cis* stereochemical relationship, and therefore it presents as a mixture of four optical isomers. Using Sharpless oxidation the chiral epoxides **32** and **33** were prepared from the allylic alcohol **34** (**Scheme 3**). Enantiomerically pure **32** and **33** were converted to the tosylates **35** and **36**, respectively, as described in our earlier publication. The reaction of **35** with **37** gave SCH 49999 (**38**) (**Scheme 4**), whereas the reaction of **36** with **37** yielded SCH 50002 (**40**) (**Scheme 5**). Alternatively, when **35** was treated with **41** it yielded SCH 50000 (**42**) (**Scheme 6**), and **36** on treatment with **41** yielded SCH 50001 (**43**) (**Scheme 7**). The phenol **37** was prepared by reacting **44** with the enantiomerically pure tosylate **45**, followed by *O*-demethylation (**Scheme 8**). Similarly, **41** was prepared by reacting **44** with the enantiomerically pure tosylate **46**, followed by *O*-demethylation (**Scheme 8**).

With all the four isomers of SCH 45012 in hand, we determined their antifungal activities. Interestingly, SCH 49999 and SCH 50000 possessing 5-(*S*) stereochemistry were inactive, whereas SCH 50001 and SCH 50002, possessing 5-(*R*) stereochemistry, were highly active as antifungals, and were both equally active. These results emphasize the importance of synthesizing possible stereoisomers to determine their biological activity in drug discovery, because there

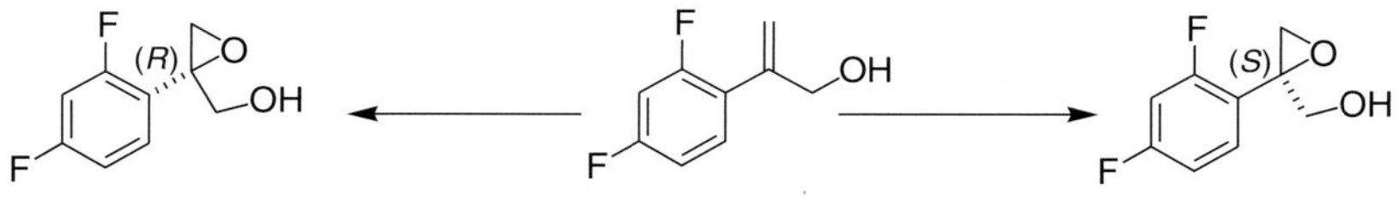

32   **34**   **33**

Scheme 3

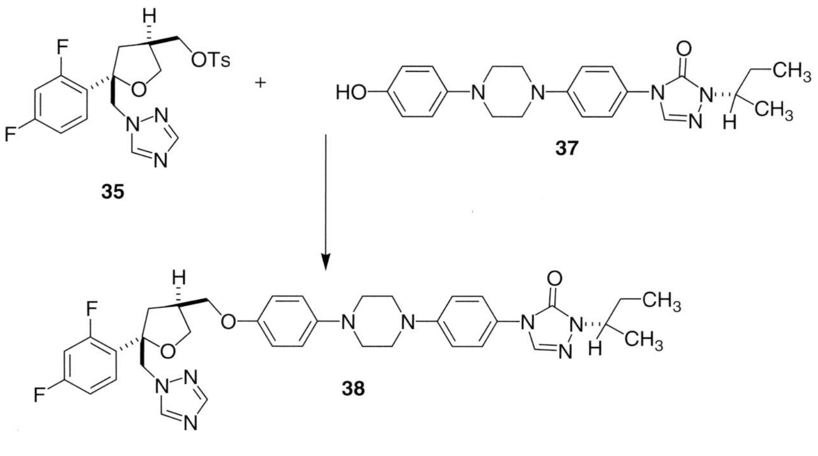

35   **37**

38

Scheme 4

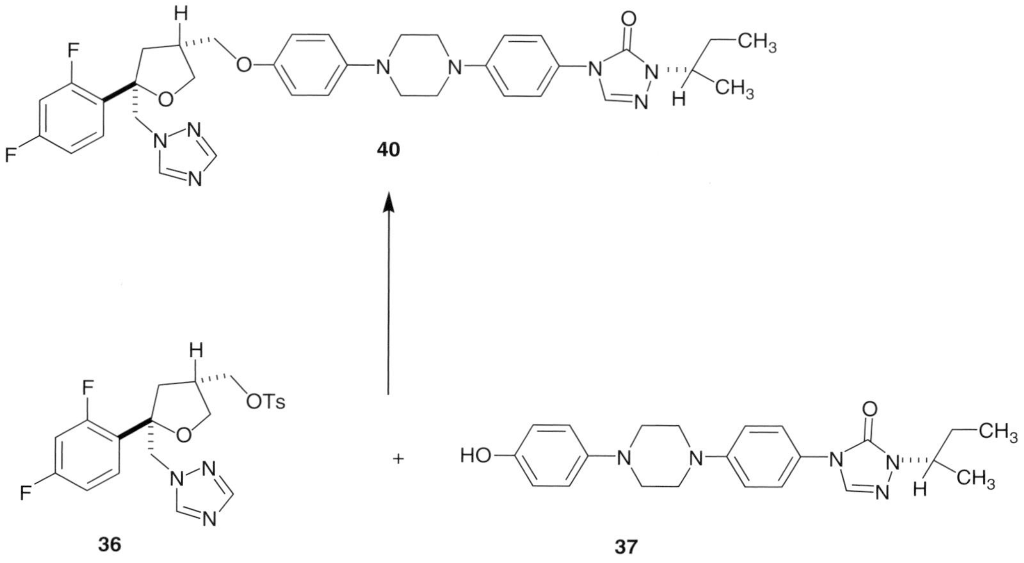

Scheme 5

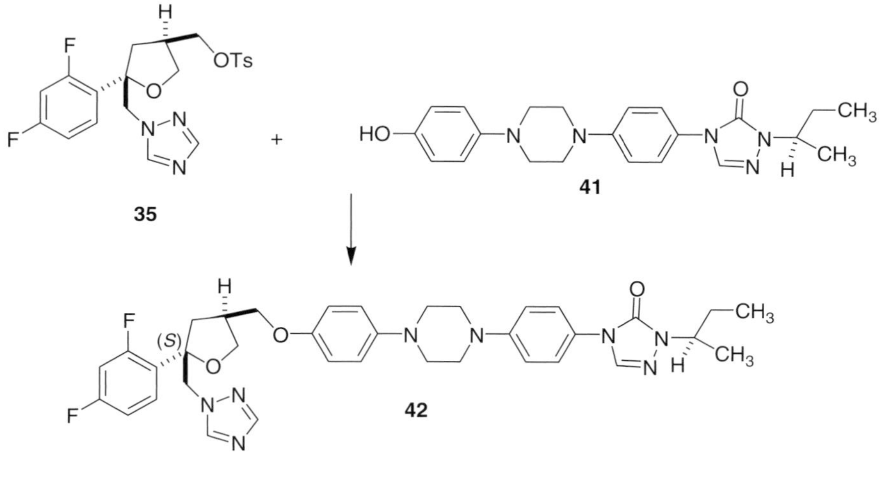

Scheme 6

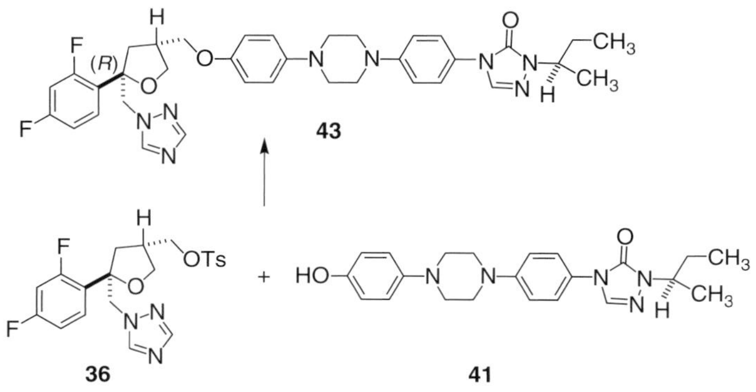

Scheme 7

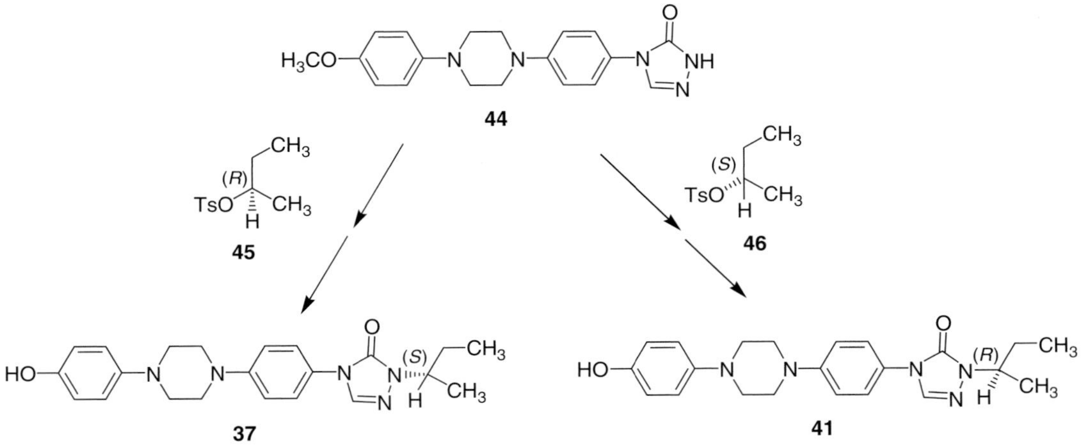

Scheme 8

is no other way at the outset to know which stereoisomer will be of importance. SCH 50001 and SCH 50002 being equiactive also suggests that the stereochemistry of the side chain is not important. With this information in hand, we synthesized SCH 51048 (**47**),[21] following similar schemes as used for the synthesis of SCH 50001 and SCH 50002, thus eliminating one asymmetric center, which of course makes the synthesis much simpler.

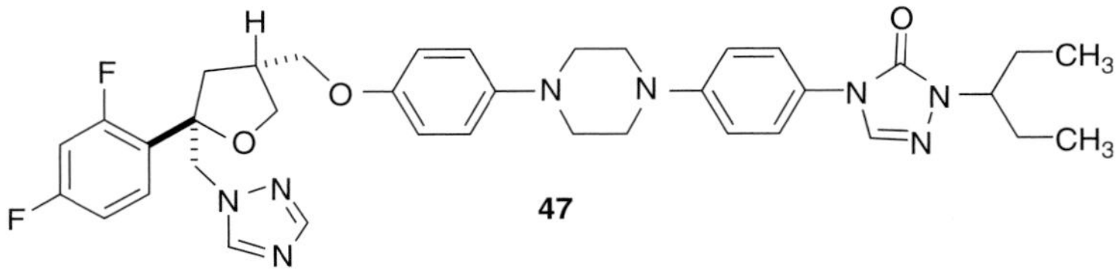

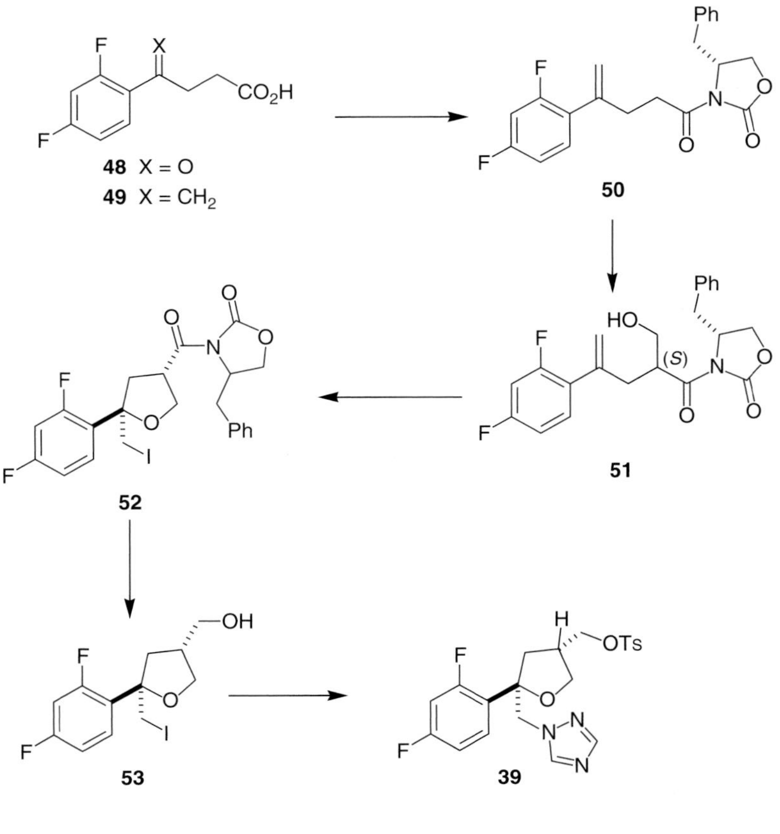

Scheme 9

SCH 51048 was much more potent than either SCH 50001 or SCH 50002, and also had a much broader spectrum of activity as an antifungal. Thus, eliminating one asymmetric center in SCH 51048 resulted in a compound that was more active and easier to synthesize.

8.04.3.3   The Improved Synthesis of SCH 51048[22]

The Friedel–Crafts reaction of 2,4-difluorobenzene with succinic anhydride yielded the ketoacid **48**, which was converted in Wittig reaction to the olefin **49** (**Scheme 9**). Stereoselective hydroxymethylation of **50** obtained from **49** yielded **51**. Iodocyclization of **51** yielded **52**, which on reduction with lithium borohydride yielded the alcohol **53**. Displacement of the iodo compound with sodium triazole followed by tosylation yielded **39**. The synthesis of SCH 51048 was completed by reacting **39** with the phenol **54**.

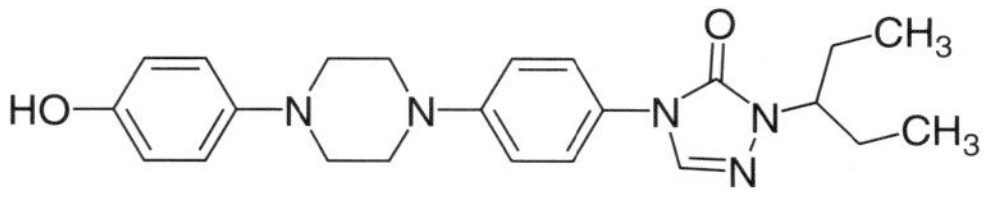

54

8.04.3.4   The Design of Posaconazole

During in vivo studies it was found that SCH 51048 produced an active metabolite, and it was concluded from mass spectrometry studies that the metabolic hydroxylation occurred at one of the carbons in the side chain of SCH 51048. As hydroxylation of the tertiary center will produce an unstable compound, the metabolic hydroxylation must have occurred either at the primary (two isomers) or secondary carbon (four isomers) center. We synthesized all the possible isomers, and determined their activities. Based on their spectrum of activity, potency, and pharmacokinetic properties, posaconazole (**21**) was selected as a clinical candidate.

8.04.3.5   The Synthesis of Posaconazole[15,22]

The synthesis of posaconazole was achieved by reacting **36** with the phenol **55** followed by deprotection of the alcohol protecting group. Compound **55** was prepared by *O*-demethylation of **56**, which in turn was prepared by reacting **44** with **57** (**Scheme 10**). Compound **57** was derived from (*S*)-lactic acid.

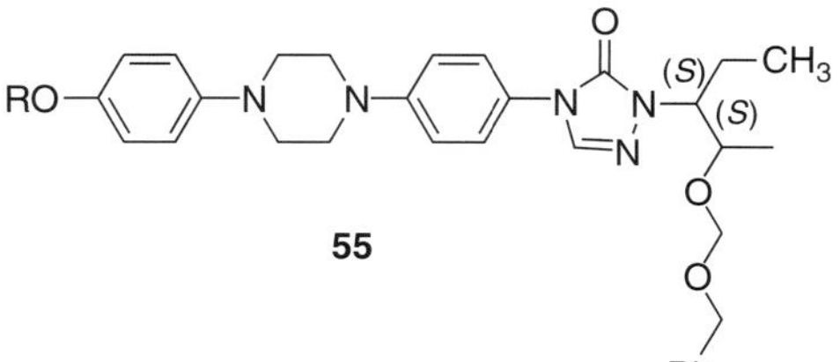

55

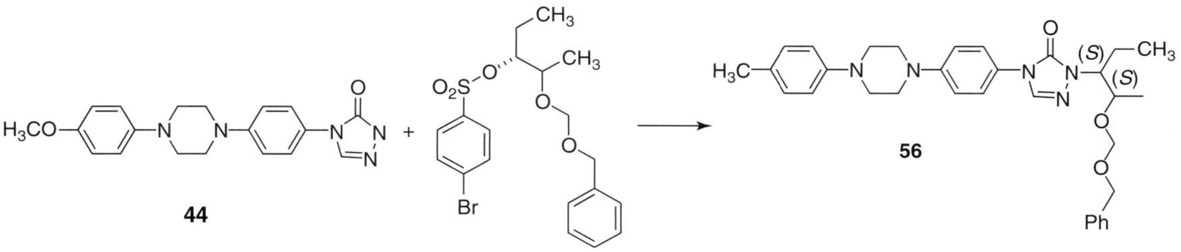

44   **57**   **56**

Scheme 10

8.04.3.6   The Antifungal Activity of Posaconazole

Posaconazole was tested against 285 strains of yeasts and filamentous fungi comprising 37 different species of fungi, including fluconazole-sensitive and -resistant strains of *Candida*, and also against *Aspergillus*. The mean minimum inhibitory concentration (MIC) against *Candida* was $0.018\,\mu g\,mL^{-1}$, and that against *Asperigillus* was $0.048\,\mu g\,mL^{-1}$. These MIC values are superior when compared with other clinically used antifungals. It has already been pointed out in this chapter that posaconazole was found in the clinic to be safe and efficacious as an antifungal – it is active against *Candida* and, particularly, serious infections caused by *Aspergillus*.

8.04.4   The Discovery of Lonafarnib

8.04.4.1   Overview

In recent years, considerable progress has been made in cancer chemotherapy, with the discovery of paclitaxel, hercepetin, imatinib, etc., yet there is a growing need for new anticancer agents with fewer side effects. Among several possible approaches, we[23,24] and others[25–27] decided to attempt to discover an anticancer agent that would work by inhibiting the enzyme FPT. In this chapter we discuss the path we took to the discovery of lonafarnib (**58**), a potent inhibitor of FPT, and discuss its anticancer activity.

58

It was well documented[28] at the outset of our work that approximately 30% of all human tumors contained a mutated Ras protein. Three mammalian Ras genes (i.e., H, K, and N), with mutations at positions 12, 13, and 61, have been identified in these tumors. Each of these genes encodes closely related proteins known as Ras P-21 proteins, containing 189 amino acids.

Ras proteins undergo a series of post-translational modifications before attaching to the cell membrane. In the 'off' state the modified Ras protein is bound to guanosine diphosphate (GDP); however, during cell activation, Ras-GDP is exchanged to the guanosine triphosphate form (Ras-GTP). In the normal cellular process, Ras-GTP (the 'on' state) is hydrolyzed by a GTPase (GTP-activating protein) to Ras-GDP, and cellular function returns to the 'off' state. However, in cancer cells the above hydrolysis step is much slower, and cellular proliferation continues in an uncontrolled manner.

At the C-terminus of the Ras protein there is a CAAX sequence (where C is cysteine, A is an aliphatic amino acid, and X is a variable such as methionine). However, when X is leucine the protein becomes geranyl geranylated, which is a more common event in a normal cellular process, and it uses a related enzyme called geranyl geranyl protein transferase (GGPT). It is therefore important that a desirable FPT inhibitor should be selective over GGPT to avoid potential toxicity problems.

The post-translation modification[29,30] of the RAS protein involves modification of the CAAX box using the following steps: (1) farnesylation of the cysteine residue using FPT, (2) cleavage of the CA bond using Ras-converting enzyme, and (c) methylation of the newly generated carboxylic acid group using carboxymethyl transferase. Thus, the transformed Ras protein; attaches to the cell membrane. It should be noted that selective inhibition of any of the above three steps will provide an anticancer agent.

Several different approaches have been taken in different laboratories to discover potent inhibitors of FPT, and as a result four compounds have entered clinical trials: L-778123 (Merck)[25] (**59**), BMS-214662 (Bristol Meyers-Squibb)[26] (**60**), R-115777 (Janssen)[27] (**61**), and lonafarnib (Schering-Plough)[23,24] (**58**). In this chapter we focus on the discovery of lonafarnib.

59   **60**

61

8.04.4.2   Screening of the SPRI Library of Compounds to Discover Initial Leads

The discovery of lonafarnib began with the screening of SPRI compound libraries. Several tricycles, including compounds **62**, **63**, and **64**, were found to possess weak to moderate activity against FPT, and these compounds had selectivity over GGPT; however, they showed poor cellular activity (**Figure 8**).

8.04.4.3   Structure–Activity Relationships and the Discovery of SCH 44342

Following the above leads, several amides were prepared. In **Table 3**, the activities of a few of these amides are summarized to highlight structure–activity relationships. SCH 44342[31,32] was one of the most active analogs synthesized.

It should be noted in **Table 3** that the aliphatic amide **65** was less active than the aromatic amides **66** and **68**, and within the aromatic amides the introduction of a CH_2 spacer group significantly improved their activities. The reason for the improvement of activity from **66** to **67**, and from **68** to **69** (SCH 44342), became clear when an x-ray analysis of

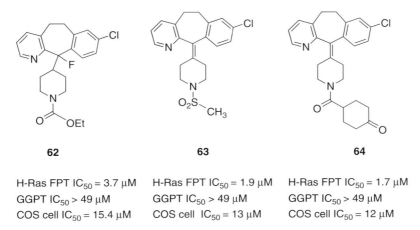

62   **63**   **64**

H-Ras FPT IC_{50} = 3.7 μM   H-Ras FPT IC_{50} = 1.9 μM   H-Ras FPT IC_{50} = 1.7 μM
GGPT IC_{50} > 49 μM   GGPT IC_{50} > 49 μM   GGPT IC_{50} > 49 μM
COS cell IC_{50} = 15.4 μM   COS cell IC_{50} = 13 μM   COS cell IC_{50} = 12 μM

Figure 8   Activities of initial lead compounds. (Reprinted with permission from Ganguly, A. K.; Doll, R. J.; Girijavallabhan, V. M. *Curr. Med. Chem.* **2001**, *8*, 1421 © Bentham Science Publishers Ltd.)

Table 3   Activities of amides

Compound	R	FPT IC$_{50}$ (μM)
65	CH_2—CH_3	8.4
66		5.3
67		0.8
68		4.7
69 (SCH 44342)		0.25
70		15.8

Reprinted with permission from Ganguly, A. K.; Doll, R. J.; Girijavallabhan, V. M. *Curr. Med. Chem.* **2001**, *8*, 1421
© Bentham Science Publishers Ltd.

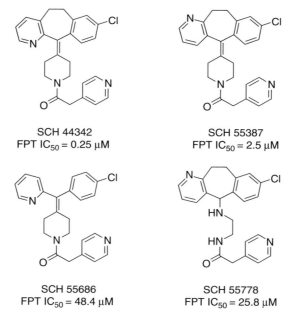

SCH 44342
FPT IC$_{50}$ = 0.25 μM

SCH 55387
FPT IC$_{50}$ = 2.5 μM

SCH 55686
FPT IC$_{50}$ = 48.4 μM

SCH 55778
FPT IC$_{50}$ = 25.8 μM

Figure 9   A change in position of the pyridine nitrogen or from ring cleavage results in inactivity.

an inhibitor bound to FPT was available, and we shall return to this point later. When the spacer group was increased in length, the resulting compounds were inactive. We also found that when the position of the pyridine nitrogen atom was changed, or when the rings B and D were cleaved, the resulting compounds were essentially inactive (**Figure 9**).

FPT is an obligatory dimer composed of α and β subunits. The crystal structure of unliganded FPT shows that the α unit is composed of seven pairs of anti-parallel α helices packed to form a crescent shape. The six pairs of helices in the β subunit are arranged as a double-walled barrrel. The active site cavity is situated near the center of the protein, and lined with residues of both the subunits. The catalytic zinc is liganded by three side chains arising from the β subunit, and a water molecule occupies the fourth site.

8.04.4.4   X-ray Crystallography Results and the Design of Future Analogs

An x-ray crystallographic study[33] of SCH 44342 bound to FPT (**Figure 10**) demonstrated an extended interaction between SCH 44342 and FPT. SCH 44342 binds in the center of the active site cavity, and the tricyclic ring A-B-C is situated deep in the cavity, and fits well in the lipophilic pocket and near the farnesyl pyrophosphate residue. Ring E is closest to the molecular surface. In three dimensions, the tricyclic ring is at a right angle to the rest of the molecule. The pyridine nitrogen atom is hydrogen bonded to a water molecule, which in turn is hydrogen bonded to the Ser99β residue. Aromatic rings A and C stack against the aromatic amino acid residues of the α and β subunits. One of the most important interactions involves the hydrogen bonding of the amide carbonyl with a water molecule, which is hydrogen bonded in turn to the Phe360β and Tyr361β residues of the protein backbone. It was also clear from x-ray studies that a few more interactions could be gained in SCH 44342 if there was a substituent at the 3-position of the pyridine ring. Thus, a number of analogs with different substituents at the 3-position (Br, Cl, F, and CH_3) were synthesized, and their activities are summarized in **Table 4**. Although **74** was highly active, as the methyl group is expected to undergo metabolic oxidation we decided to substitute bromine at the 3-position in ring A for the synthesis of future analogs.

At this stage we also investigated[34] the importance of having an sp^3 center at C-11, and our results are summarized in **Figure 11**. It was intriguing that the C-11 enantiomers **75** and **76** had equal activities, as did the enantiomers **77** and **78**. These compounds were premetabolized to the pyridine N-oxides, and, as expected, they had better pharmacokinetic properties compared to the pyridine analogs. X-ray analysis of the desbromo enantiomers **79** and **80** bound to FPT revealed that the two enantiomers bind in a similar way, and when the bound structures were superimposed they overlapped almost perfectly. The (S)-(−) enantiomer binds in the same location as the (R)-(+)

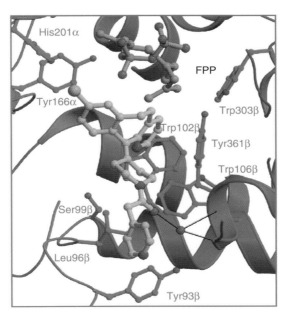

Figure 10   X-ray crystal structure of SCH 44342 bound to FPT ($IC_{50} = 250$ nM). (Reprinted with permission from Strickland, C. L.; Weber, P. C.; Windsor, W. T.; Wu, Z.; Le, H. V.; Albanese, M. M.; Alvarez, C. S.; Cesarz, D.; del Rosario, J.; Deskus, J. et al. *J. Med. Chem.* **1999**, *42*, 2125. Copyright (1999) American Chemical Society.)

Table 4   Activities of analogs of SCH 44342 with different substituents

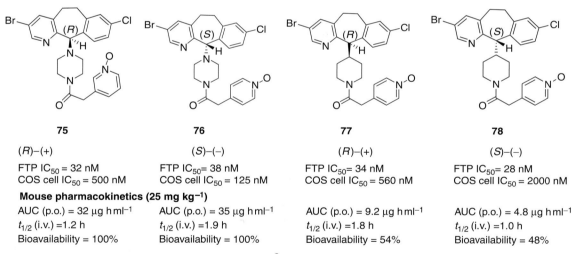

Compound	R	IC_{50} (μM)
SCH 44342 (**69**)	H	0.25
71	F	0.65
72	Cl	0.07
73	Br	0.06
74	CH₃	0.04

75

(R)–(+)

FTP IC$_{50}$ = 32 nM
COS cell IC$_{50}$ = 500 nM
Mouse pharmacokinetics (25 mg kg⁻¹)
AUC (p.o.) = 32 μg h ml⁻¹
t$_{1/2}$ (i.v.) =1.2 h
Bioavailability = 100%

76

(S)–(–)

FTP IC$_{50}$ = 38 nM
COS cell IC$_{50}$ = 125 nM

AUC (p.o.) = 35 μg h ml⁻¹
t$_{1/2}$ (i.v.) =1.9 h
Bioavailability = 100%

77

(R)–(+)

FTP IC$_{50}$= 34 nM
COS cell IC$_{50}$ = 560 nM

AUC (p.o.) = 9.2 μg h ml⁻¹
t$_{1/2}$ (i.v.) =1.8 h
Bioavailability = 54%

78

(S)–(–)

FTP IC$_{50}$= 28 nM
COS cell IC$_{50}$ = 2000 nM

AUC (p.o.) = 4.8 μg h ml⁻¹
t$_{1/2}$ (i.v.) =1.0 h
Bioavailability = 48%

Figure 11   Properties of analogs of SCH 44342 with an sp³ center at C-11 (AUC, area under the plasma concentration–time curve).

enantiomer, except that the tricyclic ring is rotated 180° relative to the active site. The increased potency of the 3-bromo analogs was due to the additional interactions with the aromatic amino acids, as mentioned above. A summary of the interactions revealed in the x-ray analysis of **79** and **80** are depicted in **Figure 12**.

8.04.4.5   Lonafarnib and Its Biological Activity

From these studies it was clear that there is enough flexibility at C-11, and that it could be either a C–C bond or a C–N bond. Several amide analogs were synthesized, and, as indicated in the x-ray studies, this portion of the molecule is exposed to the surface, and, as a result, most of them were active. Among these amides, **81** and its enantiomer were highly

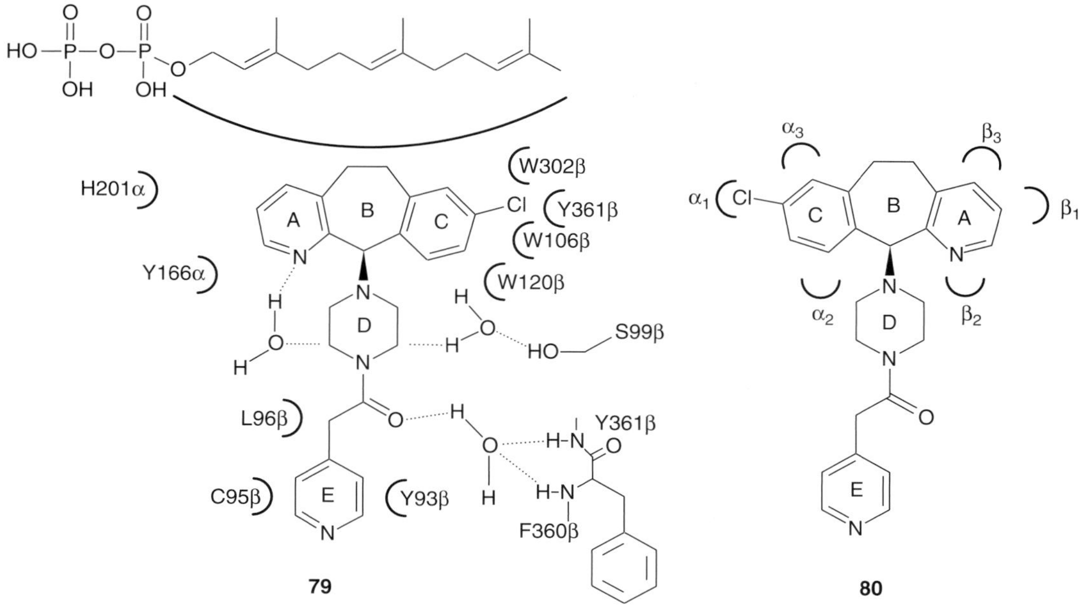

Figure 12 X-ray analysis of the des bromo enantiomers (**79**) and (**80**). (Reprinted with permission from Strickland, C. L.; Weber, P. C.; Windsor, W. T.; Wu, Z.; Le, H. V.; Albanese, M. M.; Alvarez, C. S.; Cesarz, D.; del Rosario, J.; Deskus, J. *et al. J. Med. Chem.* **1999**, *42*, 2129. Copyright (1999) American Chemical Society.)

active. X-ray studies with these enantiomers bound to FPT indicated, besides the interactions noted above, that further substitutions of these compounds at either the 7- or 10-position should be beneficial in improving their in vitro potency.

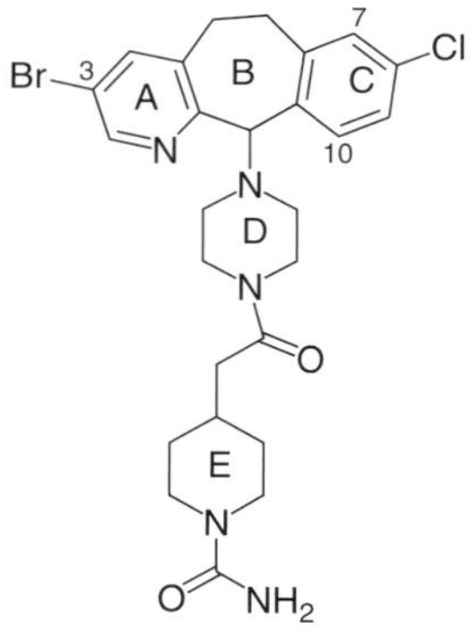

We synthesized both of these classes of compounds with different amide side chains, and C-11 substituents. Among all of the compounds synthesized, lonafarnib (SCH 66336) was one of the most potent analogs, possessing desirable pharmacokinetic properties. It also showed potent cell-based activity and antitumor activity. Unlike the enantiomers **79** and **80**, which were equally active in vitro, in the case of lonafarnib the (*R*)-(+) enantiomer was active and the (*S*)-(−) enantiomer was inactive.

The x-ray crystal structure[33] of lonafarnib bound to FPT is presented in **Figure 13**. As the (*S*)-(−) enantiomer was inactive, it could not be co-crystallized with FPT for x-ray studies.

The profile of the biological activity of lonafarnib is summarized in **Table 5**.

8.04.4.6   Synthesis of Lonafarnib

An overview of the synthesis of lonafarnib[23,24] is given in **Scheme 11**.

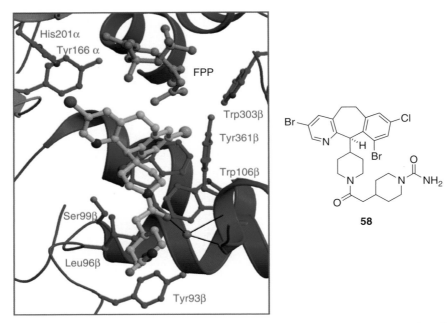

Figure 13   X-ray crystal structure of lonafarnib bound to FPT ($IC_{50} = 1.9$ nM). (Reprinted with permission from Strickland, C. L.; Weber, P. C.; Windsor, W. T.; Wu, Z.; Le, H. V.; Albanese, M. M.; Alvarez, C. S.; Cesarz, D.; del Rosario, J.; Deskus, J. *et al. J. Med. Chem.* **1999**, *42*, 2130. Copyright (1999) American Chemical Society.)

Table 5   Biological activity of lonafarnib

Inhibition
H-ras FPT $IC_{50} = 1.9$ nM
K-ras FPT $IC_{50} = 5.2$ nM
N-ras FPT $IC_{50} = 2.8$ nM
GGPT $IC_{50} >> 50\,000$ nM
COS cell $IC_{50} = 10$ nM

Soft agar
NIH-H-ras $IC_{50} = 75$ nM
NIH-K-ras $IC_{50} = 500$ nM

Mouse pharmacokinetics (25 mg kg^{-1})
$AUC = 24.1$ mg h μM^{-1}
C_{max} (p.o.) $= 8.84\,\mu M$
$t_{1/2}$ (i.v.) $= 1.4$ h
Bioavailability $= 76\%$

Monkey pharmacokinetics (10 mg kg^{-1})
$AUC = 14.7$ mg h μM^{-1}
C_{max} (p.o.) $= 1.9\,\mu M$
$t_{1/2}$ (i.v.) $= 3$ h
Bioavailability $= 50\%$

Reprinted with permission from Ganguly, A. K.; Doll, R. J.; Girijavallabhan, V. M. *Curr. Med. Chem.* **2001**, *8*, 1433 © Bentham Science Publishers Ltd.

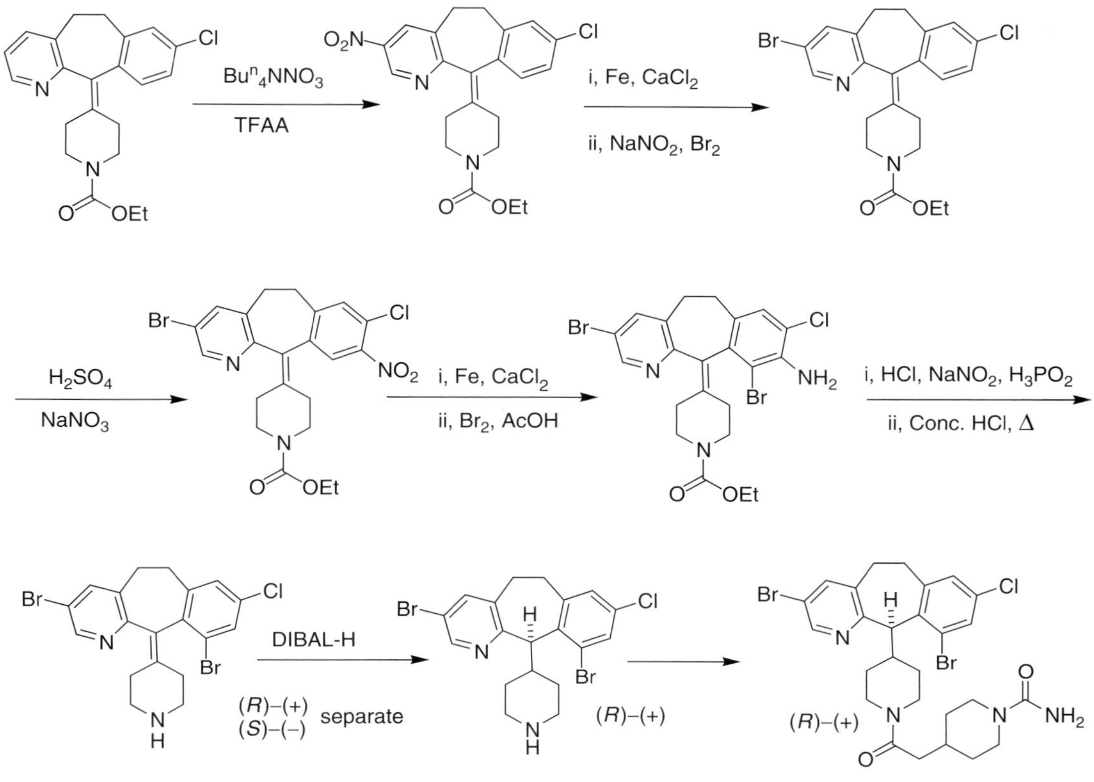

Scheme 11 Reprinted with permission from Ganguly, A. K.; Doll, R. J.; Girijavallabhan, V. M. *Curr. Med. Chem.* **2001**, *8*, 1427 © Bentham Science Publishers Ltd.

8.04.5   Summary

Based on all the above observations lonafarnib was recommended for clinical studies, and presently it is in third phase of the trial in cancer patients.

References

1. Clader, J. W. *J. Med. Chem.* **2004**, *47*, 1–9.
2. Sliskovic, D. R.; Picard, J. A.; Krause, B. R. *Prog. Med. Chem.* **2002**, *3*, 121–171.
3. Schnitzer-Polokoff, R.; Compton, D.; Boykow, G.; Davis, H.; Burrier, R. *Comp. Biochem. Physiol.* **1991**, *99*, 665–670.
4. Burnett, D. A.; Caplen, M. A.; Davis, H. R., Jr.; Burrier, R. E.; Clader, J. W. *J. Med. Chem.* **1994**, *37*, 1733–1736.
5. Clader, J. W.; Burnett, D. A.; Caplen, M. A.; Domalski, M. S.; Dugar, S.; Vaccaro, W.; Sher, R.; Browne, M. E.; Zhao, H.; Burrier, R. E. et al. *J. Med. Chem.* **1996**, *39*, 3684–3693.
6. Salisbury, B. G.; Davis, H. R.; Burrier, R. E.; Burnett, D. A.; Boykow, G.; Caplen, M. A.; Clemmons, A. L.; Compton, D. S.; Hoos, L. M.; McGregor, D. G. et al. *Atherosclerosis* **1995**, *115*, 45–63.
7. Van Heek, M.; France, C. F.; Compton, D. S.; McLeod, R. L.; Yumibe, N. P.; Alton, K. B.; Sybertz, E. J.; Davis, H. R., Jr. *J. Pharmacol. Exp. Ther.* **1997**, *283*, 157–163.
8. Rosenblum, S. B.; Huynh, T.; Afonso, A.; Davis, H. R., Jr.; Yumibe, N.; Clader, J. W.; Burnett, D. A. *J. Med. Chem.* **1998**, *41*, 973–980.
9. Thiruvengadam, T. K.; Fu, X.; Tann, C.-H.; Mcallister, T. L.; Chiu, J. S.; Colon, C. International Patent WO0034240, June 15 2000.
10. Dujovne, C. A.; Ettinger, M. P.; McNeer, J. F.; Lipka, L. J.; LeBeaut, A. P.; Suresh, R.; Yang, B. O.; veltri, E. P. *Am. J. Cardiol.* **2002**, *90*, 1092–1097.
11. Davidson, M. H.; McGarry, T.; Bettis, R.; Melani, L.; Lipka, L. J.; LeBeaut, A. P.; Suresh, R.; Sun, S.; Veltri, E. P. *J. Am. Coll. Cardiol.* **2002**, *40*, 2125–2134.
12. Gagne, C.; Bays, H. E.; Weiss, S. R.; Mata, P.; Quinto, K.; Melino, M.; Cho, M.; Musliner, T. A.; Gumbiner, B. *Am. J. Cardiol.* **2002**, *90*, 1084–1091.

13. Ballantyne, C. M.; Houri, J.; Notarbartolo, A.; Melani, L.; Lipka, L. J.; Suresh, R.; Sun, S.; LeBeaut, A. P.; Sager, P. T.; Veltri, E. P. *Circulation* **2003**, *107*, 2409–2415.
14. Altmann, S. W.; Davis, H. R., Jr.; Zhu, Li-ji.; Yao, X.; Hoos, L. M.; Tetzloff, G.; Iyer, S. P. N.; Maguire, M.; Golovko, A.; Zeng, M. et al. *Science* **2004**, *303*, 1201–1204.
15. Saksena, A. K.; Girijavallabhan, V. G.; Lovey, R. G.; Pike, R. E.; Wang, H.; Liu, Y. T.; Pinto, P.; Bennett, F.; Jao, E.; Patel, N.; et al. In *Anti-Infectives: Recent Advances in Chemistry and Structure Activity Relationships*; Bentley, P. H.; O'Hanlon, P. J. Eds.; Royal Society of Chemistry: Cambridge, UK, 1997; pp 180–199.
16. Richardson, K. In *Recent Advances in the Chemistry of Anti-infective Agents:*; Bentley, P. H.; Ponsford, R. Eds.; Royal Society of Chemistry: Cambridge, UK, 1993; Chapter 12, p 182
17. Street, S. D. A. In *Recent Advances in Chemistry and Structure Activity Relationships*; Bentley, P. H.; Ponsford, R. Eds.; Royal Society of Chemistry: Cambridge, UK, 1997; pp 141–151.
18. De Beule, K. *Int. J. Antimicrob. Agents* **1996**, *6*, 175.
19. Girijavallabhan, V. M.; Ganguly, A. K.; Saksena, A. K.; Cooper, A. B.; Lovey, R,; Rane, D. F.; Pike, R. E.; Desai, J. A.; Jao, E. In *Recent Advances in the Chemistry of Anti-infective Agents;*, Bentley, P. H.; Ponsford, R. Eds.; Royal Society of Chemistry: Cambridge, UK, 1993; p 191.
20. Saksena, A. K.; Girijavallabhan, V. M.; Lovey, R. G.; Pike, R. E.; Desai, J.; Ganguly, A. K.; Hare, R. S.; Loebenberg, D.; Cacciapuoti, A.; Parmegiani, R. M. *Bioorg. Med. Chem. Lett.* **1994**, 2023.
21. Saksena, A. K.; Girijavallabhan, V. M.; Lovey, R. G.; Desai, J. D.; Pike, R. E.; Jao, E.; Wang, H.; Ganguly, A. K.; Loebenberg, D.; Hare, R. et al. *Bioorg. Med. Chem. Lett.* **1995**, *5*, 127–132.
22. Saksena, A. K.; Girijavallabhan, V. M.; Wang, H.; Liu, Y. T.; Pike, R. E.; Ganguly, A. K. *Tetrahedron Lett.* **1996**, *37*, 5657.
23. Ganguly, A. K.; Doll, R. J.; Girijavallabhan, V. M. *Curr. Med. Chem.* **2001**, *8*, 1419–1436.
24. Njoroge, F. G.; Taveras, A. G.; Kelly, J.; Remiszewski, S.; Mallams, A. K.; Wolin, R.; Afonso, A.; Cooper, A. B.; Rane, D. F.; Liu, Y.-T. et al. *J. Med. Chem.* **1998**, *41*, 4890.
25. Britten, C. D.; Rowinsky, E.; Yao, S.-L.; Rosen, N.; Eckhardt, S. G.; Drengler, R.; Hammond, L.; Siu, L. L.; Smith, L. et al. *Proc. Am. Soc. Clin. Oncol.* **1999**, 597.
26. Hunt, J. T. Abstracts of the 219th American Chemistry Society National. Meeting, 2000.
27. Hudes, G. R.; Schol, J.; Baab, J.; Rogatko, A.; Bol, K.; Horak, I.; Langer, C.; Goldstein, L. J.; Szarka, C.; Meropl, N. J.; Weiner, L. *Proc. Am. Soc. Clin. Oncol.* **1999**, 601.
28. Bos, J. L. *Cancer Res.* **1989**, *49*, 4682.
29. Boyartchuk, V. L.; Ashby, M. N.; Rine, J. *Science* **1997**, *275*, 1796.
30. Clarke, S. *Annu. Rev. Biochem.* **1996**, *271*, 11541.
31. Bishop, W. R.; Bond, R.; Petrin, J.; Wang, L.; Patton, R.; Doll, R.; Njoroge, G.; Catino, J.; Schwartz, J.; Windsor, W. et al. *J. Biol. Chem.* **1995**, *270*, 30611.
32. Njoroge, F. G.; Doll, R. J.; Vibulbhan, B.; Alvarez, C. S.; Bishop, W. R.; Petrin, J.; Kirschmeier, P.; Carruthers, N. I.; Wong, J. K.; Albanese, M. M. et al. *Bioorg. Med. Chem.* **1997**, *5*, 101.
33. Strickland, C. L.; Weber, P. C.; Windsor, W. T.; Wu, Z.; Le, H. V.; Albanese, M. M.; Alvarez, C. S.; Cesarz, D.; del Rosario, J.; Deskus, J. et al. *J. Med. Chem.* **1999**, *42*, 2125.
34. Mallams, A. K.; Njoroge, F. G.; Doll, R. J.; Snow, M. E.; Kaminski, J. J.; Rossman, R. R.; Vilbulbhan, B.; Bishop, W. R.; Kirschmeier, P.; Liu, M. et al. *Bioorg. Med. Chem.* **1997**, *5*, 93.

Biography

A K Ganguly was born in India and educated partly in India and in England. He received his PhD degree from Imperial College, London under the supervision of Sir Derek Barton.

Prof Ganguly immigrated to the United States in 1967 and worked with Sir Derek at the Research Institute of Medicine and Chemistry, Cambridge, Massachusetts before joining the Schering-Plough Research Institute, Kenilworth, NJ in 1968 as a Senior Scientist. At Schering-Plough Research Institute he progressed to the position of Senior Vice President of Chemical Research in which capacity he directed all aspects of Chemical Research at the

institute. In September 1999 he joined Stevens Institute of Technology, Hoboken, NJ as a Distinguished Research Professor of Chemistry where he teaches Medicinal Chemistry and directs research programs for graduate students.

Prof Ganguly has made many significant contributions in drug discovery. He is associated with the discovery of Ezetimibe (Zetia), a cholesterol absorption inhibitor; Noxafil (Posaconazole), a potent antifungal and Lonafarnib (Sarasar), a highly selective farnesyl protein transferase inhibitor for the treatment of cancer. Prof Ganguly is also recognized for his many contribution toward synthesis of biologically active molecules and determining structures of complex oligosaccharide antibiotics such as Ziracin. Prof Ganguly has published 176 papers and is a coinventor of 80 patents. He has been a plenary lecturer at many international meetings and received several awards. In 2003, Prof Ganguly received the prestigious Hershberg award from the American Chemical Society for making important contributions in Medicinal Chemistry. In 2004, Prof Ganguly received Doctor of Engineering (Honoris Causa) from Stevens Institute of Technology and was also the recipient of the 'Life time achievement award in Chemical Sciences' from the Indian Chemical Society. Prof Ganguly remains as a consultant at Schering-Plough Research Institute.

Comprehensive Medicinal Chemistry II
ISBN (set): 0-08-044513-6

ISBN (Volume 8) 0-08-044521-7; pp. 29–51

6. Naesens, L.; Bischofberger, N.; Augustijns, P.; Annaert, P.; Van den Mooter, G.; Arimilli, M. N.; Kim, C. U.; De Clercq, E. *Antimicrob. Agents Chemother.* **1998**, *42*, 1568–1573.
7. Cundy, K. C.; Sueoka, C. M.; Lynch, G. R.; Griffin, L.; Lee, W. A.; Shaw, J.-P. *Antimicrob. Agents Chemother.* **1998**, *42*, 687–690.
8. Shaw, J.-P.; Sueoka, C. M.; Oliyai, R.; Lee, W. A.; Arimilli, M. N.; Kim, C. U.; Cundy, K. C. *Pharm. Res.* **1997**, *14*, 1824–1829.
9. Arimilli, M.; Kim, C.; Bischofberger, N. *Antiviral Chem. Chemother.* **1997**, *8*, 557–564.
10. Barditch-Crovo, P.; Deeks, S. G.; Collier, A.; Safrin, S.; Coakley, D. F.; Miller, M.; Kearney, B. P.; Coleman, R. L.; Lamy, P. D.; Kahn, J. O. et al. *Antimicrob. Agents Chemother.* **2001**, *45*, 2733–2739.
11. Cundy, K. C. Abstract 318. In *Clinical Anti-HIV Activity of Tenofovir Disoproxil Fumarate Correlates with Intracellular Drug Levels,* 7th European Conference on Clinical Aspects and Treatment of HIV Infection, Lisbon, Portugal, Oct 23–27, 1999; Poster 318: Lisbon, Portugal, 1999.
12. Schooley, R. T.; Ruane, P.; Myers, R. A.; Beall, G.; Lampiris, H.; Berger, D.; Chen, S.-S.; Miller, M. D.; Isaacson, E.; Cheng, A. K. *AIDS* **2002**, *16*, 1257–1263.
13. Squires, K.; Pierone, G.; Berger, D.; Steinhart, C.; Bellos, N.; Becker, S. L.; Chen, S. S.; Miller, M. D.; Coakley, D. F.; Cheng, A. et al. A 48-Week Final Analysis from a Phase III Randomized, Double Blind, Placebo Controlled Study in Antiretroviral Experienced Patients (Study 907). In *Programs and Abstracts of the 9th Conference on Retroviruses and Opportunistic Infections,* Seattle, WA, February 24–28, 2002; poster abstract 413-W: Seattle, WA, 2002.
14. Margot, N. A.; Johnson, A.; Cheng, A.; C oakley, D. F.; Miller, M. D. Final 48-Week Genotypic and Phenotypic Analyses of Study 907: Tenofovir DF (TDF) Added to Stable Background Regimens (poster abstract). In *9th Conference on Retroviruses and Opportunistic Infections,* Seattle, WA, Feb 24–28, 2002; Seattle, WA, 2002, p 209; poster abstract 414-W.
15. Kearney, B. P.; Flaherty, J. F.; Sayre, J. R.; W olf, J. J.; Coakley, D. F. A Multiple-Dose, Randomized, Crossover Drug Interaction Study Between Tenofovir DF and Lamivudine or Didanosine. In *The 1st IAS Conference on HIV Pathogenesis and Treatment,* Buenos Aires, Argentina, Jul 8–11, 2001; poster number 337: Buenos Aires, Argentina, 2001.
16. Flaherty, J. F.; Kearney, B. P.; Wolf, J. J.; Sayre, J. R.; Coakley, D. F. A Multiple-Dose, Randomized, Crossover Drug Interaction Study between Tenofovir DF and Efavirenz, Indinavir, or Lopinavir/Ritonavir (poster). In *The 1st IAS Conference on HIV Pathogenesis and Treatment,* Buenos Aires, Argentina, Jul 8–11, 2001; poster number 336: Buenos Aires, Argentina, 2001.
17. Arimilli, M.; Cundy, K.; Dougherty, J. P.; Kim, C. V.; Oliyai, R.; Stella, V. J. Antiviral Phosphonomethoxy Nucleotide Analogs Having Increased Oral Bioavailability. Patent US 6,043,230.
18. Yuan, L.-C.; Dahl, T. C.; Oliyai, R. *Pharm. Res.* **2001**, *18*, 234–237.
19. Yuan, L.-C.; Dahl, T. C.; Oliyai, R. *Pharm. Res.* **2000**, *17*, 1098–1103.

Biographies

Reza Oliyai, PhD, is the Director of Formulation and Process Development, Gilead Sciences. He is a co-inventor of Viread, Truvada, and the triple combination of Efavirenz/Emtricitabine/Tenofovir DF. Dr Oliyai has also been involved with the development of Hepsera and Tamiflu. Dr Oliyai received his PhD from the University of Kansas Pharmaceutical Chemistry and his BS in Pharmacy from Oregon State University.

Maria Fardis, PhD, has been at Gilead since 2001. Dr Fardis has been involved in multiple projects at Gilead ranging from immunology to antivirals. Prior to Gilead, Dr Fardis was at Intrabiotics Pharmaceuticals where she was involved in antibacterial programs. Dr Fardis received her PhD from University of California, Berkeley and her BS degree in Chemistry at the University of Illinois, Urbana-Champaign.

Comprehensive Medicinal Chemistry II
ISBN (set): 0-08-044513-6

ISBN (Volume 8) 0-08-044521-7; pp. 53–58

8.06 Hepsera

R Oliyai and M Fardis, Gilead Sciences Inc., Foster City, CA, USA

8.06.1	**Introduction**	**59**
8.06.2	**Synthesis and Formulation**	**60**
8.06.3	**Mechanism and Site of Bioconversion**	**61**
8.06.4	**Toxicity Issues**	**62**
8.06.5	**Conclusion**	**62**
	References	**62**

8.06.1 Introduction

According to the World Health Organization, approximately 2 billion people have been infected with hepatitis B and over 350 million have chronic hepatitis B infection.[1] Over 1 million patients have been diagnosed with hepatitis B in the US and Europe.[2] Only 15% of the infected patient population is receiving treatment for hepatitis B. Lack of treatment in patients with high hepatitis B virus (HBV) replication will lead to cirrhosis of the liver within a few years. Patients with liver cirrhosis have a short life expectancy due to liver failure and/or hepatocellular carcinoma.[3]

Chronic HBV infection commonly develops in patients who were exposed to HBV in childhood. Approval of a hepatitis B vaccine in 1982 has resulted in a decline in chronic HBV cases in recent years.

Currently available treatments for hepatitis B include interferon alfa-2b (Intron, from Schering Corporation), lamivudine (Epivir-HBV, GlaxoSmithKline), and adefovir dipivoxil (Hepsera, Gilead Sciences). Interferon alfa is administered parenterally and is associated with a number of adverse effects, such as depression, fatigue, irritability, and influenza-like symptoms, as well as bone marrow suppression. Lamivudine, a nucleoside also referred to as 3TC, is administered at a daily dose of 100 mg. Lamivudine is well tolerated and reduces the viral load significantly. However, viral resistance develops in approximately two-thirds of patients after a 3-year treatment period.[4]

Adefovir is a potent antiviral agent with activity against human immunodeficiency virus (HIV), herpes simplex virus (HSV), simian immunodeficiency virus (SIV), and HBV (**Figure 1**). Despite its broad spectrum of activity, it is not clinically useful as it suffers from poor oral bioavailability. The pharmacokinetics and bioavailability of adefovir have been studied at doses of 1.0 or 3.0 mg kg^{-1} in human clinical trials.[5] The terminal half-life of adefovir dosed by intravenous infusion is 1.6 ± 0.5 h. Over 98% of the dose is recovered unchanged in the urine within 24 h. Serum clearance of adefovir is 223 ± 53 mL h^{-1} kg^{-1}, which is similar to the renal clearance of the drug (205 ± 78 mL h^{-1} kg^{-1}). Considering the low protein binding,[6] active tubular secretion accounts for approximately 60% of the clearance of adefovir. The steady-state volume of distribution of adefovir is large (418 ± 76 mL kg^{-1}) and suggests complete distribution of the compound throughout body water. The oral bioavailability of adefovir at 3.0 mg kg^{-1} dose is variable and less than 12%. The subcutaneous bioavailability of adefovir at 3.0 mg kg^{-1} is over 100%.

The preclinical and human pharmacokinetics data in combination with the low permeability coefficient across Caco-2 cells[7] indicate that the low oral bioavailability is due to poor permeation across intestinal epithelium, rather than metabolic degradation.[8] A number of prodrugs were prepared to improve the bioavailability of adefovir by increasing the lipophilicity of the compound.[9,10] Adefovir dipivoxil demonstrated the most favorable properties and was progressed through clinical trials.

Adefovir dipivoxil is dosed orally at 10 mg day^{-1}. It is effective in the treatment of the e antigen-positive and e antigen-negative HBV patients, and in patients who are resistant to lamivudine,[11,12] with a median reduction of serum HBV DNA of 4.3 log$_{10}$ copies mL^{-1}. Following oral dosing, the prodrug is very efficiently cleaved and adefovir is released. No intact adefovir dipivoxil or monoester was detected in plasma following oral dosing in animal studies.[13]

Upon initiation of treatment of HBV patients with adefovir dipivoxil, clearance of HBV DNA is observed in a biphasic curve.[14] Initially, a sharp drop in HBV DNA levels (corresponding to clearance of viral particles from plasma) is observed with a half-life of 1 day. In the second, slower phase the infected virus-producing cells are eliminated. A single-dose pharmacokinetic study using adefovir dipivoxil 10 mg showed a $C_{max} = 18.4 \pm 6.26$ ng mL^{-1} at $T_{max} = 1.75$ h (C_{max}, the maximum observed plasma concentration; T_{max}, time to reach the maximum observed plasma concentration).[15] The

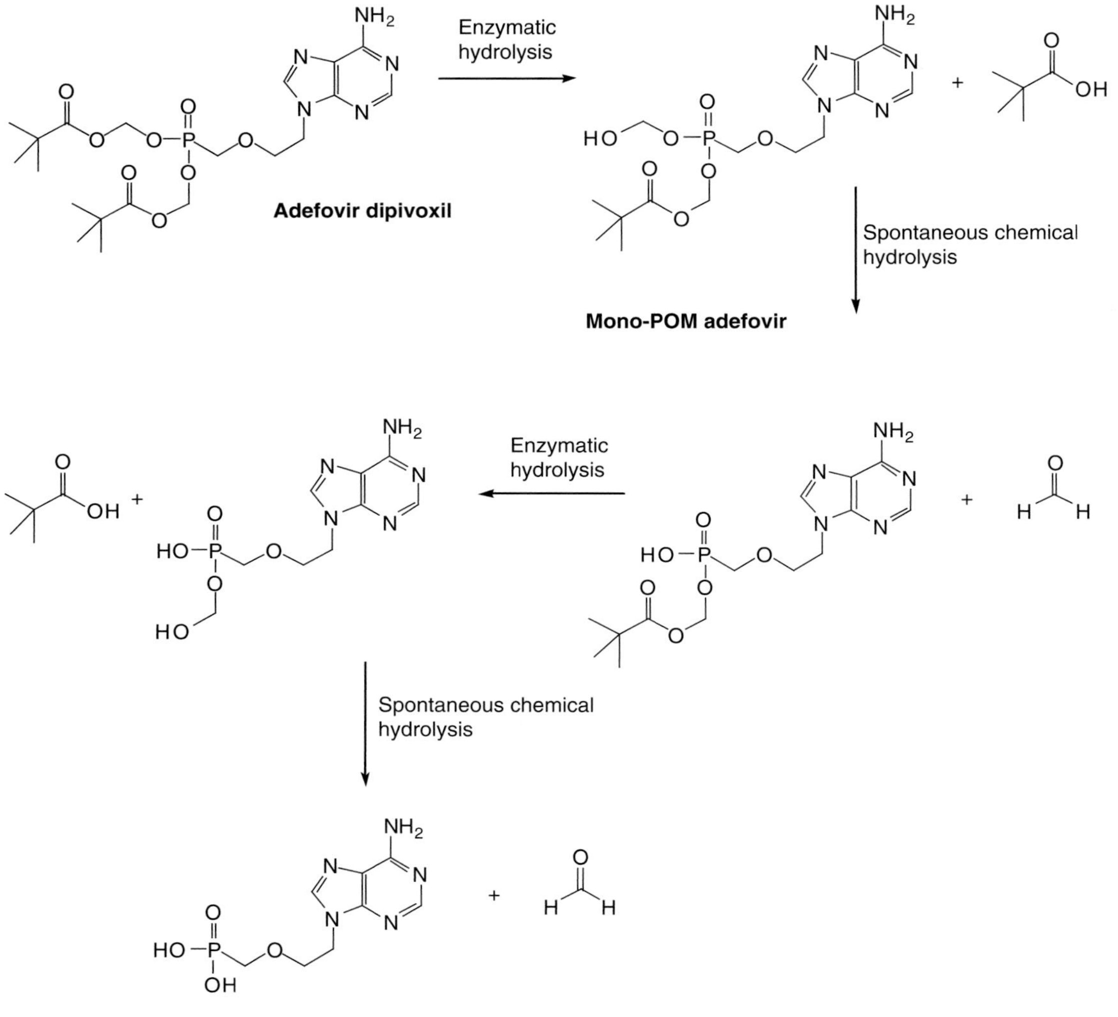

Figure 1 Structures of adefovir dipivoxil, adefovir, and the bioconversion pathway.

terminal elimination half-life of adefovir in plasma is 7.48 ± 1.65 h while the in vitro intracellular half-life is approximately 17 h. The efficacy and toxicity of adefovir dipivoxil were studied in two phase III clinical trials. The 48-week studies demonstrated improved liver histology in 53% of HBV e antigen-positive and 64% of HBV e antigen-negative patients. Adefovir dipivoxil also demonstrated efficacy in patients who were resistant to lamivudine. In addition, levels of serum alanine aminotransferase, which is a marker for biochemical response to hepatitis B treatment, were normalized.[16] Adefovir dipivoxil is well tolerated and was found to have a safety profile similar to placebo.

A low incidence of resistance is associated with administration of adefovir dipivoxil. Recently, two cases of adefovir resistance have been described in which the patients still responded to subsequent lamivudine therapy.[17] Mutant N236T in domain D of the HBV polymerase causes a reduction in susceptibility of HBV to adefovir both in vivo and in vitro.[18] The in vitro experiments demonstrated decreased replication capacity associated with these resistant viruses. Chronic dosing of adefovir dipivoxil is necessary; as with lamivudine, acute exacerbation of infection is observed upon termination of treatment with adefovir dipivoxil.[19]

8.06.2 Synthesis and Formulation

The preparation of adefovir dipivoxil was first described in the literature over 10 years ago.[20,21] Optimization of the synthesis and formation of crystalline final product was subsequently described (**Figure 2**).[22,23] The synthesis of

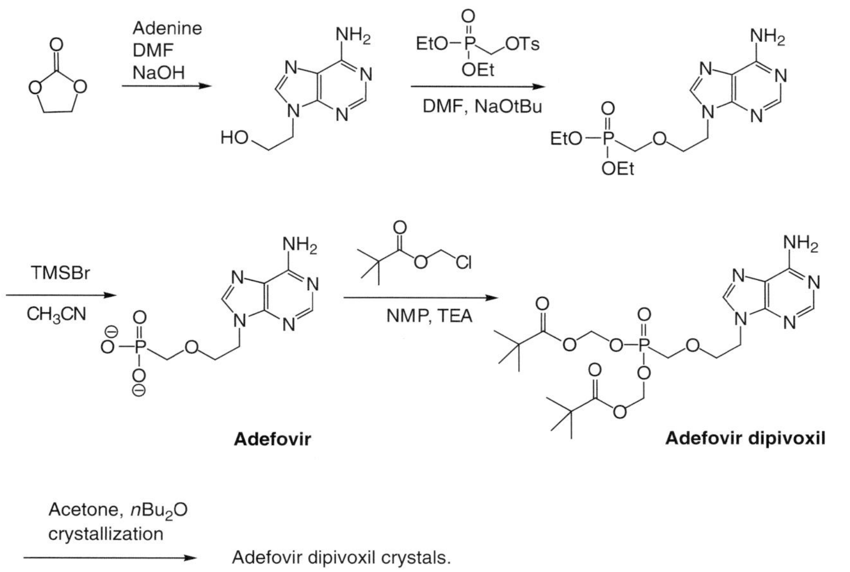

Figure 2 Synthetic scheme of crystalline adefovir dipivoxil.

diethyl p-toluenesulfonyloxymethyl-phosphonate is achieved by heating diethyl phosphite and paraformaldehyde under basic conditions at $87\,^\circ$C, followed by addition of p-toluenesulfonyl chloride at $0\,^\circ$C. The preparation of 9-(2-hydroxyethyl)adenine is accomplished through heating adenine, molten ethylene carbonate, and sodium hydroxide in DMF. The newly formed 9-(2-hydroxyethyl)adenine is further elaborated via an SN_2-type substitution in the presence of diethyl p-toluenesulfonyloxymethyl-phosphonate and a sodium alkoxide base in N,N-dimethylformamide (DMF). Conversion of the diethyl phosphonate to the diphosphonic acid is performed under standard bromotrimethylsilane/acetonitrile conditions. The synthesis of adefovir dipivoxil is completed by addition of the chloromethyl pivalate to a solution of the corresponding phosphonic acid in N-methylpyrrolidone (NMP), using triethylamine as base.

Adefovir dipivoxil is a white crystalline powder with high aqueous solubility at pH 2.0 ($19\,\mathrm{mg\,mL^{-1}}$) and lower solubility at pH 7.4 ($0.4\,\mathrm{mg\,mL^{-1}}$). The degradation kinetics of adefovir dipivoxil is governed by two distinct, but interrelated degradation pathways: (1) hydrolysis of the pivaloyloxymethyl moiety; and (2) formaldehyde-mediated dimerization of the adenine ring.[24] Hydrolysis of adefovir dipivoxil produces one equivalent each of mono-POM adefovir, pivalic acid, and formaldehyde. Formaldehyde can further react with the N^6-amine of adenine to form the corresponding carbinolamine intermediate. Dehydration of the carbinolamine intermediate leads to the formation of the reactive imine cation (Schiff base) of adefovir dipivoxil, which can react with an additional adefovir dipivoxil or mono-POM adefovir molecule to form the methylene-linked dimer. Both degradation pathways are known to be pH-dependent in solution.

Each tablet of Hepsera for oral administration contains 10 mg of the active ingredient, adefovir dipivoxil, in addition to the inactive ingredients croscarmellose sodium, lactose monohydrate, magnesium stearate, pregelatinized starch, and talc.

8.06.3   Mechanism and Site of Bioconversion

As described in the previous section, adefovir dipivoxil is an oral prodrug of adefovir in which the phosphonic acid is masked as the *bis*-(pivaloyloxymethyl) ester. The bioconversion of adefovir dipivoxil to adefovir is mediated by esterases.[25] The bioconversion mechanism involves rapid enzymatic hydrolysis of the *bis*-ester followed by spontaneous decomposition of the hydroxymethyl intermediate. The monoester most likely undergoes a similar degradation, leading to the rapid formation of adefovir. Adefovir is then transported into different cell lines by various mechanisms such as a saturable protein-mediated process in Hela cells[26] or by fluid-mediated endocytosis in human, caueasian, peripheral blood, leukemia, acute lymphoblastic (CCRF CEM) T-lymphoblastoid tissue.[27] Adefovir is phosphorylated to adefovir monophosphate by various kinases, one of which present in lymphoid cells is identified as adenylase kinase 2.[28] A

second phosphorylation of adefovir monophosphate provides adefovir diphosphate, which is incorporated into the elongating HBV DNA through HBV DNA transcriptase. Absence of the necessary 3'-hydroxyl group required for chain elongation during transcription in adefovir results in chain termination of HBV DNA. Adefovir is efficiently phosphorylated in hepatocytes yielding adefovir diphosphate, which has a half-life of 33 h in human Hep G2 cells.[29]

8.06.4   Toxicity Issues

Adefovir dipivoxil 10 mg day^{-1} has a safety profile that is similar to that of placebo[30] and is well tolerated by healthy, renally, and hepatically impaired patients, as well as lamivudine-resistant HBV patients coinfected with HIV.[31] A potential concern with adefovir dipivoxil is nephrotoxicity, which was observed at doses ≥ 30 mg daily.[32] This observation is consistent with the high clearance of adefovir which exceeds the glomerular filtration rate.[5] Monitoring of renal function may be required during treatment with adefovir dipivoxil. In addition, increase in serum creatinine (>0.5 mg dL^{-1}) has been observed at the 10 mg dose in 20% of pre- and posttransplant patients. However, no dose adjustment is required for hepatically impaired patients as no substantial changes in adefovir pharmacokinetics were observed in that patient population.

The bioconversion of adefovir dipivoxil to adefovir leads to the formation of pivalic acid and formaldehyde. The major clinical toxicological concern resulting from the pivalate released by prodrugs is related to the impact of pivalate on carnitine homeostasis. Adefovir dipivoxil is the only pivalate-containing prodrug used in chronic treatment. The daily pivalate load from adefovir dipivoxil is only 0.04 mmol day^{-1}, which is negligible compared to the total body carnitine pool (120 mmol). The daily formaldehyde load of 0.04 mmol day^{-1} is considered insignificant as well.

8.06.5   Conclusion

Adefovir dipivoxil (Hepsera) is an excellent example of a prodrug that can overcome the oral delivery problem associated with poor permeation across the intestinal mucosa. Since its launch in 2002, Hepsera has become an important agent for the treatment of hepatitis B patients with evidence of active viral replication. A safety profile similar to placebo has been observed for Hepsera, which allows it to be prescribed as an effective drug for chronic treatment of hepatitis B.

References

1. World Health Organization. Available online at: www.who.int/csr/disease/hepatitis/whocdscsrlyo20022/en/index1.html (accessed May 2006).
2. Hepatitis. http://hepatitis.about.com/od/hepatitisnews (accessed May 2006).
3. Beasley, R. P. *Cancer* **1988**, *61*, 1942–1956.
4. Papatheodoridis, G. V.; Hadziyannis, S. J. *Aliment. Pharmacol. Ther.* **2004**, *19*, 25–37.
5. Cundy, K. C.; Barditch-Crovo, P.; Walker, R. E.; Collier, A. C.; Ebeling, D.; Toole, J. J.; Jaffe, H. S. *Antimicrob. Agents Chemother.* **1995**, *39*, 2401–2405.
6. Qaqish, R. B.; Mattes, K. A.; Ritchie, D. J. *Clin. Ther.* **2003**, *25*, 3084–3099.
7. Shaw, J.-P.; Cundy, K. C. *Pharm. Res.* **1993**, *10*, S294.
8. Cundy, K. C.; Shaw, J.-P.; Lee, W. A. *Antimicrob. Agents Chemother.* **1994**, *38*, 365–368.
9. Naesens, L.; Neyts, J.; Balzarini, J.; Bischofberger, N.; De Clercq, E. *Nucleosides Nucleotides* **1995**, *14*, 767–770.
10. Shaw, J.-P.; Louie, M. S.; Krishnamurthy, V. V.; Arimilli, M. N.; Jones, R. J.; Bidgood, A. M.; Lee, W. A.; Cundy, K. C. *Drug Metab. Dispos.* **1997**, *25*, 362–366.
11. Peters, M. G.; Hann, H. W.; Martin, P.; Heathcote, J. E.; Buggisch, P.; Rubin, R.; Bourliere, M.; Kowdley, K.; Trepo, C.; Gray, D. F. et al. *Gastroenterology* **2004**, *126*, 91–101.
12. Perrillo, R.; Hann, H.-W.; Mutimer, D.; Willems, B.; Leung, N.; Lee, W. M.; Moorat, A.; Gardner, S.; Woessner, M.; Bourne, E. et al. *Gastroenterology* **2004**, *126*, 81–90.
13. Cundy, K. C.; Fishback, J. A.; Shaw, J.-P.; Lee, M. L.; Soike, K. F.; Visor, G. C.; Lee, W. A. *Pharm. Res.* **1994**, *11*, 839–843.
14. Rivkin, A. M. *Ann. Pharmacother.* **2004**, *38*, 625–633.
15. Hepsera™ (adefovir dipivoxil) tablets. US Prescribing Information. Gilead Sciences, Inc. Foster City, CA, September 2002.
16. Tong, M. J.; Tu, S. S. *Semin. Liver Dis.* **2004**, *24*, 37–44.
17. Brunetto, M. R.; Bonino, F. *Curr. Pharm. Des.* **2004**, *10*, 2063–2075.
18. Angus, P.; Vaughan, R.; Xiong, S.; Yang, H.; Delaney, W.; Gibbs, C.; Brosgart, C.; Colledge, D.; Edwards, R.; Ayres, A. et al. *Gastroenterology* **2003**, *125*, 292–297.
19. Kumar, R.; Agrawal, B. *Curr. Opin. Invest. Drugs* **2004**, *5*, 171–178.
20. Starrett, J. E.; Jr; Tortolani, D. R.; Russell, J.; Hitchcock, M. J. M.; Whiterock, V.; Martin, J. C.; Mansuri, M. M. *J. Med. Chem.* **1994**, *37*, 1857–1864.
21. Starrett, J. E., Jr; Tortolani, D. R.; Hitchcock, M. J. M.; Martin, J. C.; Mansuri, M. M. *Antiviral Res.* **1992**, *19*, 267–273.
22. Arimilli, M. N.; Lee, T. T. K.; Manes, L. V. Nucleotide Analog Compositions. WO 99/04774, 4 February, 1999.
23. Yu, R. H.; Schultze, L. M.; Rohloff, J. C.; Dudzinski, P. W.; Kelly, D. E. *Org. Process Res. Dev.* **1999**, *3*, 53–55.

24. Yuan, L.-C.; Dahl, T. C.; Iliyai, R. *Pharm. Res.* **2000**, *17*, 1098–1103.
25. Naesens, L.; Balzarini, J.; Bischofberger, N.; De Clercq, E. *Antimicrob. Agents Chemother.* **1996**, *40*, 22–28.
26. Cihlar, T.; Rosenberg, I.; Votruba, I.; Holý, A. *Antimicrob. Agents Chemother.* **1995**, *39*, 117–124.
27. Olsanska, L.; Cihlar, T.; Votruba, I.; Holy, H. *Collect. Czech. Chem. Commun.* **1997**, *62*, 821–828.
28. Robbins, B. L.; Greenhaw, J. J.; Connelly, M. C.; Fridland, A. *Antimicrob. Agents Chemother.* **1995**, *39*, 2304–2308.
29. Ray, A. S.; Vela, J. E.; Olson, L.; Fridland, A. *Biochem. Pharmacol.* **2004**, *68*, 1825–1831.
30. Hadziyannis, S. J.; Tassopoulos, N.; Heathcote, E. J.; Chang, T.-T.; Kitis, G.; Rizzetto, M.; Marcellin, P.; Lim, S. G.; Goodman, Z.; Wulfsohn, M. S. et al. *N. Engl. J. Med.* **2003**, *348*, 800–807.
31. Benhamou, Y.; Bochet, M.; Thibault, V.; Calvez, V.; Fievet, M. H.; Vig, P.; Gibbs, C. S.; Brosgart, C.; Fry, J.; Namini, H. et al. *Lancet* **2001**, *358*, 718–723.
32. Marcellin, P.; Chang, T.-T.; Lim, S. G.; Tong, M. J.; Sievert, W.; Shiffman, M. L.; Jeffers, L.; Goodman, Z.; Wulfsohn, M. S.; Xiong, S. et al. *N. Engl. J. Med.* **2003**, *348*, 808–816.

Biographies

Reza Oliyai, PhD, is the Director of Formulation and Process Development, Gilead Sciences. He is a co-inventor of Viread, Truvada, and the triple combination of Efavirenz/Emtricitabine/Tenofovir DF. Dr Oliyai has also been involved with the development of Hepsera and Tamiflu. Dr Oliyai received his PhD from the University of Kansas in Pharmaceutical Chemistry and his BS in Pharmacy from Oregon State University.

Maria Fardis, PhD, has been at Gilead since 2001. Dr Fardis has been involved in multiple projects at Gilead ranging from immunology to antivirals. Prior to Gilead, Dr Fardis was at Intrabiotics Pharmaceuticals where she was involved in antibacterial programs. Dr Fardis received her PhD from University of California, Berkeley and her BS degree in Chemistry at the University of Illinois, Urbana-Champaign.

Comprehensive Medicinal Chemistry II
ISBN (set): 0-08-044513-6

ISBN (Volume 8) 0-08-044521-7; pp. 59–63

8.07 Ezetimibe

J W Clader, Schering-Plough Research Institute, Kenilworth, NJ, USA

8.07.1	**Introduction**	**65**
8.07.2	**Discovery of the Prototype Azetidinone Cholesterol Absorption Inhibitor**	**66**
8.07.3	**Defining the Nature of the Target**	**68**
8.07.3.1	Structure–Activity Relationship on the Azetidinone Nucleus	68
8.07.3.2	Rigid Analogs	71
8.07.4	**The Discovery of Ezetimibe**	**72**
8.07.5	**Synthesis**	**77**
8.07.6	**Ezetimibe and Statins**	**77**
8.07.7	**Clinical Results**	**77**
8.07.8	**Mechanism of Action Studies**	**80**
8.07.9	**Conclusion**	**80**
References		**80**

8.07.1 Introduction

Atherosclerotic coronary artery disease remains a major cause of death and morbidity worldwide and a significant drain on healthcare resources, especially in developed countries. Cardiovascular disease claimed over 900 000 lives in the USA in 2002. The World Health Organization (WHO) estimates that worldwide 17 million people die every year from cardiovascular disease, especially heart attack and stroke. The direct and indirect costs of cardiovascular disease are estimated to reach $393 billion in 2005.[1]

Of the many risk factors for cardiovascular disease, dyslipidemia is among the best understood and one that has clearly lent itself to both lifestyle and pharmacological intervention. Several large clinical studies conducted over the last few decades have indicated that overall mortality rates from cardiovascular disease can be significantly reduced with aggressive pharmacological intervention and risk factor management. Dietary counseling, exercise, and the use of drug therapy to reduce low-density lipoprotein (LDL) levels can significantly reduce the risk of developing coronary artery disease. While statins have dominated the market for pharmacotherapy, recent changes in the target LDL levels recommended by the National Cholesterol Education Program as well as a desire to limit the required dose of statins have highlighted the need for even more effective therapies.[2–4]

Ezetimibe is the first of a new class of drugs that treat hypercholesterolemia by inhibition of absorption of cholesterol from the intestines. While some intestinal cholesterol comes from the diet, the majority of the cholesterol that is absorbed from the intestines comes from the liver and has been excreted into the intestines in bile. Ezetimibe alone reduces LDL-C 15–18%, and adding ezetimibe to a starting dose of a statin produces a reduction in cholesterol levels equivalent to that seen with an eightfold higher statin dose.[5–8] Ezetimibe has been the subject of a number of reports in the biological and medical literature.[9–12]

This case history traces the development of ezetimibe from the discovery of the first azetidinone cholesterol absorption inhibitors (CAIs) through the most recent information on the mechanism of action of this class of drugs. Beyond its significance as a new treatment for hypercholesterolemia, ezetimibe is of considerable interest because of the unusual path that led to its discovery. Although it began as a traditional medicinal chemistry effort, early in the program it became clear that ezetimibe and related azetidinone CAIs were not inhibiting cholesterol absorption via any known mechanism. Because of the absence of clear understanding of the molecular target, no in vitro assay existed to assess activity. As a result, every compound in this class had to be evaluated exclusively using in vivo models, most commonly in the cholesterol-fed hamster. Despite this limitation, these compounds displayed structure–activity relationships (SARs) that were well behaved and consistent with interaction with a structurally well-defined molecular target, although the nature of this target was unknown. Thus, the path that led to ezetimibe was arguably less

technologically driven and more biology-based than typical drug discovery programs. The process relied heavily on hypothesis and experimentation to understand both the nature of the molecular target and how the activity of these compounds could be optimized. The history that is presented attempts to capture not only the facts and chronology of the discovery but also the thoughts, hypotheses, and milestones that led to an evolving understanding of the nature of the unknown biological target and how the activity of compounds could be optimized when a more targeted approach was not possible.[13]

8.07.2   Discovery of the Prototype Azetidinone Cholesterol Absorption Inhibitor

The discovery program that led to ezetimibe began as a traditional drug discovery program to discover novel acyl-coenzyme A cholesterol acyltransferase (ACAT) inhibitors.[14] Although ACAT was known to be involved in a variety of cholesterol trafficking events including cholesterol absorption in rodents, the relevance of ACAT in nonrodent species was still unclear at the time this program began. Nonetheless, a variety of structural classes were known to be potent ACAT inhibitors in vitro and to be active in rodent animal models that reflect a potential for lowering cholesterol levels. Among these models was the cholesterol-fed hamster.[15] A high-cholesterol diet dramatically increases liver cholesteryl ester (CE) levels in these animals, making them especially sensitive to ACAT inhibition. By contrast, serum cholesterol (SC) levels are not dramatically changed by cholesterol feeding in these animals, and most ACAT inhibitors have minimal effect on serum cholesterol levels. **Figure 1** shows the in vitro and in vivo profiles of **1** and **2**,[16,17] which are typical early compounds from this effort, as well as a reference ACAT inhibitor **3** (CI-976).[18]

As would be expected with a well-defined molecular target, ACAT inhibitors displayed clearly defined SARs in these models that followed logically from structural changes. For instance, proper conformational constraint of **1** led to indane **2**, which showed both a significant increase in in vitro potency as well as a commensurate improvement in potency in the cholesterol-fed hamster.[16,17]

In addition to **2**, a number of alternate conformationally constrained analogs were prepared to probe their impact on in vitro and in vivo potency. Among these was azetidinone **4** proposed by Burnett et al.[19] These compounds were prepared by ester-enolate condensation to give azetidinone **5**, which was then deprotected by CAN oxidation, reduced, and acylated with a variety of acids (**Figure 2**).

In practice, the ester-enolate condensation gave a very modest yield of the desired azetidinone **5** accompanied by a small amount of a by-product **6** apparently derived from deprotonation of **5** followed by Claisen condensation with the ethyl phenylacetate starting material. Both the desired product **4** as well as the intermediate **5** and by-product **6** were evaluated for in vitro activity against ACAT as well as for in vivo activity in the cholesterol-fed hamster (**Figure 3**).[19]

None of the compounds was a potent ACAT inhibitor, although compound **6** did show some modest ACAT activity with an IC_{50} of about 7 µM. Despite this relatively weak activity, **6** showed moderate in vivo activity in the cholesterol-fed hamster assay. This included effects on both CE as well as a modest but reproducible effect on SC. Initial follow-up of this lead structure by Burnett and co-workers (**Figure 4**) demonstrated that several analogs of **6** displayed a similar profile of weak ACAT activity accompanied by a modest but reproducible effect on CE and SC in the hamster.[20]

Even at this early stage, elements of clear structure–activity trends were apparent, such as the loss in activity with aliphatic derivative **9**. Borrowing elements of known ACAT inhibitors such the 2,4,6-trimethoxy moiety of **3**, the ACAT activity of these compounds could be improved to give compounds such as **11**, but this had little or no impact on the

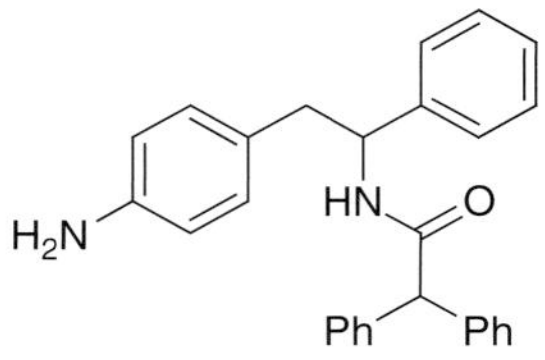

1
ACAT: IC_{50} = 900 nM
Hamster: at 50 mg kg^{-1} PO
SC: NE
CE: −80%

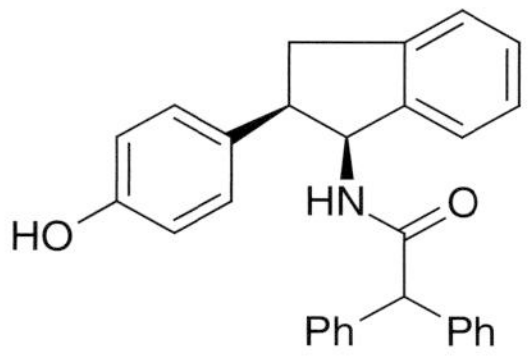

2
ACAT: IC_{50} = 40 nM
Hamster: at 50 mg kg^{-1} PO
SC: −7%
CE: −88%

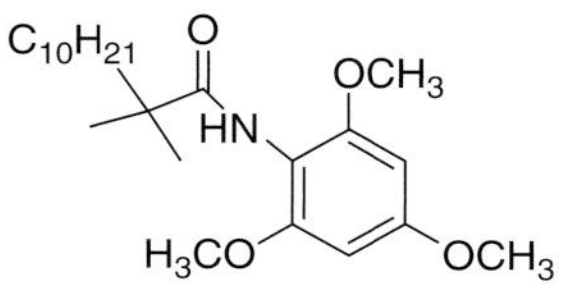

3 (CI-976)
ACAT: IC_{50} = 4200 nM
Hamster: at 50 mg kg^{-1} PO
SC: −10%
CE: −67%

Figure 1   Prototype ACAT inhibitors.

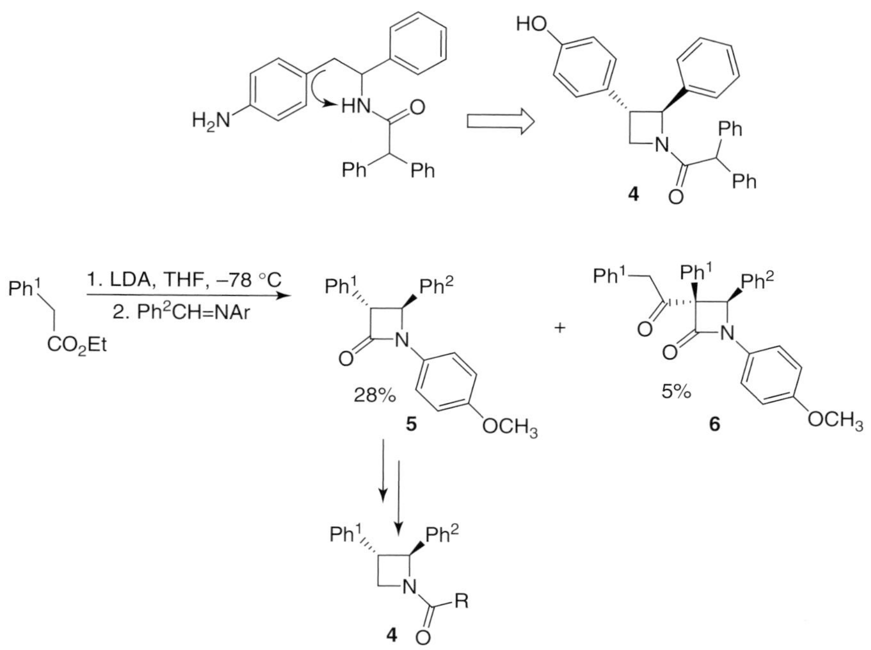

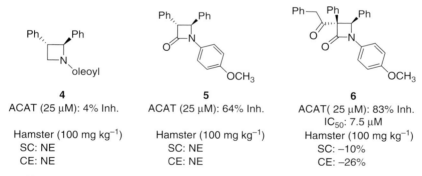

Figure 2 Design and synthesis of early azetidinones.

4
ACAT (25 µM): 4% Inh.

Hamster (100 mg kg⁻¹)
SC: NE
CE: NE

5
ACAT (25 µM): 64% Inh.

Hamster (100 mg kg⁻¹)
SC: NE
CE: NE

6
ACAT(25 µM): 83% Inh.
IC₅₀: 7.5 µM
Hamster (100 mg kg⁻¹)
SC: −10%
CE: −26%

Figure 3 In vitro and in vivo activity of first azetidinones.

in vivo activity of compounds. Based on this observation, Burnett *et al.* opted to disregard the ACAT activity of subsequent analogs and focus on optimizing the in vivo activity of compounds, guided solely by activity in the cholesterol-fed hamster. The absence of an in vitro assay clearly made this an extraordinarily challenging medicinal chemistry effort. Nonetheless, this work culminated in the discovery of azetidinone **13** and its resolved form **14**, the prototype azetidinone cholesterol absorption inhibitors and the starting point for all subsequent work. In an interesting portent of things to come, the difference between **13** and less active analogs such as **12** is the addition of a single well-placed methoxy group at the C4 phenyl. **Figure 5** shows the in vitro and in vivo profile of **14**.[21]

In addition to blocking accumulation of hepatic CEs in the hamster, **14** reduces SC levels in the cholesterol-fed rat, dog, and monkey. Of particular significance is the activity in the cholesterol-fed rhesus monkey. **Figure 6** shows the effects of **14** on rhesus monkeys fed a high-cholesterol diet for 4 weeks.[21] Control animals show a profound hyper-cholesterolemia after 3 weeks which is completely blocked by 1 mg⁻¹ kg dose of **14** administered in diet. When the control animals are then treated with 1 mg⁻¹ kg of **14**, their cholesterol levels return to nearly baseline within 1 week, while withdrawal of drug causes a gradual rise in cholesterol levels over the same time period.

While ACAT inhibitors are known to inhibit cholesterol absorption in rodents, such potent antihypercholesterolemic activity in nonrodents was unprecedented with ACAT inhibitors. In addition, Salisbury *et al.* demonstrated that ACAT

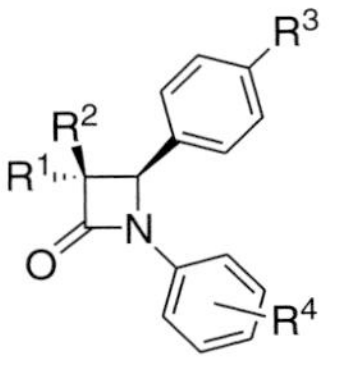

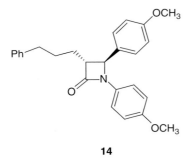

Compound	R^1	R^2	R^3	R^4	ACAT IC_{50}	Dose (mg kg^{-1})	% Change SC	% Change CE
6	PhCH$_2$CO	Phenyl	H	4-CH$_3$O	7.5 μM	50	−10	−26
7	Phenyl	Ethyl	H	4-CH$_3$O	>10 μM	50	−15	−39
8	Ph(CH$_2$)$_3$	Ethyl	H	4-CH$_3$O	>10 μM	50	−12	−29
9	n-C$_{10}$H$_{21}$	H	H	4-CH$_3$O	>10 μM	50	NE	NE
10	Ph(CH$_2$)$_3$	Phenyl	H	4-CH$_3$O	6 μM	50	−11	−16
11	Ethyl	Ph(CH$_2$)$_3$	H	2,4,6-CH$_3$O	1.7 μM	50	NE	−31
12	Ph(CH$_2$)$_3$	H	H	4-CH$_3$O	>10 μM	50	NE	−15
13	Ph(CH$_2$)$_3$	H	CH$_3$O	4-CH$_3$O	18 μM	10	−21	−60
14	(R)-Ph(CH$_2$)$_3$	H	(S)-CH$_3$O	4-CH$_3$O	26 μM	10	−43	−93

Figure 4 Early azelidinone structure–activity relationships.

ACAT: IC_{50} = 26 μM
Cholesterol-fed animal models:

ED_{50}: Hamster: 2.2 mg kg^{-1a}
Rat: 2.0 mg kg^{-1}
Monkey: 0.2 mg kg^{-1}
Dog: 0.1 mg kg^{-1}

^{a} Endpoint in hamster is hepatic cholesteryl esters, others are total plasma cholesterol.

14

Figure 5 Profile of **14** in cholesterol–fed animal models.

inhibitors and **14** have different effects on intracellular cholesteryl levels.[21] In these experiments, the ACAT inhibitor **3** blocked the accumulation of ^{14}C-cholesteryl esters but had no effect on free cholesterol in the intestinal wall of cholesterol-fed hamsters. By contrast, **14** inhibited the accumulation of both esterified and unesterified cholesterol. The reduction in cholesteryl ester levels was due entirely to reduction in free cholesterol substrate, while ACAT activity remained essentially unchanged. Thus, both the in vitro data and the in vivo pharmacology suggested that the azetidinone cholesterol absorption inhibitors act via a unique mechanism that is upstream from ACAT.

8.07.3   Defining the Nature of the Target

8.07.3.1   Structure–Activity Relationship on the Azetidinone Nucleus

Starting from the initial discovery of **14** by Burnett *et al.*, follow-up studies focused on establishing the SAR profile around the azetidinone nucleus.[20] While important in any medicinal chemistry effort, this was an especially critical early step in this program since it established that the observed in vivo activity followed predictable SARs despite the potential complications associated with use of an in vivo rather than an in vitro model. Among other things, these early studies confirmed the importance of the C4 *p*-methoxyphenyl moiety of **14** (**Figure 7**).

Activity decreases as the 4-methoxy group is moved to the 3- and 2-positions or is replaced by either a 3,4-dimethoxy or methylenedioxy group (**13–18**). On the other hand, adding a 2-hydroxyl group in addition to the

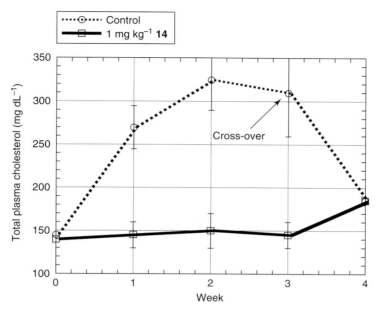

Figure 6 Effect of **14** in the cholesterol-fed rhesus monkey.

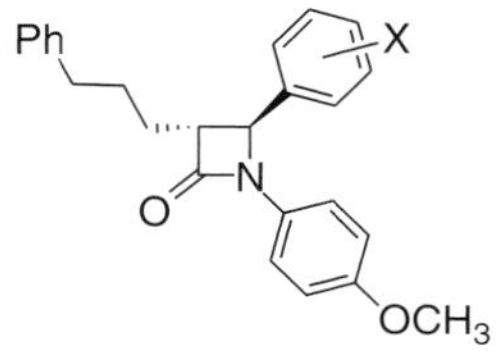

Compound	X	Dose (mg kg^{-1})	% Change CE	% Change SC
13	4-CH$_3$O	50	−78	−29
15	3-CH$_3$O	50	−24	−17
16	2-CH$_3$O	50	0	0
17	3,4-CH$_3$O	50	−33	−11
18	3,4-OCH$_2$O	50	−19	0
19	2-OH, 4-CH$_3$O	10	−95	−30
20	4-OH	50	−48	−16
21	3,4-OH	50	−39	0
22	CH$_2$OH	50	−46	−21
23	CH$_2$OCH$_3$	50	−56	−27
24	NMe$_2$	50	−60	−29
25	NH$_2$	50	−0	0
26	F	50	−12	0

Figure 7 SAR on the C4 phenyl.

4-methoxy group (**19**) increases activity.[22] Some activity is retained when the 4-methoxy group is replaced by similar hydrogen bonding moieties (**20–25**), but is essentially abolished when this group is replaced by nonpolar groups such as fluorine (**26**). These observations are consistent with subsequent work by Van Heek *et al.* described below that suggests that phenol **20** is the bioactive form of **14**. More drastic changes to the C4 phenyl itself, i.e., **27**, cause a marked reduction in activity even in the presence of a methoxy substituent (**Figure 8**).

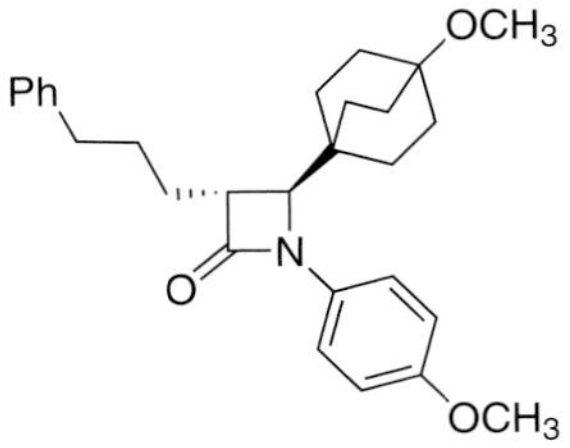

27
CE: NE at 20 mg kg^{-1}

Figure 8 Compound **27**.

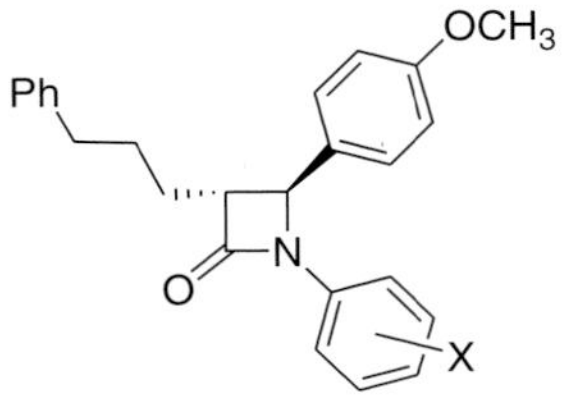

Compound	X	% Change CE	% Change SC
28	3-CH$_3$O	−91	−38
29	2-CH$_3$O	−87	−49
30	3,4-OCH$_2$O	−52	−19
31	4-OH	−90	−35
32	H	−95	−40
33	F	−95	−38
34	CH$_3$	−90	−38

Figure 9 SAR on the N1 phenyl group.

Unlike the C4 position, the N-aromatic ring, while required for activity, is tolerant of a variety of substituents or no substituent while retaining good activity in the 7-day cholesterol-fed hamster model (**28–34**) (**Figure 9**).

An aromatic moiety directly attached to the azetidinone nitrogen is also required for good activity (**Figure 10**). Simple alkyls, aralkyls, N-H, and N-acyl derivatives **35–39** are devoid of activity, although the N-cyclopropyl derivative **40** retains some activity.

The remaining aromatic ring tethered to the azetidinone 3-position via a 3-carbon chain is also an important part of the pharmacophore, in that activity can be modulated by changing the tether length (**Figure 11**) (**41–44**) or nearly abolished by removing the aromatic moiety (**45**). Activity is also reduced by disubstitution at C-3 (**46–47**), which may be related to the conformational requirements of the tethering chain. Unlike the chiral center at C4, which shows a clear preference for the 4-*S* configuration, both the 3-(*R*) and 3-(*S*) forms often show comparable activity with no consistent preference.

One remaining issue concerned the importance of the azetidinone itself. Unlike drugs such as antibiotics or elastase inhibitors which are also built on an azetidinone nucleus, there is no evidence that a reactive azetidinone is required for cholesterol absorption inhibition. Compound **14** is completely unchanged after 7 days' incubation at 37 °C with 0.1 N benzylamine, while a variety of beta-lactam antibiotics are completely consumed by these conditions. The azetidinone nucleus of **14** is also stable in vivo, although interestingly its enantiomer is rapidly hydrolyzed in vivo to the amino acid **48**.[13] Neither enantiomer of **48** has appreciable CAI activity, a fact that confuses the interpretation of the apparent preference for the 4-*S* configuration. Although the azetidinone carbonyl is required for good activity, activation of the azetidinone by N-acylation with a variety of groups (e.g., **37**) essentially abolishes the cholesterol absorption inhibitory activity. On the other hand, some activity is retained by analogous gamma-lactams and related compounds **50–51** as well as sultams **52–53**, all of which suggests that the primary role of the azetidinone nucleus is to provide a suitable scaffold (**Figure 12**).[23]

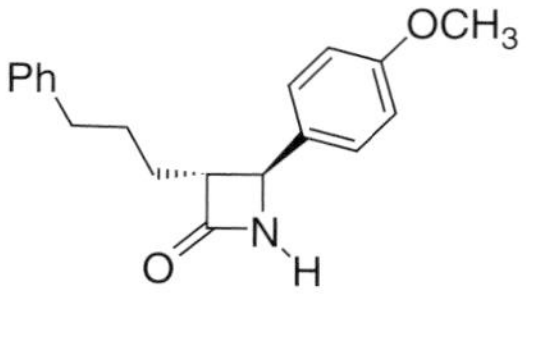

Figure 10 Modifications to the N1 phenyl group.

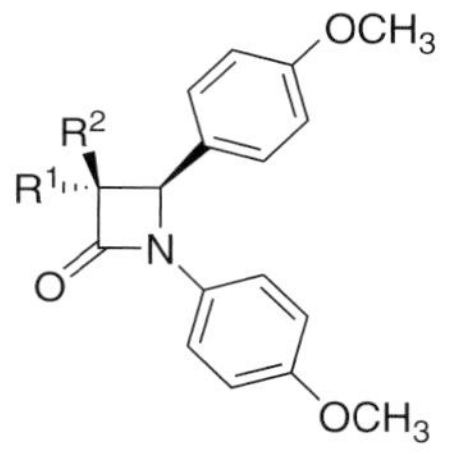

35
CE: NE at 50 mg kg^{-1}

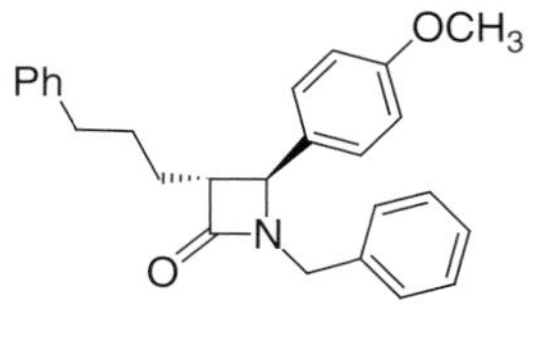

36
CE: NE at 50 mg kg^{-1}

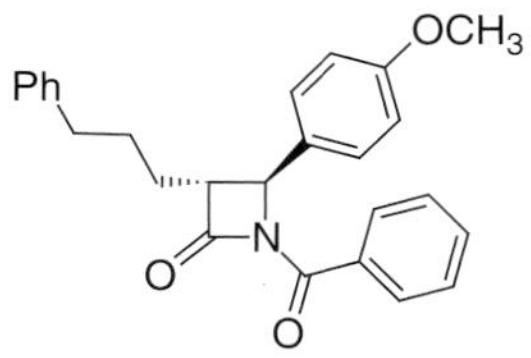

37
CE: NE at 50 mg kg^{-1}

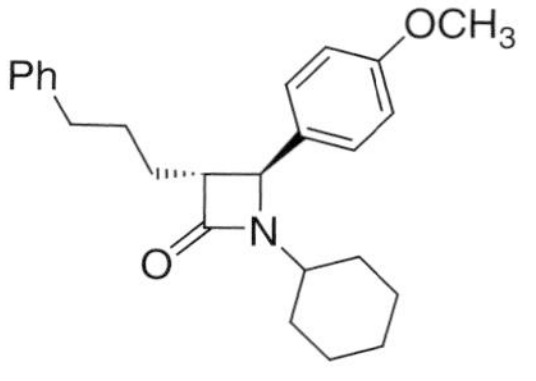

38
CE: NE at 50 mg kg^{-1}

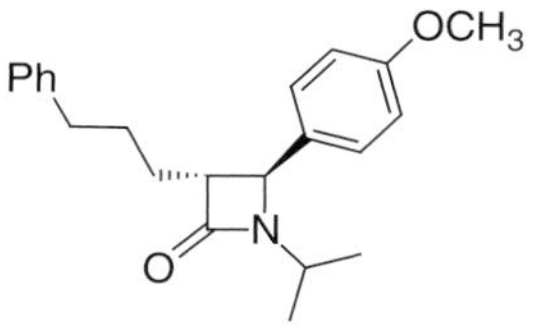

39
CE: −30% at 50 mg kg^{-1}

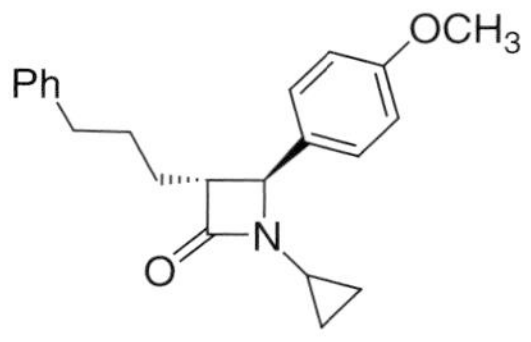

40
CE: −61% at 50 mg kg^{-1}

Compound	R^1	R^2	% Change CE	% Change SC
41	Ph(CH$_2$)$_2$	H	0	0
42	H	Ph(CH$_2$)$_2$	−19	0
43	Ph(CH$_2$)$_4$	H	−23	−8
44	H	Ph(CH$_2$)$_4$	−72	−28
45	C$_6$H$_{11}$(CH$_2$)$_3$	H	−28	−12
46	Ph(CH$_2$)$_3$	CH$_3$	−21	0
47	CH$_3$	Ph(CH$_2$)$_3$	−56	0

Figure 11 SAR at C3.

In the absence of an in vitro assay, it is impossible to unequivocally determine which of these effects on CAI activity reflect changes in intrinsic affinity for the unknown target of these compounds and which reflect purely pharmacokinetic influences. The striking difference in the metabolic stability of SCH 48461 versus its enantiomer is a clear example of the dangers of overinterpreting these data. Nonetheless, these initial studies of azetidinone CAIs demonstrated that activity in this series follows clearly defined SARs consistent with a well-defined molecular target. None of the compounds showed significant ACAT activity.

8.07.3.2 Rigid Analogs

Some follow-up studies focused on defining the active conformation of azetidinone CAIs, with the dual purpose of improving the potency of compounds as well as providing additional tools to understand the nature of the molecular target. That the conformation of the sidechain tether was important for activity was first suggested by the difference in activity of the *E*- and *Z*-propenyl derivatives **54** and **55** (**Figure 13**).[13] Additionally, the fact that potent compounds

48
CE: NE at 50 mg kg^{-1}

49
CE: NE at 50 mg kg^{-1}

50
CE: –52% at 50 mg kg^{-1}

51
CE: –47% at 50 mg kg^{-1}

52
CE: –51% at 50 mg kg^{-1}

53
CE: –43 at 50 mg kg^{-1}

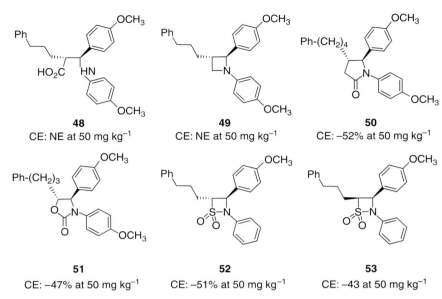

Figure 12 Nonazetidinone compounds.

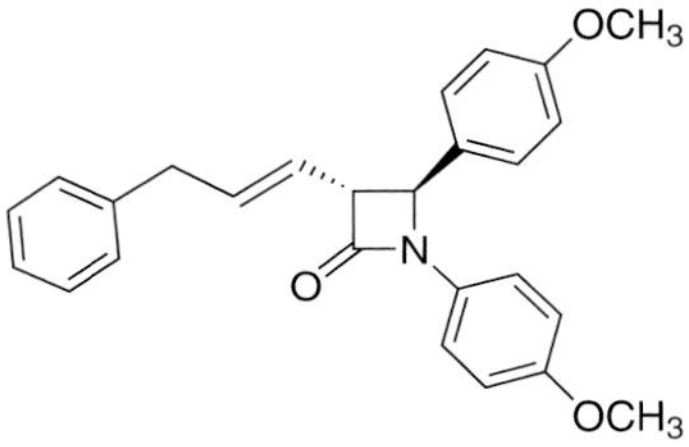

54
CE = –18% at 50 mg kg^{-1}

55
CE = –95% at 50 mg kg^{-1}

Figure 13 Activity of compounds **54** and **55**.

exist with either the 3-(*R*) or 3-(*S*) stereochemistry led Dugar *et al.* to propose that in the binding conformation, the tethered aromatic ring is located in a position in space that is equally accessible from both the 3-(*R*) and 3-(*S*) stereochemistries (**Figure 14**).[24] According to this model, the reduced potency of C3-disubstituted compounds results from steric hindrance that prevents the side chain from adopting the required conformation. Based on this model, they proposed a series of azaspirononanone derivatives which could exist in either a *syn* or *anti* form (**Figure 14**), both of which were consistent with the conformational model.

In practice, the *anti* derivative **56** (**Figure 15**) is slightly more active than **14** in the cholesterol-fed hamster model, while *syn* isomer **57** is much less active. Other aspects of the SARs in the spirocyclic series follow similar trends in the acyclic series, including the absolute stereochemistry and substitution pattern on the phenyl at the position equivalent to C4 of nonspirocyclic azetidinones. Unlike the acyclic compounds, enantiomers **60** and **61** in the spirocyclic series show reasonable metabolic stability, indicating that the stereochemical preference likely reflects differences in intrinsic activity. This provides some of the most compelling evidence that the azetidinone CAIs have a well-defined molecular target.

8.07.4 The Discovery of Ezetimibe

While the data discussed so far provided encouragement that the activity of this class of compounds could be optimized, the absence of an in vitro assay made it difficult to separate potential effects on intrinsic potency from effects on pharmacokinetics. This issue was confounded by the fact that **14** is extensively metabolized in vivo, making

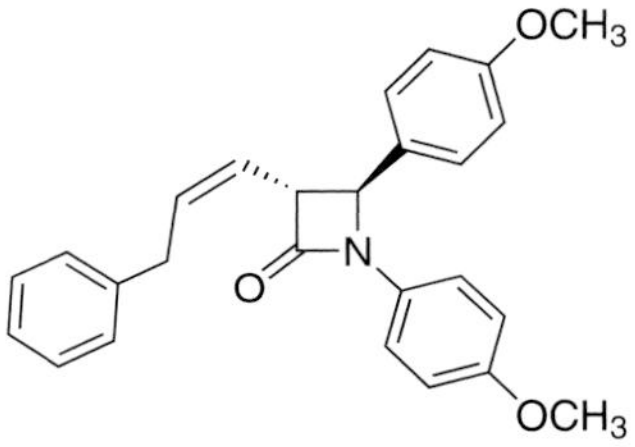

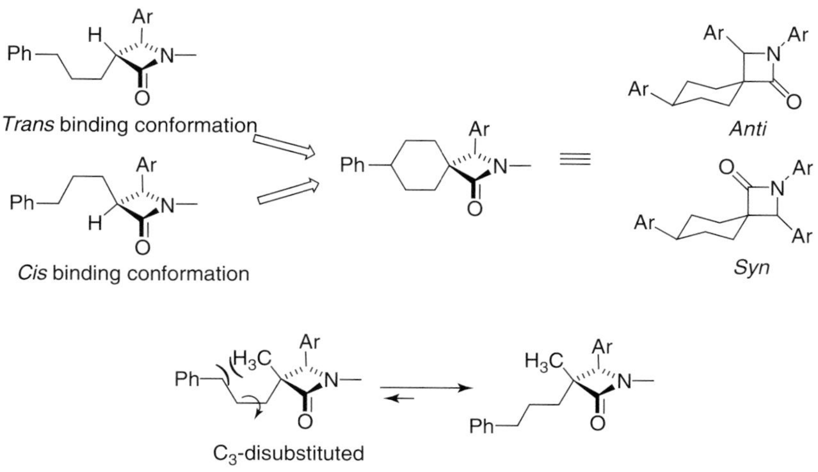

Figure 14 Conformational model leading to design of spirocyclic compounds.

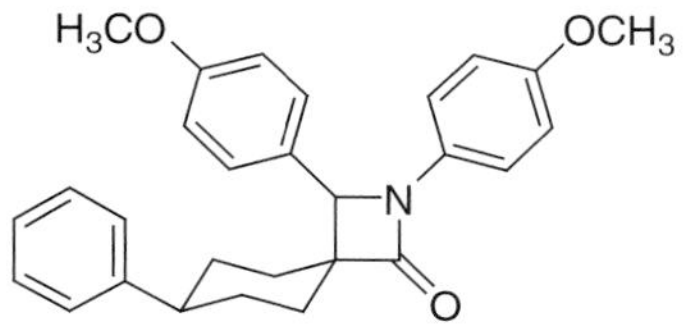

56
CE = −79% at 10 mg kg^{-1}

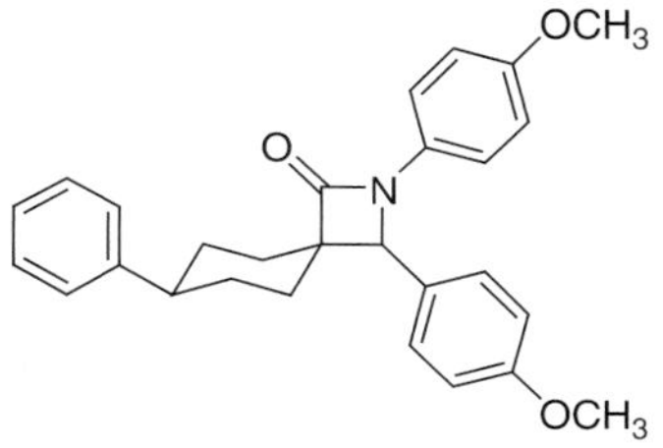

57
CE = −17% at 50 mg kg^{-1}

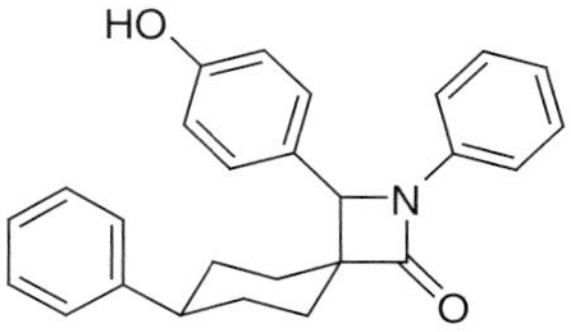

58
CE = −65% at 3 mg kg^{-1}

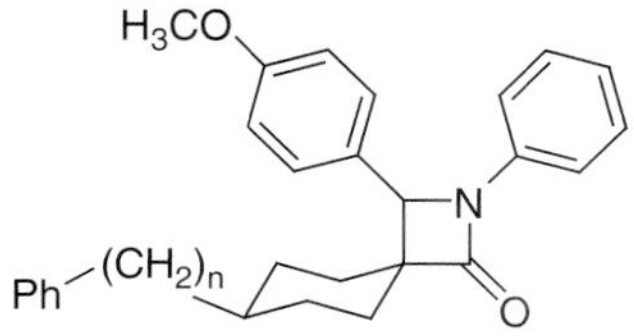

59
n = 1 or 2
CE = −43% at 50 mg kg

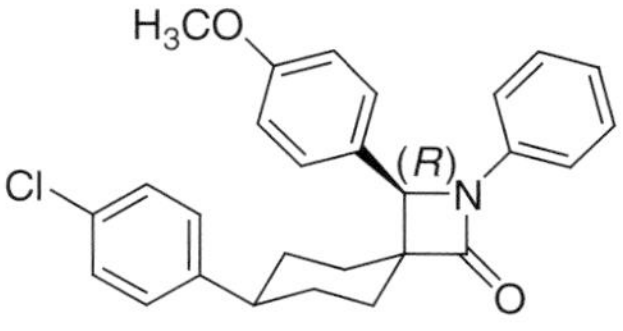

60
ED$_{50}$ = 0.66 mg kg^{-1}

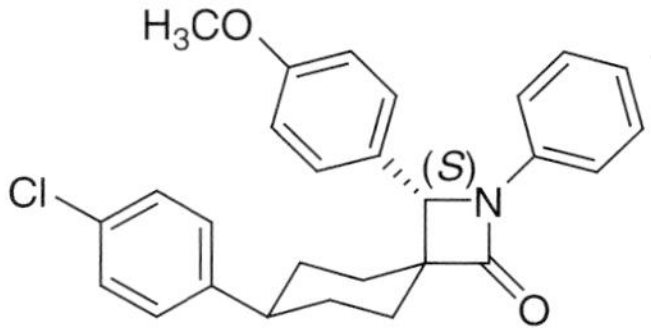

61
CE = −32% at 25 mg kg^{-1}

Figure 15 Spirocyclic derivatives.

the identity of the true active species unclear.[13] To address this latter issue, Van Heek and colleagues devised a unique experimental protocol designed to determine if there were active metabolites and how they contributed to the overall in vivo profile of **14**.[25] The experiment was divided into two parts, both of which used an intestinally cannulated, bile duct-diverted rat model (**Figure 16**).

In the first part of the experiment, animals were dosed via intraduodenal cannula with ^{3}H-**14**. Bile from these animals was collected via the bile duct cannula to give so-called 'metabolite' bile, which based on previous experiments was known to contain the majority of the metabolites of **14**. Concurrently, control animals were dosed with vehicle and their bile was also collected. ^{3}H-**14** was added directly to this bile to match the specific activity of the metabolite bile to give so-called 'parent' bile. In the second part of the experiment, both the metabolite bile and the parent bile were again dosed intraduodenally into a second group of diverted animals along with ^{14}C-cholesterol, and the counts of both ^{14}C and ^{3}H in various tissues were measured. In this way, both the pharmacological effects as well as the disposition of drug could be measured in a single experiment by following counts of ^{14}C and ^{3}H, respectively. Furthermore, because of

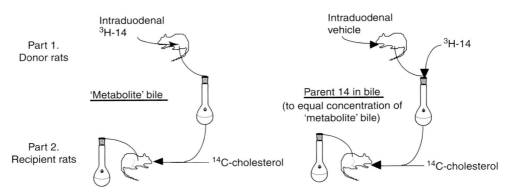

Figure 16   Intestinally cannulated bile duct-diverted rat protocol. (Reprinted with permission from Clader, J. W. *J. Med. Chem.* **2004**, *47*, 1–9. Copyright 2004 American Chemical Society.)

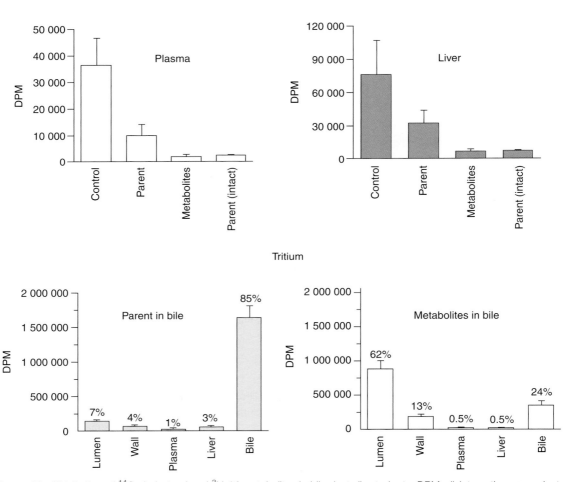

Figure 17   Distribution of ^{14}C-cholesterol and ^{3}H-**14** metabolites in bile duct-diverted rats. DPM, disintegrations per minute.

the bile duct diversion, the pharmacological effects reflected primarily the activity of the species being dosed and not the effects of any subsequently produced metabolites. In a third control arm of part 2, parent bile was dosed into intact, undiverted animals, where formation of active metabolites could still contribute to activity. **Figure 17** shows the results of these experiments.

While the parent bile reduced the appearance of ^{14}C cholesterol in both the plasma and liver, the metabolite bile was clearly more effective and reduced ^{14}C levels comparably to the parent compound **14** in intact animals. These data strongly suggested that formation of one or more active metabolites plays a significant role in the in vivo activity of **14**. Consistent with this, when animals were dosed with parent bile the majority of the tritium counts were recovered in the bile of recipient animals. However, when ^{3}H metabolite bile was dosed, the majority of the counts remained in the intestinal wall and lumen. These data suggest that not only are there active metabolites but that these metabolites localize at the putative site of action more efficiently than **14** itself.

Clearly the next important question was the identity of these active metabolites. To address this, metabolite bile was fractionated by high-performance liquid chromatographs (HPLC) and each fraction was evaluated according to the paradigm described in part 2 above. The results of this experiment are shown in **Figure 18**. Based on the appearance of ^{14}C-cholesterol in plasma, these data showed that the bulk of the activity resided in fraction 6, and subsequent analysis showed that this fraction was composed primarily of the glucuronide of compound **20**, a phenolic metabolite of **14**.

To complete the experiment, the activity of the metabolite bile, crude fraction **6**, and authentic **20** were compared in bile duct-diverted rats (**Figure 19**). These data show that the activity of **20** is identical to that of fraction 6 and both are substantially more active than the crude mixture of metabolites.

In total, these data strongly suggested that much of the in vivo activity of **14** was due to the formation of **20**. Furthermore, they suggested that metabolism of **14–20** helped to localize the compound in the intestines at the putative site of action of the compound.

The experiments by Van Heek *et al.* provided compelling evidence for the presence of at least one active metabolite of **14**. Nonetheless, there were also other less prominent metabolites whose formation could either contribute to the activity of **14** or could diminish it. To understand this, Rosenblum *et al.* prepared authentic samples of a number of known or putative metabolites which were evaluated for activity in the cholesterol-fed hamster (**Figure 20**).

Among the various putative metabolites of **14** were a variety of phenols produced by dealkylation of either or both of the methoxy groups or via aromatic hydroxylation. Both phenol **62** and bisphenol **63** showed substantial activity in the cholesterol-fed hamster, although neither was more active than **14** or **20**, suggesting that metabolism on the N-aryl moiety was not required for activity. On the other hand, phenol **64** was less active than **14**, suggesting that this route of metabolism might be detrimental to activity. In addition to the phenols, another route of metabolism involved hydroxylation of the 3-phenylpropyl side chain to produce alcohols and ketones. (*S*)-alcohol **65** was substantially more active than **14**, the (*R*)-alcohol **66** was less active, and the corresponding ketone **67** had intermediate activity. This

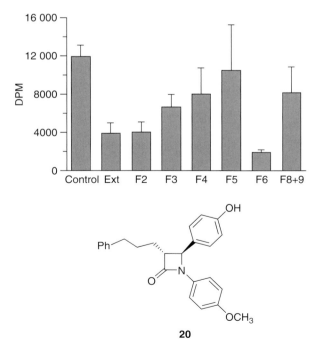

20

Figure 18 Efficacy of total bile extract (Ext) and fractionated bile.

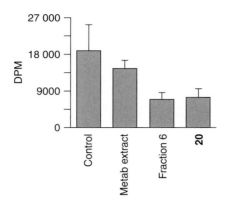

Figure 19   Efficacy of fraction 6 versus **20** in bile duct-diverted rats.

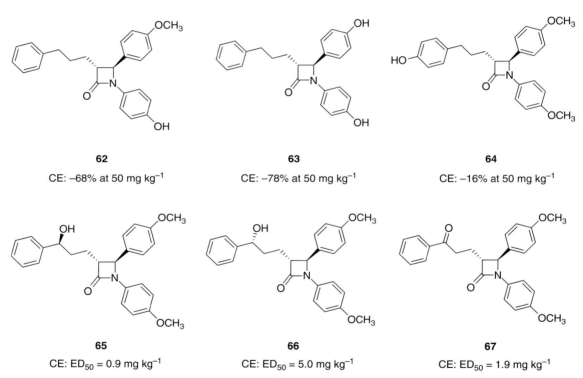

62	**63**	**64**
CE: −68% at 50 mg kg^{-1}	CE: −78% at 50 mg kg^{-1}	CE: −16% at 50 mg kg^{-1}
65	**66**	**67**
CE: ED$_{50}$ = 0.9 mg kg^{-1}	CE: ED$_{50}$ = 5.0 mg kg^{-1}	CE: ED$_{50}$ = 1.9 mg kg^{-1}

Figure 20   Activity of possible metabolites of **14** in cholesterol-fed hamster.

suggested that metabolism to the (*S*)-alcohol, if not required for activity, might improve the activity of the compound. Other combinations of these routes of metabolism were also investigated with similar results.

Based on the combined observations of experiments in bile duct-diverted rats and the activity of various putative metabolites, a strategy emerged for the design of a second-generation compound, namely:

1. Premetabolize profitable sites of metabolism on the C4 aryl and the phenylpropyl sidechain to improve activity, minimize plasma levels, and localize the compound in the intestines.
2. Block unprofitable sites of metabolism to maximize activity and limit further oxidative metabolism.

This strategy was in fact applied to a number of chemical series related to **14**,[26–29] but most successfully by Rosenblum *et al.* to give azetidinone **68**, now known as ezetimibe.[30]

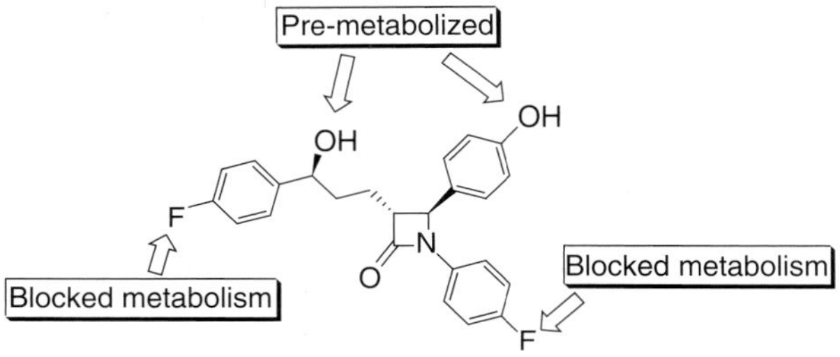

68 (Ezetimibe)

ED$_{50}$:	**14**	Ezetimibe
Hamster:	2.2 mg kg^{-1}	0.04 mg kg^{-1}
Rat:	2.0 mg kg^{-1}	0.03 mg kg^{-1}
Monkey:	0.2 mg kg^{-1}	0.0005 mg kg^{-1}
Dog:	0.1 mg kg^{-1}	0.007 mg kg^{-1}
Blood levels:		$\ll$ 14

Figure 21 Comparison of structure and in vivo profile of **14** and ezetimibe. (Reprinted with permission from Clader, J. W. *J. Med. Chem.* **2004**, *47*, 1–9. Copyright 2004 American Chemical Society.)

Figure 21 compares the activity of ezetimibe to **14** in a number of cholesterol-fed animal models.[25] In every case, but most dramatically in the monkey, ezetimibe is substantially more active than **14** and shows substantially lower plasma levels.

8.07.5 Synthesis

A number of syntheses of ezetimibe and related compounds have been reported, several of which were utilized in the course of these investigations.[31–38] Many of these are based on an Evans-type oxazolidinone condensation to establish the correct stereochemistry on the azetidinone ring. **Figure 22** shows a representative synthesis of ezetimibe. In addition to the use of the oxazolidinone, this synthesis also features a Corey oxazaborolidine reduction to set the (*S*) stereochemistry of the side chain hydroxyl group.

8.07.6 Ezetimibe and Statins

All of the in vivo data described thus far involve animals fed diets that are substantially higher in fat and cholesterol than the animals' normal chow diet. Despite the substantial activity of ezetimibe and other azetidinone CAIs in these models, none of the compounds tested significantly reduced plasma cholesterol levels in animals fed a normal chow diet. While this was initially a concern, this observation ultimately led to one of the most important aspects of the profile of ezetimibe. In considering the possible reasons for the lack of substantial effect on serum cholesterol in the absence of a high cholesterol diet, Davis reasoned that a CAI might stimulate hepatic hydroxymethylglutaryl-coenzyme A (HMG-CoA) reductase activity. This could compensate for a reduced cholesterol load due to inhibition of intestinal cholesterol absorption. If this were the case, then coadministration of a CAI and an HMG-CoA reductase inhibitor should produce an enhanced reduction in serum cholesterol at doses that were less effective or ineffective as monotherapy. To test this hypothesis, Davis *et al.* administered ezetimibe (0.007 mg kg^{-1}) or lovastatin (5 mg kg^{-1}) to chow-fed dogs over 14 days.[39] While neither compound had a substantial effect on serum cholesterol alone, the combination produced a profound reduction in serum cholesterol (**Figure 23**).

Similar experiments demonstrated a comparable effect in other species and with other statins. These data suggested that ezetimibe would be effective at reducing cholesterol levels in humans and would be particularly effective in combination HMG-CoA reductase inhibitors.

8.07.7 Clinical Results

Human clinical trials with ezetimibe supported the expectations of animal studies with ezetimibe both as monotherapy and in combination with statins.[40–43] **Table 1** shows the results of phase III human trials with ezetimibe as

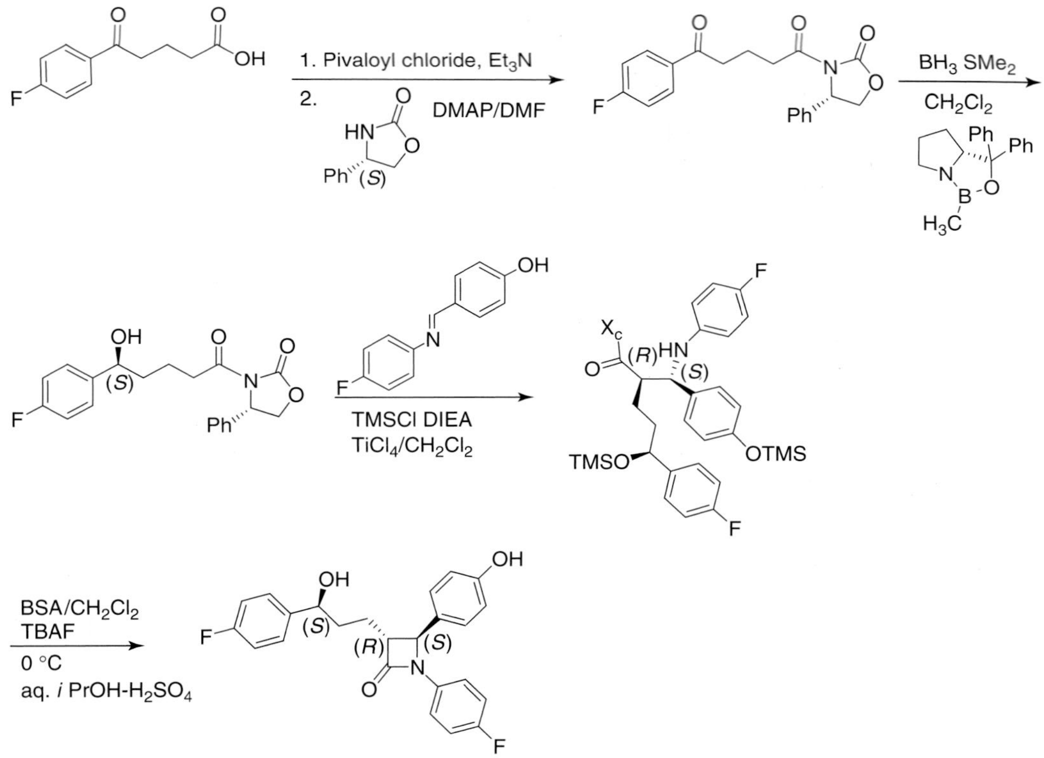

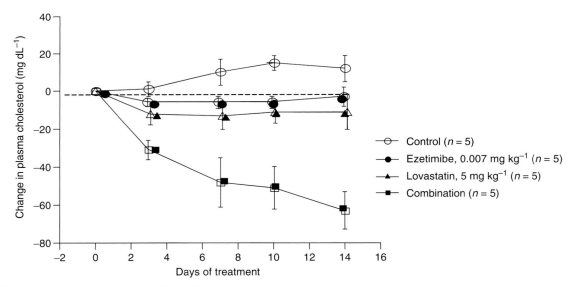

Figure 22 Representative synthesis of ezetimibe.

Figure 23 Effect of ezetimibe and lovastain in chow-fed dogs.

monotherapy. Ezetimibe produced a significant reduction in total cholesterol, LDL cholesterol, and triglycerides as well as a small but significant increase in HDL cholesterol.

Figure 24 compares the effect of simvastatin or atorvastatin either alone or when coadministered with ezetimibe on LDL cholesterol. In each case, ezetimibe produced an additional 15–18% reduction in LDL cholesterol above that

Table 1 Ezetimibe phase III monotherapy efficacy results

Treatment	Mean % change from baseline at endpoint			
	LDL cholesterol	Total cholesterol	HDL cholesterol	Triglycerides
Placebo (n = 226)	+0.4	+0.8	−1.6	+5.7
Ezetimibe 10 mg (n = 666)	−16.9*	−12.5*	+1.3*	−5.7*

*Significantly different from placebo (p < 0.01).

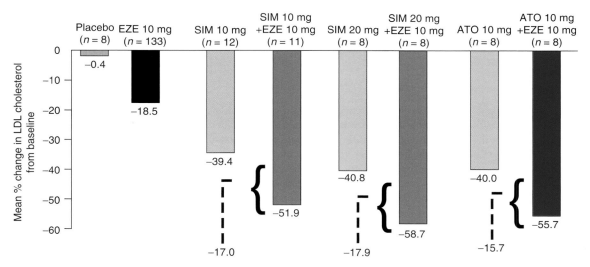

Figure 24 Ezetimibe (EZE) coadministered with simvastatin (SIM) or atorvastatin (ATO).

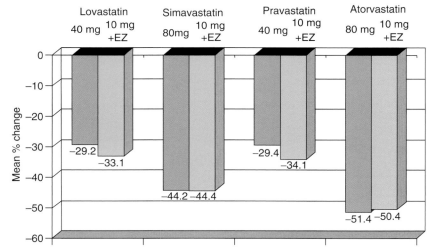

Figure 25 Effect of high and low dose coadministered with ezetimibe (EZ).

achieved by the statin alone. Finally, **Figure 25** compares the effect of a high dose of statin alone with a low dose of statin coadministered with ezetimibe. In each case, coadministration of ezetimibe and low-dose statin produced an equivalent reduction in LDL cholesterol as the high dose of statin alone. Combined, these data demonstrate that ezetimibe alone or coadministration with statins provides favorable effects on the major lipid parameters in patients with hypercholesterolemia.

8.07.8   Mechanism of Action Studies

Clearly one of the important goals in this area has been identification and characterization of the molecular target of these compounds and the establishment of an in vitro assay. Considerable progress has been made in this area since the discovery of ezetimibe, with the primary focus being on Niemann–Pick C1 Like 1 Protein (NPC1L1). Altman, Davis *et al.* have shown that NPC1L1 is critical for the uptake of cholesterol by intestinal enterocytes and is a key modulator of whole-body cholesterol homeostasis. NPC1L1-null mice were completely insensitive to ezetimibe, suggesting that this or an associated protein may be the molecular target of this class of compounds.[44,45] Recent data showing that several ezetimibe analogs bind to NPC1L1 provide compelling evidence that this protein is in fact the molecular target of ezetimibe.[46] The identification of this protein is an important scientific achievement not just for its role in promoting the development of this class of drugs but also for furthering our understanding of the mechanisms of cholesterol absorption and cholesterol homeostasis.

8.07.9   Conclusion

The past 30 years of drug discovery have seen a procession of breakthrough technologies designed to improve the efficiency and overall success rate of drug discovery programs. Ironically, the fact that drug discovery remains a difficult and risky endeavor despite the use of these technologies has caused some to question the value of these technologies compared to more traditional biologically driven approaches towards drug discovery, with ezetimibe often cited as a case in point. The technological advances of the past decades have unquestionably had a profound and positive impact on the discovery process, and new technologies such as genomics, which played a pivotal role in elucidating the mechanism of action of ezetimibe, will likely continue to shape how we discovery drugs in the future. The fact that ezetimibe was discovered without the benefit of many of these technologies is less a testament to the relative value of biological versus technological approaches than it is a reaffirmation of the continued importance of science and scientists in drug discovery.

References

1. American Heart Association. *HeartDisease and Stroke Statistics*: 2004 Update; American Heart Association: Dallas, TX, 2003.
2. Davidson, M. H.; Toth, P. P. *Prog. Cardiovasc. Dis.* **2004**, *47*, 73–104.
3. Dunbar, R. L.; Rader, D. J. *Drug Disc. Today* **2004**, *1*, 169–176.
4. Bruckert, E. *Cardiology* **2002**, *97*, 59–66.
5. Shepherd, J. In *73rd European Atherosclerosis Society Congress*, Salzburg, Austria, July 7, 2002.
6. Stein, E.; Stender, S.; Mata, P.; Ponsonnet, D.; Melani, L.; Lipka, L.; Suresh, R.; Veltri, E. In *73rd European Atherosclerosis Society Congress*, Salzburg, Austria, July 9, 2002.
7. Stein, E. A. *Am. J. Manag. Care* **2002** *8*, S36–S39; discussion S45–S47.
8. Al-Shaer, M. H.; Choueiri, N. E.; Suleiman, E. S. *Lipids in Health and Disease* **2004**, *3*, 22.
9. Wang, H. H.; Lammert, F.; Wang, D. Q.-H. *Vasc. Dis. Prevent.* **2005**, *2*, 165–175.
10. Murdoch, D.; Scott, L. J. *Am. J. Cardiovasc. Drugs* **2004**, *4*, 405–422.
11. Flores, N. A. *Curr. Opini. Invest. Drugs* **2004**, *5*, 984–992.
12. Rees, A. *Br. J. Diabetes Vasc. Dis.* **2004**, *4*, 163–171.
13. Some aspects of the case history of the discovery of ezetimibe and structure–activity relationships of this class of compounds have been reviewed previously. See Clader, J. W. *J. Med. Chem.* **2004**, *47*, 1–9; Clader, J. W. *Curr. Topics Med. Chem.* **2005**, *5*, 243–256; Burnett, D. A. *Curr. Med. Chem.* **2004**, *11*, 1873–1887.
14. Sliskovic, D. R.; Picard, J. A.; Krause, B. R. *Prog. Med. Chem.* **2002**, *39*, 121–171.
15. Schnitzer-Polokoff, R.; Compton, D.; Boykow, G.; Davis, H.; Burrier, R. *Comp. Biochem. Physiol. A* **1991**, *99*, 665–670.
16. Clader, J. W.; Berger, J. G.; Burrier, R. E.; Davis, H. R.; Domalski, M.; Dugar, S; Kogan, T. P.; Salisbury, B.; Vaccaro, W. *J. Med. Chem.* **1995**, *38*, 1600–1607.
17. Vaccaro, W.; Amore, C.; Berger, J.; Burrier, R.; Clader, J.; Davis, H.; Domalski, M.; Fevig, T.; Salisbury, B.; Sher, R. *J. Med. Chem.* **1996**, *39*, 1704–1719.
18. Sliskovic, D. R.; White, A. D. *Trends Pharmacol. Sci.* **1991**, *12*, 194–199.

19. Burnett, D. A.; Caplen, M. A.; Davis, H. R., Jr.; Burrier, R. E.; Clader, J. W. *J. Med. Chem.* **1994**, *37*, 1733–1736.
20. Clader, J. W.; Burnett, D. A.; Caplen, M. A.; Domalski, M. S.; Dugar, S.; Vaccaro, W.; Sher, R.; Browne, M. E.; Zhao, H.; Burrier, R. E.; Salisbury, B.; Davis, H. R., Jr. *J. Med. Chem.* **1996**, *39*, 3693–3694.
21. Salisbury, B. G.; Davis, H. R.; Burrier, R. E.; Burnett, D. A.; Boykow, G.; Caplen, M. A.; Clemmons, A. L.; Compton, D. S.; Hoos, L. M.; McGregor, D. G.; Schnitzer-Polokoff, R.; Smith, A. A; Weig, B. C.; Zilli, D. L.; Clader, J. W.; Sybertz, E. J. *Atherosclerosis* **1995**, *115*, 45–63.
22. Vaccaro, W. D.; Sher, R.; Davis, H. R., Jr. *Bioorg. Med. Chem.* **1998**, *6*, 1429–1437.
23. Dugar, S.; Kirkup, M.; Clader, J. W.; Lin, S.-I.; Rizvi, R.; Snow, M. E.; Davis, H. R., Jr.; McCombie, S. W. *Bioorg. Med. Chem. Lett.* **1995**, *5*, 2947–2952.
24. Dugar, S.; Clader, J. W.; Chan, T.-M.; Davis, H. R. *J. Med. Chem.* **1995**, *38*, 4875–4877.
25. Van Heek, M.; France, C. F.; Compton, D. S.; McLeod, R. L.; Yumibe, N. P.; Alton, K. B.; Sybertz, E. J.; Davis, H. R., Jr. *J. Pharm. Exper. Ther.* **1997**, *283*, 157–163.
26. McKittrick, B. A.; Ma, K.; Dugar, S.; Clader, J. W.; Davis, H., Jr.; Czarniecki, M. *Bioorg. Med. Chem. Lett.* **1996**, *6*, 1947–1950.
27. Dugar, S.; Yumibe, N.; Clader, J. W.; Vizziano, M.; Huie, K.; Van Heek, M.; Compton, D. S.; Davis, H. R., Jr. *Bioorg. Med. Chem. Lett.* **1996**, *6*, 1271–1274.
28. Kirkup, M. P.; Rizvi, R.; Shankar, B.; Dugar, S.; Clader, J. W.; McCombie, S. W.; Lin, S.-I.; Yumibe, N. *Bioorg. Med. Chem. Lett.* **1996**, *6*, 2069–2072.
29. McKittrick, B. A.; Ma, K.; Huie, K.; Yumibe, N.; Davis, H. R., Jr.; Clader, J. W.; Czarniecki, M.; McPhail, A. T. *J. Med. Chem.* **1998**, *41*, 752–759.
30. Rosenblum, S. B.; Huynh, T.; Afonso, A.; Davis, H. R., Jr.; Yumibe, N.; Clader, J. W.; Burnett, D. A. *J. Med. Chem.* **1998**, *41*, 973–980.
31. Homann, M. J.; Previte, E. Stereoselective microbial reduction for the preparation of 1-(4-fluorophenyl)-3(R)-[3(S)-Hydroxy-3-(4-fluorophenyl)propyl)]-4(S)-(4-hydroxyphenyl)-2-azetidinone. US Patent 6,133,001, Oct 17, 2000.
32. Thiruvengadam, T. K.; Fu, X.; Tann, C.-H.; Mcallister, T. L.; Chiu, J. S.; Colon, C. Process for the synthesis of azetidinones. PCT Int. Patent WO200034240, 2000.
33. Wu, G.; Wong, Y.; Chen, X.; Ding, Z. *J. Org. Chem.* **1999**, *64*, 3714–3718.
34. Wu, G.-Z.; Chen, X.; Wong, Y.-S.; Schumacher, D. P.; Steinman, M. 3-hydroxy gamma-lactone based enantioselective synthesis of azetidinones. US Patent 5,886,171, Mar 23, 1999.
35. Shankar, B. B. Process for preparing 1-(4-fluorophenyl)-3(R)-(3(S)-hydroxy-3-([phenyl or 4-fluorophenyl])-propyl)-4(S)-(4-hydroxyphenyl)-2-azetidinone. US Patent 5,856,473, June 5, 1999.
36. Wu, G.-Z.; Chen, X.; Wong, Y.-S.; Schumacher, D. P.; Steinman, M. 3-Hydroxy gammalactone based enantioselective synthesis of azetidinones. PCT Int. Patent WO199745406, 1997.
37. Shankar, B. B.; Kirkup, M. P.; McCombie, S. W.; Clader, J. W.; Ganguly, A. K. *Tetrahedron Lett.* **1996**, *37*, 4095–4098.
38. Burnett, D. A. *Tetrahedron Lett.* (**1994**) *35*, 7339-7342.
39. Davis, H. R., Jr.; Pula, K. K.; Alton, K. B.; Burrier, R. E.; Watkins, R. W. *Metab. Clin. Exp.* **2001**, *50*, 1234–1241.
40. Dujovne, C. A.; Ettinger, M. P.; McNeer, J. F.; Lipka, L. J.; LeBeaut, A. P.; Suresh, R.; Yang, B. O.; Veltri, E. P. *Am. J. Cardiol.* **2002**, *90*, 1092–1097.
41. Davidson, M. H.; McGarry, T.; Bettis, R.; Melani, L.; Lipka, L. J.; LeBeaut, A. P.; Suresh, R.; Sun, S.; Veltri, E. P. *J. Am. Coll. Cardiol.* **2002**, *40*, 2125–2134.
42. Gagne, C.; Bays, H. E.; Weiss, S. R.; Mata, P.; Quinto, K.; Melino, M.; Cho, M.; Musliner, T. A.; Gumbiner, B. *Am. J. Cardiol.* **2002**, *90*, 1084–1091.
43. Ballantyne, C. M.; Houri, J.; Notarbartolo, A.; Melani, L.; Lipka, L. J.; Suresh, R.; Sun, S.; LeBeaut, A. P.; Sager, P. T.; Veltri, E. P. *Circulation* **2003**, *107*, 2409–2415.
44. Altmann, S. W.; Davis, H. R., Jr.; Zhu, L.-J.; Yao, X.; Hoos, L. M.; Tetzloff, G.; Iyer, S. P. N.; Maguire, M.; Golovko, A.; Zeng, M.; Wang, L.; Murgolo, N.; Graziano, M. P. *Science* **2004**, *303*, 1201–1204.
45. Davis, H. R., Jr.; Zhu, L.-J.; Hoos, L. M.; Tetzloff, G.; Maguire, M.; Liu, J.; Yao, X.; Iyer, S. P. N.; Lam, M.-H.; Lund, E. G.; Detmers, P. A.; Graziano, M. P.; Altmann, S. W. *J. Biol. Chem.* **2004**, *279*, 33586–33592.
46. Garcia-Calvo, M.; Lisnock, J.; Bull, H. G.; Hawes, B. E.; Burnett, D. A.; Braun, M. P.; Crona, J. H.; Davis, H. R., Jr.; Dean, D. C.; Detmers, P. A.; Graziano, M. P.; Hughes, M.; MacIntyre, D. E.; Ogawa, A.; O'Neill, K.; Iyer, S. P. N.; Shevell, D. E.; Smith, M. M.; Tang, Y.-S.; Makarewicz, A. M.; Ujjainwalla, F.; Altmann, W. S.; Chapman, K.; Thornberry, N. A. *Proc. Natl. Acad. Sci. USA* **2005**, *102*, 8132–8137.

Biography

John W Clader received his PhD in organic chemistry in 1980 from Indiana University. After two years at the University of Notre Dame he began his career in medicinal chemistry at Hoffmann-La Roche in Nutley, New Jersey. He moved to the Schering-Plough Research Institute in 1985, where he is currently Distinguished Research Fellow in Medicinal Chemistry. In addition to ezetimibe, his research interests at Schering-Plough have included potential treatments for Alzheimer's disease and HIV infection as well as the application of chemoinformatics and other computer methods to facilitate drug discovery.

Comprehensive Medicinal Chemistry II
ISBN (set): 0-08-044513-6

ISBN (Volume 8) 0-08-044521-7; pp. 65–82

8.08 Tamoxifen

V C Jordan, Fox Chase Cancer Center, Philadelphia, PA, USA

8.08.1	**Background and Introduction**	83
8.08.2	**Nonsteroidal Antiestrogens**	83
8.08.3	**The Discovery of ICI46,474**	86
8.08.4	**Studies Published by Scientists at ICI Pharmaceuticals Division**	86
8.08.5	**Patenting Problems**	88
8.08.6	**A Conversation between the Laboratory and the Clinic**	89
8.08.7	**Twenty-First Century View of Tamoxifen as a Treatment for Breast Cancer**	91
8.08.8	**Concerns about Tamoxifen**	91
8.08.9	**Selective Estrogen Receptor Modulation**	94
8.08.10	**Tamoxifen and Breast Cancer Prevention**	94
8.08.11	**Current Chemoprevention**	96
8.08.12	**Tamoxifen's Legacy: A Menu of Medicines**	96
	References	98

8.08.1 Background and Introduction

In 1896 George Beatson[1] demonstrated that removal of the ovaries from premenopausal women could cause the regression of breast cancer. By the turn of the century it was established[2] that about one-third of all premenopausal women with advanced breast cancer could benefit from oophorectomy and from that time, a principal strategy for the treatment and prevention of breast cancer has been either to block or to restrict the action of estradiol in its target tissue, the breast. However, the successful clinical development of the antiestrogenic drug tamoxifen did not initially focus on the therapy for breast cancer but evolved to this application by drawing upon expertise in several unrelated disciplines. Most of the early interest in antiestrogens was focused on reproductive endocrinology but it was clear from the beginning of clinical studies that the effects of the drugs on cholesterol biosynthesis would play a pivotal role in assessing safety considerations for long-term therapy. Ultimately the discovery of the estrogen receptor[3,4] in the 1960s and the application of this basic knowledge to understand hormone-dependent breast cancer growth[5] focused interest on the development of tamoxifen as a targeted agent to block estrogen action in the tumor directly.

8.08.2 Nonsteroidal Antiestrogens

Tamoxifen was not the first antiestrogen but the value of the drug slowly increased as the fashions in research changed. During the late 1950s and throughout the 1960s there was a focus on the development of new contraceptives in the wake of the success of oral contraceptives. These medicines did not treat a disease but altered lifestyle so there was huge potential for widespread use. But the application of antiestrogens as contraceptives failed and the compounds were drugs looking for a disease to treat! This perspective changed following the start of the 'War on Cancer' declared on 23 December 1971. There were now incentives to conduct translational cancer research and introduce new treatments. Regrettably, the process was slow for the introduction of targeted treatments as all hopes were initially pinned on combination cytotoxic chemotherapy to cure cancer.

In 1958, Lerner and coworkers at the William S. Merrell Company reported the biological properties of the first nonsteroidal antiestrogen, MER25 (**Figure 1**).[6] The discovery of nonsteroidal antiestrogens was an example of serendipity and is best described in Lerner's own words[7]:

> As 1954 was drawing to its end a triphenylethanol compound was synthesized, not for the purpose of investigation for estrogen antagonism but for testing by the cardiovascular research section at Merrell since it had been

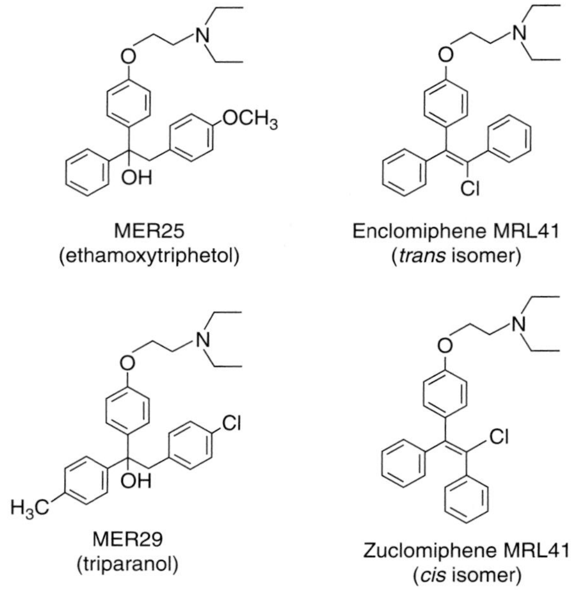

Figure 1 The formulae of the first nonsteroidal antiestrogen, MER25, and the structurally related compound MER29 used to lower circulating cholesterol. MER29 was withdrawn from the market because of the increased incidence of cataracts. The next antiestrogen, MRL41, was a mixture of the *cis* and *trans* isomers of a substituted triphenylethylene later named clomiphene.

reported that a related compound had some effect on blood flow. A request for a sample of that compound for study as a possible estrogen antagonist was answered by the cardiovascular system pharmacologist with his entire supply since it was essentially inactive in his studies. This compound, 1-(*p*-1 diethylamino-ethoxyphenyl)-2(*p*-methoxyphenyl)-1-phenylethanol was tested in immature mice at the arbitrary 3-day screening dose of 5 mg. It was administered subcutaneously twice daily for three days alone or in combination with 0.03 μg of estradiol benzoate, and the uterine weight and intraluminal fluid served as the end points to be measured on the day after the last treatment. The results of this study were highly questioned since neither the uteri of the mice administered the compound alone nor the uteri of the animals receiving the compound plus estrogen were significantly heavier than those of controls treated with olive oil vehicle alone. It was thought that this was a 'bad study.' The compound, however, was retested and the results were identical to those of the first study. The increase in uterine weight and intraluminal fluid by estradiol treatment was completely prevented by simultaneous administration of the compound that was eventually to be called MER25 or ethamoxytriphetol.

The compound was found to be an antiestrogen in all species tested and was found to have no other hormonal or antihormonal properties. However, the discovery MER25 was considered to be of importance at the time because the compound was a postcoital contraceptive in laboratory animals.[8–10] Obviously, one application could have been as a 'morning-after pill' but after clinical evaluation in numerous situations the results were disappointing. MER25 underwent initial evaluation for the induction of ovulation[11] and the treatment of chronic cystic mastitis, breast and endometrial carcinoma[12,13] but the low potency and severe side effects on the central nervous system prohibited further clinical development.[7]

It is relevant to point out that the antiestrogen MER25 is a structural derivative of the cholesterol-lowering drug triparanol (MER29) (**Figure 1**). In the late 1950s there was initial enthusiasm about the potential benefits of triparanol as a hypocholesterolemic drug.[14] However, the finding that triparanol caused an accumulation of desmosterol (an intermediate in cholesterol biosynthesis)[15–18] and the linking of this biochemical effect to cataract formation,[19–21] caused withdrawal of the drug in 1962 (**Figure 2**). Nevertheless, triparanol was first evaluated as a potential therapy for breast cancer[22] but again the results were disappointing.

A successor compound to MER25, MRL41 or clomiphene (**Figure 1**), was a more potent antiestrogen but drug development for long-term use was to be retarded because of toxicological concerns. Clomiphene is an effective antifertility agent in laboratory animals[23] but paradoxically induces ovulation in subfertile women.[24–26] Again, the prospect of developing a 'morning-after pill' for women was not realized. Although clomiphene showed some activity in

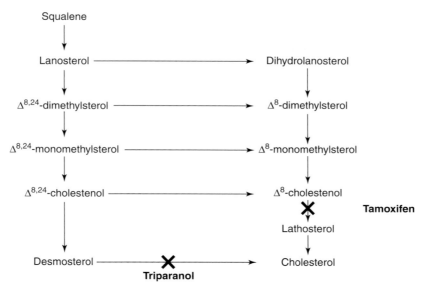

Figure 2 The mechanism of action of triparanol is to inhibit cholesterol biosynthesis but also to increase demosterol levels which is implicated in cataract formation. In contrast, tamoxifen has an alternate mechanism of action to lower cholesterol levels.

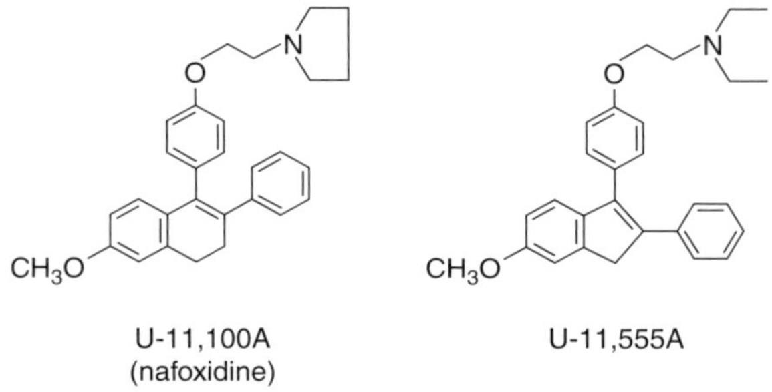

Figure 3 The formulae of compounds discovered by the Upjohn Company as antifertility agents. Nafoxidine subsequently was tested as a therapy for breast cancer but development was not pursued because of toxic side effects experienced by the majority of patients. Two observations merit comment. The fixed ring structure confirmed the idea that a 'trans'-like' structure was necessary for potent antifertility and, later, antiestrogenic activity. The compound U-11,555A became a template for numerous agents developed throughout the 1990s and into the twenty first Century.

the treatment of advanced breast cancer,[27,28] the drug was only developed for short-term use for the induction of ovulation[29] because desmosterol was noted in patient sera during prolonged treatment.[30]

Clomiphene is marketed as an impure mixture of geometric isomers (**Figure 1**) which have opposing biological activities: one isomer is an estrogen and one isomer is an antiestrogen.[31] Unfortunately the isomers were initially (1967) given the incorrect designation but this was corrected by 1976.[32] Although breast cancer clinical trials were still being reported with the impure mixture of isomers in 1974,[28] the antiestrogenic isomer eventually entered into clinical trial for breast cancer treatment but the studies were dropped by the National Institutes of Health (NIH) due to the interest in tamoxifen (**Figure 2**).[33]

In contrast, the compound nafoxidine (U-11,100A)[34,35] and U-11,555A[36] are dihydronaphthalene and indene derivatives respectively and therefore cannot isomerize (**Figure 3**). The drugs are potent antiestrogens with antifertility properties in laboratory animals.[37–39] Subsequent studies by Terenius[40] demonstrated tight and somewhat irreversible binding of nafoxidine to the estrogen receptor derived from mouse target tissues. Nafoxidine exhibits antitumor properties in laboratory models of kidney cancer in the hamster[41] and the dimethylbenzanthracene (DMBA)-induced rat mammary carcinoma model[42] but following extensive testing as a treatment for breast and renal cancer, the drug was not developed further because of unacceptable side effects experienced by all patients.[43,44]

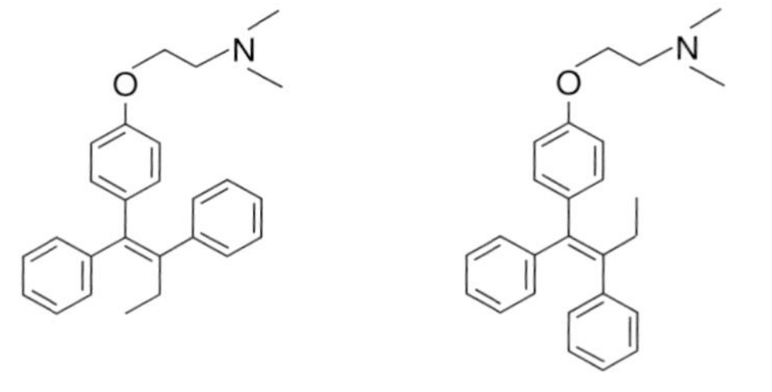

ICI46,474 (tamoxifen)   ICI47,699 (*cis* isomer)

Figure 4 The formula of the mixed estrogen/antiestrogen ICI46,474 and its estrogenic isomer ICI47,699 discovered in antifertility screens at ICI Pharmaceuticals Division. The *trans* isomer designated ICI46,474 was eventually developed as the breast cancer drug tamoxifen (Nolvadex).

8.08.3   The Discovery of ICI46,474

Although the discovery of ICI46,474 (tamoxifen) (**Figure 4**), the antiestrogenic, pure *trans* isomer of a substituted triphenylethylene, was made by Harper, Richardson, and Walpole[45–47] as part of the Fertility Control Program at ICI Pharmaceuticals (now AstraZeneca), Cheshire, UK, the study of cancer therapies was Walpole's long-term interest.[48]

In the late 1940s Walpole was a member of staff at ICI's Dyestuffs Division Biological Laboratory in Wilmslow, Cheshire. This establishment was the fledgling predecessor of the Pharmaceuticals Division Research Laboratories built in 1956–57 at Alderley Park near Macclesfield, Cheshire. Walpole was asked to establish animal models for the bioassay of potential alkylating agents[49–54] to evaluate compounds as bladder carcinogens[55] and to assess the potential health hazards for workers in the dyestuffs industry.[56] Walpole made the important discovery that tris-ethyleneimino-S-triazine (M9500) was an active anticancer agent in the Walker rat carcinoma 256[50] and conducted extensive structure–activity relationship studies with many mono and bifunctional compounds. Although tris-ethyleneimino-S-triazine is now only of historical interest, the related compound, hexamethylmelamine (M10,567) was also found to be active by Walpole.[52] Hexamethylmelamine is active in a broad spectrum of tumors including ovarian carcinoma resistant to other alkylating agents. However, Walpole's interest in alkylating agents and carcinogenesis provides only part of the background which led to his suggestion that ICI46,474 should be tested in breast cancer. Walpole was also interested in estrogens and was aware in 1963 that antiestrogens could be used for the treatment of breast cancer. The story of ICI's involvement with the hormonal treatment of breast cancer goes back to the early 1940s following the laboratory discovery by others that chlorotriphenylethylene is an orally active estrogen.[57] ICI supplied the triphenylethylenes initially used by Haddow, Watkinson, and Paterson in their landmark study of the antitumor effects of synthetic estrogens in advanced breast and prostate cancer.[58] These studies paved the way for the standard use of high-dose estrogen therapy to treat both breast and prostate cancer for the next two decades. In 1949, Walpole and Paterson[59] studied the antitumor actions of nonsteroidal estrogen therapy on breast cancer in the hope of finding the reason why some patients responded and others did not. These early studies by Walpole were unsuccessful but the subcellular mechanism of hormonal-dependent growth was eventually discovered by Jensen with his pioneering work on the estrogen receptor.[3,60]

8.08.4   Studies Published by Scientists at ICI Pharmaceuticals Division

The focus of the fertility control research program at ICI Pharmaceuticals Division was to investigate the physiology of reproduction in laboratory animals, use interesting antihormonal agents to block the reproductive process and to propose possible clinical evaluations of novel compounds.[45,46,61–77]

The work of others, testing compounds from competing pharmaceutical companies, established the activity and potential value of antiestrogens for the induction of ovulation in subfertile women.[12,13,24,26,78–80] Cancer therapy was a remote possibility as the treatment strategy at the time for advanced (metastatic) breast cancer was either endocrine ablation or additive high-dose hormonal therapy.[81] Although additive hormonal therapy was cheap, only one in three patients responded and the average response duration was 1 year. Nevertheless, a few unsuccessful attempts to test antiestrogens had occurred (**Table 1**).

Walpole's team of endocrinologists concentrated exclusively on the study of ovulation so that it could be prevented. 'No egg no pregnancy,' to quote Gregory Pincus, the Director of the Worcester Foundation, when describing how oral contraceptives worked. Additionally, the team at ICI Pharmaceuticals Division studied implantation so that the process

Table 1 Preliminary clinical trials of antiestrogens for the treatment of metastatic breast cancer

Compound	Reference	Daily dose (mg)	Total patients (% response)	Toxicities
Triparanol	Kraft (1962)[22]	250–1000	8(11)	None[a]
MER25	Kistner and Smith (1960)[12]	500–4500	4(25)	Acute psychotic episode[b]
Clomiphene	Herbst et al. (1964)[27]	100–300	56(34)	None reported or mild[c]
	Hecker et al. (1974)[28]			
Nafoxidine	Legha et al. (1976)[44]	180–240	198(31)	Bilateral cataracts, ichthyosis Cutaneous photophobia[d]
Tamoxifen	Cole et al. (1971)[87]	20–40	114(31)	Transient thrombocytopenia[e]

[a] Withdrawn from the market by the William S. Merril Co. in cooperation with the Food and Drug Administration in April 1962.
[b] Does not include patients treated by Dr Roy Hertz when therapy was stopped due to hallucinations.[7]
[c] Visual symptoms.[30]
[d] Affecting 80–100% of patients.[44]
[e] "The particular advantage of this drug is the low incidence of side effects"[87]; "The side effects were usually trivial."[88]

could be prevented to avoid pregnancy. Walpole, Harper, and subsequently Labhsetwar all published extensively throughout the 1960s on the physiology of reproduction and the use of inhibitory compounds, discovered through a screen in vivo, to dissect the mechanisms of the reproductive process in animals.

The compound ICI46,474 was discovered in the screening process for antifertility agents in the rat. Many of the laboratory studies with ICI46,474 focused on the application of an antiestrogen to confirm that the implantation of the blastocyst in the rat requires an estrogen surge on day 4 following mating.[67,68] In other words, an understanding of reproductive mechanisms was the primary goal of the research group. ICI46,474 was subsequently shown to be an antiestrogen in the rat but an estrogen in the mouse (**Figure 4**). The *cis* geometric isomer ICI47,699 (**Figure 4**) was an estrogen in the rat and the mouse.[45] Only at the end of Walpole's career at ICI Pharmaceuticals Division did interest turn to the interaction of compounds with the estrogen receptor. The geometric isomers of tamoxifen and clomiphene all inhibited the binding of tritiated estradiol to rat and mouse estrogen receptor derived from uterus and pituitary gland.[69] Unfortunately, no firm conclusion could be drawn to explain estrogen/antiestrogen action in different species. This species difference in the pharmacology of antiestrogens also raised the question of whether ICI46,474 would be an estrogen or an antiestrogen in humans. Estrogens were already used in the treatment of breast cancer but an antiestrogenic compound would be of value clinically because there may be fewer side effects. Nevertheless, based on the earlier experience at Merrill in the USA, only compounds that did not cause an increase in desmosterol could be used for long-term treatment in humans.

The reproductive endocrinology team advanced clinical testing in several areas primarily endocrinology and gynecology. Indeed in 1972, Walpole reviewed all of the progress in advancing ICI46,474 to become a clinically useful drug. These data were either just published at the time or subsequently published a few years later.

There was, at the time, an ever-expanding literature on the use of clomiphene for the induction of ovulation in subfertile women.[82] ICI46,474 was successfully tested as an agent for the induction of ovulation[83,84] and was approved for use in clinical practice in 1975 in the UK and several other countries around the world. Additionally, ICI46,474 was noted to block the uptake of ^{3}H-estradiol in the human uterus in vivo[85] and had some benefit in diminishing bleeding for patients with menometrorrhagia.[86] As noted earlier, Walpole was interested in cancer therapy and had connections at the Christie Hospital in Manchester.[59] A small preliminary study using ICI46,474 to treat late breast cancer showed some benefit in 10 of 46 patients[87] and in a preliminary dose response study[88] found that 12 out of 33 (33%) of patients receiving 10 mg twice daily and 14 out of 35 (40%) of patients receiving 20 mg twice daily had definite responses.

At that time, in 1973, there was little or no enthusiasm at ICI Pharmaceuticals Division to pursue a major program of drug development for the treatment of breast cancer. Walpole, in contrast, was optimistic about exploiting ICI46,474 as a breast cancer drug and agreed not to take early retirement if the antiestrogen was supported for clinical approval. This was achieved in the UK through the Committee for Safety of Medicines in 1973 and in December 1977 by the Food and Drug Administration (FDA) in the United States.

The reasons for the reluctance to pursue global development of ICI46,474 were the perceived vulnerability of the product in the absence of a patent in the US market and the assessment, which was correct at the beginning of the 1970s, that there was virtually no market. Only one in three patients with advanced breast cancer would respond to endocrine therapy for about a year. This market would amount to no more than £50 000 of sales per annum and the competitor products of high-dose estrogen or androgen cost only pennies per dose whereas the new antiestrogen ICI46,474 would, by necessity, have to cost 10–20 times more per dose. The only advantage for ICI,46,474 was a reduction in adverse side effects (**Table 1**).

The scientists at ICI Pharmaceuticals Division did not conduct any systematic study of the mechanism of action or antitumor properties of ICI46,474. These studies were conducted outside ICI Pharmaceuticals Division with academic collaborators (the process was advanced through the good offices of Walpole who remained as a consultant for ICI after his retirement in the early 1970s). Laboratory programs were established at the Worcester Foundation for Experimental Biology in Shrewsbury, MA (1972–74) in collaboration with ICI Americas (Stuart Pharmaceuticals) and subsequently at Leeds University, UK (1974–79) with a formal Leeds University/ICI joint research scheme. Tamoxifen was to be reinvented as a pioneering targeted antihormonal treatment for breast cancer by using appropriate laboratory models to create a scientific basis for pursuing rational clinical trials.

8.08.5   **Patenting Problems**

Adequate patent protection is required to develop an innovation in a timely manner. In 1962, ICI Pharmaceuticals Division (now AstraZeneca and formerly Zeneca) filed a broad patent in the UK.[228]

> The application stated: The alkene derivatives of the invention are useful for the modification of the endocrine status in man and animals and they may be used for the control of hormone-dependent tumours or for the management of the sexual cycle and aberrations thereof. They also have useful hypocholesterolemic activity.

This was published in 1965 as UK Patent GB 1013907 which described the innovation that different geometric isomers of substituted triphenylethylenes had either estrogenic or antiestrogenic properties. The original patent was enhanced with UK Patent GB 1064629 published 1967 which is a process for the manufacture of *cis* and *trans* isomers.

In 1973, Nolvadex, the ICI brand of tamoxifen (as its citrate salt), was approved by the Committee on the Safety of Medicines in the UK for the treatment of breast cancer. Although tamoxifen was approved for the treatment of advanced breast cancer in postmenopausal women on 30 December 1977 in the US (ICI Pharmaceuticals Division received the Queen's Award for Technological Achievement in the UK on 6 July 1978), the patent situation was unclear. ICI Pharmaceuticals Division was repeatedly denied patent protection in the USA (with an exclusion of claims for a cancer treatment) until the 1980s because of the perceived primacy of the earlier Merrill patents[229] and because no advance (that is, a safer, more specific drug) was recognized by the US Patent Office. In other words, the clinical development of tamoxifen advanced steadily for more than a decade in the USA without the assurance of exclusivity. This situation also illustrates how unlikely the usefulness of tamoxifen was considered to be by the pharmaceutical industry in general. In theory, tamoxifen could have been marketed by other companies as a generic drug. Remarkably, when tamoxifen was hailed as the adjuvant endocrine treatment of choice for breast cancer by the National Cancer Institute in 1984, the patent application, initially denied in 1984, was awarded through the court of appeals in 1985. This was granted with precedence to the patent dating back to 1965! So, at a time when worldwide patent protection was being lost, the patent protecting tamoxifen started a 17-year life in the USA. The unique and unusual legal situation did not go uncontested by generic companies but AstraZeneca rightly retained patent protection for their pioneering product, most notably, from the Smalkin Decision in Baltimore, 1996.[230] Nevertheless, one generic company (Barr Pharmaceuticals), in a separate out-of-court agreement, did distribute tamoxifen supplied by ICI Pharmaceuticals Division throughout the 1990s. The tablets were priced slightly lower than Nolvadex. Worldwide there are now dozens of generic brands of tamoxifen supplied to healthcare systems outside the USA. Most notable is the Hungarian brand of tamoxifen marketed under the name of Zitozonium. Worldwide sales probably exceeded $10 billion for AstraZeneca and its earlier founder companies which provided the resources for all subsequent drug discovery and development in cancer. Early successes were the antiandrogen Casodex and the sustained release luteinizing hormone-releasing hormone superagonist Zoladex. The actual figures of total worldwide sales for generic tamoxifen are hard to estimate.

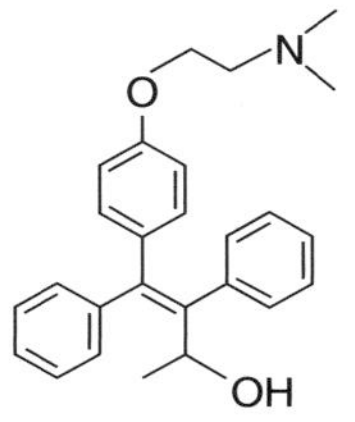

Alpha-hydroxytamoxifen

Figure 6 The metabolite of tamoxifen believed to be responsible for rat liver carcinogenesis.

carcinogenesis. There had not been a significant increase in hepatocellular carcinoma since the two initial cases reported in 1989.[143] Similarly, epidemiology studies[176] had not shown a rise in hepatocellular carcinoma in breast cancer patients since tamoxifen was approved for use in the USA in 1978. In contrast, oral contraceptives cause a tenfold increase in the risk for the development of hepatocellular carcinoma,[177] but this risk is considered to be acceptable to regulatory authorities because of the rarity of the disease. Clearly there was a problem in translating laboratory studies of the toxicology of tamoxifen to clinic experience with the drug. Unfortunately, this was not relevant to the media and those scientists promoting other products to replace tamoxifen.

During the early years of the 1990s there was intense interest in discovering the initiating event for tamoxifen-induced rat liver carcinogenesis and determining the relevance for humans. Han and Lehr[148] first noted an accumulation of DNA adducts in the liver of Sprague–Dawley rats on repeated injections of $20\,mg\,kg^{-1}$ (cf. human dosage of $0–3\,mg\,kg^{-1}$). This observation was adequately confirmed by numerous investigators and the focus of investigation turned to the identification of the actual DNA adduct. Several candidates were proposed: an epoxide,[178–180] 4-hydroxytamoxifen,[181,182] Metabolite E,[183] or alpha-hydroxytamoxifen.[184–186] Osborne *et al.*[187] prepared alpha-acetoxytamoxifen which is able to react with DNA to a greater extent (1 in 50 bases) than alpha-hydroxytamoxifen (1 in 10^5 DNA bases). The products of the reaction were identical to those isolated from DNA of rat hepatocytes or the livers of rats treated with tamoxifen. The adduct of tamoxifen and DNA was identified at the nucleoside deoxyguanosine in which the alpha position of tamoxifen is linked covalently to the exocyclic amino of deoxyguanosine.

These important observations provided a framework to study the metabolic activation of tamoxifen in human systems and to identify any DNA adducts in human tissues. The metabolic activation of tamoxifen and its metabolite alpha-hydroxytamoxifen (**Figure 6**) were compared using primary cultures of rat, mouse, and human hepatocytes.[180] Although DNA adducts were readily identified in rat and mouse hepatocytes (90 and 15 adducts per 10^8 nucleotides respectively), DNA adducts were not detected in tamoxifen-treated human hepatocytes. Additionally, human hepatocytes also appeared to produce 50-fold lower levels of alpha-hydroxytamoxifen from tamoxifen compared to rat hepatocytes. Further studies showed that if cells were treated with alpha-hydroxytamoxifen human hepatocytes had 300-fold lower levels of adducts compared to rat hepatocytes.

Studies in humans have confirmed that the human is not as susceptible as the rat to DNA adduct formation with tamoxifen. The pattern of DNA adducts found in the rat liver was not found in humans treated with tamoxifen,[173] DNA adducts were not found in lymphocytes,[188] and there is a lack of genotoxicity of tamoxifen in human endometrium.[157] In the latter studies, DNA adducts could be produced in endometrial samples with alpha-hydroxytamoxifen but not with tamoxifen. The authors proved that tissue was capable of metabolizing tamoxifen to alpha-hydroxytamoxifen but apparently it is incapable of producing adducts. Endometria from patients taking tamoxifen for up to 9 years were analyzed for DNA adducts. No evidence for any DNA adducts induced by tamoxifen was found in any of the patients examined. The authors concluded that the genotoxic events observed with tamoxifen in the rat may not apply to the human endometrium.[157] This conclusion supports the previous suggestion that tamoxifen, or indeed any new antiestrogen which has partial agonist actions, will cause the activation and detection of pre-existing disease.[174]

The results from the EBCTCG (2005) provide a current evaluation of the benefits and side effects experienced with adjuvant tamoxifen in clinical trials during the past two decades.[141] A comparison of all patients receiving adjuvant tamoxifen versus those not receiving tamoxifen reveals that there is no significant excess of deaths from any particular cause. The average non-breast cancer death rate was calculated to be 0.8% per year for women receiving tamoxifen or not. There is a small excess of deaths in women receiving tamoxifen from thromboembolism and uterine cancer (but not liver cancer) but these data are nonsignificant. Presuming there is a real excess of death for both side effects combined for the 60 000 women–years of tamoxifen exposure in the trials, this would represent an absolute risk of death of about 0.2% per decade. The EBCTG suggest this risk is small in comparison with the 10-year benefit in reducing breast cancer mortality by 5.3% (node-negative) and 12.2% (node-positive).

The trend in clinical practice of using longer and longer treatment regimens with tamoxifen stimulated the investigation of the development of drug resistance to tamoxifen. Drug resistance to tamoxifen therapy can take many forms.[189,190] Obviously, if tumors are estrogen receptor-negative there is only a small probability of a response to antiestrogen therapy. In the case of metastatic breast cancer about 10% of estrogen receptor-negative and progesterone receptor-negative patients respond to any form of endocrine modulation.[191] Similarly the overview analysis[140] of clinical trials shows that post-menopausal, node-positive patients with receptor-poor disease do not benefit from adjuvant tamoxifen.

In the laboratory, tamoxifen was found to inhibit estrogen-stimulated growth of MCF7 breast tumors implanted into athymic mice.[123] Nevertheless, continuous therapy with tamoxifen results in the emergence of tamoxifen-stimulated breast tumors that will grow in response to either estrogen or tamoxifen.[150–152,192] Since there were clinical reports of tamoxifen-stimulated tumors that have a withdrawal response to tamoxifen,[193,194] new second-line agents (or first-line agents) were becoming necessary to control tumors that grow after extended tamoxifen treatment. New nonestrogenic agents were introduced to improve response rates and reduce side effects. However, this goal was only to become successful in the twenty-first century (see Section 8.08.12). Tamoxifen was first destined to be tested as the first chemopreventive to reduce the incidence of breast cancer in high-risk women.

8.08.9   Selective Estrogen Receptor Modulation

In the 1960s and 1970s, antiestrogenicity was correlated with antitumor activity. However, the finding that nonsteroidal antiestrogens expressed increased estrogenic properties, i.e., vaginal cornification and increased uterine weight in the mouse, raised questions about the reasons for the species specificity. One obvious possibility was species-specific metabolism, i.e., the mouse converts antiestrogens to estrogens via novel metabolic pathways. However, no species-specific metabolic routes to known estrogens were identified but knowledge of the mouse model created a new dimension for study that ultimately led to the recognition of the target site-specific actions of antiestrogens. This concept was subsequently referred to as selective estrogen receptor modulation (SERM) to describe the target site-specific effects of raloxifene (see 8.09 Raloxifene), an antiestrogen originally targeted for an application in breast cancer but now used, paradoxically, as a preventive for osteoporosis. Now the whole class of drugs is known as SERMs.

The estrogen receptor-positive breast cancer cell line MCF7[110] can be heterotransplanted to immune-deficient athymic mice but the cells will only grow into tumors with estrogen support.[123] Paradoxically, tamoxifen, an estrogen in the mouse, does not support tumor growth but stimulates mouse uterine growth with the same spectrum of tamoxifen metabolites present in both the uterus and the human tumor.[195] To explain the selective actions of tamoxifen in different targets of the same host, it was suggested that the estrogen receptor complex could be interpreted as a stimulatory or inhibitory signal at different sites. The concept was consolidated with experimental evidence from two further models. First, tamoxifen and raloxifene maintain bone density in the ovariectomized rat but both compounds inhibit estradiol-stimulated uterine weight and prevent carcinogen-induced mammary tumorigenesis.[122,196] Second, the finding that tamoxifen would partially stimulate the growth of a human endometrial carcinoma transplanted into athymic mice allowed the investigation of two human tumors bitransplanted in the same mouse to determine whether tamoxifen could inhibit estrogen-stimulated growth of two tumors in the same host equally. Tamoxifen demonstrated target site specificity: breast tumor growth was controlled but endometrial tumors continued to grow.[197] Again, the spectrum of tamoxifen metabolites was consistent in all target tissues despite the contrasting biological responses, so it was concluded that the estrogen receptor complexes must be interpreted differently in different target tissues.

The laboratory principle of selective estrogen receptor modulation translated to the clinic with the findings that tamoxifen maintained bone density[198] and lowered circulating cholesterol.[199] These were extremely important findings because there were justifiable concerns that tamoxifen, an 'antiestrogen,' might prevent breast cancer but increase risks for osteoporosis and coronary heart disease. The beneficial effects of SERM action on bones and circulating cholesterol were important to advance clinical studies testing the worth of tamoxifen as a chemopreventive in high-risk women. Additionally, the recognition that tamoxifen increases the risk of endometrial cancer was an advantage for screening volunteers for trials. Nevertheless, the reports of carcinogenicity associated with tamoxifen naturally created major problems for recruitment to chemoprevention trials.

8.08.10   Tamoxifen and Breast Cancer Prevention

Thirty years ago, tamoxifen was shown to prevent the induction and promotion of carcinogen-induced mammary cancer in rats.[92,200] Similarly, tamoxifen was also shown to prevent the development of mammary cancer induced by ionizing radiation in rats. These laboratory observations, coupled with the emerging preliminary clinical observation that

adjuvant tamoxifen could prevent contralateral breast cancer in women,[201] provided a rationale for Powles to start a toxicology study at the Royal Marsden Hospital, London, UK to test whether tamoxifen would be acceptable to prevent breast cancer in high-risk women. This vanguard study opened for recruitment in 1986[202] and was to provide important toxicological and compliance data for subsequent trialists.

This toxicology and compliance study was supplemented by parallel investigations of tamoxifen as a chemopreventive in animal models of tumorigenesis[122] and the safety studies of tamoxifen to establish the effects on bone and circulating lipids (*see* Section 8.08.9).

In the decade following the Powles initiative, several studies were started to answer the question: "Does tamoxifen have worth in the prevention of breast cancer in select high-risk women?" Eventually four studies were available to evaluate the veracity of the question - the Royal Marsden study, the NSABP/NCI study, the Italian study, and the International Breast Intervention Study (IBIS). The results have been adequately summarized by Cuzick and coworkers[203] but the NSABP Study will be presented in detail because it was the only prospective study to achieve its recruitment goal.

The NSABP P-1 study opened in the USA and Canada in May of 1992 with an accrual goal of 16 000 high-risk women to be screened and recruited at 100 North American sites. It closed after accruing 13 338 in 1997 due to the high-risk status of the participants. Those eligible for entry included any woman over the age of 60 or women between the ages of 35 and 59 whose 5-year risk of developing breast cancer, as predicted by the Gail model,[204] was equal to that of a 60-year-old woman. Additionally, any woman over age 35 with a diagnosis of lobular carcinoma in situ (LCIS) treated by biopsy alone was eligible for entry to the study. In the absence of LCIS, the risk factors necessary to enter the study varied with age, such that a 35-year-old woman must have a relative risk (RR) of 5.07, whereas the required RR for a 45-year-old woman was 1.79. Routine endometrial biopsies to evaluate the incidence of endometrial carcinoma in both arms of the study were also performed.

The breast cancer risk of women enrolled in the study was extremely high with no age group having an RR of less than 4, including the over-60s group. Recruitment was also balanced with about one-third younger than 50 years, one-third between 50 and 60 years old, and one-third older than 60 years. Secondary end points of the study included the effect of tamoxifen on the incidence of fractures and cardiovascular deaths. Most importantly, the study planned to provide the first information about the role of genetic markers in the etiology of breast cancer. Unfortunately the question of whether tamoxifen has a role to play in the treatment of women who are found to carry somatic mutations in the *BRCA-1* and *BRCA-2* gene could not adequately be answered[205] because of the low incidence of women with mutations in the P-1 study overall.

The first results of the NSABP study were reported in September 1998, after a mean follow-up of 47.7 months.[206] There were a total of 368 invasive and noninvasive breast cancers in the participants; 124 in the tamoxifen group and 224 in the placebo group. A 49% reduction in the risk of invasive breast cancer was seen in the tamoxifen group and a 50% reduction in the risk of noninvasive breast cancer was observed. A subset analysis of women at risk due to a diagnosis of LCIS demonstrated a 56% reduction in this group. The most dramatic reduction was seen in women at risk due to atypical hyperplasia where risk was reduced by 86%.

The benefits of tamoxifen were observed in all age groups with a relative risk of breast cancer ranging from 0.45 in women aged 60 and older to 0.49 for those in the 50–59-year age group and 0.56 for women aged 49 and younger. A benefit for tamoxifen was also observed for women with all levels of breast cancer risk within the study, indicating that the benefits of tamoxifen are not confined to a particular lower risk or higher risk subset. Benefits were observed in women at risk on the basis of family history and those whose risk was due to other factors.

As expected, the effect of tamoxifen occurred on the incidence of estrogen receptor-positive tumors which were reduced by 69% per year. The rate of estrogen receptor-negative tumors in the tamoxifen group (1.46 per 1000 women) did not significantly differ from the placebo group (1.20 per 1000 women). Tamoxifen reduced the rate of invasive cancers of all sizes but the greatest difference between the groups was the incidence of tumors 2.0 cm or less. Tamoxifen also reduced the incidence of both node-positive and node-negative breast cancer. The beneficial effects of tamoxifen were observed for each year of follow-up in the study. After year 1 the risk was reduced by 33% and in year 5 by 69%.

Tamoxifen also reduced the incidence of osteoporotic fractures of the hip, spine, and radius by 19%. However, the difference approached, but did not reach, statistical significance. This reduction was greatest in women aged 50 and older at study entry. No difference in the risk of myocardial infarction, angina, coronary artery bypass grafting, or angioplasty was noted between the groups.

This study confirmed the association between tamoxifen and endometrial carcinoma. The relative risk of endometrial cancer in the tamoxifen group was 2.5. The increased risk was seen in women aged 50 and older whose relative risk was 4.01. All endometrial cancers in the tamoxifen group were grade 1 and none of the women on the

tamoxifen died of endometrial cancer. There was one endometrial cancer death in the placebo group. Although there is no doubt that tamoxifen increases the risk of endometrial cancer, it is important to recognize that this increase translates to an incidence of 2.3 women per 1000 per year who develop endometrial carcinoma.

More women in the tamoxifen group developed deep vein thrombosis (DVT) than in the placebo group. Again, this excess risk was confined to women aged 50 and older. The relative risk of DVT in the older age group was 1.71 (95% CI 0.85 to 3.58). An increase in pulmonary emboli was also seen in the older women taking tamoxifen, with a relative risk of approximately 3. Three deaths from pulmonary emboli occurred in the tamoxifen arm, but all were in women with significant comorbidities. An increased incidence of stroke (RR 1.75) was also seen in the tamoxifen group, but this did not reach statistical significance.

An assessment of the incidence of cataract formation was made using patient self-report. A small increase in cataracts was noted in the tamoxifen group: a rate of 24.8 women per 1000 compared to 21.7 in the placebo group. There was also an increased risk of cataract surgery in the women on tamoxifen. These differences were marginally statistically significant and observed in the older patients in the study. These findings emphasize the need to assess the patient's overall health status before making a decision to use tamoxifen for breast cancer risk reduction. These observations are also particularly interesting based on the early controversy in the 1960s (*see* Section 8.08.2) about the safety of this drug group.

An assessment of quality of life showed no difference in depression scores between groups. Hot flushes were noted in 81% of the women on tamoxifen compared to 69% of the placebo group and the tamoxifen-associated hot flushes appeared to be of greater severity than those in the placebo group. Moderately bothersome or severe vaginal discharge was reported by 29% of the women in tamoxifen group and 13% in the placebo group. No differences in occurrence of irregular menses, nausea, fluid retention, skin changes, or weight gain or loss were reported.[207,208]

8.08.11   Current Chemoprevention

Based on a thorough review of all the available data, the FDA approved tamoxifen for the reduction of breast cancer incidence in high-risk pre- and postmenopausal women in 1998. However, the report that tamoxifen caused a small but significant increase in uterine sarcoma[209] resulted in an industry request for a black box inclusion for tamoxifen from the FDA. Additionally, the IBIS-1 study noted an unacceptable increase in deaths from tamoxifen treated patients who inadvertently had surgery during the study acceptability of tamoxifen as a chemopreventive.[210] This led to the development of IBIS-2 using an aromatase inhibitor to prevent breast cancer. Aromatase inhibitors have fewer side effects than tamoxifen and it is known that during adjuvant treatment, they reduce the incidence of contralateral breast cancer even more than tamoxifen.[211–213]

Another approach is the evaluation of the SERM raloxifene as a preventive for breast cancer in high-risk postmenopausal women. The Study of Tamoxifen and Raloxifene (STAR) has reduced its recruitment goal of 19 000 volunteers and the results will be available by July 2006.

The promise of the chemoprevention for breast cancer is becoming a reality. However, there are many challenges. Tamoxifen, the pioneering medicine, is considered by many to be too controversial to be widely used as a chemopreventive. However, there are no alternatives for the premenopausal woman at high risk for breast cancer and the good news is that this risk group has the best risk–benefit ratio.[214] For postmenopausal women, where the side effects are well defined, the future depends on the results of current clinical trials with raloxifene or aromatase inhibitors. Unfortunately, there are no comparisons of a SERM with an aromatase inhibitor so the choice of a chemopreventive strategy will need to be made on a patient-by-patient basis. In other words, the options are the use of raloxifene or an aromatase inhibitor with bone monitoring and a bisphosphonate to avoid osteoporosis.

8.08.12   Tamoxifen's Legacy: A Menu of Medicines

Tamoxifen became the most investigated anticancer agent over the 40 years of its development. The success of the drug as an adjuvant therapy has been quantified: 400 000 women with breast cancer are alive because of long-term tamoxifen treatment. Most importantly, the development of tamoxifen demonstrated that there was an advantage for patients by targeting the estrogen receptor specifically. This in turn encouraged the pharmaceutical industry to invest in research to discover both safer and more effective drugs. This is best illustrated by comparing treatment options for advanced breast cancer in 1970, i.e., before tamoxifen (**Figure 7**) with the therapeutic options for all stages of breast cancer in 2005 (**Figure 8**).

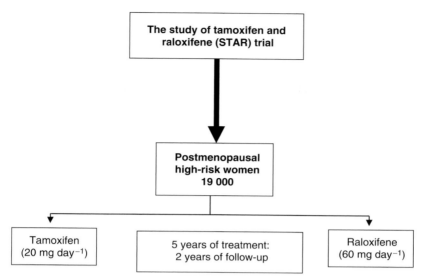

Figure 8 The design of the STAR trial. The STAR trial is a phase III, double-blinded trial that will assign eligible postmenopausal women to either daily tamoxifen (20 mg orally) or raloxifene (60 mg orally) therapy for 5 years. It is the first head-to-head trial comparing the effect of raloxifene with that of tamoxifen in reducing the incidence of invasive breast cancer in postmenopausal women at risk for the disease. Approximately 19 000 postmenopausal women 35 years of age or older having at least a 1.66% estimated Gail risk of developing breast cancer or a history of lobular carcinoma in situ (LCIS) are being enrolled. The trial is double-blinded, and study participants will be randomized to receive either 20 mg day^{-1} tamoxifen or 60 mg day^{-1} raloxifene for 5 years. The STAR trial's primary aim is to determine whether long-term therapy is effective in preventing the occurrence of invasive breast cancer in high-risk postmenopausal women. The comparison is to be made to the established drug, tamoxifen. The secondary aim is to establish the relative safety profiles of raloxifene and tamoxifen.

also complete a minimum of 2 additional years of follow-up after therapy is stopped. The primary aim of the STAR trial is to determine whether long-term raloxifene therapy is effective in preventing the occurrence of invasive breast cancer in postmenopausal women who are identified as being at high risk for the disease. It is the first head-to-head trial comparing tamoxifen with raloxifene. The secondary aim is to establish the net effect of raloxifene therapy, by comparison of cardiovascular data, fracture data, and general toxicities with tamoxifen. The results from the STAR trial are anticipated in 2006.

The significant reduction in risk of invasive breast cancer observed in the MORE trial led directly to the design of the CORE study (**Figure 7**). Results from the CORE trial[170] revealed that the incidences of invasive breast cancer and estrogen receptor-positive invasive breast cancer were reduced by 59% (HR, 0.41; 95% CI, 0.24–0.71) and 66% (HR, 0.34; 95% CI, 0.18–0.66), respectively, in the raloxifene group compared with the placebo group. Raloxifene, however, did not significantly ($P = 0.86$) alter the incidence of estrogen receptor-negative invasive breast cancer. Over the 8 years of both trials, the incidences of invasive breast cancer and estrogen receptor-positive invasive breast cancer were reduced by 66% (HR, 0.34; 95% CI, 0.22–0.50) and 76% (HR, 0.24; 95% CI, 0.15–0.40), respectively, in the raloxifene group compared with the placebo group.[170]

8.09.7   Raloxifene and Lipids

Estrogen increases HDL cholesterol levels and decreases LDL cholesterol levels in humans[39,171] as well in animal models of atherosclerosis, partly because of estrogen receptor-mediated upregulation of the hepatic LDL receptor.[172] In ovariectomized rats, raloxifene treatment has been shown to reduce serum total cholesterol concentrations,[97,173] and this reduction correlates with the extent of raloxifene binding to the estrogen receptor.[97,173] These results are not surprising for a 'nonsteroidal antiestrogen,' as the original observations for clomiphene analogs and tamoxifen show (*see* 8.08 Tamoxifen). Raloxifene may also have cardioprotective effects because of its antioxidant properties. This is important since oxidative modifications of LDL have been implicated in atherogenesisis.[174] Raloxifene also appears to have a favorable effect on lipid parameters in postmenopausal women. In the published European trial,[78] treatment with raloxifene in a dosage of 30, 60, or 150 mg day^{-1} resulted in significant decreases in the serum concentrations of total and LDL cholesterol over a 24-month period ($P < 0.05$ versus placebo). These decreases were evident during the

first 3 months of therapy and were maintained thereafter. Notably, none of the treatment groups showed any changes in serum concentrations of HDL cholesterol and triglycerides. The effect of raloxifene on serum lipid levels was also assessed in 390 healthy postmenopausal women.[175] In this study, raloxifene (60 and 120 mg day^{-1}) was compared with HRT (0.625 mg day^{-1} of conjugated estrogen and 2.5 mg day^{-1} of medroxyprogesterone given continuously) and placebo. Assessments were made at baseline, 3 months, and 6 months. Over the 6-month study period, both dosages of raloxifene lowered serum LDL cholesterol levels by about 12% compared with placebo ($P < 0.001$). This finding was similar to the 14% reduction that occurred with continuous HRT.[78] The effect of raloxifene on cardiovascular events was also examined in osteoporotic postmenopausal women from the MORE trial. In the study design, patients were randomly assigned to receive raloxifene 60 mg day^{-1} ($n = 2557$), or 120 mg day^{-1} ($n = 2572$), or placebo ($n = 2576$) for 4 years. Barrett-Connor and coworkers[176] reported that raloxifene therapy for 4 years did not significantly affect the risk of cardiovascular events in the overall cohort but did significantly reduce the risk of cardiovascular events in the subset of women with increased cardiovascular risk. In addition, there was no evidence that raloxifene caused an early increase in risk of cardiovascular events.

To address the question of whether raloxifene reduces the risk of CHD, a total of 10 101 women (at increased risk of CHD) have been recruited to receive placebo or raloxifene in the RUTH trial, with cardiovascular disease as a primary endpoint.[177] The RUTH trial (**Figure 9**) is designed to determine whether raloxifene (60 mg day^{-1}), compared with placebo, reduces the risk of coronary events and invasive breast cancer in postmenopausal women at risk for a major coronary event.

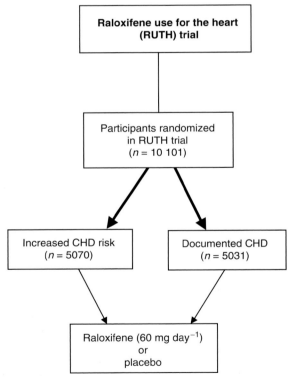

Figure 9 Study design of the Raloxifene for Use in The Heart (RUTH) trial. The RUTH trial is a double-blind, placebo-controlled, randomized clinical trial designed to evaluate whether 60 mg day^{-1} of oral raloxifene compared with placebo reduces the risk of coronary events. Between June 1998 and August 2000, 11 767 women signed an informed consent agreement to participate in RUTH at 187 sites in 26 countries. After excluding 1411 women who did not meet inclusion criteria and 255 women who met more than one exclusion criteria, a total of 10 101 women were randomized to raloxifene 60 mg day^{-1} or placebo. Of these, 5070 were at increased risk for CHD events and 5031 had documented CHD. Women were eligible for randomization if they were aged 55 years or older, at least 1 year postmenopausal, and had documented CHD, peripheral artery disease, or multiple risk factors for CHD. Breast cancer incidence will be determined by mammograms performed 2, 4, and 6 years after the qualifying mammogram. The study is planned to end after the pre-specified number of participants experience their first acute coronary event. The total duration of treatment is projected to range from 5 to 7.25 years.

8.09.8   Other Selective Estrogen Receptor Modulators

Current interest in new SERM molecules has built on the experience of the prototypes with the goal of enhancing bioavailability and selectivity and decreasing side effects (i.e., breast cancer, uterine cancer, and blood clots). All compounds under study have predominantly antiestrogenic effects in the rodent uterus with virtually no estrogen agonist properties. In order to improve upon the raloxifene pharmacophore, some groups have reported on the effect of modifying the benzothiophene nucleus. Particularly noteworthy were two discoveries made by the chemists at Eli Lilly which improved upon raloxifene.[77] One change involved the introduction of a methyl ether on either the 5-OH or the 4'OH position, which resulted in compounds with increased potency in a cholesterol reduction assay in ovariectomized rats.[178,179] The other change involved the replacement of the carbonyl 'hinge' with other atoms or groups, including N, CH_2, S, and O. The change to the oxygen atom resulted in a compound with very little or no uterine effects in preclinical rodent models as well as increased potency in preventing bone loss in the ovariectomized rat model.[180] These combined changes resulted in the development of arzoxifene.

8.09.9   Arzoxifene

Arzoxifene (LY 353,381.HCl; **Figure 2**) is a new benzothiophene analog that is structurally related to raloxifene.[180,181] Its structure differs from that of raloxifene by the replacement of a carbonyl group with oxygen. It was designed to improve the bioavailability of raloxifene and provide sustained antiestrogenic blockade in the treatment of breast cancer without any of the agonist effects seen with tamoxifen. It is classified as a second-generation SERM, based on its differential estrogenic/antiestrogenic effects in vivo on estrogen target tissues.[181] It is metabolized by demethylations and both the parent compound and the metabolite bind to the estrogen receptor with high affinity and inhibit estrogen-dependent growth of MCF-7 breast cancer cells.[182–184] Arzoxifene protects against bone loss and reduces serum cholesterol levels in ovariectomized rats with a potency that is 30–100 times greater than that of raloxifene and it has minimal uterine effects.[181,185] It is highly effective at preventing N-methyl-N-nitrosourea-induced mammary cancer in rats and is significantly more potent than raloxifene in this regard.[186] Interestingly, arzoxifene has also been shown to be only partially cross-resistant with tamoxifen in models of drug-resistant breast and endometrial cancer[187,188]; however, recent evidence indicates that it is superior to raloxifene as a chemopreventive in rat mammary carcinogenesis.[182,189,190] In a small phase I study in 32 pre- and postmenopausal women with locally advanced or metastatic breast cancer who had previously received endocrine therapy, arzoxifene (10, 20, 50, and 100 mg) did not produce any significant responses, suggesting cross-resistance between arzoxifene and tamoxifen.[191] In a phase II study in 119 pre- and postmenopausal women with advanced or metastatic breast cancer, two doses of arzoxifene (20 versus 50 mg day^{-1}) were compared in patients who had either tamoxifen-sensitive or tamoxifen-resistant disease and 20 mg arzoxifene was found to be as effective as 50 mg in the treatment of metastatic breast cancer.[191] A phase III trial was subsequently initiated comparing arzoxifene (20 mg day^{-1}) with tamoxifen (20 mg day^{-1}) in postmenopausal women with advanced disease; however, at the interim review, the trial was terminated and development of arzoxifene discontinued for this indication.

8.09.10   Bazedoxifene

Bazedoxifene (TSE-424; **Figure 10**) is a novel SERM developed by Wyeth Pharmaceuticals that is currently in phase III clinical trials for the prevention and treatment of postmenopausal osteoporosis. It is an indole-based estrogen receptor ligand that has been stringently selected to ensure an improved profile over its predecessor raloxifene. It was developed using preclinical selection parameters, which included favorable effects on the skeleton and lipid metabolism, demonstrable mammary and uterine safety, and neutral effects on hot flashes.[192] Bazedoxifene treatment maintains bone mineral density, preserves normal bone histology, increases bone compressive strength, and reduces total cholesterol levels in animal models.[192–194] It lacks uterotropic activity[194] and it blocks raloxifene-induced increases in uterine weight[192] and inhibits E_2-induced proliferation in MCF-7 breast cancer cells.[192] Based on the favorable preclinical evaluation, it is suggested that bazedoxifene has the potential to improve the SERM profile beyond that achieved by raloxifene.

8.09.11   Lasofoxifene

Lasofoxifene (**Figure 10**) is a novel nonsteroidal SERM that is in clinical trials for the prevention and treatment of osteoporosis in postmenopausal women.[195] It is a diaryltetrahydronaphthalene derivative referred to as CP336156. The structure of CP336156 is reminiscent of nafoxidine (**Figure 3**) if it were to be demethylated in vivo. There are two diastereometric salts. CP336156 is the L enantiomer that has 20 times the binding affinity of the D enantiomer. Studies

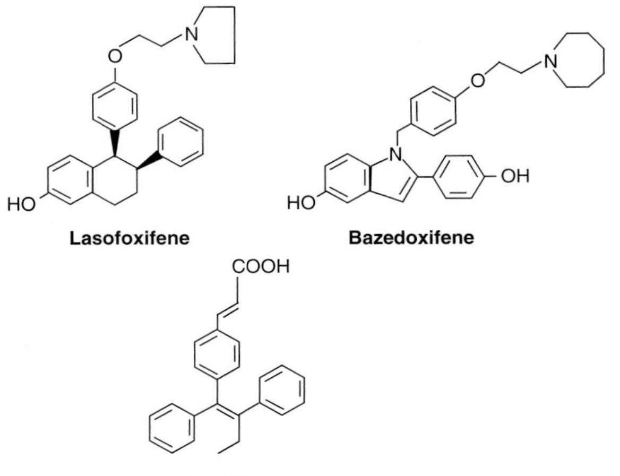

Figure 10   Bazedoxifene, lasofoxifene, and GW 5638.

demonstrated that the L enantiomer had twice the bioavailability of the D enantiomer.[195] It has a high binding affinity for estrogen receptor and preclinical studies indicate that it prevents lumbar vertebral bone loss in the ovariectomized rat model, with greatly enhanced potency relative to raloxifene.[195–197] Lasofoxifene also lowers serum cholesterol levels without induction of uterine hypertrophy in rat models,[195,196] and it inhibits breast tumor formation in mice injected with human MCF-7 breast cancer cells and blocks N-nitrosomethylurea-induced mammary carcinomas in rats.[198] In a phase III clinical trial conducted by Pfizer[199] involving 410 postmenopausal women randomly assigned to CP336156 (0.25 or 1 mg day^{-1}), raloxifene (60 mg day^{-1}), or a placebo, CP336156 increased bone mineral density at the lumbar spine by about 2% after 2 years of treatment, compared to no increase with raloxifene and a 2% decrease in the placebo group. Changes in bone turnover markers were also greater with CP336156 and the drug reduced LDL cholesterol levels by a mean of 20% versus 12% with raloxifene. Overall, the drug was well tolerated and there were no reports of increased endometrial hyperplasia or vaginal bleeding.

8.09.12   GW 5638

GW 5638 (**Figure 10**) is a tamoxifen analog (*see* 8.08 Tamoxifen) that was discovered by Willson and colleagues in 1994[200] at Glaxo Wellcome in North Carolina. It functions as a full estrogen receptor agonist in bone and the cardiovascular system in ovariectomized rats, but, unlike tamoxifen, it is a more potent antagonist in breast cancer cells and has no uterotrophic behavior.[201] GW 5638 does not have the usual tertiary amino antiestrogenic side chain but a shorter allylcarboxylic group on a triphenylethylene carrier molecule.[200,201] GW 5638 can induce a unique conformational change in ERα that is recognized by synthetic peptides selected by phage display.[202,203] These peptides recognize GW5638–ERα complexes but not tamoxifen–ERα or other ligand-bound estrogen receptor complexes,[200,201] indicating that conformational changes elicited by GW 5638 and tamoxifen are different. Recent crystallograpy studies[204] of ERα ligand binding domain (LBD) bound to GW 5638 have revealed a new LBD conformation in which AF2 H12 is repositioned by direct contacts between the carboxyl side chain of GW 5638 and the N-terminus of H12. In addition to preventing coactivator recruitment by occlusion of the AF2 cleft, GW 5638 also destabilizes ERα, although less so than the more potent estrogen receptor antagonist ICI 164,380/182,780 (fulvestrant),[204] which suggests that it is a more potent growth inhibitor than tamoxifen and raloxifene. This effect is associated with a rotation of H12, induced by the tethering of Leu-536 and Tyr-537 to the carboxyl moiety on GW 5638/7604, which leads to an increase in the surface hydrophobicity of the ERα LBD and a decrease in ERα stability. The fact that tamoxifen-resistant breast cancers are not cross-resistant to GW 5638[205] suggests that this SERM has significant potential as a therapeutic agent.

8.09.13   Conclusion

During the last 15 years there has been a revolution in understanding the multifaceted aspects of the estrogen receptor as a changeable target. Subtle three-dimensional changes in ligand structures lead to conformational changes between

ligand–receptor complexes that yield distinct physiological responses. However, despite our rapid advance in understanding these complex phenomena, it is still unclear as to exactly how these multisignalling pathways of estrogen ultimately determine certain biological endpoints. Many of the basic building blocks have been identified and they must now be assembled. What is interesting about the SERM story is that the first- and second-generation compounds were available long before they were recognized to have SERM activity.[66] The clinical development of nonsteroidal antiestrogens over the past 40 years has resulted in the first agents (clomiphene and tamoxifen) for the induction of ovulation in subfertile women, the first antiestrogen (tamoxifen) specifically for the treatment of estrogen receptor-positive breast cancer, the first chemopreventive (tamoxifen) to reduce the incidence of breast cancer in high-risk pre- and postmenopausal women, and the first SERM (raloxifene) for the treatment and prevention of osteoporosis. The potential effect of raloxifene to reduce the incidence of invasive breast cancer is currently being evaluated in three prevention trials, the CORE, RUTH, and STAR trials. Each raloxifene trial has enrolled a unique cohort and, when all trials are complete, they will provide important information about the occurrence of invasive breast cancer in diverse populations encompassing over 33 000 postmenopausal women with widely varying breast cancer risks. If the results of these raloxifene trials demonstrate a significant reduction in the incidence of invasive breast cancer, as was observed in the MORE trial, while confirming a better risk–benefit profile than tamoxifen, raloxifene will become an important therapy for the reduction of breast cancer risk among postmenopausal women.

Development of drugs of the SERM class is being advanced by numerous companies; however, the sheer cost of mounting the clinical trials necessary to prove efficacy as a breast cancer preventive, and preventive for osteoporosis and/or CHD is staggering. Nevertheless, new and improved SERMs are available for further development (arzoxifene, Eli Lilly and Company; bazedoxifene, Wyeth Laboratories; lasofoxifene, Pfizer). What is remarkable is the reinvention of the molecules for different indications over the past 50 years (*see* 8.08 Tamoxifen). The basic structures originally discovered as potential antifertility agents in the early 1960s are today continuing to be mined by medicinal chemists as advances in basic knowledge of molecular mechanisms occur.

References

1. Schwartz, N. B.; McCormack, C. E. *Annu. Rev. Physiol.* **1972**, *34*, 425–472.
2. Burger, H. G. *Int. J. Fertil.* **1981**, *26*, 153–160.
3. Sherman, B. M.; Korenman, S. G. *J. Clin. Invest.* **1975**, *55*, 699–706.
4. Sherman, B. M.; West, J. H.; Korenman, S. G. *Clin. Endocrinol. Metab. J. Clin. Endocrinol. Metab.* **1976**, *42*, 629–636.
5. Korenman, S. G.; Sherman, B. M.; Korenman, J. C. *Clin. Endocrinol. Metab.* **1978**, 7, 625–643.
6. Sarto, G. E. *Int. J. Gynaecol. Obstet.* **1977**, *15*, 189–192.
7. Nordin, B. E.; Horsman, A.; Brook, R.; Williams, D. A. *Clin. Endocrinol. (Oxf.)* **1976**, 5, 353S–361S.
8. Krolner, B.; Pors Nielsen, S. *Bone Miner. Clin. Sci. (Lond.)* **1982**, *62*, 329–336.
9. Richelson, L. S.; Wahner, H. W.; Melton, L. J., III; Riggs, B. L. *N. Engl. J. Med.* **1984**, *311*, 1273–1275.
10. Reinbold, W. D.; Genant, H. K.; Reiser, U. J.; Harris, S. T.; Ettinger, B. *Bone Miner. Radiol.* **1986**, *160*, 469–478.
11. Ettinger, B. *Int. J. Fertil.* **1986**, *31*, 15–20.
12. Nilas, L.; Gotfredsen, A.; Hadberg, A.; Christiansen, C. *Bone Miner.* **1988**, *4*, 95–103.
13. Nilas, L.; Christiansen, C. *Eur. J. Clin. Invest.* **1988**, *18*, 529–534.
14. Kannel, W. B.; Hjortland, M. C.; McNamara, P. M.; Gordon, T. *Ann. Intern. Med.* **1976**, *85*, 447–452.
15. Stampfer, M. J.; Colditz, G. A.; Willett, W. C. *Ann. NY Acad. Sci.* **1990**, *592*, 193–203, discussion 257–262.
16. Paganini-Hill, A.; Henderson, V. W. *Am. J. Epidemiol.* **1994**, *140*, 256–261.
17. Henderson, V. W.; Paganini-Hill, A.; Emanuel, C. K.; Dunn, M. E.; Buckwalter, J. G. *Arch. Neurol.* **1994**, *51*, 896–900.
18. Henderson, V. W.; Watt, L.; Buckwalter, J. G. *Psychoneuroendocrinology* **1996**, *21*, 421–430.
19. Coope, J.; Thomson, J. M.; Poller, L. *Br. Med. J.* **1975**, *4*, 139–143.
20. Campbell, S.; Whitehead, M. *Clin. Obstet. Gynaecol.* **1977**, *4*, 31–47.
21. Hailes, J. D.; Nelson, J. B.; Schneider, M.; Rennie, G. C.; Burger, H. G. *Med. J. Aust.* **1981**, *2*, 340–342.
22. Ettinger, B.; Genant, H. K.; Cann, C. E. *Ann. Intern. Med.* **1985**, *102*, 319–324.
23. The website www.premarin.com/home.asp is owned by Wyeth Pharmaceuticals. This website provides information about estrogen replacement therapy (ERT/HRT) (accessed May 2006).
24. Gordon, S. F.; Thompson, K. A.; Ruoff, G. E.; Imig, J. R.; Lane, P. J.; Schwenker, C. E. *Int. J. Fertil. Menopausal Stud.* **1995**, *40*, 126–134.
25. Gordon, S. F. *Obstet. Gynecol. Am. J. Obstet. Gynecol.* **1995**, *173*, 998–1004.
26. Castallo, M. A. *Pa Med.* **1967**, *70*, 80–81.
27. McEwen, D. C. *Can. Nurse* **1967**, *63*, 34–37.
28. May, W. J. *Am. Fam. Physician* **1977**, *16*, 109–113.
29. Genant, H. K.; Baylink, D. J.; Gallagher, J. C.; Harris, S. T.; Steiger, P.; Herber, M. *Obstet. Gynecol.* **1990**, *76*, 579–584.
30. LaRosa, J. C. *J. Reprod. Med.* **1985**, *30*, 811–813.
31. Stampfer, M. J.; Colditz, G. A.; Willett, W. C.; Manson, J. E.; Rosner, B.; Speizer, F. E.; Hennekens, C. H. *N. Engl. J. Med.* **1991**, *325*, 756–762.
32. Ross, R. K.; Paganini-Hill, A.; Mack, T. M.; Arthur, M.; Henderson, B. E. *Lancet* **1981**, *1*, 858–860.
33. Colditz, G. A.; Willett, W. C.; Stampfer, M. J.; Rosner, B.; Speizer, F. E.; Hennekens, C. H. *N. Engl. J. Med.* **1987**, *316*, 1105–1110.
34. Sullivan, J. M.; Vander Zwaag, R.; Lemp, G. F.; Hughes, J. P.; Maddock, V.; Kroetz, F. W.; Ramanathan, K. B.; Mirvis, D. M. *Ann. Intern. Med.* **1988**, *108*, 358–363.
35. Beard, C. M.; Kottke, T. E.; Annegers, J. F.; Ballard, D. J. *Mayo Clin. Proc.* **1989**, *64*, 1471–1480.

36. Stampfer, M. J.; Colditz, G. A. *Prev. Med.* **1991**, *20*, 47–63.
37. Grady, D.; Rubin, S. M.; Petitti, D. B.; Fox, C. S.; Black, D.; Ettinger, B.; Ernster, V. L.; Cummings, S. R. *Ann. Intern. Med.* **1992**, *117*, 1016–1037.
38. Blumenfeld, Z.; Aviram, M.; Brook, G. J.; Brandes, J. M. *Maturitas* **1983**, *5*, 77–83.
39. Kaplan, N. M. *J. Reprod. Med.* **1985**, *30*, 802–804.
40. Ottosson, U. B.; Carlstrom, K.; Johansson, B. G.; von Schoultz, B. *Gynecol. Obstet. Invest.* **1986**, *22*, 198–205.
41. Ziel, H. K.; Finkle, W. D. *N. Engl. J. Med.* **1975**, *293*, 1167–1170.
42. Gambrell, R. D., Jr.; Massey, F. M.; Castaneda, T. A.; Ugenas, A. J.; Ricci, C. A. *J. Am. Geriatr. Soc.* **1979**, *27*, 389–394.
43. Greenblatt, R. B.; Gambrell, R. D., Jr.; Stoddard, L. D. *Pathol. Res. Pract.* **1982**, *174*, 297–318.
44. Hemminki, E.; Kennedy, D. L.; Baum, C.; McKinlay, S. M. *Am. J. Public Health* **1988**, *78*, 1479–1481.
45. Wysowski, D. K.; Golden, L.; Burke, L. *Obstet. Gynecol.* **1995**, *85*, 6–10.
46. Keating, N. L.; Cleary, P. D.; Rossi, A. S.; Zaslavsky, A. M.; Ayanian, J. Z. *Ann. Intern. Med.* **1999**, *130*, 545–553.
47. Bergkvist, L.; Adami, H. O.; Persson, I.; Hoover, R.; Schairer, C. *N. Engl. J. Med.* **1989**, *321*, 293–297.
48. Steinberg, K. K.; Smith, S. J.; Thacker, S. B.; Stroup, D. F. *Epidemiology* **1994**, *5*, 415–421.
49. Colditz, G. A.; Hankinson, S. E.; Hunter, D. J.; Willett, W. C.; Manson, J. E.; Stampfer, M. J.; Hennekens, C.; Rosner, B.; Speizer, F. E. *N. Engl. J. Med.* **1995**, *332*, 1589–1593.
50. Colditz, G. A. *J. Natl. Cancer Inst.* **1998**, *90*, 814–823.
51. Persson, I.; Weiderpass, E.; Bergkvist, L.; Bergstrom, R.; Schairer, C. *Cancer Causes Control* **1999**, *10*, 253–260.
52. Steinberg, K. K.; Thacker, S. B.; Smith, S. J.; Stroup, D. F.; Zack, M. M.; Flanders, W. D.; Berkelman, R. L. *JAMA* **1991**, *265*, 1985–1990.
53. Hulley, S.; Grady, D.; Bush, T.; Furberg, C.; Herrington, D.; Riggs, B.; Vittinghoff, E. *JAMA* **1998**, *280*, 605–613.
54. Grady, D.; Yaffe, K.; Kristof, M.; Lin, F.; Richards, C.; Barrett-Connor, E. *Am. J. Med.* **2002**, *113*, 543–548.
55. Hulley, S.; Furberg, C.; Barrett-Connor, E.; Cauley, J.; Grady, D.; Haskell, W.; Knopp, R.; Lowery, M.; Satterfield, S.; Schrott, H. et al. *JAMA* **2002**, *288*, 58–66.
56. The Women's Health Initiative Study Group. *Control Clin. Trials* **1998**, *19*, 61–109.
57. Rossouw, J. E.; Anderson, G. L.; Prentice, R. L.; LaCroix, A. Z.; Kooperberg, C.; Stefanick, M. L.; Jackson, R. D.; Beresford, S. A.; Howard, B. V.; Johnson, K. C. et al. *JAMA* **2002**, *288*, 321–333.
58. Shumaker, S. A.; Legault, C.; Rapp, S. R.; Thal, L.; Wallace, R. B.; Ockene, J. K.; Hendrix, S. L.; Jones, B. N., III; Assaf, A. R.; Jackson, R. D. et al. *JAMA* **2003**, *289*, 2651–2662.
59. Chlebowski, R. T.; Hendrix, S. L.; Langer, R. D.; Stefanick, M. L.; Gass, M.; Lane, D.; Rodabough, R. J.; Gilligan, M. A.; Cyr, M. G.; Thomson, C. A. et al. *JAMA* **2003**, *289*, 3243–3253.
60. Rapp, S. R.; Espeland, M. A.; Shumaker, S. A.; Henderson, V. W.; Brunner, R. L.; Manson, J. E.; Gass, M. L.; Stefanick, M. L.; Lane, D. S.; Hays, J. et al. *JAMA* **2003**, *289*, 2663–2672.
61. Anderson, G. L.; Limacher, M.; Assaf, A. R.; Bassford, T.; Beresford, S. A.; Black, H.; Bonds, D.; Brunner, R.; Brzyski, R.; Caan, B. et al. *JAMA* **2004**, *291*, 1701–1712.
62. Beral, V. *Lancet* **2003**, *362*, 419–427.
63. Beral, V.; Bull, D.; Reeves, G. *Lancet* **2005**, *365*, 1543–1551.
65. The website www.nhlbi.nih.gov/whi/ is operated by the National Institutes of Health and the National Heart, Lung, and Blood Institute (accessed May 2006).
66. Jordan, V. C. *Cancer Res.* **2001**, *61*, 5683–5687.
67. Jordan, V. C. *J. Med. Chem.* **2003**, *46*, 1081–1111.
68. Fisher, B.; Costantino, J. P.; Wickerham, D. L.; Redmond, C. K.; Kavanah, M.; Cronin, W. M.; Vogel, V.; Robidoux, A.; Dimitrov, N.; Atkins, J. et al. *J. Natl. Cancer Inst.* **1998**, *90*, 1371–1388.
69. Osborne, C. K. *N. Engl. J. Med.* **1998**, *339*, 1609–1618.
70. EBCTCG. *Lancet* **1998**, *351*, 1451–1467.
71. EBCTCG. *Lancet* **2005**, *365*, 1687–1717.
72. MacDougall, M. J.; Tan, S. L.; Hall, V.; Balen, A.; Mason, B. A.; Jacobs, H. S. *Fertil. Steril.* **1994**, *61*, 1052–1057.
73. Ecochard, R.; Mathieu, C.; Royere, D.; Blache, G.; Rabilloud, M.; Czyba, J. C. *Fertil. Steril.* **2000**, *73*, 90–93.
74. Boostanfar, R.; Jain, J. K.; Mishell, D. R., Jr.; Paulson, R. J. *Fertil. Steril.* **2001**, *75*, 1024–1026.
75. Wiseman, L. R.; Goa, K. L. *Drugs* **1997**, *54*, 141–160.
76. Vogel, C. L. *Oncol. (Williston Park)* **1998**, *12*, 9–13.
77. Jones, C. D.; Jevnikar, M. G.; Pike, A. J.; Peters, M. K.; Black, L. J.; Thompson, A. R.; Falcone, J. F.; Clemens, J. A. *J. Med. Chem.* **1984**, *27*, 1057–1066.
78. Delmas, P. D.; Bjarnason, N. H.; Mitlak, B. H.; Ravoux, A. C.; Shah, A. S.; Huster, W. J.; Draper, M.; Christiansen, C. *Bone Miner. N. Engl. J. Med.* **1997**, *337*, 1641–1647.
79. Ettinger, B.; Black, D. M.; Mitlak, B. H.; Knickerbocker, R. K.; Nickelsen, T.; Genant, H. K.; Christiansen, C.; Delmas, P. D.; Zanchetta, J. R.; Stakkestad, J. et al. *JAMA* **1999**, *282*, 637–645.
80. Johnston, C. C., Jr.; Bjarnason, N. H.; Cohen, F. J.; Shah, A.; Lindsay, R.; Mitlak, B. H.; Huster, W.; Draper, M. W.; Harper, K. D.; Heath, H., III et al. *Bone Miner. Arch. Intern. Med.* **2000**, *160*, 3444–3450.
81. Delmas, P. D.; Ensrud, K. E.; Adachi, J. D.; Harper, K. D.; Sarkar, S.; Gennari, C.; Reginster, J. Y.; Pols, H. A.; Recker, R. R.; Harris, S. T. et al. *J. Clin. Endocrinol. Metab.* **2002**, *87*, 3609–3617.
82. Vogel, V. G.; Costantino, J. P.; Wickerham, D. L.; Cronin, W. M.; Wolmark, N. *Clin. Breast Cancer* **2002**, *3*, 153–159.
83. Siris, E. S.; Harris, S. T.; Eastell, R.; Zanchetta, J. R.; Goemaere, S.; Diez-Perez, A.; Stock, J. L.; Song, J.; Qu, Y.; Kulkarni, P. M. et al. *J. Bone Miner. Res.* **2005**, *20*, 1514–1524.
84. Mosca, L.; Barrett-Connor, E.; Wenger, N. K.; Collins, P.; Grady, D.; Kornitzer, M.; Moscarelli, E.; Paul, S.; Wright, T. J.; Helterbrand, J. D. et al. *Am. J. Cardiol.* **2001**, *88*, 392–395.
85. Mosca, L. *Ann. NY Acad. Sci.* **2001**, *949*, 181–185.
86. Black, L. J.; Jones, C. D.; Falcone, J. F. *Life Sci.* **1983**, *32*, 1031–1036.
87. Osborne, C. K.; Hobbs, K.; Clark, G. M. *Cancer Res.* **1985**, *45*, 584–590.
88. Thompson, E. W.; Reich, R.; Shima, T. B.; Albini, A.; Graf, J.; Martin, G. R.; Dickson, R. B.; Lippman, M. E. *Cancer Res.* **1988**, *48*, 6764–6768.
89. Poulin, R.; Merand, Y.; Poirier, D.; Levesque, C.; Dufour, J. M.; Labrie, F. *Breast Cancer Res. Treat.* **1989**, *14*, 65–76.

90. Wakeling, A. E.; Valcaccia, B.; Newboult, E.; Green, L. R. *J. Steroid Biochem.* **1984**, *20*, 111–120.
91. Gottardis, M. M.; Jordan, V. C. *Cancer Res.* **1987**, *47*, 4020–4024.
92. Anzano, M. A.; Peer, C. W.; Smith, J. M.; Mullen, L. T.; Shrader, M. W.; Logsdon, D. L.; Driver, C. L.; Brown, C. C.; Roberts, A. B. et al. *J. Natl. Cancer Inst.* **1996**, *88*, 123–125.
93. Gottardis, M. M.; Ricchio, M. E.; Satyaswaroop, P. G.; Jordan, V. C. *Cancer Res.* **1990**, *50*, 3189–3192.
94. Clemens, J. A.; Bennett, D. R.; Black, L. J.; Jones, C. D. *Life Sci.* **1983**, *32*, 2869–2875.
95. Buzdar, A. U.; Marcus, C.; Holmes, F.; Hug, V.; Hortobagyi, G. *Oncology* **1988**, *45*, 344–345.
96. Jordan, V. C.; Phelps, E.; Lindgren, J. U. *Breast Cancer Res. Treat.* **1987**, *10*, 31–35.
97. Black, L. J.; Sato, M.; Rowley, E. R.; Magee, D. E.; Bekele, A.; Williams, D. C.; Cullinan, G. J.; Bendele, R.; Kauffman, R. F.; Bensch, W. R. et al. *J. Clin. Invest.* **1994**, *93*, 63–69.
98. Turner, C. H.; Sato, M.; Bryant, H. U. *Endocrinology* **1994**, *135*, 2001–2005.
99. Sato, M.; Kim, J.; Short, L. L.; Slemenda, C. W.; Bryant, H. U. *J. Pharmacol. Exp. Ther.* **1995**, *272*, 1252–1259.
100. Lerner, L. J.; Holthaus, F. J., Jr.; Thompson, C. R. *Endocrinology* **1958**, *63*, 295–318.
101. Lunan, C. B.; Klopper, A. *Clin. Endocrinol. (Oxf.)* **1975**, *4*, 551–572.
102. Jordan, V. C. *Pharmacol. Rev.* **1984**, *36*, 245–276.
103. Jordan, V. C. *J. Med. Chem.* **2003**, *46*, 883–908.
104. Segal, S. J.; Nelson, W. O. *Proc. Soc. Exp. Biol. Med.* **1958**, *98*, 431–436.
105. Emmens, C. W. *Br. Med. Bull.* **1970**, *26*, 45–51.
106. Emmens, C. W. *Annu. Rev. Pharmacol.* **1970**, *10*, 237–254.
107. Holtkamp, D. E.; Greslin, J. G.; Root, C. A.; Lerner, L. J. *Proc. Soc. Exp. Biol. Med.* **1960**, *105*, 197–201.
108. Greenwald, G. S. *Fertil. Steril.* **1965**, *16*, 185–194.
109. Callantine, M. R.; Humphrey, R. R.; Lee, S. L.; Windsor, B. L.; Schottin, N. H.; O'Brien, O. P. *Endocrinology* **1966**, *79*, 153–167.
110. Harper, M. J.; Walpole, A. L. *J. Reprod. Fertil.* **1967**, *13*, 101–119.
111. Harper, M. J.; Walpole, A. L. *J. Endocrinol.* **1967**, *37*, 83–92.
112. Greenblatt, R. B.; Barfield, W. E.; Jungck, E. C.; Ray, A. W. *JAMA* **1961**, *178*, 101–104.
113. Huppert, L. C. *Fertil. Steril.* **1979**, *31*, 1–8.
114. Klopper, A.; Hall, M. *Br. Med. J.* **1971**, *1*, 152–154.
115. Williamson, J. G.; Ellis, J. D. *J. Obstet. Gynaecol. Br. Commonw.* **1973**, *80*, 844–847.
116. Harper, M. J.; Walpole, A. L. *Nature* **1966**, *212*, 87.
117. Palopoli, F. P.; Feil, V. J.; Allen, R. E.; Holtkamp, D. E.; Richardson, A., Jr. *J. Med. Chem.* **1967**, *10*, 84–86.
118. Jordan, V. C.; Haldemann, B.; Allen, K. E. *Endocrinology* **1981**, *108*, 1353–1361.
119. Clark, J. H.; Guthrie, S. C. *Biol. Reprod.* **1981**, *25*, 667–672.
120. Camerman, N.; Chan, L. Y.; Camerman, A. *J. Med. Chem.* **1980**, *23*, 941–945.
121. Iyer, R. N.; Gopalchari, R.; Kamboj, V. P.; Kar, A. B. *Ind. J. Exp. Biol.* **1967**, *5*, 169–170.
122. Iyer, R. N.; Gopalchari, R. *Ind. J. Pharmacol.* **1969**, *31*, 49–54.
123. Gopalchari, R.; Iyer, R. N.; Kamboj, V. P.; Kar, A. B. *Contraception* **1970**, *2*, 199–205.
124. Gopalchari, R.; Iyer, R. N. *Ind. J. Chem.* **1973**, *11*, 229–233.
125. Jones, C. D.; Suarez, T.; Massey, E. H.; Black, L. J.; Tinsley, F. C. *J. Med. Chem.* **1979**, *22*, 962–966.
126. Legha, S. S.; Carter, S. K. *Cancer Treat. Rev.* **1976**, *3*, 205–216.
127. Legha, S. S.; Slavik, M.; Carter, S. K. *Cancer* **1976**, *38*, 1535–1541.
128. Rose, D. P.; Fischer, A. H.; Jordan, V. C. *Eur. J. Cancer Clin. Oncol.* **1981**, *17*, 893–898.
129. Jordan, V. C.; Gosden, B. *Mol. Cell Endocrinol.* **1982**, *27*, 291–306.
130. Jones, C. D.; Blaszczak, L. C.; Goettel, M. E.; Suarez, T.; Crowell, T. A.; Mabry, T. E.; Ruenitz, P. C.; Srivatsan, V. *J. Med. Chem.* **1992**, *35*, 931–938.
131. Witte, R. S.; Pruitt, B.; Tormey, D. C.; Moss, S.; Rose, D. P.; Falkson, G.; Carbone, P. P.; Ramirez, G.; Falkson, H.; Pretorius, F. J. *Cancer* **1986**, *57*, 34–39.
132. Lee, R. W.; Buzdar, A. U.; Blumenschein, G. R.; Hortobagyi, G. N. *Cancer* **1986**, *57*, 40–43.
133. Jordan, V. C.; Collins, M. M.; Rowsby, L.; Prestwich, G. *J. Endocrinol.* **1977**, *75*, 305–316.
134. Borgna, J. L.; Rochefort, H. *J. Biol. Chem.* **1981**, *256*, 859–868.
135. Coezy, E.; Borgna, J. L.; Rochefort, H. *Cancer Res.* **1982**, *42*, 317–323.
136. Black, L. J.; Goode, R. L. *Life Sci.* **1980**, *26*, 1453–1458.
137. Black, L. J.; Goode, R. L. *Endocrinology* **1981**, *109*, 987–989.
138. Black, L. J.; Jones, C. D.; Goode, R. L. *Mol. Cell Endocrinol.* **1981**, *22*, 95–103.
139. Scholl, S. M.; Huff, K. K.; Lippman, M. E. *Endocrinology* **1983**, *113*, 611–617.
140. Jordan, V. C.; Gosden, B. *J. Steroid Biochem.* **1983**, *19*, 1249–1258.
141. Wakeling, A. E.; Valcaccia, B. *J. Endocrinol.* **1983**, *99*, 455–464.
142. Jordan, V. C.; Gosden, B. *Endocrinology* **1983**, *113*, 463–468.
143. Lerner, L. J.; Jordan, V. C. *Cancer Res.* **1990**, *50*, 4177–4189.
144. Love, R. R.; Mazess, R. B.; Barden, H. S.; Epstein, S.; Newcomb, P. A.; Jordan, V. C.; Carbone, P. P.; DeMets, D. L. *N. Engl. J. Med.* **1992**, *326*, 852–856.
145. Turner, R. T.; Wakley, G. K.; Hannon, K. S.; Bell, N. H. *J. Bone Miner. Res.* **1987**, *2*, 449–456.
146. Turner, R. T.; Wakley, G. K.; Hannon, K. S.; Bell, N. H. *Endocrinology* **1988**, *122*, 1146–1150.
147. Turken, S.; Siris, E.; Seldin, D.; Flaster, E.; Hyman, G.; Lindsay, R. *J. Natl. Cancer Inst.* **1989**, *81*, 1086–1088.
148. Broulik, P. D. *Endocr. Regul.* **1991**, *25*, 217–219.
149. Feldmann, S.; Minne, H. W.; Parvizi, S.; Pfeifer, M.; Lempert, U. G.; Bauss, F.; Ziegler, R. *Bone Miner.* **1989**, *7*, 245–254.
150. Smith, C. L.; Nawaz, Z.; O'Malley, B. W. *Mol. Endocrinol.* **1997**, *11*, 657–666.
151. McKenna, N. J.; Lanz, R. B.; O'Malley, B. W. *Endocrinol. Rev.* **1999**, *20*, 321–344.
152. Enmark, E.; Gustafsson, J. A. *J. Intern. Med.* **1999**, *246*, 133–138.
153. Wolf, D. M.; Jordan, V. C. *Breast Cancer Res. Treat.* **1994**, *31*, 129–138.
154. Wolf, D. M.; Jordan, V. C. *Breast Cancer Res. Treat.* **1994**, *31*, 117–127.

155. Catherino, W. H.; Wolf, D. M.; Jordan, V. C. *Mol. Endocrinol.* **1995**, *9*, 1053–1063.
156. Levenson, A. S.; Catherino, W. H.; Jordan, V. C. *J. Steroid Biochem. Mol. Biol.* **1997**, *60*, 261–268.
157. Brzozowski, A. M.; Pike, A. C.; Dauter, Z.; Hubbard, R. E.; Bonn, T.; Engstrom, O.; Ohman, L.; Greene, G. L.; Gustafsson, J. A.; Carlquist, M. *Nature* **1997**, *389*, 753–758.
158. Shiau, A. K.; Barstad, D.; Loria, P. M.; Cheng, L.; Kushner, P. J.; Agard, D. A.; Greene, G. L. *Cell* **1998**, *95*, 927–937.
159. Liu, H.; Park, W. C.; Bentrem, D. J.; McKian, K. P.; Reyes Ade, L.; Loweth, J. A.; Schafer, J. M.; Zapf, J. W.; Jordan, V. C. *J. Biol. Chem.* **2002**, *277*, 9189–9198.
160. MacGregor Schafer, J.; Liu, H.; Bentrem, D. J.; Zapf, J. W.; Jordan, V. C. *Cancer Res.* **2000**, *60*, 5097–5105.
161. Anghel, S. I.; Perly, V.; Melancon, G.; Barsalou, A.; Chagnon, S.; Rosenauer, A.; Miller, W. H., Jr.; Mader, S. *J. Biol. Chem.* **2000**, *275*, 20867–20872.
162. Beall, P. T.; Misra, L. K.; Young, R. L.; Spjut, H. J.; Evans, H. J.; LeBlanc, A. *Calcif. Tissue Int.* **1984**, *36*, 123–125.
163. Sato, M.; McClintock, C.; Kim, J.; Turner, C. H.; Bryant, H. U.; Magee, D.; Slemenda, C. W. *J. Bone Miner. Res.* **1994**, *9*, 715–724.
164. Kanis, J. A.; Johnell, O.; Black, D. M.; Downs, R. W., Jr.; Sarkar, S.; Fuerst, T.; Secrest, R. J.; Pavo, I. *Bone* **2003**, *33*, 293–300.
165. Gradishar, W.; Glusman, J.; Lu, Y.; Vogel, C.; Cohen, F. J.; Sledge, G. W., Jr. *Cancer* **2000**, *88*, 2047–2053.
166. Snyder, K. R.; Sparano, N.; Malinowski, J. M. *Am. J. Health Syst. Pharm.* **2000**, *57*, 1669–1675; quiz 1676–1678.
167. Jordan, V. C. *Breast Cancer Res. Treat.* **1983**, *3*, S73–S86.
168. Cummings, S. R.; Eckert, S.; Krueger, K. A.; Grady, D.; Powles, T. J.; Cauley, J. A.; Norton, L.; Nickelsen, T.; Bjarnason, N. H.; Morrow, M. et al. *JAMA* **1999**, *281*, 2189–2197.
169. Cauley, J. A.; Norton, L.; Lippman, M. E.; Eckert, S.; Krueger, K. A.; Purdie, D. W.; Farrerons, J.; Karasik, A.; Mellstrom, D.; Ng, K. W. et al. *Breast Cancer Res. Treat.* **2001**, *65*, 125–134.
170. Martino, S.; Cauley, J. A.; Barrett-Connor, E.; Powles, T. J.; Mershon, J.; Disch, D.; Secrest, R. J.; Cummings, S. R. *J. Natl. Cancer Inst.* **2004**, *96*, 1751–1761.
171. Vejtorp, M.; Christensen, M. S.; Vejtorp, L.; Larsen, J. F. *Acta Obstet. Gynecol. Scand.* **1986**, *65*, 391–395.
172. Ma, P. T.; Yamamoto, T.; Goldstein, J. L.; Brown, M. S. *Proc. Natl. Acad. Sci. USA* **1986**, *83*, 792–796.
173. Kauffman, R. F.; Bensch, W. R.; Roudebush, R. E.; Cole, H. W.; Bean, J. S.; Phillips, D. L.; Monroe, A.; Cullinan, G. J.; Glasebrook, A. L.; Bryant, H. U. *J. Pharmacol. Exp. Ther.* **1997**, *280*, 146–153.
174. Zuckerman, S. H.; Bryan, N. *Atherosclerosis* **1996**, *126*, 65–75.
175. Walsh, B. W.; Kuller, L. H.; Wild, R. A.; Paul, S.; Farmer, M.; Lawrence, J. B.; Shah, A. S.; Anderson, P. W. *JAMA* **1998**, *279*, 1445–1451.
176. Barrett-Connor, E.; Grady, D.; Sashegyi, A.; Anderson, P. W.; Cox, D. A.; Hoszowski, K.; Rautaharju, P.; Harper, K. D. *JAMA* **2002**, *287*, 847–857.
177. Wenger, N. K.; Barrett-Connor, E.; Collins, P.; Grady, D.; Kornitzer, M.; Mosca, L.; Sashegyi, A.; Baygani, S. K.; Anderson, P. W.; Moscarelli, E. *Am. J. Cardiol.* **2002**, *90*, 1204–1210.
178. Grese, T. A.; Sluka, J. P.; Bryant, H. U.; Cullinan, G. J.; Glasebrook, A. L.; Jones, C. D.; Matsumoto, K.; Palkowitz, A. D.; Sato, M.; Termine, J. D. et al. *Proc. Natl. Acad. Sci. USA* **1997**, *94*, 14105–14110.
179. Grese, T. A.; Cho, S.; Finley, D. R.; Godfrey, A. G.; Jones, C. D.; Lugar, C. W., III; Martin, M. J.; Matsumoto, K.; Pennington, L. D.; Winter, M. A. et al. *J. Med. Chem.* **1997**, *40*, 146–167.
180. Palkowitz, A. D.; Glasebrook, A. L.; Thrasher, K. J.; Hauser, K. L.; Short, L. L.; Phillips, D. L.; Muehl, B. S.; Sato, M.; Shetler, P. K.; Cullinan, G. J. et al. *J. Med. Chem.* **1997**, *40*, 1407–1416.
181. Sato, M.; Turner, C. H.; Wang, T.; Adrian, M. D.; Rowley, E.; Bryant, H. U. *J. Pharmacol. Exp. Ther.* **1998**, *287*, 1–7.
182. Suh, N.; Glasebrook, A. L.; Palkowitz, A. D.; Bryant, H. U.; Burris, L. L.; Starling, J. J.; Pearce, H. L.; Williams, C.; Peer, C.; Wang, Y. et al. *Cancer Res.* **2001**, *61*, 8412–8415.
183. Licun, W.; Tannock, I. F. *Clin. Cancer Res.* **2003**, *9*, 4614–4618.
184. Freddie, C. T.; Larsen, S. S.; Bartholomaeussen, M.; Lykkesfeldt, A. E. *Mol. Cell Endocrinol.* **2004**, *219*, 27–36.
185. Ma, Y. L.; Bryant, H. U.; Zeng, Q.; Palkowitz, A.; Jee, W. S.; Turner, C. H.; Sato, M. *J. Bone Miner. Res.* **2002**, *17*, 2256–2264.
186. Hale, L. V.; Layman, N. K.; Wilson, P.; Short, L. L.; Magee, D. E.; Wightman, S. R.; Fuchs-Young, R. *Lab. Anim. Sci.* **1997**, *47*, 82–85.
187. Dardes, R. C.; Bentrem, D.; O'Regan, R. M.; Schafer, J. M.; Jordan, V. C. *Clin Cancer Res.* **2001**, *7*, 4149–4155.
188. Schafer, J. M.; Lee, E. S.; Dardes, R. C.; Bentrem, D.; O'Regan, R. M.; De Los Reyes, A.; Jordan, V. C. *Clin. Cancer Res.* **2001**, *7*, 2505–2512.
189. Suh, N.; Lamph, W. W.; Glasebrook, A. L.; Grese, T. A.; Palkowitz, A. D.; Williams, C. R.; Risingsong, R.; Farris, M. R.; Heyman, R. A.; Sporn, M. B. *Clin. Cancer Res.* **2002**, *8*, 3270–3275.
190. Rendi, M. H.; Suh, N.; Lamph, W. W.; Krajewski, S.; Reed, J. C.; Heyman, R. A.; Berchuck, A.; Liby, K.; Risingsong, R.; Royce, D. B. et al. *Cancer Res.* **2004**, *64*, 3566–3571.
191. Munster, P. N.; Buzdar, A.; Dhingra, K.; Enas, N.; Ni, L.; Major, M.; Melemed, A.; Seidman, A.; Booser, D.; Theriault, R. et al. *J. Clin. Oncol.* **2001**, *19*, 2002–2009.
192. Komm, B. S.; Lyttle, C. R. *Ann. NY Acad. Sci.* **2001**, *949*, 317–326.
193. Miller, C. P.; Collini, M. D.; Tran, B. D.; Harris, H. A.; Kharode, Y. P.; Marzolf, J. T.; Moran, R. A.; Henderson, R. A.; Bender, R. H.; Unwalla, R. J. et al. *J. Med. Chem.* **2001**, *44*, 1654–1657.
194. Komm, B. S.; Kharode, Y. P.; Bodine, P. V.; Harris, H. A.; Miller, C. P.; Lyttle, C. R. *Endocrinology* **2005**, *146*, 3999–4008.
195. Rosati, R. L.; Da Silva Jardine, P.; Cameron, K. O.; Thompson, D. D.; Ke, H. Z.; Toler, S. M.; Brown, T. A.; Pan, L. C.; Ebbinghaus, C. F.; Reinhold, A. R. et al. *J. Med. Chem.* **1998**, *41*, 2928–2931.
196. Ke, H. Z.; Paralkar, V. M.; Grasser, W. A.; Crawford, D. T.; Qi, H.; Simmons, H. A.; Pirie, C. M.; Chidsey-Frink, K. L.; Owen, T. A.; Smock, S. L. et al. *Endocrinology* **1998**, *139*, 2068–2076.
197. Ke, H. Z.; Foley, G. L.; Simmons, H. A.; Shen, V.; Thompson, D. D. *Endocrinology* **2004**, *145*, 1996–2005.
198. Cohen, L. A.; Pittman, B.; Wang, C. X.; Aliaga, C.; Yu, L.; Moyer, J. D. *Cancer Res.* **2001**, *61*, 8683–8688.
199. The website www.drugdevelopment-technology.com/projects/lasofoxifene/ is operated by Pfizer and Ligand Pharmaceuticals (accessed May 2006).
200. Willson, T. M.; Henke, B. R.; Momtahen, T. M.; Charifson, P. S.; Batchelor, K. W.; Lubahn, D. B.; Moore, L. B.; Oliver, B. B.; Sauls, H. R.; Triantafillou, J. A. et al. *J. Med. Chem.* **1994**, *37*, 1550–1552.
201. Willson, T. M.; Norris, J. D.; Wagner, B. L.; Asplin, I.; Baer, P.; Brown, H. R.; Jones, S. A.; Henke, B.; Sauls, H.; Wolfe, S. et al. *Endocrinology* **1997**, *138*, 3901–3911.

202. Wijayaratne, A. L.; Nagel, S. C.; Paige, L. A.; Christensen, D. J.; Norris, J. D.; Fowlkes, D. M.; McDonnell, D. P. *Endocrinology* **1999**, *140*, 5828–5840.
203. Connor, C. E.; Norris, J. D.; Broadwater, G.; Willson, T. M.; Gottardis, M. M.; Dewhirst, M. W.; McDonnell, D. P. *Cancer Res.* **2001**, *61*, 2917–2922.
204. Wu, Y. L.; Yang, X.; Ren, Z.; McDonnell, D. P.; Norris, J. D.; Willson, T. M.; Greene, G. L. *Mol. Cell* **2005**, *18*, 413–424.
205. Dardes, R. C.; O'Regan, R. M.; Gajdos, C.; Robinson, S. P.; Bentrem, D.; De Los Reyes, A.; Jordan, V. C. *Clin Cancer Res.* **2002**, *8*, 1995–2001.

Biographies

Joan Lewis-Wambi is currently an Assistant Member at the Fox Chase Cancer Center, Philadelphia, Pennsylvania. She works in collaboration with Dr V Craig Jordan, the Alfred G Knudson Chair of Cancer Research and the Vice President and Research Director for Medical Sciences at Fox Chase. Dr Lewis-Wambi received her PhD in Biochemistry and Nutrition from Rutgers University (New Jersey) in May 2002. She is an Avon scholar and is the recipient of the Fellow-in-Training scholarship from the Center for Biomedical Continuing Education (CBCE).

V Craig Jordan received his PhD, DSc, and DMed (honoris causa) from Leeds University (UK). His work on SERMs has been recognized by the Bristol Myers Squibb Award, the NAMS Eli Lilly Award for SERM research as well as more than a dozen prizes and honorary degrees in the United States and Europe. He is currently Vice President and Research Director for Medical Sciences at the Fox Chase Cancer Center, Philadelphia, Pennsylvania.

Comprehensive Medicinal Chemistry II
ISBN (set): 0-08-044513-6
ISBN (Volume 8) 0-08-044521-7; pp. 103–121

incontinence – i.e., if the rhabdosphincter were contracting more forcibly it would be more likely to resist the increase in abdominal pressure during a cough, laugh, or sneeze.

Importantly, duloxetine's facilitatory effects on the sphincter 'disappeared' during a micturition contraction. In other words, during a micturition contraction, the sphincter activity was absent. During micturition, it is important for the bladder to contract and the sphincter to relax in synergy to allow efficient voiding. Thus, the fact that inhibition of the sphincter remained during a bladder contraction (i.e., synergy was maintained) indicated that urinary retention should not be a problem. (This has been confirmed in all the clinical trials to date.)

Similar effects were seen with another SNRI, venlafaxine.[28] Remarkably, however, we could not administer a combination of a selective serotonin reuptake inhibitor (SSRI) and a norepinephrine reuptake inhibitor to the same animal and obtain a similar effect to duloxetine and venlafaxine, despite our best efforts. In other words, combining two selective individual reuptake inhibitors was ineffective but having the combination of reuptake inhibition in a single molecule was effective. This remains an enigma to date.

Further studies indicated that the enhancement of sphincter activity was due to increased stimulation of $5HT_2$ serotonergic and alpha-1 adrenergic receptors resulting from increased levels of 5HT and norepinephrine associated with reuptake blockade.[27–29]

Because of the role of the sympathetic nervous system and norepinephrine in mediating contraction of the urethral smooth muscle, we examined the effects of selective norepinephrine reuptake inhibitors,[30] as well as duloxetine, for their ability to augment sympathetic-nerve-induced urethral contractions via the norepinephrine reuptake inhibition properties. We found that increasing synaptic levels of norepinephrine with a reuptake blocker produced no consistent facilitation of urethral contractions due to counteracting effects resulting from enhanced norepinephrine stimulation of relaxatory beta adrenergic receptor stimulation.

8.10.4 Challenges in Bringing Forward the First Therapeutic Treatment for Incontinence

In the early 1990s, therapy for stress urinary incontinence relied on pelvic floor exercises and surgery. Bringing the first drug forward to treat any indication provided a number of challenges, such as extent of medical need and clinical trial design. Unique to duloxetine's trials in incontinence were (1) the fact that urologic thought leaders' prevailing opinion at that time was that stress incontinence was 'an anatomical defect' that would be only amenable to surgery and not pharmacological therapy, and (2) doubts about whether a CNS approach to a urological problem was tenable.

8.10.4.1 Incontinence Markets

Without any historical pharmaceutical sales data for an indication, the market potential is difficult to predict because most financial models are based on sales of competitors' products. Since there were no well-marketed products for stress urinary incontinence, it was difficult to develop a financial model. In 1992, even sales of urge incontinence products were remarkably small; for example, the top UUI medicine was Ditropan, which only had 92 million days of therapy prescribed in the USA, and there were virtually no drug sales in the USA for SUI. This absence of therapy highlighted the need for new therapy with a mechanism of action that was different from previous therapy and was emphasized in 1992 by the Agency for Health Care Policy and Research (AHCPR) which released its first guideline on urinary incontinence and reported:

- 13 million Americans are incontinent; 11 million are women
- 1 in 4 women ages 30–59 have experienced an episode of urinary incontinence
- 50% or more of the elderly persons living at home or in long-term care facilities are incontinent
- $16.4 billion is spent every year on incontinence-related care: $11.2 billion for community-based programs and at home, and $5.2 billion in long-term care facilities
- $1.1 billion is spent every year on disposable products for adults.

Valuable resources for establishing characteristics of unexplored markets with great medical need are found in patient advocacy groups. One of the most prominent patient advocacy groups for incontinence was the Simon Foundation, led by Cheryl Gartley. These patient advocacy groups can provide valuable direction for understanding the patient perspective in regards to why some seek treatment versus those who do not, concerns about surgical treatments, reasons for failure with pelvic floor exercises, and motivation in regards to treating their SUI. For example,

in 1992, only about 20% of all incontinence patients sought treatment, only about half of those received medical therapy, and 90% of those stopped taking their medication within 3 months because of intolerable side effects and low efficacy. While it seemed that just about every publication on surgical procedures for stress incontinence reported 80–90% cure rates, it was not until follow-up reports were published did we become aware that these rates seen immediately after surgery, at the most prestigious academic institutions, in the hands of the best practitioners, were not reflective of long-term results of the overall surgical population and were accompanied by concerning complications.

8.10.4.2   Clinical Trial Considerations

Being the first to initiate regulatory submission quality clinical trials in a therapeutic indication also carries substantial challenges:

1. There are no physicians with regulatory submission clinical trial experience specific to your indication.
2. There are no publications indicating
    ○ which efficacy measures are best
    ○ inclusion/exclusion criteria
    ○ recruitment rates
    ○ trial duration or design
    ○ anticipated placebo response rates or intrinsic variability to allow power calculations
    ○ quality of life instruments.

Fortunately, the pharmaceutical industry and the Food and Drug Administration (FDA) worked together to develop guidelines for regulatory submissions in incontinence and laid the groundwork for the first meeting to establish 'FDA Guidance for Industry for the Development of Incontinence Drugs' in 1998.

One of the take-home messages from that public meeting was that incontinence clinical trials needed both an objective and a subjective measure of patient improvement. For example, a patient might not consider a statistically significant reduction in number of incontinence episodes from four per day to two per day clinically significant. Similarly, one must question a drug that makes the patient less bothered by incontinence if there is no statistically significant reduction in the number of incontinence episodes. Thus, both an objective measure and subjective measure were needed.

Since the American Urological Association was over 100 years old and incontinence constituted a significant portion of a urologist's practice, it was surprising that opinions about the best objective measures of the condition were so disparate. As it turned out, incontinence episode frequency recorded in the micturition diary (which one might question as being objective since it is filled out by the patient) was the most reliable and used in all clinical trials for duloxetine.

Since a psychometric instrument to assess patient perception of their incontinence was needed, Buesching of Lilly, in collaboration with Patrick of University of Washington, developed the Incontinence Quality of Life Questionnaire (IQOL).[31] This questionnaire contained domains for avoidance and limiting behaviors, psychosocial impacts, and social embarrassment and was validated for internal consistency, reproducibility, correlation with other measures, and responsivity.[32]

8.10.5   Clinical Trial Results

8.10.5.1   The First Trial – SAAA

The initial clinical trial[33] examined stress, mixed, and urge incontinent patients because of duloxetine's affects on both the rhabdosphincter and bladder capacity. This preliminary proof-of-concept trial for duloxetine's use in treating incontinence was at a dose of 20 mg q.d., which is substantially lower than that used in later trials (i.e., 40–60 mg b.i.d.). In hindsight, based on the low dose of duloxetine, the small number of SUI patients (22 on duloxetine, 11 on placebo), and the extreme variability typically found in the stress-pad test (one of our efficacy measures), it is remarkable that this trial showed a statistically significant reduction in stress-pad test weights with duloxetine treatment (**Figure 4a**) and a nonsignificant reduction in incontinence episodes (**Figure 4b**). Additionally, the trial showed that a significantly greater proportion of SUI patients on duloxetine showed at least a 30% reduction in incontinence episodes compared to placebo. Nearly 50% of the patients showed 70% improvement in incontinence episodes and stress-pad test, but this stratification was not statistically significant (**Figure 5**). No effects were seen in the mixed or urge incontinent

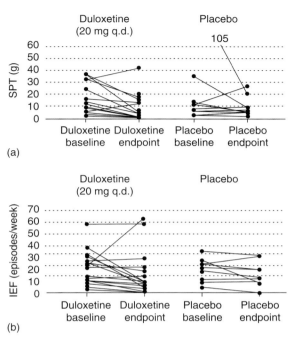

Figure 4 Line graph 'baseline to endpoint' for (a) stress-pad test (SPT) and (b) incontinence episode frequency (IEF). Each line connects an individual SUI patient's baseline and endpoint values. Duloxetine significantly decreased amount of SPT leakage ($p = 0.02$ versus placebo, ANCOVA on ranked changes with baseline values as covariate) and frequency of incontinence, but it was not statistically significant ($p = 0.34$ versus placebo, ANCOVA on ranked changes with baseline values as covariate). (Adapted from Mulcahy, J. J.; Kirkemo, A.; Rudy, D. C.; Blaivas, J. G.; Wahle, G. R.; Sirls, L. T.; Laddu, A. R.; Faries, D.; Debrota, D.; Thor, K. B. *Neurourol. Urodyn.* **1996**, *15*, 92–395.)

patients. These results led to a focus on SUI in subsequent trials. Remarkably, the placebo-treated group actually had a higher proportion of adverse events than the duloxetine-treated group, and the subsequent US trial dosed at 20, 30, and 40 mg q.d.

8.10.5.2 The Japanese Trial

This trial[34] was the first to show efficacy in neurogenic bladder (and is the only trial published for any overactive bladder condition to date). This important trial also supported the early positive results for duloxetine in SUI. This study showed a reduction from 1.7 to 0.3 incontinence episodes per day in neurogenic bladder patients at 20 mg (but no effect at 10 mg) and a reduction from 3 to 1 incontinence episodes/day at both 10 and 20 mg doses in SUI patients. Although the trial was a single-blinded study and contained no placebo group (which was traditional in Japan at that time to ensure all patients were treated with something), these results added to the early suggestions of duloxetine's efficacy.

8.10.5.3 The Second US Trial – SAAB

The next trial measured incontinence episode frequency, IQOL, stress-pad test weights, and 24-h pad weights with about 35 patients in each dose group and restricted itself to stress and mixed UI patients.[35] Unexpectedly, only the 20 mg group showed statistically significant improvement in all measures, while those at dose 40 mg showed significance in only stress-pad test and IQOL, and the 30 mg dose group only showed significance in IQOL (**Figures 6** and **8**). In hindsight, this absence of a dose response is not surprising since the dose increments of 20, 30, and 40 mg duloxetine are proportional to increments of, for example, 1, 1.5, and 2 aspirins. One might not expect to see a dose-dependent reduction in headaches in groups of 35 patients across those doses of aspirin. When all duloxetine arms were pooled, significance was retained for all measures except 24-h pad weights, which still showed twice as much reduction as placebo. As in the first trial, the overall incidence of adverse events on placebo was actually worse than any of the duloxetine groups. However, nausea (now recognized as the most prominent side effect of duloxetine) did show a higher (though not significant) increase with duloxetine compared to placebo. In a dose-related pattern similar to the efficacy described above, the highest rate of

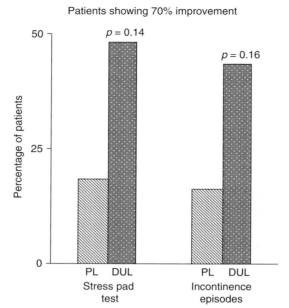

Patients showing 70% improvement

Figure 5 Responder analysis of 20 mg q.d. duloxetine (DUL) versus placebo (PL) as defined by a 70% improvement from baseline in SUI patients. (Adapted from Mulcahy, J. J.; Kirkemo, A.; Rudy, D. C.; Blaivas, J. G.; Wahle, G. R.; Sirls, L. T.; Laddu, A. R.; Faries, D.; Debrota, D.; Thor, K. B. *Neurourol. Urodyn.* **1996**, *15*, 92–395.)

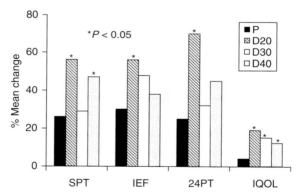

Figure 6 Effects of placebo (P) and duloxetine (D) at 20, 30, 40 mg q.d. on stress-pad test (SPT), incontinence episode frequency (IEF), 24-h pad weight (24PT), and incontinence quality of life (IQOL) in SUI patients. (Adapted from Zinner, N.; Sarshik, S.; Yalçin, I.; Faries, D.; DeBrota, D.; Riedl, P.; Thor, K.B. *Efficacy and Safety of Duloxetine in Stress Urinary Incontinent Patients: Double-Blind, Placebo-Controlled Multiple Dose Study.* International Continence Society, 28th Annual Meeting, Jerusalem, Israel, Sept. 1998.)

nausea was seen in the 20 mg group (17%), followed by the 40 mg group (14%), with the 30 mg group trailing (10%). Again, this rank order for nausea mimicked the rank order for efficacy across these doses.

8.10.5.4   Incontinence Severity Index in SAAB

The fact that all efficacy measures showed improvements with duloxetine at all doses (but lacked statistical significance for all points) suggested that patients were really improving with duloxetine but that the small number of patients coupled with the noise in each of the individual parameters were obscuring the positive signal. Therefore, a factor analysis approach, the incontinence severity index (ISI), was developed by Ilker Yalçin.[36,37]

The underlying supposition in the creation of the ISI was that incontinence is a multifaceted condition that cannot be represented with a single measure. For example, if two women leak three times a day, one at a volume of 5 g and the other at 10 g, episode frequency itself will not reflect the condition or improvement in the condition appropriately.

Biography

Karl B Thor, PhD, formed Urogenix, Inc., as a subsidiary of Astellas Pharmaceuticals, Inc., in March of 2007 and holds the position of vice-president of research. In addition, Dr Thor holds Adjunct Research Associate Professor positions in the Department of Surgery/Division of Urology and the Department of Obstetrics and Gynecology at Duke University Medical Center in Durham, North Carolina. He is Co-director of the Laboratory of Neurourology at the Veterans Affairs Medical Center, Surgical Research Services, also in Durham. The Laboratory of Neurourology receives grant support from the NIH, Veterans Administration, and Christopher Reeves Foundation for studies of neural control of the lower urinary tract under pathological conditions including spinal cord injury.

Dr Thor received his PhD in Pharmacology from the University of Pittsburgh School of Medicine where he trained under William (Chet) de Groat, PhD and was supported by a PhARMA predoctoral fellowship. He held a National Research Service Award postdoctoral fellowship from the NIH at Uniformed Services University of the Health Sciences in Bethesda, Maryland, and was a Senior Staff Fellow in the Laboratory of Neurophysiology at the NIH. He joined Eli Lilly in the Neuroscience Division in 1990, where he discovered duloxetine (Yentreve) as a treatment for stress urinary incontinence. In 1998, he formed PPD GenuPro as a subsidiary of PPD Inc., where he discovered the clinical potential of dapoxetine as a therapy for premature ejaculation. These two drugs are the first agents to be submitted to regulatory agencies for their respective indications. In 2002, he founded Dynogen Pharmaceuticals Inc., a neuroscience-based drug discovery and development company targeting genitourinary and gastrointestinal diseases, where two clinical program for overactive bladder and two clinical programs for irritable bowel syndrome were initiated.

Dr Thor has published numerous articles and book chapters on CNS control of lower urinary tract function and holds several patents for methods of treating urinary bladder and sexual dysfunction. He is a member of the Urodynamics Society, the International Continence Society, and the Society for Neuroscience. He served as a member of the FDA-PhARMA Industry Guidelines for Urinary Incontinence Trials in 1998. He has served as a peer reviewer for the NIH on the Urology Special Emphasis Study Section, the National Institutes of Neurological Disorders and Stroke Neurology B Study Section, and the Small Business Initiative Review Study Section. He has served on the World Health Organization's International Consultation on Incontinence since its inception.

Comprehensive Medicinal Chemistry II
ISBN (set): 0-08-044513-6

ISBN (Volume 8) 0-08-044521-7; pp. 123–135

8.11 Carvedilol

R R Ruffolo and G Z Feuerstein, Wyeth Research Laboratories, Collegeville, PA, USA

© 2007 Elsevier Ltd. All Rights Reserved.

8.11.1	**Historical Overview**	**137**
8.11.2	**The Pharmacology of Carvedilol: Historical Perspectives**	**138**
8.11.2.1	Adrenergic Receptors	138
8.11.2.2	The Multiple Actions of Carvedilol: Antioxidant Properties	139
8.11.3	**Enter Heart Failure**	**140**
8.11.3.1	Beta Blockers in Heart Failure: Perspectives	140
8.11.3.2	Carvedilol and Heart Failure: Banging our Heads against the Textbooks	140
8.11.4	**Science, Medicine, and Leadership Win Out: The Pendulum Swings in Carvedilol's Favor**	**141**
8.11.5	**Clinical Trials of Carvedilol in Heart Failure: The Bottom Line**	**142**
8.11.6	**The Food and Drug Administration: The Last Major Obstacle**	**142**
8.11.7	**Summary and Perspective**	**143**
	References	**144**

8.11.1 Historical Overview

Carvedilol (Coreg) is commonly referred to as a third-generation beta adrenoceptor blocker (beta blocker) with vasodilatory and antioxidant properties. These varied activities are attributed to different parts of the carvedilol molecule (**Figure 1**). Carvedilol was originally discovered in the early 1980s as a novel beta blocker for the primary therapeutic indications of angina and hypertension, which were traditional uses of drugs of this class at the time. However, the true novelty behind carvedilol is the fact that it represents the first beta blocker to be approved for the treatment of chronic congestive heart failure, which is a serious progressive disease that typically results in death. Until the time that carvedilol was approved by the Food and Drug Administration (FDA), beta blockers were 'contraindicated' in patients with heart failure because of the well-known cardiac depressant effects of this class of drug, as well as the prevailing view that such drugs would worsen heart failure and potentially increase mortality. The paradoxically beneficial effects of carvedilol in patients with heart failure changed not only our thinking about this serious disease, but resulted in a new treatment paradigm for patients with heart failure, as well as a new standard of care. Most importantly, carvedilol removed the death sentence from many patients with this invariably progressive and fatal disease.

More than 20 years ago, research commenced in the laboratories of SmithKline Beecham Pharmaceuticals (now GlaxoSmithKline) with the intent to explore the potential of carvedilol's unique pharmacological profile, consisting of beta blockade, alpha blockade, vasodilatation, and antioxidant activity, as a potential therapy for chronic heart failure.

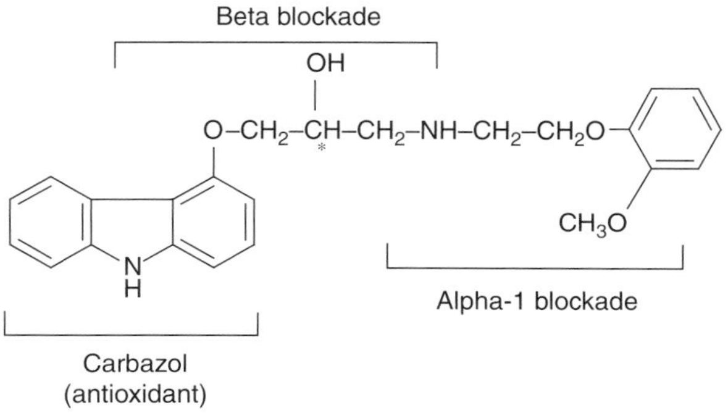

Figure 1 Chemical structure of carvedilol. The active moieties responsible for the antiadrenergic and antioxidant actions are noted. Asterisk denotes the point of asymmetry.

Over the next decade, compelling data were generated in a variety of in vitro and in vivo experimental systems that highlighted novel and unusually effective cardioprotective properties of the drug that resulted in part from the beneficial hemodynamic effects emanating from beta blockade and alpha blockade, and in part from the unique antioxidant, antiapoptotic, and antiproliferative properties of the molecule. Based on these extensive studies, the highest levels of corporate and R&D management at SmithKline Beecham took the brave, risky, controversial, and highly innovative decision to support long-term clinical trials of carvedilol in patients with congestive heart failure, for whom such drugs remained contraindicated. The importance of this decision cannot be overstated given the dogma prevalent at the time that such a drug might actually harm patients, as well as the resulting concerns related to liability and the view that the commercial value to the corporation after taking such unprecedented risks was low.

During the lengthy and costly development program for carvedilol in heart failure, clinical trials were often delayed, and the priority of the program changed regularly, as is commonly the case in the risky environment of drug development. The fact that these trials did, in the end, continue through to completion can be attributed to the vision, persistence, and courage of a few basic research and clinical scientists, including these authors among others, and the trust in them provided by corporate and R&D management, including Jan Leschly (former chief executive officer), Jean-Pierre Garnier (chief operating officer), and George Poste (former chairman of R&D). Certainly there were more doubters than supporters at the time, making the support received from SmithKline Beecham executives all the more important.

The highlight of the entire 15–20 year discovery and development effort on carvedilol was the day that an independent Data Safety and Monitoring Board (DSMB) terminated the studies of carvedilol in heart failure prematurely because of an 'unexpected beneficial effect' of the drug in heart failure patients compared to those patients receiving placebo, making it, in the judgment of the DSMB, unethical to maintain patients in the placebo part of the study. In other words, the DSMB felt that the clinical trial needed to be stopped so that the patients receiving placebo could be immediately treated with this new life-saving medicine. The reduction in mortality observed with carvedilol in patients with this fatal disease was an unprecedented 65%. In hindsight, the decision taken by SmithKline Beecham to develop carvedilol in heart failure, against the recommendations of many experts in the field at the time, represents the pharmaceutical industry at its best. This extraordinary risk taken by SmithKline Beecham, its management, and its scientists has resulted in a dramatic improvement in human health, and in this case, created a new standard of care for seriously ill patients with heart failure to extend their lives, decrease their need for hospitalization, and decrease the burden of this devastating disease on healthcare systems. The ultimate beneficiary of this extraordinary gamble taken by SmithKline Beecham is indeed the patient.

8.11.2   The Pharmacology of Carvedilol: Historical Perspectives

8.11.2.1   Adrenergic Receptors

SmithKline Beecham originally acquired rights to carvedilol in the US from Boehringer Mannhein (now part of Roche) with the obligation to develop the drug for angina and hypertension. The original pharmacological profile of carvedilol was that of a vasodilating beta blocker.[1,2] A long-standing core team of scientists at SmithKline Beecham, consisting of the authors, as well as Hieble, Nichols and Ohlstein, devoted between one and two decades of their respective lives to understanding every aspect of this drug, and to determining exactly how it might work in heart failure. Early reports suggested that the vasodilatory effects of carvedilol resulted from calcium channel blockade.[2] Through a detailed series of experiments, it was established unequivocally that the primary vasodilatory properties of carvedilol resulted from alpha-1 adrenergic receptor blockade, and not through inhibition of calcium channels.[3] This finding was initially viewed as a significant disappointment, since another combined alpha and beta blocker was known (i.e., labetalol), making carvedilol appear to be less novel. However, through other investigations, the core team subsequently discovered many additional important activities of carvedilol, and these activities made carvedilol unique from any other drug in the world.

The extreme potency of carvedilol as an alpha-1 blocker was thought to be important in the pharmacodynamic response of the drug in patients with heart failure, and would result in a beneficial reduction in afterload, making it easier for the failing heart to eject its contents, thereby improving general circulatory status. The role of alpha-1 receptor blockade in the pharmacological profile of carvedilol has proven to be critical in the ability of patients with heart failure to tolerate the beta blocking actions of the drug. Thus, the reduction in afterload produced by alpha-1 blockade helped the heart to compensate for the known cardiac depressant effects of the beta blocking actions.[4] It is also likely that alpha-1 receptor blockade contributes to the favorable metabolic profile of the drug.[5] In retrospect, establishing that the alpha-1 blocking action of carvedilol was responsible for the vasodilatory properties of the drug, and not calcium channel blockade, was a blessing in disguise inasmuch calcium channel blockers have subsequently been shown to increase risk of cardiac morbidity in certain populations of patients with heart disease.[6,7]

for further training at the National Institutes of Health (NIH), Bethesda, MD (1979–81) in the Laboratory of Clinical Sciences (Chief, I Kopin) focusing on sympathetic nervous system control of the cardiovascular system.

Giora Feuerstein holds adjunct position in academic organization (Med College of Georgia, Augusta, GA) and Jefferson Medical College, Philadelphia, PA). He also serves on editorial boards of the *Journal of Pharmacology and Experimental Therapeutics; Biochemical Pharmacology; Journal of Cerebral Blood Flow and Metabolism; Circulation Research; and Stroke*. Giora Feuerstein is the recipient of several national and international awards including Award of Excellent in Cardiovascular Research, AHA, 1987; Prix Galien Award for Drug Discover (endothelin antagonist), 1994; Conrad R Lam Award for Cardiovascular Research, Henry Ford Foundation, 2001.

Giora Feuerstein has authored and co-authored over 650 publications of which over 400 are peer review journals. He is also the coinventor on 12 patents and has edited 8 books.

Comprehensive Medicinal Chemistry II
ISBN (set): 0-08-044513-6
ISBN (Volume 8) 0-08-044521-7; pp. 137–147

8.12 Modafinil, A Unique Wake-Promoting Drug: A Serendipitous Discovery in Search of a Mechanism of Action

F Rambert, J F Hermant, and D Schweizer, Cephalon France, Maisons-Alfort, France

8.12.1	**Introduction**	149
8.12.2	**The First Step: The Discovery of Adrafinil**	149
8.12.3	**From Adrafinil to Modafinil**	150
8.12.4	**From Animal Pharmacology to a Human Therapeutic**	151
8.12.5	**In Search of a Mechanism(s) of Action**	151
8.12.6	**In Search and Discovery of Potential New Therapeutic Indications**	152
8.12.7	**Conclusions**	153
References		153

The search for additional indications for modafinil was directed toward diseases associated with wake deficit and somnolence but also to those in which symptoms could be related to cognition deficits, with modafinil showing human efficacy in attention deficit hyperactivity disorder (ADHD). Preclinical studies also showed a beneficial effect of modafinil in models of depression.

Since its discovery, modafinil has proven to be an efficient and safe therapeutic agent. Many biochemical and pharmacological studies, focused on the mechanism of action of modafinil, have failed to provide any conclusive evidence for a discrete target for this novel drug.

8.12.1 Introduction

The story of modafinil (Provigil), that began in the early 1970s, is the discovery of the unexpected activity of an NCE in classical animal models that transitioned to an effective agent that helped define and grow the field of sleep medicine. Additionally, extensive preclinical work on modafinil led to: (1) the identification of additional therapeutic indications that were not fully appreciated from the initial in vivo profile for modafinil; and (2) a continuing search, albeit unsuccessful, to define its mechanism(s) of action.

At the time that modafinil was discovered, the process of drug discovery was neither target-directed nor technology- (genome, high-throughput screening (HTS), or combinatorial chemistry) driven. The search for new drugs was thus mainly based on proven functional and empirical methods. Many of the receptors now known to be involved in sleep–wake regulation, e.g., hypocretin/orexin,[1] had not been identified, while the discrete functional role(s) of better-characterized central nervous system (CNS) neurotransmitter systems, e.g., cholinergic, noradrenergic, dopaminergic, serotonergic, histaminergic, GABAergic, in sleep function were still emerging.[2–4]

8.12.2 The First Step: The Discovery of Adrafinil

In the early 1970s, the Research Department of Laboratoire Louis Lafon, a small family-owned pharmaceutical company located in Maisons-Alfort, a suburb of Paris, had focused its research activities in three areas: (1) cardiovascular; (2) antispasmodics; and (3) nonsteroidal anti-inflammatory drugs (NSAIDs)/analgesics.

In this last field, an empirical test battery was implemented, using the best-characterized in vivo pharmacological methods generally in use at that time (some of which are still used). These included the carrageenan-induced paw edema and mechanical inflammatory pain (Randall and Selitto model) in rats, hot plate test, and the abdominal writing test in mice.

Initial screening efforts involved the investigation of many NCEs in the abdominal writing test: five compounds were simultaneously included in the same study. A trained observer, blinded to the randomized compound treatments, watched the mice for 5 min following p.o. dosing of NCEs. Typically, following administration of an agent that elicited

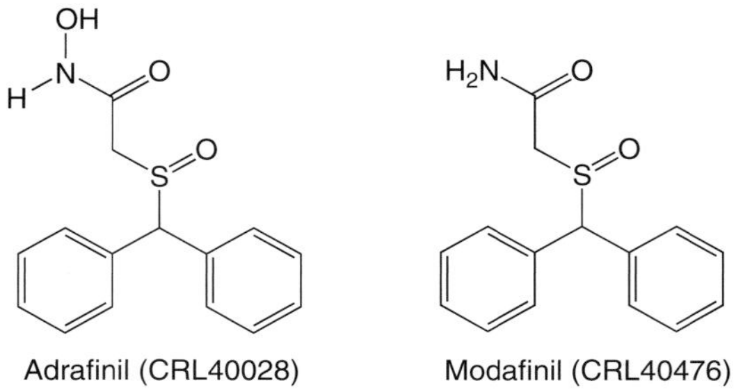

Figure 1 Structures of adrafinil and modafinil.

writhing (given i.p.), the mice were quiet and showed restricted motor activity, probably to avoid experiencing additional pain. However, during one of these experiments, the observer noticed that some mice, usually one in each group of six, did not display this behavior but instead exhibited marked locomotion, exploring their surroundings in a similar manner to control vehicle-treated mice. When the experiment was unblinded, the mice displaying this locomotor activity were all found to have received the same compound, CRL40028 ((benzhydryl sulfinyl) acetohydroxamic acid), later named adrafinil (**Figure 1**).

As a result of this unexpected finding, Louis Lafon, the founder of the company, immediately switched the objectives of the NSAID/analgesic research team to CNS therapeutic agents, more precisely to the discovery of nonamphetamine-like psychostimulant or antidepressant drugs.

Using a battery of classical behavioral tests in mice, the stimulant potential of adrafinil was confirmed, based on a dose-dependent increase in locomotor activity, antagonism of barbital-induced narcosis, and a decrease in the duration in the forced swim test.[5] Interestingly, adrafinil did not display any of the other effects normally observed with amphetamine and nonamphetamine (methylphenidate-like) stimulants: it failed to induce changes in core temperature; did not produce stereotyped or climbing behavior; and did not increase lethality in aggregated mice. Adrafinil was also devoid of other effects usually seen with classical antidepressants. It thus had no interaction with reserpine-, oxotremorine-, or apomorphine-induced hypothermia (although it slightly potentiated yohimbine-induced toxicity); lacked peripheral sympathetic effects (lack of mydriasis, salivation, piloerection, or antagonism of reserpine-induced ptosis); and lacked peripheral anticholinergic effects (lack of mydriasis or antagonism of oxotremorine-induced salivation or lacrimation).

These results led to the conclusion that, as compared to amphetaminic, anticholinergic, or antidepressant drugs, adrafinil had a unique behavioral profile in mice. This was defined by a specific stimulant activity associated with nonclassical antidepressant-like effects that did not appear to be related to a β-adrenergic mechanism, together with behavioral effects not linked to dopaminergic stimulation.

8.12.3 From Adrafinil to Modafinil

In considering adrafinil as a lead compound in search of therapeutic utility, the Chemistry Department in Laboratoire L. Lafon began to synthesize compounds chemically related to adrafinil. More than 100 compounds were then characterized using the primary behavioral screening used to determine the effects of adrafinil. In April 1976, 2 years after the first assays with adrafinil, CRL40476 was identified. This compound displayed the same pharmacological profile as adrafinil, but was more potent and longer-lasting than adrafinil. This compound, (diphenyl-methyl)-sulfinyl-2-acetamide, was named modafinil (presumably by analogy to 'modified adrafinil' or 'modulated adrafinil'?).

Because of the similar effects of adrafinil and modafinil and the available preclinical experimental data obtained with adrafinil (particularly in toxicological studies), the development of modafinil moved forward, mainly focused on putative therapeutic applications, a search for the mechanism of action, and a differentiation from amphetamine- and nonamphetamine-like stimulants. The initial publication of the unique behavioral profile of modafinil[6] aroused considerable interest from several research groups, resulting in many subsequent animal studies that confirmed, in a variety of species – mice,[7] rats,[8,9] cats,[10] narcoleptic dogs,[11] monkeys,[12,13] and even fruitfly[14] – the stimulant and awakening effects of the compound. Simultaneously, every effort was made to generate additional data to advance knowledge related to modafinil to the same level as that which was known regarding adrafinil, in order to provide a choice between the two compounds for further development. At the same time, studies on the metabolism of adrafinil indicated that this compound was primarily inactivated by conjugation but was, to a slight extent, metabolized to an active metabolite, modafinil.

invaluable insight and knowledge concerning the vastly different toleration in a rodent species of two closely related ketones – one cyclic, the other an acetophenone – which clearly established that an STR existed for this series.

The outcome of this study had a major impact on our project – one that would definitively play a significant role in our eventual successful discovery of eperezolid and linezolid. The importance of Piper's role in that success cannot be over-emphasized. First, a principal initial objective of the study was met in substantiating the rumored toxicological problems with the lead oxazolidinone. Second, the protocol had identified an equipotent analog of DuP-721 of close structural similarity with a clean toxicological profile. Third, Piper's study design had clearly demonstrated the utility of the drug-sparing protocol, thereby allowing us to plan on using this for expedited subsequent lead oxazolidinone toxicology evaluations. Finally, it shaped what would become our enabling strategy, allowing us to proceed. That strategy pivoted on the need to establish an understanding of STRs as a means of acquiring confidence in the selection of our advanced leads for further progression to drug candidate status. The necessity of conducting multiple, early, multiday toleration studies presented obvious additional hurdles for our team, obstacles that were considerably above those normally encountered in a typical SAR-driven program at that time. The strategy we instituted may arguably be one of the earliest of the few projects in the industry that succeeded in delivering a first-in-class drug to market, from a research program heavily reliant on early toxicological evaluation as a means to vet numerous, promising lead compounds.

Another series of compounds we focused on following the successful outcome with U-82965 would add substantially to our understanding of the STRs, and led to the identification of structural features that would eventually be incorporated into our drug candidates. We had chosen to examine various fused-ring heterocyclic oxazolidinone derivatives[22,23] for the explicit purpose of replacing the indanone ketone of U-82965, while retaining a five-membered benzo-fused ring. This interest included a series of active indazoles[22] and 5'-indolines, among others. With the principle of locating a carbonyl with an orientation proximate to that of the ketone in U-82965 or DuP-721, we prepared a series of amides attached to a 3-(5'-indolinyl)-5-acetamidomethyl-2-oxazolidinone core. This series of 5'-indolinyl amides[23] had superior activity to the isomeric 6'-indolinyl analogs.

Two of these active 5'-indoline amides, U-97456 (6) and U-85910 (7), were tolerated extremely well in the 30 day rat toxicology protocol.[28] Those results thereby established for the first time that a nitrogen atom substituted at the oxazolidinone phenyl *para* position could lead to compounds with improved toleration profiles. The thiophene amide 7 was of interest in that it was determined by James Kilburn and Suzanne Glickman at the Centers for Disease Control and Prevention (CDC) to have potent in vitro activity against *Mycobacterium tuberculosis*.[29] While the Gram-positive antibacterial potency of U-97456 was slightly below our targeted profile desired for a clinical candidate, the excellent safety profile of this compound and 7 laid the foundation for the eventual synthesis of many other oxazolidinone series similarly substituted with nitrogen-containing heterocycles at the *para*-phenyl position. This includes our first drug candidate, the piperazinyl fluorophenyl eperezolid, and the morpholinyl fluorophenyl analog linezolid.

6   7

U-97456 realized another important SAR finding. When the optimal *N*-hydroxyacetyl substituent found on U-97456 was incorporated into the piperazinyl fluorophenyl oxazolidinone (giving eperezolid), it also engendered very high potency, and excellent oral in vivo efficacy comparable to vancomycin subcutaneously (s.c.), as well as an excellent 30 day toleration profile. This same moiety has subsequently been singled out by other researchers as bringing about optimal activity in additional oxazolidinone and oxazolidinone-surrogate series (i.e., where other heterocycles replace the oxazolidinone).[30,30a,31a]

8.13.5   Insights Gained from Rigid Fused-Ring Oxazolidinones

An area of major interest in my laboratory involved a series of novel tricyclic-fused oxazolidinones. Contrary to prior SAR conclusions that had been reported in the literature,[24] we demonstrated that substitution could indeed be tolerated at both the *ortho*-phenyl and the oxazolidinone C-4 positions, provided these loci were connected with a short alkyl bridge in a *trans* orientation, relative to the 5-acetamidomethyl oxazolidinone side chain. My associate Peter Manninen initially synthesized the racemic *trans*-[6,5,5]-tricyclic-fused oxazolidinone analog (±)-8a corresponding to DuP-721.[32] Later,

Mark Gleave, a postdoctoral researcher in my laboratory, carried out an asymmetric Sharpless epoxidation followed by an intramolecular version of the Manninen cyclization (vide infra), to give (+)-**8a** in 98.8% ee, and this compound demonstrated good potency, with only a twofold reduction in in vitro and in vivo activity compared with DuP-721.[33]

Based on a report from DuPont researchers[34] in which the methyl ketone of DuP-721 was replaced by aromatic or heteroaromatic rings to good effect, we proceeded to examine this type of substitution on our [6,5,5]-tricyclic oxazolidinones.[35] Many of those compounds demonstrated compelling in vitro Gram-positive activity. The most active derivative was the (±)-3-pyridyl-[6,5,5]-tricyclic-fused oxazolidinone U-92300 (**8b**),[25,35] which demonstrated excellent in vitro potency, and in vivo oral efficacy commensurate with vancomycin, dosed s.c. Upon advancement of this compound into the rat toxicology protocol, however, it was disappointing to find that U-92300 elicited toxic effects in the rat when dosed twice daily at 100 mg kg^{-1} day^{-1}. Thus, this particular example is illustrative of the value afforded the team in conducting these early multiday toxicology studies.

Based on the high potency of U-92300, Gleave[35] also went on to synthesize the racemic des-fluoro-tricyclic-fused version of linezolid **8c**. Surprisingly, compound **8c** was found to be 16- to 64-fold less active in vitro than linezolid. In a similar fashion, the complexities of correctly predicting the suitability of such rigid analogs will be also illustrated by the following example of earlier exploits at attempting to design more potent compounds, based on the hypothesis-driven modification of these rigid frameworks. On the basis of computational considerations of a conformational feature we believed could influence the activity of these tricyclic compounds, our colleague Douglas Rohrer had predicted that the corresponding ethylene-bridged [6,6,5]-tricyclic fused analogs would be more active, in that this structural motif more closely approached the three-dimensional structural arrangement of the two ring systems found in the lowest-energy conformation of DuP-721.[36] Debra Allwine completed the synthesis of the 3-pyridyl-[6,6,5]-tricyclic oxazolidinone **9**, corresponding to the highly potent homologue [6,5,5] analog **8b**, only for us to find that **9** demonstrated an eightfold reduction in in vitro activity relative to **8b**.[35] It was hypothesized that the disappointing weak activity of **9** could be indicative that the [6,6,5] template, with its larger central ring than that in the [6,5,5] congener, is not as well accommodated within the oxazolidinone binding site.

8a R = COMe
8b R = 3-pyridyl
8c R = morpholinyl

9

This hypothesis would be in keeping with generalizations put forth by Brittelli and co-workers,[24] which they derived from studies of multi-substitution about the oxazolidinone phenyl ring. From that work it was suggested that the oxazolidinone binding site was narrow, with limited space in what would be the general region of our tricyclic template central ring. They also noted that this narrow binding site appeared to possess a small pocket on one (undetermined) side of the linear axis (extending through the aryl and oxazolidinone rings) that could accommodate a small *meta* substituent.[24] We subsequently prepared a series of six different rigid [6,6,5,5]-tetracyclic-fused oxazolidinones,[25] which we considered might add further insights and refinement to this DuPont model of the (then completely unknown) binding site, by taking advantage of the fixed orientation of *meta* substituents affixed by nature of the rigidity of these templates. Examining the biological activity of the pairs of congeners, each designed to fit into one of the two possible *meta*-positioned clefts, we observed modest, but significant, differences in the measured in vitro activity favoring isomers such as **10**. These data best supported the positioning of the cleft in the DuPont binding site model at the *meta* position, which is proximal to the oxazolidinone carbonyl.

10

One of the several synthetic pathways we used to construct our various tricyclic-fused oxazolidinones would come to play an important role in finding a viable solution to the need for an alternative route to optically active oxazolidinones.

This chemistry involved a route to the racemic [6,5,5]-tricyclic oxazolidinone nucleus that was developed with a summer student, Kristine Lovasz. We demonstrated that in mixtures of the *threo* and *erythro* indolinyl benzylcarbamates **11**, having an appendant alcohol at C-2, only the *threo* isomer would rapidly (0.5 h) undergo smooth intramolecular cyclization by simply treating with the mild base K_2CO_3 in refluxing CH_3CN, giving *trans*-**12**. Reaction of the *threo* isomer required 24 h to cyclize to the *cis* tricyclic oxazolidinone.[35]

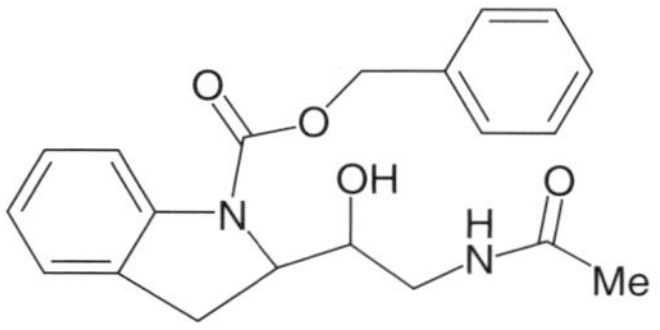

8.13.6   The Development of a Viable Synthesis of Oxazolidinones with High Enantiomeric Purity

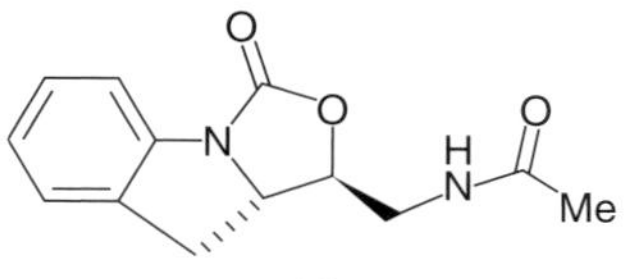

At the beginning of our oxazolidinone exploratory project, we had elected to work with racemic compounds as a means of enabling the rapid SAR exploration and discovery of proprietary oxazolidinone series. The route we chose to the racemic oxazolidinones exploited an iodocyclocarbamation reaction, first developed by Fraser-Reid,[37] and subsequently employed by others[38,39] in a more closely related sense to our work. Our use of this iodine-mediated cyclization of *N*-allyl carbamates was the first to apply it to the preparation of 3-aryl- and 3-heteroaryl-oxazolidinones. For the heteroaryl analogs, we found addition of excess pyridine and iodine was needed to prevent an undesired side reaction caused by a Friedel–Crafts-type alkylation of the heteroaryl ring.[36] Overall, the iodocyclocarbamation served us well in allowing rapid generation of a very wide range of racemic novel aryl- and heteroaryl-oxazolidinones, and was used widely by our oxazolidinone chemistry team.

However, our success in finding compounds of great interest soon required us to identify a viable approach to the preparation of optically active 5-(*S*)-acetamidomethyl-2-oxazolidinones for our more extensive PK, pharmacological, and toxicological profiling of the best leads. Additionally, we would need a robust route to high optical purity material to support an investigational new drug application and manufacture clinical bulk supplies.

The DuPont group[40] had chosen to employ a cyclization that had first been described in a racemic fashion decades earlier,[41,42] and involved the high-temperature cyclization of an optically enriched epoxide, (*R*)-glycidyl butyrate (**13**), with an aryl isocyanate, and provided an oxazolidinone butyrate ester intermediate **14** that was used for the synthesis of DuP-721 (**Scheme 1**). For our purposes, we recognized there would be some substantial limitations to our use of this approach. For the broad variety of oxazolidinone series in which we had interest, that sequence would have required our in-house preparation of many noncommercially available isocyanates. This typically involves treatment of an aryl amine with phosgene, and is complicated by the low extent of conversion to the isocyanate, as the reaction is stalled by concomitant formation of the aniline hydrochloride salt. Our most overriding concern was the significant potential safety hazards represented by frequently working with substantial quantities of phosgene in a discovery research laboratory setting. On that basis, my laboratory began searching for a new approach in earnest.

In thinking about the broader utility of exploiting the carbamate as an internal acylating agent as seen in the tricyclic-fused work above, the general idea developed to an approach for a possible new enantioselective synthesis of 5-(*S*)-acetamidomethyl oxazolidinones. This was envisioned as involving deprotonation of an aryl-NHCBZ (**15**) with a suitable base, alkylation of the resulting carbamate anion with epoxide **13**, and then cyclization of the nascent alkoxide by closure onto the CBZ carbonyl. When I first attempted this sequence using NaH as the base, the reaction proceeded to afford a prolific mixture of products – it clearly appeared that it was time to go back to the drawing board. Unbeknownst to me, Peter Manninen had proceeded, completely on his own initiative, to get this failed transformation

Scheme 1

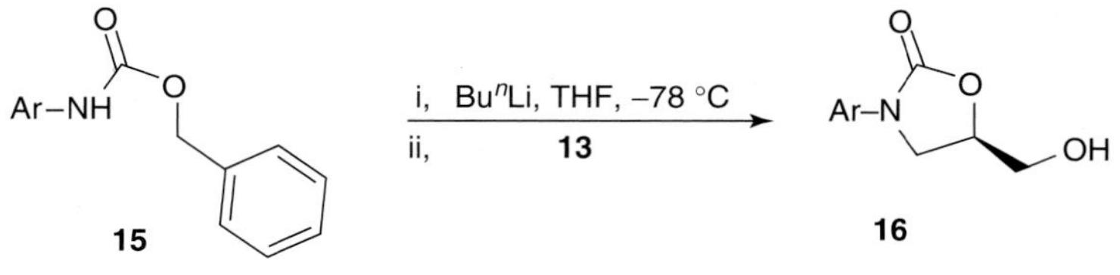

Scheme 2

to work. There followed a memorable 'eureka' moment in July 1992, when he brought forth the news that he had discovered the right conditions: the key was in using Bu^{n}Li as the base. We subsequently demonstrated that the presence of lithium was absolutely critical in this reaction.[43] As an added bonus, the product he isolated in very high yield was the 5-(R)-hydroxymethyl oxazolidinone **16**, not its butyrate ester **14** (as obtained in the isocyanate route). This was ideal, as the alcohol is a key intermediate for conversion to the active 5-(S)-acetamidomethyl oxazolidinones, and Manninen's reaction conditions provided this compound in very high enantiomeric excess, typically >99%.[44] The entire expanded chemistry team (vide infra) subsequently found the Manninen reaction was very reliable and general in scope, as has now been validated by its widespread use by many of the numerous researchers who have reported work in this field (**Scheme 2**).

This innovative contribution by Manninen had an invaluable impact on the success of our program, both in enabling the ready synthesis of oxazolidinones in high yield and optical purity, and by greatly facilitating the scale-up of initial multikilogram bulk quantities of eperezolid and linezolid. Having a viable route to the synthesis of a large range of optically active 3-aryl oxazolidinones accelerated not only our discovery efforts but also the entire time-line of progression to human trials for our two clinical candidates. It was gratifying to see that essentially the same route developed in our discovery laboratory[45] was used during the first 100 kg scale preparation of good manufacturing practices (GMP) clinical supplies of eperezolid and linezolid, carried out with only minor variations, by David Houser. After considerable efforts by a number of colleagues, the current production route for linezolid employs a more significant variation of the Manninen reaction, where instead of (R)-glycidyl butyrate, 2-(S)-3-chloro-1,2-propane-diol is employed.[28,46]

8.13.7   Formation of the Oxazolidinone Working Group

The successful outcomes of the U-82965, U-97456, and U-85910 toxicology studies (as well as the poor performance of U-92300), gave us considerable confidence in our ability to use the 30 day toxicology protocol to identify oxazolidinone series most worthy of continued pursuit – and those unworthy of advancement due to an unacceptably low therapeutic index. The establishment of this knowledge basis contributed directly to the eventual increased support of the oxazolidinone project with the allocation of additional chemistry and biology resources. In May of 1990, the chemistry laboratory headed by Douglas Hutchinson, and comprising his associates Raymond Reid and Stuart Garmon, later joined by Robert Reischer, was assigned to this project. In early 1991, Michael Barbachyn also joined the effort, along with his associates Kevin Grega and Dana Toops, and, later, Susan Hendges. With the tripling of the size of our original chemistry team, I was appointed the oxazolidinone chemistry team leader, and asked to chair a larger, interdisciplinary team of scientists called the 'Oxazolidinone Working Group.' The latter team added considerably more biology resources, to allow in-depth study of the then poorly understood mechanism of action (MOA), as well as other pharmacology of the oxazolidinones, and provide designated absorption, distribution, metabolism, excretion and toxicology (ADMET) evaluation resources. Robert Yancey's laboratory also helped profile the in vivo activity of the oxazolidinones.

The MOA studies were led initially by Keith Marotti, working with Jerry Buysse, and Dean Shinabarger, together with their associates William Demyan and Donna Dunyak. Following the PK evaluation of early compounds by John Greenfield, Mary Lou Sedlock and Robert Anstadt were added to the team in this capacity. Later, all PK and ADME work on this series was transferred to a group of colleagues led by Robert Ings at the Upjohn laboratories in Crawley, UK, including the laboratories of Martin Howard, Iain Martin, Peter Daley-Yates, Phil Jeffries, William Speed, Mark Ackland, and Neil Duncan. In Michigan, the toxicology studies on eperezolid were conducted by Richard Piper's laboratory, along with those of John Lund and Robert Denlinger; and colleagues in Upjohn's Tsukuba, Japan, laboratories would carry out the toxicological evaluation of linezolid. This latter group consisted of S Koike, H Miura, R Nakamura, and K Chiba, and James Moe.

Linezolid also has an excellent aqueous solubility of $3.7\,mg\,mL^{-1}$, which, like eperezolid, greatly facilitated the i.v. formulation development; as Zyvox, it is available as a $2.0\,mg\,mL^{-1}$ solution. The PK characteristics have made the switch from i.v. to oral therapy with linezolid particularly easy for physicians, with no dosage adjustments necessary. This capability can provide advantages over competing agents. It also is advantageous in allowing patients continuing on oral therapy to leave the hospital earlier than if they were maintained on i.v. therapy, which can result in cost benefits.[69]

The editors explicitly requested that case histories be written as personal perspectives. In that respect, the most singularly fulfilling aspect of this work – one I am certain must surely resonate with my colleagues – has come from hearing the uplifting testimonials of patients who benefited from treatment with linezolid during the clinical trials, particularly those for whom all other therapies had failed. From the period late 1997 to mid-2000, several hundred clinical investigators enrolled over 700 particularly ill patients in open-label, compassionate-use basis trials with linezolid, for the treatment of significant, antibiotic-resistant Gram-positive bacterial infections. One of the more striking accounts to come from one of these studies concerned a very seriously ill infant struggling to survive an infection with an MDR strain of VRE. As the attending physician had exhausted all treatment options, the prospects for the patient's survival were dim. Emergency arrangements were made, with the FDA's approval, to acquire linezolid from Pharmacia & Upjohn in order to treat this child on a compassionate-use basis. Because an i.v. access could not be established at that point, the drug powder was formulated with saline and dosed orally. Within 48 h after initiation of linezolid therapy, the infant was alert and sitting up, and subsequently fully recovered.[70] The reported assessment of the overall cure rate with linezolid (dosed in adults at 600 mg b.i.d. for a minimum of 10 days) observed in the entire compassionate-use basis program was 90.5%.[71]

On April 17, 2000, the FDA approved linezolid for treatment of susceptible and resistant Gram-positive infections in an initial set of indications that subsequently was broadened following additional approvals. The indications currently approved in the USA are for the treatment of hospital- and community-acquired pneumonia caused by *S. aureus* (methicillin-susceptible or MRSA) or *S. pneumoniae* (penicillin-susceptible or MDR strains), and VRE *E. faecium* (including concurrent bacteremias). Linezolid is the only approved drug for treatment of hospital-acquired MDR *S. pneumoniae*, and the first oral agent ever approved for the treatment of VRE infections.[72,72a] Linezolid has also been approved for treatment of complicated skin and skin structure infections, including diabetic foot infections without concomitant osteomyelitis, which are caused by methicillin-susceptible *S. aureus* and MRSA, *S. pyogenes*, or *Streptococcus agalactiae*. Linezolid has been approved for use in children and newborns against Gram-positive infections.

In summary, for the foreseeable future, the significant medical need presented by growing multidrug resistance will remain the impetus to continue the search for new antibacterial agents. Novel drugs such as linezolid that have a unique MOA can avoid cross-resistance to agents already in use, and may slow the rate of resistance development. The FDA approval of linezolid was the culmination of a 12 year research program that faced significant hurdles with the need to establish STRs, as well as SARs. Linezolid has established itself as an important antibiotic for the treatment of susceptible and MDR Gram-positive infections in the practice of medicine.

References

1. Culbertson, J.; Cowan, M. C. *Living Agents of Disease*; GP Putnam's Sons: New York, 1952.
2. Barber, M.; Garrod, L. P. *Antibiotic and Chemotherapy*; Williams and Wilkins: Baltimore, MD, 1963.
3. Franklin, T. J.; Snow, G. A. *Biochemistry of Antimicrobial Action*; Academic Press: New York, 1971.
4. Lesher, G. Y.; Froelich, E. J.; Gruett, M. D.; Bailey, J. H.; Brundage, R. P. *J. Med. Pharm. Chem.* **1962**, *5*, 1063–1068.
5. Walsh, C.; Wright, G. *Chem. Rev.* **2005**, *105*, 391–393.
6. Alder, J. D. *Drugs Today* **2005**, *41*, 81–90.
7. Livermore, D. M. *Clin. Infect. Dis.* **2003**, *36*, S11–S23.
8. Roberts, M. C. *Mol. Biotech.* **2004**, *28*, 47–62.
9. Weigel, L. M.; Clewell, D. B.; Gill, S. R.; Clark, N. C.; McDougal, L. K.; Flannagan, S. E.; Kolonay, F. F.; Shetty, J.; Killgore, G. E.; Tenover, F. C. *Science*, **2003**, *302*, 1569–1571.
10. Tiemersma, E. W.; Bronzwaer, S. L. A. M.; Lyytikäinen, O.; Degener, J. E.; Schrijnemakers, P.; Bruinsma, N.; Monen, J.; Witte, W.; Grundmann, H. *Emerg. Infect. Dis.* **2004**, *10*, 1627–1634.
11. Cardo, D.; Horan, T.; Andrus, M.; Dembinski, M.; Edwards, J.; Peavy, G.; Tolson, J.; Wagner, D. *Am. J. Infect. Control*. **2004**, *32*, 470–485.
12. Diekema, D. J.; Pfaller, M. A.; Schmitz, F. J.; Smayevsky, J.; Bell, J.; Jones, R. N.; Beach, M. *Clin. Infect. Dis.* **2001**, *32*, S114–S132.
13. Chambers, H. F. *Emerg. Inf. Dis.* **2001**, *7*, 178–182.
14. Slee, A. M.; Wuonola, M. A.; McRipley, R. J.; Zajac, I.; Zawada, M. J.; Bartholomew, P. T.; Gregory, W. A.; Forbes, M. *Abstracts of Papers*, Proceedings of the 27th Interscience Conference on Antimicrobial Agents and Chemotherapy 1987, Abstract No. 244; American Society for Microbiology: Washington, DC.
15. Uttley, A. H.; Collins, C. H.; Naidoo, J.; George, R. C. *Lancet* **1988**, *i*, 57–58.
16. Edmond, M. B.; Ober, J. F.; Dawson, J. D.; Weinbaum, D. L.; Wenzel, R. P. *Clin. Infect. Dis.* **1996**, *23*, 1234–1239.
17. Leclercq, R.; Derlot, E.; Duval, J.; Courvalin, P. *N. Engl. J. Med.* **1988**, *319*, 157–161.

18. Centers for Disease Control and Prevention. *Morb. Mortal. Wkly. Rep.* **1993**, *42*, 597–599.

18a. Centers for Disease Control and Prevention. *Morb. Mortal. Wkly. Rep.* **2002**, *51*, 565–567.

18b. Centers for Disease Control and Prevention. *Morb. Mortal. Wkly. Rep.* **2002**, *51*, 902.

18c. Centers for Disease Control and Prevention. *Morb. Mortal. Wkly. Rep.* **2004**, *53*, 322–323.

19. Menichetti, F. *Clin. Microbiol. Infect.* **2005**, *11*, 22–28.

20. Ranger, L. Linezolid and the Oxazolidinones – A New Class of Antibiotics. In *Creation of a Novel Class: The Oxazolidinone Antibiotics*; Batts, D. H.; Kollef, M. H.; Lipsky, B. A.; Nicolau, D. P., Weigelt, J. A. (Eds.); Innova Institute for Medical Education: Tampa, FL, 2004; Chapter 3, p 30.

21. Brickner, S. J. Antibacterial 3-(fused-ring substituted)phenyl-5.beta.-amidomethyloxazolidin-2-ones. US Patent 5,225,565, July 6, 1993.

22. Brickner, S. J. Substituted 3(5′indazolyl) oxazolidin-2-ones. US Patent 5,182,403, January 26, 1993.

23. Brickner, S. J. 5′-Indolinyl-5β-amidomethyloxazolidin-2-ones. US Patent 5,164,510, November 17, 1992.

24. Park, C. H.; Brittelli, D. R.; Wang, C. L.-J.; Marsh, F. D.; Gregory, W. A.; Wuonola, M. A.; McRipley, R. J.; Eberly, V. S.; Slee, A. M.; Forbes, M. *J. Med. Chem.* **1992**, *35*, 1156–1165.

25. Brickner, S. J. *Curr. Pharm. Des.* **1996**, *2*, 175–194.

26. Piper, R. C.; Platte, T. F.; Palmer, J. R. Unpublished data (Pharmacia Corporation).

27. Wang, C.-L. J.; Wuonola, M. A. Preparation of 5-(aminomethyl)-3-phenyl-2-oxazolidinone Derivatives and Antibacterial Pharmaceuticals Containing Them. US Patent 4,801,600, January 31, 1989.

28. Ford, C. W.; Zurenko, G. E.; Barbachyn, M. R. *Curr. Drug Targets Infect. Dis.* **2001**, *1*, 181–199.

29. Kilburn, J.; Glickman, S.; Brickner, S. J.; Manninen, P. R.; Ulanowicz, D. A.; Lovasz, K. D.; Zurenko, G. E. *Abstracts of Papers*, Proceedings of the 33rd Interscience Conference on Antimicrobial Agents and Chemotherapy, New Orleans, October 17–20, 1993. American Society for Microbiology: Washington, DC.

30. Weidner-Wells, M. A.; Boggs, C. M.; Foleno, B. D.; Melton, J.; Bush, K.; Goldschmidt, R. M.; Hlasta, D. J. *Bioorg. Med. Chem.* **2002**, *10*, 2345–2351.

30a. Gravestock, M. B.; Acton, D. G.; Betts, M. J.; Dennis, M.; Hatter, G.; McGregor, A.; Swain, M. L.; Wilson, R. G.; Woods, L.; Wookey, A. *Bioorg. Med. Chem. Lett.* **2003**, *13*, 4179–4186.

31. Barbachyn, M. R.; Cleek, G. J.; Dolak, L. A.; Garmon, S. A.; Morris, J.; Seest, E. P.; Thomas, R. C.; Toops, D. S.; Watt, W.; Wishka, D. G. et al. *J. Med. Chem.* **2003**, *46*, 284–302.

31a. Johnson, P. D.; Aristoff, P. A.; Zurenko, G. E.; Schaadt, R. D.; Yagi, B. H.; Ford, C. W.; Hamel, J. C.; Stapert, D.; Moerman, J. K. *Bioorg. Med. Chem. Lett.* **2003**, *13*, 4197–4200.

32. Brickner, S. J. Tricyclic [6.5.5]-fused Oxazolidinone Antibacterial Agents. US Patent 5,231,188, July 27, 1993.

33. Gleave, D. M.; Brickner, S. J. *J. Org. Chem.* **1996**, *61*, 6470–6474.

34. Carlson, R. K.; Gregory, W. A.; Park, C. H. 5-(Aminomethyl)-3-(4-arylphenyl)oxazolidin-2-one Derivatives Useful as Antibacterial Agents. European Patent Application 352781 A2 (*Chem. Abstr.* **1990**, *113*, 172003).

35. Gleave, D. M.; Brickner, S. J.; Manninen, P. R.; Allwine, D. A.; Lovasz, K. D.; Rohrer, D. C.; Tucker, J. A.; Zurenko, G. E.; Ford, C. W. *Bioorg. Med. Chem. Lett.* **1998**, *8*, 1231–1236.

36. Brickner, S. J.; Manninen, P. R.; Ulanowicz, D. A.; Lovasz, K. D.; Rohrer, D. C. *Abstracts of Papers*, 206th National Meeting of the American Chemical Society, Chicago, IL, August, 1993. American Chemical Society: Washington, DC, 1993.

37. Pauls, H. W.; Fraser-Reid, B. *J. Am. Chem. Soc.* **1980**, *102*, 3956–3957.

38. Takano, S.; Hatakeyama, S. *Heterocycles* **1982**, *19*, 1243–1245.

39. Cardillo, G.; Orena, M. *Tetrahedron* **1990**, *46*, 3321–3408.

40. Wang, C.-L. J.; Gregory, W. A.; Wuonola, M. A. *Tetrahedron* **1989**, *45*, 1323–1326.

41. Herweh, J. E.; Kauffman, W. J. *Tetrahedron Lett.* **1971**, 809–812.

42. Speranza, G. P.; Peppel, W. J. *J. Org. Chem.* **1958**, *23*, 1922–1924.

43. Manninen, P. R.; Little, H. A.; Brickner, S. J. *Book of Abstracts*, 212th ACS National Meeting, Orlando, FL, August 25–29, 1996. American Chemical Society: Washington, DC, 1996.

44. Manninen, P. R.; Brickner, S. J. *Org. Syn.* **2004**, *81*, 112–120.

45. Brickner, S. J.; Hutchinson, D. K.; Barbachyn, M. R.; Manninen, P. R.; Ulanowicz, D. A.; Garmon, S. A.; Grega, K. C.; Hendges, S. K.; Toops, D. S.; Ford, C. W. et al. E.; *J. Med. Chem.* **1996**, *39*, 673–679.

46. Pearlman, B. A.; Perrault, W. R.; Barbachyn, M. R.; Manninen, P. R.; Toops, D. S.; Houser, D.; Fleck, T. J. Process to Prepare Oxazolidinones. US Patent 5,837,870, November 17, 1998.

47. Carlson, R. K.; Park, C.-H.; Gregory, W. A. Aminomethyloxooxazolidinyl Arylbenzene Derivatives Useful as Antibacterial Agents. US Patent 4,948,801, August 14, 1990.

48. Barbachyn, M. R.; Ford, C. W. *Angew. Chem. Int. Ed.* **2003**, *42*, 2010–2023.

49. Hutchinson, D. K.; Brickner, S. J.; Barbachyn, M. R.; Taniguchi, M.; Munesada, K.; Yamada, H. Phenyloxazolidinone Antimicrobials. US Patent 5,883,093, March 16, 1999.

50. Chu, D. T. W.; Fernandes, P. B. *Antimicrob. Agents Chemother.* **1989**, *33*, 131–135.

51. Barbachyn, M. R.; Toops, D. S.; Grega, K. C.; Hendges, S. K.; Ford, C. W.; Zurenko, G. E.; Hamel, J. C.; Schaadt, R. D.; Stapert, D.; Yagi, B. H. et al. *Bioorg. Med. Chem. Lett.* **1996**, *6*, 1009–1014.

52. Tucker, J. A.; Brickner, S. J.; Ulanowicz, D. A. Oxazolidinone Antibacterial Agents Having a Six-membered Heteroaromatic Ring. US Patent 5,719,154, February 17, 1998.

52a. Tucker, J. A.; Allwine, D. A.; Grega, K. C.; Barbachyn, M. R.; Klock, J. L.; Adamski, J. L.; Brickner, S. J.; Hutchinson, D. K.; Ford, C. W.; Zurenko, G. E. et al. *J. Med. Chem.* **1998**, *41*, 3727–3735.

53. Zurenko, G. E.; Yagi, B. H.; Schaadt, R. D.; Allison, J. W.; Kilburn, J. O.; Glickman, S. E.; Hutchinson, D. K.; Barbachyn, M. R.; Brickner, S. J. *Antimicrob. Agents Chemother.* **1996**, *40*, 839–845.

54. Ford, C. W.; Hamel, J. C.; Wilson, D. M.; Moerman, J. K.; Stapert, D.; Yancey, R. J., Jr.,; Hutchinson, D. K.; Barbachyn, M. R.; Brickner, S. J. *Antimicrob. Agents Chemother.* **1996**, *40*, 1508–1513.

55. Piper, R. C.; Lund, J. E.; Denlinger, R. H.; Platte, T. F.; Brown, W. P.; Brown, P. K.; Palmer, J. R.; *Abstracts of Papers*, Proceedings of the 35th Interscience Conference on Antimicrobial Agents and Chemotherapy, San Francisco, CA 1995; Paper F223.

56. Eustice, D. C.; Feldman, P. A.; Zajac, I.; Slee, A. M. *Antimicrob. Agents Chemother.* **1988**, *32*, 1218–1222.

57. Murray, R. W.; Schaadt, R. D.; Zurenko, G. E.; Marotti, K. R. *Antimicrob. Agents Chemother.* **1998**, *42*, 947–948.

58. Kloss, P.; Xiong, L.; Shinabarger, D. L.; Mankin, A. S. *J. Mol. Biol.* **1999**, *294*, 93–101.

59. Lin, A. H.; Murray, R. W.; Vidmar, T. J.; Marotti, K. R. *Antimicrob. Agents Chemother.* **1997**, *41*, 2127–2131.
60. Shinabarger, D. L.; Marotti, K. R.; Murray, R. W.; Lin, A. H.; Melchior, E. P.; Swaney, S. M.; Dunyak, D. S.; Demyan, W. F.; Buysse, J. M. *Antimicrob. Agents Chemother.* **1997**, *41*, 2132–2136.
61. Colca, J. R.; McDonald, W. G.; Waldon, D. J.; Thomasco, L. M.; Gadwood, R. C.; Lund, E. T.; Cavey, G. S.; Mathews, W. R.; Adams, L. D.; Cecil, E. T. et al. *J. Biol. Chem.* **2003**, *278*, 21972–21979.
62. Zhou, C. C.; Swaney, S. M.; Shinabarger, D. L.; Stockman, B. J. *Antimicrob. Agents Chemother.* **2002**, *46*, 625–629.
63. Barbachyn, M. R.; Hutchinson, D. K.; Brickner, S. J.; Cynamon, M. H.; Kilburn, J. O.; Klemens, S. P.; Glickman, S. E.; Grega, K. C.; Hendges, S. K.; Toops, D. S. et al. *J. Med. Chem.* **1996**, *39*, 680–685.
64. Cynamon, M. H.; Klemens, S. P.; Sharpe, C. A.; Chase, S. *Antimicrob. Agents Chemother.* **1999**, *43*, 1189–1191.
65. Koike, S.; Miura, H.; Nakamura, R.; Chiba, K.; Moe, J. B. *Abstracts of Papers*, Proceedings of the 35th Interscience Conference on Antimicrobial Agents and Chemotherapy, San Francisco, CA, 1995. American Society for Microbiology: Washington, DC; Paper F22.
66. Rattan, A. *Drugs Future* **2003**, *28*, 1070–1077.
67. Stalker, D. J.; Jungbluth, G. L. *Clin. Pharmacokinet.* **2003**, *42*, 1129–1140.
68. Wienkers, L. C. *Drug Metab. Dispos.* **2000**, *28*, 1014–1017.
69. Wood, M. J. *J. Antimicrob. Chemother.* **1996**, *37*, 209–222.
70. Nachman, S. A. Personal communication, 2005.
71. Birmingham, M. C.; Rayner, C. R.; Meagher, A. K.; Flavin, S. M.; Batts, D. H.; Schentag, J. *J. Clin. Infect. Dis.* **2003**, *36*, 159–167.
72. Bain, K. T.; Wittbrodt, E. T. *Ann. Pharmacother.* **2001**, *35*, 566–575.
72a. Burleson, B. S.; Ritchie, D. J.; Micek, S. T.; Dunne, W. M. *Pharmacotherapy* **2004**, *24*, 1225–1231.

Biography

Steven J Brickner, PhD, is a Research Fellow in Antibacterials Chemistry, at Pfizer Global Research and Development in Groton, CT. He graduated from Miami University (OH) with a BS in chemistry with honors in 1976. He received his MS and PhD in organic chemistry from Cornell University (NY), and was an NIH Postdoctoral Research Fellow at the University of Wisconsin-Madison. Brickner is a medicinal chemist with over 20 years of research experience in the pharmaceutical industry, all focused on the discovery and development of novel antibacterial agents. He is an inventor/co-inventor on 21 US patents, and has published numerous scientific papers within the areas of oxazolidinones and novel azetidinones. Since 1997, Brickner has been a member of the Forum on Microbial Threats at the Institute of Medicine (National Academy of Sciences), and is a member of the Editorial Advisory Board for *Current Pharmaceutical Design*. He was named the 2002–03 Outstanding Alumni Lecturer, College of Arts and Science, Miami University, and was a co-recipient of the 2003 American Chemical Society's 31st Northeast Regional Industrial Innovation Award. Prior to joining Pfizer in 1996, he led a team at Pharmacia and Upjohn that discovered and developed Zyvox (linezolid).

Comprehensive Medicinal Chemistry II
ISBN (set): 0-08-044513-6

ISBN (Volume 8) 0-08-044521-7; pp. 157–171

8.14 Copaxone

D Teitelbaum, R Arnon, and M Sela, Weizmann Institute of Science, Rehovot, Israel

8.14.1	**Introduction**	**173**
8.14.2	**Preclinical Studies**	**174**
8.14.2.1	The Beginning	174
8.14.2.2	The Chemistry of Copaxone	175
8.14.2.3	Studies in Experimental Animal Models	175
8.14.3	**Clinical Investigations**	**175**
8.14.3.1	Early Clinical Trials	176
8.14.3.2	Clinical Studies Leading to Food and Drug Administration Approval	176
8.14.3.2.1	Bornstein study	176
8.14.3.2.2	Phase III studies	176
8.14.3.3	Recent Clinical Studies	176
8.14.3.3.1	Open-label extension of the American phase III trial	176
8.14.3.3.2	Meta-analysis of the double-blind, placebo-controlled clinical trials	177
8.14.3.3.3	Magnetic resonance imaging studies	178
8.14.3.3.4	Study of primary progressive multiple sclerosis	178
8.14.3.3.5	Oral study	178
8.14.3.3.6	Comparative studies	178
8.14.3.4	Safety and Tolerability	179
8.14.4	**The Immunopharmacology of Glatiramer Acetate**	**179**
8.14.4.1	Immunological Properties of Glatiramer Acetate	179
8.14.4.1.1	Binding to major histocompatibility complex molecules	179
8.14.4.1.2	Inhibition of T-cell responses by glatiramer acetate	179
8.14.4.1.3	Induction of antigen-specific T-regulatory cells	179
8.14.4.1.4	Effect of glatiramer acetate on antigen-presenting cells	180
8.14.4.1.5	Neuroprotective effects of glatiramer acetate	180
8.14.4.2	Immunological Effects of Glatiramer Acetate in Multiple Sclerosis Patients	181
8.14.4.2.1	Antibody response	181
8.14.4.2.2	T-cell response to glatiramer acetate in naive multiple sclerosis patients	181
8.14.4.2.3	T-cell response to glatiramer acetate in treated multiple sclerosis patients	181
8.14.4.2.4	Proposed mechanism of action of glatiramer acetate-specific immunomodulation	182
8.14.5	**Concluding Remarks**	**182**
References		**183**

8.14.1 Introduction

Multiple sclerosis (MS) is an inflammatory disease of the central nervous system (CNS) that leads to myelin destruction and axonal loss. It is the most common, nontraumatic, disabling neurological disorder in young adults. While the etiology of MS remains unknown, its pathogenesis involves autoimmune reactivity to various myelin antigens such as myelin basic protein (MBP), proteolipid protein (PLP), myelin oligodendrocyte glycoprotein (MOG), and other myelin minor components. MS is often characterized by relapsing episodes of neurological impairment followed by remissions. This type of disease is termed relapsing-remitting (RR). In about one-third of MS patients this disease evolves into a progressive course, termed secondary progressive MS.[1]

Currently approved immunomodulatory therapies for the treatment of RR-MS include glatiramer acetate (GA; Copaxone) and the recombinant β-IFNs, Avonex (IFN-β1a), Rebif (IFN-β1a), and Betaseron (IFN-β1b). All modify the course of the disease, reduce the number of relapses, and slow the accumulation of disability.[2] The clinical efficacy

of the four drugs appears to be similar; however, GA is distinct from the IFNs as it is a more specific immunomodulator and also combines in its effect neuroprotection.[3] It also has milder side effects and better tolerability.

Copaxone (GA), formerly known as Copolymer 1 or Cop1, is a synthetic polymer of four amino acids L-alanine, L-lysine, L-glutamic acid, and L-tyrosine.[4] This is a novel and unique drug: it is the first drug based on antigen-specific suppression of an autoimmune disease.[5] It is also the first case in which a synthetic polymeric substance comprises the main ingredient of a drug.[6] We are familiar with the use of biopolymers, for packing a drug, for slow and controlled release, and for many other uses, but never as an active ingredient against a disease.

In the following we intend to describe the path of discovery of this drug and its development into a drug against RR-MS. We will discuss its mechanism of action, making it a prototype for therapeutic vaccines against autoimmune diseases.[7]

8.14.2 Preclinical Studies

8.14.2.1 The Beginning

It all began as basic research into the mechanisms involved in the induction and suppression of EAE, which is the primary animal model for MS. EAE is an acute neurological autoimmune disease, induced by the injection in complete Freund's adjuvant of brain- or spinal cord-derived substances which constitute the encephalitogenic antigens. These include several proteins such as MBP, PLP, MOG, and others. The disease is mediated by CD4 + autoreactive T cells, which recognize the encephalitogenic antigen(s) in association with major histocompatibility complex (MHC) class II molecules. These autoreactive cells migrate into the CNS and mediate the pathogenic process.[8] When we started our research in 1967, the only encephalitogenic material identified in the CNS was the MBP, and the only information available about it was its overall amino acid composition. Also the basic understanding of immunology and its role in EAE was in its infancy.

Our approach to the study of EAE was the synthetic one, using copolymers of amino acids. Research with amino acid polymers was pioneered by Ephraim Katchalski.[9] These synthetic protein-like molecules were shown to be useful models to study the structure–function relationship of proteins, the structural basis of antigenicity, and other immunological processes.[6,10] Employing these amino acid polymers, immune response to a large variety of antigenic determinants including nonprotein moieties such as sugars and lipids could be induced. Of special interest was the immune response to lipid components, which was not easy either to elicit or investigate because of solubility problems. However, conjugates, in which synthetic lipid compounds were attached on to synthetic copolymers of amino acids, elicited a specific response to lipids such as cytolipin H,[11] which is a tumor-associated glycolipid, or sphingomyelin.[12] Furthermore, we demonstrated that both the sugar and lipid components of such molecules contributed to their immunological specificity. The resultant antilipid antibodies were capable of detecting the corresponding lipids both in water-soluble systems and in their physiological milieu. This was fascinating because it gave us a glimpse into some disorders involving lipid-containing tissue and consequently led to our interest in demyelinating diseases, namely, disorders in which the myelin sheath, which constitutes the lipid-rich coating of all axons, is damaged, resulting in various neurological dysfunctions.

Our working hypothesis was that EAE induced by MBP might actually be caused by demyelinating antilipid immune response and that the positively charged MBP might serve only as a carrier for an acidic lipid (e.g., phospholipids). In order to test this hypothesis we synthesized several positively charged copolymers of amino acids whose composition resembled to a certain extent that of natural MBP, of which Cop1 was the simplest one, and tested their capacity to simulate the MBP ability to induce EAE. However, efforts over the course of more than a year led to negative results. None of these synthetic polymers possessed any encephalitogenic activity.[4] Furthermore, even the conjugation of sphingolipid moiety – which could potentially enhance the antisphingomyelin response and consequently the demyelination process – did not endow these polymers with any encephalitogenic activity whatsoever. At that time we became aware of the work of Elizabeth Roboz Einstein, who was among the first to show that MBP, its modifications as well as other nonencephalitogenic basic proteins, can inhibit EAE.[13] We proceeded therefore to test the copolymers for a possible inhibitory effect on the induction of EAE. The results of the inhibition experiments showed that all the tested copolymers (Cop1, Cop2, and Cop3) showed efficacy in suppressing EAE in guinea-pigs, the most active among the series being Cop1. Thus, we had started out by trying to design a molecule that could cause EAE and ended up with one that suppressed EAE. Over the next years the study of Cop1 has proceeded along two tracks: (1) its clinical development for the treatment of MS patients; and (2) the scientific research to understand how it affects the immune system. The latter advanced in parallel with the increased knowledge of immunology and the increased sophistication of the research tools and methods.[14]

8.14.2.2   The Chemistry of Copaxone

Copaxone (Copolymer 1) is a synthetic amino acid copolymer which is prepared by polymerization of the monomers, N-carboxy α-amino acid anhydrides. These anhydrides are obtained by reacting the respective amino acids with phosgene in dioxane. The anhydrides may be readily polymerized to form amino acid polymers.[9] The polymerization is usually carried out in inert solvent such as dioxane, in the presence of suitable initiator amines or strong bases. In this type of polymerization a growing chain always reacts with a monomer, leading to a narrow, Poissonian distribution of molecular weights.

The length of the polymer depends on the ratio between the monomer and the initiator, which is usually a primary or secondary amine. Keeping this ratio constant leads to high reproducibility of molecular size in different batches of the polymers. Furthermore, the rate of polymerization is an intrinsic property of the different N-carboxyanhydride derivatives, and hence, different samples of a polymer with the same composition of amino acids, although of random sequence in their nature, will be very similar in their physical and chemical properties.

In the case of Copolymer 1, the N-carboxyanhydrides used for polymerization were those of tyrosine, alanine, γ-benzyl glutamate, and ε, N-trifluoroacetyl lysine. The γ-benzyl and ε,N-trifluoroacetyl protective groups were deblocked after polymerization, yielding a water-soluble polymer with a residue molar ratio of 4.2 L-alanine : 3.4 L-lysine : 1.4- L-glutamic acid : 1.0 L-tyrosine. The original polymer was of average molecular weight 23 kDa.[4] Currently, the molecular weight range of the polymer constituting Copaxone and termed GA by the Food and Drug Administration (FDA) is of 4700–13 000 Da.

8.14.2.3   Studies in Experimental Animal Models

GA was demonstrated to suppress EAE induced by MBP in a variety of species: guinea-pigs, rabbits, mice, and two species of monkeys – rhesus monkeys and baboons. In contrast to rodents, where GA inhibits the onset of the disease, in primates it was treatment of the ongoing disease. A remarkable degree of suppression of EAE by GA was demonstrated in all species studied, even though different encephalitogenic determinants of MBP are involved in disease induction in the different species. Furthermore, GA was effective in suppressing the chronic relapsing EAE, a disease which shows a closer resemblance to MS, that can be induced by either spinal cord homogenate or encephalitogenic peptides derived from PLP and MOG.[5] Thus, the suppressive effect of GA in EAE is a general phenomenon and is not restricted to a particular species, disease type, or the encephalitogen used for EAE induction. More recent studies have demonstrated that, in addition to the parenteral route of administration used in all the studies described so far, oral administration of GA is also effective in suppressing EAE in rats, mice, and primates. Furthermore, oral GA was more effective than oral MBP in suppressing the disease.[15,16]

The suppressive effect of GA in EAE is a specific one, since GA lacked any suppressive effect on the immune response in several systems – humoral and cellular immune responses to a variety of antigens and vaccination against various induced infections. GA treatment also did not suppress other experimental autoimmune diseases, including myasthenia gravis, thyroiditis, diabetes, and systemic lupus erythematosus.[5,17] However, it has been reported to inhibit another autoimmune disorder, namely experimental uveoretinitis,[18] a disease interrelated with MBP and EAE. Recently, GA was also shown to be effective in the case of experimental colitis.[19] In addition, GA also had an effect on a murine model for graft-versus-host disease, as well as in three systems of graft rejection.[20]

The specific effect of GA in EAE may be explicable in terms of immunological specificity. Indeed, marked cross-reactivity was demonstrated between GA and MBP, both at the cellular and the humoral levels of the immune response. Thus, using monoclonal antibodies, we could demonstrate clearly that several monoclonal anti-MBP antibodies reacted with GA and vice versa.[21] At the cellular level, cross-reaction was observed both in vitro and in vivo.[22] Of interest is the very good correlation between the extent of cellular immunological cross-reactivity and the suppressive effect on EAE of various synthetic copolymers, and of particularly interest is the observation that a polymer resembling GA in all parameters, except that it is built of D-amino acids rather than L-amino acids, does not cross-react with MBP and has no EAE-suppressing activity whatsoever.[23]

8.14.3   Clinical Investigations

Several comprehensive review articles[24–28] dedicated almost exclusively to this subject have described in detail the various clinical trials that led to the approval of GA as a drug for the treatment of MS, and its evaluation. In the following we will relate to these clinical studies briefly and focus on additional findings that were reported more recently.

8.14.3.1   Early Clinical Trials

In view of the putative resemblance between EAE and MS[8] and based on the efficacy demonstrated by GA in suppressing EAE in all species including primates, both rhesus monkeys and baboons, the next step was testing it in MS patients. We first conducted some basic toxicological studies in our laboratory which included acute and subchronic administration to mice, rats, rabbits, and beagle dogs, uptake studies and Ames test (mutagenicity test). GA was found to be nontoxic and eligible for phase I clinical trial. Two early clinical trials were conducted, one in Israel[29] and the other in the US.[30] The former, in which only 4 patients participated, receiving the same, relatively low dose (2–3 mg, 2–3 times a week for 6 months), indicated possible slight improvement in disability, but mainly no apparent adverse affect of GA. The latter, conducted in 16 patients with RRMS or chronic progressive MS, was actually a phase I trial, using increasing dosage, and led to the definition of the optimal dose, 20 mg GA daily, administered subcutaneously. While efficacy could not be evaluated in this early trial, GA treatment was well tolerated in all patients, with no toxicity noted and no adverse effects recognized in the clinical disease.

8.14.3.2   Clinical Studies Leading to Food and Drug Administration Approval

8.14.3.2.1   Bornstein study

The results of the phase I trials paved the way for a phase II double-blind, randomized, placebo-controlled pilot trial conducted by Bornstein et al.[31] The whole trial was executed without the backup of a pharmaceutical company. It was a National Institutes of Health-supported trial and the GA batches used in this trial were prepared and characterized in our laboratory. The study involved 50 patients with RRMS treated for 2 years by daily subcutaneous injections of either 20 mg GA or placebo. The results demonstrated a remarkable effect on two primary outcome measures: (1) the relapse rate (75% reduction); and (2) the proportion of relapse-free patients.

8.14.3.2.2   Phase III studies

Following the publication of these results, in 1985 TEVA Pharmaceutical Industries of Israel licensed the rights to produce and market GA. TEVA undertook a drug development program and started producing the copolymer in a chemically defined manner with consistent performance in bioassays. This substance was used in two phase III clinical studies. The first was an open-label trial, involving 271 patients conducted in four medical centers in Israel.[32] The clinical results obtained were similar to those reported in the double-blind phase II trial, namely 73% reduction in relapse rate. Since this was an open-label study, the results could not be used for regulatory purposes. Another double-blind phase III clinical trial was required to get FDA approval. To this end, a multicenter study involving 11 centers in the USA and 251 patients was designed in which patients were treated with either 20 mg GA or placebo for 2 years (core study). Results at 24 months[33] showed 29% reduction relative to placebo in relapse rate (the primary endpoint) in favor of GA ($P = 0.007$). The original core study has been extended for a totally blinded and placebo-controlled observation period up to 35 months.[34] By the end of this phase there was a 32% reduction in mean relapse rate. Secondary endpoint results showed that the proportion of patients improved by $\geqslant 1$ expanded disability status (EDSS) steps from entry favored GA (27.2% versus 12.0%; $P = 0.001$) and the mean EDSS score improved by -0.11 in the GA group and worsened by $+0.34$ in the placebo group ($P = 0006$). On the basis of the above-described results, the FDA approved GA (Copaxone) for the treatment of patients with RRMS. Copaxone is now approved in 44 countries worldwide, including the US, Canada, Australia, Israel, and all the European countries.

8.14.3.3   Recent Clinical Studies

8.14.3.3.1   Open-label extension of the American phase III trial

The American phase III trial had an additional phase – an open-label extension in which patients who received placebo crossed over to active treatment with GA and patients who received GA during the double-blind phase continued to receive GA. The open-label extension phase is ongoing and now in its 13th year, and data are available from the 6-, 8-, and 10-years time points.[35–37] The annualized relapsed rate of patients treated from the beginning of the study dropped each year and was 0.23 in the sixth year compared to a 1.52 pretrial relapse rate. This low rate was also maintained after 8 and 10 years of GA treatment. In patients who were on placebo and switched to GA, although their relapse rate was significantly higher during the placebo controlled phase, it began to equalize to that of the GA group in the third year. However, EDSS analysis showed that mean EDSS levels during 10 years increased from randomization by 0.48 steps for patients always on GA and 0.8 steps for those switching from placebo to active treatment (**Figure 1**). In addition, comparing the proportion of patients with confirmed disability progression showed that patients treated

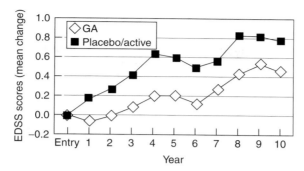

Figure 1 Yearly EDSS change from randomization for patients always on GA versus placebo/active patients. Yearly data are derived from the number of patients starting each specific year. (Adapted from Ford, C.; Johnson, K.; Brooks, B.; Goodman, A.; Kachuck, N.; Lisak, R.; Myers, L. W.; Panich, H. S.; Pruitt, A.; Rizzo, M. *et al. Mult. Scler.* **2003**, *9*, S120.)

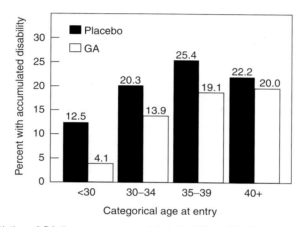

Figure 2 Effect of age at initiation of GA therapy on accumulated disability in RRMS. Patients were stratified by age roughly into quartiles and the proportion with progression defined as ≥ 1 unit change in EDSS sustained ≥ 90 days calculated for each age category. (Reproduced from Wolinsky, J. S. *Exp. Opin. Pharmacother.* **2004**, *5*, 75–891.)

with GA since randomization had significantly lower risk to progress than those taking placebo during the double-blind phase of the trial.[37] It may thus be concluded that GA continues to be effective and safe after a decade of use in a large proportion of RRMS patients. The results validate the importance of long-term compliance with GA therapy, and they also support the growing conviction that early and extended treatment offers the best outcomes in RRMS.[28]

8.14.3.3.2   Meta-analysis of the double-blind, placebo-controlled clinical trials

A meta-analysis using pooled data from 540 patients in three randomized, double-blind, placebo-controlled trials was recently published.[38] It was designed to investigate whether the treatment effect varied according to disease-related variables at baseline. Regression models were developed to estimate the annualized relapse rate, total number of on-trial relapses, and time to first relapse. Also explored were the effect of GA on accumulated disability and the potential role of clinical variables as predictors of relapse rate and treatment efficacy. There was a 28% reduction in the average annualized relapse rate in the GA-treated group; a 36% reduction in on-trial relapses occurred in the GA group. Drug assignment ($P = 0.04$), baseline EDSS score ($P = 0.02$), and the number of relapses during the 2 years prior to study entry ($P = 0.002$) were significant predictors of on-trial relapse rate. A beneficial effect of GA on slowing accumulated disability was also found (risk ratio: 0.6; $P = 0.02$).[28] In GA-treated patients the time to confirmed progression was also doubled when compared to placebo patients (ratio estimate 1.88; $P = 0.02$). Two other factors unfavorably influencing the accumulation of disability were the number of on-trial relapses and patients' age (**Figure 2**). The analysis suggests that the risk of accumulating new disability increases with age and that this risk is best curtailed by the early initiation of GA therapy. This finding underscores the importance of early initiation of MS therapy to maximize its effectiveness.

8.14.3.3.3 Magnetic resonance imaging studies

MRI offers a noninvasive estimate of some of the pathological changes that are ongoing within the CNS. Preliminary retrospective or single-center studies reported a reduction in the frequency of gadolinium (Gd)-enhancing lesions in GA-treated patients compared to treatment-naive subjects. MRI evaluation made 6 years after randomization in the open-label extension phase of the US trial showed that the odds of finding Gd-enhancing lesion were 2.5 times higher for the group originally randomized to placebo compared with the GA group.[28]

A large European–Canadian multicenter double-blind, placebo-controlled study was conducted specifically to address the onset, magnitude, and durability of the effect of GA on MRI-monitored disease activity.[39]

For the primary outcome measure, patients in the GA-treated group demonstrated 29% fewer Gd-enhancing CNS lesions (areas of acute inflammation representing disruption of the blood–brain barrier) than patients in the placebo group. For secondary MRI outcomes, GA showed significantly greater lesion reductions (ranging from 30% to 82.6%) than placebo. Although this 9-month trial period was considered too short to demonstrate a significant reduction in the volume of hypointense T1 lesions (representing areas of demyelination and axonal loss), further analysis of these scans has shown that, after 8 months, the proportion of new T2 lesions evolving into these hypointense T1 lesions (black holes) in patients receiving GA was half that shown in patients receiving placebo (P value < 0.002). These results are of significance since black holes are indicators of more severe and permanent tissue disruption and strong correlations have been found between the extent of black holes in the brain and MS-related disability. A recent post hoc analysis of this trial showed some nonsignificant reduction in brain tissue loss in the GA group during the double-blind phase, which became significantly lower in the open-label phase for patients treated with GA since randomization.[40]

8.14.3.3.4 Study of primary progressive multiple sclerosis

In addition to the phase II clinical trial in RRMS, Bornstein and co-workers[41] also conducted a double-blind trial, in two centers, in New York and Texas, which included 106 patients suffering from chronic progressive MS. The primary outcome measure of this trial was confirmed progression of disability by full-grade change in the EDSS. Out of 23 patients that fulfilled this criterion, 9 were in the GA-treated group and 14 in the placebo group, which did not manifest a statistically significant difference. Progression rates at 12 and 24 months were higher for the placebo group, with a 2-year probability of 29.5% compared to 20.4% for the treated groups ($P = 0.088$). The difference in the 2-year progression of 0.5 EDSS units ($P = 0.03$) was significant.

The patients in this trial categorized as chronic progressive included patients that now would have been considered to have secondary progressive as well as primary progressive disease types.

A double-blind, placebo-controlled study of primary progressive MS patients was started in 1999 in the US, France, and the UK. More than 900 patients in more than 50 centers participated in this trial, that was given the code name Promise. The primary endpoint of the study was to determine whether GA slows confirmed disease progression in the absence of relapses. A data safety-monitoring committee for the trial in interim analysis concluded that it was improbable that the study would reach statistical significance. All patients were taken off study medication in an organized fashion and offered entry into a natural history study. A full intent-to-treat analysis of all trial data is in progress.[28]

8.14.3.3.5 Oral study

Studies in laboratory models of EAE showed that oral administration of GA is effective in suppressing EAE in both rodents and primates in acute as well as chronic relapsing disease.[15,16] In view of these results a double-blind and placebo-controlled study was initiated in RRMS patients. In this trial, given the code name Coral, two doses of oral GA, 5 and 50 mg respectively, or placebo, were given daily for 14 months, to 1650 MS patients enrolled from 158 sites in 18 different countries. The results of this trial showed that, even though the drug was safe when administered as an oral formulation, it failed to show clinical or MRI evidence of an effect at either dose. It is not clear whether the discrepancy between the animal model and humans reflects some aspect of trial design, e.g., dosing, formulation, or site of drug release. A limited clinical trial testing a high dose of oral GA is now in progress.

8.14.3.3.6 Comparative studies

The relative efficacy of the disease-modifying therapies currently approved for use in RRMS, the three β-IFNs and GA, is a matter of great interest. Direct randomized controlled trials comparing these agents pose substantial methodological logistical and cost problems and so far no such trials have been performed. Few open-label prospective and retrospective observational studies have been conducted[28] that suggest differences in efficacy among the different treatments and somewhat larger treatment responses for GA. A study that just appeared[42] describes a 24-month

comparison of immunomodulatory treatments – a retrospective open-label study in 308 patients treated with β-IFNs or GA. The reduction of relapse rate was significantly higher for patients treated with GA compared with all β-IFNs (−0.71, $P < 0.001$). In addition, the discontinuation rate within 24 months was significantly lower for GA (8.9% versus 24%).

8.14.3.4   Safety and Tolerability

Safety data accumulated from > 3500 MS patients treated with GA in controlled and uncontrolled studies indicate withdrawal from therapy for adverse experience in 8.4%.[28] The most frequent reasons recorded for treatment withdrawals were dyspnea and vasodilation (2% for each). The most commonly reported adverse experiences are local injection site reactions which generally decline over time. They consist of erythema, pain, inflammation, pruritus, and swelling, but no skin necrosis. Localized lipoatrophy occurs in some areas after ≥ 1 year of GA treatment.

Approximately 10–15% of GA-treated patients report a postinjection systemic reaction that includes flushing, chest tightness, palpitations, dyspnea, tachycardia, and anxiety. Symptoms were generally transient and resolved spontaneously without sequelae. Controlled studies demonstrated that GA does not provoke hematological abnormalities, elevation of hepatic enzymes, flu-like symptoms, depression, or abnormalities of blood pressure.

It may thus be concluded that GA has a favorable side-effect profile, with excellent patient compliance and long-term acceptance of therapy. Based on the above it was concluded in several review articles that GA is a valuable first-line treatment option for RRMS patients.[26,28]

8.14.4   The Immunopharmacology of Glatiramer Acetate

Extensive studies conducted during the last decade in both the animal model of EAE and in humans have demonstrated several immunological properties of GA and elucidated its mechanism of action. These studies were recently summarized in several review articles.[20,25,27,43] In the following we will relate to these studies briefly and describe more recent findings.

8.14.4.1   Immunological Properties of Glatiramer Acetate

Several immunological properties of GA are thought to contribute to the effects of GA.

8.14.4.1.1   Binding to major histocompatibility complex molecules

GA exhibits a very rapid, high, and efficient binding to many different MHC class II haplotypes on living murine and human antigen-presenting cells (APCs).[44] GA was also shown to interact with purified human leukocyte antigen (HLA)-DR molecules – DR1, DR2, and DR4 – with high affinity.[45] As a result of its high and efficient binding to MHC class II molecules, GA is capable of competing for binding with MBP and their myelin associated proteins, such as PLP and MOG, and even displace them from the MHC binding site.

8.14.4.1.2   Inhibition of T-cell responses by glatiramer acetate

It has been demonstrated that GA can competitively inhibit the immune response to MBP of diverse MBP-specific murine and human T-cell lines (TCLs) and clones, which have different MHC restrictions and respond to different epitopes of MBP.[46–48] GA also inhibited the response of TCLs reactive with PLP and MOG peptides. These results suggest that the observed inhibition was due to competition between GA and nominal antigen for the MHC peptide-binding site. This mechanism may be less specific, and indeed GA was shown also to inhibit in vitro some other immune responses.[48,49] In addition to the relatively nonspecific MHC-blocking, GA was shown to inhibit the response to the immunodominant epitope of MBP peptide 82–100 in a strictly antigen-specific manner by acting as T-cell receptor (TCR) antagonist.[50]

8.14.4.1.3   Induction of antigen-specific T-regulatory cells

In vivo studies have demonstrated that GA-treated animals (either by subcutaneous injections or by oral administration) develop GA-specific T suppressor (Ts) cells in the peripheral immune system. These cells can adoptively transfer protection against EAE.[15,51] Furthermore, Ts cell hybridomas and lines that inhibited EAE in vivo could be isolated from spleen cells of mice rendered unresponsive to EAE by GA.[52] These Ts cells were characterized

as Th2/3-type cells secreting anti-inflammatory cytokines such as interleukin (IL)-4, IL10, and transforming growth factor (TGFβ), but not Th1 cytokines, in response to both GA and MBP. Other myelin antigens such as PLP, MOG and αβ crystalline could not activate the GA Ts cells to secrete Th2 cytokines. Yet the disease induced by PLP and MOG can be suppressed by these Ts cells, probably due to a bystander suppression mechanism.[53,54] More recently, it has been demonstrated that these GA-specific Th2 suppressor T cells which were induced in the periphery by either injection or oral treatment accumulate in the brain.[55,56]

The GA-specific cells accumulated in the CNS demonstrated in situ extensive expression of the anti-inflammatory cytokines IL10 and TGFβ and the brain-derived neurotrophic factor (BDNF), but not the inflammatory cytokine IFN-γ. Furthermore, the GA-induced cells infiltrating the brain induced bystander expression of IL10 and TGFβ by resident astrocytes and microglia.[57] These findings clearly indicate that the GA-specific cells which penetrate the CNS function in vivo as regulatory cells and mediate the therapeutic effect of GA in the target organ.

It was recently suggested that, in addition to the induction of GA-specific Th2 cells, GA also led to the conversion of CD4 + CD25− T cells to CD4 + CD25 + regulatory T cells through activation of transcription factor Foxp3.[58] The induction of Foxp3 by GA was mediated through its ability to produce IFN-γ and, to a lesser extent, TGFβ These findings were demonstrated both in MS patients treated with GA and in wild-type B6 mice, but not in IFN-γ knockout mice.

8.14.4.1.4 Effect of glatiramer acetate on antigen-presenting cells

Several groups have recently reported on the effects of GA on various types of APC. Thus GA blocked lipopolysaccharide-mediated induction of several activation markers of human monocytes and the release of tumor necrosis factor (TNF-α) and IL12. On the other hand, it induced increased production of IL10 by the monocytes.[59,60] Similarly, GA inhibited production of IL2 and TNF-α by in vitro-generated human dendritic cells (DC). DC exposed to GA induced IL4-secreting Th2 cells and enhanced the level of IL10.[61] There is also evidence that GA treatment modifies in vivo the properties of APC. Thus, the spontaneous and triggered release of IL10 was enhanced in monocytes from GA-treated patients whereas the stimulated secretion of IL12 was reduced.[60] It is not clear, however, whether in vivo GA affects the monocytes directly or indirectly by TH2 cytokines secreted by GA-specific T cells. It seems that APC deviation into APC favoring Th2 differentiation may be an additional contributing factor to the therapeutic effect of GA.

8.14.4.1.5 Neuroprotective effects of glatiramer acetate

Recent studies have revealed an additional aspect of GA activity – neuroprotective effects that might also be relevant to MS. It was demonstrated that, similarly to MBP, active immunization with GA as well as adoptive transfer of T cells reactive to GA can inhibit the progression of secondary degeneration after crush injury of the rat optic nerve.[62] The GA-specific T cells secreted significant amounts of BDNF,[62] a neurotrophin that plays a major role in neuronal survival. Furthermore, vaccination with GA protected neurons against glutamate cytotoxicity,[63] and aggregated beta-amyloid-induced toxicity.[64]

GA treatment also increased survival time and improved motor function in a mouse model of amyotrophic lateral sclerosis.[65] Adoptive transfer of GA-specific T cells was effective in protecting dopaminergic neurons in a mouse model of Parkinson disease.[66] Taken together, these results show that GA may have neuroprotective functions in human neurodegenerative diseases.

Several lines of evidence suggest that GA also has a neuroprotective effect in EAE and MS. The effect of GA was studied in MOG-induced EAE, which is considered to be a model that simulates neurodegeneration more than inflammation.[67] It was demonstrated that GA immunization attenuates both inflammation and associated neuronal axonal damage. In the murine model of Theiler's virus-induced demyelinating disease, it was demonstrated that polyreactive antibodies to GA promoted myelin repair.[68]

As indicated before, we have demonstrated that adoptively transferred GA reactive cells migrate to the CNS and also produce in situ BDNF in addition to anti-inflammatory cytokines.[57] In this regard it should be noted that the BDNF receptor trkB is expressed in neurons and astrocytes in MS lesions.[69] Therefore, BDNF secreted by GA-specific cells in the CNS could exert neurotrophic effects directly in the MS target tissue.

Human GA-specific T cells, of both TH1 and Th2 type, are capable of producing BDNF.[70] Studying BDNF production by 73 GA and 33 MBP reactive short-term TCLs, it was found that the mean BDNF level for the GA cell lines was significantly higher than that for MBP lines.[71]

There are also limited clinical data pointing to the neuroprotective effects of GA therapy. Thus, in the European Canadian MRI study, it was demonstrated that GA produced a 50% reduction in the proportion of new MS lesions evolving into persistent black holes[72] (i.e., lesions where severe tissue disruption has occurred). In another study, N-acetylaspartate (NAA), which is a reliable marker of neuronal and axonal injury, was measured using magnetic

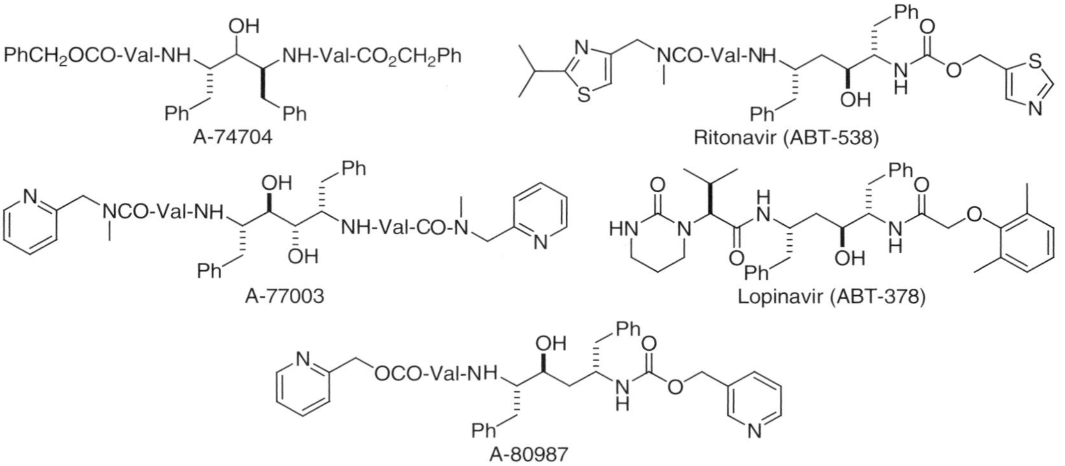

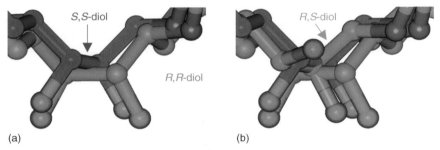

Figure 2 Structures of symmetry-based HIV protease inhibitors A-74704, A-77003, A-80987, ritonavir (ABT-538), and lopinavir (ABT-378).

Figure 3 Binding of diastereomeric diols in the HIV protease active site. (a) Overlap of the central portions of S,S-diol (A-76928, green) and R,R-diol (A-76889, magenta). Hydroxyl groups (red) are projected down toward the catalytic aspartates. (b) Overlap of the above inhibitors with the R,S-diol (A-77003, blue), showing the shift from symmetry to allow the R-hydroxyl group to interact with both aspartates. (Reprinted with permission from Hosur, M. V.; Bhat, T. N.; Kempf, D. J.; Baldwin, E. T.; Liu, B.; Gulnik, S.; Wideburg, N. E.; Norbeck, D. W.; Appelt, K.; Erickson, J. W. *J. Am. Chem. Soc.* **1994**, *116*, 847–855 © American Chemical Society.)

Although the proof-of-concept study with A-77003 was unsuccessful in demonstrating antiviral activity, this endeavor, along with other preclinical results with this series of inhibitors, provided valuable insights. Most notably, careful analysis of the x-ray crystal structures of A-77003 (R,S-diol core) and its two diasteromers (A-76889, R,R-diol and A-76928, S,S-diol) bound to HIV protease revealed three different modes of binding.[12] Whereas the R,R-diol adopted one gauche orientation across the carbon–carbon bond adjacent to the central carbon–carbon bond, the S,S-diol assumed the alternate gauche orientation, which allowed both isomers to project both hydroxyl groups toward the two enzyme aspartate residues (**Figure 3a**). Interestingly, while the S,S-diasteromer A-76928 bound perfectly symmetrically in the active site, the R,R-diol A-76889 was slightly shifted from a symmetric position. The R,S-diol A-77003 adopted a gauche orientation similar to the R,R-diol but was observed to shift even more from a symmetric binding mode (approximately one-half bond length) to allow the R-hydroxyl group to lie on the symmetry axis and hydrogen bond to both catalytic aspartate residues (**Figure 3b**). Because of the relatively weak interaction of the S-hydroxyl group of this isomer with enzyme active site residues, removal of this hydroxyl to provide the 'deoxy-diol' core analog (A-78791) resulted in increased affinity compared to A-77003. The deoxy-diol bound in an identical fashion to the R,S-diol, shifted in position from a purely symmetrical orientation with respect to the axis of symmetry of the enzyme.[12]

The implications of the above structural observations to the ongoing discovery process for this series were several fold. First, it was apparent from both structural analysis and ongoing SAR studies that the carbon framework constituting the diol cores placed adjoining amino acids groups in an optimal position for binding the P2–P3 and P2′–P3′ subsites, compared to the 'mono-ol' core (**Figure 1**). This observation is consistent with the fact that in the asymmetric substrates of HIV protease, amino acid α-carbons are separated by two atoms. Second, the asymmetric orientation of the R,S-diol and deoxy-diol with respect to the enzyme C_2-axis suggested that the contributions of

adjacent groups attached to the two ends of these core groups may differ, a prediction borne out by the activities of pairs of inhibitors functionalized with nonidentical adjacent acyl groups. Most importantly, the increased potency of the deoxy-diol core, compared to the diols, allowed the investigation of truncated compounds containing functionality binding to only five subsites (P3–P2' or P2–P3') rather than six.[13] Optimization of the initial series of truncated inhibitors provided A-80987 (**Figure 2**), with similar in vitro potency to the longer inhibitor A-77003. Importantly, whereas the oral bioavailability of A-77003 in rats was 0.7%,[10] A-80987 provided significant plasma levels in rats and dogs after oral dosing (26% and 23% bioavailability, respectively).[14] A-80987 was the second compound in this series to be advanced to human studies and, while demonstrating oral bioavailability in HIV-infected subjects, still displayed high clearance due to rapid metabolism.

8.15.4   The Discovery of Ritonavir (Norvir)

A significant improvement on the rapid clearance of A-80987 was achieved in subsequent SAR studies. In vitro metabolism studies in human liver microsomes indicated that N-oxidation of the pyridyl groups of both A-77003 and A-80987 occurred rapidly to produce the major metabolites. Systematic studies in which each of the two pyridyl groups were independently replaced with thiazolyl groups, which were more stable to oxidation, suggested that further improvements in the pharmacokinetics of A-80987 were possible. Coincidentally, alkyl groups on the P3-thiazolyl group were shown to improve in vitro potency.[14] Subsequent crystallography studies revealed a hydrophobic contact between this alkyl group and the side chain of the valine at position 82 in the enzyme active site.[15] Combining the above observations led to the discovery of ritonavir (ABT-538), which represented a substantial improvement over A-80987.[14] Thus, in MT4 cells (an immortalized T-cell-derived cell line amenable to HIV infection) the average EC_{50} of ritonavir against a panel of typical wild-type laboratory strains of HIV was 23 nM (approximately 10-fold more potent than A-80987 and A-77003). Furthermore, the oral bioavailability of ritonavir in rats, dogs, and monkeys exceeded 70%, and plasma levels remained above the in vitro EC_{50} for >6–8 h after a 10 mg kg^{-1} oral dose in all three species. On the basis of these attributes, ritonavir was advanced into human trials, and in single-dose studies, plasma concentrations >14-fold higher than those observed with A-80987 were observed.[14]

Significant advances in the synthetic routes to this series of symmetry-based HIV protease inhibitors allowed the clinical examination of the above three inhibitors. The initial synthesis of the protected diol core proceeded via a cumbersome McMurray pinacol coupling of Boc-phenylalaninal to give a mixture of R,R-, S,S-, and R,S-isomers, which were subsequently separated and identified.[7] Significant improvements were realized using a Pedersen coupling of Cbz-phenylalaninal, which produced almost exclusively the R,R-diol isomer.[16] Selective protection of one hydroxyl group and activation of the other as the corresponding mesylate, followed by stereochemical inversion via internal cyclization of one of the Cbz carbonyl groups provided the R,S-diol core required for A-77003 (**Figure 4**). Activation with α-acetoxyisobutyryl bromide produced the corresponding inverted bromoacetate, which could be debrominated to yield the deoxy-diol core. Although this route enabled the discovery of A-80987 and ritonavir, both of which contain this core unit, it was clearly too inefficient for production of supplies for toxicological and clinical studies. A key improvement of the synthesis was realized in a sequence in which the enaminoketone intermediate derived from the sequential addition of acetonitrile anion and benzyl Grignard to protected phenylalanine benzyl ester was stereoselectively reduced in a one-pot set of reactions (initial 1,4-reduction with sodium borohydride/methanesulfonic acid followed by carbonyl reduction with sodium trifluoroacetoxyborohydride) to the protected deoxy-diol core.[17] This synthesis has been scaled up to produce metric tons of both ritonavir and lopinavir, the latter of which also contains this common symmetry-based core.

Initial single-dose studies of ritonavir in healthy human volunteers confirmed its excellent pharmacokinetic profile, and four doses of ritonavir were studied as monotherapy in HIV-infected subjects.[18] In stark contrast to A-77003 and A-80987, plasma HIV RNA (initially measured with a relatively insensitive assay with a 10 000 copies mL^{-1} lower limit of quantitation) immediately plummeted upon the initiation of therapy. The rapid decline in viral load was unprecedented in clinical studies of antiretroviral agents (which up to that point consisted primarily of nucleoside reverse transcriptase inhibitors), and enabled the first quantitative estimates of HIV production and turnover in vivo.[19] The startling results, which indicated that, on average, 1–10 billion HIV particles per day are produced in an untreated, infected individual, overturned the prevailing notion of a 'latent' phase of HIV infection prior to the appearance of the opportunistic infections that define AIDS as a syndrome. These findings also began to redefine the goal of antiretroviral therapy to not only delay the progression of symptomatic AIDS, but to lower viral load to undetectable levels.

Subsequently, the clinical efficacy of ritonavir was established in an innovative phase III study. In a double-blind, placebo-controlled study, ritonavir was added to existing standard-of-care therapy (consisting of combinations of nucleosides) in individuals at high risk for developing AIDS (the median CD4 level in this patient population was 18 cells mm^{-3}). Within a few months, ritonavir treatment was shown to produce a highly statistically significant

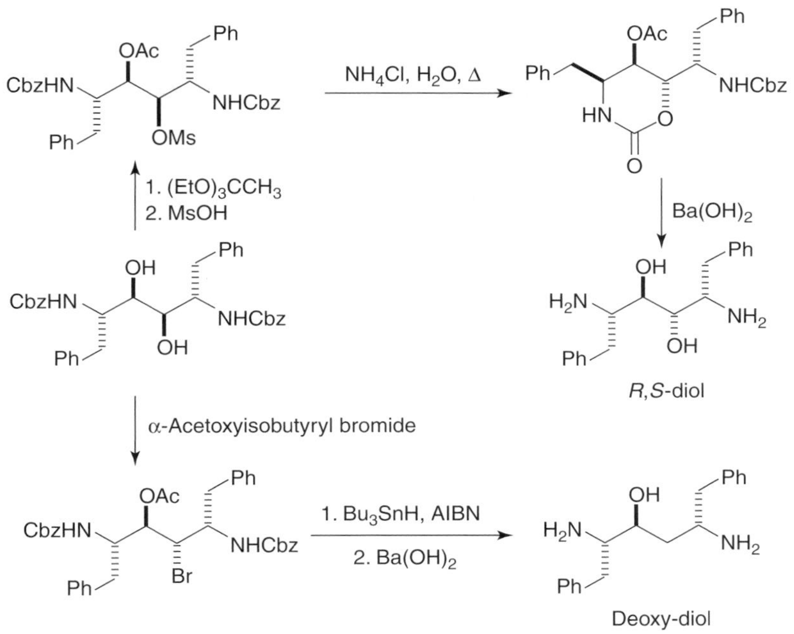

Figure 4 Initial syntheses of the *R,S*-diol and 'deoxy-diol' core diamines of A-77003 and A-80987. (Reprinted from Kempf, D. J.; Marsh, K. C.; Fino, L. C.; Bryant, P.; Craig-Kennard, A.; Sham, H. L.; Zhao, C.; Vasavanonda, S.; Kohlbrenner, W. E.; Wideburg, N. E. *Bioorg. Med. Chem.* **1994**, *2*, 847–858, with permission from Elsevier, and Kempf, D. J.; Sowin, T. J.; Doherty, E. M.; Hannick, S. M.; Codacovi, L.; Henry, R. F.; Green, B. E.; Spanton, S. G.; Norbeck, D. W. *J. Org. Chem.* **1992**, *57*, 5692–5700 © American Chemical Society.)

prolongation of the time to either death or AIDS-defining illness.[20] These results established ritonavir as the first protease inhibitor to demonstrate clinical efficacy, and the compound was licensed under the brand name Norvir in early 1996. This study also influenced the acceptance of plasma HIV RNA quantitation (viral load) as a surrogate marker for clinical efficacy, enabling the rapid development of all subsequent antiretroviral drugs.

8.15.5 Lessons from the Ritonavir Development Program

Several key findings from the development and clinical programs with ritonavir had a major impact on the subsequent HIV protease inhibitor discovery effort, leading ultimately to the discovery of lopinavir. The first was the characterization of drug resistance during ritonavir monotherapy in phase II studies. Longitudinal assessment of plasma samples from patients who initially responded to therapy with a drop in viral load, but whose plasma HIV RNA rebounded over time, revealed the stepwise accumulation of specific mutations in the HIV protease gene. Viruses isolated from the blood of these patients displayed reduced phenotypic susceptibility to ritonavir, as well as to some other protease inhibitors. Importantly, the rate at which the mutations appeared inversely correlated with the plasma trough levels of ritonavir in different patients.[21] Thus, those patients with lower drug levels experienced the evolution of drug resistance at a higher rate than those with higher drug levels. The rate at which rebound occurred also inversely correlated to the degree of suppression of viral load, indicating that ongoing replication allows the production and emergence of resistant variants.[22] These key findings led to the articulation of a hypothetical pharmacokinetic–pharmacodynamic (PK/PD) model for the emergence of drug resistance to protease inhibitors. Since protease inhibitors are reversible inhibitors of HIV protease, and, in general, penetrate into and egress from cells relatively quickly (in contrast to nucleosides, which are trapped intracellularly in mono-, di-, and/or triphosphate forms), protease inhibitor trough plasma levels are likely to be reasonable temporal surrogates for minimum intracellular drug levels. If, during a dosing cycle (prior to the next dose), drug concentrations decline to a level that is incompletely suppressive, allowing significant viral replication to begin, preexisting mutants in the HIV quasispecies have a replication advantage in the presence of drug. Preferential replication of these mutants results in the accumulation of additional mutations. With reduced susceptibility to drug, these multiple mutants begin replicating at even higher drug concentrations, providing increased opportunity for the evolution and selection of even more mutations. Ultimately, combination mutants with

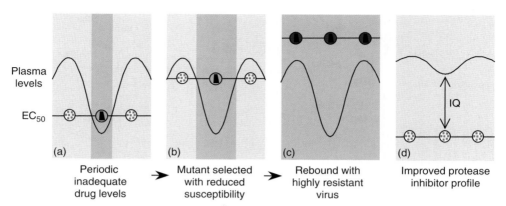

Figure 5 (a)–(c) PK/PD model for resistance development for HIV protease inhibitors; (d) idealized PK/PD profile with high inhibitory quotient.

sufficient resistance to overcome all drug concentrations encountered in a dosing cycle are produced, the drug loses its antiviral effect, and viral load rebounds to pretherapy levels (**Figure 5a–c**). This PK/PD resistance model suggested that resistance could be delayed or prevented if (a) the drug was substantially more potent (i.e., much lower EC_{50}), and/or (b) the drug maintained higher plasma concentrations (particularly trough concentrations: **Figure 5d**). Both of these goals were realized with the development of lopinavir/ritonavir (see below).

A second key finding from the ritonavir development program was the realization that this compound, like many protease inhibitors, is highly bound to human serum in vivo, both to serum albumin and α-acid glycoprotein. The effect of this high protein binding was assessed in vitro by adapting the HIV tissue culture system to tolerate the presence of 50% human serum. Upon the addition of human serum, the EC_{50} of ritonavir increased approximately 20-fold, suggesting that its potency in vivo was substantially compromised from that observed in standard tissue culture assays.[23] This finding had several important consequences. First, the average protein-adjusted EC_{50} for ritonavir for several wild-type viral strains was approximately 1 μM (as opposed to approximately 50 nM in the absence of human serum). At full dose, average trough plasma concentrations of ritonavir are only four to five times higher than this value. Thus, the PK/PD resistance model described above, wherein replication occurs to select resistance as drug levels decline to the trough, is consistent with the development of resistance to ritonavir by most patients receiving monotherapy (particularly since many patients experiencing rapid evolution were assigned to lower, investigational doses of ritonavir and had trough levels significantly lower than the full-dose average). This ratio between plasma trough levels and the human serum-adjusted EC_{50}, later to become known as the inhibitory quotient (**Figure 5**), was estimated to be four or less for all first-generation protease inhibitors. Subsequent studies reveal that a fourfold or greater decrease in viral susceptibility (i.e., $\geq$fourfold higher EC_{50} and thus average inhibitory quotient of one or less) significantly impacted the virologic response to those protease inhibitors, even in combination with other antiretrovirals, suggesting that the inhibitory quotient model as articulated above has clinical relevance.

One enigma remained with respect to the serum binding of ritonavir. While the EC_{50} increased by 20-fold upon the addition of 50% human serum, the free fraction of ritonavir in human plasma was found to be <1% (suggesting that serum binding should have an even greater effect on the EC_{50}). This issue was clarified in a recent study showing that even in the absence of human serum, ritonavir (as well as lopinavir) is relatively highly bound to the 10% fetal bovine serum present in the tissue culture antiviral assay media. In fact, the EC_{50} and free fraction in the tissue culture media were proportional under a variety of low- and high-serum conditions, allowing the calculation of both serum-free and 100% human serum-adjusted EC_{50} values.[24] The latter value for both ritonavir and lopinavir is closely approximated by the EC_{50} determined in the presence of 10% fetal bovine serum plus 50% human serum. The recognition of the importance of serum binding led to the routine screening of all new protease inhibitor analogs both in the absence and presence of 50% human serum, a change in paradigm that enabled the discovery of lopinavir (see below).

The third key finding from the ritonavir development program was the recognition of its potential as a pharmacokinetic booster by virtue of potent inhibition of the 3A isozyme of cytochrome P450 (CYP3A). CYP3A, the most predominant metabolizing enzyme in the liver and intestine, is the primary route of metabolic transformation and clearance of virtually all HIV protease inhibitors. Ritonavir was found to produce a Type II spectral perturbation in the CYP absorbance spectrum in human liver microsomes, suggesting direct binding of its unhindered 5-thiazolyl group to the CYP heme iron atom.[25] In vitro, ritonavir potently inhibited not only standard CYP3A substrates, but also the metabolism of other protease inhibitors in both rat and human liver microsomes. In rats, coadministration with ritonavir

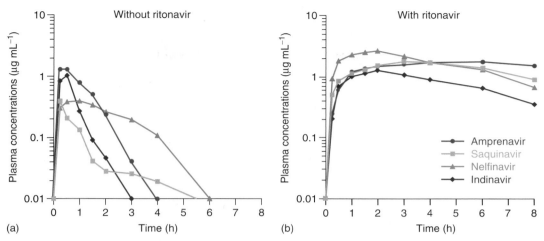

Figure 6 Pharmacokinetic boosting by ritonavir in rats. HIV protease inhibitors were dosed orally at $10\,mg\,kg^{-1}$ without or with a concomitant $10\,mg\,kg^{-1}$ dose of ritonavir. (Reprinted with permission from Kempf, D. J.; Marsh, K. C.; Kumar, G.; Rodrigues, A. D.; Denissen, J. F.; McDonald, E.; Kukulka, M. J.; Hsu, A.; Granneman, G. R. *Antimicrob. Agents Chemother.* **1997**, *41*, 654–660 © American Society for Microbiology.)

increased the plasma levels of other protease inhibitors by 8- to 46-fold and substantially increased the serum half-life (**Figure 6**). Similar enhancements were observed in humans. Given the relatively low inhibitory quotient values for these first-generation protease inhibitors, enhancements of plasma trough levels by ritonavir codosing significantly improved potency, and the use of ritonavir-boosted protease inhibitors is now recommended as the preferred method for use of this drug class in most HIV treatment guidelines. Importantly, a combination of ritonavir and saquinavir was shown to durably suppress viral replication in most patients even without the concomitant use of nucleoside therapy, providing an example of a potent class-sparing regimen.[26]

One study using ritonavir enhancement is worthy of special note with respect to validation of the inhibitory quotient PK/PD model. Although the inhibitory quotient model is normally based on trough levels, there is generally high correlation between peak levels (C_{max}), overall exposure (area under curve, AUC), and minimum concentrations; thus, the most relevant pharmacokinetic parameter to be utilized in a PK/PD model had not been established. In this study, patients failing therapy with indinavir (800 mg three times daily) plus nucleosides were switched to indinavir/ritonavir (400 mg twice daily each) without a change in backbone nucleoside therapy. Viral isolates from most patients were at least partially resistant to both indinavir and ritonavir prior to the switch. Because of the 67% decrease in total indinavir dose, the C_{max} of indinavir in combination with ritonavir was lower than that produced by indinavir 800 mg three times daily alone (the AUC was approximately the same). However, indinavir C_{trough} increased by 6.5-fold due to the increase in half-life from ritonavir boosting. Three weeks after the change from indinavir to indinavir/ritonavir, 58% of patients experienced an incremental virologic response.[27] The indinavir inhibitory quotient following the change to indinavir/ritonavir (based on indinavir C_{trough}) was the best predictor of response, providing further validation of the inhibitory quotient PK/PD model and, in particular, indicating that the C_{trough} (or C_{min}) rather than the C_{max} or AUC is the most appropriate PK parameter to include in the calculation of inhibitory quotient.

8.15.6 The Discovery of Lopinavir and the Development of Kaletra (Lopinavir/Ritonavir)

As mentioned previously, the evaluation of the antiviral potency of new protease inhibitors in the discovery program was expanded to include assays in the presence of 50% human serum, to best estimate 'in vivo potency.' Following the observation of substantial boosting by ritonavir, the preclinical pharmacokinetic screening protocol was also modified to include evaluation in rats and dogs, both alone and following codosing with ritonavir. The goal for an advanced generation protease inhibitor was twofold: improved potency in the presence of human serum and improved pharmacokinetics through ritonavir boosting. A third key element of the design of lopinavir was the incorporation of structural data on the resistant mutants isolated during ritonavir monotherapy. The primary mutation, which occurred in most individuals early after viral rebound, was an amino acid change from valine at position 82 to alanine (V82A), phenylalanine (V82F), or threonine (V82 T). As mentioned earlier, valine 82 is positioned in the active site of HIV

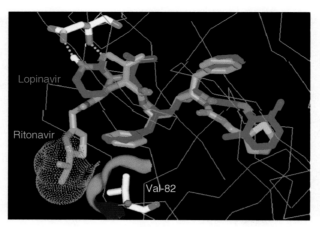

Figure 7 Overlay of ritonavir (orange) and lopinavir (green) in the HIV protease active site, illustrating the interaction of the P3-isopropylthiazolyl group of ritonavir with the side chain of valine 82. (Reprinted from Stoll, V.; Qin, W. Y.; Stewart, K. D.; Jakob, C.; Park, C.; Walter, K.; Simmer, R. L.; Helfrich, R.; Bussiere, D.; Kao, J. *et al. Bioorg. Med. Chem.* **2002**, *10*, 2803–2806, with permission from Elsevier.)

protease (P3 and P3′ subsites) and interacts via nonbonded hydrophobic interactions with the isopropyl substituent on the P3-thiazolyl group of ritonavir. Modeling of the V82A, V82F, and V82T mutant proteases suggested that in each case this hydrophobic interaction would be lost upon viral mutation, lowering the affinity of ritonavir for the enzyme. In order to minimize loss of binding to the position-82 mutants, the ritonavir structure was truncated to remove the P3 isopropylthiazolyl group. Initial analogs were much less potent, but the incremental decrease in activity upon addition of 50% human serum was less than threefold. A key finding was that the terminal urea following truncation could be cyclized, affording > 10-fold improved potency.[28] Finally, systematic studies to replace the 5-thiazolyl group remaining from ritonavir produced lopinavir (ABT-378) (**Figure 2**), which, in the presence of human serum, was 10-fold more active than ritonavir.[23] Crystallographic analysis[15] confirmed that the interaction with valine-82 was lessened, compared to ritonavir (**Figure 7**), and the K_i of lopinavir against the V82A, V82F and V82T protease increased only slightly ($\leq$ fourfold), whereas the K_i of ritonavir was increased by up to 50-fold. The average resistance of multiply mutant clinical HIV isolates to lopinavir was also substantially lower (approximately threefold) than to ritonavir. Combined with the 10-fold higher potency of lopinavir, the EC_{50} of lopinavir against highly resistant isolates remained similar to that of ritonavir against wild-type HIV.[28]

The metabolism of lopinavir occurred almost exclusively via CYP3A in rat and human liver microsomes, and was inhibited at very low concentrations of ritonavir (IC_{50} 0.036 and 0.073 µM, respectively). The concentrations required to inhibit saquinavir metabolism were significantly higher, suggesting that lopinavir would be exquisitely sensitive to ritonavir boosting. In rats, oral dosing of lopinavir alone produced very low exposures, and in dogs and monkeys, no plasma concentrations were detected due to exceedingly rapid metabolic clearance. By contrast, coadministration of lopinavir with ritonavir produced high and sustained plasma levels.[28] In dogs, the AUC increased by > 350-fold, and lopinavir levels remained stable for > 12 h at concentrations > 64-fold above the human serum-adjusted antiviral EC_{50}. Thus in lopinavir, the dual goals of improving potency and pharmacokinetics over ritonavir had been achieved, along with improved activity against resistant virus. Lopinavir, enhanced by ritonavir, was advanced into clinical studies as the first protease inhibitor regimen designed to be pharmacokinetically boosted.

Initial pharmacokinetic studies of lopinavir/ritonavir in humans confirmed the exquisite sensitivity of lopinavir to ritonavir pharmacokinetic enhancement, providing high plasma levels of lopinavir. At steady-state, a twice-daily regimen of 400 mg lopinavir with low-dose ritonavir (100 mg) produced trough levels of lopinavir > 75-fold above its serum adjusted EC_{50} (inhibitory quotient > 75). This regimen eventually became the approved clinical dose and has been written lopinavir/r to signify that the low-dose ritonavir is present merely as a pharmacokinetic booster and is unlikely to elicit significant antiviral activity, in contrast with earlier dual protease inhibitor regimens using higher, efficacious doses of ritonavir. In the initial Phase II study, lopinavir/r was studied as monotherapy for 3 weeks prior to the addition of nucleosides. A mean decline in plasma HIV RNA of 1.85 log copies mL^{-1} was observed at week 3,[29] and after nearly 7 years, 95% of patients remaining on study had < 50 copies mL^{-1} of HIV RNA in their plasma.[30] In the same group of patients, the average increase in CD4 levels was 511 cells mm^{-3}, demonstrating substantial and prolonged immune restoration. Because of its high inhibitory quotient, lopinavir/r was also active in patients who had

previously failed therapy with other protease inhibitors and whose viruses were drug-resistant. In one study in multiple protease inhibitor-experienced patients, activity indistinguishable from that in treatment-naive patients was observed in subjects whose baseline (study entry) viruses displayed up to 10-fold reduced susceptibility to lopinavir in vitro[31] and up to five mutations associated with reduced susceptibility to lopinavir.[32] Evidence of partial activity in patients with baseline strains with up to 40- to 60-fold reduced lopinavir susceptibility provided confirmation of the high inhibitory quotient erected by this regimen.[33] The statistically significant correlation between virologic response and individual inhibitory quotient values in these patients also served to validate the inhibitory quotient as an appropriate PK/PD model for protease inhibitor efficacy.[34]

In a large phase III study, lopinavir/r was compared in a placebo-controlled, double-blind fashion to another protease inhibitor (nelfinavir) in combination with two nucleosides (stavudine and lamivudine). The virologic response in lopinavir/r-treated patients was statistically significantly superior to the response in nelfinavir-treated subjects.[35] In addition, analysis of the viral isolates from patients in both study arms with HIV RNA >400 copies mL^{-1} revealed a startling difference in the evolution of resistance.[36] Thus, 43/96 (45%) of nelfinavir-treated patients demonstrated genotypic resistance to nelfinavir and 79/96 (82%) displayed resistance to lamivudine. In contrast, none of the 51 patients treated with lopinavir/ritonavir for whom genotypes were available demonstrated resistance to lopinavir (or any other protease inhibitor). Furthermore, the rate of lamivudine resistance was also significantly lower (19/51, 37%) than in nelfinavir-treated patients. This study revealed a substantial barrier to resistance erected by lopinavir/r in previously untreated patients that was not present with earlier protease inhibitors. On the basis of this phase III study, lopinavir/r was licensed in the USA in 2000 under the brand name Kaletra. Only recently have the first cases of evolution of resistance to lopinavir/r in treatment-naive patients been documented, attesting to the high barrier to resistance provided by this boosted protease inhibitor regimen.[37,38]

The high pharmacological barrier to resistance of lopinavir/r is consistent with its unique pharmacokinetic profile as a boosted protease inhibitor. Because of its high inhibitory quotient, drug concentrations are unlikely to enter the 'zone of highest selective pressure' (the concentration range just above the EC_{50} for wild-type HIV where any preexisting mutants in the HIV quasispecies with low-level reduced susceptibility have a maximal replication advantage over the major susceptible population) with normal dosing frequency.[36] Furthermore, if doses are missed, the clearance of lopinavir increases over time as drug concentrations continue to fall due to the decline in the inhibitory effects of ritonavir on hepatic CYP3A. Consequently, there is a large difference in lopinavir plasma half-life between the first 12 h following dosing ($t_{1/2}$ 8 h) compared to 24 h after a dose ($t_{1/2}$ 2.2 h), the time at which drug levels are estimated to reach the zone of highest selective pressure (**Figure 8**).[39] Since lopinavir passes rapidly through this zone and decays further to concentrations that are no longer selective for resistance, overall evolution of the multiple mutations required for resistance is disfavored, even during periods of imperfect adherence when significant viral replication is expected to commence. This non-log-linear decay is not observed for protease inhibitors unboosted by ritonavir, since hepatic clearance remains relatively constant.

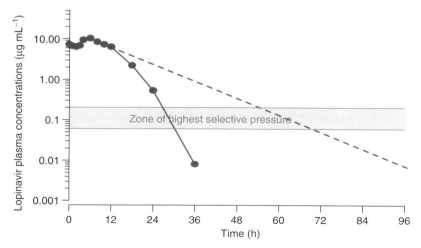

Figure 8 Estimated decay of lopinavir plasma levels through the zone of highest selective pressure following missed doses of lopinavir/r (dosed at 400 mg twice daily in healthy human volunteers at steady state). Blue line: mean plasma levels; green dashed line: extrapolated plasma concentration based on the half-life observed between 6 and 12 h following the final dose.

8.15.7   Conclusion

The high potency and generally good tolerability of lopinavir/r has prompted its wide use, particularly in the most difficult-to-treat patient populations (e.g., children, patients presenting with very high viral loads or failing other regimens, individuals coinfected with hepatitis C), and lopinavir/r has become the most widely prescribed protease inhibitor worldwide since its launch in 2000. Meanwhile, ritonavir, while no longer widely used as an active protease inhibitor for inhibiting HIV, is universally employed as a pharmacokinetic booster for the protease inhibitor class. A key element of the ritonavir and lopinavir/r discovery programs was the iterative integration of lessons learned in the development phase back into the discovery process. Structural information on resistant mutants allowed the design of analogs retaining high potency against these mutants and contributed to the erection of a high barrier to resistance with lopinavir/r. The development of a quantitative PK/PD understanding of this class enabled the optimization of characteristics contributing to high virologic efficacy. Finally, the opportunistic use of the profound drug–drug interactions of ritonavir, normally viewed unfavorably in drug development, to enhance efficacy created a new paradigm for the use of the protease inhibitor class and spurred the development of lopinavir/r. Each of these innovations ultimately contributed to improved patient care, allowing persons with HIV to live normal productive lives and providing hope for the future.

References

1. World Health Organization. http://www.who.int/hiv/facts/focus/en/index1.html (accessed May 2006).
2. Pearl, L. H.; Taylor, W. R. *Nature* **1987**, *329*, 351–354.
3. Kohl, N. E.; Emini, E. A.; Schleif, W. A.; Davis, L. J.; Heimbach, J. C.; Dixon, R. A. F.; Scolnick, E. M.; Sigal, I. S. *Proc. Natl. Acad. Sci. USA* **1988**, *85*, 4686–4690.
4. Navia, M. A.; Fitzgerald, P. M. D.; McKeever, B. M.; Leu, C.-T.; Heimbach, J. C.; Herber, W. K.; Sigal, I. S.; Darke, P. L.; Springer, J. P. *Nature* **1989**, *337*, 615–620.
5. Greenlee, W. J. *Med. Res. Rev.* **1990**, *10*, 173–236.
6. Vacca, J. P. *Methods Enzymol.* **1994**, *241*, 311–334.
7. Kempf, D. J.; Norbeck, D. W.; Codacovi, L.; Wang, X. C.; Kohlbrenner, W. E.; Wideburg, N. E.; Paul, D. A.; Knigge, M. F.; Vasavanonda, S.; Craig-Kennard, A. et al. *J. Med. Chem.* **1990**, *33*, 2687–2689.
8. Erickson, J.; Neidhart, D. J.; VanDrie, J.; Kempf, D. J.; Wang, X. C.; Norbeck, D. W.; Plattner, J. J.; Rittenhouse, J. W.; Turon, M.; Wideburg, N. et al. *Science* **1990**, *249*, 527–533.
9. Kempf, D. J. *Methods Enzymol.* **1994**, *241*, 334–354.
10. Kempf, D. J.; Marsh, K. C.; Paul, D. A.; Knigge, M. F.; Norbeck, D. W.; Kohlbrenner, W. E.; Codacovi, L.; Vasavanonda, S.; Bryant, P.; Wang, X. C. et al. *Antimicrob. Agents Chemother.* **1991**, *35*, 2209–2214.
11. Reedijk, M.; Boucher, C. A. B.; Vanbommel, T.; Ho, D. D.; Tzeng, T. B.; Sereni, D.; Veyssier, P.; Jurriaans, S.; Granneman, R.; Hsu, A. et al. *Antimicrob. Agents Chemother.* **1995**, *39*, 1559–1564.
12. Hosur, M. V.; Bhat, T. N.; Kempf, D. J.; Baldwin, E. T.; Liu, B.; Gulnik, S.; Wideburg, N. E.; Norbeck, D. W.; Appelt, K.; Erickson, J. W. *J. Am. Chem. Soc.* **1994**, *116*, 847–855.
13. Kempf, D. J.; Marsh, K. C.; Fino, L. C.; Bryant, P.; Craig-Kennard, A.; Sham, H. L.; Zhao, C.; Vasavanonda, S.; Kohlbrenner, W. E.; Wideburg, N. E. et al. *Bioorg. Med. Chem.* **1994**, *2*, 847–858.
14. Kempf, D. J.; Marsh, K. C.; Denissen, J. F.; McDonald, E.; Vasavanonda, S.; Flentge, C. A.; Green, B. E.; Fino, L.; Park, C. H.; Kong, X.-P. et al. *Proc. Natl. Acad. Sci. USA* **1995**, *92*, 2484–2488.
15. Stoll, V.; Qin, W. Y.; Stewart, K. D.; Jakob, C.; Park, C.; Walter, K.; Simmer, R. L.; Helfrich, R.; Bussiere, D.; Kao, J. et al. *Bioorg. Med. Chem.* **2002**, *10*, 2803–2806.
16. Kempf, D. J.; Sowin, T. J.; Doherty, E. M.; Hannick, S. M.; Codacovi, L.; Henry, R. F.; Green, B. E.; Spanton, S. G.; Norbeck, D. W. *J. Org. Chem.* **1992**, *57*, 5692–5700.
17. Stuk, T. L.; Haight, A. R.; Scarpetti, D.; Allen, M. S.; Menzia, J. A.; Robbins, T. A.; Parekh, S. I.; Langridge, D. C.; Tien, J.-H. J.; Pariza, R. J. et al. *J. Org. Chem.* **1994**, *59*, 4040–4041.
18. Markowitz, M.; Saag, M.; Powderly, W. G.; Hurley, A. M.; Hsu, A.; Valdes, J. M.; Henry, D.; Sattler, F.; La Marca, A.; Leonard, J. M. et al. *N. Engl. J. Med.* **1995**, *333*, 1534–1539.
19. Ho, D. D.; Neumann, A. U.; Perelson, A. S.; Chen, W.; Leonard, J. M.; Markowitz, M. *Nature* **1995**, *373*, 123–126.
20. Cameron, D. W.; Heath-Chiozzi, M.; Danner, S.; Cohen, C.; Kravcik, S.; Maurath, C.; Sun, E.; Henry, D.; Rode, R.; Potthoff, A. et al. *Lancet* **1998**, *351*, 543–549.
21. Molla, A.; Korneyeva, M.; Gao, Q.; Vasavanonda, S.; Schipper, P. J.; Mo, H.-M.; Markowitz, M.; Chernyavskiy, T.; Niu, P.; Lyons, N. et al. *Nature Med.* **1996**, *2*, 760–766.
22. Kempf, D. J.; Rode, R. A.; Xu, Y.; Sun, E.; Heath-Chiozzi, M. E.; Valdes, J.; Japour, A. J.; Danner, S.; Boucher, C.; Molla, A. et al. *AIDS* **1998**, *12*, F9–F14.
23. Molla, A.; Vasavanonda, S.; Kumar, G.; Sham, H. L.; Johnson, M.; Grabowski, B.; Denissen, J. F.; Kohlbrenner, W.; Plattner, J. J.; Leonard, J. M. et al. *Virology* **1998**, *250*, 255–262.
24. Hickman, D.; Vasavanonda, S.; Nequist, G.; Colletti, L.; Kati, W. M.; Bertz, R.; Hsu, A.; Kempf, D. J. *Antimicrob. Agents Chemother.* **2004**, *48*, 2911–2917.
25. Kempf, D. J.; Marsh, K. C.; Kumar, G.; Rodrigues, A. D.; Denissen, J. F.; McDonald, E.; Kukulka, M. J.; Hsu, A.; Granneman, G. R. et al. *Antimicrob. Agents Chemother.* **1997**, *41*, 654–660.
26. Cameron, D. W.; Japour, A. J.; Xu, Y.; Hsu, A.; Mellors, J.; Farthing, C.; Cohen, C.; Poretz, D.; Markowitz, M.; Follansbee, S. et al. *AIDS* **1999**, *13*, 213–224.

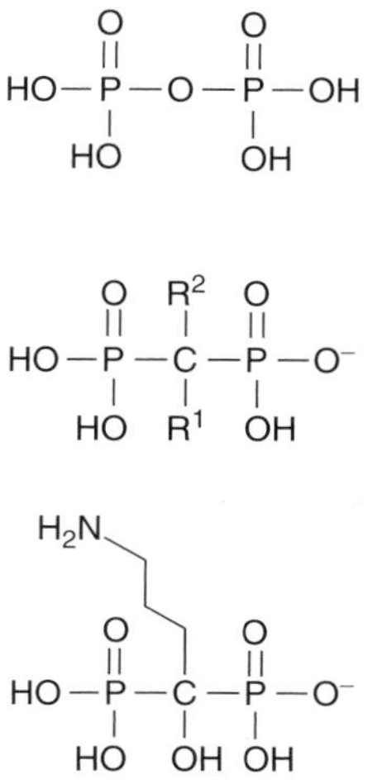

Figure 1 Structures of pyrophosphate, a basic bisphosphonate, and alendronate. Note that in the basic structure, substitutions at R^1 are generally considered to affect binding to bone, while substitutions at R^2 affect potency.

hydrophilic phosphonate moieties limit the penetration of BP molecules through cellular lipid bilayer membranes to undetectable levels, so that distribution is limited to an extracellular compartment and, one day after a dose, is essentially limited to the surface of bone. The bioavailable ALN is thus rapidly cleared from the circulation with an end result of about 50% binding to the hydroxyapatite bone mineral and the remainder being excreted in the urine.[21] This half-life in the circulation is approximately 1–2 h. In humans, the bulk of the ALN not retained on the bone surface is excreted within the first 24 h during the first elimination phase. Intermediate elimination phases exist, whereby the calculated half-lives are days to weeks. In the final elimination phase, ALN has a half-life of about 10 years in humans. Both the intermediate and the terminal half-lives exclusively represent the ALN that is released from the bone.

On bone ALN binds to the mineral surface with no known interactions with the protein matrix. Because the resting surfaces of bone are covered with cells (osteoblasts and lining cells), the most exposed sites are those undergoing active bone resorption. It was shown that these are the preferential sites for ALN uptake in bone at pharmacologically relevant doses.[38–40] At suprapharmacological doses (far above those used to treat humans), the ALN is more or less evenly distributed over the bone surface. Before bone formation is initiated at a given site, the ALN can be readily released from the surface as a result of bone resorption. Before ALN or any other BP can inhibit resorption, it must be ingested by the osteoclast. ALN release from bone is facilitated by acidification of the surface, which takes place during resorption.[38] The removal of mineral and protein from the resorption lacuna beneath the osteoclast occurs through a process of transcytosis.[41,42] This is hypothesized to release not only calcium and phosphate into the blood stream, but also ALN.[43] As time proceeds, the concentration of ALN on the surface of bone increases to steady-state level related to its half-life on the surface of bone, and the ALN inhibition of osteoclast function would gradually increase. This has been documented to occur over 3–6 months in osteoporotic women. Meanwhile, bone formation would proceed to sequester the ALN, as discussed below. Based on these facts, one would predict that, following a single dose, osteoclast-mediated release of bone-associated ALN into the bloodstream would be quite rapid at first, but it should then decline over time. Consistent with this model, the first three of four half-lives for ALN release in clinical testing were calculated as 0.80 days (days 4–7), followed by 6.6 days (days 9–16), and then 35.6 days (days 30–180).

The calculated terminal half-life for ALN is measured in years rather than days. The ALN released during this phase mostly includes BP previously buried beneath the bone surface. This is because ALN preferentially labels the bone resorption surface, and the resorption cycle is always followed by a subsequent cycle of bone formation at the same site. The ALN localized on the resorption surface is therefore covered by de novo synthesized bone, as has been demonstrated in the rat.[40] The bone formation process itself takes 3–4 months, and it can be years before a new resorption cycle reinitiates at a given site. The buried ALN remains pharmacologically inert until it is released back into the circulation as a result of normal bone turnover. Rates of turnover from both cortical and cancellous bone determine not only the subsequent release of BPs but also the relative uptake and distribution of BPs when administered. The cancellous bone takes up a relatively larger proportion of the absorbed BP than the cortical bone,

since cancellous bone is subject to substantially higher turnover. Accurate assessments of terminal half-life in pharmacokinetic analysis therefore required a substantial follow-up, since the curve for elimination from bone is nonlinear for many months.[21] The average terminal elimination half-life of ALN from the skeleton, estimated by urinary excretion in an 18-month follow-up study, is about 10 years. A similar half-life is estimated from modeling of bone turnover at the various compartments, and the total body burden of ALN after 10 years of treatment with an averaged daily dose of 10 mg orally is 75 mg.[44] Although no other BP has been studied in a clinical pharmacokinetic trial long enough to establish a terminal elimination half-life, all other BPs should theoretically exhibit a similar half-life after incorporation into bone. Clinical benefits of bone retention of ALN can be seen after discontinuation whereby bone loss is gradual in comparison to the rapid loss seen after estrogen therapy withdrawal.[45]

8.16.3   Mechanism of Action

8.16.3.1   Alendronate Action at the Molecular Level

Although tested for clinical use since the mid-1980s, the molecular target for ALN, along with other N-BPs, was not identified until 1999. Over the years, BPs were shown to affect several biochemical pathways. For example, ALN and numerous other BPs were found to inhibit the activity of several protein tyrosine phosphatases.[46–50] These actions occurred usually at the upper range of pharmacologically relevant concentrations and failed to correlate with the pharmacological potency of these agents. Although these phosphatase inhibitory activities could be involved in the mechanism of action of some BPs, more compelling proof was obtained for a different molecular target responsible for BP inhibition of osteoclastic bone resorption, as described below.

8.16.3.2   Nitrogen-Containing Bisphosphonate Inhibition of the Cholesterol Biosynthetic Pathway

Over 15 years ago, it was shown that certain BP derivatives (isoprenoid (phosphinylmethyl) phosphonates) weakly inhibit the cholesterol biosynthetic enzyme squalene synthase.[51] The search for more potent inhibitors that might block cholesterol production revealed that the N-BPs incadronate (YM175) and ibandronate potently inhibit squalene synthase.[52] Subsequent studies examined the structure–activity relationship (SAR) for inhibition of squalene synthase.[53–55] In vivo testing showed that certain compounds suppressed serum cholesterol in rodents.[53] Other cholesterol-lowering bisphosphonates were shown to trigger degradation of hydroxymethylglutaryl coenzyme A (HMG-CoA).[56–58] In the same context, utility of squalene synthase inhibition by bisphosphonate was also used for the development of an assay to measure zoledronate levels in animals and clinical serum samples.[59] Although cholesterol itself is important for osteoclast signaling and survival, the osteoclast relies on low-density lipoprotein (LDL) as an external source rather than synthesis through internal pathways.[60,61] Therefore, although ALN, like another N-BP, pamidronate, has been shown to inhibit cholesterol synthesis, this is through inhibition of an enzyme other than squalene synthase. Restoration of cholesterol in the ALN-treated osteoclast does nothing to interfere with its inhibitory action on bone resorption.[62] This then lead to a search for other possible enzymes that could account for its antiresorptive effects.

8.16.3.3   Farnesyl Diphosphate Synthase as the Molecular Target of Alendronate

The ability of ALN to inhibit sterol biosynthesis upstream of squalene synthase[52] suggested inhibition of an enzyme upstream of squalene synthase in the mevalonate pathway,[63] as was indeed identified (**Figure 2**). In subsequent studies, the key enzyme inhibited by ALN was found to be farnesyl diphosphate (FPP) synthase.[60] The reason for continuing to search within the cholesterol biosynthetic pathway for a target enzyme, despite lack of evidence that ALN's effect on cholesterol synthesis would be important to its effects on the osteoclast, was based on the observation that restoration of a branch pathway (leading to protein geranylgeranylation) was sufficient to block all effects of ALN or other N-BPs on osteoclastic bone resorption,[62,64] as discussed below. Modeling of the interaction between ALN and FPP synthase suggests binding to the geranyl diphosphate site,[65] where it acts as a transition-state analog. Enzymological studies suggest that inhibition of FPP synthase is indeed complex.[66] Both competitive and noncompetitive inhibition is reported, depending on the substrate used in the assay, isopentenyl diphosphate or geranyl diphosphate, respectively. Other studies have centered on the SAR for N-BP inhibition of FPP synthase. Modeling using the N-BP risedronate showed that modifications (e.g., addition of a methyl group) to the structure of the side chain can give rise to analogs with markedly less potent inhibition of FPP synthase, making them less effective inhibitors of bone resorption in vivo.[67]

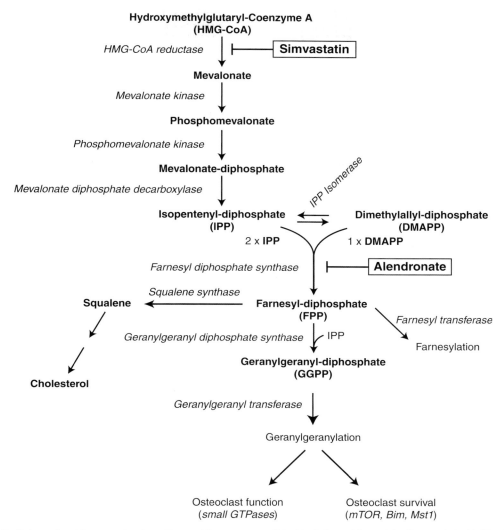

Figure 2 Schematic of the mevalonate pathway. All enzymes are listed in *italics*, while metabolites are in bold. The target of inhibition (farnesyl diphosphate synthase) for alendronate is enclosed within a box, as is the target for the statin, simvastatin, which is shown for reference.

The variable that confers potency against FPP synthase relates to the position of the nitrogen group relative to the phosphonate groups. Interestingly, a modification in one of the phosphonate groups of risedronate, while drastically reducing FPP synthase inhibition, gave rise to a new compound with new activity against type II geranylgeranyl transferase.[68] This derivative has substantially less antiresorptive activity than risedronate in vivo, likely due to reduced binding to bone.[69] Other modifications of risedronate can confer specificity for isopentenyl diphosphate isomerase in addition to FPP synthase.[70] It remains unclear, as yet, whether equivalent modifications to ALN or other N-BPs would confer similar changes in enzyme specificity.

8.16.3.4   Inhibition of Farnesyl Diphosphate Synthase Blocks Protein Isoprenylation and Sterol Synthesis

FPP synthase is responsible for the production of isoprenoid lipids FPP (15 carbon) and geranylgeranyl diphosphate (GGPP) (20 carbon). While FPP, formed by the condensation of three isopentenyl diphosphates (or isomers), is primarily used to synthesize cholesterol, it also can be used for protein isoprenylation. FPP can also be condensed with a fourth isopentenyl diphosphate to form GGPP. The blockade in synthesis of GGPP, albeit through indirect effects on

FPP synthase, is critical for N-BP effects on suppressing osteoclastic bone resorption[62] and inducing osteoclast apoptosis.[71] GGPP, like FPP, is a substrate for protein isoprenylation, and both isoprenoids exhibit specificity in the proteins to which they can be coupled. Isoprenylation involves the transfer of a farnesyl or geranylgeranyl lipid group onto a cysteine amino acid residue in characteristic C-terminal (e.g., CAAX) motifs.[72,73] Most of the isoprenylated proteins identified to date are small guanosine triphosphatases (GTPases) that are geranylgeranylated, and specific CAAX motifs are responsible for directing which lipids are attached to each respective protein.[72] Geranylgeranylated signaling proteins are important for the regulation of a variety of cell processes required for osteoclast function, including cytoskeletal regulation, formation of the ruffled border and regulation of apoptosis.[74–77]

The ability of ALN and other N-BPs to inhibit the cholesterol biosynthetic pathway and protein isoprenylation was actually first demonstrated in J774 macrophages.[52,63] The relevance of this model relates to the fact that J774 cells come from the same lineage as osteoclasts, and that these cells, like osteoclasts, undergo apoptosis in response to N-BP treatment. In these early studies, it was recognized that N-BP inhibition of the cholesterol biosynthetic pathway and isoprenylation was important.[63] Using a more relevant system, it was later discovered that ALN inhibits incorporation of [^{14}C]mevalonate into either isoprenylated proteins or sterols in purified murine or rabbit osteoclasts.[60,78] The relevance of this effect was proven through the ability of a GGPP precursor, geranylgeraniol (GGOH) to block ALN effects on the osteoclast,[62,71,79] as discussed in detail below.

8.16.3.5   Evidence for Molecular Mechanisms In Vivo

The molecular actions of the N-BPs, described above, have been confirmed in vivo using surrogate markers.[80,81] In one study, the well-documented feedback regulation of HMG-CoA reductase expression by cholesterol biosynthetic intermediates was examined.[80] ALN and other N-BPs, but not those lacking a nitrogen, suppressed expression of HMG-CoA reductase in osteoclasts from the proximal tibia. While ALN induced changes in HMG-CoA reductase expression in the osteoclast, no changes were seen in other bone- or marrow-associated cells, which is consistent with the observed targeting of ALN to the osteoclast.[38,40] This effect in the osteoclast appeared to be mediated by the accumulation of metabolites upstream of FPP synthase rather than those lying downstream. The loss of HMG-CoA reductase expression along with inhibition of FPP synthase in the osteoclast could potentially have additive effects on the mevalonate–cholesterol biosynthetic pathway. It may also prevent accumulation of too many upstream metabolites that might otherwise compete with ALN binding to FPP synthase. In the second study, osteoclasts were examined for the in vivo actions of ALN on protein geranylgeranylation.[81] In osteoclasts purified (by immunoadsorption onto magnetic beads) after ALN treatment, geranylgeranylation of the small GTPase Rap1A was suppressed. In this regard, Rap1A was used as a marker for ALN action, although there was no specific link made between the inhibition of this small GTPase and any loss of osteoclast function. For comparative purposes, clodronate was also tested, and this nitrogen-free BP had no effect on geranylgeranylation.

8.16.3.6   Mechanism of Action at the Cellular Level

The relationship between molecular action and antiresorptive effects has been documented for BPs lacking and containing nitrogen. For the non-N-BPs, which are intracellularly metabolized to form toxic analogs, the mechanism is accepted based on the ability of the toxic analogs to reproduce the effects of the parent BPs when administered to the osteoclast.[82] Perhaps the best documentation for a cause–effect relationship has been established for ALN and the other N-BPs, where inhibition of FPP synthase and consequential effects on the osteoclast (loss of resorption, induction of apoptosis) can be overcome simply by reintroducing the critical lost metabolite. Among the downstream metabolites that could specifically restore the three major processes leading to cholesterol synthesis, farnesylation or geranylgeranylation, only geranylgeraniol (GGOH), a lipid alcohol that can replenish GGPP, prevents the ALN effect.[62] Other metabolites downstream of FPP synthase that feed into farnesylation or sterol synthesis are without effect. The observation that farnesol, which is readily metabolized to form FPP, cannot restore osteoclast survival or function was unexpected.[62,83] FPP, like GGPP, is sufficient to block N-BP-induced macrophage apoptosis.[63] The reasons for farnesol not being metabolized to GGPP during BP (or statin) treatment remain to be elucidated. Interestingly, the upstream metabolite, mevalonate, can also partially rescue inhibition of resorption, although this effect disappears with increasing concentration of ALN.[62,63] This is consistent with very recent data suggesting, in part, competitive inhibition of FPP synthase by N-BPs.[66] By this token, lower concentrations of N-BP may show a disproportionate loss of activity, since upstream metabolite accumulation could result in more effective competition for binding sites within FPP synthase. In the context of the in vivo finding that ALN can also suppress HMG-CoA reductase expression,[80] this

5. Adachi, J. D. *Calcif. Tissue Int.* **1996**, *59*, 16–19.

6. Ravn, P.; Rix, M.; Andreassen, H.; Clemmesen, B.; Bidstrup, M.; Gunnes, M. *Calcif. Tissue Int.* **1997**, *60*, 255–260.

7. Hochberg, M. C.; Greenspan, S.; Wasnich, R. D.; Miller, P.; Thompson, D. E.; Ross, P. D. *J. Clin. Endocrinol. Metab.* **2002**, *87*, 1586–1592.

8. Bone, H. G.; Hosking, D.; Devogelaer, J. P.; Tucci, J. R.; Emkey, R. D.; Tonino, R. P.; Rodriguez-Portales, J. A.; Downs, R. W.; Gupta, J.; Santora, A. C. *N. Engl. J. Med.* **2004**, *350*, 1189–1199.

9. Adachi, J. D.; Saag, K. G.; Delmas, P. D.; Liberman, U. A.; Emkey, R. D.; Seeman, E.; Lane, N. E.; Kaufman, J. M.; Poubelle, P. E. E.; Hawkins, F. *Arthritis Rheum.* **2001**, *44*, 202–211.

10. Black, D. M.; Cummings, S. R.; Karpf, D. B.; Cauley, J. A.; Thompson, D. E.; Nevitt, M. C.; Bauer, D. C.; Genant, H. K.; Haskell, W. L.; Marcus, R. *Lancet* **1996**, *348*, 1535–1541.

11. Black, D. M.; Thompson, D. E.; Bauer, D. C.; Ensrud, K.; Musliner, T.; Hochberg, M. C.; Nevitt, M. C.; Suryawanshi, S.; Cummings, S. R. *J. Clin. Endocrinol. Metab.* **2000**, *85*, 4118–4124.

12. Devogelaer, J. P.; Broll, H.; Correa Rotter, R.; Cumming, D. C.; De Deuxchaisnes, C. N.; Geusens, P.; Hosking, D.; Jaeger, P.; Kaufman, J. M.; Leite, M. *Bone* **1996**, *18*, 141–150.

13. Ensrud, K. E.; Black, D. M.; Palermo, L.; Bauer, D. C.; Barrett-Connor, E.; Quandt, S. A.; Thompson, D. E.; Krapf, D. B. *Arch. Intern. Med.* **1997**, *157*, 2617–2624.

14. Ho, Y. V.; Frauman, A. G.; Thomson, W.; Seeman, E. *Osteoporos. Int.* **2000**, *11*, 98–101.

15. Kushida, K.; Shiraki, M.; Nakamura, T.; Kishimoto, H.; Morii, H.; Yamamoto, K.; Kaneda, K.; Fukunaga, M.; Inoue, T.; Nakashima, M. *J. Bone Miner. Metab.* **2004**, *22*, 462–468.

16. Ringe, J. D.; Dorst, A.; Faber, H.; Ibach, K. *Rheumatol. Int.* **2004**, *24*, 110–113.

17. Minne, H. W.; Pollhane, W.; Karpf, D. B. *Int. J. Clin. Pract. Suppl.* **1999**, *101*, 36–39.

18. Nevitt, M. C.; Thompson, D. E.; Black, D. M.; Rubin, S. R.; Ensrud, K.; Yates, A. J.; Cummings, S. R. *Arch. Intern. Med.* **2000**, *160*, 77–85.

19. Adami, S.; Mian, M.; Gatti, P.; Rossini, M.; Zamberlan, N.; Bertoldo, F.; Lo Cascio, V. *Bone* **1994**, *15*, 415–417.

20. Siris, E.; Weinstein, R. S.; Altman, R.; Conte, J. M.; Favus, M.; Lombardi, A.; Lyles, K.; McIlwain, H.; Murphy, W. A. Jr.; Reda, C. *J. Clin. Endocrinol. Metab.* **1996**, *81*, 961–967.

21. Khan, S. A.; Kanis, J. A.; Vasikaran, S.; Kline, W. F.; Matuszewski, B. K.; McCloskey, E. V.; Beneton, M. N.; Gertz, B. J.; Sciberras, D. G.; Holland, S. D. *J. Bone Miner. Res.* **1997**, *12*, 1700–1707.

22. Walsh, J. P.; Ward, L. C.; Stewart, G. O.; Will, R. K.; Criddle, R. A.; Prince, R. L.; Stuckey, B. J.; Dhaliwal, S. S.; Bhagat, C. I.; Retallack, R. W. *Bone* **2004**, *34*, 747–754.

23. Khan, A. A.; Bilezikian, J. P.; Kung, A. W.; Ahmed, M. M.; Dubois, S. J.; Ho, A. Y.; Schussheim, D.; Rubin, M. R.; Shaikh, A. M.; Silverberg, S. J. *J. Clin. Endocrinol. Metab.* **2004**, *89*, 3319–3325.

24. Guaraldi, G.; Orlando, G.; Madeddu, G.; Vescini, F.; Ventura, P.; Campostrini, S.; Mura, M. S.; Parise, N.; Caudarella, R.; Esposito, R. *HIV Clin. Trials* **2004**, *5*, 269–277.

25. Aris, R. M.; Lester, G. E.; Caminiti, M.; Blackwood, A. D.; Hensler, M.; Lark, R. K.; Hecker, T. M.; Renner, J. B.; Guillen, U.; Brown, S. A. *Am. J. Respir. Crit. Care Med.* **2004**, *169*, 77–82.

26. Hassager, C.; Colwell, A.; Assiri, A. M.; Eastell, R.; Russell, R. G.; Christiansen, C. *Clin. Endocrinol.* **1992**, *37*, 45–50.

27. Slemenda, C.; Hui, S. L.; Longcope, C.; Johnston, C. C. *J. Clin. Invest.* **1987**, *80*, 1261–1269.

28. Meunier, P. J.; Boivin, G. *Bone* **1997**, *21*, 373–377.

29. Liberman, U. A.; Weiss, S. R.; Broll, J.; Minne, H. W.; Quan, H.; Bell, N. H.; Rodriguez Portales, J.; Downs, R. W., Jr.; Dequeker, J.; Favus, M. *N. Engl. J. Med.* **1995**, *333*, 1437–1443.

30. Boivin, G. Y.; Chavassieux, P. M.; Santora, A. C.; Yates, J.; Meunier, P. J. *Bone* **2000**, *27*, 687–694.

31. Roschger, P.; Rinnerthaler, S.; Yates, J.; Rodan, G. A.; Fratzl, P.; Klaushofer, K. *Bone* **2001**, *29*, 185–191.

32. Karpf, D. B.; Shapiro, D. R.; Seeman, E.; Ensrud, K. E.; Johnston, C. C., Jr.; Adami, S.; Harris, S. T.; Santora, A. C., II.; Hirsch, L. J.; Oppenheimer, L. *JAMA* **1997**, *277*, 1159–1164.

33. Levis, S.; Quandt, S. A.; Thompson, D.; Scott, J.; Schneider, D. L.; Ross, P. D.; Black, D.; Suryawanshi, S.; Hochberg, M.; Yates, J. *J. Am. Geriatr. Soc.* **2002**, *50*, 409–415.

34. Cocquyt, V.; Kline, W. F.; Gertz, B. J.; Van Belle, S. J.; Holland, S. D.; DeSmet, M.; Quan, H.; Vyas, K. P.; Zhang, K. E.; De Greve, J. *J. Clin. Pharmacol.* **1999**, *39*, 385–393.

35. Porras, A. G.; Holland, S. D.; Gertz, B. J. *Clin. Pharmacokinet.* **1999**, *36*, 315–328.

36. van Beek, E.; Hoekstra, M.; van de Ruit, M.; Lowik, C.; Papapoulos, S. *J. Bone Miner. Res.* **1994**, *9*, 1875–1882.

37. Lin, J. H. *Bone* **1996**, *18*, 75–85.

38. Sato, M.; Grasser, W.; Endo, N.; Akins, R.; Simmons, H.; Thompson, D. D.; Golub, E.; Rodan, G. A. *J. Clin. Invest.* **1991**, *88*, 2095–2105.

39. Azuma, Y.; Sato, H.; Oue, Y.; Okabe, K.; Ohta, T.; Tsuchimoto, M.; Kiyoki, M. *Bone* **1995**, *16*, 235–245.

40. Masarachia, P.; Weinreb, M.; Balena, R.; Rodan, G. A. *Bone* **1996**, *19*, 281–290.

41. Salo, J.; Lehenkari, P.; Mulari, M.; Metsikko, K.; Vaananen, H. K. *Science* **1997**, *276*, 270–273.

42. Palokangas, H.; Mulari, M.; Vaananen, H. K. *J. Cell Sci.* **1997**, *110*, 1767–1780.

43. Reszka, A. A.; Rodan, G. A. *Mini Rev. Med. Chem.* **2004**, *4*, 711–719.

44. Rodan, G.; Reszka, A.; Golub, E.; Rizzoli, R. *Curr. Med. Res. Opin.* **2004**, *20*, 1291–1300.

45. Wasnich, R. D.; Bagger, Y. Z.; Hosking, D. J.; McClung, M. R.; Wu, M.; Mantz, A. M.; Yates, J. J.; Ross, P. D.; Alexandersen, P.; Ravn, P. *Menopause* **2004**, *11*, 622–630.

46. Schmidt, A.; Rutledge, S. J.; Endo, N.; Opas, E. E.; Tanaka, H.; Wesolowski, G.; Leu, C. T.; Huang, Z.; Ramachandaran, C.; Rodan, S. B. *Proc. Natl. Acad. Sci. USA* **1996**, *93*, 3068–3073.

47. Endo, N.; Rutledge, S. J.; Opas, E. E.; Vogel, R.; Rodan, G. A.; Schmidt, A. *J. Bone Miner. Res.* **1996**, *11*, 535–543.

48. Opas, E. E.; Rutledge, S. J.; Golub, E.; Stern, A.; Zimolo, Z.; Rodan, G. A.; Schmidt, A. *Biochem. Pharmacol.* **1997**, *54*, 721–727.

49. Murakami, H.; Takahashi, N.; Tanaka, S.; Nakamura, I.; Udagawa, N.; Nakajo, S.; Nakaya, K.; Abe, M.; Yuda, Y.; Konno, F. *Bone* **1997**, *20*, 399–404.

50. Skorey, K.; Ly, H. D.; Kelly, J.; Hammond, M.; Ramachandran, C.; Huang, Z.; Gresser, M. J.; Wang, Q. *J. Biol. Chem.* **1997**, *272*, 22472–22480.

51. Biller, S. A.; Forster, C.; Gordon, E. M.; Harrity, T.; Scott, W. A.; Ciosek, C. P., Jr. *J. Med. Chem.* **1988**, *31*, 1869–1871.

52. Amin, D.; Cornell, S. A.; Gustafson, S. K.; Needle, S. J.; Ullrich, J. W.; Bilder, G. E.; Perrone, M. H. *J. Lipid Res.* **1992**, *33*, 1657–1663.

53. Ciosek, C. P., Jr.; Magnin, D. R.; Harrity, T. W.; Logan, J. V.; Dickson, J. K., Jr.; Gordon, E. M.; Hamilton, K. A.; Jolibois, K. G.; Kunselman, L. K.; Lawrence, R. M. *J. Biol. Chem.* **1993**, *268*, 24832–24837.

54. Magnin, D. R.; Biller, S. A., Jr.; Dickson, J. K.; Logan, J. V.; Lawrence, R. M.; Chen, Y.; Slusky, R. B.; Ciosek, C. P., Jr.; Harrity, T. W.; Jolibois, K. G. *J. Med. Chem.* **1995**, *38*, 2596–2605.
55. Amin, D.; Cornell, S. A.; Perrone, M. H.; Bilder, G. E. *Arzneimittelforschung* **1996**, *46*, 759–762.
56. Berkhout, T. A.; Simon, H. M.; Patel, D. D.; Bentzen, C.; Niesor, E.; Jackson, B.; Suckling, K. E. *J. Biol. Chem.* **1996**, *271*, 14376–14382.
57. Berkhout, T. A.; Simon, H. M.; Jackson, B.; Yates, J.; Pearce, N.; Groot, P. H.; Bentzen, C.; Niesor, E.; Kerns, W. D.; Suckling, K. E. *Atherosclerosis* **1997**, *133*, 203–212.
58. Jackson, B.; Gee, A. N.; Guyon-Gellin, Y.; Niesor, E.; Bentzen, C. L.; Kerns, W. D.; Suckling, K. E. *Arzneimittelforschung* **2000**, *50*, 380–386.
59. Risser, F.; Pfister, C. U.; Degen, P. H. *J. Pharm. Biomed. Anal.* **1997**, *15*, 1877–1880.
60. Bergstrom, J. D.; Bostedor, R. G.; Masarachia, P. J.; Reszka, A. A.; Rodan, G. *Arch. Biochem. Biophys.* **2000**, *373*, 231–241.
61. Luegmayr, E.; Glantschnig, H.; Wesolowski, G. A.; Gentile, M. A.; Fisher, J. E.; Rodan, G. A.; Reszka, A. A. *Cell Death Differ.* **2004**, *11*, S108–S118.
62. Fisher, J. E.; Rogers, M. J.; Halasy, J. M.; Luckman, S. P.; Hughes, D. E.; Masarachia, P. J.; Wesolowski, G.; Russel, R. G.; Rodan, G. A.; Reszka, A. A. *Proc. Natl. Acad. Sci. USA* **1999**, *96*, 133–138.
63. Luckman, S. P.; Hughes, D. E.; Coxon, F. P.; Graham, R.; Russell, G.; Rogers, M. J. *J. Bone Miner. Res.* **1998**, *13*, 581–589.
64. van Beek, E.; Lowik, C.; van der Pluijm, G.; Papapoulos, S. *J. Bone Miner. Res.* **1999**, *14*, 722–729.
65. Martin, M. B.; Arnold, W.; Heath, H. T., III.; Urbina, J. A.; Oldfield, E. *Biochem. Biophys. Res. Commun.* **1999**, *263*, 754–758.
66. Dunford, J. E.; Ebetino, F. H.; Rogers, M. J. *Bone* **2002**, *30*, 40S.
67. Dunford, J. E.; Thompson, K.; Coxon, F. P.; Luckman, S. P.; Hahn, F. M.; Poulter, C. D.; Ebetino, F. H.; Rogers, M. J. *J. Pharmacol. Exp. Ther.* **2001**, *296*, 235–242.
68. Coxon, F. P.; Helfrich, M. H.; Larijani, B.; Muzylak, M.; Dunford, J. E.; Marshall, D.; McKinnon, A. D.; Nesbitt, S. A.; Horton, M. A.; Seabra, M. C. *J. Biol. Chem.* **2001**, *276*, 48213–48222.
69. van Beek, E. R.; Lowik, C. W.; Ebetino, F. H.; Papapoulos, S. E. *Bone* **1998**, *23*, 437–442.
70. Thompson, K.; Dunford, J. E.; Ebetino, F. H.; Rogers, M. J. *Biochem. Biophys. Res. Commun.* **2002**, *290*, 869–873.
71. Reszka, A. A.; Halasy-Nagy, J. M.; Masarachia, P. J.; Rodan, G. A. *J. Biol. Chem.* **1999**, *274*, 34967–34973.
72. Zhang, F. L.; Casey, P. J. *Annu. Rev. Biochem.* **1996**, *65*, 241–269.
73. Sinensky, M. *Biochim. Biophys. Acta* **2000**, *1484*, 93–106.
74. Ridley, A. J.; Paterson, H. F.; Johnston, C. L.; Diekmann, D.; Hall, A. *Cell* **1992**, *70*, 401–410.
75. Ridley, A. J.; Hall, A. *Cell* **1992**, *70*, 389–399.
76. Zhang, D.; Udagawa, N.; Nakamura, I.; Murakami, H.; Saito, S.; Yamasaki, K.; Shibasaki, Y.; Morii, N.; Narumiya, S.; Takahashi, N. *J. Cell Sci.* **1995**, *108*, 2285–2292.
77. Clark, E. A.; King, W. G.; Brugge, J. S.; Symons, M.; Hynes, R. O. *J. Cell Biol.* **1998**, *142*, 573–586.
78. Coxon, F. P.; Helfrich, M. H.; Van't Hof, R.; Sebti, S.; Ralston, S. H.; Hamilton, A.; Rogers, M. J. *J. Bone Miner. Res.* **2000**, *15*, 1467–1476.
79. Halasy-Nagy, J. M.; Rodan, G. A.; Reszka, A. A. *Bone* **2001**, *29*, 553–559.
80. Fisher, J. E.; Rodan, G. A.; Reszka, A. A. *Endocrinology* **2000**, *141*, 4793–4796.
81. Frith, J. C.; Monkkonen, J.; Auriola, S.; Monkkonen, H.; Rogers, M. J. *Arthritis Rheum.* **2001**, *44*, 2201–2210.
82. Frith, J. C.; Monkkonen, J.; Blackburn, G. M.; Russell, R. G.; Rogers, M. J. *J. Bone Miner. Res.* **1997**, *12*, 1358–1367.
83. Reszka, A. A.; Halasy-Nagy, J. M.; Masarachia, P. J.; Rodan, G. A. *J. Biol. Chem.* **1999**, *274*, 34967–34973.
84. Rogers, H. L.; Marshall, D.; Rogers, M. J. *Bone* **2002**, *30*, 43S.
85. Alakangas, A.; Selander, K.; Mulari, M.; Halleen, J.; Lehenkari, P.; Monkkonen, J.; Salo, J.; Vaananen, K. *Calcif. Tissue Int.* **2002**, *70*, 40–47.
86. Hughes, D. E.; Wright, K. R.; Uy, H. L.; Sasaki, A.; Yoneda, T.; Roodman, G. D.; Mundy, G. R.; Boyce, B. F. *J. Bone Miner. Res.* **1995**, *10*, 1478–1487.
87. Luckman, S. P.; Coxon, F. P.; Ebetino, F. H.; Russell, R. G.; Rogers, M. J. *J. Bone Miner. Res.* **1998**, *13*, 1668–1678.
88. Shipman, C. M.; Croucher, P. I.; Russell, R. G.; Helfrich, M. H.; Rogers, M. J. *Cancer Res.* **1998**, *58*, 5294–5297.
89. Benford, H. L.; Frith, J. C.; Auriola, S.; Monkkonen, J.; Rogers, M. J. *Mol. Pharmacol.* **1999**, *56*, 131–140.
90. Glantschnig, H.; Rodan, G. A.; Reszka, A. A. *J. Biol. Chem.* **2002**, *277*, 42987–42996.
91. Glantschnig, H.; Fisher, J. E.; Wesolowski, G.; Rodan, G. A.; Reszka, A. A. *Cell Death Differ.* **2003**, *10*, 1165–1177.
92. Sugatani, T.; Hruska, K. A. *J. Biol. Chem.* **2005**, *280*, 3583–3589.
93. Graves, J. D.; Gotoh, Y.; Draves, K. E.; Ambrose, D.; Han, D. K.; Wright, M.; Chernoff, J.; Clark, E. A.; Krebs, E. G. *EMBO J.* **1998**, *17*, 2224–2234.
94. Graves, J. D.; Draves, K. E.; Gotoh, Y.; Krebs, E. G.; Clark, E. A. *J. Biol. Chem.* **2001**, *276*, 14909–14915.
95. Lee, K. K.; Ohyama, T.; Yajima, N.; Tsubuki, S.; Yonehara, S. *J. Biol. Chem.* **2001**, *276*, 19276–19285.
96. Seedor, J. G.; Quartuccio, H. A.; Thompson, D. D. *J. Bone Miner. Res.* **1991**, *6*, 339–346.
97. Bikle, D. D.; Morey Holton, E. R.; Doty, S. B.; Currier, P. A.; Tanner, S. J.; Halloran, B. P. *J Bone Miner Res* **1994**, *9*, 1777–1787.
98. Schenk, R.; Merz, W. A.; Muhlbauer, R.; Russell, R. G.; Fleisch, H. *Calcif. Tissue Res.* **1973**, *11*, 196–214.
99. Sato, M.; Grasser, W. *J. Bone Miner. Res.* **1990**, *5*, 31–40.
100. Selander, K.; Lehenkari, P.; Vaananen, H. K. *Calcif. Tissue Int.* **1994**, *55*, 368–375.
101. Murakami, H.; Takahashi, N.; Sasaki, T.; Udagawa, N.; Tanaka, S.; Nakamura, I.; Zhang, D.; Barbier, A.; Suda, T. *Bone* **1995**, *17*, 137–144.
102. Kim, T. W.; Yoshida, Y.; Yokoya, K.; Sasaki, M. *Am. J. Orthod. Dentofacial. Orthop.* **1999**, *115*, 645–653.
103. Zimolo, Z.; Wesolowski, G.; Rodan, G. A. *J. Clin. Invest.* **1995**, *96*, 2277–2283.
104. Ito, M.; Amizuka, N.; Nakajima, T.; Ozawa, H. *Bone* **2001**, *28*, 609–616.
105. Rogers, M. J.; Xiong, X.; Ji, X.; Monkkonen, J.; Russell, R. G.; Williamson, M. P.; Ebetino, F. H.; Watts, D. J. *Pharm. Res.* **1997**, *14*, 625–630.
106. Frith, J. C.; Rogers, M. J. *J. Bone Miner. Res.* **2003**, *18*, 204–212.
107. Boyce, R. W.; Wronski, T. J.; Ebert, D. C.; Stevens, M. L.; Paddock, C. L.; Youngs, T. A.; Gundersen, H. J. *Bone* **1995**, *16*, 209–213.
108. Balena, R.; Toolan, B. C.; Shea, M.; Markatos, A.; Myers, E. R.; Lee, S. C.; Opas, E. E.; Seedor, J. G.; Klein, H.; Frankenfield, D. *J. Clin. Invest.* **1993**, *92*, 2577–2586.
109. Guy, J. A.; Shea, M.; Peter, C. P.; Morrissey, R.; Hayes, W. C. *Calcif. Tissue Int.* **1993**, *53*, 283–288.
110. Lafage, M. H.; Balena, R.; Battle, M. A.; Shea, M.; Seedor, J. G.; Klein, H.; Hayes, W. C.; Rodan, G. A. *J. Clin. Invest.* **1995**, *95*, 2127–2133.
111. Mosekilde, L.; Thomsen, J. S.; Mackey, M. S.; Phipps, R. J. *Bone* **2000**, *27*, 639–645.
112. Toolan, B. C.; Shea, M.; Myers, E. R.; Borchers, R. E.; Seedor, J. G.; Quartuccio, H.; Rodan, G.; Hayes, W. C. *J. Bone Miner. Res.* **1992**, *7*, 1399–1406.

113. Mashiba, T.; Turner, C. H.; Hirano, T.; Forwood, M. R.; Johnston, C. C.; Burr, D. B. *Bone* **2001**, *28*, 524–531.
114. Mori, S.; Harruff, R.; Ambrosius, W.; Burr, D. B. *Bone* **1997**, *21*, 521–526.
115. Lanza, F.; Schwartz, H.; Sahba, B.; Malaty, H. M.; Musliner, T.; Reyes, R.; Quan, H.; Graham, D. Y. *Am. J. Gastroenterol.* **2000**, *95*, 3112–3117.
116. Lanza, F. L.; Hunt, R. H.; Thomson, A. B.; Provenza, J. M.; Blank, M. A. *Gastroenterology* **2000**, *119*, 631–638.
117. de Groen, P. C.; Lubbe, D. F.; Hirsch, L. J.; Daifotis, A.; Stephenson, W.; Freedholm, D.; Pryor Tillotson, S.; Seleznick, M. J.; Pinkas, H.; Wang, K. K. *N. Engl. J. Med.* **1996**, *335*, 1016–1021.
118. Peter, C. P.; Handt, L. K.; Smith, S. M. *Dig. Dis. Sci.* **1998**, *43*, 1998–2002.
119. Blank, M. A.; Ems, B. L.; Gibson, G. W.; Myers, W. R.; Berman, S. K.; Phipps, R. J.; Smith, P. N. *Dig. Dis. Sci.* **1997**, *42*, 281–288.
120. Elliott, S. N.; McKnight, W.; Davies, N. M.; MacNaughton, W. K.; Wallace, J. L. *Life Sci.* **1998**, *62*, 77–91.
121. Peter, C. P.; Kindt, M. V.; Majka, J. A. *Dig. Dis. Sci.* **1998**, *43*, 1009–1015.
122. Reszka, A. A.; Fisher, J. E.; Rodan, G. A. *Bone* **2001**, *28*, S95.
123. Reszka, A. A.; Halasy-Nagy, J.; Rodan, G. A. *Mol. Pharmacol.* **2001**, *59*, 193–202.
124. Suri, S.; Monkkonen, J.; Taskinen, M.; Pesonen, J.; Blank, M. A.; Phipps, R. J.; Rogers, M. J. *Bone* **2001**, *29*, 336–343.
125. Fisher, J. E.; Rodan, G. A.; Reszka, A. A. *J. Bone Miner. Res.* **2001**, *16*, S218.
126. Cryer, B.; Bauer, D. C. *Mayo Clin. Proc.* **2002**, *77*, 1031–1043.
127. Leu, C. T.; Luegmayr, E.; Freedman, L. P.; Rodan, G. A.; Reszka, A. A. *Bone* **2006**, *38*, 628–636.

Biographies

Alfred A Reszka earned his PhD degree from the University of Illinois, Champaign–Urbana in 1992. His thesis research focused on mapping out the sites of interaction between β1 integrins and the actin cytoskeleton. He then went on to conduct his postdoctoral research at the University of Washington as a recipient of a Jane Coffin Childs fellowship, where he characterized the physical and regulatory interactions between MAP kinases, ERKs 1 and 2, and the microtubule cytoskeleton. Dr Reszka joined Merck & Co. in 1997 where he focused on elucidation of the mechanism of action of Fosamax. His scientific contributions at Merck include the identification of farnesyl diphosphate synthase as the target enzyme inhibited by Fosamax in the osteoclast and the identification of the geranylgeranylation pathway as the critical path in the osteoclast inhibited by Fosamax. He has also identified key osteoclast signaling pathways downstream of prosurvival cytokines that are required for osteoclast survival (mTOR and S6 kinase) as well as characterizing Mst1 as a proapoptotic kinase involved in osteoclast cell death. He is the author of many peer-reviewed scientific articles, book chapters, and reviews. He is a member of the American Society for Bone and Mineral Research and the Endocrine Society.

Gideon A Rodan received his MD and PhD degrees from the Hebrew University and the Weizman Institute of Science, respectively. He was professor and head of the Department of Oral Biology at the University of Connecticut, moving to Merck & Co. in 1985 where he established the Department of Bone Biology and Osteoporosis Research. His scientific contributions include establishing transformed osteoblastic cell lines, regulation of bone cell metabolism by hormones and cytokines, mediation of mechanical stimulation by cAMP, cloning of alkaline phosphatase, establishing the role of $\alpha v \beta 3$ integrin in osteoclast function, and coauthoring a hypothesis that osteoblasts activate osteoclasts. He was responsible for fostering the development of the first bisphosphonate, alendronate (Fosamax) for the treatment of osteoporosis providing insight into alendronate inhibition of bone resorption in vivo and elucidating the mechanism of action of bisphosphonates. He was an author of numerous peer-reviewed scientific articles, book chapters, invited editorials, reviews, and perspectives and the coeditor of *Principles of Bone Biology*. He served as President of the Gordon Conference on Bones and Teeth, the American Society for Bone and Mineral Research, and the International Bone and Mineral Society and is the recipient of numerous awards. He was an Adjunct Professor at the University of Pennsylvania School of Medicine. He died on January 1, 2006.

Comprehensive Medicinal Chemistry II
ISBN (set): 0-08-044513-6

ISBN (Volume 8) 0-08-044521-7; pp. 199–212

the NMR shifts for the various analogs moved when going from the sulfide to the corresponding adduct, I suddenly realized that our internally well accepted theory of how the structure of the important mercapto-adduct (understood by us at that time to be a model for the enzyme–inhibitor complex) must be wrong, and that the pyridine moiety must be more heavily involved than we earlier thought. I poured out a 'wee dram' (of Scotch whisky), started to draw new possible reactions and structures, and became more and more excited. After a couple of hours (and drams) late at night, I thought I had the solution and would refer to this as 'the wonderful night.' The next morning, I started to write all the reactions on my white board and when Arne Brändström came into my room (as usual), I went through it and he accepted it. Fine! However, it subsequently turned out to be wrong again. This was not discovered until 6 months, and then my 'spiro' intermediate proposal (later published by others[37]) happened to be close enough for us to be able to circumvent the vast stability problems with the crystals of the intermediate, so we were eventually able to obtain crystals good enough for an x-ray study.

8.17.3.2 Creative Climate

My experiences of working with a number of creative people have led me to speculate about the determining factors that are the most important for cultivating a creative climate within a group of people. In my opinion, these are a suitable leadership (if there is a group leader), direct information, no hierarchy, no 'stolen' ideas, a high degree of openness, honesty, generosity, lack of prestige, and, finally, a sense of humor. Thus, it is important that all relevant information reaches the team members simultaneously, so that no one feels handicapped compared with the others. The leader of the group should convey enthusiasm for new ideas coming from the other team members, have a rewarding attitude, and provide room for mistakes. The innovator should always be named the first time his or her idea is mentioned by somebody else to an uninitiated person or a group. The leader may even teach this need for openness and generosity, in order to prevent any feelings that ideas can be 'stolen.' A flat organization within the group, without hierarchy, will facilitate direct information flow and ensure that ideas are dealt with positively. This provides a good climate for immediate release of new ideas, which is important for speeding up cross-fertilization within the group. It is therefore possible that the modern Anglo-American bonus model, which prioritizes only personal benefits, may have a negative impact on group creativity and provide incentives to conceal brilliant ideas for some time. The old Swedish model, with equal reward-sharing across the group, may have advantages here.

A creative climate is, of course, advantageous in all research organizations but it would be no exaggeration to state that creativity is compulsory for medicinal chemists in the pharmaceutical industry, where those involved must take full responsibility for generating new ideas concerning chemical structures. As the compounds they generate must be patentable, they also need to be inventive. Most of the ideas and creative work leading to new inventions are focused on problem-solving. This is reflected in the way a patent application is built up: a problem is presented, followed by information on the invention that provides a solution to the problem. A more difficult creative act, however, is to define the problem, and this can often be the great invention! Mats Sundgren has recently completed his thesis work on organizational creativity in pharmaceutical research and development.[38] Based on the interviews he conducted at pharmaceutical industries in Sweden, the UK, and the US, creativity is the most important thing for securing success, but there is hardly anyone who talks about this. The demands concerning effectiveness in drug development projects have increased markedly, and this has led to more and more detailed project planning. Sundgren concluded, however, that planning for what is going to happen comes from the generation of ideas, and the effectiveness comes from creativity. He therefore stressed the need for a balance between effectiveness and organizational creativity.

8.17.3.3 Enthusiasm

In the pharmaceutical industry a core activity is to synthesize compounds and test them in biological models. While there are many novel techniques available today compared with 10–20 years ago, these are most relevant to the very early phases of discovery. When progressing towards registration, there are few shortcuts. Indeed, the demands during the late phases are much tougher than before and, ultimately, it is the patent-protected compounds that are registered as new drugs that count, not the techniques that were used. If we are not developing compounds and testing them, there will be no new drugs. The laboratory work, which is very time-consuming, has to be done, but is also sensitive to disturbances.

In the pharmaceutical industry of today, you can spend much of your time reading and sending e-mails, informing yourself via the intranet and internet, and going to meetings, seminars, symposia, and courses. As a result, many people feel stressed and frustrated, yet most of these activities are, in themselves, useful and necessary. We cannot blame the surrounding world, but must decide for ourselves how our time should be used. We all know this, but what can we do?

In my view, the only thing that can meet and counteract this 'development' is increased enthusiasm to achieve the required goals. Prohibitions, rules, and limitations belong to the past.

A prerequisite for the generation of enthusiasm is that your manager (or evaluating person) shows interest in the work that you do and notices the results you achieve. Other important factors include the degree of participation, immediate information on results without having to wait for a meeting, the feeling of belonging to a group, and a feeling of urgency. For scientists, participation in scientific symposia may be a positive factor in generating enthusiasm, enabling you to meet with your competitors and perhaps become aware that you know things that they don't.

8.17.3.4   Patents

Scientists, especially medicinal chemists, should also take an enthusiastic interest in the patent work. Patents can be as important for the company as the compounds and drugs, and I believe that the effort put into the 'patent work' is not always as optimal as it should be.

Patent work in pharmaceutical research has two quite different aspects: the patent professional (attorney) and the scientific. A comparison can be made with the structure–activity relationship work, which also involves two sides: the biological and the chemical. In both cases, a close collaboration between the two, involving a 'bridging over' of competence, is important in determining a positive outcome. From the chemical side, this has even led to the creation of the special discipline of medicinal chemistry. As far as patent work is concerned, the attorney's basic education has generally included science, but the scientists involved do not have the corresponding education about patents. My impression is that there is normally a nonoptimal balance of bridged input in patent work in pharmaceutical research. It is therefore important for scientists, mainly medicinal chemists, to increase their patent knowledge and to show enthusiasm for patents. Furthermore, patent work and initiatives should be driven from the scientific side, which would, no doubt, provide more optimal discussion partners for the attorneys.

8.17.4   Future Perspectives of Medicinal Chemistry
8.17.4.1   Visions for the Future

Unfortunately, I am not optimistic about the future of industrial pharmaceutical research. As well as escalating demands on safety and the introduction of reference price systems, which result in increased costs and decreased profitability, there is a continuing decline in productivity, measured as new drugs coming to the market. Moreover, if a drug is successful in getting to the market, there is an increased threat of challenges to, and invalidations of, the patents. It also seems likely that reference price systems for drugs will change the basis of research in the pharmaceutical industry, which has for many years been dominated by analog-based drug design that aims to improve on already-existing medical principles. Effort will now need to switch to pioneer or first-in-class drug research, which requires a high level of innovation. However, I believe that there is a general decline in the innovative climate in the major pharmaceutical companies.

I fear that the general decrease in productivity may be related to an increasing focus on molecular biology-driven, target-oriented drug discovery, in which valuable and important techniques are frequently being used out of perspective. The approach has been heavily criticised, as highlighted by the following citations:

> The elegance of these techniques is seductive – so much so that I believe they are taken too readily as valid models of disease for evaluating drugs[39]
>
> …we are whole animals and if you do your experiments on isolated cells and the tissues can't talk, then you tend to get results that may not be representative of the whole animal[40]
>
> …the complexity of the in vivo situation cannot be mimicked. Also, cell phenotypes that develop outside the body (i.e., in vitro) might exist exclusively in the test tube[41]
>
> But what the in vitro system cannot do is construct a functional and valid in vivo biochemistry. And that is potentially a fatal flaw. For in most human diseases it is the functional biochemistry and not the anatomical biochemistry which goes wrong[42]
>
> A receptor molecule is a very dynamic molecule, built to undergo conformational change with lightning speed. That it can look like the same after removal, grinding and suspension of the tissue is a preposterous belief.[43]

The industry has invested heavily in high-throughput chemistry, computational chemistry and various enabling in vitro techniques, including HTS, supported by vast substance libraries with special delivery facilities. Furthermore, there has been a concomitant reorganization of scientists into specialist functions along the generally accepted drug discovery

time-line, not unlike assembly lines in car production years ago. I believe this is a research trap which it is not easy to escape and, in the meantime, resources are probably being misspent.

During my recent discussions with Arvid Carlsson, he presented convincing arguments in favor of in vivo screening in drug research. Carlsson Research Company has synthesized compounds, such as (–)-OSU 6162 (**5**) and ACR 16 (**6**), which are so-called dopamine stabilizers. If the release of dopamine is too high, these compounds will decrease it, and if the release is too low, they will increase it to an acceptable level. Thus the same compounds can either block or stimulate dopamine receptors in the CNS. In test studies in patients seriously ill with Parkinson's disease and Huntington's chorea, (–)-OSU 6162 has shown close to miraculous effect, as well as clear positive effects in schizophrenic patients. However, a fundamental characteristic of these compounds is that they have low, or even zero, affinity for the receptors in classical binding models, and would therefore be considered inactive and uninteresting in an HTS screen!

These findings have more general implications for in vitro versus in vivo testing. Cloning may be used to provide a chemically homogeneous form of a particular dopamine receptor, for example, although this receptor may not be functionally homogeneous in vivo. According to Arvid Carlsson, receptors in vivo have a fabulous ability to adapt themselves functionally to the concentration of their relevant transmitter substances, in this case dopamine, which may, perhaps, vary 1000-fold within the synaptic area. This adaptation is not a matter of receptor density but is functional and may have an evolutionary background, when the original concentration of the transmitter may have been zero. In the presence of different concentrations of the transmitter, the same receptor is now able to act either as a stimulator or as an inhibitory autoreceptor. Binding of antagonists to these receptors with different functions also appears to be different.

In my view, it would therefore not be surprising to find that results from HTS screening to identify micromolar affinities for a particular dopamine receptor may not transfer well to an in vivo situation. Similarly, gene knockout animals and cell lines as models for diseases have frequently proven to create misleading artifacts. Furthermore, there are known paradoxical pharmacological effects, which may have similar background, that result in the same compound acting as a full and partial agonist and antagonist at different functional states of the 'same' receptor. For example, from my own experience, the potent 5-HT$_{1A}$ agonist 8-OH-DPAT (**3**) showed a strong stimulatory effect on the sexual behavior of rats.[44] This was surprising, since stimulation of serotonin neurons normally inhibits sexual behavior, as in the case of selective serotonin reuptake inhibitors, for example, where sexual disturbances are among the common side effects.

Some of the most successful medicines are remarkably weak or nonselective, and many are active on several receptors. I believe there is a low probability that we will identify optimal drugs that are active only on one receptor, particularly for CNS-active compounds. Such compounds are, of course, excellent pharmacological tools, but most diseases are related to unbalanced multilocalized abnormal biochemical patterns. However, HTS screening of today would not rate weakly binding, nonspecific compounds as 'hits.' Ironically, omeprazole, one of the world's biggest selling drugs, is probably one of the best examples of a drug that would not have been discovered by using HTS screening. As a targeted prodrug that is only slowly converted to the active species at about neutral pH, omeprazole would not have been potent enough to become a 'hit' when applied to an H$^+$,K$^+$-ATPase enzyme-screening model.

8.17.4.2 Innovation

The risk is that the current decline in productivity in the big pharma industry is putting more and more pressure on effectivity, with research dominated by targets for delivery of numbers of synthesized compounds, CDs, and patent applications at predestined deadlines. Although high productivity in the laboratory is important, such benchmarking may be at the expense of new thinking and innovation. I believe that organization of drug discovery work according to time-lines (e.g., hit identification, hit to lead, lead optimization, etc.), with different people becoming specialized in each phase, is detrimental for the development of chemists and may lead to decreased enthusiasm. Broad competence in medicinal chemistry has, in the past, been important not only in the pioneering structure discovery and optimization of potency and selectivity, but also for the generation of new, important, 'simple' ideas, and in the difficult tasks of improving pharmacokinetics, metabolism, and bioavailability and avoiding toxicological problems. Most experienced medicinal chemists know that the efforts required to progress from the first CD in a project to a final, useful CD may be tremendous, and that the timeframe cannot easily be scheduled. For example, in 1984, we were already aware that SCH 28080, mentioned above, was an H$^+$,K$^+$-ATPase inhibitor. Thus, we knew its target, site of action, mechanism of action, and that it had a potent and desired effect in humans. In addition, we had several lead compounds and a full setup of test models, both in vivo and in vitro. Despite all these enabling prerequisites, it has taken 20 years to get to where we are today, with a CD under development.

My hope for the future is that soon the pendulum will swing back, and we will once again focus on biological activity and in vivo screening. In the meantime, we need to safeguard the classical medicinal chemistry discipline and to improve it with better knowledge about patents and increased understanding of the factors of importance for nurturing creativity and enthusiasm.

References

1. Lindberg, P. PhD Thesis, Lund Institute of Technology, Lund, Sweden, **1977**.
2. Lindberg, P.; Bergman, R.; Wickberg, B. *J. Chem. Soc. Perkin Trans. 1* **1977**, 684–691. http://www.rsc.org/publishing/journals/p1 (accessed July 2006).
3. Carlsson, A.; Henning, M.; Lindberg, P.; Martinson, P.; Trolin, G.; Waldeck, B.; Wickberg, B. *Acta Pharmacol. Toxicol.* **1978**, *42*, 292–297.
4. Marchner, H. PhD Thesis, University of Uppsala, Uppsala, Sweden, **1979**.
5. Tottmar, O.; Lindberg, P. *Acta Pharmacol. Toxicol.* **1997**, *40*, 476–481.
6. Jonsson, M.; Lindquist, N. G.; Ploen, L.; Ekvarn, S.; Kronevi, T. *Toxicology* **1979**, *12*, 89–100.
7. Arvidsson, L. E.; Hacksell, U.; Nilsson, J. L.; Hjorth, S.; Carlsson, A.; Lindberg, P.; Sanchez, D.; Wikstrom, H. *J. Med. Chem.* **1981**, *24*, 921–923.
8. Hjorth, S.; Carlsson, A.; Wikstrom, H.; Lindberg, P.; Sanchez, D.; Hacksell, U.; Arvidsson, L.; Svensson, U.; Nilsson, J. L. *Life Sci.* **1981**, *28*, 1225–1238.
9. Tedroff, J.; Torstenson, R.; Hartvig, P.; Sonesson, C.; Waters, N.; Carlsson, A.; Neu, H.; Fasth, K. J.; Langstrom, B. *Synapse* **1988**, *28*, 280–287.
10. Nilsson, M.; Carlsson, A.; Markinhuhta, K. R.; Sonesson, C.; Pettersson, F.; Gullme, M.; Carlsson, M. L. *Prog. NeuroPsychopharmacol. Biol. Psychiatry* **2004**, *28*, 677–685.
11. Lambrecht, N.; Munson, K.; Vagin, O.; Sachs, G. *J. Biol. Chem.* **2000**, *275*, 4041–4048.
12. Lindberg, P.; Nordberg, P.; Alminger, T.; Brändström, A.; Wallmark, B. *J. Med. Chem.* **1986**, *29*, 1327–1329.
13. Brändström, A.; Lindberg, P.; Bergman, N.-A.; Alminger, T.; Anker, K.; Junggren, U.; Lamm, B.; Nordberg, P.; Erickson, M.; Grundevik, I. et al. *Acta Chem. Scand.* **1989**, *43*, 536–548.
14. Lindberg, P.; Brändström, A.; Wallmark, B. *Trends Pharmacol. Sci.* **1987**, *8*, 399–402.
15. Lindberg, P.; Brändström, A.; Wallmark, B.; Mattsson, H.; Rikner, L.; Hoffman, K-J. *Med. Res. Rev.* **1990**, *10*, 1–54.
16. Wallmark, B.; Carlsson, E.; Larsson, H.; Brändström, A.; Lindberg, P. *3rd SCI-RSC Medicinal Chemistry Symposium*, Cambridge, England, September 15–18, **1985**, abstract S16.
17. Lambert, R. W., Ed.; *Proceedings*, Whitstable, Kent, **1986**, pp 293–311.
18. Figala, V.; Klemm, K.; Kohl, B.; Krüger, U.; Rainer, G.; Schaefer, H.; Senn-Bilfinger, J.; Sturm, E. *3rd SCI-RSC Medicinal Chemistry Symposium*, Cambridge, England, September 15–18, **1985**, abstract p20.
19. Lindberg, P. Presented at the Royal Society of Chemistry Fine Chemicals and Medicinals Group, *2nd Bath International Symposium on Medicinal Chemistry*, September 25–27 Bath, England, **1994**.
20. Carlsson, E.; Lindberg, P.; von Unge, S. *Chem. Britain* **2002**, *38*, 42–45.
21. Olbe, L.; Carlsson, E.; Lindberg, P. *Nature (Lond.) Rev.* **2003**, *2*, 132–139.
22. Lindberg, P.; Keeling, D.; Frycklund, J.; Andersson, T.; Lundborg, P.; Carlsson, E. *Aliment. Pharmacol. Ther.* **2003**, *17*, 481–488.
23. Andersson, T.; Hassan-Alin, M.; Hasselgren, G.; Röhss, K.; Weidolf, L. *Clin. Pharmacokinet.* **2001**, *40*, 411–426.
24. Andersson, T.; Röhss, K.; Bredberg, E.; Hassan-Alin, M. *Aliment. Pharmacol. Ther.* **2001**, *15*, 1563–1569.
25. Hassan-Alin, M.; Andersson, T.; Niazi, M.; Röhss, K. *Eur. J. Clin. Pharmacol.* **2005**, 779–784.
26. Lind, T.; Rydberg, L.; Kylebäck, A.; Jonsson, A.; Andersson, T.; Hasselgren, G.; Holmberg, J.; Röhss, K. *Aliment. Pharmacol. Ther.* **2000**, *14*, 861–867.
27. Kahrilas, P.; Falk, G.; Johnson, D.; Schmitt, C.; Collins, D.; Whipple, J.; D'Amico, D.; Hamelin, B.; Joelsson, B. *Aliment. Pharmacol. Ther.* **2000**, *14*, 1249–1258.
28. Richter, J.; Kahrilas, J.; Johanson, J.; Maton, P.; Breiter, J.; Hwang, C.; Marino, V.; Hamelin, B.; Levine, J. *Am. J. Gastroenterol.* **2001**, *96*, 656–665.
29. Castell, D.; Kahrilas, P.; Richter, J.; Vakil, N.; Johnson, D.; Zuckerman, S.; Skammer, W.; Levine, J. *Am. J. Gastroenterol.* **2002**, *97*, 575–583.
30. Labenz, J.; Armstrong, D.; Katelaris, P.; Schmidt, S.; Adler, J.; Eklund, S. *Gut* **2004**, *53*, A108.
31. Labenz, J.; Armstrong, D.; Lauritsen, K.; Katelaris, P.; Schmidt, S.; Schutze, K.; Wallner, G.; Juergens, H.; Preiksaitis, H.; Keeling, N. et al. *Aliment. Pharmacol. Ther.* **2005**, *21*, 739–746.
32. Lauritsen, K.; Devière, J.; Bigard, M.-A.; Bayerdörffer, E.; Mózsik, G.; Murray, F.; Kristjánsdóttir, S.; Savarino, V.; Vetvik, K.; de Freitas, D. et al. *Aliment. Pharmacol. Ther.* **2003**, *17*, 333–341.
33. Wallmark, B.; Keeling, D.; Lindberg, P. *Des. Enzyme Inhib. Drugs* **1994**, *2*, 672–691.
34. Holstein, B.; Holmberg, A.; Florentzson, M.; Holmberg, A. A.; Andersson, M.; Andersson, K. *Gastroenterology* **2004**, *126*, A–469.
35. Lindberg, P.; Wikström, H.; Sanchez, D.; Arvidsson, L.-E.; Hacksell, U.; Nilsson, J.; Hjorth, S.; Carlsson, A. Dopamine Receptor Agonists 2, Carlsson, A., Nilsson, J. L. G., Eds.; In *Proceedings of the Swedish Academy of Pharmaceutical Sciences Symposium* April 20–23, 1982; Swedish Pharmaceutical Press: Stockholm, **1983**, pp 48–55.
36. Wikstrom, H.; Sanchez, D.; Lindberg, P.; Hacksell, U.; Arvidsson, L.; Johansson, A.; Thorberg, S. O.; Nilsson, J. L.; Svensson, K.; Hjorth, S. *J. Med. Chem.* **1984**, *27*, 1030–1036.
37. Rackur, G.; Bickel, M.; Fehlhaber, H-W.; Herling, A.; Hitzel, V.; Lang, H. J.; Rösnerand, M.; Weger, R. *Biophys. Res. Commun.* **1985**, *128*, 477–484.
38. Sundgren, M. PhD thesis, Chalmers University of Technology, Göteburg, Sweden, **2004**.
39. Higgs, G. *Drug Disc. Today* **2004**, *9*, 727–729.
40. Carey, J.; Barton, S.; Campbell, S. *Drug Disc. Today* **2005**, *10*, 233–236.
41. Persson, C.; Erjefält, J.; Uller, L.; Andersson, M.; Grieff, L. *Trends Pharmacol. Sci.* **2001**, *22*, 538–541.
42. Horrobin, D. *Nat. (Lond.) Rev.* **2003**, *2*, 151–154.
43. Jönsson, B. *Kemivärlden Med. Kemisk Tidskr.* **2000**, *11*, 22–24.
44. Ahlenius, S.; Larsson, K.; Svensson, L.; Hjorth, S.; Carlsson, A.; Lindberg, P.; Wikstrom, H.; Sanchez, D.; Arvidsson, L. E.; Hacksell, U. et al. *Pharmacol. Biochem. Behav.* **1981**, *15*, 785–792.

Biography

Per Lindberg graduated in organic chemistry from the University of Technology Lund, Sweden, in 1977 and was appointed Associate Professor of Organic Chemistry at the Chalmers Institute of Technology, Gothenburg, Sweden, in 1982. After working for several years with Prof Arvid Carlsson (Nobel Prize winner in Medicine, 2000) in the Department of Pharmacology, University of Gothenburg, he joined Hässle AB (today AstraZeneca R&D Mölndal, Sweden) in 1982 as Head of Medicinal Chemistry GI and held that position until 1993. He was involved in several projects, including the development of the first proton pump inhibitor omeprazole (Losec) and elucidation of its mechanism of action, as well as the follow-up project to develop esomeprazole (Nexium). From 1993 to 1998 he was Director of the Preclinical Alliances Group. In 1998, he became Scientific Advisor in the Scientific Patent Support Team directed to patent litigation of omeprazole and, since April 2002, has been Senior Scientific Advisor within the Gastrointestinal Therapeutic Area. He has been the Scandinavian representative in the IUPAC Medicinal Chemistry Section since 1990, and is author of numerous publications in medicinal chemistry and many inventorships, including co-inventorship of esomeprazole.

8.18 Calcium Channel α_2–δ Ligands: Gabapentin and Pregabalin

A J Thorpe and C P Taylor, Pfizer Global Research and Development, Ann Arbor, MI, USA

8.18.1	**Early Preclinical Work Leading to the Development of Gabapentin**	**227**
8.18.1.1	Selection of γ-Amino-Butyric Acid as a Mimetic Compound to Target (GABA$_B$ Receptors)	228
8.18.1.2	The Discovery and Structure–Activity Relationship (SAR) of Gabapentin: Serendipity Leads to the Discovery of a New Drug Target	228
8.18.2	**Additional Structure–Activity Relationship at the Calcium Channel α_2–δ Site**	**231**
8.18.2.1	Studies of Gabapentin Derivatives	231
8.18.2.2	A Critical Piece of the Puzzle: Transport of Amino Acids Across Membrane Barriers	234
8.18.2.3	The Discovery and Structure–Activity Relationship of Pregabalin: A Second Fortuitous Discovery	235
8.18.2.4	Beta Amino Acids	236
8.18.2.5	Aliphatic Side-Chain Replacements	237
8.18.3	**Clinical Development of Gabapentin**	**238**
8.18.3.1	Early Clinical Studies	238
8.18.3.2	Development of Gabapentin for Epilepsy	239
8.18.3.3	Development of Gabapentin for Neuropathic Pain	239
8.18.4	**Studies of the Mechanism of Action of Gabapentin and Pregabalin**	**240**
8.18.4.1	Discovery of the Calcium Chemical α_2–δ Binding Site	240
8.18.4.2	Drug-Induced Changes in Calcium Channel Function?	240
8.18.4.3	Studies Relating to γ-Amino-Butyric Acid Concentrations and GABA$_B$ Receptors	240
8.18.4.4	Mutation and Deletion Studies of α_2–δ Protein	241
8.18.4.5	Mechanism of Action: Beyond the Drug-Binding Site	241
8.18.4.6	Decreases in Neurotransmitter Release	242
8.18.4.7	Mechanisms of Action – Summary	242
References		**242**

8.18.1 Early Preclinical Work Leading to the Development of Gabapentin

Gabapentin was conceived as part of a drug discovery program to treat neurological diseases, including epilepsy, spasticity, multiple sclerosis, and other central nervous system (CNS) disorders. This program began in the early 1970s at the German company, Goedecke, A.G., in Freiburg, Germany, which was a part of Warner-Lambert (now incorporated into Pfizer). The history of this project included chemical attempts to inhibit γ-amino-butyric acid (GABA) degradation in brain with compounds that inhibited the catalytic pyridoxylphosphate of GABA-transaminase. It had already been known for some time that GABA was a key inhibitory neurotransmitter, and that experimental chemical impairment of GABA systems could cause seizures in experimental animals. The GABA transaminase project at Goedecke had progressed a compound to phase I clinical trials, but these were halted because of safety concerns. The chemical matter developed within the GABA transaminase project had no direct relationship to the chemical matter that led to gabapentin, although both had a similar conceptual approach based on GABA.

8.18.1.1   Selection of γ-Amino-Butyric Acid as a Mimetic Compound to Target (GABA_B Receptors)

In early 1973, Gerhardt Satzinger of Goedecke conceived a series of about 25 derivatives of GABA that were later synthesized. This project was an attempt to design GABA mimetic drugs acting at GABA receptors that (unlike GABA) would penetrate the blood–brain barrier. At this early time, GABA was only recently acknowledged as an inhibitory neurotransmitter, and there were only a few known GABA agonists, of which baclofen (a selective GABA_B agonist) was one example.[1] Most of the compounds that were later synthesized turned out to be inactive in a GABA_B radioligand-binding assay, but nevertheless several prevented seizures in mice when given systemically against chemical challenge with the GABA synthesis inhibitors mercaptopropionic acid or thiosemicarbazide. The mechanism of action of these anticonvulsant drugs was not known at the time, but there were already several known differences between the Goedecke 3-GABA derivatives and baclofen. Many of the original pharmacological studies compared the effects of gabapentin and related compounds to those of baclofen[2] but there were many differences, particularly a complete lack of gabapentin effects on GABA_B receptors.

8.18.1.2   The Discovery and Structure–Activity Relationship (SAR) of Gabapentin: Serendipity Leads to the Discovery of a New Drug Target

As it is sometimes the case in drug discovery, the discovery of gabapentin was made through a rather indirect path. Since GABA does not penetrate the blood–brain barrier, Satzinger and colleagues directed attention to increasing the oral bioavailability of GABA, by raising the log P of its analogs through the incorporation of lipophilic groups on the carbon backbone.[1–3] Although compounds described in Satzinger's original work did possess anticonvulsant activity (as did centrally administered GABA), it was ultimately found that none of these compounds affected either metabotropic or ionotropic GABA receptors, nor fluxed through the blood–brain barrier by passive diffusion. However, it was from these fortuitous studies that (1-aminomethyl-cyclohexyl)-acetic acid (**2**) (gabapentin, Neurontin) emerged as a potent and efficacious anticonvulsant in the thiosemicarbazide-induced tonic convulsion model in mice (**Table 1**). Compared to other analogs in this series, gabapentin clearly had greater activity. Several years later, its anticonvulsant properties were confirmed in the low-intensity mouse electroshock model, where again it proved to be the most potent member within this series. Notably, this early SAR study, driven largely through in vivo work, hinted at the optimal size of the cycloalkyl ring for anticonvulsant activity, with the cyclohexyl (**2**) and cycloheptyl (**3**) moieties most preferred.

As interest in the anticonvulsant properties of gabapentin grew, efforts to identify the molecular target in the CNS were initiated by several research groups. This culminated in the discovery of the calcium channel α_2–δ subunit as the molecular target of this compound.[4] A retrospective study of Satzinger's original work identified gabapentin as the most potent amino acid with regard to affinity for this protein. When studied much later in a radioligand-binding assay,[5] the cyclopentyl (**1**) and cyclooctyl (**4**) analogs showed a two- and 15-fold drop respectively in binding affinity at pig brain membrane α_2–δ sites (measured with [^{3}H]-gabapentin) when compared to unlabeled gabapentin. Thus, this

Table 1   Effect of varying ring size on α_2–δ binding affinity and anticonvulsant action

Compound	n	TSCZ ED_{50} (mg kg^{-1} IP)	Low-intensity electroshock ED_{50} (mg kg^{-1} IP)	α_2–δ binding affinity (IC_{50}, nM)
1	1	<31	NT	260
2	2	5	4.5	140
3	3	<63	~60	110
4	4	>250	NT	1810

TSCZ, thiosemicarbazide.

Table 14 Cyclic β-amino acids and their affinity for α_2–δ and system L

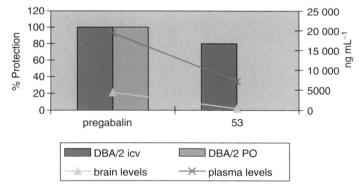

Compound	R_1	R_2	α_2–δ binding affinity (IC$_{50}$, nM)	System L (IC$_{50}$, μM)
5	n/a	n/a	80	158
50	CH(CH$_2$CH$_3$)$_2$	H	56	>100
51	(CH$_2$)$_2$CH$_3$	(CH$_2$)$_2$CH$_3$	200	>100
52	–(CH$_2$)$_4$		23	>100
53	–(CH$_2$)$_5$		13	>100
54	CH(CH$_3$)$_2$	H	340	>100
55	CH$_2$CH(CH$_3$)$_2$	H	330	>100
56	c-C$_3$H$_5$	H	200	>100
57	H	CH$_2$CH(CH$_3$)$_2$	37	>100
58	H	CH(CH$_2$CH$_3$)$_2$	630	>100
59	H	H	>10 000	>100

Figure 3 Biological and exposure data for amino acid **53** compared to that of pregabalin **5**. Compound **53** potentially inhibits [^{3}H]gabapentin binding but is not active in the DBA/2 seizure model with oral dosing (blue bars). This is related to the lack of activity of compound **53** at the system L amino acid transporter, resulting in very low exposures of **53** to brain (yellow symbols). Despite this, direct administration of **53** by ICV injection prevents seizures in DBA/2 mice (red bar). (The graph is from [27].)

DBA/2 mouse test. However, once the blood–brain barrier was circumvented by direct intracerebroventricular administration, the spirocarboxylic acid (**53**) gave 80% protection against seizures. Further analysis of the plasma exposure of both pregabalin and the acid (**53**) after oral administration to mice (**Figure 3**) showed a clear difference, indicating low absorption of **53** from the gut. Furthermore, extremely low brain levels were measured with **53**, further indicating poor brain penetration. Presumably, the levels of **53** obtained in the brain were not enough to elicit antiseizure activity, despite potent binding at α_2–δ.

8.18.2.5 Aliphatic Side-Chain Replacements

Schelkun and coworkers examined the effect of replacement of the isobutyl side chain of pregabalin with heteroaromatic, aromatic, and heterocyclic groups in order to expand the scope of SAR (**Table 15**).[28] Binding affinity in this series trended with the following rank order: 2-furan $\sim$ 3-furan > 3-thiophene > 2-thiophene. Overall, though, modifications in this study

Table 15

$$R \underset{}{\overset{}{\bigvee}} \overset{CO_2H}{\underset{NH_2}{}}$$

Compound	R	Stereochemistry	α_2–δ binding affinity (IC$_{50}$, nM)	System L (IC$_{50}$, μM)	DBA/2 (30 mg kg^{-1}, PO; % of mice protected)
5	iPr	S	80	158	100
60	(tetrahydrofuranyl)	–	>10 000		0
61	(2-furyl)	–	421	10 000	20
62	(3-furyl)	–	518	1422	0
63	(3-thienyl)	–	2140	2936	40
64	(2-thienyl)	–	832	7034	60
65	(2-furyl)	S	178	10 000	40
66	(2-furyl)	R	1460	10 000	0

led to compounds with lower affinity for α_2–δ measured by displacement of [^{3}H]-gabapentin from pig brain membranes than pregabalin. The 2-furyl analog **61** was the most potent ligand in this class. Enantiomers of **61** were stereoselectively synthesized using the oxazolidinone strategy. Consistent with the stereodivergence seen with enantiomers of pregabalin **5** and 3-Me gabapentin **22**, the (S)-enantiomer **65** was eightfold more potent at α_2–δ than the corresponding (R)-enantiomer **66** and only the more potently binding enantiomer was active in vivo. Besides reducing affinity for α_2–δ binding, incorporation of heteroatoms in the amino acid framework resulted in decreased affinity for blocking [^{3}H]-leucine uptake into CHO cells. It is thought that reduced affinity for both the system L amino acid carrier and the α_2–δ protein led to diminished in vivo activity as measured by both the Vogel conflict model for anxiolytic-like action (not shown) and also the DBA/2 seizure model with this heterocyclic class. However, DBA/2 activity with compounds **63**, **64** and **65** suggests the untested idea that these compounds traverse membrane barriers by mechanisms other than System L transport.

8.18.3   Clinical Development of Gabapentin

8.18.3.1   Early Clinical Studies

Gabapentin was studied in animal toxicology at Goedecke from about 1980 to 1982, and was first studied for tolerance, safety, and pharmacokinetics in healthy human subjects in a study contracted from Goedecke in 1982.[29] Clinical phase I single-dose and multiple-dose tolerance studies showed elimination with a plasma half-life of about 6 h and dose-proportional absorption.[29]

Clinical phase II studies with gabapentin began in 1983 and continued in 1984 and 1985 with several quite small studies against Huntington's disease, hemiplegia, and spasticity (single-blind placebo-controlled) with dosages of 200 and 400 or 600 and 900 mg day^{-1} given q.i.d. or t.i.d. In early 1984, spasticity was studied in open-label clinical trials at three medical centers, each with relatively low dosages of gabapentin and very small numbers of patients (total $n = 10$–65

patients per study; one study used rising dosages up to 1800 mg day^{-1} given t.i.d. for 1 week). Unfortunately, due to the low number of patients, lack of parallel placebo groups, and heterogeneity of spasticity diagnoses and severity, no conclusions could be drawn from these studies other than a relatively benign side-effect profile for gabapentin.

8.18.3.2 Development of Gabapentin for Epilepsy

The first clinical study of gabapentin in epilepsy showing efficacy was conducted with 25 patients in 1985 as a placebo-controlled add-on crossover study in refractory partial seizures (Bernd Schmidt *et al.*, Goedecke, unpublished data), unpublished data. All patients were maintained on their prior medications, but were then crossed over between four different study groups with the addition of placebo, 300, 600, or 900 mg gabapentin per day. The 900 mg dose group showed a significant difference from placebo in weekly seizure frequency (overall 45% decrease compared to placebo). In addition, there was a 43% responder rate (number of patients with more than 50% decline in seizure frequency) in this study with 900 mg day^{-1}, versus 14% responder rate at 300 mg day^{-1}. These results started the serious clinical development of gabapentin, which was done by a planning team at first chaired by Schmidt, and run from Goedecke.

Later (in mid-1986) a global planning team for gabapentin was formed, and although still run from Goedecke, it included regulatory and clinical personnel from Parke-Davis Research in Ann Arbor, Michigan. Subsequent to 1986, the pivotal clinical studies of gabapentin for epilepsy were mainly coordinated by Jan Wallace and Kent Schellenberger (Parke-Davis, Ann Arbor) and after 1990 by Elizabeth Garofalo and colleagues[30,31] (Parke-Davis, Ann Arbor). Positive results in the pivotal clinical efficacy studies with gabapentin treatment as add-on for refractory partial seizures were eventually filed with the US Food and Drug Administration for the New Drug Application (NDA) for gabapentin in 1992. These studies were coordinated by many outside investigators, including Dennis Chadwick in the UK,[32,33] and Michael McLean,[34] Eugene Ramsey,[35] BJ Wilder, Ahmad Beydoun[36] (and others) in the US. Three months before the US Food and Drug Administration (FDA) advisory committee meeting for the epilepsy approval in December 1992, Mark Pierce (Parke-Davis, now Pfizer Ann Arbor, recently retired) joined the clinical group from his prior position at Abbott, and defended the clinical efficacy and safety information before the FDA epilepsy advisory committee. Approval of the gabapentin epilepsy NDA was achieved after final label wording negotiations on December 31 of 1993. Neurontin was launched by Parke-Davis for the US epilepsy market in February 1994.

After the gabapentin product launch, gabapentin was used extensively by physicians for treating epilepsy. Because of a relatively benign adverse event profile and few drug–drug interactions, it was also prescribed for off-label indications, including neuropathic pain, anxiety, and other psychiatric indications, essential tremor, spasticity, postsurgical pain, and prevention of postmenopausal hot flashes. None of these additional indications were supported in the gabapentin product labeling or approved by regulatory agencies until the FDA approved a supplemental NDA for gabapentin to treat postherpetic neuralgia in July 2001 (*see* Section 8.18.3.3, below).

8.18.3.3 Development of Gabapentin for Neuropathic Pain

Lakhbir Singh and Mark Field performed initial studies of gabapentin for use as an analgesic in animal models in 1992–1993 at the Parke-Davis Cambridge (UK) Research Centre. They found that gabapentin was active in several rat models of antihyperalgesic action (formalin test, carrageenan test),[37,38] although at rather high dosages. Gary Bennett obtained a sample of gabapentin in 1993, and tested it in his model of neuropathic pain from sciatic nerve ligation in rats.[39] These results from animal models were presented by Bennett at a national meeting in 1994, and at about the same time, the first case reports of gabapentin use for neuropathic pain appeared in the literature.[40,41] Subsequently, a large number of investigators found gabapentin to be active in animal models of pain states[14,38,42–61] and also in several clinical studies.[62–65]

In 1995 and 1996, Parke-Davis began two large placebo-controlled parallel group studies of gabapentin for treating neuropathic pain. Clinical trials for diabetic peripheral neuropathy were supervised by Elizabeth Garofalo (Parke-Davis, Ann Arbor), and postherpetic neuralgia studies by Leslie Magnus-Miller (Parke-Davis, Morris Plains). Both studies were done with virtually identical protocols. These studies were both resoundingly positive in comparison to placebo treatment, and the results were published side by side in the *Journal of the American Medical Association* in 1998.[66,67]

In 1998 and 1999, based on the published *Journal of the American Medical Association* clinical studies, the first regulatory approvals were obtained for gabapentin to treat neuropathic pain in Asia and Latin America, then later in Europe. In early 2001, it was decided to submit a supplementary NDA (sNDA) application in the US with gabapentin for neuropathic pain, based upon the recently obtained clinical data. The submission was completed in August 2001 and the FDA approved the sNDA and provided revised labeling for gabapentin for treatment of postherpetic neuralgia on May 24, 2002. The commercial launch of gabapentin for postherpetic neuralgia was in August of 2002.

8.18.4   Studies of the Mechanism of Action of Gabapentin and Pregabalin

The original animal pharmacology studies of gabapentin from Goedecke (1981–1984: W. Reimann, G. Bartoszyk, and others, unpublished data) included studies of anticonvulsant action in mice, rats, and monkeys and antispasticity effects in animal models with mice, rats, and anesthetized cats. Early mechanism of action work centered on electrophysiological changes in spinal reflexes (W. Steinbrecher, 1982, unpublished data) and reduced monoamine neurotransmitter release from brain tissue slices in vitro (W. Reimann, E. Schlicker).[2,68] Also in 1984, a group of pharmacologists at Parke-Davis (P. Boxer, R. Anderson *et al.*, Ann Arbor, MI, unpublished data) studied gabapentin in several preclinical models of spasticity in mice. Gabapentin reduced muscle rigidity and increased locomotor agility (apparently by reducing spinal polysynaptic reflexes) at dosages of 10–100 mg kg^{-1} IP. The results with gabapentin compared favorably with other experimental treatments for spasticity (e.g., baclofen, diazepam). Slightly later, mechanism studies focused on transport of gabapentin by the system L amino acid transporter (which proved to be the main mechanism of gabapentin entry across the gut and also across the blood–brain barrier[69,70]). However, the lack of pharmacological activity with several system L transporter substrates that lacked high affinity for α_2–δ (data not shown) soon turned attention to other potential sites of action.

8.18.4.1   Discovery of the Calcium Chemical α_2–δ Binding Site

Important mechanism of action work began in about 1991 in the Parke-Davis Cambridge Research Centre. This work lead to the identification of a specific [^{3}H]gabapentin-binding site in rat brain tissues by David Hill, Nirmala Suman-Chauhan, and colleagues.[5,71] They showed that the site was distributed heterogeneously in rat brain, with high densities of binding in regions that were also rich with synaptic endings. Later, Nicholas Gee and Jason Brown of the Cambridge Unit used protein biochemistry techniques to purify solubilized protein fractions from pig brain that bound with high affinity to [^{3}H]gabapentin. After four stages of column purification, and sequencing of a short peptide fragment from the purified protein, the high-affinity site was identified as the α_2–δ type 1 protein, a subunit of voltage-gated calcium channels.[4] Recombinant production of α_2–δ protein in mammalian cells showed [^{3}H]gabapentin-binding properties essentially identical to those of pig or rat brain membranes, confirming the identity of the binding site.[4]

8.18.4.2   Drug-Induced Changes in Calcium Channel Function?

Building upon previous findings, David Dooley and colleagues (Pfizer Ann Arbor) showed that gabapentin reduces calcium influx and neurotransmitter release from rat and human brain tissues.[72–78] Alexander McKnight and YP Maneuf (Parke-Davis Cambridge) found that gabapentin reduced glutamate release in a manner suggesting relevance for analgesia.[79,80] Electrophysiologists in Cambridge and at Aberdeen University in the UK demonstrated changes in calcium channel and synaptic function that were presumed to be caused by gabapentin action at the α_2–δ site.[81–85] These findings together with SAR described in Sections 8.18.2.3–8.18.2.5 provided a strong circumstantial argument that binding at to the α_2–δ site was important to the pharmacology of gabapentin and pregabalin.

8.18.4.3   Studies Relating to γ-Amino-Butyric Acid Concentrations and GABA$_B$ Receptors

At about the same time as the α_2–δ publications, reports appeared indicating that gabapentin in humans caused an elevation of brain GABA concentrations measured by molecular resonance spectroscopy (MRS; O.A. Petroff, R. Mattson *et al.*, Yale University.[86–91] Other studies suggested that gabapentin increased nonsynaptic release of GABA from brain tissues (Jeffery Kocsis, George Richerson, Yale University.[92–95] Furthermore, with prolonged treatment, both pregabalin and gabapentin increased the number of GABA transporters present on neuronal cell membranes (Michael Quick, University of Alabama Birmingham).[96] However, the changes in whole-brain GABA concentration reported with gabapentin were smaller in magnitude than after treatment with the GABA-transaminase inhibitor vigabatrin.[88,89,97] Furthermore, other reports showed increases in MRS brain GABA signals caused by topiramate and lamotrigine, two very different antiepileptic drugs that are not known to alter brain GABA metabolism or degradation.[97–99] In addition, neither gabapentin nor pregabalin altered GABA concentrations in rat brain,[100] despite the robust pharmacological action of both gabapentin and pregabalin in rat models of seizures, pain, and antianxiety-like actions. It is somewhat confusing to reconcile these apparently conflicting data. However, MRS studies of whole-brain GABA measure almost exclusively the large pool of GABA contained within inhibitory neurons, and not the much smaller extracellular GABA

pool that is available to bind to inhibitory neuronal receptors. Therefore, the relevance of the whole-brain GABA findings to drug pharmacology is unclear. Finally, electrophysiological studies of both gabapentin and pregabalin in anesthetized rats suggest that rapid GABA synaptic action in rat brain is reduced, rather than enhanced, by drug treatment.[101] Therefore, any potential contribution of changes in brain GABA concentration to the anticonvulsant action of gabapentin and pregabalin remains far from clear.

To complicate the picture further, several publications from a group at the Merck-Frosst Research Center in Canada[102–104] implicated certain subtypes of the $GABA_B$ receptor as targets for gabapentin. However, several other laboratories were unable to replicate these findings,[105–110] and therefore, actions of gabapentin at $GABA_B$ receptors are not generally accepted to be relevant. Furthermore, neither pregabalin nor gabapentin displace radioligand binding to subtypes of $GABA_B$ receptors, including recombinant heteromeric $GABA_{BR1a}/GABA_{BR2}$ receptors, heteromeric $GABA_{BR1b}/GABA_{BR2}$, and $GABA_B$ receptors from native rat brain membranes (unpublished data, not shown). Therefore, consistent and reproducible actions of this class of drugs at $GABA_B$ receptors have not been observed.

8.18.4.4   Mutation and Deletion Studies of α_2-δ Protein

More recent studies of gabapentin and pregabalin mechanisms relating to the α_2-δ binding site were initiated based on findings by Ti-Zhi Su and James Offord (Pfizer Molecular Sciences, Ann Arbor) and Jason Brown and Nicolas Gee (Parke-Davis Research Centre, Cambridge) in about 1998–1999. This work began with experimental deletions and mutations to the DNA sequence for α_2-δ type 1, expressed in recombinant cell systems in vitro.[111,112] Work with mutant mice lacking α_2-δ proteins was hindered by the fact that a global knockout of α_2-δ type 1 caused lethality (J. Offord, unpublished data) and mouse mutations that cause a functional knockout of α_2-δ type 2 protein caused extreme behavioral abnormalities and early lethality.[113–115]

Subsequently, Offord and colleagues began work to produce genetically altered mice that selectively expressed a single amino acid mutation within the coding sequence of α_2-δ type 1. This single change to recombinant α_2-δ proteins in vitro had previously been shown to diminish greatly drug binding affinity for [^{3}H]gabapentin and [^{3}H]pregabalin.[112] When homozygous genetically altered mice incorporating the same single amino acid change were obtained (called R217A mutants because of a substitution of alanine for the wild-type arginine at position 217 of the α_2-δ type 1 protein), it was confirmed that these mice had reduced drug binding for gabapentin and pregabalin selectively to portions of brain and spinal cord.[116–118] Furthermore, this mutation did not cause untoward effects such as ataxia or seizures in whole animals. Finally, studies of pregabalin and gabapentin given systemically to wild-type and R217A mutant mice showed that drug actions in models of analgesia, anticonvulsant actions, and antianxiety-like effects were selectively reduced in R217A mutant mice, while the actions of other drugs, such as morphine, amitriptyline, and phenytoin, remained unchanged.[118,119] These findings strongly support the independent findings from structure–activity studies (see Section 8.18.2, above) to indicate that high-affinity binding to the α_2-δ type 1 protein is required for pharmacological effects of both pregabalin and gabapentin in vivo. It is interesting that preliminary findings with R217A mice showed a less marked effect of the mutation on anticonvulsant action of pregabalin than on analgesic or antianxiety-like actions of pregabalin. This suggests the possibility that pregabalin binding to α_2-δ type 2 proteins (that were not altered by the R217A mutation) may be important for anticonvulsant actions. However, additional experiments will be required to test this idea.

8.18.4.5   Mechanism of Action: Beyond the Drug-Binding Site

As can be appreciated from the preceding paragraphs, the mechanism of action of gabapentin (and also pregabalin) has been the subject of some debate. Despite that, quite a lot is known about their pharmacology. Neither compound is active at a wide variety of radioligand-binding sites associated with the actions of some other antiepileptic drugs, including batrachotoxinin sites on voltage-gated calcium channels, GABA receptors ($GABA_A$, $GABA_B$, $GABA_C$, benzodiazepine, or GABA transporter sites), or glutamate receptors (N-methyl-d-aspartate, strychnine-insensitive glycine, alpha-amino-3-hydroxy-5-methyl-4-isoxazolepropionic acid (AMPA), kainate or mGluR1, mGluR5 glutamate receptors). Furthermore, both compounds are remarkably silent even at high concentrations in a wide array of other radioligand-binding assays for common drug and neurotransmitter receptors. This is probably due to the unusual chemical nature of the molecules, that are both relatively small, hydrophilic, and possess strong amine and acid moieties. Both compounds, in contrast, are high-affinity ligands for α_2-δ type 1 and α_2-δ type 2 proteins, labeled with [^{3}H]gabapentin[5] or [^{3}H]pregabalin[117,120] or [^{3}H]L-leucine.[121]

8.18.4.6   Decreases in Neurotransmitter Release

How does binding at the α_2–δ site produce anticonvulsant effects? Numerous studies have examined voltage-gated calcium channel function with gabapentin or pregabalin, and these studies have produced conflicting results. Several studies report that gabapentin reduced current through voltage-clamped calcium channels on neuronal cell body membranes.[82,83,85,122–125] However, other studies with neuronal cell bodies[126,127] or recombinant cell systems expressing calcium channels with α_2–δ subunits[128] have shown no change in calcium currents after application of gabapentin. Despite these conflicting data, studies with calcium influx measured with fluorescent probes in synaptosomes from rat or human neocortex[75,78,127] show that both gabapentin and pregabalin reduce calcium influx measured at presynaptic terminals. In addition, studies measuring the synaptic release of glutamate,[73,79,80,129–133] norepinephrine, serotonin, or dopamine from neocortical tissues,[73,74,77] or the release of sensory peptide neurotransmitters substance P or calcitonin gene-related peptide (CGRP)[134] have shown subtle but reproducible reductions in calcium-dependent neurotransmitter release. In some systems, pretreatment in vivo with inflammatory agents or in vitro with neuropeptides or protein kinase activators are required before clear drug effects are seen.[80,134]

Recently, several studies have indicated that not only calcium-dependent release of neurotransmitters, but also asynchronous (miniature potentials – calcium-independent) release of vesicles is reduced by drug treatment. These experiments include the release of vesicles from glutamate synapses in spinal cord slices,[81,129] entorhinal cortex slices,[135] cultured rat hippocampal neurons,[141] and the release of cholinergic vesicles from mouse neuromuscular junction.[142] Miniature synaptic potentials or asynchronous release is reduced in each of these preparations by treatment with gabapentin or pregabalin. These results suggest that the actions of α_2–δ ligands to reduce neurotransmitter release may not always require inhibition of calcium influx. And so, drug actions might also be mediated by an interaction between α_2–δ and synaptic proteins in addition to calcium channel α subunits that are involved in the release or trafficking of synaptic vesicles.

8.18.4.7   Mechanisms of Action – Summary

The question remains whether these actions of pregabalin and gabapentin on α_2–δ proteins to reduce neurotransmitter release can fully account for all of the pharmacological actions of these drugs. Could other pharmacological sites of action contribute to the pharmacological actions of pregabalin? It is impossible to rule out contributions from other unknown drug targets. However, based on the great similarity in pharmacology in many different animal models between gabapentin, pregabalin, and several other α_2–δ ligand compounds that are earlier in development, it does not appear necessary to postulate additional sites of action to account for the pharmacology of this drug class. Furthermore, no other candidate molecular drug targets for these compounds have yet been identified. The potential drug targets of glutamic acid dehydrogenase,[136] branched chain amino acid aminotransferase,[137–139] GABA-transaminase,[140] and the system L transporter[19] have mostly been ruled out. This is because they are not affected similarly by gabapentin and pregabalin, and none of these sites show sufficient affinity for the known compounds to account for their clinically relevant drug actions. Conversely, for GABA transaminase and system L transport mechanisms, there are known compounds that act at these sites that do not share the pharmacology of gabapentin and pregabalin. Therefore, because of the lack of other potentially relevant molecular targets, and with known actions of both gabapentin and pregabalin ascribed to binding at the α_2–δ site, other potential sites of drug action remain speculative.

References

1. Satzinger, G. *Arznei. Forsch.* **1994**, *44*, 261–266.
2. Reimann, W. *Eur. J. Pharmacol.* **1983**, *94*, 341–344.
3. Bryans, J. S.; Wustrow, D. J. *Med. Res. Rev.* **1999**, *19*, 149–177.
4. Gee, N. S.; Brown, J. P.; Dissanayake, V. U.; Offord, J.; Thurlow, R.; Woodruff, G. N. *J. Biol. Chem.* **1996**, *271*, 5768–5776.
5. Suman-Chauhan, N.; Webdale, L.; Hill, D. R.; Woodruff, G. N. *Eur. J. Pharmacol. Mol. Pharmacol. Sect.* **1993**, *244*, 293–301.
6. Lim, J.; Stock, N.; Pracitto, R.; Boueres, J. K.; Munoz, B.; Chaudhary, A.; Santini, A. M.; Orr, K.; Schaffhauser, H.; Bezverkov, R. E. et al. *Bioorg. Med. Chem. Lett.* **2004**, *14*, 1913–1916.
7. Stearns, B. A.; Anker, N.; Arruda, J. M.; Campbell, B. T.; Chen, C.; Cramer, M.; Hu, T.; Jiang, X.; Park, K.; Ren, K. K. *Bioorg. Med. Chem. Lett.* **2004**, *14*, 1295–1298.
8. Hu, T.; Stearns, B. A.; Campbell, B. T.; Arruda, J. M.; Chen, C.; Aiyar, J.; Bezverkov, R. E.; Santini, A.; Schaffhauser, H.; Liu, W. et al. *Bioorg. Med. Chem. Lett.* **2004**, *14*, 2031–2034.
9. Bryans, J. S.; Davies, N.; Gee, N. S.; Horwell, D. C.; Kneen, C. O.; Morrell, A. I.; O'Neill, J. A.; Ratcliffe, G. S. *Bioorg. Med. Chem. Lett.* **1997**, *7*, 2481–2484.
10. Bryans, J. S.; Horwell, D. C.; McGuffog, J.; Morrell, A. I.; O'Niell, J. A.; Ratcliffe, G. S. *Med. Chem. Res.* **1998**, *8*, 153–162.
11. Bryans, J. S.; Horwell, D. C.; Ratcliffe, G. S.; Receveur, J.-M.; Rubin, J. R. *Bioorg. Med. Chem.* **1999**, *7*, 715–721.
12. Receveur, J.-M.; Bryans, J. S.; Field, M. J.; Singh, L.; Horwell, D. C. *Bioorg. Med. Chem. Lett.* **1999**, *9*, 2329–2334.

13. Bryans, J. S.; Davies, N.; Gee, N. S.; Dissanayake, V. U. K.; Ratcliffe, G. S.; Horwell, D. C.; Kneen, C. O.; Morrell, A. I.; Oles, R. J.; O'Toole, J. C. et al. *J. Med. Chem.* **1998**, *41*, 1838–1845.
14. Field, M. J.; Singh, L.; Hughes, J. *Br. J. Pharmacol.* **2000**, *131*, 282–286.
15. Urban, M. O.; Ren, K.; Park, K. T.; Campbell, B.; Anker, N.; Stearns, B.; Aiyar, J.; Belley, M.; Cohen, C.; Bristow, L. *J. Pharmacol. Exp. Ther.* **2005**, *313*, 1209–1216.
16. Patani, G. A.; LaVoie, E. J. *Chem. Rev.* **1996**, *96*, 3147–3176.
17. Burgos-Lepley, C. E.; Thompson, L. R.; Kneen, C. O.; Osborne, S. A.; Bryans, J. S.; Capiris, T.; Suman-Chauhan, N.; Dooley, D. J.; Donovan, C. M.; Field, M. J. et al. *Bioorg. Med. Chem. Lett.* **2006**, *16*, 2333–2336.
18. Herr, R. J. *Bioorg. Med. Chem.* **2002**, *10*, 3379–3393.
19. Su, T. Z.; Feng, M. R.; Weber, M. L. J. *Pharmacol. Exp. Ther.* **2005**, *313*, 1406–1415.
20. Andruszkiewicz, R.; Silverman, R. B. *Synthesis* **1989**, *12*, 953–955.
21. Andruszkiewicz, R.; Silverman, R. B. *J. Biol. Chem.* **1990**, *265*, 22288–22291.
22. Silverman, R. B.; Andruszkiewicz, R.; Nanavati, S. M.; Taylor, C. P.; Vartanian, M. G. *J. Med. Chem.* **1991**, *34*, 2295–2298.
23. Taylor, C. P.; Vartanian, M. G.; Andruszkiewicz, R.; Silverman, R. B. *Epilepsy Res.* **1992**, *11*, 103–110.
24. Yuen, P. W.; Kanter, G. D.; Taylor, C. P.; Vartanian, M. G. *Bioorg. Med. Chem. Lett.* **2003**, *4*, 823–826.
25. Taylor, C. P.; Vartanian, M. G.; Yuen, P. W.; Bigge, C.; Suman, C. N.; Hill, D. R. *Epilepsy Res.* **1993**, *14*, 11–15.
26. Belliotti, T.; Ekhato, I. V.; Capiris, T.; Kinsora, J.; Vartanian, M. G.; Field, M.; Meltzer, L. T.; Heffner, T.; Schwarz, J. B.; Taylor, C. P. et al. *J. Med. Chem.* **2005**, *48*, 2294–2307.
27. Schwarz, J. B.; Gibbons, S. E.; Graham, S. R.; Colbry, N. L.; Guzzo, P. R.; Le, Van-Duc; Vartanian, M. G.; Kinsora, J. J.; Lotarski, S. M.; Li, Z. et al. *J. Med. Chem.* **2005**, *48*, 3026–3035.
28. Schelkun, R. M.; Yuen, P.-W.; Wustrow, D. J.; Kinsora, J.; Su, T.-Z.; Vartanian, M. G. *Bioorg. Med. Chem. Lett.* (in press).
29. Vollmer, K. O.; vonHodenberg, A.; Koelle, E. U. *Arzneim. Forsch. Drug Res.* **1986**, *36*, 830–839.
30. Trudeau, V.; Myers, S.; Lamoreaux, L.; Anhut, H.; Garofalo, E.; Ebersole, J. *J. Child Neurol.* **1996**, *11*, 470–475.
31. Burgey, G. K.; Morris, H. H.; Rosenfeld, W.; Blume, W. T.; Penovich, P. E.; Morrell, M. J.; Liederman, D. B.; Crockatt, J. G.; Lamoreaux, L.; Garofalo, E. et al. *Neurology* **1997**, *49*, 739–745.
32. Andrews, J.; Chadwick, D.; Bates, D. *Lancet* **1990**, *335*, 1114–1117.
33. Chadwick, D. Gabapentin: Clinical Use. In *Antiepileptic Drugs*, 4th ed.; Levy, R. H., Mattson, R. H., Meldrum, B. S., Eds.; Raven Press: New York, 1995, pp 851–856.
34. McLean, M. J.; Ramsey, R. E.; Leppik, I.; Rowan, A. J.; Shellenberger, M. K.; Wallace, J.; Group, U. G. S. *Neurology* **1993**, *43*, 2292–2298.
35. Ramsay, R. E. *Neurology* **1994**, *44*, S23–S30, discussion S31–S32.
36. Beydoun, A.; Fischer, J.; Labar, D. R.; Harden, C.; Cantrell, D.; Uthman, B. M.; Sackellares, J. C.; Abou-Khalil, B.; Ramsay, R. E.; Hayes, A. et al. *Neurology* **1997**, *49*, 746–752.
37. Singh, L.; Field, M. J.; Ferris, P.; Hunter, J. C.; Oles, R. J.; Williams, R. G.; Woodruff, G. N. *Psychopharmacology* **1996**, *127*, 1–10.
38. Field, M. J.; Holloman, E. F.; McCleary, S.; Hughes, J.; Singh, L. J. *Pharmacol. Exp. Ther.* **1997**, *282*, 1242–1246.
39. Xiao, W.-H.; Bennett, G. J. *Analgesia* **1996**, *2*, 267–273.
40. Mellick, G. A.; Mellicy, L. B.; Mellick, L. B. *J. Pain Symptom Manage.* **1995**, *10*, 265–266.
41. Mellick, G. A.; Mellick, L. B. *Arch. Phys. Med. Rehabil.* **1997**, *78*, 98–105.
42. Hwang, J. H.; Yaksh, T. L. *Reg. Anesth.* **1997**, *22*, 249–256.
43. Shimoyama, N.; Shimoyama, M.; Davis, A. M.; Inturrisi, C. E.; Elliott, K. J. *Neurosci. Lett.* **1997**, *222*, 65–67.
44. Hunter, J. C.; Gogas, K. R.; Hedley, L. R.; Jacobson, L. O.; Kassotakis, L.; Thompson, J.; Fontana, D. J. *Eur. J. Pharmacol.* **1997**, *324*, 153–160.
45. Stanfa, L. C.; Singh, L.; Williams, R. G.; Dickenson, A. H. *Neuroreport* **1997**, *8*, 587–590.
46. Feng, Y.; Cui, M.; Willis, W. D. *Anesthesiology* **2003**, *98*, 729–733.
47. Jones, D. L.; Sorkin, L. S. *Brain Res.* **1998**, *810*, 93–99.
48. Chapman, V.; Suzuki, R.; Chamarette, H. L.; Rygh, L. J.; Dickenson, A. H. *Pain* **1998**, *75*, 261–272.
49. Abdi, S.; Lee, D. H.; Chung, J. M. *Anesth. Analg.* **1998**, *87*, 1360–1366.
50. Gillin, S.; Sorkin, L. S. *Anesth. Analg.* **1998**, *86*, 111–116.
51. Hulsebosch, C. E.; Xu, G.-Y.; Perez-Polo, J. R.; Westlund, K. N.; Taylor, C. P.; McAdoo, D. J. *J. Neurotrauma* **2000**, *17*, 1205–1217.
52. Field, M. J.; McCleary, S.; Hughes, J.; Singh, L. *Pain* **1999**, *80*, 391–398.
53. Yoon, M. H.; Yaksh, T. L. *Anesth. Analg.* **1999**, *89*, 434–439.
54. Lu, Y.; Westlund, K. N. *J. Pharmacol. Exp. Ther.* **1999**, *290*, 214–219.
55. Cesena, R. M.; Calcutt, N. A. *Neurosci. Lett.* **1999**, *262*, 101–104.
56. Dixit, R.; Bhargava, V. K.; Kaur, N. *Methods Find. Exp. Clin. Pharmacol.* **1999**, *21*, 481–482.
57. Hulsebosch, C. E.; Xu, G.-Y.; Perez-Polo, J. R.; Westlund, K. N.; Taylor, C. P.; McAdoo, D. J. *J. Neurotrauma* **2000**, *17*, 1205–1217.
58. Cheng, J. K.; Pan, H. L.; Eisenach, J. C. *Anesthesiology* **2000**, *92*, 1126–1131.
59. Kaneko, M.; Mestre, C.; Sanchez, E. H.; Hammond, D. L. *J. Pharmacol. Exp. Ther.* **2000**, *292*, 743–751.
60. Kayser, V.; Christensen, D. *Pain* **2000**, *88*, 53–60.
61. Takahashi, I.; Andoh, T.; Nojima, H.; Shiraki, K.; Kkuraishi, Y. *J. Pharmacol. Exp. Ther.* **2001**, *296*, 270–275.
62. Rosner, H.; Rubin, L.; Kestenbaum, A. *Clin. J. Pain* **1996**, *12*, 56–58.
63. Portenoy, R. K. Management of Cancer Pain: Opioid and Adjuvant Pharmacotherapy. In *Real Patients, Real Problems: Optimal Assessment and Management of Cancer Pain*; Portenoy, R. K., Ed.; American Pain Society: Glenview, Illinois, USA, 1997, pp 5–15.
64. Serpell, M. G. *Pain* **2002**, *99*, 557–566.
65. Mueller, M. E.; Gruenthal, M.; Olson, W. L.; Olson, W. H. *Arch. Phys. Med. Rehabil.* **1997**, *78*, 521–524.
66. Backonja, M.; Beydoun, A.; Edwards, K. R.; Schwartz, S. L.; Fonseca, V.; Hes, M. S.; Lamoreaux, L.; Garofalo, E. *JAMA* **1998**, *280*, 1831–1836.
67. Rowbotham, M. C.; Harden, N.; Stacey, B.; Bernstein, P.; Magnus-Miller, L. *JAMA* **1998**, *280*, 1837–1842.
68. Schlicker, E.; Reimann, W.; Gothert, M. *Arzneim. Forsch. Drug Res.* **1985**, *35*, 1347–1349.
69. Thurlow, R. J.; Brown, J. P.; Gee, N. S.; Hill, D. R.; Woodruff, G. N. *Eur. J. Pharmacol. Mol. Pharmacol. Sect.* **1993**, *247*, 341–345.
70. Su, T. Z.; Lunney, E.; Campbell, G.; Oxender, D. L. *J. Neurochem.* **1995**, *64*, 2125–2131.
71. Hill, D. R.; Suman, C. N.; Woodruff, G. N. *Eur. J. Pharmacol.* **1993**, *244*, 303–309.
72. Dooley, D. J.; Suman, C. N.; Madden, Z. *Soc. Neurosci. Abstr.* **1996**, *22*, 1992–1995.

73. Dooley, D. J.; Donovan, C. M.; Pugsley, T. A. *J. Pharmacol. Exp. Ther.* **2000**, *296*, 1086–1098.
74. Dooley, D. J.; Mieske, C. A.; Borosky, S. A. *Neurosci. Lett.* **2000**, *280*, 107–110.
75. Fink, K.; Meder, W.; Dooley, D. J.; Gothert, M. *Br. J. Pharmacol.* **2000**, *130*, 900–906.
76. Meder, W. P.; Dooley, D. *J. Brain Res.* **2000**, *875*, 157–159.
77. Dooley, D. J.; Donovan, C. M.; Meder, W. P.; Whetzel, S. Z. *Synapse* **2002**, *45*, 171–190.
78. Fink, K.; Dooley, D. J.; Meder, W. P.; Suman-Chauhan, N.; Duffy, S.; Clusmann, H.; Gothert, M. *Neuropharmacology* **2002**, *42*, 229–236.
79. Maneuf, Y. P.; Hughes, J.; McKnight, A. T. *Pain* **2001**, *93*, 191–196.
80. Maneuf, Y. P.; McKnight, A. T. *Br. J. Pharmacol.* **2001**, *134*, 237–240.
81. Patel, M. K.; Gonzalez, M. I.; Bramwell, S.; Pinnock, nR. D.; Lee, K. *Br. J. Pharmacol.* **2000**, *130*, 1731–1734.
82. Sutton, K. G.; Scott, R. H.; Lee, K.; Pinnock, R. D. *Soc. Neurosci. Abstr.* **2000**, *26*, Abstract 234.4.
83. Sutton, K. G.; Martin, D.; Pinnock, R. D.; Lee, K.; Scott, R. H. *Br. J. Pharmacol.* **2001**, *135*, 257–265.
84. Sutton, K. G.; Snutch, T. P. *Drug Dev. Res.* **2002**, *54*, 167–172.
85. Martin, D. J.; McClelland, D.; Herd, M. B.; Sutton, K. G.; Hall, M. D.; Lee, K.; Pinnock, R. D.; Scott, R. H. *Neuropharmacology* **2002**, *42*, 353–366.
86. Petroff, O. A.; Rothman, D. L.; Behar, K. L.; Lamoureux, D.; Mattson, R. H. *Ann. Neurol.* **1996**, *39*, 95–99.
87. Mattson, R. H.; Rothman, D. L.; Behar, K. L.; Petroff, O. A. C. *Epilepsia* **1997**, *38*, 65–66.
88. Petroff, O. A.; Hyder, F.; Rothman, D. L.; Collins, T. L.; Mattson, R. H. *Epilepsia* **1998**, *39*, 71.
89. Petroff, O. A.; Rothman, D. L. *Mol. Neurobiol.* **1998**, *16*, 97–121.
90. Petroff, O. A.; Hyder, F.; Rothman, D. L.; Mattson, R. H. *Epilepsia* **2000**, *41*, 675–680.
91. Errante, L. D.; Williamson, A.; Spencer, D. D.; Petroff, O. A. *Epilepsy Res.* **2002**, *49*, 203–210.
92. Kocsis, J. D.; Honmou, O. *Neurosci. Lett.* **1994**, *169*, 181–184.
93. Honmou, O.; Oyelese, A. A.; Kocsis, J. D. *Brain Res.* **1995**, *692*, 273–277.
94. Honmou, O.; Kocsis, J. D.; Richerson, G. B. *Epilepsy Res.* **1995**, *20*, 193–202.
95. Richerson, G. B.; Wu, Y. *Adv. Exp. Med. Biol.* **2004**, *548*, 76–91.
96. Whitworth, T. L.; Quick, M. W. *Biochem. Soc. Transact.* **2001**, *29*, 736–741.
97. Petroff, O. A.; Hyder, F.; Collins, T.; Mattson, R. H.; Rothman, D. L. *Epilepsia* **1999**, *40*, 958–964.
98. Kuzniecky, R. I.; Hetherington, H. P.; Ho, S.; Pan, J. W.; Martin, R.; Gilliam, F.; Hugg, J.; Faught, E. *Neurology* **1998**, *51*, 627–629.
99. Kuzniecky, R. I.; Ho, S.; Pan, J. W.; Martin, R.; Gilliam, F.; Faught, E.; Hetherington, H. P. *Neurology* **2002**, *58*, 368–372.
100. Errante, L. D.; Petroff, O. A. C. *Seizure* **2003**, *12*, 300–306.
101. Stringer, J. L.; Taylor, C. P. *Epilepsy Res.* **2000**, *41*, 155–162.
102. Ng, G. Y. K.; Bertrand, S.; Sullivan, R.; Ethier, N.; Wang, J.; Yergey, J.; Belley, M.; Trimble, L.; Bateman, K.; Alder, L. et al. *Mol. Pharmacol.* **2001**, *59*, 144–152.
103. Bertrand, S.; Ng, G. Y. K.; Purisai, M. G.; Wolfe, S. E.; Severidt, M. W.; Nouel, D.; Robitaille, R.; Low, M. J.; O'Neill, G. P.; Metters, K. et al. *J. Pharmacol. Exp. Ther.* **2001**, *298*, 15–24.
104. Bertrand, S.; Nouel, D.; Morin, F.; Nagy, F.; Lacaille, J. C. *Synapse* **2003**, *50*, 95–109.
105. Lanneau, C.; Green, A.; Hirst, W. D.; Wise, A.; Brown, J. T.; Donnier, E.; Charles, K.; Wood, M.; Davies, C. H.; Pangalos, M. N. *Neuropharmacology* **2001**, *41*, 965–975.
106. Jensen, A. A.; Mosbacher, J.; Elg, S.; Lingenhoehl, K.; Lohmann, T.; Johansen, T. N.; Abrahamsen, B.; Mattsson, J. P.; Lehmann, A.; Bettler, B. et al. *Mol. Pharmacol.* **2002**, *61*, 1377–1384.
107. Vartanian, M. G.; Donevan, S. D.; Weber, M. L.; Stoehr, S. J.; Dooley, D. J.; Taylor, C. P.; Donevan, S. D. *Soc. Neurosci. Abstr.* **2002**, 603.3.31, 603.3.
108. Shimizu, S.; Honda, M.; Tanabe, M.; Ono, H. *J. Pharmacol. Sci.* **2004**, *96*, 444–449.
109. Cheng, J.-K.; Lee, S. Z.; Yang, J.-R.; Wang, C.-H.; Liao, Y.-Y.; Chen, C. C.; Chiou, L.-C. *J. Biomed. Sci.* **2004**, *3*, 346–355.
110. Dang, K.; Bowery, N.; Sugiyura, T.; Urban, L. *Am. Pain Soc. Abstr. Poster* **2003**, poster 782.
111. Brown, J. P.; Gee, N. S. *J. Biol. Chem.* **1998**, *273*, 25458–25465.
112. Wang, M.; Offord, J.; Oxender, D. L.; Su, T.-Z. *Biochem. J.* **1999**, *342*, 313–320.
113. Barclay, J.; Balaguero, N.; Mione, M.; Ackerman, S. L.; Letts, V. A.; Brodbeck, J.; Canti, C.; Meir, A.; Page, K. M.; Kusumi, K. et al. *J. Neurosci.* **2001**, *21*, 6095–6104.
114. Brill, J.; Klocke, R.; Paul, D.; Boison, D.; Gouder, N.; Klugbauer, N.; Hofmann, F.; Becker, C.-M.; Becker, K. *J. Biol. Chem.* **2004**, *279*, 7322–7330.
115. Ivanov, S. V.; Ward, J. M.; Tessarollo, L.; McAreavey, D.; Sachdev, V.; Fananapazir, L.; Banks, M. K.; Morris, N.; Djurickovic, D.; Devor-Henneman, D. E. et al. *Am. J. Pathol.* **2004**, *165*, 1007–1018.
116. Li, Z.; Piechan, J.; Schwarz, R. D.; Taylor, C. P.; Wong, E. *Soc. Neurosci. Abstr.* **2003**, *33*, 791.2.
117. Bian, F.; Li, Z.; Offord, J. D.; Davis, M. D.; McCormick, J. A.; Taylor, C. P.; Walker, L. C. *Brain Res.* **2006**, *1075*, 68–80.
118. Bramwell, S.; Cox, P. J.; Ackley, M.; Offord, J.; Stott, E.; Wain, L.; Su, T.-Z.; Williams, D. C.; Field, M. *J. Soc. Neurosci. Abstr.* **2004**, *523*, 19.
119. Taylor, C. P. *CNS Drug Rev.* **2004**, *10*, 159–164.
120. Piechan, J. L.; Donevan, S. D.; Taylor, C. P.; Dickerson, M. R.; Li, Z. *Soc. Neurosci. Abstr.* **2004**, *115*, 11.
121. Brown, J. P.; Dissanayake, V. U.; Briggs, A. R.; Milic, M. R.; Gee, N. S. *Anal. Biochem.* **1998**, *255*, 236–243.
122. Stefani, A.; Spadoni, F.; Giacomini, P.; Lavaroni, F.; Bernardi, G. *Epilepsy Res.* **2001**, *43*, 239–248.
123. Alden, K. J.; Garcia, J. *J. Pharmacol. Exp. Ther.* **2001**, *297*, 727–735.
124. Stefani, A.; Spadoni, F.; Bernardi, G. *Neuropharmacology* **1998**, *37*, 83–91.
125. Alden, K. J.; Garcia, J. *Am. J. Physiol.-Cell Physiol.* **2002**, *283*, C941–C949.
126. Schumacher, T. B.; Beck, H.; Steinhauser, C.; Schramm, J.; Elger, C. E. *Epilepsia* **1997**, *39*, 355–363.
127. van Hooft, J. A.; Dougherty, J. J.; Endeman, D.; Nichols, R. A.; Wadman, W. J. *Eur. J. Pharmacol.* **2002**, *449*, 221–228.
128. Canti, C.; Davies, A.; Dolphin, A. C. *Curr. Neuropharmacol.* **2004**, *1*, 209–217.
129. Shimoyama, M.; Shimoyama, N.; Hori, Y. *Pain* **2000**, *85*, 405–414.
130. Bayer, K.; Seifollah, A.; Zeilhofer, H. U. *Neuropharmacology* **2004**, *46*, 743–749.
131. Kumar, N.; Coderre, T. J. *J. Pain* **2004**, *5*, S27–S28.
132. Brown, J. T.; Randall, A. D. *Synapse* **2005**, *55*, 262–269.
133. Coderre, T. J.; Kumar, N.; Lefebvre, C. D.; Yu, J. S. C. *J. Neurochem.* **2005**, *94*, 1131–1139.

Abraham descriptors
  computational oral absorption
        prediction 5:685–686
  molecular descriptors 5:544
Abraxane, as fast-tracked drug
        1:760*T*
ABS-103 6:291
Absence (petit-mal) seizures 6:283
  calcium channels 2:852
Absolute average error (AAE), solubility
        models 5:631
Absolute configuration determination,
        chiral drug discovery 2:720
Absolute oral bioavailability (*F*)
        5:700–701
AbSolv 5:679*T*
  aqueous solubility models 5:644*T*
Absorption 1:52
  affecting factors 5:532, 5:532*F*,
        5:702*T*
  bioavailability, effects on 5:702
  bioavailability *vs. see* Bioavailability
  Biopharmaceuticals Classification
        System 5:702, 5:703*T*
  blood–brain barrier 1:52
  carriers/transporters 5:703
  cell-based assays 3:628
  dielectric heating 3:840
  dissolution 5:702
  dose 5:703
  flexibility 5:704
  formulation effects 5:703
  hydrogen bonding 5:704
  intraspecies differences 5:40
    bioavailability 5:40, 5:41*F*
    fraction absorbed 5:41, 5:42*F*
    gastrointestinal tract length
        5:40–41
    gastrointestinal tract
        physiology 5:40
    humans *vs.* dogs 5:41, 5:42
    humans *vs.* monkey 5:41–42,
        5:42*F*
    humans *vs.* rats 5:40–41, 5:42*F*,
        5:43*F*
    intestinal tract secretions 5:41
    lipophilic drug solubility 5:41
    physicochemical factors 5:40
  ionizable drugs 5:383
  limitation, BCS *see* Biopharmaceutical
        Classification Scheme
        (BCS)
  lipophilicity 5:703
    measurement 5:703–704
  model comparisons, VolSurf
        descriptors *see* VolSurf
  nonoral routes *see* Nonoral absorption
  parameter (π), in vitro oral absorption
        prediction *see* Oral absorption
        prediction, in vitro
  permeability 5:702, 5:703
  pharmacokinetics *see*
        Pharmacokinetics

polarity 5:704
  amphiphilicity 5:704
  hydrophilic–lipophilic balance
        (HLB) 5:704
  polar surface area 5:704
prediction
  in vitro oral prediction 5:676
  pH stability testing 5:496
prediction in drug discovery 5:260
quantification, pulmonary
        administration *see* Pulmonary
        administration
quantum mechanical
        calculations 4:398
salt form effects 5:519*F*, 5:523
size/shape 5:704
solubility 5:703
  affecting factors 5:703
  formulations 5:703
structure–absorption relationships
        (SABRs) 1:52
Absorption/emission- based optical
        assays 3:578
Absorption enhancers
  nasal administration 5:285*F*, 5:286
  oral absorption 5:672
Absorption rate constants (*k*a)
  inhibition, in vitro studies 5:248
  metabolic drug–drug
        interactions 5:835–836
  oral absorption prediction *see* Oral
        absorption prediction,
        physiologically based models
ABT-229, as prokinetic agent 6:681
ABT-239 6:205–206
  narcolepsy treatment 6:156–157
ABT-255 7:589
ABT-378, structure 2:302*F*
ABT-418
  Alzheimer's disease treatment 6:205
  structure 6:181*F*
ABT-431, Parkinson's disease
        treatment 6:215
ABT-594 6:340
  discovery/development 1:380
  structure 1:379*F*; 6:341*F*
  quantum mechanical
        calculations 4:391, 4:391*F*
ABT-702, structure 6:344*F*
ABT-719 7:584–586, 7:589
ABT-751 7:95–96
ABT-769 6:291
  structure 6:292*F*
ABT-874, IBD treatment 6:623–624
Abundant protein depletion,
        proteomics 3:31–32
Academia
  collaborations
    with pharmaceutical companies *see*
        Collaborations, industry/academia
    tamoxifen discovery 8:88
  corporate conflicts of interest 1:711
  SMEs 1:491, 1:521*T*

Acamprosate
  alcohol addiction 6:180*T*, 6:183
  structure 6:181*F*
Acarbose 6:439*T*, 6:445
  advantages 6:446
  clinical use 6:446
  disadvantages 6:446
  mechanisms of action 6:445
  oral absorption 5:680*T*
  pharmacokinetics 6:445
  side effects/contraindications 6:446
  structure 6:440*F*
Acardi *see* Pimobendan
ACAT inhibitors 6:487, 6:488*F*
Accelerated mass spectrometry (AMS),
        metabolism studies 5:34
Accelrys database 1:517; 3:307
  BIOSTER 3:307
  Biotransformations 3:307
  Catalyst 3:295
  Kinase ChemBioBase 3:307
  Metabolism 3:307
Acceptable criteria, statistical
        methods 5:563
Acceptable risk, postmarketing
        surveillance 1:764
Acceptor–donor distances, fluorescence
        resonance energy
        transfer 3:605
Accessibility, drug metabolism 5:810
Accession number, UniProtKB
        flatfile 3:360
'Accessory proteins'
  mitochondria complex I 2:1070–1071
  nicotinic acetylcholine
        receptors 2:886
Accolate *see* Zafirlukast
Accumulation risk, inhibition, in vitro
        studies 5:248
Accuracy
  aqueous solubility models *see*
        Solubility models, aqueous
  comparative computational
        modeling 3:557
  fluorescence screening assays
        3:600–601
  homology-based protein structure
        prediction 3:326
  ionization prediction, in silico tools *see*
        Ionization prediction, in silico
        tools
  NMR structures, problems 3:542
  physiologically based
        pharmacokinetics 5:877
ACD
  aqueous solubility models 5:644*T*
  log D calculation program 5:662*T*
  log P calculation program 5:661*T*
  oral absorption prediction, molecular
        descriptors 5:683
  p*K*a, ionization prediction 5:616*T*,
        5:620
  p*K*aDB, ionization prediction 5:623*T*

Acebutol
  bioavailability 5:712T
  oral absorption 5:468T, 5:680T
Acecol see Temocapril hydrochloride
ACE inhibitors
  see also specific drugs
  active metabolites 6:721
  active-site modeling 4:17
  aldosterone receptor blocker
      combination 6:590
  analog design 2:651
  bioavailability 6:719T
  clinical use 1:594–595, 1:606–607
    age-related macular degeneration
        treatment 6:315T
    congestive heart failure
        treatment 6:698
    hypertension treatment 1:532T,
        6:585, 6:590, 6:699–700, 6:718
    renal dysfunction 6:585
  discovery/development 1:16
    protein crystallography, rational
        drug design 3:876, 3:877F
    second generation 1:17
  dose 6:719T
  half-life 6:719T
  in vitro potency 6:719T, 6:721
  mechanism of action 6:588, 6:590,
      6:721
  me-too prescribing 1:713–714
  peptidomimetics 2:613, 2:614F
  pharmacokinetics 6:721
  side effects 6:727
  structure 6:587F, 6:719T
Acemannan 1:236
Acetamide, structure 5:359
5-Acetamidomethyl derivatives,
    oxazolidinones see
    Oxazolidinones
Acetaminophen
  ADME 5:912
  bioavailability 5:712T
  glucuronic acid conjugation 5:149F,
      5:150
  glutathione conjugation 5:160,
      5:160F
  migraine treatment 6:373–374,
      6:373T
  oral absorption 5:468T, 5:680T
  structure 2:9F, 5:370F, 6:373–374
  sulfation 5:148, 5:149F
  toxicity 5:911, 5:912F
    glutathione lack 5:911–912
[11]C Acetate, as metabolic probe 3:665
Acetazolamide
  central sleep apnea
      treatment 6:162–163
  development 1:530
  historical perspective 6:287
Acetic acid
  microwave heating, loss
      tangent 3:840T
  structure 5:369F

Acetixime, structure 5:371F
Acetohexamide 6:438, 6:442T
  advantages 6:443
  clinical use 6:443
  disadvantages 6:443
  mechanisms of action 6:440
  pharmacokinetics 6:440
  reduction 5:127
  side effects/contraindications 6:443
  structure 6:440F
Acetone, microwave heating 3:840T
Acetonitrile, microwave
    heating 3:840T
Acetylation see Conjugation, drug
    metabolism
N-Acetylation, solid support
    synthesis 2:466F, 2:467,
    2:467F
O-Acetylation, solid support
    synthesis 2:466F, 2:467, 2:468F
2-Acetylchlortetracycline
  biosynthesis 7:602
  structure 7:601F
Acetylcholine
  gastrointestinal tract 6:645–646
  receptors, molecular probes 3:664,
      3:664T
  sleep 6:146
Acetylcholine M receptor
    blockers 6:748
  see also specific drugs
  effectiveness 6:748
  mechanism 6:748
  toxicity 6:748
Acetylcholinesterase
  Alzheimer's disease 6:202
  drug metabolism 5:203
  nicotinic acetylcholine receptor
      binding 2:885
Acetylcholinesterase inhibitors
  Alzheimer's disease
      treatment 6:203–204
  autoimmune neuromuscular
      disorders 6:12
  dynamic combinatorial
      libraries 3:970–971, 3:972F,
      3:973
  myasthenia gravis treatment 6:12,
      6:249
  as 'old' target 2:274–275
  ring equivalent bioisosterism 2:656,
      2:657F
Acetylcholinesterase-releasing
    modulators, Alzheimer's disease
    treatment 6:204
Acetyl-CoA, tetracyclines
    biosynthesis 7:600–601
Acetyl-CoA carboxylase inhibitors,
    antifungal drugs see Antifungal
    drugs
N-Acetyl-L-glutamate inhibitors,
    schizophrenia treatment
    6:38–39

N-Acetylmuramic acid analogs, solid-
    phase pool libraries 3:705F
3-Acetylpyridine
  ionization constant
      calculation 5:379F, 5:394
  structure 5:360
Acetylsalicylic acid (aspirin) 6:338
  bioavailability 5:712T
  discovery/development 1:322–323,
      1:530
  gastric damage
      phosphodiesterase-5
          inhibitors 2:949
      rolipram 2:948
  heart disease 1:323
  hydrolysis 5:491, 5:491F
  mechanism of action 1:323
  metabolic syndrome
      management 6:385
  migraine treatment 6:373–374,
      6:373T
  molecular representation 3:236,
      3:237F
  oral absorption 5:468T, 5:680T
  structure 2:9F, 5:369F, 6:336F,
      6:373–374, 6:776F
  thrombolytic drugs 6:775
  use 1:323
N-Acetyltransferase 1 (NAT-1), drug
    metabolism 5:208
N-Acetyltransferase 2 (NAT-2)
  drug metabolism 5:208
  polymorphisms 5:208
Acetyltransferases
  aminoglycoside resistance 7:642,
      7:643
  drug metabolism 5:208
N-Acetyltransferases (NAT) 5:184
  characteristics 5:169T
  deficiency, adverse drug
      effects 1:282T
  genes 5:184
  human liver cytosolic fraction 5:234
  mechanism of action 5:184
  molecular modeling 5:820
  species distribution 5:184
Achromycin see Tetracyclines
Aciclovir see Acyclovir
Acid–base reaction equilibria, ionization
    see Ionization
Acid–base titration method,
    solubility 5:401
α_1-Acid glycoprotein (AGP) 5:849
  binding prediction models 5:854,
      5:859
    model associated with Hervé
        5:859, 5:859F
  biological function 5:849
  composition 5:849
  drug binding 5:849
  pharmacokinetics 5:14
  plasma protein binding 5:322,
      5:325–326

Acid hydrazone exchange, dynamic combinatorial libraries 3:970–971
Acidic compounds, PAMPA 5:471
Acid microclimate, intestinal absorption 5:262
Acid prodrugs *see* Prodrugs
Acids
  definition 5:358, 5:360
  weak *see* Weak acids
Acid-sensitive ion channels (ASICs) 2:879, 2:910; 6:321, 6:322, 6:345
  antagonists 6:322, 6:345
  characteristics 2:881*T*
  diseases 2:912
    pain perception 2:911
  expression 2:879
  molecular biology 2:910
  nomenclature 2:910
  pharmacology 2:911
    neuropeptides 2:911
    toxins 2:911
  physiological functions 2:911
    neuronal distribution 2:911
    pain 2:911
    retinal activity 2:911
  regulation 2:911
  structure 2:880*F*, 2:910
  therapeutics 2:911
    ischemia 2:911
Acitretin, psoriasis treatment 7:978*T*
ACLARA Biosciences, microfluidic devices 2:350*F*, 2:359
Acne 7:935, 7:957
  see also Acne rosacea
  androgens 7:936*T*, 7:937, 7:958
    dehydroepiandrosterone sulfate 7:937
  classification 7:957–958
  follicular hyperkeratinization 7:936, 7:936*T*, 7:958, 7:959*F*
  inflammation 7:960
  neuroendocrine factors 7:937
    corticotropin-releasing hormone 7:937
    melanocortins 7:937
    α-melanocyte-stimulating factor 7:937
    peroxisome proliferating receptors 7:937
    substance P 7:937
  pathogenesis 7:936, 7:936*T*
  prevalence 7:935
  Propionibacterium acnes 7:936, 7:936*T*, 7:937, 7:958, 7:959, 7:959*F*
    inflammation 7:938
  sebum production 7:936, 7:936*T*, 7:958, 7:958*F*
  severity definition 7:960
  treatment *see* Acne treatment

Acne rosacea 7:957, 7:960
  antibacterial agents
    benzoyl peroxide 7:961
    doxycycline 7:963
    erythromycin 7:963
    metronidazole 7:961, 7:963
    minocycline 7:963
    sodium sulfacetamide 7:961
    tetracycline 7:963
  causes 7:961
  Helicobacter pylori 7:961
  intensity 7:960–961
  treatment 7:961
Acne treatment 7:938, 7:938*F*, 7:957–968
  antibacterial agents 7:938, 7:961
    azelaic acid (AZA) 7:939, 7:966
    azithromycin 7:962–963
    benzoyl peroxide 7:938–939, 7:961
    clindamycin 7:938–939, 7:961, 7:962–963
    dapsone 7:961
    erythromycin 7:961, 7:962–963
    macrolide antibiotics 7:939
    minocycline 7:611–612
    nadifloxacin 7:961
    oxytetracycline 7:611–612
    sodium sulfacetamide 7:961
    sulfamethoxazole 7:962–963
    tetracyclines 7:611–612, 7:939
    topical antibiotics 7:961
    trimethoprim 7:962–963
  combinations 7:966
  future treatments 7:966
    azithromycin 7:966
    clindamycin 7:966
    doxycycline 7:966
    PSK-3841 7:966
    XMP.629 7:966
  glycolic acid 7:966
  hormonal agents 7:939
    finasteride 7:940
    oral contraceptives 7:939–940
  lactic acid 7:966
  light therapy 7:940, 7:940*T*, 7:966
    laser therapy 7:941
    photodynamic therapy 7:940
    Propionibacterium acnes 7:940
  pathogenesis targets 7:960, 7:960*F*
  radiofrequency treatment 7:941
  retinoids 7:963
    isotretinoin 7:963
    oral 7:939
    tazarotene 7:964
    topical 7:939
    tretinoin 7:963
  sales 7:967
  salicylic acid 7:966
  steroids 7:965
    desogestrel 7:965
    epristeride 7:965

    finasteride 7:965
    spironolactone 7:965
    systemic treatments 7:962
Acne vulgaris *see* Acne
Acquired long QT syndrome (LQTS) 5:935
Acquisitions, pharmaceutical companies 2:101
Acridines 7:118
  see also specific drugs
Acromegaly
  diabetes mellitus 6:423
  drug introductions 1:108, 1:231
Acrylamides, PASP 3:792–793
ACSLXtreme, metabolic drug–drug interaction software 5:841
Act 496 automatic synthesizer, PASP 3:806
Actaea racemosa (black cohosh) 1:424
Actarit 1:212
ACTH receptors, in human disease 2:775*T*
Actimmune 1:499
Actinomadura, tetracyclines synthesis 7:600
Actinomadura brunnea, Sch 33256 biosynthesis 7:602
Actinomycetes, tetracycline biosynthesis 7:601
Action Alliance 1:648
Action potentials
  cardiac ion channels 6:734, 6:734*F*
  potassium channels 6:734*F*
  sodium channels 6:734*F*
Activating protein 1 (AP-1)
  cDNA functional screening 3:15–16
  GPCR functional assay screening 2:812
  node network theory 3:4
Active amines, amides of 5:1026
Active analog approach, pharmacophore detection 4:127*T*, 4:128, 4:128*F*, 4:129*T*, 4:133
Active-analog drug design *see* Computer-assisted drug design (CADD)
Active efflux systems, β-lactam antibiotic resistance 7:487
Active ingredients, traditional medicine 1:420
Active ligand copying, de novo drug design 4:284
Active molecules, similarity measures, comparison of 4:179–180
Active oxygen species, flavin monooxygenases (FMO) 5:112
Active permeation, membrane transport proteins 5:52, 5:53*F*
Active secretion (CL$_{rs}$), plasma protein binding 5:323–324
Active site, phosphodiesterases 4:758, 4:758*F*
Active-Site Mapping 4:17

Active-site modeling *see* Computer-assisted drug design (CADD)

Active targeting, drug–polymer conjugates **5**:1046

Active transfer, pharmacokinetics **5**:13

Active transport
  biliary clearance **5**:47
  cell-based assays **3**:627
    Caco-2 cell permeability assays **5**:265–266
  GRIND descriptors **5**:541
  intestinal absorption *see* Intestinal absorption
  intestinal membrane **5**:458
  membrane transport proteins **5**:52–54, **5**:53*F*
  peptide/protein drugs **2**:591
  prediction, permeability assays *see* Permeability assays
  renal clearance **5**:46

Active transporters, blood–placenta transfer modeling **5**:737

Active tubular secretion, renal excretion **5**:16

Activity, definition **5**:358

Activity-based protein profiling, proteomics **3**:44–45

Activity cluster analysis, sourcing development **2**:216, **2**:216*F*, **2**:217*F*

Activity determination *see* Sourcing development

Activity interdependency, sourcing development **2**:214

Activity modeling, biological fingerprints *see* Biological fingerprints

Activix **1**:514

Actonel *see* Risedronate sodium

Actos *see* Pioglitazone hydrochloride

ACT-UP **1**:719
  drug discovery **2**:270–271

Acumen Explorer **3**:686*T*, **3**:687

Acute depression, treatment **6**:71

Acute infections, parasitic infections **7**:750

Acute lymphoblastic leukemia (ALL)
  Bcr-Abl kinases **2**:972
  pegaspargase **1**:148
  receptor tyrosine kinase FLT3 **3**:5

Acute mania, treatment **6**:71

Acute myelogenous leukemia (AML)
  gemtuzumab ozogamicin **1**:139
  histone deacetylases (HDACs) **7**:224
  new drugs **1**:107
  PKC-412 **2**:971
  retinoic acid receptor-α (RAR-α) **2**:1027
  sodium phenyl butyrate (buphenyl) **1**:274

Acute myocardial infarction (AMI), quantum mechanical calculations **4**:384, **4**:392, **4**:395, **4**:397–399

Acute pain
  chemical pain models **6**:332
  definition **6**:328
  mechanical models **6**:332
    pinprick tests **6**:332
    Randall Selitto test **6**:332
    von Frey monofilaments **6**:332
  models **6**:331
    see also specific models
  thermal models **6**:331
    cold water baths **6**:332
    fixed-temperature water baths **6**:332
    hot-plate assay **6**:332
    radiant heat source **6**:332
    tail-flick test **6**:331–332
  treatment
    analgesics **6**:337*T*
    differential efficacy **6**:337–338

Acute promyelocytic leukemia, histone deacetylase (HDAC) inhibitors **1**:273–274

Acute stress disorder **6**:5
  treatment **6**:6

Acute systemic toxicity, animal experimentation alternatives *see* Animal experimentation alternatives

Acute *vs.* chronic treatment, drug development productivity **2**:128

ACV-1, nicotinic acetylcholine receptor binding **2**:888

Acyclic nucleosides
  see also specific drugs
  hepatitis B antivirals **7**:318

Acyclic X, structure **6**:88*F*

Acyclovir **7**:297
  absorption studies **5**:42
  alpha-herpesvirus infections **7**:257–258
  bioavailability **7**:299
  cytomegalovirus **7**:297
  discovery/development **1**:372
  Epstein–Barr virus **7**:297
  herpes simplex viruses **7**:297
  herpes simplex virus type 1 **7**:257–258, **7**:299
  herpes simplex virus type 2 **7**:257–258, **7**:299
  intestinal absorption, active transport **5**:266
  ionization constant (pK_a) **5**:366*T*
  mechanism of action **7**:297
  oral absorption **5**:468*T*
  phosphorylation **7**:297
  prodrugs of, amino acid esters **5**:1023
  products of **5**:1023*F*

SAR studies **7**:298
structure **1**:70*F*, **1**:373*F*
target information **4**:750*T*
varicella zoster virus **7**:257–258, **7**:297, **7**:299

Acyclovir triphosphate, hepatitis B **7**:318

Acylation, hydrophobic depoting **2**:586

N-Acylation, aminoglycosides **7**:635

Acyl-cyanamides, carboxylic acids, bioisosterism **2**:659

Acyl esters, prodrugs **5**:1022

Acyl-glucuronides
  glucuronic acid conjugation **5**:151, **5**:151*F*
  glutathione transferases **5**:161–162, **5**:162*F*

2-Acylglycerol *O*-acyltransferase, coenzyme A conjugation **5**:157

Acyl halides, glutathione transferases **5**:161

Acyl hydrazones
  dynamic combinatorial libraries **3**:968
  exchange, dynamic combinatorial libraries (DCLs) **3**:969

Acylides **7**:553
  clinical indication **7**:553–555
  SAR **7**:553–555

Acyloxymethyl groups, prodrugs **5**:1020

Acyl resorcinol carbamates, solid-phase discrete libraries **3**:709*T*

Acyl-sulfonamides, carboxylic acids, bioisosterism *see* Bioisosterism, carboxylic acids

AD237 **1**:504

AD337 **1**:504

AD452 **1**:504

AD923 **1**:504

AD-9677, structure **6**:406*F*

Adalimumab **1**:212, **1**:334*T*, **1**:336, **1**:433*T*, **1**:440, **1**:517; **6**:240*T*
  brands, global sales **1**:605*T*
  cancer development **1**:442
  dosing **1**:441
  immunogenicity **1**:442
  infections **1**:442
  introduction of **1**:108
  psoriasis treatment **7**:975
  rheumatoid arthritis treatment **1**:433–434; **7**:858
  skin inflammation treatment **7**:975
  tuberculosis development **1**:441
  tumor necrosis factor-α binding **1**:441

ADAM33, COPD **7**:880

Adapalene
  acne treatment **7**:939
  development **7**:964

ADAPT, computational oral absorption prediction **5**:686

AG331 4:74
structure 4:74F
AG1024 7:190, 7:197
structure 7:197F
AG1296 7:198
structure 7:198F
AG013736 7:203
structure 7:202F
AG-13958, wet age-related macular degeneration treatment 6:306T
AG14361, IBD treatment 6:628–629
Agalsidase alfa 1:247
Age effects
metabotyping 5:1000, 5:1000F
obesity 6:391
Agency for Health Care Policy and Research, urinary incontinence markets 8:127
Agenerase see Amprenavir
Age-related macular degeneration (AMD)
see also Dry age-related macular degeneration; Wet age-related macular degeneration
blindness 6:299
clinical trials 6:305
disease basis 6:300
amphiphilic pyridinium ion (A2E) 6:301, 6:301F
cigarette smoking 6:302
complement 6:302
infections 6:302
inflammation 6:302
lipofuscin 6:301
lipoprotein/cholesterol metabolism 6:302
mitochondria 2:1054–1055
oxidative stress 6:302
photoreceptor function 6:300
retinaldehyde 6:300
disease state 6:299
development 6:299
models 6:303
in vitro models 6:303
new research areas 6:314, 6:315T
prevalence 6:298
treatment 6:308
unmet medical needs 6:311
vascular endothelial growth factor 3:202
Agglomerative testing, microarrays 3:93
Aggrastat see Tirofiban hydrochloride
Aggrecanase inhibitors, arthritis treatment 7:865–866
Aggregate data use, postmarketing surveillance 1:770
Aggregate-forming compounds, promiscuous ligands 2:738

Aggregation inhibitors, Huntington's disease treatment 6:221
Aggregation reactions
indomethacin 5:407, 5:408F, 5:409F
2-naphthoic acid 5:407, 5:408F, 5:409F
phenazopyridine 5:409F, 5:410
piroxicam 5:408F, 5:410, 5:411F
solubility see Solubility
Aggregators see Promiscuous ligands, computational prediction
AGI-1067 6:491
structure 6:489F
Agilent, microarrays 3:89
Aging
Caenorhabditis elegans model system 3:8
mitochondria dysfunction 2:1060
Agitoxin complex, K+ ion channel–toxin interactions 4:719
AGN29
as farnesoid X receptor ligand 2:1024
structure 2:1025F
AGN31
as farnesoid X receptor ligand 2:1024
structure 2:1025F
AGN34
as farnesoid X receptor ligand 2:1024
structure 2:1025F
Agomelatine 6:76–77
primary insomnia treatment 6:160
Agonist/antagonist screening, GPCRs 3:637
Agonists, peptide/protein drugs see Peptide/protein drugs
Agonist state modeling
GPCRs 4:683
G protein-coupled receptor models 4:683
Agonist substitution therapies, addiction 6:179
Agouron 2:761T
Agrylin see Anagrelide hydrochloride
AGT depletion, triazene resistance 7:138, 7:139F
AH-1058 6:753
AHN 2-005, structure 6:187F
AIDA 6:108
structure 6:107F
AIDS 5:223
see also HIV infection
clinical definition 7:330
highly active antiretroviral therapy 8:53
historical perspective 1:68; 4:599–600
immunosuppression 5:223
incidence 8:53
orphan drugs approvals and designations 1:665–666, 1:667
pathogenesis 1:68
prevalence and mortality, USA 1:595
research history 1:528

social infrastructure/organization effects 1:634
vaccine 1:666
AIDS Action Council 1:665–666
AIDS dementia 6:10
Airflow limitation, asthma 7:873–874
Air pollution, cancer risk factors 7:4
Airway disorders, potassium channels 2:863
Airway inflammation, asthma 7:874–875
Airway remodeling
asthma see Asthma
COPD see Chronic obstructive pulmonary disease (COPD)
AK-37
discovery/development 1:382
structure 1:382F
AK295 6:272
structure 6:272F
AKAP79/250, GPCR interaction 2:797T
AKT2, PDB entries 2:965T
Akt/PDK1 7:208
in cancer 7:208
GPCR interaction 2:786T
Akt/PDK1 inhibitors 7:209
structure 7:208F
AL-108, Alzheimer's disease 6:212
AL-438 2:1016–1017
structure 2:1017F
AL-2512, dry eye treatment 6:317
ALADDIN program
pharmacophore 3D database searching 4:139T
substructure searching 4:467
Alanex 2:760T
D-Alanyl-D-peptidase (Streptomyces R61) 3:532–533
structure 3:534F
Ala scan, peptide/protein agonists 2:582
Albany Molecular Research 2:760T
Albendazole
ionization constant (pKa) 5:366T
nematode infection treatment 7:760
trematode infection treatment 7:761
Albright hereditary osteodystrophy, Gα3 subfamily mutations 2:780–782
Albuferon-α, hepatitis C 7:267, 7:390
Albumin 5:849
biological functions 5:849
crystal structure 5:849, 5:851T
fatty acid binding 5:849
depletion, proteomics 3:31–32
diffusion, tissue slices 5:329
drug binding sites 5:849, 5:853T
drug–polymer conjugates 5:1051
structure 5:849, 5:850F
PDB 5:851T
synthesis 5:849

Albumin binding **5**:322, **5**:325–326
  hydrophobic depoting **2**:586
  plasma protein binding *vs.*
    5:859
  prediction models **5**:854,
    5:855
    apparent association constant
      (log$K^{app}HSA$) **5**:857–858
    association constant
      (logK_{HSA}) **5**:856–857
    future work **5**:864–865
    general regression neural
      networks **5**:858–859
    Kier and Hall associated
      topographical index
      5:856
    multilayer feedforward neural
      networks (MFNN)
      5:858–859
    multiple linear regression-based
      model **5**:858
    tetracyclines **7**:618
Albuterol
  asthma treatment **7**:883
  Comer, William, work by
    8:2–3
  library similarity searches **4**:563,
    4:564*T*
  mechanism of action, GPCR
    targeting **2**:773*T*
Albuterol sulfate **6**:240*T*
  structure **6**:238*F*
Alcohol
  breast cancer **7**:151
  cancer risk factors **7**:4
Alcohol addiction
  comorbidities **6**:184*T*
  counseling **6**:179–182
  medications for **6**:179, **6**:180*T*,
    6:184*T*
    acamprosate **6**:180*T*, **6**:183
    antalarmin **6**:184*T*
    baclofen **6**:184*T*
    CEE-03-310 **6**:184*T*
    diazepam **6**:180*T*
    disulfiram **6**:179–182, **6**:180*T*,
      6:183
    lorazepam **6**:180*T*
    nalmefene **6**:184*T*
    naltrexone **6**:180*T*, **6**:183
    rimonabant **6**:184*T*
    topiramate **6**:184*T*
    valproate **6**:184*T*
  physical dependence **6**:178
  treatment **6**:9
    naltrexone **6**:9
Alcohol dehydrogenases
    (ALDs) **5**:121, **5**:177
  drug metabolism **5**:202
  steroid metabolism **5**:199*F*,
    5:202
  yeast selection markers
    3:117

Alcohol/fatty acid conjugation,
    acetylation **5**:154
Alcohol oxidase 1 (AOX1), yeast
    selection markers **3**:117
Alcohol/phenol prodrugs *see* Prodrugs
Alcohols
  ezetimibe, metabolite
    formation **8**:75–76
  UDP-glucuronyltransferases **5**:150*F*,
    5:151, **5**:151*F*, **5**:153*F*
  ionization constants (pK_a) **5**:372
  sulfotransferases **5**:148–149, **5**:148*F*
Alcohol sulfotransferase **5**:147
Alcophenac, hydrolysis **5**:142, **5**:144*F*
Aldara *see* Imiquimod
Aldehyde dehydrogenases
    (ALDH) **5**:120, **5**:177
  categories **5**:120
  characteristics **5**:169*T*
  deficiency, adverse drug
    effects **1**:282*T*
  disulfiram inhibition **5**:120
  mechanism of action **5**:121*F*
Aldehyde oxidase (AO) **5**:119, **5**:176
  mechanism of action **5**:119, **5**:119*F*
  multiplicity **5**:176
  reductions **5**:125
  structure **5**:177
  tissue distribution **5**:176–177
Aldehyde reductases **5**:177–178
Aldesleukin **1**:501
  toxicity **7**:46*T*
Aldo/keto reductases (AKRs) **5**:125,
    5:178
  characteristics **5**:169*T*
  isoforms **5**:125
  mechanism of action **5**:126*F*
Aldol formation, dynamic combinatorial
    library generation **3**:962*F*
Aldose reductase inhibitors,
    pharmacophores **4**:517*F*
Aldose reductases **5**:177–178
  quantum mechanical
    calculations **4**:404
Aldosterone **2**:1017–1018
  structure **2**:1018*F*
Aldosterone receptor antagonists
  ACE inhibitor combination **6**:590
  hypertension treatment **6**:590
  mechanism of action **6**:590
  structure **6**:587*F*
Aldurazyme *see* Laronidase
ALE0540, structure **6**:346*F*
ALEC *see* Pumactant
Alefacept **1**:235, **1**:500
  introduction of **1**:108
  labeling changes **1**:758*T*
  psoriasis treatment **7**:946, **7**:946*T*,
    7:972
Alemtuzumab **1**:135, **1**:334*T*, **1**:433*T*,
    1:502; **6**:240*T*
  antibody-directed enzyme prodrug
    therapy **5**:1034–1035

chronic lymphocytic leukemia
    1:135
  clinical indication **6**:245
  leukemia treatment **1**:444
  toxicity **7**:46*T*
Alendronate **6**:512; **8**:199–212
  clinical trials **8**:200
  clinical use **8**:200
    postmenopausal
      osteoporosis **8**:200
  efficacy **8**:200
  mechanism of action **8**:201, **8**:202
    see also Bone resorption/remodelling
    cellular level **8**:204
    cholesterol synthesis
      inhibition **8**:202
    farenyl diphosphate synthetase
      inhibition **8**:202, **8**:203*F*
    in vivo evidence **8**:204
    protein isoprenylation **8**:203
    protein tyrosine phosphatase
      inhibition **4**:640–641
    sterol synthesis **8**:203
    tissue level **8**:207
  oral absorption **5**:1026–1027
  pharmacokinetics **8**:200
    half-life **8**:201–202
  structure **5**:1027*F*; **6**:510*F*; **8**:200,
    8:201*F*
  target information **4**:750*T*
  toxicology **8**:207
    esophagus **8**:207–208
Alesion *see* Epinastine hydrochloride
Alfa 1-antitrypsin (AAT) *see*
    α_1-Antitrypsin
Alfacalcidol
  structure **2**:1018*F*
  as vitamin D receptor ligand
    2:1019
Alfadolone **1**:506–507
Alfentanil **6**:353
  absorption, pH dependency **5**:383
  anesthetic adjuvants **6**:362
  oxidation **5**:110, **5**:111*F*
  structure **6**:363*F*
Alfuzosin
  benign prostatic hyperplasia
    treatment **6**:556*T*
  prostatitis treatment **6**:571
  structure **6**:565*F*
Algorithms
  see also specific types
  clustering **4**:185
  core molecular modeling **4**:88,
    4:101
  dissimilarity-based compound
    selection **4**:186
  ionization prediction, in silico
    tools **5**:610
  pharmacophore detection *see*
    Pharmacophores, detection
  postmarketing surveillance **1**:764,
    1:764*T*

screening library selection 4:501
sequence alignment,
bioinformatics 3:319–320
Ali *see* Loteprednol etabonate
Alignment methods
3D-QSAR *see* 3D quantitative
structure–activity relationship
(3D-QSAR)
regulatory regions,
bioinformatics 3:322
Alignment scores, homology-based
protein structure
prediction 3:326
Alinidine 6:747, 6:753
Aliphatic alcohols, ionization constants
(pK_a) 5:372
Aliphatic amidene multiatom probe,
Molecular Interaction
Fields 4:246, 4:246F
Aliphatic bases, ionization constants
(pK_a) 5:377
Aliphatic side-chain replacements,
gabapentin 8:237
Aliskiren, structure 6:724F
Alitretinoin, **350**
characteristics 2:1012T
Kaposi's sarcoma 1:136
Alkane functionalization, PASP 3:796,
3:796F
Alkane/water partitioning,
lipophilicity 5:433, 5:682
Alkene oxides, epoxide hydrolases *see*
Epoxide hydrolases
Alkene substitution, peptides 2:604
Alkoxycarbonyloxyalkyl groups,
prodrugs 5:1020
Alkoxyprolines, solid-phase discrete
libraries 3:709T
Alkylaminobenzanilides, solid-phase
pool libraries 3:705F
Alkylating compounds 7:33,
7:129–148
see also specific drugs
future work 7:144
pharmacogenetics 3:61
toxicity 7:46T
N-Alkylation, solid support
synthesis 2:466F, 2:467, 2:467F
O-Alkylation, solid support
synthesis 2:466F, 2:467, 2:468F
Alkylpolyamines
SAR 7:828F, 7:829F
trypanosomiasis treatment
7:826–827
Alkyl-tethered diisopropylarylsilane
linker, solid-phase
libraries 3:738F, 3:740
2-Alkylthiobenzimidazoles,
PASP 3:805
Allegra *see* Fexofenadine
Allele-specific chain extension *see* Single
nucleotide polymorphisms
(SNPs)

Allele-specific polymerase chain
reaction, single nucleotide
polymorphisms 2:365
Allergens, atopic dermatitis 7:947–948
Allergic asthma, omalizumab 1:206
Allergic conjunctivitis 6:316
models 6:316
Allergic inflammation, STAT (signal
transducer and activator of
transcription) 3:208
Allergic rhinitis
betotastine besilate 1:204
epinastine hydrochloride 1:204
ramatroban 1:206
Allergies 5:217–218
see also specific types
cephalosporins 7:489
penicillin-induced 5:220–221
Alliance for Cellular Signaling, cell-
based assays 3:643–644
Alligator Bioscience 1:515
ALLN 7:235–236
Allobarbital, hydrolysis 5:142, 5:144F
Allodynia
definition 6:328–329
transient receptor potential ion
channels 2:909–910
Allometric scaling
biliary clearance models 5:872
bioavailability models *see*
Bioavailability models
in vivo ADME studies 5:47
Allometry
renal clearance models 5:871–872
volume of distribution (V_{Dss}) *see*
Volume of distribution (V_{Dss})
Allopurinol
bioavailability 5:712T
oral absorption 5:468T
Allosteric effects
comparative QSAR 4:55
potassium channels 2:862
Allosteric modulation, GPCRs *see* G
protein-coupled receptors
(GPCRs)
Alloxan, diabetes mellitus type 1 animal
models 6:433
All-*trans*-retinoic acid (ATRA)
as retinoic acid receptor ligand 2:1020
structure 2:1020F
Allwine, Debra, oxazolidinone
development 8:159–160
Allylamines
as antifungals 7:430T
structure 5:359
17-Allylaminogeldanamycin 7:230–231
discovery/development 1:386
structure 1:387F
Allylic carbon oxidation, cytochrome
P450 5:92, 5:93F, 5:94F
Allylic sulfur compounds 1:449–450
Alma-Ata conference (1978)
1:638–639

Almogran *see* Almotriptan
ALMOND software
GRIND descriptors 5:541, 5:542F
library design 4:363
Almotriptan 1:181
migraine treatment 6:373T
AlogP98, computational oral absorption
prediction 5:688
AlogP method, log P calculation 5:655
ALGOPS
aqueous solubility models 5:644T
log D calculation program 5:662T
log P calculation program 5:661T
ALOGPS program, self-learning
feature 5:664
Alosetron hydrochloride 1:201
animal studies 6:653
5-HT$_3$ receptor 6:653
introduction of 1:107–108
irritable bowel syndrome
treatment 6:655T, 6:657
withdrawal of 6:653
Aloxi *see* Palonosetron
Alpha 1-antitrypsin (AAT) *see*
α_1-Antitrypsin
Alpha adrenergic receptors *see*
α-Adrenergic receptors
α1-antitrypsin (AAT) *see* α_1-Antitrypsin
α-subunits, G protein structure 2:780
$\alpha_2\delta$ binding site
gabapentin mechanism of
action 8:240
mutation studies, gabapentin
mechanism of action 8:241
Alphadolone
GABA$_A$ receptors 2:893
structure 6:353F
Alpha erythropoietin, brands, global
sales 1:605T
Alphagen *see* Brimonidine
Alphaglobulins, plasma protein
binding 5:322
Alpha helices, scaffold-based design *see*
Scaffold-based design
Alpha-herpesviruses, antiviral
drugs 7:254T, 7:257–258
A-5021 7:259
acyclovir 7:257–258
brivudine 7:257–258
cyclohexenylguanine 7:259
famciclovir 7:257–258
idoxuridine 7:257–258
penciclovir 7:257–258
synguanol 7:259
trifluridine 7:257–258
valacyclovir 7:257–258
Alpha-naphthoflavone, reaction
phenotyping studies 5:243T
AlphaScreen cAMP assay, GPCR
functional assay
screening 2:809
Alpha-synuclein, NMR structure
determination 3:490, 3:491F

Alpha tocopherol, structure **6**:260*F*
Alpha Tocopherol and Beta Carotene
    (ATBC) study **1**:455
Alphaxalone **6**:361
    GABA receptor interactions **6**:357
    structure **6**:353*F*
Alprazolam
    bioavailability **5**:712*T*
    development **6**:98
    oral absorption **5**:680*T*
    structure **2**:652*F*; **6**:99*F*
Alprenolol
    bioavailability **5**:712*T*
    oral absorption **5**:468*T*, **5**:680*T*
ALS Functional Rating Scale
    (ALSFRS) **6**:223
Alteplase **1**:499; **6**:769
    adverse events **6**:770
        intracranial hemorrhage **6**:770
    clinical trials **6**:772
        COBALT study **6**:773
        Fibrinolytic Therapy Trialists
            (FTT) meta-analysis **6**:773
        GISSI-2 **6**:772
        GUSTO-1 **6**:772
        ISIS-3 **6**:772
    structure **6**:770
Alteration of function (neomorphic),
    chemical mutagenesis **3**:13–14
Alternative medicine
    see also Complementary medicine
    definition **1**:408
Alternative RNA splicing, STAT (signal
    transducer and activator of
    transcription) **3**:204
Alumina, microwave-assisted
    chemistry **3**:853
Alumoxane-activated metallocene
    catalysts, olefin
    polymerization **1**:791
Alveolar macrophages,
    tetracyclines **7**:620
Alvimopan **1**:509
    irritable bowel syndrome
        treatment **6**:658
Alvocidib *see* Flavopiridol
Alzheimer's disease **6**:9, **6**:196
    clinical trial issues **6**:201, **6**:202*F*
        assessment scales **6**:201
    cognitive function **6**:195
    Comer, William, work by **8**:4
    competence and consent
        issues **1**:698
    cultural attitudes **1**:629
    disease basis **6**:197
        amyloid deposits **6**:10, **6**:209
        amyloid hypothesis **6**:198
        cholinergic hypothesis **6**:10, **6**:198
        genetic risk factors **6**:198, **6**:199*T*
        inflammatory cascade
            hypothesis **6**:198
        mitochondria **2**:1054–1055
        tau hypothesis **6**:198

disease state/diagnosis **6**:196
    biomarkers **6**:197
    blood–brain barrier-permeable
        molecules **6**:197
    Clinical Dementia Rating
        (CDR) **6**:197
    computed tomography **6**:197
    Consortium to Establish a Registry
        for Alzheimer's Disease
        (CERAD) **6**:197
    DSM-IV **6**:9, **6**:196
    exclusionary diagnosis **6**:196
    magnetic resonance imaging **6**:197
    National Institute for Aging
        (NIA) **6**:197
    post-mortem diagnosis **6**:197
epidemiology **2**:52–53
    gender differences **1**:630
    inheritance **6**:195
    prevalence **6**:9, **6**:196
knockout animal models **3**:160
models **6**:199
    genetic models **6**:201
    natural models **6**:201
    pharmacological/lesion
        models **6**:199
new research areas **6**:10, **6**:203
    acetylcholinesterase-releasing
        modulators **6**:204
    β-amyloid aggregation
        blockers **6**:209
    anti-inflammatory
        compounds **6**:211
    GABA receptor antagonists
        6:206
    histamine receptor
        antagonists **6**:205
    hormone replacement
        therapy **6**:207
    5-HT receptor antagonists **6**:207
    muscarinic acetylcholine receptors
        (mAChRs) **6**:204
    neuroprotective drugs **6**:211
    nicotinic acetylcholine receptors
        (nAchRs) **6**:205
    prolyl endopeptidase
        inhibitors **6**:207
    secretase inhibitors **6**:207, **6**:208*F*
    statins **6**:211
    symptomatic treatment
        6:203–204
    tau kinase inhibitors **6**:209
    thyrotropin-releasing hormone
        agonists **6**:207
positional cloning **2**:307
prognosis **6**:11
proteomics **2**:388
symptoms **6**:9, **6**:196, **6**:199*T*
treatment **6**:10, **6**:202
    see also specific drugs
    phosphodiesterase-8
        inhibitors **2**:949
    tetracyclines **7**:620

unmet medical needs **6**:203
    disease-modifying therapies **6**:203
    Women's Health Initiative Memory
        Study **2**:53
Alzheimer's Disease Assessment Scale
    (ADAS) **6**:201
AM80
    as retinoic acid receptor
        ligand **2**:1020–1021
    structure **2**:1020*F*
AM-112 **7**:512
    structure **7**:512*F*
AM-1241 **6**:342–343
    structure **6**:343*F*
AMA-1, malaria **7**:767
Amantadine
    adverse effects **7**:377
    bioavailability **5**:712*T*
    licensing **7**:377
    mechanism of action **7**:377
    orthomyxoviruses **7**:274–275
    Parkinson's disease
        treatment **6**:214–215
    resistance **7**:377
    structure **7**:377*F*
Amaryl *see* Glimepiride
Amaryllidaceae alkaloids, PASP **3**:815,
    3:816*F*, **3**:819
AMBER
    docking, scoring functions **4**:483
    K$^+$ ion channels **4**:710
    molecular dynamics
        simulations **4**:652
    quantum mechanical
        calculations **4**:385
    regression-based **4**:484
Amberlyst 15 ion exchange resin,
    PASP **3**:821
Ambient temperature stability, drug
    reference standard **2**:167,
    2:168*T*
AmBisome **1**:501
Ambrisentan **1**:508–509
Ambuic acid **1**:361
    structure **1**:360*F*
AMD-3100 **7**:360
    structure **7**:361*F*
Amebiasis **7**:751*T*
    biology **7**:758
    clinical manifestations **7**:758
    drug treatment **7**:753*T*, **7**:758
    epidemiology **7**:758
    tetracyclines **7**:613
Amedis Pharmaceuticals **1**:492*T*
American Academy of Pediatrics
    (AAP) **6**:128–129
American Association of Colleges of
    Pharmacy **2**:85
American Chemistry Council Long-
    Range Research
    Initiative **1**:57*T*
American Heart Association (AHA),
    hypertension **6**:699

Angiotensin AT$_1$ receptor antagonists
  see also specific drugs
  nonpeptide drugs 2:618–619, 2:620F
Angiotensin AT$_1$ receptors, in human
    disease 2:775T
Angiotensin AT$_2$ receptor antagonists
  see also specific drugs
  development,
    pharmacophores 4:533F
  nonpeptide drugs 2:619–621,
    2:621F, 2:622F
Angiotensin AT$_2$ receptors
  activation 2:785T
  desensitization 2:801
  in human disease 2:775T
Angiotensin-converting enzyme (ACE)
  crystallization 3:876, 3:882–883
  inhibitors see ACE inhibitors
  lisinopril complex structure 4:17,
    4:18F
Angiozyme (RPI-4610) 7:200
AngloScandinavian Coronary Outcomes
    Study (ASCOT) 6:474, 6:474T
Angola, functional public health
    system 1:642
Anhydrotetracyclines
  see also specific drugs
  formation 7:604–605, 7:605F
  Tet repressor binding 7:615
Anhydrous to monohydrate transition,
    solid-state
    physicochemistry 5:523, 5:523F
Anidulafungin 7:430T
  clinical profiles 7:462
    candidosis 7:462
  medicinal chemistry 7:458–459,
    7:459T
  pharmacokinetics 7:461–462
Aniline
  quantum mechanical
    calculations 4:387, 4:387T
  structure 5:359
bis-Anilinopyrimidines, cyclin-
    dependent kinase
    inhibition 7:206
Animal experimentation
    alternatives 1:463–487
  see also Toxicogenomics
  acute systemic toxicity 1:467T, 1:476
    MEIC study 1:476
  advantages 1:464
  carcinogenicity 1:467T, 1:477
    chemical biotransformation 1:478
    databases 1:477
    gap junction intercellular
      communication (GJIC)
      assay 1:478
    in vivo testing regulations 1:477
    morphological cell
      transformation 1:467T, 1:477
    predictive performance 1:477
    Syrian hamster embryo (SHE) cell
      assay 1:477

chronic toxicity 1:467T, 1:476
  co-culture systems 1:476–477
  single cell systems 1:476–477
disadvantages 1:464
drug metabolism and
    pharmacokinetics 5:33
embryotoxicity 1:467T, 1:475
  frog embryo teratogenesis assay
    (FETAX) 1:467T, 1:475
endocrine toxicity 1:467T, 1:475
  mitogenesis assays 1:467T, 1:475
  subcellular hormone receptor
    binding 1:467T, 1:475
  transactivation systems 1:467T,
    1:475
eye irritation 1:466, 1:467T
  bovine corneal opacity and
    permeability (BCOP)
    assay 1:466, 1:467T
  chicken enucleated eye test
    (CEET) 1:466, 1:467T
  chorioallantoic membrane vascular
    assay (CAMVA) 1:466, 1:467T
  co-culture systems 1:466, 1:467T
  fluorescein leakage assay
    (FLL) 1:466, 1:467T
  hen's egg test-chorioallantoic
    membrane (HET-CAM) 1:466,
    1:467T
  isolated chicken eye (ICE)
    assay 1:470
  isolated rabbit eye (IRE)
    assay 1:470
  rabbit enucleated eye test
    (REET) 1:466, 1:467T
  red blood cell (RBC) assay 1:467T
  three-dimensional tissue
    constructs 1:466, 1:467T
genetically engineered cells 1:479
  biotransforming potential
    cells 1:479
  reporter gene constructs 1:479
genotoxicity 1:467T, 1:470
  Ames test 1:467T, 1:470
  bacterial gene mutation
    assays 1:467T, 1:470
  chromosomal damage 1:467T, 1:470
  COMET assay 1:467T, 1:470–471
  gene mutations 1:470–471
  International Conferences on
    Harmonization 1:471
  in vitro chromosomal
    aberrations 1:467T
  micronuclei 1:467T, 1:471
  mouse lymphoma assay 1:471
human cell use 1:479
information sources 1:465
  British Toxicology Society 1:465
integrated tier-testing
    strategies 1:482
limitations 1:481
  biotransformation 1:481
  in vivo data risk assessment 1:481

neurotoxicity 1:467T, 1:474
photogenotoxicity 1:467T, 1:472
  photo-Ames test 1:467T, 1:472
  photo-chromosome aberration
    test 1:467T, 1:472
  photo-micronucleus test
    (PMNT) 1:467T, 1:472
  photo-thymidine kinase
    test 1:467T, 1:472
photoirritancy 1:467T, 1:471
  human 3-D skin model
    phototoxicity test (H3D-
    PT) 1:471–472
  RBC phototoxicity test (RBC-
    PT) 1:467T, 1:471–472
  3T3 neutral red uptake
    phototoxicity test (3T3-NRU-
    PT) 1:471
photosensitization 1:467T, 1:471
phototoxicity 1:467T, 1:471
prospective human studies 1:478
  Declaration of Helsinki 1:479
  UK regulations 1:479
proteomics 1:481
pyrogenicity 1:465, 1:467T
  human whole-blood test
    (WBT) 1:466, 1:467T
  limulus amebocyte lysate (LAL)
    test 1:465–466, 1:467T
  lipopolysaccharides 1:465
reproductive toxicity 1:467T, 1:475
screening 1:464
skin corrosivity 1:467T, 1:472
  Corrositex 1:467T, 1:472
  Epidermal Skin Test (EST-
    1000) 1:467T, 1:473
  Episkin 1:467T, 1:472
  transepidermal resistance
    (TER) 1:467T, 1:472
  ZK1350 corrosivity test 1:467T,
    1:472
skin irritation 1:467T, 1:473
  EpiDerm 1:467T, 1:473
  Episkin 1:467T, 1:473
  excised skin 1:467T, 1:473
  Prediskin 1:467T, 1:473
  sequential testing strategy 1:473
  SkinEthic 1:473
  skin integrity function tests 1:473
skin permeability 1:467T, 1:472
skin sensitization 1:467T, 1:473
  antigen presentation 1:474
  cytokine release 1:474
  DEREK 1:473–474
  METEOR 1:474
stem cells 1:480
  adult 1:480
  embryonic 1:480
  hepatocyte derivation 1:480
  human umbilical cord blood
    (HUCB) 1:480
toxicity testing 1:464
  EU Cosmetics Directive 1:464

Animal experimentation alternatives
    (*continued*)
  toxicokinetics   **1**:474
    physiologically based biokinetic
      (PBPK) modeling   **1**:474
  validation   **1**:465, **1**:476*F*, **1**:483*F*
    definition   **1**:465
    European Centre for the Validation
      of Alternative Methods
      (ECVAM)   **1**:465
    Interagency Co-ordinating
      Committee on the Validation of
      Alternative Methods
      (ICCVAM)   **1**:465
    Organization for Economic Co-
      operation and Development
      (OECD)   **1**:465
Animal models   **1**:533
  alternatives *see* Animal
    experimentation alternatives
  aminoglycosides   **7**:724
  antimicrobial agents *see* Antimicrobial
    drugs
  anxiety disorders *see* Anxiety disorders,
    animal models
  atherosclerosis *see* Atherosclerosis
  BCS   **5**:978
  benign prostatic hyperplasia
    6:558–559, **6**:558*F*
  bone diseases/disorders   **6**:507
    metastatic bone tumors   **6**:508
    osteoporosis *see* Osteoporosis
  Comer, William (views on)   **8**:3
  diabetes mellitus type 2 *see* Diabetes
    mellitus type 2
  disadvantages   **3**:617–618
  drug discovery *see* Drug discovery
  experimental autoimmune
    neuritis   **6**:239–242
  hepatitis C   **7**:389
  lead optimization   **2**:139
  microarrays   **3**:93
  modafinil   **8**:150
  narcolepsy *see* Narcolepsy
  nervous system autoimmune diseases
    see Nervous system autoimmune
    diseases
  neurodegenerative diseases *see*
    Neurodegenerative diseases
  phenotype-oriented screening *see*
    Phenotype-oriented screening
  proteomics   **3**:43
  psoriasis *see* Psoriasis
  regulation   **1**:533
  systemic lupus erythematosus
    6:239
  target identification   **2**:312
  toxicity   **1**:533
  toxicogenomics   **3**:56
  vaginal administration   **5**:290
Animal nasal mucosa, nasal
    administration model *see* Nasal
    administration

Animals (Scientific Procedures) Act
    (1986)   **1**:533
Animal skin, dermal/transdermal
    administration   **5**:293
Animal studies, principles   **1**:812
Animal toxicity studies
  drug development   **2**:28–29, **2**:29*T*
  resistance   **5**:912
Anion aggregation, solubility   **5**:407
Anionic lipid interactions, KcsA   **4**:712,
    4:712*F*
Aniracetam   **2**:903
Anisotropic diffraction patterns, protein
    crystallography   **3**:441
Anisotropic etching *see* Lithography/
    photolithography
Anisotropic network model (ANM), K$^+$
    ion channels   **4**:711
Anisotropic systems
  lipophilicity measurement *see*
    Lipophilicity measurement
  lipophilicity measurement, ions/
    zwitterions *see* Lipophilicity
    measurement, ions/zwitterions
Anisoylated purified streptokinase
    activator complex
    (APSAC)   **6**:767, **6**:769
  mechanism of action   **6**:769
  pharmacokinetics   **6**:769
Anistreplase *see* Anisoylated purified
    streptokinase activator complex
    (APSAC)
Annotation
  Genome Reviews   **3**:356
  PDB   **3**:376
  UniProtKB/Swiss-Prot   **3**:359–360
Annual Reports of Medicinal Chemistry,
    drug introductions   **1**:105
Anomalous dispersion *see* X-ray
    crystallography, phase problem
    solutions
Anomalous scattering, protein
    crystallography, rational drug
    design   **3**:890
Anorexiants *see* Obesity treatment; *see
    specific drugs*
ANOVA analysis, microarrays   **3**:101
Ansamycins   **1**:386
  see also specific drugs
  heat shock protein 90 (hsp90)
    inhibitors   **7**:230
  mechanism of action   **1**:387
  stereochemistry   **1**:387
Antacids
  gastric/mucosal ulcer
    treatment   **6**:604
  gastroesophageal reflux disease
    treatment   **1**:439
Antalarmin
  alcohol addiction   **6**:184*T*
  irritable bowel syndrome
    treatment   **6**:661*T*
  structure   **6**:103*F,* **6**:181*F*

Antebate *see* Betamethasone butyrate
    propionate
Anterior fibromuscular stroma, urinary
    incontinence   **6**:571
Antevate *see* Betamethasone butyrate
    propionate
Anthocyanidins, nutraceuticals   **1**:452*T*
Anthocyanins,
    pharmacokinetics   **1**:454*T*
Anthracyclines   **7**:116
  see also specific drugs
  discovery/development   **1**:381–382
  mechanism of action   **2**:1040
  structure   **1**:382*F*
Anthralin, psoriasis treatment   **7**:944
Anthraquinone derivatives,
    reduction   **5**:123, **5**:123*F*
Anthraquinones   **7**:117
  see also specific drugs
Anti-AIDS drugs
  see also specific drugs
  development   **1**:68
  monotherapies   **1**:69
    resistance development   **1**:69
  structure-based design   **1**:68
  viral entry inhibitors   **1**:79
    see also specific drugs
Anti-allergy drugs
  see also Histamine H$_1$ receptor
    antagonists; *specific drugs*
  drug introductions   **1**:204
  ophthalmic diseases   **6**:316
Anti-Alzheimer's disease drugs, drug
    introductions   **1**:192
Anti-apoptotic drugs
  see also specific drugs
  carvedilol   **8**:137–138, **8**:141
  Huntington's disease
    treatment   **6**:221
Anti-apoptotic pathways, NF-κB
    pathway   **3**:196
Antiarrhythmic drugs   **6**:729–762
  see also specific drugs
  class 1 *see* Sodium channel blockers
  class 1a   **6**:742
    effectiveness   **6**:742
    mechanism   **6**:742
    toxicity   **6**:742
  class 1b   **6**:743
    effectiveness   **6**:743
    mechanism   **6**:743
    toxicity   **6**:743
  class 1c   **6**:743
    effectiveness   **6**:743
    mechanism   **6**:743
    toxicity   **6**:743
  class 2, β–adrenergic receptor
    antagonists
  class 3 *see* K$^+$ ion channel blockers
  class 4 *see* Calcium-channel blockers
  class 5 (pace maker current
    blockers)   **6**:747
  new research areas   **6**:753

classification systems **6:**742
  see also specific classes
  Sicilian gambit **6:**742
  Vaughan Williams–Singh **6:**742
clinical trials **6:**740, **6:**741*T*
  endpoints **6:**740
  surrogates **6:**740
future work **6:**758
  example **6:**758
  mechanistic action
    definition **6:**758
  model generation **6:**758
  target identification **6:**758
hERG blocker modeling **5:**942
introductions **1:**162
mechanism of action **6:**735
  potassium channels **2:**862
new research areas **6:**750
  class 1/3/4 blockers **6:**755
  class 1-5 combinations **6:**753
  class 1b/3 blockers **6:**754
  class 3/4 blockers **6:**754
  fatty acids/fish oils **6:**756
  gap junction activators **6:**755
  sodium–calcium exchanger
    blockers **6:**756
  sodium–hydrogen exchanger
    blockers **6:**756
  stretch activated channel
    blockers **6:**755
rational design **6:**756
  arrhythmogens **6:**757
  channel-specific blockage **6:**757
  combination blockers **6:**757
  mutated ion channels **6:**757
  pathology **6:**757
  selective channel blocking **6:**757
regional differences **6:**741
Antiasthmatic drugs
  see also specific drugs
  introductions **1:**108, **1:**207
Antibacterial drugs
  see also specific drugs
  acne treatment *see* Acne treatment
  antimycobacterial agents *see*
    Antimycobacterial drugs
  cell wall targets **7:**733
  definition **7:**470
  DNA synthesis inhibitors **7:**733
  experimental models **7:**677
    in vitro models **7:**677–678
    minimum inhibitory
      concentration **7:**677–678
    rodent models **7:**677–678
    safety assays **7:**677–678
  future work **7:**475–476
  genomic identification *see* Genomics,
    antibacterial target identification
  history **7:**732
    pharmaceutical companies lack of
      interest **7:**732
  natural products **1:**368
  protein synthesis inhibitors **7:**733

RNA synthesis inhibitors **7:**733
  target identification, genomics *see*
    Genomics, antibacterial target
    identification
2′.4.4′-trichloro-2′-
    hydroxydiphenylether **2:**1068
Antibiotic multidrug resistance
    8:158
Antibiotic resistance
  gene expression vectors **3:**107
  history **7:**732
  tetracyclines *see* Tetracyclines
Antibiotics
  see also specific drugs
  atopic dermatitis treatment **7:**950
  bioethics, drug pricing/
    profitability **1:**689
  cell culture **3:**622
  definition **7:**470, **7:**470*T*
  drug resistance **1:**691
  Helicobacter pylori eradication
    6:607–609
  Hirschmann, R F and **1:**20
  introduction **1:**107, **1:**108
  prostatitis treatment **6:**571
Antibodies
  affinity chromatography **3:**422
  development, biomarker
    identification **3:**72
  generation **5:**217
  protein kinase measurement **3:**72
  SMEs **1:**516
  specificity, ELISA **3:**580
Antibody arrays, proteomics **3:**40
Antibody-based purification,
    proteomics **3:**32
Antibody capture assays,
    ELISA **3:**579–580, **3:**579*F*
Antibody-directed enzyme prodrug
    therapy (ADEPT) **5:**1032,
    5:1033
  antibody characteristics **5:**1033
    Campath-1H **5:**1034–1035
    ING-1 **5:**1034–1035
  benzoic acid mustards **5:**1034,
    5:1035*F*
  polyfluorinated analogs **5:**1036
  linkers **5:**1036
  mechanism of action **5:**1033,
    5:1035*F*
  peptidases **5:**1034
    carboxypeptidase G2 **5:**1034
    intrinsic enzymes **5:**1034
  prodrugs **5:**1035*F*
  promoieties **5:**1036
Antibody libraries
  phage display **1:**435
  ribosome display **1:**435
Antibody response
  chemical *vs.* biological **2:**117
  glatiramer acetate **8:**181
Antibody staining, HCS labeling
    techniques **3:**689

Anticancer drugs **7:**6, **7:**8*T,* **7:33–53**
  see also Antineoplastic drugs; Cancer
    chemotherapy; *specific drugs*
  alkylating compounds *see* Alkylating
    compounds
  antimetabolites *see* Antimetabolites
  attrition rate **7:**222
  bortezomib **2:**577
  clinical trials **7:**222
    market forces **7:**222–223
  conjugation, drug metabolism **5:**149
  DNA topoisomerase inhibitors *see*
    DNA topoisomerase inhibitors
  endocrine-modulating *see* Anticancer
    drugs, endocrine-modulating
  evaluation **7:**34
    HTS **7:**34, **7:**35*F,* **7:**36*F*
    hollow-fiber assay **7:**34
  introductions
    adjuvants **1:**152
    antiemetic agents **1:**107, **1:**149
    antineoplastic drugs **1:**143
    chemotherapy-induced
      leukopenia **1:**152
    cytoprotective agents **1:**151
    hypercalcemia **1:**153
    malignant melanoma **1:**107
    5α-reductase inhibitors **1:**151
  in vitro cell panels **7:**34
  mechanism of action **7:**6–7
  models *see* Anticancer drugs, tumor
    models
  monoclonal antibodies **1:**444
  new compounds **7:221–251**
  new research areas **7:**245
  P-glycoprotein interactions **5:**769*T*
  pharmacodynamics **7:**34
  pharmacokinetics **7:**34
  pharmacology **7:**34
  phosphodiesterase-4 inhibitors **2:**948
  platinum antitumor compounds *see*
    Platinum antitumor compounds
  resistance **7:**44
    ABCC transporters **7:**44–45
    ABC transporters **7:**44–45, **7:**45*T*
    circumvention **7:**44
    P-glycoprotein **7:**44–45
  targeted therapies **7:**9
  toxicity **7:**44, **7:**46*T*
  unmet medical needs **7:**245
Anticancer drugs, endocrine-
    modulating **7:149–181**
  see also Breast cancer; *specific agents*;
    specific drugs
  biphenyls/diaryls **7:**155, **7:**155*F*
  coumestrol-based
    compounds **7:**156–157, **7:**157*F*
  2,3-diaryl piperazine series **7:**156,
    7:156*F*
  1,2-diphenylenediamine platinum
    compounds **7:**155–156, **7:**155*F*
  pyrroloindolizine ring
    structure **7:**156, **7:**156*F*

Anticancer drugs, endocrine-
   modulating (*continued*)
   clinical trials   7:152
      Study of Raloxifene and Tamoxifen
         (STAR)   7:152
   disease state   7:149
   estrogen synthesis inhibition   7:175
      see also Aromatase inhibitors;
         Gonadotropin-releasing hormone
         (GnRH) receptor ligands
   new research   7:177
   structures   7:153*F*
   triphenylethylenes *see*
      Triphenylethylene anticancer
      drugs
   unmet medical needs   7:177
Anticancer drugs, microtubule-
   targeting   **7:81–110**
   see also specific drugs
   characterization   7:81
   mechanism of action   7:81, 7:82*F*
   microtubule assembly
      inhibition   7:81, 7:82
      colchicine binding domain
         inhibitors   7:88
      vinca alkaloid binding   7:82
   microtubule assembly
      promoters   7:99
   microtubule stabilizers   7:81, 7:99
      synthetic   7:103, 7:104*F*
   toxicity   7:46*T*
Anticancer drugs, tumor models   7:34,
   7:222
   immunocompromised animals
      7:34–35
   in vivo gene targeting   7:37
   knockout animals   7:37
   leukemia/lymphoma models
      7:34–35
   liquid/hematologic models   7:36
   orthotopic/metastatic models   7:36
   predictive value   7:37, 7:39*T*
      cell line integrity   7:42–43
      comparative genomic hybridization
         (CGH)   7:42–43
      design errors   7:37
      experimental errors   7:37
      immortalized cell lines   7:37
      limitations   7:37
      model numbers   7:38
      model suitability   7:38
      'significant activity'
         definition   7:38
   solid tumors   7:34–35
   subcutaneous/flank tumor
      models   7:35, 7:39*T*, 7:43
      limitations   7:44
      predictive value   7:39*T*
   transgenic animals   7:37
Anticholinergic drugs
   see also specific drugs
   gastric/mucosal ulcer
      treatment   6:604

irritable bowel syndrome
   treatment   6:657
   market size   6:599*T*
   as 'soft drug'   **5:**1011, **5:**1011*F*
Anticoagulant drugs
   see also specific drugs
   introductions   1:169
Anti-cocaine vaccine, cocaine
   abuse   6:181*T*, 6:185
Anticonvulsant drugs   6:280, 6:287,
   6:288
   see also specific drugs
   bipolar affective disorder
      treatment   6:71
   clinical trials   6:287
   gabapentin   8:228–229
   historical perspective   6:287
   mechanisms   6:284
      understanding of   6:292
   newer types   6:290
   NINDS screening program   6:287
   QSAR   4:162, 4:162*F*
   screening   6:293
   targets   6:280, 6:280*T*
Antidepressant drugs   2:62; 6:65
   see also specific drugs
   ADHD treatment *see* Attention deficit
      hyperactivity disorder treatment
   analgesia   6:339
   characteristics   2:64*T*
   drug profitability   1:692–693
   efficacy   6:66*T*
   introductions   1:176
   irritable bowel syndrome
      treatment   6:655*T*, 6:659
   market size   6:599*T*
   modafinil enhancement   8:153
   primary insomnia treatment   6:157
   side effects, transporter binding   2:62
   suicidal behavior association   1:687,
      1:713
   suicide in children/adolescents   6:62
Antidiabetic drugs
   see also specific drugs
   drug introductions   1:108, 1:227
Antidiarrheal drugs
   see also specific drugs
   irritable bowel syndrome
      treatment   6:655*T*, 6:658
Antidigoxin polyclonal antibody   **1:243**
Antidotes, introductions   1:243
Anti-double-stranded DNA antibodies,
   systemic lupus
   erythematosus   6:233
Antiemetic drugs   6:672, 6:681
   see also specific drugs
   cannabinoid receptor
      antagonists   6:684
   cholinergic neurotransmission
      modulators   6:683
   clinical trials   6:676
   dopamine D_2/D_3 receptor
      antagonists   6:683

5-HT$_3$ receptor antagonists   6:681
   future work   6:688
   histamine H$_1$ receptor
      antagonists   6:683
   5-HT$_3$ ligand gated ion channel
      blockage   3:633–634
   introductions   1:107, 1:149
   market   6:600*T*
   new research areas   6:687
      dopamine D$_3$ receptor
         antagonists   6:687
      GABA$_B$ receptor agonists   6:687
      NK$_1$ receptor antagonists   6:682
Antiepileptic drugs
   see also specific drugs
   analgesia   6:339
   calcium-channel blockers   2:849–850
   introductions   1:107, 1:187
   side effects   6:339
   sodium channels   2:839–840
Antiestrogens   7:154
   see also specific agents; specific drugs
   ICI Pharmaceuticals fertility control
      research program   8:86, 8:87*T*
   toxicity   7:46*T*
Antifertility agent, clomiphene
   8:84–85
Antifolates   7:62*T*, 7:69
   see also specific drugs
   active transport   7:58
   clinical trials   7:72
   structure-based drug design   4:72
Antifungal drugs   7:429, 7:430*T*,
   7:445–468
   see also specific drugs
   acetyl-CoA carboxylase
      inhibitors   7:439, 7:439*F*
      candidosis   7:439
      soraphen A   7:439, 7:439*F*
   biological fingerprints   4:809–810,
      4:810*F*
   discovery/ developments   7:432,
      7:433
      cell wall targeting   7:433
      echinocandins   7:433
      pneumocandins   7:433
      pradamicins   7:433
   ergosterol-affecting   7:446
      ergosterol biosynthesis   7:446,
         7:446*F*
      lanosterol-14α-demethylase
         inhibitors   7:449
      squalene epoxidase inhibitors *see*
         Squalene epoxidase inhibitors
   fungal cell wall disrupters   7:456
      polyenes *see* Polyenes
   fungal cell wall synthesis
      inhibitors   7:457
      chitin synthase inhibitors   7:462
      glucan synthase inhibitors *see*
         Glucan synthase inhibitors
      inositol phosphoryl ceramide synthase
         inhibitors   7:463

introductions **1:**107, **1:**131
isoleucyl-tRNA synthetase inhibitors (ITRS) **7:**433
isophosphorylceramide synthase inhibitors **7:**434, **7:**435*F*
lanosterol-14α-demethylase inhibitors *see* Lanosterol-14α-demethylase inhibitors
limitations **7:**429
mechanism of action **7:**431*F*
N-myristoyltransferase inhibitors **7:**436, **7:**464
  benzofuran inhibitors **7:**465
  candidosis **7:**436
  co-crystallographic studies **7:**436, **7:**437*F*, **7:**438*F*
  CP-123,457 **7:**436, **7:**437*T*
  CP-123457 **7:**465
  HTS **7:**465
  UK-370.485 **7:**436, **7:**437*T*
  UK-362,091 **7:**436, **7:**437*T*
natural products **1:**369
  extremophiles **1:**361
new directions **7:**463
preclinical discovery **7:**434
protein biosynthesis inhibitors **7:**439
  candidosis **7:**439–440
  elongation factors **7:**439–440
  GR135402 **7:**439–440, **7:**440*F*
resistance **7:**425–426
RNA polymerase III inhibitors **7:**436
  aspergillosis **7:**436–439
  candidosis **7:**436–439
  HTS **7:**436–439
  UK-118,005 **7:**436–439, **7:**438*T*
sphingolipid biosynthesis inhibitors **7:**434, **7:**435*F*
superficial mycoses **7:**429
systemic mycoses **7:**431
2′.4.4′-trichloro-2′-hydroxydiphenylether **2:**1068
Antifusion drugs
see also specific drugs
influenza viruses, antivirals **7:**385, **7:**386*F*
Antigen-antibody reactions, nanotechnology **2:**361
Antigen capture assays, ELISA **3:**579–580, **3:**579*F*
Antigen Express **1:**516
Antigenic drift, influenza viruses **7:**374–376
Antigenic shift, influenza viruses **7:**374–376
Antigen presentation **5:**219; **7:**846
asthma **7:**876
CD1 molecules **5:**222
intracellular *vs.* extracellular proteins **5:**219–220
major histocompatibility complex **5:**219

non-animal skin sensitization testing **1:**474
to T cells **5:**218–219
Antigen-presenting cells **5:**218–219
glatiramer acetate **8:**180
Antigen processing, ubiquitin - proteasome pathway **3:**190
Antiglaucoma drugs
see also specific drugs
drug introductions **1:**108, **1:**238
Anti-GPIIb/IIIa antibody, animal models **6:**767
Anti-Gram-positive natural products **7:653–671**
see also specific drugs
glycopeptides **7:**654
  clinical indication **7:**656–657
  mechanism of action **7:**655*F*
  resistance **7:**655–656
  SAR **7:**656–657
lipopeptides **7:**665
pseudomonic acid **7:**658
targets **7:**654, **7:**654*F*
Antihistamines (H₁ antagonists) *see* Histamine H₁ receptor antagonists
Antihypertensive drugs
see also specific drugs
ADHD treatment *see* Attention deficit hyperactivity disorder treatment
drug actions **1:**532–533, **1:**532*T*
renal dysfunction **6:**585
Anti-infective drugs
see also specific drugs
definition **7:**470*T*
introductions **1:**106*T*, **1:**107, **1:**108
  antibiotics **1:**107, **1:**108
  antifungals **1:**107, **1:**131
  antimalarials **1:**106–107, **1:**134
  antisepsis **1:135**
  antivirals **1:**107, **1:**119
  natural products **1:**367
  ophthalmic diseases **6:**317
Anti-inflammatory actions
glucocorticoid receptor **2:**1006
liver X receptor (LXR) **2:**1010
Anti-inflammatory cytokines
see also specific drugs
nervous system autoimmune diseases, treatment **6:**244
Anti-inflammatory drugs
see also specific drugs
Alzheimer's disease treatment **6:**211
amyotrophic lateral sclerosis treatment **6:**224
atherosclerosis **6:**490
central nervous system injury **6:**272, **6:**273*F*
Huntington's disease treatment **6:**221
introductions **1:**216
ophthalmic diseases **6:**317
prostatitis treatment **6:**571

Anti-interleukin-2 antibodies
Crohn's disease treatment **6:**244
multiple sclerosis treatment **6:**244
Anti-interleukin-6 receptor antibodies, systemic lupus erythematosus treatment **6:**248
Antileukotriene drugs
see also specific drugs
atopic dermatitis treatment *see* Atopic dermatitis treatment
skin inflammation treatment **7:**979
Antimalarial drugs **1:**370; **7:**752, **7:**753*T*, **7:**754, **7:765–814**, **7:**800*T*
see also specific drugs
antimicrobial agents **7:**474, **7:**795
apicoplast as target **7:**792
  fatty acid biosynthesis inhibitors **7:**792, **7:**792*F*, **7:**793*T*
  isoprenoid biosynthesis inhibitors **7:**794, **7:**795*F*
artemisinin *see* Artemisinin
chalcones **7:**798
classification **7:**772
combination therapies **7:**799
complicated malaria **7:**755
de novo pyrimidine synthesis inhibitors **7:**788, **7:**790*T*
doxycycline **7:**612
farnesyltransferase inhibitors **7:**791
future prospects **7:**803
  resistance problems **7:**803–804
glycolytic pathway inhibitors **7:**785
  fructose-1,6-bisphosphate aldolase **7:**785–786
  lactate dehydrogenase **7:**785–787
  triosephosphate isomerase **7:**785–786
introductions **1:**106–107, **1:**134
natural products **7:**365–366
nucleotide mechanism **7:**787
peptide deformylase inhibitors **7:**796
phospholipid metabolism inhibitors **7:**797
  bis-quaternary ammonium salts **7:**797–798
  monoquaternary ammonium salts **7:**797
prophylaxis **7:**755
protease inhibitors **7:**778
  aspartic proteases (plasmepsins) inhibitors **7:**781, **7:**783*T*, **7:**786*F*
  cysteine protease (falcipains) inhibitors **7:**781, **7:**785*T*, **7:**786*F*
  falcilysin inhibitors **7:**784
purine salvage pathway inhibitors **7:**787, **7:**788*F*
quinolines *see* Quinolines
tetracyclines **7:**610, **7:**612
uncomplicated malaria **7:**754

Antimetabolites 7:6–7, 7:33, **7:55–79**
see also specific drugs
clinical trial issues 7:60
adverse events 7:60, 7:61*T*
efficacy doses 7:61
experimental models 7:60
cell proliferation 7:60
in vivo models 7:60
tumor type 7:60
xenografts 7:60
intracellular transformations 7:59,
7:59*F*
folylpolyglutamate
synthetase 7:59
new research areas 7:74
combination treatments 7:74–75,
7:75*T*
expression profiling 7:75
toxicity reduction 7:74
nucleoside mimics 7:62*T*, 7:63
clinical trials 7:67
delivery 7:68
mechanism of action 7:63, 7:64*T*
toxicity 7:46*T*
unmet medical needs 7:74
Antimicrobial drugs
see also specific drugs
animal models 7:472
half-life differences 7:472–473
antimalarial drugs 7:795
classification 7:471, 7:471*T*
by target site 7:471
clinical needs 7:473, 7:474*T*
developing countries 7:474
diarrheal disease 7:474
drug resistance 7:473
HIV 7:474
industrial countries 7:473,
7:474–475
malaria 7:474
morbidity/mortality *vs.*
burden 7:473
prosthetic implants 7:474–475
tuberculosis 7:474
definition 7:470*T*
doxycycline *see* Doxycycline
evaluation process 7:472
toxicity 7:472
future work 7:475
antibacterial agents 7:475–476
antivirals 7:475
methicillin-resistant *Staphylococcus
aureus* 7:475
natural products 7:475
resistance 7:475
historical perspective 7:469,
7:470–471
Ehrlich, Paul 7:469
natural products 7:469,
7:470–471
synthetics 7:470–471
HTS 2:397
in vitro activity 7:473

mechanism of action, ribosomal RNA
(rRNA) interactions 2:1041
screening 7:471
'indicator organisms' 7:472
in vitro screening 7:472
pharmacology 7:472
as 'soft drug' 5:1011
tetracycline C4 derivatives 7:604
therapeutic development 7:471
Antimicrobial peptides
see also specific drugs
psoriasis 7:942
Antimicrotubule drugs *see* Anticancer
drugs, microtubule-targeting
Antimigraine drugs
see also specific drugs
introductions 1:107, 1:181
Anti-MRSA cephalosporins *see*
Cephalosporins
Antimuscarinics, urinary incontinence
treatment 6:543–544
Antimycin
mitochondria complex III 2:1075
structure 2:1076*F*
Antimycobacterial drugs 7:474, **7:699–
730**, 7:705*T*
see also specific drugs
clinical trials 7:703
diagnostic tools 7:703–704
duration 7:703–704
latent infection 7:703
multidrug-resistance 7:703
patient numbers 7:703–704
current 7:704
directly observed treatment, short
course 7:704
mechanisms of action 7:704, 7:704*F*
new research areas 7:720, 7:720*F*,
7:720*T*
persistence-specific 7:726
unmet medical needs 7:719
Antinausea drugs
see also specific drugs
market 6:600*T*
Antineoplastic drugs
see also Anticancer drugs; *specific drugs*
apoptosis induction 1:382
caspase activation 1:382
introductions 1:143
natural products 1:380
toxicity 1:380
Anti-NS5b nonnucleosides *see* Hepatitis
C virus, RNA polymerase
inhibitors
Anti-NS5b nucleosides *see* Hepatitis C
virus, RNA polymerase inhibitors
Anti-obesity, drug introductions 1:242
Antioxidants
atherosclerotic plaque
formation 6:469
carvedilol *see* Carvedilol
central nervous system injury 6:266,
6:267*F*

coronary heart disease
treatment 6:478
hypertension treatment, mechanism
of action 6:590–591
obesity management, mechanism of
action 6:590–591
oxytetracycline 7:619
renal dysfunction
management 6:590
spinal cord injury treatment 6:260*F*,
6:266
tetracyclines 7:619
traumatic brain injury
treatment 6:260*F*, 6:266
Antioxidant Supplementation in
Atherosclerosis Prevention
(ASAP) 6:478
Antipain 6:272
Antiparkinsonian drugs
see also specific drugs
introductions 1:107, 1:179
Anti-phosphoserine antibodies,
biomarker detection 3:72
Anti-phosphotyrosine antibodies
biomarker detection 3:72
protein kinase substrate
purification 3:72
Antiplatelet therapy
see also specific drugs
abciximab 1:434
monoclonal antibodies 1:434
Antiproliferative properties,
carvedilol 8:137–138, 8:141
Antipsoriasis, drug
introductions 1:108, 1:238
Antipsychotic drugs
see also specific drugs; specific drugs
development,
pharmacophores 4:516*F*
first generation 6:29
dopamine D2 receptor
antagonism 6:31
second generation *vs.* 6:33
market size 6:599*T*
second generation 6:29
first generation *vs.* 6:33
mechanisms of action 6:32
serotonin receptor/D2 receptor
interactions 6:33
side effects 6:33
Antipyrine
BCS 5:973*T*
blood–placenta transfer
modeling 5:737
effective permeability (P_{eff})
5:673*T*
oral absorption 5:468*T*, 5:680*T*
Antirheumatic drugs
see also specific drugs
introductions 1:212
Antirhinoviral drugs
see also specific drugs
PASP 3:804, 3:804*F*

Antisecretory compounds, market
    size 6:598
Antisense-based therapeutics
    see also Messenger RNA (mRNA);
        specific therapies
    development 1:601, 1:602
    drug discovery 2:35, 2:35F
    drug–polymer conjugates 5:1045
    vascular endothelial growth
        factor 3:203
Antisense oligonucleotides, transgenic
    animal generation 3:152
Antisense phosphorodiamidate
    morpholino oligonucleotides,
    Danio rerio model system 3:10
Antisense RNA, target selection 2:761
Antisepsis drugs
    see also specific drugs
    definition 7:470T
    introductions 1:135
Antispasmodic drugs
    see also specific drugs
    irritable bowel syndrome
        treatment 6:659
    market size 6:599T
Antithrombotic drugs
    see also specific drugs
    introductions 1:164
Anti-TNF-α therapies
    see also specific drugs
    Crohn's disease treatment 6:244
    rheumatoid arthritis treatment 6:244
Antitrust liability, market
    regulation 1:730
α₁-Antitrypsin
    deficiency, COPD 7:875
    drug delivery, orphan
        designation 1:658
Antitrypsin depletion,
    proteomics 3:31–32
Antitumor antibiotics
    see also specific drugs
    toxicity 7:46T
Antitumor antibodies 7:33
    see also specific antibodies
Anti-tumor necrosis-α drugs
    see also specific drugs
    IBD treatment 6:623–624
Antityrosinemia, drug
    introductions 1:247
Antiulcerative drugs
    see also specific drugs
    introductions 1:198
    market size 6:599T
Anti-VEGF antibodies, psoriasis
    treatment 7:979
Antiviral drugs 7:253–293, 7:254T
    see also specific drugs; specific viruses
    conjugation, drug metabolism 5:149
    introductions 1:107, 1:119
    lipodystrophy syndrome,
        mitochondrial effects 2:1082
    mitochondrial effects see Mitochondria

molecular connectivity indices
    (MCIs) 4:563
natural products 1:372
2'.4.4'-trichloro-2'-
    hydroxydiphenylether 2:1068
Antiviral mononucleotides, as
    prodrugs 5:1021
Anti-VLA therapies, multiple sclerosis
    treatment 7:926
Antixerostomia, drug
    introductions 1:246
Antizol see Fomepizole
Antocin see Atosiban
Anton Paar GmbH, microwave-assisted
    chemistry 3:851
    pressure measurement 3:852
    Synthos 3000 3:851, 3:852F
    temperature measurement 3:852
Anxiety disorders 6:5, 6:85–115
    see also specific diseases/disorders
    classification 6:86
    corticotrophin-releasing factor
        receptor 3:93–94
    diagnosis 6:5–6
    incidence 6:5
    neurochemistry 6:90
        cholecystokinin 6:90
        corticotropin-releasing factor 6:90
        dopamine 6:90
        GABA 6:90
        galanin 6:90
        glutamate 6:90
        5-hydroxytryptamine 6:90
        neuropeptide Y 6:90
        neurotrophic factors 6:90
        norepinephrine 6:90
        substance P 6:90
    neurocircuitry 6:87, 6:88F
        amygdala 6:87
        bed nucleus of the stria terminalis
            (BNST) 6:89
        behavioral components 6:89
        medial frontal cortex
            (mFC) 6:87–89
        sensory information 6:87
        threat information 6:87–89
    neuroimaging 6:89
        functional magnetic resonance
            imaging (fMRI) 6:89
        magnetic resonance imaging
            (MRI) 6:89
        positron emission tomography
            (PET) 6:89
        single photon emission computed
            tomography (SPECT) 6:89
    prognosis 6:6
    treatment 6:6
        benzodiazepines 6:6
Anxiety disorders, animal models 6:92
    conflict models 6:93
        four-plate test 6:93
        Geller–Seifter (GS) test 6:93
        Vogel conflict test (VCT) 6:93

defensive burying 6:93
defensive test battery 6:94
    benzodiazepines 6:94
    chlordiazepoxide 6:94
    diazepam 6:94
    5-HT₁A agonists 6:94
    fluoxetine 6:94
    imipramine 6:94
    mice rats 6:94
    phenelzine 6:94
elevated plus maze model 6:92
fear potentiated startle models 6:93
genetic basis 6:90
light-dark box 6:92
marmoset human threat test 6:95
    diazepam 6:95
open field model 6:94
social interaction test 6:95
stress-induced hyperthermia 6:95
ultrasonic vocalization 6:95
Anxiety disorders, genetic basis
    animal models 6:90
    GABA 6:90
        β₂ subunit gene deletion 6:91
        GAD67 knockout models 6:90
    human genetics 6:91
        phenotype definition 6:91
    noradrenergic 6:91
        alpha₂-adrenoceptor deletion 6:91
        COMT 6:91
        monoamine oxidase knockout
            models 6:91
        norepinephrine transporter 6:91
    serotonergic 6:91
        5-HT₁A receptor deletion 6:91
        5-HT₁B receptor deletion 6:91
        5-HT₁C receptor deletion 6:91
        serotonin transporter
            deletion 6:91
Anxiety disorder treatment 6:97
    see also specific drugs
    benzodiazepines 6:97
    clinical trials 6:96
    corticotropin-releasing factor
        antagonists 6:102
    5-HT₁A agonists 6:101
    GABA_A receptor modulators, subtype-
        specific 6:103
    gabapentinoids 6:108
    melanin-concentrating
        hormone 6:109
    melanocortins 6:109
    metabotropic glutamate receptor
        agonists 6:106
    monoamine transport inhibitors
        6:99
    neurokinin receptor antagonists see
        Neurokinin receptor antagonists
    new research areas 6:102
Anxiolytic drugs
    see also specific drugs
    introductions 1:197
Anzemet see Dolasetron mesylate

AP6 **2**:900
AP22161
  SH2 inhibitor **2**:983
  structure **2**:983*F*
AP22408
  SH2 inhibitor **2**:983
  structure **2**:983*F*
AP23451
  analogs **2**:977
  Src kinase inhibitor **2**:976–977
  structure **2**:978*F*
AP23464 **7**:190, **7**:192
  Src kinase inhibitor **2**:976–977
  structure **2**:974*F*, **2**:978*F*, **7**:190*F*
AP23573
  characteristics **2**:984*T*
  PI-3K–Akt/PKB–aTOR pathway
    kinase inhibitors **2**:980–981
  structure **2**:981*F*
AP23644, Bcr-Abl kinase
    inhibitor **2**:973
AP23848
  KIT inhibitor **2**:972
  structure **2**:971*F*
Apaziquone **1**:514
APC-366
  asthma treatment **7**:882
  clinical trials **7**:882
APC mutations, cancer
    differentiation **7**:12
APD-125, primary insomnia
    treatment **6**:159*T*
APD356 (lorcaserin
    hydrochloride) **6**:414
  structure **6**:414*F*
Apex-3D, pharmacophore
    searching **4**:135*T*
APFSiA, as prokinetic agent **6**:686
Apical cell membrane, passive
    transcellular transport **5**:264
Apical sodium-dependent bile acid
    transporter (ASBT), 3D
    QSAR **5**:595
Apicidin **1**:371–372
  structure **1**:371*F*
Apigenin **1**:452*T*
API stability, drug reference
    standard **2**:167
Aplasmomycin, carbon–boron
    bioisosterism **2**:698
Aplidine **1**:394
  analogs **1**:395
  matrix metalloproteinase
    blocking **1**:395
  structure **1**:394*F*
Apolar surface area (ASA), molecular
    descriptors **5**:582
Apolipoprotein-A (apoA), plasma
    levels **6**:466
Apolipoprotein A-I (ApoA-I), coronary
    heart disease **6**:471
Apolipoprotein-B (apoB), in
    lipoproteins **6**:463

Apolipoprotein-E (apoE) knockouts,
    atherosclerosis animal
    model **6**:472
Apomorphine
  as antiemetic agent **6**:683
  Parkinson's disease treatment **6**:11,
    6:215
Apopain *see* Caspase-3
Apoptolidin, structure **2**:725*F*
Apoptosis **7**:13
  Bcl-2 overexpression **2**:1076–1077
  benign prostatic hyperplasia **6**:556–
    557
  Caenorhabditis elegans model
    system **3**:8
  cell-based assays **3**:629
  cellular characteristics **3**:195–196
  cytochrome c *see* Cytochrome c
  defects in cancer **7**:13
    apoptosome defects **7**:14
    Bak **7**:14
    Bax **7**:14
    Bcl-2 superfamily **7**:14
    evasion **7**:18
    inhibitors of apoptosis **7**:14
    Warburg effect **7**:13
  evasion, metastasis *see* Metastasis
  extrinsic pathway (death
    receptors) **3**:195–196; **7**:13
    apoptosis-inducing factor
      (AIF) **3**:196
    Bak **3**:196
    Bax **3**:196
    cytochrome *c* **3**:196
  genes, immunosuppression
    5:223–224
  imaging, [^{111}In]-DTPA-PEG
    annexin **3**:668
  induction
    antineoplastic drugs **1**:382
    retinoblastoma protein (Rb) **7**:11
    tetracyclines **7**:620
    tumor necrosis factor-α (TNF-
      α) **3**:196–197
  initiation, mitochondria complex
    I **2**:1068–1069
  intrinsic pathway **3**:195–196; **7**:13
    see also Mitochondria
  mitochondria *see* Mitochondria
  NF-κB pathway **3**:192, **3**:195
  small animal imaging **3**:668
Apoptosis-inducing factor (AIF) **3**:196
Apoptosis inhibitors
  see also specific drugs
  age-related macular degeneration
    treatment **6**:315*T*
Apoptosome defects, in cancer **7**:14
Apoptotic neurodegeneration, central
    nervous system injury **6**:273
Apparent association constant
    (log$K^{app}HSA$) **5**:857–858
Apparent membrane permeability
    coefficient ($P_{m,app}$) **5**:465

Apparent permeability (P_{app}) **5**:309
Appendix, DCN **2**:155
Appetite suppressants
  see also specific drugs
  safety **1**:686
Applicability domain, QSAR *see*
    Quantitative structure–activity
    relationship (QSAR)
Application limits, toxicity,
    computational models **5**:916
Applied Biosystems, microarrays **3**:89
Approval systems, R&D
    management **2**:250
APPROVe trial **1**:684
Apraclonidine
  side effects **6**:311
  structure **6**:309*F*
Apramycin **7**:632
  mechanism of action **7**:640
  structure **7**:633*F*
Aprepitant **1**:149
  as antiemetic agent **6**:682, **6**:685
Aprovel *see* Irbesartan
Aptamers
  biomarker identification **3**:73
  conjugation, 5-bromo-
    deoxyuridine **3**:73
Aptiganel **6**:259
AQUA **3**:547*T*
AQUASOL database **5**:632
Aqueous environment, Molecular
    Interaction Fields **4**:238
Aqueous free energy of dissociation,
    ionization **5**:605
Aqueous solubility
  electrotopological indices (E-state
    index) *see* Electrotopological
    indices (E-state index)
Aqueous solutions
  chemical stability **5**:490–491
  microwave-assisted chemistry **3**:854
  solubility definition **5**:628
Aqulpa *see* Nedaplatin
Ara-A *see* Arabinosyladenine
Arabic culture, natural
    products **1**:356–357
Arabidopsis thaliana, genome
    sequencing **3**:317*T*
Arabinosyladenine
  discovery/development **1**:372
  structure **1**:373*F*
Arachidonic acid
  hypertension **6**:582, **6**:582*F*
  structure **2**:909*F*
Arachidonic acid pathway, leukotriene
    biosynthesis inhibition **7**:904
Arachnoid epithelium, central nervous
    system **5**:302
Arakis **1**:504
Aralkylamine *N*-acetyltransferase,
    acetylation **5**:153
Aranidipine **1**:153
  SAR **6**:716*T*

Arava *see* Leflunomide

Archangel syndicate, Scotland **1**:564

Archive of Obsolete PDB
  Entries **3**:535–536

Arcoxia *see* Etoricoxib

Area postrema, emesis **6**:672–673

Area under plasma concentration curve,
  definition **5**:869*T*

Area under the curve (AUC)
  bioavailability **5**:701
  pharmacokinetics **5**:18

Arena Pharmaceuticals **1**:514

Arenaviruses, antiviral drugs **7**:254*T*,
  7:278

Arene oxides
  epoxide hydrolases **5**:141, **5**:143*F*,
    5:144*F*
  glutathione transferases **5**:161,
    5:161*F*

Arexis **1**:511

Argenta Discovery **1**:492*T*, **1**:513

Arglabin **1:136**
  mechanism of action **2**:851–852

Aricept *see* Donepezil; Donepezil
  hydrochloride; Duloxetine

Ariëns, Eef, collaborations, Timmerman,
  Henk **8**:8

Arimidex *see* Anastrozole

Arimoclomol
  amyotrophic lateral sclerosis
    treatment **6**:224

A-ring chemistry, tetracyclines *see*
  Tetracyclines

Aripiprazole **1:184**
  mechanism of action **6**:34
  schizophrenia treatment **6**:34
  side effects **6**:33

Arixtra *see* Fondaparinux sodium

Armodafinil
  narcolepsy treatment **6**:155*T*, **6**:156
  structure **6**:181*F*

Arndt, Ulie, x-ray
  crystallography **4**:66–67

Aromasin *see* Exemestane; Formestane

Aromatase inhibitors **7**:175, **7**:176*F*
  see also specific drugs
  breast cancer treatment **8**:97*F*
  introduction of **1**:107

Aromatic amine/amide activation,
  cytochrome P450 1A2 **5**:918,
    5:919*F*

Aromatic bases, ionization constants
  (pK_a) **5**:378

Aromatic dicarboxylic acid derivatives,
  hydroxamic acids **7**:226

Aromatic hydrolamines,
  sulfotransferases **5**:148*F*, **5**:149

Aromatic-hydroxylamine *O*-
  acetyltransferase,
  acetylation **5**:153

Aromatic moiety, azetidinone
  cholesterol absorption
  inhibitors **8**:70, **8**:71*F*

Aromatic para-amino substitutes,
  bioisosterism **2**:674

Aromatic polyamine amides, solid-phase
  discrete libraries **3**:709*T*

Aromatic residues, hERG **5**:938–939,
  5:939*F*

Aromatic rings
  azetidinone cholesterol absorption
    inhibitors **8**:70, **8**:71–72, **8**:71*F*,
    8:73*F*
  oxidation, cytochrome P450s **5**:101,
    5:102*F*
  quinolone antibiotics **7**:583, **7**:583*T*

Aromatic sp^2 hydroxyl probe, molecular
  interaction fields **4**:239–240,
    4:239*F*

Aroyl-pyrrole-hydroxy-amides (APHAs),
  hydroxamic acids **7**:225–226

ARQ-501 **1**:512

Arqule **1**:512; **2**:760*T*

Array Biosciences **1**:512; **2**:760*T*

Arrayed DNA-based siRNA libraries,
  RNA interference **3**:18

ArrayExpress, cDNA microarrays
  3:331

Array formats, cDNA functional
  screening **3**:15–16, **3**:16*F*

ArrayScan II system, high-content
  screening (HCS) **2**:427–428

ArrayScan VTi **3**:686*T*, **3**:687

β-Arrestin *see* Beta-arrestin

Arrhythmias **6**:694, **6**:700, **6**:730
  see also specific types
  atrial **6**:730
    mechanistic models **6**:738, **6**:739*T*
    premature contractions **6**:731
    unmet medical needs **6**:750
  cardiac surgery **6**:749
    Maze technique **6**:749
  diagnosis **6**:730
  disease basis **6**:733
    see also Cardiac ion channels
    cardiac pathologies **6**:733
    conduction velocity **6**:733
    endogenous chemical
      substances **6**:733
    excitability **6**:733
    genetics **6**:733
    iatrogenic causes **6**:733
    ion channels **6**:733
    pacemaking **6**:733
    refractoriness **6**:733
  electrical treatment **6**:741, **6**:749
    ablation treatment **6**:749
    DC conversion to sinus
      rhythm **6**:749
    implantable cardioverter
      defibrillators **6**:749
    pacing **6**:749
  mechanistic macro model **6**:737,
    6:739*T*
    ectopic foci **6**:737
    re-entry **6**:737, **6**:738*F*

models **6**:739
  computer modeling **6**:739
  electrophysiological
    properties **6**:739
  hearts **6**:739
  intracellular action
    potentials **6**:739
  single ion channels **6**:739
  whole cell currents **6**:739
nodal **6**:730
  AV disturbances **6**:730
  mechanistic models **6**:738, **6**:739*T*
  paroxysmal supraventricular nodal
    tachycardia **6**:730
  SA nodal tachycardia/
    bradycardia **6**:730
prevalence **6**:695*T*, **6**:700
treatment **6**:700
unmet medical needs **6**:749
ventricular **6**:731
  mechanistic macro model **6**:739*T*
  unmet medical needs **6**:750

Arrhythmogens, antiarrhythmic drugs,
  rational design **6**:757

Arroe Therapeutics **1**:507

Arsenic trioxide, orphan approval/
  designation **1**:658

Arsenite, immunosuppressive
  effects **5**:224
  cytokine production **5**:224
  JAK-STAT pathway **5**:224

Arsphenamine **1**:530
  development **1**:530

Arteflene **7**:782*T*

Artemether **1**:134; **7**:777
  absorption **7**:777–778
  discovery/development **1**:370
  introduction of **1**:106–107
  structure **1**:371*F*

Artemin, knock-in animal
  models **3**:160

Artemisinin **7**:777
  biological target **7**:778
  derivatives **1**:370–371
  discovery/development **1**:327, **1**:370
  mechanism of action **1**:370–371;
    7:778, **7**:781*F*
  neurotoxicity **7**:778
  quantum mechanical
    calculations **4**:406, **4**:406*F*
  SAR **7**:778, **7**:779*T*
  structural modifications **7**:778
  structure **1**:371*F*

Artemisinin-based combination
  treatment (ACT) **7**:755, **7**:799

Artemotil *see* Artemether

Arterial Biology for the Investigation of
  the Treatment Effects of
  Reducing cholesterol
  (ARBITER-2) trial **6**:479–480

Arterioles, blood–brain barrier
  permeation **5**:302

Artesian Therapeutics **1**:508

Artesunate **7**:777
  malaria treatment **7**:755
Artherectomy **6**:696
Arthritis **7**:846
  see also Osteoarthritis (OA);
    Rheumatoid arthritis (RA)
  animal models **6**:508
  pathophysiology **7**:857–858
  proinflammatory cytokines
    7:857–858
  prevalence **7**:846–847
Arthritis treatment **7**:849
  see also specific drugs
  arachidonic acid cascade **7**:849,
    7:850*F*
  nonsteroidal anti-inflammatory
    drugs **7**:849
  cyclooxygenase inhibitors **7**:849,
    7:852
    type1 *vs.* type 2 **7**:852
  cytokine blocking **7**:857
  disease-modifying anti-inflammatory
    drugs **7**:856
  joint space narrowing
    therapies **7**:863
  leukotriene inhibitors **7**:854
  lipoxygenase inhibitors **7**:849
  media influence on medical need
    evaluation **1**:629
  methotrexate **7**:847
  nonsteroidal anti-inflammatory
    drugs **7**:846–847
  phospholipase A2 inhibitors **7**:849,
    7:850
    cytosolic **7**:850–855
    secretory **7**:850–851
    substrate-based **7**:851
  'pyramid of treatment' **7**:847,
    7:847*F*
  steroids **7**:854
    estrogen receptor **7**:856
    glucocorticoid mimetics **7**:855
  structure-modifying anti-inflammatory
    drugs **7**:857
    aggrecanase inhibitors **7**:865–866
    IKK blockers **7**:861–862
    interleukin-1β blocking **7**:860
    matrix metalloproteinase-13
      blockers **7**:863
    p38 kinase **7**:861
    proteoglycans **7**:865–866
    tumor necrosis factor-α
      blocking **7**:857–858
    tumor necrosis factor-α converting
      enzyme inhibitors **7**:858
Articaine **6**:361
  structure **6**:354*F*
Artifact correction, fluorescence
    screening assays **3**:613
Artificial chromosomes, gene
    therapy **1**:313
Artificial hearts, medical needs *vs.*
    individual desires **1**:628

Artificial membranes **5**:740
  in vitro oral absorption prediction *see*
    Oral absorption prediction, in
    vitro
  low-density polyethylene **5**:742
  mechanistic pharmacokinetic
    studies **5**:38
  polydimethylsiloxane **5**:740
    artificial neural networks **5**:741
    compound restriction **5**:740–741
    hydrogen bonding **5**:741
    modeling **5**:741
  silicone **5**:741
    human epidermis *vs.* **5**:742
    hydrogen bonding **5**:741–742
    lipophilicity **5**:741–742
Artificial neural networks (ANN)
  bioavailability models **5**:709, **5**:719,
    5:720*F*
  blood–breast milk transfer
    modeling **5**:738
  computational oral absorption
    prediction **5**:688–689
  dermal penetration models **5**:731
  electrotopological indices (E-state
    index) **5**:570
  ionization prediction, in silico
    tools **5**:616
  log P calculation **5**:659
  polydimethylsiloxane **5**:741
  protein binding, E-state
    index **5**:566–567, **5**:567*F*
  QSAR **4**:57, **4**:156
  statistical methods, E-state
    index **5**:562
Artificial pancreas, diabetes mellitus
    management **6**:455
Aryl 1,5-enediols, solid-phase discrete
    libraries **3**:709*T*
Aryl alcohols, quinolines **7**:772, **7**:775
Arylamine *N*-acetyltransferase,
    acetylation **5**:153
Arylamines
  acetylation **5**:154
  solid-phase discrete libraries **3**:709*T*
2-Arylaminothiazoles, solid-phase
    discrete libraries **3**:707*F*
N-Arylation, solid support
    synthesis **2**:466*F*, **2**:467, **2**:467*F*
O-Arylation, solid support
    synthesis **2**:466*F*, **2**:467, **2**:468*F*
Aryl benzothiophenes, estrogen
    receptor-β-selective
    ligands **7**:166–167, **7**:167*F*
Aryl boronic acid synthesis,
    PASP **3**:813
Aryl bromide synthesis, PASP **3**:813
Aryl derivatives, carboxylic acid
    bioisosterism **2**:660, **2**:662*T*
Aryl diamidines, solid-phase discrete
    libraries **3**:709*T*
Aryl hydrocarbon receptor, protein
    kinase inhibitors **1**:390–391

5-Arylidine 4-thiazolidinones,
    PASP **3**:806–807, **3**:807*F*
Arylindoles, solid-phase pool
    libraries **3**:705*F*
Arylmethyl amidines, solid-phase
    discrete libraries **3**:709*T*
Arylmethyl cinnamic acids, solid-phase
    discrete libraries **3**:709*T*
Aryl monophosphates, as
    prodrug **5**:1025
N-Aryloxazolidin-2-ones, solid-phase
    discrete libraries **3**:709*T*
Aryloxoisobutyric acid, solid-phase
    discrete libraries **3**:709*T*
Aryloxyprolines, solid-phase pool
    libraries **3**:705*F*
Aryloxypropanolamines, as 'soft
    drug' **5**:1011, **5**:1011*F*
Aryl phosphate-binding site, protein
    tyrosine phosphatase-1B **4**:639,
    4:640*F*
Arylpiperazine derivatives, solid-phase
    discrete libraries **3**:709*T*
N-Arylpiperazines, solid-phase discrete
    libraries **3**:709*T*
Arylpyrimidine derivatives, solid-phase
    discrete libraries **3**:709*T*
Arylsulfonamide hydroxamates
  solid-phase discrete libraries **3**:709*T*
  solid-phase libraries **3**:725–726,
    3:726*F*
Arylsulfonamides
  solid-phase discrete libraries **3**:707*F*
  structure **7**:96*F*
Aryl tertiary amides, solid-phase
    discrete libraries **3**:709*T*
Arylthioisobutyric acid, solid-phase
    discrete libraries **3**:709*T*
Aryl vinyl sulfones, solid-phase discrete
    libraries **3**:709*T*
Aryplase, as fast-tracked drug **1**:760*T*
Arzoxifene **7**:154; **8**:115
  as estrogen receptor ligand **2**:1014
  structure **2**:1013*F*; **7**:153*F*; **8**:105*F*
AS-9705, as antiemetic agent
    6:687–688
Ascariasis **7**:751*T*
  biology **7**:759
  clinical manifestations **7**:759
  drug treatment **7**:753*T*
  epidemiology **7**:759
Ascaricides, mitochondria complex
    I **2**:1068–1069
Ascaris lumbricoides see Ascariasis
Ascomycin, discovery/
    development **1**:327
Ascorbic acid (vitamin C)
  coronary heart disease
    treatment **6**:478
  discovery, Szent-Gyorgi,
    Albert **2**:258
  oral absorption **5**:680*T*
  structure **5**:370*F*

Asenapine **6**:32
Asimadoline, irritable bowel syndrome
    treatment **6**:658–659, **6**:661*T*
Asoprisnil
    progesterone receptor ligands
        2:1015
    structure **2**:1015*F*
Asparagine deaminations, heterogeneous
        protein crystallization **3**:436
Aspartate aminotransferase, structure,
        monomer oddity **3**:540
Aspartic acid proteases, HIV protease
        inhibitors **7**:347–349
Aspartic proteases
    see also specific enzymes
    members **2**:625
    nonpeptide drugs **2**:625
    peptidomimetics **2**:625
    pharmacological target space **4**:422*T*
    three-dimensional structures **2**:626*T*
Aspartic proteases (plasmepsins)
        inhibitors
    see also specific drugs
    antimalarial drugs **7**:781, **7**:783*T*,
        7:786*F*
Aspergillosis **7**:427
    aureobasidin A **7**:434
    caspofungin **7**:462
    as emerging disease **7**:425–426
    micafungin **1**:133
    naftifine **7**:447
    opportunistic infections **7**:425*F*,
        7:425*T*
    RNA polymerase III
        inhibitors **7**:436–439
    virulence factors **7**:427
    voriconazole **1**:133–**1**:134;
        7:432–433
Aspergillus fumigatus, structure **7**:421*F*
Aspergillus niger, BMS-192548
        biosynthesis **7**:602
Asperlicin, structure **2**:36*F*
Aspirin *see* Acetylsalicylic acid (aspirin)
Assays
    development, HTS *see* High-
        throughput screening (HTS)
    false positives **2**:737
    HTS automation **3**:682
    quality **3**:91
        evaluation *see* High-throughput
            screening (HTS)
        readout interference, promiscuous
            ligands **2**:746
        throughput, GPCR functional assay
            screening **2**:814
        validation effects, screening library
            selection **4**:506
        variation, HTS **2**:403
Assembly algorithms, genome
        sequencing **3**:317
ASSENT-1, tenecteplase **6**:774
ASSENT-3, tenecteplase **6**:774
Assessment, project teams **2**:146

Assessment responsibilities, best in
        class *vs.* first in class **2**:119–120
Assessment of the Safety of a New
        Thrombolytic: TNK-tPA *see*
        ASSENT-1; ASSENT-3
Association constant (logK_{HSA}), albumin
        binding prediction
        models **5**:856–857
Association for University Research and
        Industry Links (AURIL)
        1:582
Association of British Pharmaceutical
        Industries (ABPI) **1**:584
Association of Laboratory
        Automation **3**:680
Association rate constant (on rate: k_{on}),
        ligand–receptor
        interactions **4**:194
Assumed dimensionality, topological
        descriptors **5**:560
Assumed dose effects, intestinal
        absorption **5**:261
Astat *see* Lanoconazole
Astbury, W T, protein
        crystallography **3**:450
Astellas, mergers and
        acquisitions **1**:530
Astemizole
    IK$_r$ blockade **6**:746
    structure **5**:934*F*
Astex Molecules **2**:761*T*
    characteristics **2**:226*T*
AstexViewer **4**:796, **4**:797*F*
Asthma **7**:848, **7**:873
    airway remodeling **7**:878
        angiogenesis **7**:878
        epithelial cells **7**:878
        epithelial damage **7**:878
        extracellular matrix **7**:878
        mast cells **7**:878, **7**:897
        mucus secretion **7**:878
        smooth muscle **7**:878
    COPD *vs.* **7**:873–874, **7**:874*T*, **7**:875,
        7:880
    disease basis **7**:881
        airflow limitation **7**:873–874
        airway inflammation **7**:874–875
        bronchoconstriction **7**:874–875
        eotaxin (CCL11) **7**:876–877
        granulocyte-macrophage colony-
            stimulating factor
            (GM-CSF) **7**:876–877
        immune cells **7**:875
        inflammation **7**:875
        interferon-γ **7**:876–877
        interleukin-1 (IL-1) **7**:876–877
        interleukin-2 (IL-2) **7**:876–877
        interleukin-3 (IL-3) **7**:876–877
        interleukin-4 (IL-4) **7**:876–877
        interleukin-5 (IL-5) **7**:876–877
        interleukin-13 (IL-13) **7**:876–877
        NF-κB pathway **3**:197
        STATs **3**:208

disease status **7**:875
    forced expiratory lung volume
        (FEV$_1$) **7**:875
    peak expiratory flow (PEF) **7**:875
    pulmonary function tests **7**:875
    severity classification **7**:875
    symptom severity/
        frequency **7**:874–875
epidemiology **7**:874*T*
    genetics **1**:286; **7**:848
    incidence **7**:873
    prevalence **7**:848
leukocyte activation/
        trafficking **7**:875
    antigen presentation **7**:876
    B cells **7**:876
    CD4+ T cells **7**:877
    cytokines **7**:876–877
    eosinophils **7**:876–877, **7**:888
    immunoglobulin E **7**:875
    interleukin-5 **7**:902–903
    mast cells **7**:876–877
    prostaglandin D$_2$ **7**:888–889
    T cells **7**:876
    Th2 cells **7**:877
models **7**:880
    advantages/disadvantages **7**:881
    BLT$_1$ knockout animals
        7:880–881
    eosinophil-negative transgenic
        animals **7**:880–881
    knockout animals **7**:880–881
    montelukast **7**:880
    protein antigens **7**:880
    rodents **7**:880–881
    transgenic animals **7**:880–881
    zileuton **7**:880
pulmonary administration **5**:286–287
Asthma treatment **7**:848, **7**:883,
        7:883*T*
see also specific drugs
albuterol **7**:883
beclomethasone **7**:883
β-adrenergic receptor agonists **7**:848
clinical trials **7**:881
    APC-366 **7**:882
    BMS-639623 **7**:882
    cathepsin S inhibitors **7**:882
    CCR3 antagonists **7**:882
    GSK-766994 **7**:882
    inflammation **7**:881–882
    leukotriene inhibitors **7**:882
    matrix metalloproteinase-12
        inhibitors **7**:882
    mepolizumab **7**:882
    montelukast **7**:881–882
    serine protease inhibitors **7**:882
    zileuton **7**:882
epinastine hydrochloride **1**:204
formoterol **7**:883
glucocorticoids **7**:848
imatinib mesylate **3**:3
inhaled corticosteroids **7**:883

Asthma treatment (*continued*)
  interleukin-5 receptor inhibitors *see*
      Interleukin-5 receptor inhibitors
  leukotriene modifiers 7:883
  long-acting β-adrenergic receptor
      agonists 7:883
  montelukast 7:883
  new research areas 7:884
      see also specific areas
      chemokine receptor CCR3 *see*
          CCR3
      CRTH2/DP-1 receptor *see* CRTH2/
          DP-1 receptor
      interleukin-5 receptor inhibitors *see*
          Interleukin-5 receptor inhibitors
      leukotriene biosynthesis inhibition
          see Leukotriene biosynthesis
          inhibition
      mast cell tryptase/chymase *see* Mast
          cell tryptase/chymase
      phosphodiesterase type 4 *see*
          Phosphodiesterase-4
      pharmacogenetics *see*
          Pharmacogenetics
  phosphodiesterase 4 inhibitors 7:893
  phosphodiesterase-4
      inhibitors 2:935–936, 2:948
  salmeterol 7:883
  short-acting β-adrenergic receptor
      agonists 7:883
  theophylline 7:883
  triamcinolone 7:883
  unmet medical needs 7:884
  zileuton 7:883
AstraZeneca
  see also ICI Pharmaceuticals
  academia collaborations 1:577*T*
  Integrative Pharmacology
      Fund 1:586
  pipeline of products 1:542
  R&D spending 2:101*T*
  sales/R&D expenditure 1:542*T*
  SME acquisition 1:496*T*
Astressin
  development 2:582
  irritable bowel syndrome
      treatment 6:659, 6:661*T*
Astrocytic glial cells, blood–brain barrier
      permeation 5:302
Asymmetric units
  crystals 3:451
  protein crystallization 4:762
  x-ray structures, problems 3:540
Atacand *see* Candesartan
Atazanavir 1:120; 2:579*T*; 7:279–282
  discovery/development 1:76*F*;
      7:353
  dosing regimens 7:357*T*
      with ritonavir 7:358*T*
  structure 1:76*F*; 7:354*F*
Atazoa 1:492*T*
ATC-0065 6:109
  structure 6:108*F*

ATC-0175 6:109
  structure 6:108*F*
Atelec *see* Clinidipine
Atenolol
  absorption 5:42, 5:468*T*, 5:680*T*
  BCS 5:973*T*
  bioavailability 5:712*T*, 6:712*T*
  bioisosterism 2:675, 2:677*F*
  discovery/development 1:328–329,
      1:374
  dose 6:712*T*
  effective permeability (P_{eff}) 5:673*T*
  half-life 6:712*T*
  mechanism of action, GPCR
      targeting 2:773*T*
  paracellular transport 5:265
  structure 1:374*F*; 2:677*F*; 6:587*F*,
      6:712*T*
Atevirdine 7:344
Athelas 1:507
Atherosclerosis 6:459–494, 6:461*F*
  see also Atherosclerotic plaques
  animal models 6:472
      higher species 6:472
      murine models 6:472
  clinical trial issues 6:472
  disease basis 6:462
      see also Lipoproteins
      plaque rupture 6:461
      reactive oxygen species 6:466
      vessel wall remodeling 6:460
  imaging methods 6:461
  new research areas 6:487
      anti-inflammatory agents 6:490
  treatment 6:480
      see also specific drugs
      inflammatory mediators 6:466
  unmet medical needs 6:486
      high-density lipoprotein
          cholesterol-elevating
          agents 6:486
      low-density lipoprotein cholesterol
          reduction 6:486
Atherosclerosis Risk in Communities
      (ARIC) study 6:470
Atherosclerotic plaques
  formation 6:461*F*, 6:466, 6:468*F*,
      6:765
      antioxidants 6:469
      chemokines 6:468
      fatty acid oxidation 6:467–468
      hydroperoxidases 6:467–468
      inflammation 6:468–469
      LDLs 6:467
      5-lipoxygenase 6:467
      15-lipoxygenase 6:467
      mast cells 6:469
      matrix metalloproteinases 6:468
      myeloperoxidase 6:467
      T cells 6:469
  rupture 6:461
Atipamezole, glucuronic acid
      conjugation 5:152–153, 5:153*F*

ATM gene, cell cycle 7:11–12
Atom–atom relationships,
      electrotopological indices
      (E-state index) 5:556
Atom-based methods
  3D-QSAR 4:582
  de novo drug design *see* Drug design,
      de novo
  whole-molecule properties, similarity
      searching 4:169–170
Atom-by-atom substructures,
      CSD 3:390–391
Atom exclusion, ligand–receptor
      interaction models 4:206
Atom hybridization, quantum
      mechanical calculations *see*
      Structure, quantum mechanical
      calculations
Atomic *B*-factors, protein 3D
      structure 3:546
Atomic coordinate reliability *see* X-ray
      crystallography, problems
Atomic descriptors, ionization
      prediction, in silico tools 5:613
Atomic distance method, hERG blocker
      modeling 5:943–944, 5:944*F*
Atomic electronegativity, specific
      consideration of
      hydrogen 4:548
Atomic fragmental system,
      lipophilicity 5:586, 5:586*F*
Atomic increment projection, hydrogen
      bonding 5:586–587
Atomic model derivation *see* X-ray
      crystallography
Atomic occupancies, protein 3D
      structure validation 3:512
Atomic order, protein crystallography *see*
      X-ray crystallography
Atomic states, electrotopological indices
      (E-state index) 5:556
Atom-layer descriptors, reagent-based
      library design 4:342–344
Atomoxetine 1:195
  ADHD treatment 6:129, 6:129*T*,
      6:132
  cataplexy treatment 6:154
  mechanism of action 6:132
  side effects 6:132
Atom pair descriptors
  fingerprint/dataprint
      descriptors 4:470, 4:470*F*
  library design, ligand-based 4:315,
      4:315*F*
  QSAR 4:153, 4:154*F*
  3D QSAR 4:155
  two-dimensional descriptors, similarity
      searching 4:171
ATOM record, protein 3D structure
      validation 3:513, 3:514*F*
Atom sigma orbital electronegativity,
      electrotopological indices
      (E-state index) 5:557

luteinizing hormone
   antagonists   6:568
new targets   6:568
phosphodiesterase-5 inhibitors
   2:948
phytotreatment   6:568
progression   6:568
Proscar Long-Term Efficacy and Safety
   Study (PLESS)   6:568
prostate-specific antigen
   levels   6:568
receptors   6:563
   α_1-adrenergic receptors   6:563,
      6:564F
   estradiol receptors   6:564
   vitamin D receptors   6:564
5α-reductase inhibitors   6:564, 6:566
testosterone replacement
   therapy   6:537
transurethral incision of the prostate
   (TIRP)   6:543
transurethral microwave hyperthermia
   (TUMT)   6:543
transurethral needle ablation
   (TUNA)   6:543
transurethral resection of the prostate
   (TURP)   6:541, 6:543
vitamin D receptor analogs   6:568
Benoxaprofen, amino acid-coenzyme A
   conjugation   5:156F, 5:157
Benoxinate
   bioisosterism   2:677F
   structure   2:677F
Benserazide, effective permeability
   (P_{eff})   5:673T
Benzaldehyde, structure   5:371F
Benzamide
   derivatives, solid-phase discrete
      libraries   3:709T
   histone deacetylase (HDAC)
      inhibitors   7:227
   structure   5:359
Benzamidoxime reductase   5:124
   mechanism of action   5:125F
Benzazapine carboxamides, respiratory
   syncytial virus (RSV), fusion
   inhibitors   7:409F, 7:411
Benzazepine-3-one synthesis, solution
   phase library synthesis
   3:782–783, 3:784F
Benzazepine antagonists, phenol
   bioisosterism   2:671, 2:672F
Benzene-sulfonamide,
   structure   5:371F
Benzenethiol, structure   5:371F
Benzidazole
   structure   7:823F
   trypanosomiasis treatment   7:758,
      7:822–823
Benzidine, structure   5:359
Benzimadazole ribosides,
   cytomegalovirus
   antivirals   7:308–309

Benzimidazole compounds
anti-NS5b nonnucleosides   7:401,
   7:402, 7:406F
estrogen receptor-β-selective
   ligands   7:169, 7:169F
hepatitis C virus inhibition   2:1047,
   2:1048F
PASP   3:804
ring analogs, rolipram   7:894
solution phase library
   synthesis   3:780, 3:782F
Benzimidazolin-2-ones, PASP   3:805,
   3:805F
Benzimidazolones, solid-phase
   libraries   3:705F, 3:735, 3:736F
Benzo[a][1,2,3]triazinones, solid-phase
   discrete libraries   3:709T
Benzocaine, structure   5:359
Benzo compound splitting (benzo
   cracking), ring system
   reorganization
   (bioisosterism)   2:678, 2:679F
Benzodiazepines
see also specific drugs
abuse potential   6:99
anesthetic adjuvants   6:362
anesthetics   6:365
anxiety disorder treatment   6:6, 6:97
chemical variation   3:925
defensive test battery   6:94
GABA receptor interactions   6:357
   GABA$_A$ receptors   2:893
   GABA$_C$ receptors   2:893
generalized anxiety disorder
   treatment   6:99
historical perspective   6:287
irritable bowel syndrome
   treatment   6:655T
mechanism-based PK/PD
   modeling   5:896, 5:897, 5:897F
mechanism of action   6:98
origins   6:97–98
panic disorders treatment   6:99
PASP   3:799–800, 3:800F
phosphodiesterase-2 inhibitors
   2:933
posttraumatic stress disorder (PTSD)
   treatment   6:99
primary insomnia treatment   6:158
receptor ligands, carboxylic ester
   bioisosterism   2:663, 2:667F
receptors, molecular probes   3:664,
   3:664T
scaffold-based design   4:32–33, 4:33F
solid-phase libraries   3:703, 3:706F,
   3:736–737, 3:736F
discrete libraries   3:709T
1,4-Benzodiazepines, solid support
   synthesis   2:479, 2:481F
Benzodiazepinones, solid-phase pool
   libraries   3:705F
Benzofuran derivatives, solid-phase
   discrete libraries   3:709T

Benzofuran inhibitors,
   N-myristoyltransferase
   inhibitors   7:465
Benzofurans
2,3-substituted, solid-phase pool
   libraries   3:705F
bioisosterism   2:680F
ring system reorganization
   (bioisosterism)   2:679
structure   2:680F
Benzoic acids
antibody-directed enzyme prodrug
   therapy *see* Antibody-directed
   enzyme prodrug therapy
   (ADEPT)
coenzyme A conjugation   5:155
structure   5:369F
Benzopiperazinones, solid support
   synthesis   2:479, 2:481F
Benzopyrans
solid-phase discrete libraries
   3:709T
solid-phase pool libraries   3:707F
structure   1:364F
variation/analogs   1:365
Benzopyrones, ring system
   reorganization
   (bioisosterism)   2:679
Benzothiadiazine, anti-NS5b
   nonnucleosides   7:403F, 7:404
Benzothiazepines, as calcium channel
   blocker   2:849–850
Benzothienol[3,2-*a*]thiadiazine 2,2-
   dioxides, phosphodiesterase-7
   inhibitors   2:943
Benzothiophenes   8:104–106
Benzotriazole, structure   5:371F
5-Benzotriazole, bioisosterism   2:673T
Benzoxazines, GABA receptors   2:893
5-Benzoxazolone, bioisosterism   2:673T
Benzoyl-CoA synthetase, coenzyme A
   conjugation   5:155
Benzoyl peroxide
acne rosacea treatment   7:961
acne treatment   7:938–939, 7:961
side effects   7:961
Benzphetamine
obesity treatment   6:397
   contraindications   6:397
   indications   6:397
   mechanism of action   6:397
   side effects   6:397
   structure   6:395T
Benztropine, Parkinson's disease
   treatment   6:214–215
Benzydamine, oxidation   5:113–114
Benzylamine, structure   5:359
Benzylguanidine derivatives, solid-phase
   discrete libraries   3:709T
O^6-Benzylguanine, triazene
   resistance   7:138
Benzylic carbons, oxidation, cytochrome
   P450   5:92, 5:93F

N3-Benzyl nirvanol, reaction phenotyping studies **5**:243*T*

7-Benzyloxy-4-trifluoromethycoumarin (BFC), in vitro inhibition studies **5**:247*T*

Benzyloxyquinoline (BQ), in vitro inhibition studies **5**:247*T*

Benzyl piperidine motif, CCR3 inhibitors **7**:885–886

5-(Benzylsulfonyl)-4-bromo-2-methyl(2H)-pyridazione (BBMP) **2**:1061, **2**:1061*F*

Bepridil
  bioavailability **5**:712*T*
  IK$_r$ blockade **6**:746

Berberine **2**:1048
  as antimicrobial **2**:1048
  Chinese medicine **2**:1048
  cholesterol-lowering properties **2**:1048
  low-density lipoprotein receptor stabilization **2**:1048
  structure **2**:1048*F*

Berenil
  structure **7**:831*F*
  trypanosomiasis treatment **7**:829–830

Berkeley DB **3**:294

Berkeleydione **1**:361
  structure **1**:360*F*

Berkeley Drosophila Genome Project, *Drosophila melanogaster* model system **3**:9

Berkeley Madonna, metabolic drug–drug interaction software **5**:841

Berkeley Structural Genomics Center (BSGC) **3**:552*T*

Berkeleytrione **1**:361
  structure **1**:360*F*

Bermexane **2**:903

Bernal, J D, protein crystallography **3**:433–434, **3**:450

Beromun *see* Tasonermin

Berzelius, Jons Jakob, proteomics **3**:28

Best, Charles, insulin discovery/development **1**:585

Best in class *vs.* first in class *see* Philosophy (of drug discovery/development)

Best-match searches, similarity searching **4**:167–168

Best volumes, VolSurf descriptors **5**:539

Beta adrenergic receptors *see* β-Adrenergic receptors

Beta amino acids **8**:236, **8**:237*F*, **8**:238*T*
  proteolysis resistance **2**:584, **2**:585

Beta-arrestin
  GPCR desensitization **2**:801
  GPCR interaction **2**:788, **2**:797*T*

phosphodiesterase-4 association **2**:926

β$_2$ subunit gene deletion, anxiety disorders, genetic basis **6**:91

β-cell replacement therapy *see* Diabetes mellitus management

βγ-regulation, G proteins **2**:781*T*, **2**:784, **2**:786*T*

β-blockers *see* β-Adrenergic receptor antagonists

β-turn mimetics, peptidomimetics **2**:608

Beta defensins, psoriasis **7**:942

Beta-endorphin ligand generation, dynamic combinatorial libraries **3**:967–968, **3**:968*F*

Beta erythropoietin, brands, global sales **1**:605*T*

Beta-galactosidase, genetically engineered animals **3**:153–154

βγ-subunits, G protein structure **2**:780

Betaglobulins, plasma protein binding **5**:322

Beta-herpesviruses, antiviral drugs **7**:254*T*, **7**:259, **7**:260

Beta interferon *see* Interferon-β

β-lactam antibiotics **7**:479–518
  see also specific drugs
  activity spectrum **7**:481
  characteristics **7**:481
  discovery **7**:480
  dual-action **7**:509
    drawbacks **7**:509
    structure **7**:510*F*
  gentamicin combination **7**:637
  market **7**:481
  mechanism of action **7**:481, **7**:485*F*
    see also Bacterial cell walls
    molecular rationalization **7**:482
    SAR **7**:482, **7**:487*F*
    transpeptidase targeting **7**:483
  nomenclature **7**:480
  resistance **7**:483
    see also β-lactamases
    active efflux systems **7**:487
    outer membrane permeability **7**:487
    transpeptidase modification **7**:487
  safety **7**:481
  structure **7**:480*F*, **7**:481*F*
  synthesis, PASP **3**:829, **3**:829*F*
  tobramycin combination **7**:637

β-lactamase
  Molecular Interaction Fields **4**:247
  optical assays **3**:580–581
  promiscuous ligands **2**:740, **2**:741*F*
    enzyme concentration **2**:741
    as model system **2**:742
  tetraiodophenolphthalein, inhibition by **2**:740, **2**:741*F*

β-lactamase inhibitors **7**:510
  see also specific drugs

bridged β-lactams **7**:511
  derivation **7**:511
  clavulanic acid **7**:510, **7**:512
  6-hydroxyl penams **7**:511
  mechanism of action **7**:493*F*
  7-methylidene cephems **7**:511
  6-methylidene penams **7**:511
  methylidene penems **7**:511
  oxapenems **7**:512
  penam sulfones **7**:511
  2β′-penam sulfones **7**:512
  structure **7**:511*F*, **7**:512*F*

β-lactamases **7**:483, **7**:485*F*
  biostructure information **7**:485, **7**:485*F*
  conserved structural motifs **7**:486, **7**:486*F*
  classification **7**:485, **7**:486*T*
  GPCR functional assay screening **2**:812
  quantum mechanical calculations *see* Quantum mechanical calculations, enzyme mechanisms
  transpeptidases *vs.* **7**:487

Betamethasone butyrate propionate **1**:216
  bioavailability **5**:712*T*
  psoriasis treatment **7**:944

Betamipron **1**:116

β-oxidation, coenzyme A conjugation *see* Coenzyme A conjugation

Beta-secretase, knockout animal models **3**:160

Betaseron **1**:501

Beta-sheets/beta-turns, scaffold-based design *see* Scaffold-based design

Beta-turn peptidomimetics, GPCR library screening **4**:693

Betaxalone
  GABA receptor interactions **6**:357
  structure **6**:353*F*

Betaxolol
  bioavailability **5**:712*T*
  oral absorption **5**:680*T*

Bethanechol, as prokinetic agent **6**:680

Betotastine besilate **1**:204
  allergic rhinitis **1**:204

Beuhler test, skin sensitization, computational models **5**:923

Bevacizumab **1**:334*T*, **1**:433*T*, **1**:499; **7**:20, **7**:200
  brands, global sales **1**:605*T*
  cancer treatment **3**:202
  clinical use **1**:606–607
  colorectal cancer treatment **1**:444; **3**:202; **7**:8
  indications **7**:200
  postmarketing surveillance **1**:761, **1**:762*T*
  toxicity **7**:46*T*

Bevantolol, bioavailability **5**:712*T*

Bexarotene **1**:136
  characteristics **2**:1012*T*
Bextra *see* Valdecoxib
Bexxar *see* Tositumomab
Bezafibrate **2**:1021
  structure **2**:1022*F*; **6**:479*F*
BHK cells, protein expression
    systems **3**:117
BIAcore *see* Surface plasmon resonance
    (SPR)
Biapenem **1**:109; **7**:507
  activity **7**:506*T*
1,5-Biaryl pyrazoles, PASP **3**:808,
    3:809*F*
Biaryls, solid-phase pool
    libraries **3**:705*F*
Biaryl substituted pyrimidines, hepatitis
    B antivirals **7**:321
Bibliographic databases *see* Literature
    databases
Bibliographic references, databases *see*
    Databases
BIBN4096, migraine treatment
    6:374
BIBO 3304, structure **6**:407*F*
BIBP3226, structure **6**:407*F*
BIBW22, P-glycoprotein
    interaction **5**:781*T*
BIBX 1382BS **7**:195
  structure **7**:195*F*
Bicalutamide **1**:143
  characteristics **2**:1012*T*
  prostate cancer **1**:144
  structure **2**:1016*F*
  target information **4**:750*T*
Bicistronic approach, *Escherichia coli*
    protein expression
    systems **3**:113
Bicuculline, structure **2**:893*F*
Bicyclic amines, solid-phase
    libraries **3**:706–708, **3**:708*F*
Bicyclic guanidines
  solid-phase pool libraries **3**:705*F*
  solid support synthesis **2**:481*F*
Bicyclic pyrimidines, solid-phase pool
    libraries **3**:705*F*
Bicyclic systems, ring system
    reorganization
    (bioisosterism) **2**:675
BIDD Databases **3**:305
  Drug ADME Associated Protein
    Database **3**:305
  Drug Adverse Reaction Target
    database **3**:305
  Kinetic Data of Biomolecular
    Interaction database **3**:305
  National University of Singapore
    Bioinformatics Drug Design
    Group **3**:305
  Therapeutically Relevant Multiple
    Pathways Database **3**:305
  Therapeutic Target Database **3**:305
BiDil, ethnic targeting **1**:703

Bifeprunox
  mechanism of action **6**:35
  schizophrenia treatment **6**:35
Bifonazole, structure **5**:359
Big Blue Mouse **3**:161
Big Blue Rat **3**:161
Biguanides **6**:444
  see also specific drugs
  advantages **6**:445
  clinical use **6**:445
  disadvantages **6**:445
  mechanism of action **6**:444
  pharmacokinetics **6**:444
  side effects/contraindications **6**:445
    lactic acidosis **6**:445
  structure **6**:440*F*
Biheterocyclic imidazoles, solid-phase
    pool libraries **3**:705*F*
Biheterocyclic imidazolidinones, solid-
    phase pool libraries **3**:705*F*
Biheterocyclic imidazolines, solid-phase
    pool libraries **3**:705*F*
BIIB-513 **6**:756
Bilberry (*Vaccinium myrtillus*) **1**:424
Bilder, molecular graphs/graphics **4**:68
Bile, intestinal membrane **5**:458
Bile acid-CoA: amino acid *N*-
    acyltransferase, coenzyme A
    conjugation **5**:155
Biliary clearance
  active transport **5**:47
  clearance models *see* Clearance models
  elimination studies **5**:46
BILN 2061 *see* Hepatitis C virus,
    protease inhibitors
BILN 2061 (Ciluprevir), hepatitis
    C **7**:267
BILS 179BS, herpes viruses **7**:259,
    7:310
Bimatoprost **1**:238
  introduction of **1**:108
  mechanism of action **6**:308
  structure **6**:309*F*
BIMU1, as prokinetic agent **6**:678
Binary formats, two-dimensional
    representations **3**:298
Binary quantitative structure–activity
    relationship **4**:54
Binase, NMR structure
    determination **3**:498, **3**:499*F*
BIND (Biomolecular Interaction
    Network Database)
  blueprint databases **3**:305–306
  metalloproteomics **3**:556–557
  optical biosensors **3**:595
  protein function prediction,
    bioinformatics **3**:337
  regulatory networks **3**:330
Binding affinity
  fragment screening **4**:443–444
  ligand–receptor interactions **4**:194
Binding conformations, gabapentin
    derivatives **8**:231, **8**:231*F*

Binding diversity, ligand–receptor
    interaction models **4**:205
Binding energies
  estimation, drug design, de
    novo **4**:284
  quantum mechanics **4**:413
Binding ligand location, NMR, drug
    discovery **3**:917
Binding promiscuity, hERG **5**:940
Binding site
  ligand–receptor interaction
    models **4**:207
  macrolide antibiotics **7**:527–528
  SH2 domain **4**:627, **4**:628*F*
Binomial distribution, adverse drug
    event frequency
    estimation **1**:767
BIO
  discovery/development **1**:390
  structure **1**:390*F*
BIO-176 **2**:903
Bioactive conformations
  pharmacophore detection **4**:127*T*
    matching points **4**:129*T*
    matching points/QSAR **4**:135*T*
  quantum mechanical
    calculations **4**:392
Bioactivity databases **3**:293–313
  bibliographic references **3**:296,
    3:300, **3**:302*F*
    author names **3**:301
    BIOSIS **3**:300
    Chemical Abstracts Service **3**:300
    design **3**:300
    Digital Object Identifier
      (DOI) **3**:300–301
    entrance point **3**:300–301
    hyperlinks **3**:300
    internet references **3**:301
    journal descriptors **3**:301
    MEDLINE **3**:300
    patent descriptors **3**:301
  biological activity **3**:301
    A_2 **3**:302
    EC_{50} **3**:302
    ED_{50} **3**:302
    IC_{50} **3**:301
    K_D **3**:302
    K_d **3**:302
    multiple targets **3**:302
    pharmacokinetic data **3**:302
    standard error of mean **3**:302
  biological activity databases **3**:301*F*,
    3:304
    BIDD Databases **3**:305
    Blueprint databases **3**:305
    Chembank **3**:306
    Chmoogle **3**:307
    *Ki*Bank **3**:306
    Ligand Info **3**:306
    PDSP K_i **3**:306
    PubChem **3**:307
    ZINC **3**:307

Bioactivity databases (*continued*)
  biological information
      databases **3**:304
    Enzyme Structures Database
        (EC-PDB) **3**:304–305
    European Bioinformatics Institute
        (EBI) **3**:304–305
    GPCRDB **3**:305
    MEROPS **3**:305
    NucleaRDB **3**:305
    Representivity of Target Families in
        the Protein data Bank
        (PDBRTF) **3**:304–305
    TCDB Database **3**:305
    TransportDB **3**:305
    Transporter Classification (TC)
        System **3**:305
    UniProtKB/Swiss-Prot Protein
        Knowledge database (Swiss-
        Prot) **3**:304–305
    UniProtKB/TrEMBL **3**:304–305
  chemical information **3**:296,
      3:297
    InChI **3**:298
    molecular fingerprinting **3**:298
    molecular similarity **3**:298
    one-dimensional
        representations **3**:298
    SMILES **3**:298
    three-dimensional
        representations **3**:298
    two-dimensional
        representations **3**:298
  commercially available
      databases **3**:307
    Accelrys **3**:307
    Aureus Pharma **3**:308
    CambridgeSoft **3**:308
    Eidogen-Sertanty **3**:308
    Elsevier MDL **3**:308
    GVK Biosciences **3**:309
    Jubilant Biosys **3**:309
    Sunset Molecular Discovery
        Databases **3**:309
  definition **3**:294
  design guidelines **3**:295
    see also specific topics
    biological *vs.* chemical data
        3:296
    searchability **3**:296
  examples **3**:303
    see also specific examples
  external resource linking **3**:301*F*
  hierarchical *vs.* relational **3**:295,
      3:297*F*
  homology models **3**:544
  integration **3**:302
    CABINET **3**:303
    CORBA **3**:303
    external identifiers **3**:303
    heterogeneity **3**:303
    HTML **3**:303
    Java **3**:303

    supplementary linking
        tables **3**:303, **3**:304*F*
    XML **3**:303
  management systems **3**:295
    Catalyst **3**:295
    ChemFinder **3**:295
    ChemoSoft **3**:295
    definition **3**:294
    JChem Base **3**:295
    MDL Isentris framework **3**:295
    MDL ISIS/Base **3**:295
    THOR **3**:295
    UNITY **3**:295
  open-source DBMS **3**:294
  options **3**:294
    desktop databases **3**:294
    server databases **3**:294
    web-enabled databases **3**:294
  protocol information **3**:296, **3**:299
    variation in **3**:300
  specialized fields **3**:303–304
  structured query language **3**:294
    American National Standards
        Institute (ANSI) **3**:294
    Data Definition Language
        (DDL) **3**:294
    Data Manipulation Language
        (DML) **3**:294
    relational database management
        system (RDBMS) **3**:294
  target information **3**:296, **3**:299
    basic data **3**:299
    full-text searching **3**:300
    keywords **3**:299
    synonyms **3**:299–300
    Uniform Resource Locators **3**:299
Bioadhesive substances
  buccal administration **5**:283–284
  nasal administration **5**:286
Bioassays, structure-based fragment
    screening **4**:788, **4**:789*F*, **4**:791
Bioavailability (*F*) **5**:22
  see also Oral bioavailability
  absorption, effects of **5**:702
  absorption *vs.* **5**:700
    see also Absorption
    absolute bioavailability (*F*)
        5:700–701
    area under the curve (AUC) **5**:701
  acyclovir **7**:299
  adefovir dipivoxil **7**:316
  affecting factors **5**:701, **5**:701*F*
  amount reaching circulation **5**:22
  calculation of **5**:700
  clearance (CL$_{int}$) in vitro
      screens **5**:959
  definition **5**:869*T*, **5**:974
  famciclovir **7**:300–301
  HIV protease inhibitors **4**:604
  individual variability **5**:705
    drug concentration **5**:705
  intravenous administration **5**:22,
      5:23*F*

  metabolism **5**:705
    CYPs **5**:705
    route of **5**:705
  models *see* Bioavailability models
  nutraceuticals **1**:455
  oral administration **5**:22
  phenolic nutraceuticals **1**:450
  QSAR **5**:700
  quantitative structure–
      pharmacokinetic relationship
      (QSPkR) **5**:700
  quantum mechanical
      calculations **4**:398
  quinolone antibiotics **7**:587
  rate of absorption **5**:22
  salt form effects **5**:523, **5**:524
  traditional medicine **1**:422
Bioavailability models **5**:**699–724**
  absorption and **5**:710
    dose/solubility ratios **5**:710
    membrane permeability **5**:710
  allometric scaling **5**:706
    drawbacks **5**:706
    preclinical data
        extrapolation **5**:706
  classification models **5**:710, **5**:712*T*
    adaptive fuzzy partitioning (AFP)
        study **5**:711
    fuzzy adaptive least squares (FALS)
        analysis **5**:710
    isotopic field orientation
        (IFO) **5**:711–718, **5**:719*T*
    ORMUCS **5**:710
    self-organizing map **5**:711
    3D-structure **5**:711–718
  data quality/type **5**:707
    difficulties **5**:707, **5**:707*T*
    heterogenous data sets **5**:707–708
    uncertainty *vs.* variability **5**:707,
        5:707*F*
  data sources **5**:708, **5**:708*T*
    AurSCOPE database **5**:708
    bias **5**:709
    molecular structures **5**:708
  data usage **5**:706
    difficulties **5**:706
    physiologically-based
        pharmacokinetic (PBPK) **5**:706
  druglikeness **5**:700
  in vitro data **5**:706
    hepatic intrinsic clearance
        data **5**:706
  pharmacokinetics **5**:700
  quantitative models **5**:718
    artificial neural networks **5**:719,
        5:720*F*
    regression models **5**:718
  scoring system **5**:721
  software **5**:708*T*, **5**:709
    ADME Boxes **5**:708*T*, **5**:709
    ADME/Tox **5**:708*T*, **5**:709
    artificial neural networks **5**:709
    AurSCOPE **5**:708*T*

BioPrint 5:708*T*
  comparative molecular field
    analysis 5:709
  ECLiPS libraries 5:708*T*
  genetic algorithms 5:709
  KnowItAll 5:708*T*, 5:709
  soft independent modeling of class
    analog (SIMCA) 5:709
  truPK bioavailability
    predictor 5:708*T*, 5:709
Biochamin A 1:452*T*
Biochemical assays
  fragment-based drug discovery 4:446
  promiscuous ligands *see* Promiscuous
    ligands, identification
Biochemical pathways
  see also specific pathways
  diagrams, Cellomics
    Knowledgebase 2:435
  proteomics 2:330
Bioequivalence 5:23
  measurement 5:23, 5:24*F*
  pharmaceutical form of drug 5:23
  pharmacophore identification 3:242
Bioethics 1:681–708
  drug allocation/access 1:691
    diseases of affluence 1:691
    enhancement and lifestyle
      drugs 1:701
    global disease burden 1:692,
      1:693*T*
  drug pricing/profitability 1:688,
    1:689
    antibiotics 1:689
    me-too drugs 1:688
  drug safety 1:683
    antidepressants and youth suicidal
      behavior 1:687
    COX-2 inhibitors,
      regulation 1:684
    Fen Phen 1:686
    hormone replacement
      therapy 1:686
    Propulsid (cisapride) 1:685
  drug treatment 1:682
  embryo research 1:559
  enhancement and lifestyle
    drugs 1:698, 1:699
    allocation/access 1:701
    birth control 1:701–702
    definition 1:699
    erectile dysfunction 1:701–702
    memory enhancement 1:700
    menopausal symptoms 1:701–702
    Ritalin and ADHD 1:700
    smoking cessation 1:701–702
  mental illness 1:695
    competence 1:696
    informed consent 1:696
    treatment 1:695
  pharmacogenomics/
    pharmacogenetics 1:702
  R&D, developing countries 1:647

regulation and
    recommendations 1:690
  stem cell research 1:559, 1:682–683
Biofluids, biomarkers 3:45
Biogen-IDEC 1:499, 1:499*T*, 1:500
  early stage finance 1:556
  formation 2:102
Biogeny 1:492*T*
Bioinformatics 3:315–347
  definition 2:317; 3:316
  domain mapping 3:879
  drug toxicity markers 3:78
  expression data 3:316, 3:330
    mRNA transcript 3:330–331
  future work 3:340
  genetic variation analysis *see* Genetic
    variation analysis
  genome sequencing 3:316, 3:317
    see also Genome sequencing
  HTS *see* High-throughput screening
    (HTS)
  molecular networks *see* Molecular
    networks, bioinformatics
  molecular sequence analysis *see*
    Molecular sequence analysis,
    bioinformatics
  molecular structure prediction 3:324
    see also Protein structure prediction;
      RNA structure prediction,
      bioinformatics
  protein function prediction *see* Protein
    function prediction
  receptor binding 1:42–43
    fuzzy coordinates 1:42–43
  target identification *see* Target
    identification
Bioisosteres, soft drugs 5:1010
Bioisosterism 2:649–711
  see also specific drugs
  analog design 2:651
    market value 2:651
  anomalies 2:688
    ether–oxygen/methylene
      groups 2:692, 2:693*F*
    fluorine–hydrogen *see* Bioisosterism,
      fluorine–hydrogen
    Friedman's paradox 2:692
  carbon–boron *see* Bioisosterism,
    carbon–boron
  carbon–silicon *see* Bioisosterism,
    carbon–silicon
  carboxamides 2:666
    cyclic amidines 2:668, 2:670*T*
    histamine H$_3$ receptor
      antagonists 2:668
    HIV-1 protease inhibitors 2:668
    hydroxyethylureas 2:668, 2:669*F*
    1,2,4-oxadiazoles 2:668, 2:670*T*
    1,3,4-oxadiazoles 2:668, 2:670*T*
    oxazole rings 2:668, 2:670*T*
    phenylimidazole 2:668, 2:670*T*
    proteolytic enzyme
      inhibitors 2:668

    pyrrole rings 2:668, 2:670*T*
    renin inhibitors 2:668
    serotonin receptor
      antagonists 2:668
    sulfonamide conversion 2:666,
      2:669*F*
    vinyl fluorides 2:668, 2:669*F*
  carboxylic acids *see* Bioisosterism,
    carboxylic acids
  carboxylic esters 2:663, 2:668*F*
    benzodiazepine receptor
      ligands 2:663, 2:667*F*
    bioavailability 2:663–665,
      2:667*F*
    bradykinin B$_1$ receptor
      antagonists 2:663
    ester to amide 2:663
    N-methoxy imidoyl nitrile
      2:663–665, 2:667*F*
    monoamine transporter
      antagonists 2:663
    muscarinic receptor ligands 2:663,
      2:667*F*
    opiate receptor antagonists
      2:663
    1,2,4-oxadiazoles 2:663
    serotonin antagonists 2:663
    1,2,4-thiadiazoles 2:663
  core structure reorganization
    2:675
    ring system reorganization *see*
      Bioisosterism, ring system
      reorganization
    scaffold hopping *see* Bioisosterism,
      scaffold hopping
  definition 2:650, 2:651
  fluorine–hydrogen *see* Bioisosterism,
    fluorine–hydrogen
  'functional analogs' 2:659
    carboxamides *see* Bioisosterism,
      carboxamides
    carboxylic acids *see* Bioisosterism,
      carboxylic acids
    carboxylic esters *see* Bioisosterism,
      carboxylic esters
    catechols 2:672, 2:674*F*, 2:675*F*
    definition 2:651
    phenols *see* Bioisosterism, phenols
    sulfonamides *see* Bioisosterism,
      sulfonamides
    thiourea 2:668
    urea 2:668
  functional group reversal 2:673
    aromatic para-amino
      substitutes 2:674
    atenolol 2:675, 2:677*F*
    enzyme resistance 2:673–674,
      2:677*F*
    meperidine 2:675
    1-methyl-4-phenyl-4-propionoxy
      piperidine 2:675, 2:677*F*
    neoorthoforms 2:674, 2:677*F*
    serotonin 2:675, 2:677*F*

Bioisosterism (*continued*)
'me-to drugs' **2**:650
minor metalloids **2**:694
carbon–boron *see* Bioisosterism,
carbon–boron
carbon–silicon *see* Bioisosterism,
carbon–silicon
selenium *see* Bioisosterism, selenium
modification analysis **2**:685
electronic parameters **2**:687,
2:688*F*
solubility **2**:687, **2**:688*F*
structural parameters **2**:686,
2:687*F*
phenols **2**:670
amide groups **2**:670–671
benzazepine antagonists **2**:671,
2:672*F*
dimethylamino-
sulfonamide **2**:670–671
dopamine receptor
antagonists **2**:671, **2**:674*F*
hydroxyisopropyl **2**:670–671
hydroxymethyl **2**:670–671
methanesulfamide **2**:670–671
N-methyl-D-aspartate (NMDA)
receptor antagonists **2**:671,
2:673*T*
quantum mechanical
calculations **4**:386–387
ring equivalents *see* Bioisosterism, ring
equivalents
ring system reorganization *see*
Bioisosterism, ring system
reorganization
scaffold hopping *see* Bioisosterism,
scaffold hopping
selenium *see* Bioisosterism, selenium
'structural analogs,' definition **2**:651
sulfonamides **2**:672
sumatriptan **2**:672–676
Bioisosterism, carbon–boron **2**:698
aplasmomycin **2**:698
applications **2**:698
boranophosphate salts **2**:699, **2**:700*F*
bond lability **2**:699–700
boromycin **2**:698
boronic chalcones **2**:698–699, **2**:699*F*
Boron Neutron Capture Therapy
(BNCT) **2**:698, **2**:700, **2**:701*F*
compounds **2**:701*F*
carboxyboranes **2**:698–699
compounds **2**:699*F*
tolboxane **2**:698
toxicity **2**:698
riboflavin deficiency **2**:698
Velcade **2**:698–699, **2**:699*F*
Bioisosterism, carbon–silicon **2**:694
biological activity **2**:694–695, **2**:695*F*
bond length **2**:694
clinical studies **2**:697, **2**:697*F*
karenitecin **2**:697
silperisone **2**:697

diols **2**:695, **2**:696*F*
diphenylpiperidine analogs **2**:696,
2:697*F*
germanium **2**:697, **2**:698*F*
hydrolysis sensitivity **2**:694
lipophilicity **2**:694
metabolic attacks **2**:695
reactivity **2**:695, **2**:696*F*
trimethylsilyl group **2**:695
silicon diols **2**:696
toxicology **2**:694
Bioisosterism, carboxylic acids **2**:659
acyl-cyanamides **2**:659
acyl-sulfonamides **2**:659
angiotensin receptor
agonists **2**:659
beta-adrenergic receptor
antagonists **2**:659
chemokine receptor 1 (CXCR1)
inhibitors **2**:659
hepatitis C inhibitors **2**:659
aryl derivatives **2**:660, **2**:662*T*
diamino-cyclobutene-diones **2**:663,
2:666*F*
direct derivatives **2**:659, **2**:661*T*
hydroxamic acids **2**:659, **2**:663*F*
bufexamac **2**:659
histone deactylase
inhibitors **2**:659
ibuproxam **2**:659
matrix metalloproteinase
inhibitors **2**:659
oxametacin **2**:659
TNF-α converting enzyme
inhibitors **2**:659
nonclassical isosterism **2**:659
nonplanar sulfur-derived acidic
functions **2**:661, **2**:662*T*,
2:669*F*
as phosphonate surrogates **2**:663,
2:667*F*
phosphorus-derived acidic
functions **2**:661, **2**:662*T*, **2**:664*T*,
2:669*F*
CCK antagonists **2**:661
GABAB antagonists **2**:661
glutamate antagonists **2**:661
p*k*a/logP values **2**:661, **2**:666*F*
planar acidic heterocycles **2**:659,
2:660, **2**:662*T*
acidity **2**:661
3,5-dioxo-1,2,4-
oxadiazolidine **2**:661
3-hydroxy-γ-pyrones **2**:661
3-hydroxy-1,2,5-thiadiazoles **2**:661
therapeutic fields **2**:660–661
thiazolidinediones **2**:661
Bioisosterism, fluorine–hydrogen **2**:688
d orbital absence **2**:689, **2**:690*F*,
2:690*T*
electronic aspects **2**:688
bond energies **2**:688–689, **2**:689*F*
stabilities **2**:689, **2**:689*F*

lipophilicity **2**:690, **2**:690*T*, **2**:691*T*
acids **2**:668–669, **2**:690
alcohols **2**:690, **2**:691*T*
alkanes **2**:690, **2**:691*T*
ketones **2**:690
propranolol analogs **2**:668,
2:692*T*
roflumilast analogs **2**:692, **2**:693*F*
steric aspects **2**:688
bond length **2**:688, **2**:689*T*
tripelennamine analogs **2**:691,
2:692*F*
Bioisosterism, ring equivalents **2**:652
best-selling drugs **2**:652, **2**:653*T*,
2:660*T*
heterocycles **2**:657
cyclooxygenase-2 (COX-2)
inhibitors **2**:658, **2**:658*F*
uncertainty of results **2**:658–659,
2:659*F*
imidazo[1,2-a]pyridines **2**:656
bradykinin receptor
antagonists **2**:656, **2**:657*F*
pyridazines **2**:655*F*, **2**:656
acetylcholinesterase
inhibitors **2**:656, **2**:657*F*
boiling points **2**:656–657,
2:658*F*
N-(6-chloronaphthalen-2-)
sulfonylpipazine
derivatives **2**:655–656, **2**:656*F*
minaprine **2**:656, **2**:657*F*
ridogrel **2**:655
thromboxane A$_2$ synthetase
inhibitors **2**:655
pyridine **2**:654
sulfonamide antibacterial
agents **2**:654, **2**:654*F*
Bioisosterism, ring system
reorganization **2**:675
benzo compound splitting (benzo
cracking) **2**:678, **2**:679*F*
bicyclic systems **2**:675
dissociation **2**:679
amiodarone **2**:679, **2**:680*F*
benzofurans **2**:679
benzopyrones **2**:679
chromonar **2**:679
khellin **2**:679
3-methyl-chromone **2**:679
restructuring **2**:678
2-methoxybenze-[e]-hexahydo-
isoindole **2**:678–679, **2**:679*F*
SKF 38 393 **2**:678–679, **2**:679*F*
tetramizole **2**:678, **2**:679*F*
spiro derivatives **2**:675
gabapentin **2**:676, **2**:678*F*
guanethidine analogs **2**:675–676,
2:678*F*
tricyclic systems **2**:675
Bioisosterism, scaffold hopping **2**:651,
2:652*F*, **2**:680
chlorpromazine **2**:680, **2**:681*F*

computational approaches
2:681
abl tyrosine kinase ligands 2:685,
2:686*F*
BIOSTER software 2:685
estrogen receptor ligands 2:683,
2:684*F*, 2:685*F*
fragment replacement
2:681–683
pharmacophore searching
2:681–683, 2:683*F*
selective optimization offside
activities (SOSA) 2:683
shape matching 2:681–683
similarity searching 2:681–683
Skelgen software 2:683
software 2:683
topological similarities 2:683–685,
2:685*F*
cyclooxygenase-2 inhibitors 2:680,
2:681*F*
GABA_A ligands 2:681, 2:682*F*
imipramine 2:680, 2:681*F*
promethazine 2:680, 2:681*F*
Bioisosterism, selenium 2:700
applications 2:701
nitric oxide production
inhibition 2:703
photodynamic treatment 2:702*T*,
2:703
1,3-selenazol-4-one
derivatives 2:703
toxicity 2:701
Biolistics, gene therapy 1:312
Biological Abstracts 3:274
Biological activity
brain uptake prediction 5:746
databases *see* Databases
determination, cell-based
assays 3:617
monitoring, career guidelines
2:19
Biological activity space,
definition 4:807
Biological conception, as
bottleneck 1:38*T*
Biological connectivity, target
validation 3:4, 3:4*F*
Biological fingerprints 4:807–818
see also Fingerprint/dataprint
descriptors
activity modeling 4:808
compound training set 4:808
cyclooxygenase-1
inhibitors 4:808–809
reference panel 4:808
specific binding sites 4:809
chemical fingerprint
correlation 4:808
chemogenomics 4:815
definition 4:815
National Cancer Institute
(NCI) 4:816

'computational' 4:809
DOCK-generated scores 4:809
generalized neighborhood behavior
(GNB) models 4:809
selectivity/differentiation
searching 4:809
antifungal agents 4:809–810,
4:810*F*
BioPrint 4:811
lead differentiation 4:811, 4:811*F*
off-target activity 4:809–810,
4:810*F*
sequence clustering 4:813
GPCR 4:813–815
protein kinases 4:813–815
structural clustering 4:811, 4:812*F*
chemical similarity
definitions 4:812–813
correlation 4:813, 4:814*F*
dopamine inhibitors 4:813
'fuzzy pharmacophores' 4:812–813
National Cancer Institute
(NCI) 4:813
Biological information databases *see*
Databases
Biological macromolecules 1:431–447
see also Monoclonal antibodies; *specific
macromolecules*
current status 1:434
FDA approval 1:434
revenue 1:434
definition 1:431–432, 1:432*F*
discovery 1:434
drug discovery 2:47
future work 1:443
efficacy 1:444
generic competition 1:444
merger effects 1:444
refinement 1:443
safety 1:444
target identification 1:443
history 1:432
molecular biology 1:432
ionization *see* Ionization
Molecular Interaction Fields 4:238
pharmacokinetics 1:438
pharmacology 1:438
production 1:434
hybridoma technology *see*
Hybridoma technology
phage display *see* Phage display
ribosome display *see* Ribosomes
transgenic animals *see* Transgenic
animals
QSAR 4:59
routes of administration 1:438
safety/tolerability 1:438
SAR 1:437
modifications 1:437
small molecule drugs *vs.* 1:438,
1:439*T*
cost 1:438
patient populations 1:438

therapeutic areas 1:432
hemostasis 1:432–433
Biological matrix, stability testing *see*
Stability testing
Biological membranes 5:455
see also Cell membranes; Lipid
bilayers; *specific types*
components 5:455, 5:455*F*
ceramides 5:456, 5:456*F*
phosphatidylcholine
5:455–456
phosphatidylethanolamine
5:455–456
phosphatidylglycerol 5:455–456
phosphatidylinositol 5:455–456
phosphatidylserine 5:455–456
depolarization, potassium
channels 2:860
mitochondria *see* Mitochondria
perturbation, tetracyclines 7:613*F*,
7:614
protein analysis, 1D-SDS PAGE
proteomics 3:39
proteins
baculovirus expression vector
systems 3:115*F*, 3:119
cell-free expression systems *see*
Cell-free expression systems
Escherichia coli protein expression
systems *see* *Escherichia coli* protein
expression systems
protein analysis 3:429
Semliki Forest Virus expression
systems 3:115*F*, 3:119
targets, cell-based assays 3:618
transport proteins *see* Membrane
transport proteins
Biological model, cell-based
assays 3:623
Biological Process, PDB 3:377
Biological production, chemical *vs.*
biological 2:117
Biological products 1:334
drug discovery, decline in 1:623
small molecules *vs.*, drug discovery
productivity 2:126,
2:128*F*
Biological variability, metabolic drug–
drug interactions 5:842
Biologics Control Act (1904),
postmarketing
surveillance 1:756
Biologics license applications
(BLAs) 2:561, 2:561*F*
Biology
discipline changes 2:88
pharmaceutical R&D 2:105
BioLoom database, ionization
prediction, in silico
tools 5:623*T*
Bioluminescence energy transfer
(BRET), GPCRs 2:798–799,
2:813

Bioluminescent assays   3:578, 3:590
    see also Green fluorescent protein
        (GFP)
    ATP measurement   3:590, 3:591F
    drug development   2:193
    hit validation   3:591, 3:592F
    microfluidic format   3:590, 3:591T
    optical assays see Optical assays
    small animal imaging   3:660T, 3:661,
        3:668
BioMarker (database)   3:309
Biomarkers   3:69–85
    advantages   3:82
    affective disorders see Affective
        disorders
    Alzheimer's disease   6:197
    bipolar affective disorder   6:49,
        6:50T, 6:51T
    cancer see Cancer biomarkers
    central nervous system injury   6:261
    clinical trials see Biomarkers, clinical
        trials
    definition   2:185; 3:70
        Biomarkers and Surrogate Endpoint
            Working Group   3:70
    depression   6:49, 6:50T, 6:51T
    downstream biological effects   3:73
        definition   3:73
        pathway characterization   3:73
        protein amounts   3:73
    drug toxicity markers   3:78
        ADME-related tissue   3:78
        bioinformatics   3:78
        genomics   3:78
        imaging technology   3:78
        luminescent imaging   3:78
        metabonomics   3:78
        oligonucleotide arrays   3:78
        proteomics   3:78
    examples   3:70
    fluorescent imaging   3:77
        fluorescent resonance energy
            transfer (FRET)   3:77
        green fluorescent protein
            (GFP)   3:77
        limitations   3:77
    functional imaging   3:76
        magnetic resonance imaging   3:77
        NMR   3:76
        positron emission
            tomography   3:77
    gene expression profiling   3:74
        experimental vs. clinical   3:75
        gene copy changes   3:75
        genomic alterations   3:74
        genomic amplification   3:75
        genomic deletion   3:75
        individual transcripts   3:74–75
        luminescent imaging   3:77–78
        quantitative reverse transcription
            polymerase chain reaction   3:75
    genetically engineered
        animals   3:154

history   3:70
IBD   6:620
identification/validation
    high-performance liquid
        chromatography-NMR (HPLC-
        NMR)   5:999
    microarrays   3:104
    proteomics   3:43
    RNAi (RNA interference)   3:172
lead identification   3:69
luminescent imaging   3:77–78
    drug toxicity markers   3:78
    gene expression regulation
        3:77–78
    luciferase   3:77–78
metabonomics   3:76
    drug toxicity markers   3:78
    low weight components   3:76
    NMR   3:76
    thin layer chromatography   3:76
    toxicology   3:76
multiple sclerosis   6:232
Parkinson's disease   6:212
personalized medicine   1:349
pharmacogenetics see
    Pharmacogenetics
preclinical models   3:76
protein detection methods   3:72
    affibodies   3:73
    antibody development   3:72
    aptamers   3:73
    protein kinases   3:72
    surface plasmon resonance   3:73
proteomics see Proteomics
signal transduction   3:73
target identification   3:69, 3:71
target modulation   3:73
    definition   3:73
    flow cytometry   3:74
    immunoassays   3:73–74
    immunohistochemistry   3:74
target selection   2:763
target validation   3:70, 3:71,
    3:71F
    controls   3:70–71
    identified target   3:70–71
    novel target   3:70–71
tissue microarrays   3:74
    biomarker identification   3:75
    immunohistochemistry   3:75
    pharmacodynamics   3:74–75
    reverse phase   3:75
    in situ hybridization   3:75
    structure   3:75
2D-PAGE   3:45–46
utility of   3:71, 3:72T
    reverse engineering   3:71–72
Biomarkers, clinical trials   2:30–31,
    2:185; 3:79
assays   3:79–80
    ELISA   3:79–80
    immunostaining   3:79–80
endpoints   3:79

pharmacodynamics   3:74–75, 3:79,
    3:80, 3:81F
    definition   3:80
    dosage   3:80–81
pharmacokinetics   3:80, 3:81F
    definition   3:80
regulatory implications   3:81
    surrogate endpoints   3:81–82
surrogate endpoints   3:80
    clinical validity   3:81–82
    definition   3:80
Biomarkers and Surrogate Endpoint
    Working Group   3:70
BioMart, proteome analysis   3:367,
    3:368
Biomimetics, PAMPA   5:474, 5:475F
Biomolecular probes
    ^{123}I   3:666
    ^{124}I   3:666
    ^{125}I   3:666
    ^{131}I   3:666
    ^{111}In   3:666
Biomolecular simulations,
    history   4:651
Biomolecule spectra, NMR   2:509
BioPath, metabolism
    prediction   5:797T, 5:799
Biopharmaceutical Classification
    Scheme (BCS)   5:970–987,
    5:972F
    absorption   5:702, 5:703T
    absorption limitation   5:978
        low-permeability drugs
            5:979
        prodrugs   5:979
    advantages   5:972–973
    aims/goals   5:971
    cost reduction   5:971–972
    drug discovery/preclinical
        development   5:976
        animal models   5:978
        class I drugs   5:976
        in vitro ADME processes   5:978
        permeability/solubility   5:975F,
            5:977
        single-pass perfusion
            technique   5:975F, 5:977
        theoretical descriptors   5:978
    ER drug delivery   5:979, 5:980F
        class II drugs   5:980
        low-solubility compounds   5:981
    food–drug interactions   5:981
        class I drugs   5:982
        class II drugs   5:982
        class III drugs   5:982
        mechanisms   5:982–983
    future work   5:983
        class I drugs   5:983
        class II drugs   5:983
        class III drugs   5:983
        dissolution rate   5:983–984
        expansion   5:984
        membrane transporters   5:984

in vitro/in vivo correlation **5**:981
class II drugs **5**:981
class III drugs **5**:981
in vitro oral absorption
prediction **5**:674
membrane transport proteins *see*
Membrane transport proteins
product development **5**:978
regulatory applications **5**:972–973,
5:974
biowaivers **5**:974–975
dissolution rates **5**:975, **5**:975*F*,
5:976*F*
formulation excipients **5**:975
in vitro dissolution testing
5:974–975
relevant pharmacokinetic
definitions **5**:974
solubility **5**:1013
Biopharmaceuticals, drug
discovery **1**:594
Biopharmaceutic Drug Disposition
Classification System
(BDDCS) **5**:984
Biophysical compartment models, oral
absorption *see* Oral absorption
prediction, physiologically based
models
Biopolymers, ionization prediction, in
silico tools *see* Ionization
prediction, in silico tools
Bioprecursors, prodrugs *see* Prodrugs
BioPrint **5**:679*T*
bioavailability models **5**:708*T*
biological fingerprints **4**:811
Bioprospecting *see* Traditional medicine
Bioreactors
cell culture **3**:622–623
genetically engineered
animals **3**:151
Biosafety role, innovation **1**:731–732
Bioscience Innovation and Growth
Team (BIGT) **1**:592
Biosensors *see* Surface plasmon
resonance (SPR)
GPCR functional assay screening *see* G
protein-coupled receptors
(GPCRs), assays
BIOSIS (Biological Abstracts)
3:271*T*
databases **3**:300
BIOSTER software
Accelrys databases **3**:307
scaffold hopping **2**:685
Biotage AB, microwave-assisted
chemistry **3**:844, **3**:851
Emrys Advancer batch reactor **3**:851,
3:851*F*
Emrys Knowledge database **3**:844
Initiator reactor **3**:844, **3**:845*F*
Biotechnology **1**:591–613
clinical practice impact **1**:603
scale of change **1**:604, **1**:605*T*

scope of change **1**:604
speed of change **1**:606
definition **1**:592, **1**:593; **2**:225
developments
patenting problems **1**:807
risk factors for venture capitalists *see*
Venture capital
distributed innovation
system **1**:554–555
drug development **1**:600
scale of change **1**:601, **1**:602*T*
scope of change **1**:600
speed of change **1**:602, **1**:603*T*
drug discovery **1**:33, **1**:36, **1**:36*T*,
1:549, **1**:593
context of change **1**:593
scale of change **1**:596, **1**:596*F*
scope of change **1**:594
speed of change **1**:598
future work **1**:81
Ganguly's views on **8**:18
governments, in drug
discovery **1**:742
history **1**:554
investment trends **1**:557
see also Venture capital
myth, evidence supporting **1**:593,
1:608
orphan drugs R&D **1**:664, **1**:664*F*
policy issues **1**:591
venture capital *see* Venture capital
Biotechnology and Bioengineering
Abstracts **3**:271*T*
Biotechnology and Biological Sciences
Research Council (BBSRC),
industry/academia
collaborations **1**:576
Biotechnology companies **2**:99, **2**:102
see also specific companies
automation **2**:103
characteristics **2**:225, **2**:226*T*
development **2**:22, **2**:23*T*
failure of **2**:22, **2**:24*T*
future problems **2**:55
history **2**:99
human genome sequence, effects
of **2**:103
manufacturing **2**:102–103
partnering **2**:102–103
products *vs.* technology
platforms **2**:231
collaborations **2**:231
contracts **2**:231
form/formulation
enhancement **2**:232
innovation decrease **2**:231
mechanization **2**:231
protein therapeutics **2**:102
R&D finances **2**:225, **2**:226*F*
size **2**:110–111
small molecule drugs **2**:102
target selection **2**:753–754, **2**:757*T*
venture capital funded **2**:8

Biotech Validation Suite **3**:547*T*
Biotica technology **1**:512
Biotransformation
animal experimentation
alternatives **1**:481
conjugation, drug metabolism
5:159
genetically engineered cells **1**:479
Biotransformations, Accelrys
databases **3**:307
Biowaivers, BCS **5**:974–975
Biphasic systems, lipophilicity
measurement **5**:432
Biphenylalanines, solid-phase discrete
libraries **3**:709*T*
Biphenylamines, solid-phase discrete
libraries **3**:709*T*
Biphenyls/diaryls, endocrine-modulating
anticancer agents *see* Anticancer
drugs, endocrine-modulating
Biplots, microarrays **3**:92, **3**:93–94,
3:94*F*
Bipolar affective disorder (BAPD) **6**:2,
6:5, **6**:46
see also Affective disorders
biomarkers **6**:49, **6**:50*T*, **6**:51*T*
classification **6**:49
disease state/diagnosis **6**:5, **6**:46
problems **6**:58
environmental causes **6**:53
genetics **6**:53
5-HT$_3$ receptors **2**:891
new research **6**:73
treatment **6**:71, **6**:71*T*
see also specific drugs
acute depression **6**:71
acute mania **6**:71
unmet medical needs **6**:72
BIRB796, arthritis treatment **7**:860
Birefringent crystals, protein
crystallography **3**:441
Biricodar, structure **5**:770*F*
BIRT-377, IBD treatment **6**:633
Birth control *see* Contraceptives
Bisacodyl, irritable bowel syndrome
treatment **6**:654
Bisarylthio *N*-hydroxypropionamides,
solid-phase discrete
libraries **3**:709*T*
N-N-Bis (2-chloroethyl)-*N*-nitrosourea
see BCNU
Bisoprolol
bioavailability **5**:712*T*
GPCR targeting **2**:773*T*
Bisperidine ring system, CCR3
inhibitors **7**:886
Bisphonal *see* Incandronic acid
Bisphosphonates **6**:510, **6**:511; **8**:200
see also specific drugs
bioavailability **6**:511–512
cell entry **6**:511
indications **6**:512
osteogenesis imperfecta **6**:512

Bisphosphonates (*continued*)
  osteoporosis **6**:512
    Paget's disease **6**:504, **6**:512
  leishmaniasis treatment **7**:838
  mechanism of action **6**:511, **6**:511*F*
  oral absorption **5**:1026–**5**:1027;
    6:511–512
  as prodrugs **5**:1021
  protein tyrosine phosphatase
    inhibition **4**:640–641
  structure **6**:510*F*; **8**:200, **8**:201*F*
Bis-quaternary ammonium salts,
    antimalarial drugs **7**:797–798
Bisulfate sequencing,
    epigenetics **1**:267*T*
Bitmaps, pharmacophore
    detection **4**:127–128
Bivalirudin **1**:164; **2**:579*T*
BL21 Origami strains, disulfide bond
    expression **3**:116
Black, James **1**:328–329
  β-adrenergic receptor
    antagonists **6**:711
  cimetidine, invention of **2**:253,
    2:254
  collaborators, Ganellin, C R
    8:42–43
  medicinal chemistry, definition
    of **2**:92
  Nobel Prize **2**:253
  propranolol, invention of **2**:253,
    2:254
  training **2**:253
Black cohosh (*Cimicifuga racemosa/Actaea
    racemosa*) **1**:424
Black lipid membrane, PAMPA
    vs. **5**:467–469
Black piedra **7**:421*T*
Bladder
  activity inhibition,
    duloxetine **8**:126–127, **8**:126*F*
  benign prostatic hyperplasia
    effects **6**:554, **6**:555*F*, **6**:559
  dysfunction **6**:542
Bladder cancer
  nedaplatin **1**:147
  valrubicin **1**:143
  xenograft models **7**:39*T*
Bladder outlet obstruction
    (BOO) **6**:543
  benign prostate hyperplasia **6**:559,
    6:560*F*
  characteristics **6**:543
BLAST database
  comparative genomics **7**:738
  sequence alignment,
    bioinformatics **3**:320
  sequence searching **2**:344–345
  siRNA design **3**:176
Blastomycosis **7**:424*T*
  amphotericin **7**:457
BLEEP program, docking, scoring
    functions **4**:485

Bleomycin
  see also AVBD regimen
  toxicity **7**:46*T*
Blessed Dementia Scale (BDS) **6**:201
Blessed Information Memory
    Concentration (BIMC) **6**:201
Blindness
  age-related macular
    degeneration **6**:299
  Danio rerio model system **3**:10
Blockade prediction, hERG blocker
    modeling **5**:941
'Blockbuster orphan'
    phenomenon **1**:658
Block copolymers, drug–polymer
    conjugates **5**:1046*F*
BlogP method, log P calculation
    5:658
Blood **5**:848
  cell-based assays **3**:620
  composition **5**:848
Blood barriers, membrane transport
    protein pharmacokinetics **5**:78,
    5:79*F*
Blood–brain barrier (BBB) **5**:302,
    5:745
  see also Brain drug uptake
  absorption **1**:52
  amino acid transport **8**:234
  automation **5**:346*T*
  cellular anatomy **5**:302, **5**:304*F*
    arterioles **5**:302
    astrocytic glial cells **5**:302
    microglia **5**:302
    pericytes **5**:302
  endothelial tight junctions
    5:302–303
  in vitro surrogate models **5**:314,
    5:315*F*
    coculture models **5**:314
    gastrointestinal models *vs.* **5**:314
    labor-intensive **5**:314
    permeability *vs.* lipophilicity **5**:314
  lipid composition **5**:459*T*
  morphology **5**:458
  permeability models **5**:313
    bovine model **5**:313
    dynamic in vitro BBB model (DIV-
      BBB) **5**:314
    intraluminal flow **5**:314
    porcine models **5**:313
    tight junctions **5**:313
    tightness assessment **5**:313–314
    tracer molecules **5**:313–314
    transendothelial electrical
      resistance (TEER) **5**:313–314
  permeation
    see also Brain drug uptake; Brain
      uptake prediction
    chemoinformatics, ADME property
      prediction **3**:259
    4D-QSAR **5**:598
    in vivo prediction problems **5**:467

mechanistic pharmacokinetic
    studies *see* Mechanistic
    pharmacokinetic studies
  molecular descriptors **5**:587
  QSAR **4**:57
  quantum mechanical
    calculations **4**:398–399
  routes across **5**:305
  see also Brain drug uptake
    ABCG2 transporter **5**:79
    ABC transporters **5**:79
    adsorptive-mediated transcytosis
      (AMT) **5**:302–303,
      5:304*F*
    carrier-mediated efflux **5**:304*F*
    carrier-mediated influx **5**:304*F*
    cell migration **5**:304*F*
    lipophilicity effects **5**:302–303,
      5:305*F*, **5**:308
    OATP (*SLC21/SLCO*)
      transporters **5**:67, **5**:79
    passive diffusion **5**:304*F*
    P-glycoprotein (P-gp) **5**:79
    receptor-mediated transcytosis
      (RMT) **5**:302–303, **5**:304*F*
    SLC transporters **5**:79
    transcellular route **5**:302–303,
      5:304*F*
  structure **5**:458, **5**:460*F*
  surface activity parameters
    5:482
Blood–brain barrier-permeable
    molecules, Alzheimer's
    disease **6**:197
Blood–brain partitioning **5**:747
  brain perfusion *vs.* **5**:755, **5**:756*T*
  calculations **5**:747, **5**:749*T*,
    5:750*T*
  comparisons **5**:751*T*
  data sets **5**:747–748
  multiple linear regression
    analysis **5**:750–751
  neural network **5**:751
  partial least squares (PLS)
    method **5**:748
  root mean square error
    (RMSE) **5**:751–752
  TOPS-MODE **5**:752
  definition **5**:747
  electrotopological indices *see*
    Electrotopological indices
    (E-state index)
  general rules **5**:760
    passive diffusion **5**:760
  influencing factors **5**:752
    compound size **5**:752
    ionization **5**:753–754
    passive transport **5**:753
    plasma water to brain partition
      coefficient **5**:754
    polar nature **5**:752
    standard free energy of
      hydration **5**:753

Bumetanide **6**:708
  bioavailability **5**:712*T*; **6**:709*T*
  dose **6**:709*T*
  half-life **6**:709*T*
  oral absorption **5**:680*T*
  structure **6**:586*F*, **6**:709*T*
Bump detection, protein 3D structure validation **3**:517
Bunyaviruses, antiviral drugs **7**:254*T*, **7**:278
Buparvaquone
  prodrugs of **5**:1014*T*
  water solubility **5**:1013*F*, **5**:1014*T*
Bupidine
  sila analog **2**:697*F*
  structure **2**:697*F*
Bupivacaine **6**:355, **6**:361
  as anesthetic **6**:353
  structure **6**:354*F*
Buprenorphine
  discovery/development **1**:379
  heroin addiction **6**:9, **6**:180*T*, **6**:183
  structure **1**:379*F*; **6**:181*F*
Bupropion
  ADHD treatment **6**:133–134
  narcolepsy treatment **6**:156
  nicotine addiction **6**:180*T*, **6**:183
  oral absorption **5**:468*T*, **5**:680*T*
  structure **6**:131*F*, **6**:181*F*
  transporter inhibition **6**:66*T*
Bureau of Home Economics (BHE/Cdb) rat, diabetes mellitus type 2 animal models **6**:435
Bureau of Nutritional Sciences, functional food definition **1**:408
Burimamide
  discovery/development **6**:597–598, **6**:605
  Ganellin, C R and **8**:42–43
Burroughs Wellcome, history **2**:100–101
Buserelin, nasal administration **5**:284
Business angels
  SME funding **1**:494
  venture capital provision **1**:556, **1**:564
Business for Social Responsibility **1**:721
Business model, pharmaceutical industry **2**:99
Business plans, venture capital deals *see* Venture capital
Business unit, pharmaceutical industry organization **2**:110
Buspirone **6**:101–102
  structure **5**:359; **6**:101*F*
Busulfan, toxicity **7**:46*T*
1-Butanol, microwave heating **3**:840*T*
2-Butanol, microwave heating **3**:840*T*
Butenafine **7**:430*T*
  medicinal chemistry **7**:448

Butenolides, solid-phase discrete libraries **3**:709*T*
Butirosins **7**:630–631
  structure **7**:631*F*
tert-Butylamodiaquine **7**:775
Butyric acid, structure **5**:369*F*
Butyrophenones **6**:29–31
Butyrylcholinesterase, Alzheimer's disease **6**:202
BVT-933 **6**:414
BW-12C **4**:70
BX-320 **7**:209
  structure **7**:208*F*
BX-471
  IBD treatment **6**:633–635
  transplant rejection **7**:921
BYK 36399, structure **6**:610*F*
Byproduct scavenging, PASP **3**:826
Bz-423, mitochondria complex V **2**:1079
BzATP, structure **2**:906*F*

C

^{11}C
  PET **3**:664*T*
  pharmacokinetic measurements **3**:672–673
C2 modifications, tetracyclines **7**:603
C4-dedimethylamino tetracycline **7**:604
C4 derivatives, tetracyclines *see* Tetracyclines
C6-hydroxyl group
  chlortetracycline **7**:604–605, **7**:605*F*
  oxytetracycline **7**:604–605, **7**:605*F*
  tetracycline **7**:604–605, **7**:605*F*
C57BL/6J mice, diabetes mellitus type 2 animal models **6**:435–436
CA-15-3, cancer biomarker **7**:49
CA-125
  cancer biomarker **7**:49
  cancer screening/diagnosis **7**:5
CAAT box, mammalian protein expression systems **3**:117
CAB-175 **7**:497
  activity **7**:496*T*
  structure **7**:495*F*
Cabergoline
  cocaine abuse **6**:184*T*
  structure **6**:181*F*
CABINET, database integration **3**:303
Caco-2 cell models **5**:671
  automation **5**:350–351
  cell-based assays **3**:628
  definition **5**:958
  flux prediction **5**:690
    multiple linear regression **5**:690
    partial least squares **5**:691
    permeability *vs.* lipophilicity **5**:690
    3D-QSAR approaches **5**:691
    2D topological descriptors **5**:691

permeability assays **5**:260, **5**:268, **5**:309, **5**:314
  active drug transport prediction **5**:271
  active transport **5**:265–266
  compound ranking **5**:261–262, **5**:269–270
  conditions **5**:268, **5**:269*F*
  continuous-discontinuous system **5**:267
  differentiation effects **5**:263–264
  FDA approval **5**:269
  inconsistencies **5**:269
  miniaturization **5**:269
  other systems *vs.* **5**:270–271
  stirred *vs.* unstirred **5**:261, **5**:262*F*
  VolSurf descriptors *vs.* **5**:540
Cacospongionolide, arthritis treatment **7**:851
Cadaver skin, dermal/transdermal administration **5**:292–293
CADD *see* Computer-assisted drug design (CADD)
Cadherins
  biological role **7**:15
  cancer, invasive growth **7**:15
Cadmium
  immunosuppression **5**:223
  protein crystallization **3**:885
Caenorhabditis elegans model system **2**:312–313; **3**:8, **3**:652
  advantages **3**:8
  affective disorder models **6**:54*T*
  aging **3**:8
  anesthetic target identification **3**:8
  apoptosis **3**:8
  BCL-2 transgenes **3**:652
  cell-cell signaling **3**:8
  gastrointestinal motility models **6**:675
  genome sequencing **3**:647–649, **3**:648*T*, **3**:652
  history **3**:647, **3**:652
  muscular dystrophy model **3**:8
  neurogenesis **3**:8
  RNA interference technology *see* RNAi (RNA interference)
  small-molecule screening **3**:652
Caffeic acid, pharmacokinetics **1**:454*T*
Caffeic acid phenethyl ester (CAPE), mitoprotective drugs **2**:1064, **2**:1064*F*
Caffeine
  bioavailability **5**:712*T*
  discovery/development **1**:322, **1**:406
  narcolepsy treatment **6**:154
  oral absorption **5**:468*T*, **5**:680*T*
  structure **6**:401*F*
Calanolide, structure **1**:74*F*
Calblock *see* Azelnipidine
Calcein AM influx assay, P-glycoprotein models **5**:776
Calcidiol, structure **6**:499*F*

Calcineurin inhibitors, transplant
    rejection 7:919
Calcipotriol
    psoriasis treatment 7:976, 7:977T
    structure 2:1018F; 7:977T
    as vitamin D receptor ligand 2:1019
Calcitonin 6:500
    bone resorption/remodelling 6:498
    eel 2:579T
    human 2:579T
    nasal administration 5:284
    salmon 2:579T
        intranasal administration
            2:594–595
        nasal administration 5:281T
        PEGylation 2:589
        as peptide/protein drug 2:580
Calcitonin gene-related peptide
    (CGRP)
    gastrointestinal tract 6:645–646
    glaucoma treatment 6:313T
    inhibition, triptans 6:374
    migraine treatment 6:377
Calcitonin gene-related peptide
    antagonists
    see also specific drugs
    migraine treatment 6:376
Calcitonin receptors
    constitutively active 2:790T
    in human disease 2:775T
Calcitriol
    bone resorption/remodelling
        6:498–499
    characteristics 2:1012T
    psoriasis treatment 7:976, 7:977T
    structure 2:1018F; 6:499F; 7:977T
    synthesis 6:499
Calcium
    antagonists, as 'soft drug' 5:1011
    bone disease treatment 6:513
    homeostasis
        central nervous system
            injury 6:271–272
        vitamin D receptor (VDR) 2:1007
    mitochondria inner membrane
        binding 2:1059
    mitochondrial permeability transition
        (MPT) 2:1060
    phosphodiesterase-1
        regulation 2:923
    tetracycline binding 7:617–618
    toleration, mitochondria
        diversity 2:1067
Calcium 3 assay, GPCR functional assay
    screening 2:811
Calcium-based screening
    assays 3:625–626, 3:626F
Calcium-channel blockers 2:849–850;
    6:715T, 6:747
    see also specific drugs
    bioavailability 6:715T
    characteristics 2:829T
    clinical trials 6:259

clinical use 1:594–595
    aneurysmal subarachnoid
        hemorrhage treatment 6:263
    cardiac arrhythmias 2:851
    cardiovascular drugs 2:851
    epilepsy 2:851
    glaucoma treatment 6:312–314,
        6:313T
    hypertension treatment 1:532T;
        6:585, 6:588, 6:699–700,
        6:715
    irritable bowel syndrome
        treatment 6:657
    pulmonary artery hypertension
        (PAH) treatment 6:725
    renal dysfunction 6:585
development, selection of 2:73
effectiveness 6:747
half-life 6:715T
mechanism of action 6:588,
    6:715–718
    glycine receptors 2:896
    nitric oxide enhancement 6:588
    P-glycoprotein
        interactions 5:769T
neuronal hyperexcitability 2:851
new research areas 6:753
pharmacokinetics 6:718
pharmacology 6:718
safety data interpretation 1:713
safety evaluation 6:715–718
SAR 6:718
side effects 6:718
    toxicity 6:747
structure 6:587F
third-generation 6:716T, 6:718
Calcium channels 2:841; 6:734,
    6:735
action potentials 6:734F
genetic disease 2:852
    absence epilepsy 2:852
    episodic ataxia 2:852
    familial hemiplegic migraine
        (FHM) 2:852
    hypokalemic periodic
        paralysis 2:852
    malignant hyperthermia 2:852
    spinocerebellar ataxia 2:852
    visual impairments 2:853
GPCR interaction 2:786T
L-type 6:735–736
molecular biology 2:844
    auxiliary subunits 2:844
    cardiac tissue 2:844
    muscle cells 2:844
    neurons 2:844–845
    tissue expression 2:844
nomenclature 2:841, 2:843T
    IUPHAR-approved scheme 2:842,
        2:843T
    N-type 2:842
    subfamily 1 2:842–844
    T-type 2:842

pain modulators 6:342
pharmacology 2:838T, 2:849
    see also specific drugs
    inorganic metal ions 2:849
    peptide toxins 2:849
    small molecules 2:849–850
physiological functions 2:846
    cardiac muscle contraction 2:847
    neurons 2:847–848
    N-type channels 2:847–848
    P-type channels 2:847–848
    second messenger effects
        2:846–847
    skeletal muscle 2:847
    smooth muscle cells 2:847
    transmembrane voltage 2:846–847
regulation 2:848
    GPCR 2:848
structure 2:828F, 2:841–842, 2:845
    $\alpha_2\delta$ subunit 2:845
    α-interaction domain (AID) 2:845
    β subunit 2:846
    cation selectivity 2:845
    γ subunit 2:846
    hinged lid mechanism 2:845
    subunits 2:841–842
target selection 2:73
therapeutics 2:838T, 2:849–850
    see also specific drugs
Calcium sensing G protein-coupled
    receptors
    allosteric modulators 2:790T
    constitutively active 2:790T
    human disease 2:792T
    in human disease 2:775T
Calcium-sensitive dyes, GPCR
    functional assay
    screening 2:811
Calculated descriptor model, hERG
    blocker modeling 5:942
Caliper Technologies, microfluidic
    devices 2:359, 2:360T
Calmodulin, GPCR interaction 2:786T,
    2:797T
Calmodulin–C20W peptide complex,
    NMR structure
    determination 3:492, 3:493F,
    3:497–498, 3:498F, 3:500F, 3:501
Calnexin, GPCR interaction 2:797T
Calpain
    activation 6:272
    central nervous system injury 6:271,
        6:272, 6:272F
    inhibition 6:271, 6:272F
    neuroprotective drug
        biomarker 6:262–263
Calpeptin 6:272
    structure 6:272F
Calsed see Amrubicin hydrochloride
Calu-3 lines, in vitro nasal
    administration models
    5:284–285
Camazepam, oral absorption 5:680T

Cambridge Antibody
　　Technology **1**:516
　characteristics **2**:226*T*
Cambridge Crystallographic data Centre
　　(CCDC) *see* Cambridge
　　Structural Database (CSD)
CambridgeSoft Corporation
　ChemFinder **3**:295
　commercially available
　　databases **3**:308
Cambridge Structural Database
　　(CSD) **3**:**389**–**410**
　advantages **3**:399
　applications **3**:398
　　bond length values **3**:400
　　conformational analysis **3**:400
　　conformational preferences **3**:400
　　conformational variation **3**:400*F*,
　　　3:401
　　crystal structure relevance **3**:399
　　intermolecular interactions *see below*
　　intramolecular geometries **3**:400
　　metal coordination **3**:401
　　MIMUMBA **3**:401
　　valence angle values **3**:400
　chemoinformatics, three-dimensional
　　structure **3**:240
　computer-assisted drug design
　　(CADD) **4**:14
　ConQuest Search **3**:393, **3**:394*F*
　conformational analysis **3**:394*F*,
　　3:400, **3**:400*F*
　pharmacophore 3D database
　　searching **4**:139*T*
　query combinations **3**:393–395
　　3D geometrical parameters **3**:393
　　3D substructures **3**:393
　　2D substructures **3**:393
　data acquisition **3**:391
　　Crystallographic Information File
　　　(CIF) **3**:391–392
　　personal communications
　　　3:391–392
　drug design/development **3**:399
　　conformational information **3**:399
　　intermolecular interactions
　　　3:399–400
　　structure correlation
　　　principles **3**:399
　information content **3**:390, **3**:390*T*,
　　3:391*F*
　　atom-by-atom
　　　substructures **3**:390–391
　　bond-by-bond
　　　substructures **3**:390–391
　　PreQuest **3**:391, **3**:393
　　2D structural
　　　representations **3**:390–391
　　validation **3**:391
　intermolecular interactions **3**:399–
　　400, **3**:401, **3**:406
　　computational paradigms **3**:402
　　halogens **3**:404

　　hydrogen bonds **3**:401–402
　　IsoStar, *see above*
　　nonbonded interactions **3**:402
　　non-hydrogen bonds **3**:404
　　prediction **3**:406
　　strong hydrogen bonds **3**:402,
　　　3:403*F*, **3**:404*F*
　　SuperStar program **3**:406,
　　　3:407*F*
　　weak hydrogen bonds **3**:404
　　X-SITE program **3**:406
　IsoStar (intermolecular
　　interactions) **3**:395, **3**:398,
　　3:398*F*
　　applications **3**:405
　　contoured surface
　　　generation **3**:405
　　scatterplots **3**:405
　　'unlikelihood' **3**:403*F*, **3**:405,
　　　3:405*F*
　knowledge-based libraries **3**:395
　Mercury Visualizer **3**:395, **3**:395*T*,
　　3:396*F*
　　hydrogen bonds **3**:395
　　nonbonded interactions **3**:402,
　　　3:402*F*
　Mogul (intramolecular
　　geometry) **3**:395, **3**:396
　　geometric features **3**:396,
　　　3:396*F*
　　hyperlinking **3**:396–398
　　search tree generation **3**:396
　protein–ligand docking
　　programs **3**:406
　　FlexX **3**:406
　　GOLD **3**:406
　　LUDI **3**:407
　rational drug design **3**:399
　site analysis, drug design **4**:286
　software components **3**:393
　statistical overview **3**:392, **3**:392*T*
　　growth rate **3**:393*F*
　　number of entries **3**:392
　Vista data analysis **3**:394*F*, **3**:395
　　conformational analysis **3**:394*F*,
　　　3:400, **3**:400*F*
　WebCite **3**:398
Camellia (green tea) **1**:425
CAMK1, PDB entries **2**:965*T*
cAMP
　phosphodiesterase(s) specific
　　to **2**:920
　phosphodiesterase substrate
　　recognition **4**:758–759
cAMP assays
　EIA, G protein-coupled receptors
　　(GPCRs) **3**:634–636
　GPCRs *see* G protein-coupled
　　receptors (GPCRs), assays
Campath *see* Alemtuzumab
cAMP dependent protein kinase, three-
　　dimensional structures **2**:636*T*
Camphor, structure **2**:909*F*

cAMP response element-binding
　　protein (CREB)
　identification, cDNA functional
　　screening **3**:15–16
　nicotinic acetylcholine receptor
　　regulation **2**:886–887
cAMP responsive element (CRE),
　　GPCR functional assay
　　screening **2**:812
Campto *see* Irinotecan hydrochloride
Camptothecin (CPT) **7**:113, **7**:121
　analogs **1**:381–382; **7**:123
　discovery/development **1**:64–65,
　　1:381–382
　DNA interactions, quantum
　　mechanical calculations **4**:410
　drug–polymer conjugates **5**:1048,
　　5:1049*F*
　SAR **1**:64–65, **1**:65*F*
　structure **1**:382*F*; **3**:700*F*; **4**:410*F*
Campylobacter infections, irritable bowel
　　syndrome **6**:650
Campylobacter jejuni infections, Guillain–
　　Barré syndrome **6**:234, **6**:235
Canada
　IPA law **1**:560
　malaria infections 1994-2004 **1**:666
　medicine marketing system **1**:632
　orphan drugs legislation **1**:662
　traditional medicine **1**:420
Cancer **7**:1–**31**, **7**:55
　see also Carcinogenicity; Tumors
　angiogenesis **7**:19
　　fibroblast growth factor **7**:21
　　platelet derived growth
　　　factor **7**:21
　　vascular endothelial growth factor
　　　see Vascular endothelial growth
　　　factor (VEGF)
　animal models **7**:222
　apoptosis *see* Apoptosis
　biomarkers *see* Cancer biomarkers
　cell cycle *see* Cell cycle
　characterization, proteomics **3**:43
　classes, microarrays **1**:268
　definition **7**:55
　development, adalimumab **1**:442
　differentiation **7**:12
　　APC mutations **7**:12
　　cessation **7**:12
　　dedifferentiation **7**:12
　　defective antitumor
　　　response **7**:12–13
　　epigenesis **7**:12
　disease basis **7**:56, **7**:184, **7**:222
　　EGFR family kinases **2**:967
　　faulty genomic imprinting **1**:271
　　genetic basis **7**:2
　　nuclear receptors **2**:1025
　disease state **7**:222
　epidemiology **7**:2
　　prevalence **7**:2
　epigenetics *see* Epigenetics

Cancer (*continued*)
 growth factor signal
  transduction **7**:22
  c-kit activation **7**:22
  deregulated cytosolic signaling
   pathways **7**:23
  deregulated receptor tyrosine
   kinases **7**:22
  epidermal growth factor
   receptors **7**:22
  PI-3K-PDK1-PKB pathway **7**:23,
   7:24*F*
  Ras-Raf-MEK-MAPK
   pathway **7**:23, **7**:23*F*
 incidence **7**:33, **7**:55,
  7:56*T*
 invasive growth *see* Cancer, invasive
  growth
 metastasis *see* Metastasis
 molecular epidemiology **7**:3
  molecular markers **7**:3
  risk factor identification **7**:3
 mortality **7**:55, **7**:56*T*, **7**:221
 normal cells *vs.* **7**:33
 proteomics **2**:388
 risk factors **7**:3
  air pollution **7**:4
  alcohol **7**:4
  cancer chemotherapy **7**:4
  carcinogens **7**:3, **7**:5
  diet **7**:3
  family history **7**:3–4
  identification **7**:3
  ionizing radiation **7**:4
  obesity **7**:3
  occupations **7**:3
  reproductive factors **7**:4
  sunlight exposure **7**:4
  tobacco smoking **7**:3, **7**:4
  virus infections **7**:4
 screening/diagnosis **7**:5
  biomarkers *see* Cancer biomarkers
  chromosome aberrations **7**:5
  genetic markers **7**:5
  mutation frequency **7**:5
 signal transduction **2**:960–962;
  7:184
 staging **7**:2
 suppression, gene therapy **1**:303
 target selection **2**:73
 treatment **7**:6, **7**:7
  see also Anticancer drugs; Cancer
   chemotherapy
  angiogenesis inhibition **3**:202
  bevacizumab **3**:202
  cilostamide **2**:947–948
  lifestyle changes **7**:4
  monoclonal antibodies **1**:434
  photodynamic treatment **7**:7
  radiotreatment **7**:7
  rituximab **1**:434
  surgery **7**:6
  trastuzumab **1**:434

  vascular endothelial growth factor *see*
   Vascular endothelial growth factor
   (VEGF)
Cancer, invasive growth **7**:14,
  7:185
 adhesion molecules **7**:14
 cadherins **7**:15
 extracellular matrix
  remodeling **7**:15–16
 integrins **7**:15
  angiogenesis **7**:16
  epidermal growth factor
   (EGF) **7**:15–16
  extracellular matrix
   remodeling **7**:16
  focal adhesion kinase (FAK)
   7:15–16
 matrix metalloproteinases **7**:16–17,
  7:241–242
  MMP-7 **7**:17
  as targets **7**:17
  tissue inhibitors of metalloproteinases
   (TIMPs) **7**:17
  upregulation **7**:17
 plasminogen activators **7**:16–17
  tissue-type **7**:17–18
  urokinase-type **7**:17–18
 proteolytic enzymes **7**:16
  see also specific types
 tissue remodeling **7**:14
Cancer biomarkers **7**:5, **7**:44,
  7:48*F*
 biopsies **7**:6
 CA-125 **7**:5
 clinical endpoints **7**:48
 discovery **7**:49
  genomics **7**:49
  metabolomics **7**:49
  proteomics **7**:49
 future work **7**:49
  normal *vs.* patient population
   differentiation **7**:49–50
  preclinical models **7**:50
  technology **7**:50
 gene profiling **7**:5–6
 Her2/neu expression **7**:48
 molecular profiling **7**:48
 pharmacodynamic markers **7**:49
 population studies **7**:47–48
 prostate-specific antigen **7**:5
 protein profiling **7**:5–6
 surrogate endpoints **7**:48
 types **7**:44
  histological markers **7**:44–47
  serum markers **7**:44–47
Cancer cell targeting, RNAi (RNA
  interference) **3**:184
Cancer chemotherapy
 see also Anticancer drugs; *specific drugs*
 cancer risk factors **7**:4
 current status **1**:63
 definition **7**:470*T*
 induced leukopenia **1**:152

 kinase inhibitors *see* Protein kinase
  inhibitors
 neuropathic pain models **6**:335
 new compounds **7**:221–251
Cancerlit **3**:271*T*
Cancer-related pain, animal
  models **6**:335
Cancidas *see* Caspofungin acetate
Cand5, wet age-related macular
  degeneration treatment
  6:306*T*
Candesartan **1**:154
 bioavailability **6**:722*T*
 drug action **1**:532*T*
 half-life **6**:722*T*
 hypertension treatment **6**:721–723
 structure **6**:587*F*, **6**:722*T*
Candesartan and Lisinopril
  Microalbuminuria (CALM)
  study, renal dysfunction **6**:585
Candida, naftifine effects **7**:447*T*
Candida albicans, structure **7**:421*F*
Candidate genes
 hypertension **6**:707
 pharmacogenomics **1**:345
Candidate Ligand Identification
  Program (CLIP), feature-based
  alignment **4**:474
Candidate parallel development,
  HTS **2**:401, **2**:402*F*
Candidate seeking, lead
  optimization **2**:205
Candidosis **7**:426
 complications **7**:426
 cutaneous **7**:423*T*
 diagnosis, PCR **7**:429
 diversity of disease **7**:426
 as emerging disease **7**:425–426
 opportunistic infections **7**:425*F*,
  7:425*T*
 organ transplants **7**:426
 treatment
  acetyl-CoA carboxylase
   inhibitors **7**:439
  amphotericin **7**:457
  anidulafungin **7**:462
  aureobasidin A **7**:434, **7**:463–464
  caspofungin **7**:426, **7**:462
  fluconazole **7**:426, **7**:432, **7**:455
  icofungipen **7**:433–434
  itraconazole **7**:455
  ketoconazole **7**:455
  micafungin **1**:133; **7**:462
  N-myristoyltransferase
   inhibitors **7**:436
  naftifine **7**:447
  protein biosynthesis
   inhibitors **7**:439–440
  RNA polymerase III
   inhibitors **7**:436–439
  voriconazole **1**:133–**1**:134; **7**:426,
   7:432–433
Candins, as antifungals **7**:430*T*

Canicular multispecific organic ion transporter *see* ABCC2 (MRP2) transporter
Canine model, drug-induced QT interval prolongation 5:902
Cannabimimetics, as analgesic 6:342–343
Cannabinoid CB$_1$ receptors
  constitutively active 2:790T, 2:793T
  in human disease 2:775T
Cannabinoid CB$_2$ receptors
  constitutively active 2:790T
  in human disease 2:775T
Cannabinoid receptor agonists 6:405F
  IBD treatment 6:629
Cannabinoid receptor antagonists
  see also specific drugs
  antiemetic drugs 6:684
  obesity treatment 6:404
Cannabinoid receptor modulators, irritable bowel syndrome treatment 6:661T, 6:665
Cannabinoids
  glaucoma treatment 6:313T
  pain 6:342
Cannabis, discovery/development 1:322
Canolide A, discovery/development 1:74
Canonical activation pathway, NF-κB pathway 3:192, 3:193F
Canonical representations, structure/substructure searching 3:238
Canon Medicinae, natural products 1:356–357
Cantharidin, structure 2:982F
CAOLD, database interfaces 3:283
Capability improvement, outsourcing 2:207
Capacity factors, VolSurf descriptors 5:539
Capacity issues, screening library selection 4:500
Capecitabine 1:144; 7:62T, 7:64
  activation, thymidine phosphorylase 5:1015
  adverse events 7:61T
  approval 7:62T
  clinical indication 7:62T, 7:68
    breast cancer 1:144
  formulations 7:68
  mechanism of action 7:64T
  as prodrug 5:1015
  structure 5:1016F
  toxicity 7:46T
Caperitide 1:167
Capillary electrophoresis
  epigenetics 1:267T
  ionization 5:350
  ionization constants (pK$_a$) measurement 5:392

plasma protein binding measurement 5:328
  proteomics 2:347, 2:347F
Capillary electrophoresis-mass spectrometry (CE-MS) 5:996
Capillary isoelectric focusing, proteomics 2:347
Capital investment, pharmaceutical industry 2:111
CAplus 3:271T, 3:273
  database interfaces 3:283
CAPON, schizophrenia 6:24
Capoten *see* Captopril
Capravirine, structure 1:74F
Capreomycin
  mycobacterium treatment 7:705T, 7:713, 7:714–715
  mechanism of action 7:714
  resistance 7:714
  structure 7:714F
CAPRICORN *see* Carvedilol Post Infarction Survival Control in Left Ventricular Dysfunction (CAPRICORN)
Capromorelin, as prokinetic agent 6:685–686
Capsaicin
  mitochondria complex I 2:1068–1069
  structure 2:909F
Capsaicin-derived products, transient receptor potential ion channels 2:909
Capsazepine
  IBD treatment 6:637
  structure 2:909F
Captive offshoring, definition 2:206
Captopril 6:718
  analog design 2:651
  bioavailability 5:712T; 6:719T
  design/development 1:16, 1:375; 2:625; 3:876–877, 3:877F, 3:929, 3:930F
    rational drug design 1:594
    structure-based design 4:685
  dose 6:719T
  drug action 1:532T
  half-life 6:719T
  in vitro potency 6:719T
  me-too prescribing 1:713–714
  number of patients needed to treat 8:144T
  oral absorption 5:680T
  structure 1:378F; 6:719T
  structure-based drug design 4:77T
  target information 4:750T
D-Captopril, structure 7:493F
Carbachol derivatives, rolipram 7:894
Carbamate esters, prodrugs 5:1023
Carbamates
  microtubule targeting agents 7:96
  prodrugs 5:1027

Carbamazepine 6:280–281, 6:288
  analgesia 6:339
  BCS 5:973T
  bioavailability 5:712T
  bipolar affective disorder treatment 6:71
  characteristics 2:829T
  effective permeability (P$_{eff}$) 5:673T
  glucuronic acid conjugation 5:143F, 5:151–152
  hydrolysis 5:142–144, 5:144F
    species differences 5:144
  immunotoxicity 5:221T
  oral absorption 5:468T
  structure 2:829T; 6:281F, 6:336F
Carbapenems 7:503
  activity 7:506T
  mechanism of action 7:508
  oral 7:508
    activity 7:509T
    bioavailability 7:508
    spectrum 7:508
    stability 7:508, 7:510F
    structure 7:509F
  parenteral anti-MRSA 7:508
  parenteral anti-*Pseudomonas* 7:503
    SAR 7:504
  SAR 7:504, 7:505F
  spectrum 7:508
  structure 7:507F
Carbazol moiety
  antioxidant properties 8:137F, 8:139
  carvedilol 8:137F, 8:139
Carbendazim
  ionization constant (pK$_a$) 5:366T
  structure 5:359
Carbenin *see* Betamipron; Panipenem
Carbidopa, GPCR targeting 2:773T
Carbohydrate-based dynamic combinatorial libraries *see* Dynamic combinatorial libraries (DCLs)
Carbohydrate metabolism, insulin 6:429–430
Carbó index, 3D molecular alignment 4:112
Carbomycin 7:556–557
Carbonate esters, prodrugs 5:1023
Carbon atoms α to heteroatom, oxidation, cytochrome P450 5:94
Carbon–boron bioisosterism *see* Bioisosterism, carbon–boron
Carbon–carbon bond-forming reactions, solid support synthesis *see* Solid support synthesis
^{14}C-labeled compounds, metabolism studies 5:34
Carbon–heteroatom coupling reactions, solid support synthesis 2:466, 2:466F

Carbonic anhydrase
  de novo drug design **4:**297
  mutations, osteopetrosis **6:**507
  structure, coordinate errors **3:**532
Carbonic anhydrase inhibitors
  see also specific drugs
  docking applications **4:**276
  dynamic combinatorial
    libraries **3:**969–970
  glaucoma treatment **6:**313*T*
Carbon–silicon bioisosterism *see*
    Bioisosterism, carbon–silicon
Carbon tetrachloride,
    reduction **5:**121–122, **5:**122*F*
Carbonyl carbon atoms, oxidation,
    cytochrome P450 **5:**103
Carbonyl groups, dynamic combinatorial
    library generation **3:**961–962,
    3:962*F*
Carbonyl oxygen probe, Molecular
    Interaction Fields **4:**239,
    4:239*F*
Carbonyl reductases (CBRs) **5:**127,
    5:178
  see also specific enzymes
  characteristics **5:**169*T*
  cytoplasmic **5:**178
  ontogenic development **5:**179
  SDR superfamily **5:**178
  tissue distribution **5:**179
Carboplatin **7:**142
  activation **7:**142–143
  bioavailability **7:**143
  clinical indication **7:**143
  mechanism of action **7:**143
  side effects **7:**143
    toxicity **7:**46*T*
  structure **7:**143*F*
  xenograft models **7:**39*T*
Carborane, structure **2:**701*F*
Carboxamides
  glucuronic acid conjugation **5:**143*F*,
    5:151–152
  oxazolidinones **7:**683–684
  phosphodiesterase-5 inhibitors **2:**940
N-Carboxyanhydrides, Hirschmann,
    R F **1:**8
Carboxyboranes, carbon-boron
    bioisosterism **2:**698–699
Carboxyhemoglobin levels, COPD
    clinical trials **7:**881–882
Carboxylesterases **5:**181
  characteristics **5:**169*T*
  hydrolases **5:**135
Carboxylic acids
  amine prodrugs **5:**1020
  bioisosterism *see* Bioisosterism,
    carboxylic acids
  ester prodrugs **5:**1018, **5:**1019*F*
  ionization constants (pK_a) *see*
    Ionization constants (pK_a)
Carboxylic derivatives,
    hydrolases **5:**135

Carboxylic ester bioisosterism,
    benzodiazepine receptor
    ligands **2:**667*F*
Carboxylic ester hydrolases, drug
    metabolism **5:**203
16α-Carboxylic esters/acids, steroid-
    derived modulators **7:**172–173,
    7:172*F*
Carboxy PEGylation *see* Polyethylene
    glycol (PEG) conjugates
Carboxypeptidase-A, three-dimensional
    structures **2:**626*T*
Carboxypeptidase G2, antibody-directed
    enzyme prodrug
    therapy **5:**1034
Carcinogenicity
  animal experimentation alternatives
    see Animal experimentation
    alternatives
  computational models **5:**920
    Carcinogenic Potency Database
      (CPDB) **5:**920
    DNA interaction *vs.* **5:**920
    future work **5:**921
    International Agency for Cancer
      Research **5:**920
    in vivo tests **5:**920
    predicted *vs.* actual results **5:**921,
      5:921*T*, **5:**922*T*
    rodent testing, relevance of
      5:920
    SAR-based models **5:**921
    TD$_{50}$ **5:**920
  electrotopological indices (E-state
    index) **5:**572
  genetically engineered animals *see*
    Genetically engineered animals
Carcinogenic Potency Database
    (CPDB) **5:**920
Carcinogens **7:**3, **7:**5
Carcinoma, definition **7:**2, **7:**55
Cardiac action potentials, sodium
    channels **2:**836
Cardiac arrhythmias
  calcium-channel blockers **2:**851
  sodium channels **2:**840–841
Cardiac disorders
  arrhythmias **6:**733
  potassium channels **2:**863
Cardiac expression, peroxisome
    proliferator-activated receptor-γ
    (PPAR-γ) **2:**1009
Cardiac hypertrophy, heart
    failure **8:**141
Cardiac ion channels **6:**733
  see also specific channels
  action potentials **6:**734, **6:**734*F*
  depolarizing currents **6:**734
  pacing currents **6:**734
  rational drug design **6:**735
  repolarizing currents **6:**734
  structure **6:**735
  voltage-gated currents **6:**735

Cardiac muscle contraction, calcium
    channels **2:**847
Cardiac myocytes, sodium
    channels **2:**834
Cardiac protection, carvedilol **8:**139
Cardiac remodeling, heart
    failure **8:**141
Cardiac tissue
  calcium channels **2:**844
  phosphodiesterase-3 **2:**924
Cardiolipin
  cytochrome *c* interactions **2:**1077
  mitochondria membranes
    inner **2:**1059
Cardiome **1:**507
Cardiostimulants
  see also specific drugs
  introductions **1:**168
Cardiotonic drugs
  see also specific drugs
  introductions **1:**175
Cardiotoxicity
  doxorubicin **7:**123
  hERG related *see* hERG
  quinolone antibiotics **7:**588
Cardiovascular disease (CVD)
    6:693–704
  see also specific diseases/disorders
  biomarkers **3:**70
  characterization, proteomics **3:**43
  economic impact **6:**693
  epidemiology **6:**694
  market **6:**694
  metabolic syndrome **6:**383, **6:**420
  metabonomics **5:**1003
  mortality **6:**461–462, **6:**693, **6:**694
  obesity **6:**420
  prevalence **6:**694, **6:**695*T*
    ethnicity **6:**694
  risk factors **6:**700, **6:**701*T*
    see also specific risk factors
  SMEs **1:**508
  tamoxifen **7:**153–154
Cardiovascular disease treatment
  see also specific drugs
  antiarrhythmics *see* Antiarrhythmic
    drugs
  calcium-channel blockers **2:**851
  market **6:**694
    development spending **6:**694
  monoclonal antibodies **1:**444
  natural products **1:**374
  thrombolytics *see* Thrombolytic drugs
  vascular endothelial growth
    factor **3:**203
Cardiovascular drugs
  see also specific drugs
  introductions **1:**106*T*, **1:**107, **1:**153
    antiarrhythmic **1:**162
    anticoagulants **1:**169
    antithrombotic agents **1:**164
    cardiostimulants **1:**168
    cardiotonic **1:**175

fibrinolytic **1**:175
heart failure **1**:167
hematological **1**:174
hemophilia **1**:175
hypocholesteremic agents **1**:170
iron chelation **1**:173
subarachnoid hemorrhage **1**:174
Cardiovascular homeostasis, estrogen
receptors (ERs) **2**:1005
Career guidelines **2**:17
gender issues **2**:244
mission **2**:17
responsibilities **2**:18
biological activity monitoring **2**:19
compound submission **2**:18
compound synthesis **2**:18
experimental records **2**:18
goal awareness **2**:19
independence/initiative **2**:18
invention records **2**:18
laboratory equipment
treatment **2**:18
literature work **2**:18
management **2**:19
oral/written reports **2**:18
personal interactions **2**:19
safe laboratory conduct **2**:18
timeliness **2**:19
salaries **2**:19
success keys **2**:17
Carfecillin, oral absorption **5**:680T
Cariporide **6**:756
study monitoring **1**:717
Carlson, Arvid, collaborators, Lindberg,
P **8**:214
Carmustine *see* BCNU (carmustine)
Carnitine, structure **2**:67F
Carotene and Retinol Efficacy Trial
(CARET) **1**:455
Carotenoid-processing enzyme
monotropic membrane protein
crystallization **3**:435, **3**:435F
protein crystallization **3**:435
Carotenoids **1**:449–450, **1**:455
clinical trials **1**:455
lung cancer **1**:455
structure **1**:455
Carotid artery injection technique,
mechanistic pharmacokinetic
studies **5**:39
Carpanone, PASP **3**:818, **3**:819F
Car–Parrinello molecular dynamics
(CP-MD) **4**:385
Carrier-linked prodrugs **5**:1017,
5:1018
Carrier-mediated efflux, blood–brain
barrier permeation **5**:304F
Carrier-mediated influx
blood–brain barrier
permeation **5**:304F
permeability **5**:672
Carrier-mediated transport, pulmonary
administration **5**:287

Carrier proteins, hydrophobic
depoting **2**:585
Carroll Depression Scale **6**:64
Carteolol, bioavailability **5**:712T
Cartesian coordinates
pharmacophore detection **4**:126–127
x-ray crystallography **4**:763
Carumonam **7**:489
discovery **7**:480
Carvedilol **8:137–148**
adrenergic receptors **8**:138
afterload reduction **8**:138
α-adrenoceptor antagonists **8**:138
antiapoptotic properties **8**:137–138,
8:141
antioxidant properties **8**:137–138,
8:139
carbazol moiety **8**:137F, **8**:139
in heart failure **8**:141
oxygen free radical
inhibition **8**:139
structure **8**:139
antiproliferative properties **8**:137–
138, **8**:141
bioavailability **6**:712T
clinical applications
cardiac protection **8**:139
hypertension treatment
6:587–588, **6**:711
myocardial infarction **8**:139
clinical trials **8**:137–138, **8**:142
COMET **8**:142
early termination **8**:138, **8**:142
phase I **8**:142
phase II **8**:142
phase III **8**:142
dose **6**:712T
FDA **8**:142
approval **8**:137
GPCR targeting **2**:773T
half-life **6**:712T
in heart failure **8**:140
antioxidant properties **8**:141
commercial doubts **8**:141
contraindications **8**:137, **8**:140
historical overview **8**:137
Boerhinger Mannheim, acquisition
from **8**:138
Garnier, Jean-Pierre **8**:138
Leschly, Jan **8**:138
new science **8**:141
Poste, George **8**:138
SmithKline Beecham
Pharmaceuticals **8**:137–138
ionization constant (pK_a) **5**:362,
5:363F
number of patients needed to
treat **8**:143, **8**:144T
pharmacology **8**:138
see also specific mechanisms
portal hypertension
treatment **6**:725–726
reprofiling **1**:349

structure **6**:587F, **6**:712T, **8**:137F
antioxidant properties **8**:139
x-ray diffraction **8**:139
vasodilation **8**:137
Carvedilol Post Infarction Survival
Control in Left Ventricular
Dysfunction
(CAPRICORN) **1**:349
number of patients needed to
treat **8**:144T
Case-control studies, phase IV clinical
trials **2**:191
CASE program, computational oral
absorption prediction **5**:688
Casodex *see* Bicalutamide
CASP (Critical Assessment of
Techniques for Protein Structure
Prediction)
computer-assisted drug design
(CADD) **4**:22
homology models reliability **3**:544
protein structure prediction
3:324–325
Caspase, activation, antineoplastic
drugs **1**:382
Caspase-3
central nervous system injury
6:273–274
fragment-based drug discovery **3**:950
optical assays **3**:581, **3**:581F
Caspase inhibitors **1**:382
see also specific drugs
Huntington's disease
treatment **6**:221
mechanism of action **2**:577
structure **1**:383F
Caspofungin **7**:430T, **7**:432–433
clinical profiles **7**:462
aspergillosis **7**:462
candidosis **7**:426, **7**:462
medicinal chemistry **7**:460
pharmacokinetics **7**:461
Caspofungin acetate **1:131**
CASREACT **3**:281–282, **3**:282T
CAS registry
de novo drug design **4**:297
structure databases **3**:280
Cassette dosing, drug metabolism and
pharmacokinetics **5**:33
Catalog databases **3**:279
Catalyst/HipHop, pharmacophore
searching **4**:129T
Catalyst/Hypogen, pharmacophore
searching **4**:135T
CATALYST program **5**:581T
3D-QSAR **4**:582–583
Accelrys **3**:295
database management systems **3**:295
homology models **4**:487–488
5-HT$_{1A}$ receptor structure
4:686–687
P-glycoprotein models **5**:784, **5**:785
pharmacophore detection **4**:134

CATALYST program (*continued*)
  pharmacophore 3D database
    searching 4:139*T*
  structure-based fragment
    screening 4:792
  3D structure generation 4:465
  virtual screening 4:790–791
Catalytic domain structure,
    phosphodiesterases 4:757,
    4:757*F*
Catalytic screening dynamic
    combinatorial libraries *see*
    Dynamic combinatorial libraries
    (DCLs)
Cataplexy 6:148
  treatment 6:154
Cataracts
  atopic dermatitis 7:948–949
  formation, tamoxifen 8:96
Catch-and-release principle,
    PASP 3:795, 3:796*F*, 3:799,
    3:800*F*, 3:826
Catechin 1:452*T*
Catecholamine sulfotransferase, 3D
    QSAR 5:593
Catechol derivatives,
    phosphodiesterase-4
    inhibitors 2:936
Catechol *O*-methyltransferase
    (COMT) 5:145, 5:184,
    5:819–820
  anxiety disorders, genetic basis 6:91
  polymorphisms, narcolepsy 6:149
  schizophrenia 6:23
  3D QSAR 5:593, 5:594*F*
Catechol-*O*-methyltransferase
    inhibitors 6:70
  see also specific drugs
  Parkinson's disease treatment 6:11,
    6:217
Catechols, 'functional analogs' 2:672,
    2:674*F*, 2:675*F*
β-Catenin inhibitors, docking
    applications 4:273, 4:273*F*
CATH (Class, Architecture, Topology
    and Homologous
    superfamily) 3:374
  protein function prediction,
    bioinformatics 3:336–337
  search capabilities 3:376–377
Cathelicidin, psoriasis 7:942
Cathepsin-B, three-dimensional
    structures 2:626*T*
Cathepsin D, disease role 3:808
Cathepsin D inhibitors
  see also specific drugs
  PASP 3:808
Cathepsin K inhibitors, bone disease
    treatment 6:515,
    6:515*F*
Cathepsin S inhibitors
  see also specific drugs
  asthma treatment 7:882

clinical trials 7:882
  COPD treatment 7:882
Cation aggregation, solubility 5:410
Cat's claw (*Uncaria tomentosa*) 1:424
CATS descriptors, library design 4:361
Caudal trigeminal nucleus,
    migraine 6:370
Causative-acting drugs,
    pharmacogenetics 3:54
CAVEAT, pharmacophore 3D database
    searching 4:138, 4:139*T*
Caveolin-1, GPCR interaction 2:797*T*
Cavernous angioma/malformation,
    epilepsy 6:285
CBNB (Chemical Business News
    base) 3:271*T*, 3:275
CC (comments), UniProtKB
    flatfile 3:360, 3:361*F*
CC-4047 2:759
  structure 2:759*F*
CC-11006 2:759
  structure 2:759*F*
CCDs *see* Charge-coupled devices
    (CCDs)
CCI-779, characteristics 2:984*T*
CCNU, structure 7:139*F*
CCR1 antagonists, transplant
    rejection 7:921
CCR2
  constitutively active 2:790*T*
  in human disease 2:775*T*
  IBD 6:633–635
CCR3
  constitutively active 2:790*T*
  eosinophil expression 7:884
  eotaxin (CCL11) 7:884
  in human disease 2:775*T*
  knockout models 7:884
  virtual screening 4:689
CCR3 antagonists
  see also specific drugs
  asthma treatment 7:882, 7:884
    benzyl piperidine motif 7:885–886
    bisperidine ring system 7:886
    3,4-dichloro substituted benzyl
      piperidines 7:888
    pyrrolidino-
      hydroquinazolines 7:887
    SAR 7:884
  atopic dermatitis treatment 7:981
  clinical trials 7:882
  skin inflammation treatment 7:981
CCR5
  constitutively active 2:790*T*
  HIV life cycle 7:331, 7:358–359
  in human disease 2:775*T*
  IBD 6:633–635
  pharmacogenetics 3:61
CCR5 antagonists
  IBD treatment 6:633–635
  solid-phase libraries 3:727–735,
    3:734*F*
CCR5R4, HIV life cycle 7:360

CCR9, IBD 6:635–636
CCT018159 7:232
CD1 molecules, antigen
    presentation 5:222
CD4 receptors, HIV life cycle 7:331,
    7:358–359
CD4+ T cells
  see also Helper T cells
  asthma 7:877
  HIV infection 7:330
CD18 reduced expression, psoriasis
    animal models 7:971
CDA-54 6:343–344
  structure 6:344*F*
Cdc-42 GTP hydrolysis, quantum
    mechanical calculations 4:404
cDNA expressed systems, in vitro
    expression studies 5:246
cDNA functional screening 3:14
  angiogenesis 3:15–16
  array formats 3:15–16, 3:16*F*
  'gain-of-function' screens 3:15,
    3:15*F*
  JAK gene isolation 3:14
  limitations 3:15*F*
  Mammalian Gene Collection
    3:15–16
  metastasis 3:15–16
  NF-κB signaling pathways 3:14,
    3:15–16
  osteogenesis 3:15–16
  overexpression pools 3:15
  p53 3:15–16
  signal transduction pathways
    3:15–16
  STAT gene isolation 3:14
  TORC1 identification 3:15–16
  Wnt signaling pathway 3:15–16
cDNA microarrays 3:88–89, 3:331
  advantages 3:90
  definition 3:331
  disadvantages 3:90
  experiment configuration 3:331
    ArrayExpress 3:331
    Gene Expression Omnibus
      (GEO) 3:331
    MIAME standard 3:331
    National Center for Computational
      Biology (NCBI) 3:331
    normalization 3:331
    quality control 3:331
  low-level analysis 3:331
  probe classification 3:332
    gene co-expression 3:332
    metabolic pathways 3:332
    regulatory pathways 3:332
  probe numbers 3:331
  sample classification 3:332
    decomposition 3:332
    methods 3:332
  single-channel *vs.* multi-
    channel 3:89
cDNA precursors, siRNAs 3:172

CDP-571, IBD treatment **6**:620–621
CDPPB, schizophrenia treatment **6**:38
CE, protein function prediction,
  bioinformatics **3**:336–337
CEA, cancer biomarker **7**:49
CEABA-VTB **3**:271*T*
CEE-03-310
  alcohol addiction **6**:184*T*
  structure **6**:181*F*
Cefaclor **7**:500
  activity **7**:502*T*
  structure **7**:500*F*
Cefadroxil **7**:500
  structure **7**:500*F*
Cefazolin
  renal elimination **5**:46
  structure **5**:371*F*; **7**:492*F*
Cefcapene pivoxil **1**:109; **7**:500
  activity **7**:502*T*
  structure **7**:501*F*
Cefdinir **7**:500
  activity **7**:502*T*, **7**:509*T*
  structure **7**:500*F*
Cefditoren
  activity **7**:502*T*
  structure **7**:501*F*
Cefditoren pivoxil **1**:110; **7**:500
Cefepime
  activity **7**:493*T*
  development **7**:493
  structure **7**:494*F*
Cefetamet **7**:500
  activity **7**:502*T*
  structure **7**:501*F*
Cefetecol **7**:492–493
  activity **7**:493*T*
  structure **7**:494*F*
Cefixime **7**:500
  activity **7**:502*T*
  structure **7**:500*F*
Cefluprenam
  activity **7**:493*T*
  development **7**:493
  structure **7**:494*F*
Cefmaliten
  activity **7**:502*T*
  structure **7**:500*F*
Cefoperazone
  renal elimination **5**:46
  structure **7**:492*F*
Cefoselis **1**:110; **7**:495
  activity **7**:493*T*
  development **7**:493
  structure **7**:494*F*
Cefotaxime
  clinical indication **7**:492
  structure **7**:493*F*
Cefoxitin **7**:489–492
  structure **7**:492*F*
Cefozopran hydrochloride **1**:111
  activity **7**:493*T*
  development **7**:493
  structure **7**:494*F*

Cefpirome
  activity **7**:493*T*
  development **7**:493
  structure **7**:494*F*
Cefpodoxime **7**:500
  activity **7**:502*T*
  structure **7**:501*F*
Cefprozil **7**:500
  structure **7**:500*F*
Ceftazidime
  activity **7**:493*T*
  clinical indication **7**:492
  structure **7**:493*F*
Ceftibuten **7**:500
  activity **7**:502*T*
  structure **7**:500*F*
Ceftizoxime **7**:500
  activity **7**:502*T*
  structure **7**:501*F*
Ceftobiprole **7**:497
  activity **7**:496*T*, **7**:497*T*
  structure **7**:495*F*
  synthesis **7**:497, **7**:498*F*
Ceftriaxone
  activity **7**:493*T*, **7**:497*T*
  clinical indication **7**:492
  immunotoxicity **5**:221*T*
  oral absorption **5**:468*T*
  repositioning **2**:276
  structure **7**:493*F*
Cefuroxime
  activity **7**:502*T*
  discovery/development **1**:325–326
  oral absorption **5**:468*T*
  structure **7**:492*F*
Celastrol, mitoprotective
  drugs **2**:1064, **2**:1064*F*
Celebrex *see* Celecoxib
Celebrix *see* Celecoxib
Celecoxib **1**:213; **6**:339
  amyotrophic lateral sclerosis
    treatment **6**:224
  as analgesic **6**:337*T*
  arthritis treatment **7**:853
  introduction of **1**:108
  safety **1**:684
  structure **2**:9*F*, **2**:658*F*, **2**:681*F*;
    3:808; **6**:307*F*, **6**:336*F*
  target information **4**:750*T*
  wet age-related macular degeneration
    treatment **6**:306*T*
Celecoxib Long-Term Arthritis Safety
  Study (CLASS) **1**:684
Cell(s)
  differentiation
    immunosuppression **5**:223
    phenotype-oriented
      screening **3**:136
  extraction/disruption,
    proteomics **3**:30, **3**:31
  fusion, cell-based assays
    3:618–619
  HCS labeling techniques **3**:688

  immortalization, Epstein–Barr virus
    (EBV) **3**:620
  infection, influenza viruses
    7:377–378
  lysis, bacterial expression
    systems **3**:420
  metabolism, small animal imaging *see*
    Small animal imaging
  nitrogen mustard resistance **7**:132,
    7:133
  permeability assays *see* Permeability
    assays
  potassium channel
    distribution **2**:857, **2**:860
  proliferation
    angiotensin II **6**:584
    imaging **3**:671
    peroxisome proliferator-activated
      receptor-δ (PPAR-δ) **2**:1010
    small animal imaging **3**:671
  siRNA targeting **3**:182, **3**:183
  sources, cell-based assays *see* Cell-
    based assays
  survival, HCS **3**:691
  tissue binding **5**:850–852
  triazene resistance **7**:138
Cell-based assays **3**:617–646
see also specific assays
  active transport **3**:627
  ADME **3**:623, **3**:628
    absorption **3**:628
    excretion **3**:629
    metabolism **3**:628, **3**:629
  advantages **3**:618, **3**:623
    agonists *vs.* antagonists **3**:618
    biological model **3**:623
    endpoint measurements **3**:623
    membrane-bound targets **3**:618
    real-time measurements **3**:618
    signal pathways **3**:618
    test protocols **3**:623
  biological activity
    determination **3**:617
  calcium-based **3**:625–626, **3**:626*F*
  cell culture **3**:622
    antibiotics **3**:622
    bioreactors **3**:622–623
    extracellular matrices **3**:622–623
    media **3**:622
    organ system mimics **3**:623
    supplements **3**:622
  cell migration/chemotaxis **3**:627,
    3:628
    Boyden chamber assays **3**:627–628
    membrane inserts **3**:627–628
  cell sources/selection **3**:619,
    3:620*T*
    blood **3**:620
    cell fusion **3**:618–619
    collections **3**:622
    history **3**:618–619
    immortalization techniques **3**:620
    insect cells **3**:619

Cell-based assays *(continued)*
  naturally proliferating cells 3:620–621
  primary cells 3:619–620
  transgenic lines 3:621–622
  tumors 3:619
  validation 3:620–621
  variation of responses 3:620–621, 3:621F
  Xenopus oocytes 3:619
 cell surface transmitters 3:624T
 co-culture, primary cells 3:619–620
 drug discovery productivity 2:124–125
 early drug development 3:617–618
 EIA 3:631
  competitive assays 3:631
  ELISPOT 3:632
  enzymes used 3:632
  fluorescence resonance energy transfer (FRET) 3:632
  methodology determinants 3:632
  radioimmunoassays *vs.* 3:632
  sandwich assays 3:631
  transcription factor production 3:632
 electrophysiology 3:624
  rubidium-86 3:625
  throughput constraints 3:624–625
  voltage clamp 3:624–625
 examples 3:633
  see also specific assays
 flow cytometry 3:629, 3:630F
  intracellular protein staining 3:630
 fluorescence correlation spectroscopy-related confocal fluorimetric techniques (FCS+) 2:411F
 fluorescent dyes 3:625
  agonists *vs.* antagonists 3:627
  drug interference 3:625
  false positives 3:625
  fluorescence resonance energy transfer (FRET) 3:618, 3:625
  GPCRs 3:627
  hERG potassium channel 3:625–626, 3:626F
 future work 3:641
  Alliance for Cellular Signaling 3:643–644
  microenvironment changes 3:642
  precursor cells 3:643
  signaling proteins 3:643–644
  stem cells 3:641–642, 3:643
  technology changes 3:641
 GPCRs 3:624T
 high-throughput screens 3:618
 ion fluxes 3:624, 3:624T
  example *see* 5-HT3 receptor
 limitations
  ADME 3:623
  whole-organism 3:623

multiple outputs (multiplexing) 3:630
 antibody-based 3:631
 disadvantages 3:639
 DNA-binding proteins 3:631
 electrophoretic mobility shift assay (EMSA) 3:631
 enzyme immunoassays 3:631
 example 3:638
 liquid assays 3:631
 mRNA profiles 3:630–631
 nucleic acid quantification 3:631
 positional (flat) assays 3:631
 protein expression 3:630–631
 RNA-binding proteins 3:631
 transcription factor activity 3:631
nuclear hormone receptors 3:624T
phenotype-oriented screening 3:136
proliferation assays 3:628
 cell migration 3:628
 contact inhibition 3:628
 differentiation 3:628
 nucleic acid quantification 3:628
protein kinases 3:624T
readouts 3:623, 3:624T
 see also specific readouts
receptor binding 3:627
 radioligands 3:627
 sodium concentrations 3:627
regulatory systems 3:617–618
reporter genes 3:633
 examples 3:633
 response element sequence 3:633
technology improvements 3:618
toxicity 3:629
 apoptosis 3:629
 necrosis 3:629
Cell-based models
 brain drug uptake *see* Brain drug uptake
 in vitro oral absorption prediction 5:675T, 5:677
 library design, ligand-based *see* Library design, ligand-based
 pharmacogenomics *vs.* 3:56
Cell-cell signaling, *Caenorhabditis elegans* model system 3:8
Cellcept *see* Mycophenolate mofetil
CellChip System *see* High-content screening (HCS)
Cell compartments, drug discovery 2:41–43
Cell culture
 cell-based assays *see* Cell-based assays
 HIV infection 7:336
 oseltamivir, structure-based design 4:620
 QSAR *see* Quantitative structure–activity relationship (QSAR)
Cell cycle 7:9, 7:10F
 control, *Saccharomyces cerevisiae* model system 3:5–6
 cyclin-dependent kinases 7:9

 G1/S transition 7:9
  retinoblastoma protein (Rb) 7:9
 G2/M transition 7:11
  ATM gene 7:11–12
  ATR 7:11–12
  Chk2 7:12
  DNA duplication 7:11–12
  DNA repair 7:11–12, 7:11F
  p53 protein 7:12
  TP53 gene 7:12
 high-content screening *see* High-content screening
 large-scale virus vector libraries 3:18–19
 microtubules 7:81
 proteomics 3:30–31
 ubiquitin -proteasome pathway 3:190
Cell-free expression systems 3:120, 3:120F, 3:415
 advantages 3:416
 components 3:121
 design 3:121
  batch mode reactions 3:121
  continuous-exchange 3:121
  continuous-flow 3:121
  transcription–translation systems 3:121
 disadvantages 3:416
 disulfide-bridged proteins 3:123
 extract preparation 3:122
  bacterial sources 3:122
  commercially available systems 3:122
  protein synthesis using recombinant elements (PURE) system 3:122
  reticulocytes 3:122
  RNAse-deficient *Escherichia coli* 3:122
  wheat germs 3:122
 isotope labeling 3:120–121
 membrane proteins 3:115F, 3:122
  GPCRs 3:122–123
  multidrug resistance transporters 3:122–123
 NMR protein labeling 3:428
 other systems *vs.* 3:113T
 reaction parameters 3:120
Cell lab IC 100 3:686T
Cell lines
 baculovirus expression systems 3:417, 3:417T
 Drosophila melanogaster 3:651
 drug metabolism, in vitro studies 5:233
 integrity 7:42–43
 P-glycoprotein models 5:777
Cell lysis
 mammalian cell expression systems 3:420
 protein purification 3:420
 yeast expression systems 3:420

Cell-mediated immune responses, multiple sclerosis 6:233
Cell membranes
composition, passive transcellular transport 5:264
depolarization, sodium channels 2:836
excitability, potassium channels 2:859
hydrophobic depoting 2:585
Cell migration/chemotaxis
blood–brain barrier permeation 5:304F
cell-based assays see Cell-based assays
Cell models, hepatitis C 7:388–389
Cellomics Knowledgebase see High-content screening (HCS)
Cell proliferation assays
antimetabolites 7:60
cell-based assays see Cell-based assays
Cell signaling, phenolic nutraceuticals 1:451
Cell-substrate adhesion, Ultra-High-Throughput Screening System (Aurora Biosciences) 2:431, 2:441F
Cell surface receptors
dynamic combinatorial libraries 3:973
siRNAs 3:183
small animal imaging see Small animal imaging
Cell surface transmitters, cell-based assays 3:624T
Celltech, academia collaborations 1:577T
Cellular Genomics 1:515
Cellular immune response 7:846
Cellular proteomics see Proteomics
Cellular retinoic-acid-binding protein type II (CRABP), structure, deliberately incorrect 3:535
Cellulose, PASP 3:793
Cell uptake measurements, brain drug uptake see Brain drug uptake
CEM Corporation, microwave-assisted chemistry 3:850
MARS Microwave Synthesis System 3:850, 3:850F
power outputs 3:851
CEM Discover platform, microwave-assisted chemistry 3:845, 3:846F
automation upgrades 3:845–846
flow-through system 3:846, 3:846F
high-throughput synthesis 3:846, 3:847F
pressure regulation 3:847
Raman spectroscopy 3:845–846
solid-phase peptide synthesis 3:846–847, 3:847F
sub-ambient chemistry 3:847–848
temperature measurements 3:847

Centaureidin 7:93
SAR 7:94F
structure 7:94F
Center for Drug Evaluation and Research (CDER) 1:546–547
Center for Epidemiological Studies Depression Scale (CESID) 6:64
Center for Eukaryotic Structural Genomics 3:552T
Centers for Health Research 1:57T
Centers of excellence in drug discovery (CEDDS) 2:109–110
Centers of excellence proximity, pharmaceutical industry organization 2:110
Centoxin, FDA failure 1:557, 1:559
Central alveolar hypoventilation syndrome 6:149
Central domain, nuclear receptors 2:995–997
Central dopaminergic systems, modafinil 8:151
Central Drug Research Institute, raloxifene development 8:106–107
Centralization, compound storage/management see Compound storage/management
Central nervous system (CNS)
barriers 5:302, 5:303F
see also Blood–brain barrier
arachnoid epithelium 5:302
imaging, [^{18}F]fallypride 3:667
nicotinic acetylcholine receptors 2:884
phosphodiesterase-4 2:926–927
small animal imaging 3:664T, 3:667
sodium channels 2:833–834
Central nervous system diseases/disorders
see also specific diseases/disorders
diagnosis 6:2
drug introductions 1:196
DSM-IV-TR 6:2, 6:3T
drugs 2:53T
future prospects 6:14
GABA$_A$ receptor 5:896
market 6:1
neurodegenerative diseases see Neurodegenerative diseases
phosphodiesterase-4 2:925–926
phosphodiesterase-10 2:930–931
preclinical research 2:52
psychiatric disorders see Psychiatric diseases/disorders
quality of life 6:1
SMEs 1:509
social cost 6:1
Central nervous system drugs 6:1–16
see also specific drugs

Central nervous system injury 6:253–277
see also Neuroprotective drugs; Spinal cord injury (SCI); Stroke; Traumatic brain injury (TBI)
clinical trial issues 6:259, 6:261, 6:262T
biomarkers 6:261
dose regimens 6:261
dose–response analysis 6:254
drug interactions 6:262
endpoint imprecision 6:262
multiple injury models 6:262
pharmacokinetics 6:261
preclinical efficacy testing 6:261
therapeutic window 6:262
time course 6:261
disease basis 6:254
experimental models 6:256, 6:257T
new research areas 6:266
anti-inflammatory drugs 6:272, 6:273F
antioxidants 6:266, 6:267F
apoptotic neurodegeneration prevention 6:273
calpain inhibition 6:271, 6:272F
glutamate-mediated excitotoxicity 6:266
mitochondrial dysfunction prevention 6:269, 6:271F
neurorestorative drugs see Neurorestorative drugs
thyrotropin-releasing hormone analogs 6:275
pathogenesis 6:254, 6:255F
apoptotic neurodegeneration 6:273
calcium homeostasis 6:271–272
calpain activation 6:272
caspase-3 activation 6:273–274
depolarization 6:254–255
mitochondrial dysfunction 6:254–255, 6:269
nitric oxide production 6:254–255
peroxynitrile production 6:255, 6:266–267, 6:267F, 6:268F, 6:270, 6:270F
secondary 6:254
treatment 6:263
unmet medical need 6:265
Central nervous system uptake 5:747, 5:756
affecting factors 5:757
active vs. inactive molecules 5:757
anatomical therapeutic chemical classification (ATC) 5:758
lipophilicity 5:757
molecular sum over all fragments (MSOF) 5:758
molecule polarity 5:758
physicochemical descriptors 5:757

Central nervous system uptake
(*continued*)
recursive feature elimination
(RFE) **5:**758–759
surface active properties **5:**757
assessment **5:**756
data lack **5:**756–757
lamivudine **7:**314
nasal administration **5:**286
Central sensitization, chronic
pain **6:**330–331
Central sleep apnea (CSA) **6:**149
disease basis **6:**152
treatment **6:**162
acetazolamide **6:**162–163
CEP-701 **7:**199
acute myeloid leukemia **2:**971
characteristics **2:**984*T*
structure **2:**971*F,* **7:**199*F*
CEP-1347, Parkinson's disease
treatment **6:**218
CEP-7033, structure **7:**202*F*
CEP-7055 **2:**969–970; **7:**203
characteristics **2:**984*T*
structure **2:**970*F*
VEGFR family kinase
inhibitors **2:**969
Cephalexin **7:**500–501
atopic dermatitis treatment
7:950
BCS **5:**973*T*
oral absorption **5:**468*T*
structure **7:**500*F*
Cephalon **1:**499, **1:**499*T*
Cephalosporins **7:**489
see also specific drugs
allergies to **7:**489
anti-MRSA **7:**495, **7:**495*F*
activity **7:**496*T,* **7:**497*T*
SAR **7:**495
catechol, structure **7:**494*F*
discovery **7:**480
discovery/development **1:**325–326
first-generation **7:**489, **7:**492*F*
fourth-generation **7:**492, **7:**494*F*
activity **7:**493*T*
SAR **7:**493–495
spectrum **7:**493
structure **7:**494*F*
oral **7:**500
absorption **7:**500
administration **7:**500–502
classification **7:**500
spectrum **7:**502–503, **7:**502*T*
structure **7:**500*F*
parenteral **7:**489, **7:**491*T*
protein binding, E-state index
5:566
second-generation **7:**489, **7:**492*F*
structure **7:**480
third-generation **7:**492
activity **7:**492, **7:**493*T,* **7:**502*T*
clinical indication **7:**492

SAR **7:**492
structure **7:**493*F,* **7:**501*F*
variety development **7:**733
Cephalothin, structure **7:**492*F*
Ceramides
biological membranes **5:**456, **5:**456*F*
solid-phase libraries **3:**746*F,* **3:**751–752
discrete libraries **3:**707*F*
stratum corneum **5:**291, **5:**456, **5:**456*F,* **5:**460
CEREP/BioPrint models, oral
absorption prediction **5:**693
Cerebral infarct, oxidative
phosphorylation
uncoupling **2:**1079–1080
Cerebral ischemia, tetracyclines **7:**620
Cerebrospinal fluid (CSF)
psychotropic drug
development **6:**51*T*
sampling, mechanistic
pharmacokinetic studies **5:**39
Cerebrovascular disease, mortality/
morbidity, developing
countries **1:**642*T*
Cerebyx *see* Fosphenytoin sodium
Ceredist *see* Taltirelin
Cerestat, nervous system trauma
treatment **6:**13
Cerestem **1:**492*T*
Cerezyme *see* Imiglucerase
CERIUS², log P calculation
program **5:**661*T*
Cerius² ADME/Tox **5:**679*T*
plasma protein binding,
prediction **5:**856*T*
Cerivastatin **1:**170; **6:**473–474, **6:**476, **6:**483
adverse effects **6:**477
discovery/development **1:**375
doses **6:**484
structure **1:**376*F;* **6:**469*F*
withdrawal **1:**757*T*
Cervical cancer
irinotecan hydrochloride **1:**147
nedaplatin **1:**147
Cestode (tapeworm)
infections **7:**751*T,* **7:**761
biology **7:**761
clinical manifestations **7:**761
drug treatment **7:**753*T,* **7:**761
see also specific drugs
epidemiology **7:**761
Cethromycin **7:**541, **7:**550
activity **7:**550
development **7:**540–541, **7:**550, **7:**550*F*
pharmacokinetics **7:**551
variations **7:**551, **7:**552*F,* **7:**553*F*
Cetirizine
acid/base reaction equilibria **5:**608, **5:**609*F*
GPCR targeting **2:**773*T*

ionization constant (pK_a) **5:**366*T*
structure **5:**360, **5:**444*F*
CETP inhibitors **6:**488
see also specific drugs
Cetrorelix **1:**232; **2:**579*T*
controlled release
formulations **2:**593
hydrolysis **5:**140, **5:**141*F*
Cetrotide *see* Cetrorelix
Cetus Corporation
formation **2:**102
mergers and acquisitions **1:**541–542
PCR invention **2:**256
Cetuximab **1:**136, **1:**334*T,* **1:**433*T;* **7:**194
colorectal cancer **1:**137
toxicity **7:**46*T*
Cevimeline hydrochloride **1:**246
Cf 172 **7:**259
Cf 1742, varicella zoster virus **7:**260
Cf 1743, varicella zoster virus **7:**260
cFLIP, NF-κB pathway
regulation **3:**196–197
CG-53135 **1:**504
CGI-17341
mycobacterium treatment **7:**721
structure **7:**721*F*
cGMP, phosphodiesterases **2:**920; **4:**758–759
allosteric binding sites **2:**928
CGP-36742, structure **2:**893*F*
CGP-40215
structure **7:**831*F*
trypanosomiasis treatment **7:**829–830
CGP-41251 *see* PKC-412
CGP-53716 **7:**198
characteristics **2:**984*T*
structure **7:**198*F*
CGP-62464 **7:**191–192
structure **7:**191*F*
CGP-71683A, structure **6:**408*F*
CGP-76030
Bcr-Abl kinase inhibitor **2:**973
Src kinase inhibitor **2:**976–977
structure **2:**974*F,* **2:**978*F*
CGP-77675, structure **7:**191*F*
CGP-79787 *see* PTK-787
CGS-27023A, arthritis
treatment **7:**863
CGX-1007, discovery/
development **1:**380
CGX-1160, discovery/
development **1:**380
Chain identifiers, protein 3D structure
validation **3:**513, **3:**514*F*
Chairpersons, project
management **2:**14
Chalcones **7:**96
antimalarial drugs **7:**798
mitochondria, drug targeting **2:**1068
SAR **7:**96, **7:**97*F*
structure **7:**97*F*

Chameleon effect, molecular descriptors **5**:588, **5**:588*F*
Channel gating mechanisms, KcsA **4**:712
Channel-specific blockage, antiarrhythmic drugs, rational design **6**:757
CHAP-31
  discovery/development **1**:386
  structure **1**:385*F*
Chaperone co-expression, *Escherichia coli* protein expression systems *see Escherichia coli* protein expression systems
Chaperonins, structure, problems with **3**:543
CHAPS, promiscuous ligands **2**:740
Characterization studies, protein purification **3**:555
Charge-coupled devices (CCDs)
  Flash Plate assays **3**:589
  GPCR functional assay screening **2**:813–814
  HCS image detection **3**:685, **3**:687
  scintillation proximity assay **3**:584
  two-dimensional gel electrophoresis **2**:335
  x-ray detection **3**:468
Charged aggregates, solubility **5**:408–409
Charge densities, molecular descriptors **5**:583–584, **5**:584*F*
Charge desolvation, ligand–receptor interactions **4**:197
Charged groups, scaffold-based design **4**:24
Charge distribution
  force-field molecular modeling **4**:92
  x-ray–crystal interaction **3**:452–453
Charged partial surface area (CPSAs), molecular descriptors **5**:547
Charged substrates, lipophilicity measurement **5**:434
Charge separation, electrolyte diffusion **5**:413
CHARMM
  docking, scoring functions **4**:483
  K$^+$ ion channels **4**:710
  molecular dynamics simulations **4**:652
Chartreusin analogs **7**:123
Chaste berry (*Vitex agnus-castus*) **1**:424
Checklist exchange, DDBJ/EMBL/GenBank **3**:351–352
Checkpoint kinase-1 (Chk1) inhibitors, docking applications **4**:276
CHEK1, PDB entries **2**:965*T*
Chelocardin
  biosynthesis **7**:602
  Nocardia sulfurea **7**:602
  structure **7**:601*F*, **7**:603
ChemAxon Ltd, JChem Base **3**:295
Chembank **3**:306

Chembridge Corporation **2**:760*T*
ChemBridge DIVERSet, promiscuous ligands, computational prediction **2**:749
ChemDBS-3D program
  pharmacophore searching **4**:129*T*
  3D databases **4**:139*T*
  3D structure generation **4**:465
ChemFinder
  CambridgeSoft Corporation **3**:295
  database management systems **3**:295
ChemGPS, adaptive in combo strategy **5**:966
Chemical Abstracts **3**:271*T*, **3**:273
  databases **3**:300
  searching **3**:274
Chemical assessment, solid-phase libraries *see* Solid-phase libraries
Chemical biology **3**:129–149, **3**:921
  combinatorial libraries *see* Combinatorial libraries
  compound collections (libraries) **3**:132
    see also Libraries
    commercially available **3**:132
    federal sources **3**:132
    Molecular Libraries screening Center Network **3**:132
    National Cancer Institute (NCI) **3**:132
    National Institute of Neurological Disorders and Stroke (NINDS) **3**:132
    natural products **3**:133
    NIH Chemical Genomics Center (NCGC) **3**:132
  definition **3**:129, **3**:130, **3**:921
  drug discovery **3**:146
    academia/industry collaboration **3**:146
    human genome project **3**:146
  forward chemical genetics **3**:130, **3**:131*F*
  HTS **3**:130
    historical perspective **3**:130
  Molecular Initiatives Roadmap **3**:129
  phenotype-oriented screens *see* Phenotype-oriented screening
  profiling/computational analysis **3**:144
    cytologic profiling **3**:144, **3**:145*F*
    genetic profiling **3**:144, **3**:144*F*
  reverse chemical genetics **3**:130, **3**:131*F*
  target-oriented screening *see* Target-oriented screening
Chemical biotransformation, non-animal carcinogenicity testing **1**:478
CHEMICALC-2, log P calculation program **5**:661*T*
Chemical complexity, fragment-based drug discovery **4**:449

Chemical destruction of β-cells, diabetes mellitus type 1 animal models **6**:433
Chemical development, predevelopment **2**:141
Chemical Diversity Labs Inc, ChemoSoft **3**:295
Chemical diversity space, definition **4**:807
Chemical Effects in Biological Systems (CEBS) database **2**:569, **2**:570*F*
Chemical environment, Molecular Interaction Fields **4**:239
Chemical exchange effects *see* Nuclear magnetic resonance (NMR), drug discovery
Chemical fingerprints, biological fingerprint correlation **4**:808
Chemical footprinting, aminoglycosides **7**:638
Chemical functionality, reagent-based library design **4**:342–344
Chemical genetics **3**:923
  definition **3**:921, **3**:923
  drug discovery **3**:923, **3**:923*F*, **3**:923*T*
  natural products **1**:366
  target identification **3**:700, **3**:701, **3**:701*F*
Chemical–genomic profiling, genomics **3**:5
Chemical information databases *see* Databases
Chemical instability
  decomposition reactions **5**:491
    see also specific reactions
  decreasing of **5**:490
  hydration **5**:492
  hydrolysis **5**:491, **5**:491*F*, **5**:492*F*
  in vivo **5**:491
  prodrugs **5**:491
  oxidation **5**:492
    dimerization **5**:492, **5**:493*F*
    pH **5**:492
    photolysis **5**:492
    temperature **5**:492
    trace metal ions **5**:492
  stereochemical conversion **5**:493, **5**:493*F*
Chemical ionization, gas chromatography-mass spectrometry (GC-MS) **5**:993–994, **5**:994*F*
Chemical labeling, LC-MS (peptide-based) proteomics **3**:38
Chemical libraries
  property space **5**:598
  target selection *see* Target selection
Chemically advanced template searches (CATS), two-dimensional descriptors, similarity searching **4**:171

Chemically-induced diabetes mellitus **6**:423
Chemically-induced seizures, epilepsy, models **6**:286
Chemical Markup Language (CML), two-dimensional representations **3**:298–299
Chemical modification, protein crystallization **3**:888
Chemical mutagenesis, mouse models *see* Mouse models
Chemical properties
 Molecular Interaction Fields **4**:239
 toxicity, computational models **5**:914
Chemical proteomics *see* Proteomics
Chemical reactions, dissolution *see* Dissolution
Chemical reactivity, promiscuous ligands, identification **2**:746
Chemical representation, pharmacophore identification **3**:242
Chemical shifts
 NMR spectroscopy **3**:945
 perturbations *see* Nuclear magnetic resonance (NMR), drug discovery
Chemical space
 drug discovery **2**:49, **2**:51*F*
 fragment-based drug discovery **3**:943
Chemical space interpolation, ionization prediction, in silico tools **5**:619–620
Chemical space mapping, adaptive in combo strategy *see* Adaptive in combo strategy
Chemical stability **5**:489–507
 see also Stability testing
 laboratory conditions **5**:490
 solid-state physicochemistry *see* Solid-state physicochemistry
 storage conditions **5**:490
  aqueous solutions **5**:490–491
  light **5**:490
  oxygen removal **5**:490
  temperature **5**:490
  water removal **5**:490
Chemical tractability, library design **4**:339–340
Chemical *vs.* biological *see* Philosophy (of drug discovery/development)
Chemiluminescence nitrogen detection (CLND), high throughput purification **3**:863
Chemiluminescent assays **3**:578, **3**:590
 electrochemiluminescence **3**:592
 ELISA **3**:592
 GPCR functional assay screening **2**:809
ChemInformRX **3**:281–282, **3**:282*T*
Chemistry
 DCN **2**:154
 discipline changes **2**:88

Chemistry limitations, solution phase parallel chemistry **3**:767
'Chemistry space,' definition **4**:308
CHEMLIST **3**:277
ChemoCentryx **1**:515
Chemogenomics **2**:567; **3**:921–3:937; **4**:421–433
 advantages **2**:567
 agonists/antagonists **3**:923
  GPCRs **3**:923, **3**:924*F*
 biological fingerprints *see* Biological fingerprints
 definition **3**:921
 druggable genome
  identification **4**:429
  known targets **4**:429
  protein numbers **4**:429
 drug/lead chemical properties **4**:422
  'druglikeness' **4**:423
  hydrophobic properties **4**:422
  physicochemical property limits **4**:422, **4**:424*T*
  'rule-of-five' **4**:422
  solubility **4**:422
 GPCR drug design **3**:932; **4**:368
 integrin ligands **3**:933, **3**:933*F*
 kinase inhibitors **3**:930, **3**:931*F*
 molecular property trends **4**:423
  'ligand efficiency' **4**:423
  molecular weight **4**:423, **4**:428*F*
  screening data **4**:423, **4**:427*F*
 nuclear receptor ligands **3**:932, **3**:933*F*, **3**:934*F*
 phosphodiesterase inhibitors **3**:930, **3**:931*F*
 physiological-based pharmacokinetic models **2**:567, **2**:568*F*
 privileged structures **3**:925, **3**:925*F*, **3**:926*T*
  chemical modifications **3**:925–926, **3**:927*F*
 protease inhibitors **3**:929
 protein structure analysis **4**:430
  decision-tree approach **4**:431
  a priori analysis **4**:431
 regulatory implications **2**:567
 SOSA approach **3**:925*F*, **3**:926
 target family-directed master keys **3**:927, **3**:929*F*
 transporter ligands **3**:934*F*, **3**:935
Chemoinformatics **1**:40; **3**:235–264
 ADME *see* Chemoinformatics, ADME property prediction
 combinatorial library design **3**:256
  definition **3**:256, **3**:256*F*
  optimization techniques **3**:257
  product-based selection **3**:256–257
  reactant- *vs.* product-based selection **3**:256
 definition **3**:235
 diversity analysis *see* Diversity analysis
 historical perspective **3**:235

 molecular conformation assessment **1**:40
 molecule representation *see* Chemoinformatics, molecule representation
 multiobjective optimization *see* Multiobjective optimization
 prioritization **1**:40
 three dimensional structure *see* Chemoinformatics, three-dimensional structure; 3D-structure
 virtual screening *see* Chemoinformatics, virtual screening
Chemoinformatics, ADME property prediction **3**:257
 computational filters **3**:258
  drug-like classifications **3**:258
  genetic algorithms **3**:258
  lead-likeness prediction **3**:258
  'rule-of-five' **3**:258
 physicochemical properties **3**:259
  blood–brain barrier penetration prediction **3**:259
  hydrophobicity prediction **3**:259
  polar surface area prediction **3**:259
  solubility prediction **3**:259
 throughput **3**:257–258
Chemoinformatics, molecule representation **3**:236
 example **3**:236, **3**:237*F*
 SMILES (simplified molecular input line entry specification) **3**:236
 structure/substructure searching **3**:236, **3**:239*F*
  Augmented Connectivity Molecular Formula **3**:238
  canonical representations **3**:238
  connection tables **3**:238
  fragment bitstrings **3**:238, **3**:238*F*
  graph theory **3**:236–237
  hashed fingerprints **3**:239
  molecular graphs **3**:236–237, **3**:237*F*
  subgraph isomorphism algorithms **3**:239
  Ullmann algorithm **3**:239–240
Chemoinformatics, three-dimensional structure **3**:240
 Cambridge Structural Database **3**:240
 compound numbers **3**:240
 Concord program **3**:240
 Corina program **3**:240
 PDB **3**:240
 pharmacophore identification *see* Pharmacophores, detection
 searching **3**:241
  connection tables **3**:241
  rigid searching **3**:242
  subgraph isomorphs **3**:242
  substructure searching **3**:241–242

target binding **3**:241, **3**:241*F*
3D graphs **3**:241
structure generation programs **3**:240
Chemoinformatics, virtual
screening **3**:244
computational techniques **3**:245
definition **3**:244–245
machine learning methods **3**:249
classification techniques **3**:250
recursive partitioning **3**:250
support vector machine
(SVM) **3**:250
similarity searching *see* Similarity
searching
structure-based *see* Structure-based
virtual screening
3D similarity searching **3**:247
alignment-based **3**:248
3D-vector **3**:248
2D similarity searching **3**:246
Extended Connectivity Fingerprints
(ECFPs) **3**:246
fragment methods **3**:246
Functional Connectivity
Fingerprints (FCFPs) **3**:246
graph-based similarity **3**:246
reduced graph similarity **3**:247
Chemokine receptors
antagonists
IBD treatment **6**:633
transplant rejection **7**:920
peptide/protein drugs, N-terminus
modification **2**:583
SMEs **1**:515
Chemokines
atherosclerotic plaque
formation **6**:468
metastasis **7**:19
structure-based drug design,
peptidomimetics **4**:672*F*, **4**:693
Chemometric methods
pharmacophore detection **4**:132
pharmacophore searching **4**:129*T*
ChemoSoft
Chemical Diversity Labs Inc **3**:295
database management systems **3**:295
Chemotaxis *see* Cell migration/
chemotaxis
Chemotherapeutics *see* Cancer
chemotherapy
CHEMSCORE program, regression-
based **4**:483–484
Chenodeoxycholic acid (CDCA)
as farnesoid X receptor ligand **2**:1024
structure **2**:1025*F*
CHF-1301 **1**:179
CHF-2819 **6**:203–204
CHF-3381 **2**:903
χ-angle, protein 3D structure
validation **3**:517
Chicken enucleated eye test (CEET),
non-animal eye irritation
testing **1**:466, **1**:467*T*

Chicken ovalbumin upstream promoter-
transcription factor,
nomenclature **2**:996*T*
Childhood diseases, mortality/morbidity,
developing countries **1**:642*T*
Children
human research ethics **1**:715, **1**:718
off-label drug use **1**:547
Chile, healthcare systems
reform **1**:641
Chimera agglomerates, buccal
administration **5**:283–284
'Chimeric' amino acids,
peptides **2**:604
Chimeric G proteins *see* G protein-
coupled receptors (GPCRs),
assays
Chimeric monoclonal
antibodies **1**:335–336, **1**:436
China
epidemiological transition
model **1**:693
IPA law **1**:560
market dominance **2**:94
market growth **1**:540
natural products **1**:356
Chinese hamster ovary cells *see* CHO
cells
Chinese medicine **1**:356
berberine **2**:1048
Chiral 1,3-oxazolidines, solid-phase
libraries **3**:722–724, **3**:723*F*
Chiral drug discovery **2**:713–736
see also specific drugs
chiral switches **2**:728
scalable methodologies **2**:729–730
'significant' improvements **2**:728–
729
success of **2**:729, **2**:729*F*
clinical rationales **2**:713
future work **2**:731
historical perspective **2**:713
'isometric ballast' **2**:714–715
market delivery **2**:718
available compounds **2**:719,
2:720*F*
examples **2**:718–719, **2**:719*F*
stereoisomer synthesis **2**:715,
2:716*F*, **2**:717*F*
diastereomer resolution **2**:715,
2:718*F*
large scale synthesis **2**:718
purity **2**:715
stereoselective syntheses **2**:715–
718, **2**:719*F*
stereospecific analyses **2**:720
see also specific methods
absolute configuration
determination **2**:720
Chiral inversion, coenzyme A
conjugation *see* Coenzyme A
conjugation
Chiral switches *see* Chiral drug discovery

Chirocaine *see* Levobupivacaine
Chiron Corporation **1**:499, **1**:499*T*,
1:501
formation **2**:102
market status **1**:541–542
mergers and acquisitions **1**:541–542
Chiron Vaccines, mergers and
acquisitions **1**:541–542
Chitin synthase inhibitors **7**:462
see also specific drugs
Chk2, cell cycle **7**:12
Chlamydia infection treatment **7**:612
tetracyclines **7**:612
Chlomethiazole, nervous system trauma
treatment **6**:13
Chlomipramine
cataplexy treatment **6**:154
structure **2**:1062*F*
Chlorambucil
bioavailability **5**:712*T*
clinical use **7**:132
development **7**:130
structure **7**:131*F*
toxicity **7**:46*T*
Chloramphenicol
bioavailability **5**:712*T*
classification, by target site **7**:471*T*
hydrolysis **5**:491, **5**:492*F*
mechanism of action **7**:688
ribosome interactions **2**:1042–1043
oral absorption **5**:468*T*, **5**:680*T*
structure **2**:1043*F*
Chlorazepate
development **6**:98
structure **6**:99*F*
Chlordiazepoxide
bioavailability **5**:712*T*
defensive test battery **6**:94
origins **6**:97–98
SAR **6**:98
structure **6**:88*F*
Chloride channel inhibitors
see also specific drugs
glaucoma treatment **6**:313*T*
Chloride channel openers
see also specific drugs
irritable bowel syndrome
treatment **6**:661*T*, **6**:665
4-Chloro-3,5-dimethylphenol,
structure **5**:360
Chlorobenzene, microwave
heating **3**:840*T*
Chloroform **6**:358
microwave heating **3**:840*T*
oxidation **5**:100
structure **6**:352*F*
Chlorogenic acid,
pharmacokinetics **1**:454*T*
Chloromuconate cycloisomerase
structure
incorrect structure **3**:533–534
Ramachandran plots **3**:533–534
R-factor **3**:533–534

Chloromycetin, patent
      protection **1**:534
N-(6-Chloronaphthalen-2-)
      sulfonylpiperazine derivatives,
      ring equivalent
      bioisosterism **2**:655–656,
      2:656*F*
3-Chlorophenol, structure **5**:370*F*
4-Chlorophenol
      ionization constant (pK_a) **5**:372
      structure **5**:370*F*
m-Chlorophenylpiperazine **6**:413
      structure **6**:414*F*
Chloroprocaine **6**:361
      structure **6**:354*F*
Chloroquine
      bioavailability **5**:712*T*
      blood to plasma ratios **5**:325*T*
      blood volume of distribution
            ($V_{SS,b}$) **5**:325*T*
      malaria prophylaxis **7**:754,
      7:755
            at-risk individuals **7**:755–756
      plasma clearance (CL_p) **5**:325*T*
      resistance **7**:752–754, **7**:755, **7**:773,
      7:777
      SARS **7**:273–274
      structure–activity
            relationship **7**:773, **7**:774*T*
Chlorothiazide
      bioavailability **5**:712*T*; **6**:709*T*
      dose **6**:709*T*
      half-life **6**:709*T*
      oral absorption **5**:468*T*
      structure **6**:586*F*, **6**:709*T*
Chlorotriphenylethylene, tamoxifen
      discovery **8**:86
Chlorpheniramine
      bioavailability **5**:712*T*
      primary insomnia treatment
            6:157–158
Chlorpromazine **6**:29–31, **6**:442*T*
      adverse effects **6**:29–31
      bioavailability **5**:712*T*
      discovery/development **3**:925*F*,
            3:926–927
      nonulcer dyspepsia treatment
            6:600
      oral absorption **5**:468*T*
      P-glycoprotein interaction **5**:781*T*
      scaffold hopping **2**:680, **2**:681*F*
      structure **2**:681*F*, **2**:687*F*; **5**:359;
            6:88*F*, **6**:440*F*
Chlorpropamide **6**:438
      advantages **6**:443
      bioavailability **5**:712*T*
      clinical use **6**:443
      disadvantages **6**:443
      mechanisms of action **6**:440
      pharmacokinetics **6**:440
      side effects/contraindications **6**:443
Chlorprothixene **6**:29–31
      structure **2**:687*F*

Chlortetracycline **7**:599
      biosynthesis **7**:600–601, **7**:600*F*
      C6-hydroxyl group **7**:604–605,
            7:605*F*
      discovery **7**:599
      structure **7**:599*F*, **7**:600*F*
      synthesis, *Streptomyces
            aureofaciens* **7**:599
Chlorthalidone **6**:708
      bioavailability **5**:712*T*; **6**:709*T*
      dose **6**:709*T*
      half-life **6**:709*T*
      structure **6**:709*T*
Chmoogle database **3**:307
CHO cells
      active drug transport
            prediction **5**:271–272
      cell-based assays **3**:620–621
      hERG blocker screening **5**:937
      mammalian cell expression
            systems **3**:418
      protein expression systems **3**:117
Cholate-CoA ligase, coenzyme A
      conjugation **5**:155
Cholebine *see* Colestimide
Cholecalciferol, structure **6**:499*F*
Cholecystokinin
      anxiety disorders **6**:90
      Ganellin, C R and **8**:48
Cholecystokinin-1 (CCK1) antagonists
      see also specific drugs
      prokinetic drugs **6**:686
Cholecystokinin-2 (CCK-2) antagonists
      see also specific drugs
      development **4**:521*F*
Cholecystokinin agonists, obesity
      treatment **6**:411
Cholecystokinin antagonists, carboxylic
      acid bioisosterism
      2:661
Cholecystokinin-A receptor agonists
      see also specific drugs
      nonpeptide drugs **2**:620*F*, **2**:621
Cholecystokinin-A receptor antagonists
      see also specific drugs
      nonpeptide drugs **2**:618–619,
            2:620*F*
      peptidomimetics **2**:608–609, **2**:611*F*
Cholecystokinin-A receptors, in human
      disease **2**:775*T*
Cholecystokinin-B receptor antagonists
      see also specific drugs
      nonpeptide drugs **2**:618–619, **2**:620*F*
      peptidomimetics **2**:608–609, **2**:611*F*
Cholecystokinin-B receptors
      drug binding **5**:382
      in human disease **2**:775*T*
Cholecystokinin receptor-2 (CCKR₂),
      constitutively active **2**:790*T*
Cholecystokinin receptor antagonists
      see also specific drugs
      irritable bowel syndrome
            treatment **6**:661*T*, **6**:664

Cholecystokinin receptors
      activation **2**:785*T*
      constitutively active **2**:790*T*
      as targets **2**:300–301
Cholesterol **6**:701
      as biomarker, cardiovascular disease
            and **3**:70
      C-reactive protein **6**:702
      high-density lipoprotein **6**:464,
            6:701–702
      intestinal membrane **5**:458
      low-density lipoprotein **6**:464,
            6:701–702
      management **6**:702
      removal, peroxisome proliferator-
            activated receptor-α (PPAR-
            α) **2**:1009
      stratum corneum **5**:291, **5**:460
      structure **5**:455*F*
      synthesis
            alendronate inhibition **8**:202
            Comer, William **8**:1
      tagging, siRNAs **3**:183
Cholesterol absorption
            inhibitors **6**:485, **6**:485*F*
      see also specific drugs
      clinical trials **6**:477
      statin combination **6**:477
Cholesterol and Recurrent Events
      (CARE) trial **6**:474*T*
      statins **6**:474
Cholesterol esters, coenzyme A
      conjugation *see* Coenzyme A
      conjugation
Cholesterol-fed dog model **8**:67, **8**:68*F*
Cholesterol-fed hamster model **8**:66,
      8:66*F*, **8**:67*F*, **8**:68*F*
Cholesterol-fed monkey model **8**:67,
      8:68*F*
Cholesterol-fed rat model **8**:67,
      8:68*F*
Cholesterol-lowering drugs
      see also specific drugs
      berberine **2**:1048
      market size **6**:599*T*
      natural products **1**:374
      tamoxifen **8**:94
'Cholinergic hypothesis,' Alzheimer's
      disease **6**:10, **6**:198
Cholinergic neurotransmission
      modulators
      see also specific drugs
      antiemetic drugs **6**:683
      prokinetic drugs **6**:680
Cholinesterase inhibitors **6**:202
      see also specific drugs
      dementia with Lewy bodies
            treatment **6**:36
Cholinesterases
      bambuterol activation **5**:1015
      hydrolases **5**:135
CHOP regimen **7**:134
      xenograft models **7**:39*T*

factor VII **6**:764
factor VIII **6**:764
factor IX **6**:764
factor X **6**:764
factor XI **6**:764
factor XII **6**:764
factor XIII **6**:764
intrinsic pathway **6**:764
kallikrein **6**:764
platelets **6**:764
thromboplastin **6**:764
tissue factor pathway
inhibitor **6**:764–765
Cloxacillin, oral absorption **5**:468*T*
Clozapine **6**:32
analog design **2**:651, **2**:652*F*
bioavailability **5**:712*T*
cognition effects **6**:36
mechanism of action **6**:32
muscarine receptor
interactions **6**:37
α7-neuronal nicotinic receptor
agonist **6**:39–40
as multitarget drug **2**:275–276
nervous system disorders **6**:15
oral absorption **5**:468*T*
pharmacology **6**:32
side effects **1**:696; **6**:33, **6**:35
Clozapine hypothesis,
schizophrenia **6**:22
Clozapine receptors, imaging **3**:667
Cluster analysis, microarrays **3**:93
Cluster-based compound selection *see*
Diversity analysis
Clustered descriptors, parallel library
design **4**:325
Clustering **3**:253, **3**:254*F*; **4**:184
algorithms **4**:185
ligand-based library design **4**:312,
4:313*F*
fragment/ligand selection drug
design **4**:296
intermolecular similarities **3**:255
Jarvis–Patrick clustering **3**:254;
4:185
Kelley measure **4**:185
K-means clustering **4**:185
library design **4**:363–364, **4**:366
microarrays **3**:93
screening library selection **4**:498,
4:502–503
statistical methods, E-state
index **5**:562
Ward's clustering **3**:254–255; **4**:185
Cluster of orthologous genes (COGs),
protein function prediction,
bioinformatics **3**:334
CluSTr database, Integr8 **3**:367
proteome analysis **3**:367–368
CMF regimen **7**:134
cMOAT *see* ABCC2 (MRP2) transporter
CMV 423 **7**:261
c-myc, RNAi **3**:180–181

CNI-1493, IBD treatment **6**:627–628
CNQX **2**:900
structure **2**:901*F*
CNS 5161 **2**:903
CNTO 1275 **6**:240*T*
IBD treatment **6**:623–624
CNT (*SLC28*) transporters **5**:69*T*, **5**:72
distribution **5**:69*T*, **5**:72
members **5**:72
purine/pyrimidine transport **5**:72
⁵⁵Co, PET **3**:664*T*
Coadministered drugs
clinical trials **2**:187
pharmacokinetics **5**:22
Coal tar, psoriasis treatment **7**:944
Coatec *see* Loprinone hydrochloride
Cobalt, tetracycline binding **7**:618
COBALT study, alteplase clinical
trials **6**:773
COBRA program, 3D structure
generation **4**:460–461
Cocaine
abuse *see* Cocaine abuse
as anesthetic **6**:353, **6**:361
discovery/development **1**:322
dopamine transporter
interaction **2**:63
ethanol interactions **5**:135
hydrolysis **5**:135, **5**:138*F*
aromatic hydroxylation **5**:135
N-demethylation **5**:135
hCE-1 **5**:135
hCE-2 **5**:135
N-oxygenation **5**:135
structure **2**:65*F*; **6**:173*F*, **6**:354*F*
toxicity **5**:135
vaccine **6**:184*T*
structure **6**:181*F*
Cocaine abuse **6**:8–9, **6**:173*T*
comorbidities **6**:184*T*
dopamine release **6**:171
medications for **6**:179, **6**:181*T*,
6:184*T*
adrogolide **6**:184*T*
anti-cocaine vaccine **6**:181*T*, **6**:185
baclofen **6**:181*T*, **6**:184*T*
BP 897 **6**:184*T*
cabergoline **6**:184*T*
citicoline **6**:184*T*
cocaine vaccine **6**:184*T*
dextroamphetamine **6**:184*T*
disulfiram **6**:181*T*, **6**:184*T*
enadoline **6**:184*T*
gabapentin **6**:184*T*
isradipine **6**:184*T*
lobeline **6**:184*T*
memantine **6**:184*T*
modafinil **6**:184*T*; **8**:153
ondansetron **6**:184*T*
quetiapine **6**:184*T*
selegiline **6**:184*T*
sertraline **6**:184*T*
tiagabine **6**:181*T*, **6**:184*T*

tolcapone **6**:184*T*
topiramate **6**:181*T*, **6**:184*T*
vanorexine **6**:184*T*
vigabatrin **6**:184*T*
treatment **6**:9
Cocaine and amphetamine-related
transcript (CART), obesity
treatment **6**:409
Coccidiomycosis **7**:424*T*
Cocktailing, structure-based fragment
screening *see* Structure-based
fragment screening
Co-crystallization **3**:891
adducts/complexation/
formulation **5**:526
de novo drug design **4**:297
N-myristoyltransferase
inhibitors **7**:436, **7**:437*F*, **7**:438*F*
rational drug design **3**:891
Coculture models
blood–brain barrier
permeation **5**:314
brain drug uptake **5**:313
non-animal chronic toxicity
testing **1**:476–477
non-animal eye irritation
testing **1**:466, **1**:467*T*
Codeine
bioavailability **5**:712*T*
discovery/development **1**:322,
1:379
irritable bowel syndrome
treatment **6**:656
oral absorption **5**:680*T*
structure **1**:379*F*
Codelink microarray **3**:89
CODESSA PRO software, organic
solvent solubility
models **5**:645–646
Codon usage, gene expression vectors
see Gene expression vectors
CoE model, compound storage/
management (Pfizer) **3**:569,
3:570*F*
Coenzyme A conjugation **5**:134*T*,
5:155, **5**:156*F*
amino acids **5**:155
benzoic acids **5**:155
bile acid-CoA: amino acid
N-acyltransferase **5**:155
glutamine *N*-acyltransferase **5**:155
glutamine *N*-
phenylacetyltransferase **5**:155
glycine **5**:155, **5**:156*F*
glycine *N*-acyltransferase **5**:155
glycine *N*-
benzoyltransferase **5**:155
taurine **5**:155
β-oxidation/chiral inversion **5**:157
fatty acid catabolism **5**:158
cholesterol esters/hybrid
triacylglycerides **5**:156*F*, **5**:157,
5:158*F*

Coenzyme A conjugation (*continued*)
  2-acylglycerol *O*-acyltransferase
    5:157
  diacylglycerol *O*-acyltransferase
    5:157
  sterol *O*-acyltransferase **5**:157
  endogenous pathway
    interference **5**:155
  enzymes **5**:155
    benzoyl-CoA synthetase **5**:155
    cholate-CoA ligase **5**:155
    long-chain acyl-CoA
      synthetase **5**:155
    medium-chain acyl-CoA
      synthetase **5**:155
    phenacyl-CoA synthetase **5**:155
    propionate-CoA synthetase **5**:155
    short-chain fatty acyl-CoA
      synthetase **5**:155
  substrate activation **5**:155
Coenzyme Q, Huntington's disease
    treatment **6**:222*T*
Coenzyme Q10 **1**:658
Cofactors
  see also specific cofactors
  UDP-glucuronyltransferases **5**:150,
    5:151*F*
  heterogeneous protein
    crystallization **3**:437
  methyltransferases **5**:145
  nuclear receptors **4**:726
  sulfotransferases **5**:147
Cognitive disorders/dysfunction
  Alzheimer's disease **6**:195
  animal models **2**:926–927
  neurodegenerative diseases **6**:195
  phosphodiesterase-5 inhibitors **2**:949
  quetiapine effects **6**:36
  risperidone effects **6**:36
  schizophrenia **6**:18
    modafinil **8**:152–153
Cohort trials, definition **2**:30*T*
Co-immunoprecipitation,
    proteomics **3**:41–42
Coincubation models, metabolic
    stability **5**:237
Colchicine **7**:88
  P-glycoprotein interaction **5**:781*T*
  SAR **7**:88, **7**:89*F*
  structure **7**:88*F*
  synthesis **7**:88
Colchicine binding domain inhibitors,
    anticancer agents, microtubule-
    targeting **7**:88
Cold water baths, acute pain
    models **6**:332
Cole, Jonathan, universities in
    commerce **2**:93
Colesevelam hydrochloride **1**:171
Colestimide **1**:171
Colforsin daropate
    hydrochloride **1**:175
  introduction of **1**:107

Collaborations
  biotechnology companies **2**:231
  drug discovery, decline in **1**:623
  functional discipline
    organization **2**:109
  industry/academia collaborations *see*
    Collaborations, industry/academia
  insular culture *vs.* **2**:121–122
  pharma *vs.* biotech, drug
    discovery **2**:230–231, **2**:230*F*
  project management *see*
    Collaborations, project
    management
  start-up companies **2**:765
Collaborations, industry/
    academia **1:571–590**
  *see also specific academic institutions; specific
    companies*
  academic cost effects **1**:577
  analytical techniques **1**:574
  case studies **1**:585
    atracurium **1**:588
    Bradford Particle Design **1**:587
    Dundee Kinase Consortium **1**:587
    Imperial College London/
      Hammersmith Hospital/ GSK
      Imaging Centre **1**:588
    Integrative Pharmacology Fund
      (UK) **1**:586
    Oxford University/Oxford
      Instruments **1**:588
    University of Toronto-Eli Lilly
      collaboration **1**:585
  collaborative training awards **1**:577
    Cooperative Awards in Science and
      Engineering (CASE) **1**:579
    Cooperative Awards in Science and
      Technology (CAST) **1**:579
    GlaxoSmithKline **1**:577
    intellectual property
      agreements **1**:577–578
    Knowledge Transfer
      Partnership **1**:578
    purpose **1**:577–578
  contract research *vs.* **1**:579
  criteria for **1**:576
    academic researcher reasons
      1:576
    company reasons **1**:576
  drug discovery **3**:146
  education **1**:574
  globalization effects **1**:577
  history **1**:584
    Duncan, Kennedy **1**:584
    Germany **1**:584
    MIT **1**:585
    Mulford, H K **1**:584
    unacceptability **1**:585
    US **1**:584
  long-term research objectives
    1:576
  mutual benefits **1**:576
  reasons for **1**:572

research databases **1**:573
    Human Genome Project **1**:573
    NIH Roadmap Initiative **1**:574
    Structural Genomics
      Consortium **1**:574
  specialist facilities/equipment **1**:574
  specific projects **1**:574
  startup companies **1**:580
    LINK **1**:580
  trends **1**:583
    Association of British
      Pharmaceutical Industries
      (ABPI) **1**:584
    merger effects **1**:583
    UK Science Parks
      Association **1**:584
    university cultural changes **1**:584
  types **1**:574, **1**:575*T*
    precompetitive research **1**:575
  UK **1**:581
    Association for University Research
      and Industry Links
      (AURIL) **1**:582
    Biotechnology and Biological
      Sciences Research Council
      (BBSRC) **1**:576
    consultancy limitations **1**:582
    Engineering and Physical Sciences
      Research Council
      (EPSRC) **1**:576
    industry interchange
      program **1**:581
    intellectual property **1**:583
    Lambert review **1**:582, **1**:583,
      1:583*T*
    legal framework **1**:582
    Medical Research Council **1**:581
    numbers **1**:576, **1**:577*T*
    Research Council
      restructuring **1**:581
    Teaching Company scheme **1**:581
  US **1**:580, **1**:584
    Bayh-Dole Act (1980) **1**:581
    Cooperative Research and
      Development Agreements
      (CRADA) **1**:580
    legal framework **1**:581
    Marie Curie fellowship
      program **1**:581
    National Research Council
      1:580
    National Research Fund **1**:580
    NIH Roadmap **1**:580–581
    technology transfer offices **1**:582
Collaborations, project
    management **2**:151
  initiation **2**:151
    communication **2**:151
    role/responsibility
      definitions **2**:151
  logistical considerations **2**:151
    communication **2**:151
    security **2**:151

oversight **2**:152
  joint management team **2**:152
  responsibilities **2**:152
Collaborative training awards, industry/
    academia collaborations *see*
    Collaborations, industry/
    academia
Collagenase, three-dimensional
    structures **2**:626*T*
Collagenase inhibitors
  see also specific drugs
  peptidomimetics **2**:613, **2**:614*F*
Collagen sandwich model
  CYP induction, in vitro studies **5**:251
  drug metabolism, in vitro
    studies **5**:233
Collander equation, log P
    calculation **5**:652
Collection nature, screening library
    selection **4**:501
Collections, cell-based assays **3**:622
Collip, James, insulin discovery/
    development **1**:585
Colony-formation assays, HCS **3**:692
Colorectal cancer
  CpG island methylation **1**:272
  incidence **7**:56*T*
  mortality **7**:56*T*
  new drugs **1**:107
  raltitrexed **1**:140
  Src kinases **7**:191
  treatment **7**:8
    bevacizumab **1**:444; **3**:202
    cetuximab **1**:137
    infliximab **1**:444
    xenograft models **7**:39*T*
Colorimetric assays
  promiscuous ligands,
    identification **2**:740
  protein concentration
    determination **3**:426
Colpaert, Francis, risperidone, invention
    of **2**:257–258
Colterol, library similarity
    searches **4**:564*T*
CombiBUILD program **4**:321
  library design **4**:358–359
CombiCHEM system **2**:760*T*
  Milestone s.r.l. microwave-assisted
    chemistry **3**:849, **3**:850*F*
CombiDOCK **4**:321
CombiGlide **4**:321, **4**:322
Combination blockers, antiarrhythmic
    drugs, rational design
    6:757
Combination treatments
  antimetabolites **7**:74–75, **7**:75*T*
  HIV infection *see* HIV infection
Combinatorial chemistry **2**:460;
    3:697–760
  see also Libraries
  definition **2**:460; **5**:957
  development **3**:762, **3**:763

discrete compound synthesis **3**:763
  compound design **3**:763
  compound quality **3**:763
drug discovery *see* Drug discovery
genomics **2**:44, **2**:46*F*
high throughput purification
    3:861
historical perspective **1**:32; **4**:337
history **2**:461; **5**:1
  'divide–couple–recombine' **2**:461–
    462
  Frank, Ronald **2**:461
  Geysen, Mario **2**:461
  Houghten, Richard **2**:461
  Merrifield, Bruce **2**:461
  'portioning–mixing' **2**:461–462
  small organic molecules **2**:462
  techniques **2**:462
HTS automation **3**:681
invention of **2**:256
  Hobbs DeWitt, Sheila **2**:260
  Kiely, John **2**:260
  Pavia, Mike **2**:260
natural products *see* Natural products
natural products *vs.* **3**:699, **3**:700*F*
  as models **3**:699
pharmaceutical productivity
    effects **2**:132
reaction types **3**:762
scaffold-based design **4**:31
strategies **2**:464
  see also specific techniques
success of **2**:52
target identification **3**:700
  chemical genetics **3**:700, **3**:701,
    3:701*F*
  fast reading methods **3**:702
terminology **3**:763
Timmerman, Henk (views on)
    8:13
usefulness **4**:443
Combinatorial docking, library
    design **4**:309
Combinatorial libraries
  chemical biology **3**:133
    diversity-oriented **3**:133, **3**:134*F*
    privileged structure **3**:133
  chemoinformatics *see*
    Chemoinformatics
  drug design, de novo **4**:289
  drug discovery **2**:300, **2**:300*T*
  multiobjective optimization **4**:769
  3D database pharmacophore
    searching **4**:138
Combinatorial modeling, QSAR *see*
    Quantitative structure–activity
    relationship (QSAR)
Combinatorial phytochemistry, drug
    research, current status **1**:60,
    1:61*F*
Combinatorial problems, ligand
    assembly drug design *see* Ligand
    assembly drug design

Combinatorial search/optimization
    algorithms
  see specific programs
  ligand assembly drug design *see* Ligand
    assembly drug design
Combinatories, drug design, de
    novo **4**:284–285
Combined bisulfite restriction analysis
    and amplification (COBRA),
    epigenetics **1**:266–268, **1**:267*T*
Combined fractional diagonal
    chromatography (COFRADIC),
    LC-MS (peptide-based)
    proteomics **3**:38
Combined lipophilicity descriptors,
    ionization effects **5**:428
Combined 2D/3D approach, hERG
    blocker modeling **5**:943,
    5:943*F*
COMBINE program, target-based
    tuning **4**:262
Combivir **7**:340
Combrestatins **7**:91
  analogs **7**:93*F*
  discovery/development **1**:380–381
  SAR **7**:91, **7**:92–93, **7**:93*F*, **7**:94*F*
  solubility **7**:91–92, **7**:93*F*
  structure **1**:381*F*; **7**:92*F*
Combrestatin A-4, [^{18}F]FDG
    imaging **3**:668–669
Combrestatin A$_4$ phosphate
    structure **6**:307*F*
  wet age-related macular degeneration
    treatment **6**:306*T*
Comer, William **8**:1–6
  albuterol **8**:2–3
  Alzheimer's disease research **8**:4
  on animal models **8**:3
  β-adrenergic receptor agonists **8**:2
  β-adrenergic receptor
    antagonists **8**:1–2, **8**:2*F*
  cholesterol synthesis **8**:1
  on SAR **8**:4
  on structure optimization **8**:3
  therapeutic area teams **8**:3
  on toxicity **8**:3
COMET assay
  carvedilol clinical trials **8**:142
  non-animal genotoxicity
    testing **1**:467*T*, **1**:470–471
CoMFA (comparative molecular
    interaction field
    analysis) **5**:581*T*, **5**:589, **5**:590*F*
  androgen receptor **4**:741
  computer-assisted drug design
    (CADD) **4**:18
  electrostatic field values **5**:589
  estrogen receptors **4**:741
  field value hardness **5**:590
  historical perspective **4**:68
  ionization prediction, in silico
    tools **5**:617
  ligand orientation **5**:590

CoMFA (comparative molecular
    interaction field analysis)
    (*continued*)
    methyltransferase molecular
        modeling 5:820
    molecular interaction field (MIF)
        derived 5:537
    monoamine oxidase 5:593
    nuclear receptors, drug design 4:741
    P-glycoprotein models 5:782
    problems 5:589–590
    QSAR 4:156
    regression analysis 5:589
    statistics 5:589
    steric field values 5:589
    3D-QSAR *see* 3D quantitative
        structure–activity relationship
        (3D-QSAR)
    toxicity, computational models 5:915
    transporter interaction
        modeling 5:775–776
Comfrey (*Symphytum officinale*), safety
    issues 1:421, 1:422
CoMISA (comparative molecular
    similarity indices
    analysis) 5:581*T*
    3D-QSAR alignment 4:578
    development 5:590, 5:596*F*
    hERG blocker modeling 5:941–942
    P-glycoprotein models 5:782, 5:783
    steric fields 5:582
    transporter interaction
        modeling 5:775–776
CoMMA (comparative molecular
    moment analysis) 5:581*T*
    3D-QSAR 4:585
    3D-QSAR 5:591
Commercial activities
    outsourcing 2:209
    universities 2:93
Commercial doubts, carvedilol in heart
    failure 8:141
Commercial fits, drug development
    productivity 2:130
Commercially available compound
    collections 3:132
    see also specific collections
Commercially available databases *see*
    Databases
Commercially-available drugs,
    PASP 3:801
Commercially available systems, cell-
    free expression systems 3:122
Commercial models
    see also specific models
    ionization prediction, in silico
        tools 5:615
Commercial reasons, project
    failure 1:342–343
Commercial software
    see specific programs
    aqueous solubility models *see*
        Solubility models, aqueous

Commercial systems, genotoxicity,
    computational models 5:917
Commission on Health Research for
    Development 1:692
Commission on Intellectual Property
    Rights 1:694
Commission on Social Determinants of
    Health 1:650
Commitment, project teams 2:145
Committee for Proprietary Medical
    Products (CPMP), obesity
    clinical trials 6:394, 6:394*T*
Committee on Orphan Medicinal
    Products (COMP) 1:547
Committee on Safety of Medicines
    postmarketing surveillance 1:756
    tamoxifen support 8:87
Common complex diseases,
    pharmacogenetics *see*
    Pharmacogenetics
Common Rule 1:682
Common scaffolds, adaptive in combo
    strategy 5:963
Common substrate binding sites, drug
    design, de novo 4:286
Common technical document
    (CTD) 2:161
    see also Development chemistry
    drug nomenclature 2:161–162
    in-process controls (IPCs) 2:163
    manufacturer details 2:163
    packaging materials 2:163
    starting material definition 2:163
    synthesis steps 2:163
Communication
    collaborative project
        management 2:151
    diversity issues 2:244
    leadership *see* Leadership
    offshore outsourcing 2:215
    outsourcing risks 2:220
    project team leaders 2:142–143
        assessment 2:146
Community-acquired methicillin-
    resistant *Staphylococcus aureus*
    (CAMSRA) 7:481
Community-acquired respiratory
    infections, macrolide
    antibiotics 7:521–522
Comorbid mood disorders,
    schizophrenia 6:19
COMPA, 3D-QSAR alignment 4:578
Compactin
    discovery/development 1:18, 1:374–
    375
    structure 1:376*F*
Comparable company analysis,
    definition 1:567
Comparative computational modeling,
    structural genomics *see* Structural
    genomics
Comparative databases 3:363
    see also specific databases

Comparative experiments,
    microarrays 3:90
Comparative genomic hybridization
    (CGH), anticancer agents,
    tumor models 7:42–43
Comparative genomics 7:738
    antibacterial target
        identification 7:738
    in vitro growth 7:738
    virulence-related targets 7:738,
        7:743
    computer tools 7:738
        BLAST 7:738
        concordance analyses 7:739
        FASTA 7:738
        GeneMark 7:738–739
        genome to genome 7:738
        GLIMMER 7:738–739
        HOBACGEN database 7:739
        Markov model-based
            systems 7:738–739
        multiple pathogens 7:738, 7:739
        123-Genomics 7:739
        TIGR assembler 7:738
    HTS 2:401
    hybridization, microarrays 3:90
    lead identification 7:739
    repetitive elements,
        bioinformatics 3:324
    target identification 7:739
Comparative modeling 4:215–236
    accuracy 4:220
    alignment errors 4:219
    applications 4:224
    automation 4:220, 4:228
    errors 4:222
        distortions/shifts 4:223
        fold assessment 4:224
        misalignments 4:222
        missing template 4:222
        model error prediction 4:224
        self-consistency 4:224
        side-chain packing 4:223
        template selection 4:222
    evaluation 4:224
    experimental structures *vs.* 4:226
    flexibility 4:220
    fold assignment 4:218
    future work 4:228
    loop modeling 4:221
        database-based 4:221
        optimization-based
            methods 4:221
    MODELLER 4:220
    novel drug leads 4:227
    refinement 4:221
        see also specific methods
    resources 4:231
    sequence–structure alignment 4:218
        iterative sequence–structure
            alignment 4:219
        methods 4:218
        profile–profile methods 4:219

sequence–profile methods **4**:219
sequence–structure threading methods **4**:219
similarity levels **4**:218
sequence–structure gap **4**:216
side-chain modeling **4**:222
fixed backbone **4**:222
methods **4**:222
rotamers **4**:222
structural genomics **4**:217
template selection **4**:220
Comparative molecular interaction field analysis *see* CoMFA (comparative molecular interaction field analysis)
Comparative molecular moment analysis *see* CoMMA (comparative molecular moment analysis)
Comparative molecular similarity indices analysis *see* CoMISA (comparative molecular similarity indices analysis)
Comparative performances, adaptive in combo strategy **5**:962*F*, **5**:965*T*
Comparative receptor surface analysis *see* CoRSA (comparative receptor surface analysis)
Comparative sequence analysis, RNA structure prediction, bioinformatics **3**:327
Comparative studies, glatiramer acetate **8**:178
Comparison methods, 2D-PAGE protein quantitation **3**:34*F*, **3**:35–36, **3**:36*F*
Comparison Trial of Saruplase and Streptokinase *see* COMPASS trial
Compartmentalization, dynamic combinatorial libraries **3**:967–968
COMPASS program
3D-QSAR **4**:583
pharmacophore searching **4**:129*T*
COMPASS trial, saruplase **6**:774
Compatible dual-vector approach, *Escherichia coli* protein expression systems **3**:113
Competence
human research **1**:715
mental illness, bioethics **1**:696
Competing disciplines, governments, in drug discovery **1**:734–735
Competition
DCN **2**:154
intellectual property, effects of **1**:746
phosphodiesterase(s) regulation **2**:923
sourcing development **2**:213
Competition based ^{9}F screening, NMR, drug discovery **3**:908*T*, **3**:915, **3**:915*F*

Competition binding equilibria, NMR, drug discovery **3**:903
Competition experiments, NMR spectroscopy **3**:946
Competitive collaborations, culture **2**:122
Competitive enzyme immunoassay, cell-based assays **3**:631
Competitive grants, governments, in drug discovery **1**:744
Competitive inhibition, in vitro studies **5**:246
Competitiveness *see* Governments, in drug discovery
Competitive screening, NMR, drug discovery **3**:908*T*
Complement, age-related macular degeneration **6**:302
Complementary medicine, definition **1**:408
Complement C5a receptor, activation **2**:785*T*
Complexes, protein crystallography, rational drug design **3**:891
Complexity, oral absorption prediction **5**:692–693, **5**:693*F*
'Complexity-generating' reactions, solid support synthesis **2**:479–480, **2**:483*F*
Complex mixtures, peptide fingerprint searching **2**:344
Complex molecules, drug development **2**:724–725, **2**:725*F*
Complex multistep synthesis, solid support synthesis **2**:475
Compliance
antivirals **7**:336, **7**:362
ideal drug characteristics **2**:564*T*
tamoxifen studies **8**:95
COMPLIB, library design **4**:340
Component/vendor selection, customized automation **5**:347
Composite scoring, docking, scoring functions **4**:264
Composition effects, transmembrane permeation, solubility-diffusion model **5**:463
Composition of matter, intellectual property **2**:130
Compound collections *see* Libraries
Compounds
in development, licensing **2**:234, **2**:234*F*
drug discovery **1**:50*T*
fluorescence screening assays *see* Fluorescence screening assays
impurity, false positives **2**:737
libraries *see* Libraries
lipophilicity classification **5**:438
management *see* Compound storage/management
materials management **2**:531

metabolism prediction **5**:797
numbers
chemoinformatics, three-dimensional structure **3**:240
solution phase library synthesis **3**:769–770
pharmacophore identification **3**:244
physiologically based pharmacokinetics *see* Physiologically based pharmacokinetics (PBPK)
quality, combinatorial chemistry **3**:763
ranking
Caco-2 cell permeability assays **5**:261–262, **5**:269–270
metabolism studies **5**:33–34
receptor binding, free energy calculation **4**:654–655
resynthesis, outsourcing **2**:211, **2**:213
screening, mouse models **3**:653
selection, chemoinformatics, diversity analysis **3**:253
size, blood–brain distribution **5**:752
stability, HTS library composition **3**:681
stability testing **5**:502
storage *see* Compound storage/management
submission **2**:18
synthesis **2**:18
Compound space **5**:960
adaptive in combo strategy **5**:962–963
Compound storage/management **3**:561–576
acquisition/processing **3**:561, **3**:563
external suppliers **3**:563–564
in-house synthesis **3**:563–564
outsourcing **3**:564
quality control **3**:563–564
registration **3**:564
audit trails **3**:561
centralization **3**:562
local sample handing **3**:562
size of collection **3**:562
dispensing **3**:561
distribution **3**:561
example *see* Compound storage/management (Pfizer)
future developments **3**:573
liquid sample collections
media **3**:567
storage conditions **3**:567
temperature **3**:567
processing for use **3**:569
HTS **3**:569
Compound storage/management (Pfizer) **3**:562
CoE model **3**:569, **3**:570*F*
compressed sets **3**:570, **3**:572*F*

Compound storage/management (Pfizer) (*continued*)
  hardware selection 3:572
    flexibility 3:573
    human *vs.* automated 3:573
    radio frequency identification tags 3:573
  liquid sample collection 3:563, 3:566
    automation 3:566
    current practice 3:566
    customer interaction 3:568
    dispensing configuration 3:566, 3:566F
    multiple replicates 3:566–567
    quality control 3:568
  primary screen supply 3:569, 3:570F
  registration 3:564
    Global electronic Compound Registration system 3:564
    in-house synthesis 3:564
  solid samples 3:565, 3:565F
    automation 3:565
    capacity 3:565
  specific activity compounds 3:571
  subsets 3:570
    Global Diverse Representative Subset (GDRS) 3:570–571
    random sets 3:570–571
Compound training set, biological fingerprints 4:808
Comprehensive Medicinal Chemistry (CMC), information on 3:308–309
Compressed sets, compound storage/management (Pfizer) 3:570, 3:572F
Compromise, conflict management 2:248
Computational filters *see* Chemoinformatics, ADME property prediction
Computational techniques, fragment-based drug discovery 4:449
Computed Ligand Binding Energy Database, BIDD Databases 3:305
Computed tomography (CT)
  Alzheimer's disease 6:197
  applications 3:659
  coronary heart disease diagnosis 6:696
  drug development 2:192
  functional small animal imaging 3:661
  heart disease 6:461
  small animal imaging 3:660T, 3:661, 3:662
Computer-assisted drug design (CADD) 2:49, 2:50F, 4:1–11, 4:13–41
  see also specific methods/techniques topics

active-analog approach 4:15
  Gaussian representations 4:16–17
  theory 4:16
  thyrotropin-releasing hormone 4:17, 4:17F
active-site modeling 4:17, 4:18F
  ACE inhibitors 4:17
  Active-Site Mapping 4:17
approaches 4:3F
definition 4:13
de novo drug design 4:4
enalapril 4:14
factor Xa inhibitors 4:9–10, 4:9F
future work 4:10, 4:35
historical perspective 4:14, 4:68
  Cambridge Structure Database 4:14
  fragments 4:15
  Massachusetts Institute of Technology 4:14–15
  PDB 4:14
  physical models 4:14–15
  QSAR 4:15
impact of 4:2, 4:3T, 4:5T
large scale simulations 4:4
molecular mechanics 4:21
molecular simulations 4:21
  force fields 4:21
necessity 4:13–14
nuclear receptors, drug design 4:739
pharmacological profiling partnership 4:10
predictive models 4:3–4
protein structure prediction 4:19
  contact matrix of predicted folds 4:19, 4:20F
  CRMS value 4:19–21
  Levinthal paradox 4:19
  ProVal 4:19–21, 4:20F
SAR 4:14
scaffold-based *see* Scaffold-based design
SMEs 1:518
statistical modeling 4:18
  3D-QSAR 4:18
  CoMFA 4:18
  partial least squares methods 4:18
structural biology partnership 4:6
  x-ray crystallography 4:6–8
structural flexibility 4:22
  critical assessment of predicted interactions (CAPRI) 4:22
  critical assessment of structure prediction (CASP) 4:22
  HIV protease 4:22, 4:23F
  static blocking 4:22
structural targets 4:19
  known *vs.* unknown 4:19
unknown receptors 4:15
Computer-based technologies
  ADMET 2:566, 2:569, 2:570F
  arrhythmias 6:739

biological fingerprints *see* Biological fingerprints
chemoinformatics, virtual screening 3:245
comparative genomics *see* Comparative genomics
design software, HTS 2:398, 2:398F
ensemble models, P-glycoprotein models 5:784
GPCRs *see* G protein-coupled receptors (GPCRs)
mechanistic pharmacokinetic studies 5:39
scaffold hopping *see* Bioisosterism, scaffold hopping
sequence searching, protein identification 2:341, 2:343F
solution phase library synthesis 3:770
toxicity *see* Toxicity, computational models
Computerized axial tomography (CAT), coronary heart disease diagnosis 6:696
Computing LLC, molecular fingerprinting 3:298
COMT *see* Catechol *O*-methyltransferase (COMT)
Comtess *see* Entacapone
Concanavalin A library, dynamic combinatorial libraries 3:969
Concealment of use, traditional medicine 1:422
Concentration
  definition 5:358
  dissolution 5:401, 5:414, 5:414F
  microsome binding 5:332
Concept 1:510
Concerta, ADHD treatment 6:129T
Concordance analyses, comparative genomics 7:739
CONCORD program
  chemoinformatics, three-dimensional structure 3:240
  conformational analysis/searching 4:110–111
  ligand–receptor interaction models 4:208
  pharmacophore 3D database searching 4:141
  3D structure generation 4:465
  2D- to 3D structure 4:462
'Concurrent validation,' process variable statistical evaluation 2:164–165
Conditional knockouts, knockout mouse model 3:13
Conditional lethals, genomics, antibacterial target identification 7:741
'Conditional transgenic animals,' transgenic animal generation 3:152

Conditioned avoidance, schizophrenia models **6**:25

Conditioned place preference, addiction models **6**:176, **6**:177*F*

Condition optimization, PASP **3**:827–828

Conduction anesthesia, sodium channels **2**:839

Conduction velocity, arrhythmias **6**:733

Conductivity, ionization constants (pK_a) measurement **5**:386

Confidentiality, corporate ethics **1**:711

Configuration changes, core molecular modeling **4**:99–100

CONFIRM program, 3D structure generation **4**:464, **4**:465, **4**:466*F*

Conflict management, leadership *see* Leadership

Conflict of commitment policies **1**:712

Conflicts of interest, corporate ethics **1**:711

Confocal FLA, fluorescence lifetime assays **3**:610, **3**:611*F*

Confocal fluorescence, promiscuous ligands, identification **2**:745, **2**:745*F*

Confocal microscopy
fluorescence fluctuation techniques **3**:607
GPCR functional assay screening **2**:813–814
HCS *see* High-content screening (HCS)

Confocal TLA, fluorescence lifetime assays **3**:611, **3**:612*F*

Conform 500 program, 3D structure generation **4**:465

Conformational analysis/ searching **4**:104
comparisons of **4**:109
experimental data **4**:110
CSD **3**:394*F*, **3**:400, **3**:400*F*
distance geometry **4**:105
evaluation of **4**:109
published **4**:109–110
genetic/evolutionary algorithms **4**:106
definition **4**:106
process **4**:106, **4**:107*F*
goal **4**:104
molecular dynamics **4**:105
random search methods **4**:105
low-mode searches **4**:105
Metropolis Monte Carlo methods **4**:105
representative conformation generation **4**:107
distance estimation **4**:108
root mean square distance (RMSD) **4**:108, **4**:108*F*
two-stage process **4**:108, **4**:109*F*

rule-based methods **4**:107
definition **4**:107
knowledge base **4**:107
simulated annealing **4**:107
systemic search methods **4**:104
grid search **4**:104
large molecules **4**:104
Systematic Unbounded Multiple Minimum Method (SUMM) **4**:104
3D structure generation **4**:110
CONCORD **4**:110–111
CORINA **4**:110–111
software **4**:110
Vista data analysis **3**:394*F*, **3**:400, **3**:400*F*

Conformational analyzer with molecular dynamics and sampling (CAMDAS), 3D-QSAR alignment **4**:581–582

Conformational changes
GPCR desensitization **2**:800
heterogeneous protein crystallization **3**:436
K^+ ion channels **4**:703–704
ligand–receptor interactions **4**:194
nuclear receptors **4**:726, **4**:731
P-glycoprotein **5**:774
structure-based fragment screening **4**:799

Conformation-mining, pharmacophore searching **4**:129*T*

Conformation numbers, pharmacophore identification **3**:243

Conformation-on-the-fly, pharmacophore identification **3**:244

Conformations
CSD **3**:399
flexibility
pharmacophore detection *see* Pharmacophores, detection
pharmacophore identification **3**:242
4D-QSAR **5**:597
ionization **5**:606
preferences, CSD applications **3**:400
superimposition, pharmacophores **4**:134
three-dimensional descriptors, similarity searching **4**:172
variation, CSD applications **3**:400*F*, **3**:401

CONFORT program, 3D structure generation **4**:464, **4**:465

Confounding factors, genetically engineered animals **3**:155*T*, **3**:156

Confrontation, conflict management **2**:249

Congenital 5α-reductase deficiency, male hypogonadism **6**:525

Congenital long QT syndrome (LQTS) **5**:935

Congenital osteoporosis **6**:507

Congenital rickets **6**:507

Congenital rubella infection, diabetes mellitus type 1 **6**:430–431

Congestive heart failure **6**:694, **6**:698
development **6**:698
hypertension **6**:698
prevalence **6**:695*T*, **6**:698
treatment **6**:698

Congo red, Huntington's disease treatment **6**:222*T*

Conjugate acids, definition **5**:358

Conjugate bases, definition **5**:358

Conjugate gradients method, core molecular modeling **4**:102–103

Conjugation, drug metabolism **5**:134, **5**:144
acetylation **5**:153
see also Coenzyme A conjugation
alcohol/fatty acid conjugation **5**:154
aralkylamine *N*-acetyltransferase **5**:153
aromatic-hydroxylamine *O*-acetyltransferase **5**:153
arylamine *N*-acetyltransferase **5**:153
diamine *N*-acetyltransferase **5**:153
O-acetylation **5**:154
substrates **5**:154, **5**:154*F*
Coenzyme A *see* Coenzyme A conjugation
definition **5**:145
glucuronic acids
see also UDP-Glucuronyltransferases
anticancer agents **5**:149
antiviral agents **5**:149
glutathione **5**:158
see also Glutathione-*S*-transferase (GST)
biotransformation **5**:159
nucleophilic substitution **5**:161, **5**:162*F*
methylation **5**:145
see also Methyltransferases
classification **5**:145*F*, **5**:147
phases *see* Phase 1; Phase 2
phosphate conjugates **5**:149
sulfation **5**:147
see also Sulfotransferases
transferases **5**:145

Conjunctivitis, treatment **6**:317

Connection tables
chemoinformatics, three-dimensional structure **3**:241
structure/substructure searching **3**:238

Conner Adult ADHD rating Scale (CAARS) **6**:119

Conotoxins **1**:379

ConQuest Search *see* Cambridge
    Structural Database (CSD)
Consensus-based methods, regulatory
    regions, bioinformatics
    3:322
Consensus (ensemble) modeling
    aqueous solubility models *see*
        Solubility models, aqueous
    de novo drug design **4**:299
Consensus scoring *see* Docking, scoring
    functions
Consent
    free **1**:715
    informed *see* Informed consent
Conservation moment, transmembrane
    helices **4**:674
Conserved domains, GPCRs **4**:671
Conserved structural motifs,
    β-lactamases **7**:486, **7**:486*F*
Consistency, drug discovery **2**:283
Consortium for Metabonomic Toxicity
    (COMET) **5**:1000–1001
Consortium to Establish a Registry for
    Alzheimer's Disease
    (CERAD) **6**:197
Constitutive androstane receptor
    (CAR) **2**:1011
    nomenclature **2**:996*T*
    pharmacology **2**:1024
    see also specific drugs
Constitutive receptors, nuclear
    receptors **4**:726
Constrained conformational searches,
    pharmacophore detection **4**:133
Constrained peptides, proteolysis
    resistance **2**:584, **2**:585
Constraint propagation, constraint
    satisfaction programming
    (CSP) **4**:771
Constraint satisfaction programming
    (CSP) *see* Multiobjective
    optimization
Construction algorithm, atom-based
    drug design **4**:288
Constructionist approach, lipophilicity
    prediction, rule-based
    systems **5**:650
Construct validity
    schizophrenia models **6**:25
    schizophrenia social isolation
        models **6**:28
Consumer advocacy groups, drug
    discovery *see* Drug discovery
Contact analysis *see* Protein 3D
    structure validation
Contact inhibition, cell-based
    assays **3**:628
Contact matrix of predicted folds,
    protein structure
    prediction **4**:19, **4**:20*F*
Contact number, protein 3D structure
    validation **3**:525*F*, **3**:526,
    3:526*F*

Contents sensors, gene detection,
    bioinformatics **3**:321–322
Continuing Outcomes Relevant to
    Evista (CORE) trial
    bone effects **8**:111, **8**:112*F*
    breast cancer prevention **8**:112*F*,
        8:113
Continuous-exchange, cell-free
    expression systems **3**:121
Continuous-flow, cell-free expression
    systems **3**:121
Continuous flow scale up, microwave-
    assisted chemistry *see*
    Microwave-assisted chemistry
Continuous flow-trough diffusion cells,
    vaginal administration **5**:289–
    290
Continuous Infusion Versus Double-
    Bolus Administration of
    Alteplase *see* COBALT
Continuous performance test (CPT),
    ADHD models **6**:118
Continuous positive airway pressure
    (CPAP), obstructive sleep apnea
    (OSA) treatment **6**:162
Contoured surface generation, IsoStar
    (intermolecular
    interactions) **3**:405
Contraceptives **6**:548
    see also specific drugs
    bioethics **1**:701–702
    introductions **1**:234
    nonsteroidal antiestrogens **8**:83
    side effects **1**:715
Contract research **1**:579
    industry/academia collaborations
        vs. **1**:579
Contract research organizations
    (CROs) **2**:226
    changing roles **2**:236
    by country **2**:228*T*
    growth of **2**:205, **2**:227, **2**:228*T*
    history **2**:102
    importance of **2**:104
    virtual companies **2**:766
Contracts
    biotechnology companies **2**:231
    pharma *vs.* biotech, drug
        discovery **2**:230, **2**:230*F*, **2**:230*T*
Contraindications, carvedilol in heart
    failure **8**:137, **8**:140
Contrast drugs, small animal
    imaging **3**:660
Control groups
    clinical trials **2**:181
    microarrays **3**:100–101
Controlled cortical impact (CCI)
    model, traumatic brain
    injury **6**:258
Controlled pore glass, PASP **3**:793
Controlled release formulations
    drug metabolism and
        pharmacokinetics **5**:32–33

drug–polymer conjugates *see*
    Drug–polymer conjugates
    nonoral absorption **5**:279
    peptide/protein drugs *see* Peptide/
        protein drugs, injection
    peptide/protein drugs, injection
        hydration effects **2**:592
    lactic acid/glycolic acid
        polymers **2**:592
    polymer matrix **2**:592
    side chain urea functional
        groups **2**:593
    soluble formulations **2**:592
Control probes, microarrays **3**:90
Controls
    homogeneous time-resolved
        fluorescence **2**:425
    HTS data management **2**:453
Control spectra, NMR **5**:991, **5**:991*F*
Convective diffusion mass transport,
    dissolution **5**:413
Convective diffusion with simultaneous
    chemical reaction model (CDR),
    dissolution **5**:401
Convenience, ideal drug
    characteristics **2**:564*T*
Conventional thermal heating,
    microwaves *vs. see* Microwaves
CONVERTER program
    3D structure generation **4**:465
    2D- to 3D structure **4**:463
Coomassie blue G-250 **2**:333
Coomassie blue R-250 **2**:333
Coomassie stains
    2D-PAGE protein quantitation **3**:35,
        3:35*T*
    two-dimensional gel
        electrophoresis **2**:333
Cooperative Awards in Science and
    Engineering (CASE), industry/
    academia collaborations **1**:579
Cooperative Awards in Science and
    Technology (CAST), industry/
    academia collaborations **1**:579
Cooperative Research and Development
    Agreements (CRADA), industry/
    academia collaborations **1**:580
Coordinate data, drug design, de
    novo **4**:285
Coordinated/vectorial transports,
    membrane transport protein
    expression **5**:74
Coordinate errors
    carbonic anhydrase structure **3**:532
    cycloamylase structure **3**:532
    estimate, protein 3D structure **3**:546
    rubredoxin structure **3**:532
    streptavidin structure **3**:532
Coordinate sources, 3D database
    pharmacophore searching *see*
    Pharmacophores, 3D database
    searching
Copaxone *see* Glatiramer acetate

Diabetes mellitus   **6**:417–458, **6**:702
see also Insulin
animal models   **6**:433, **6**:434*T*
neuropathic pain   **6**:335
nitric oxide function   **6**:581
superoxide dismutase   **6**:591
biomarkers
glucose   **3**:70
hemoglobin A1c   **3**:70
cardiovascular disease   **6**:420
clinical trials   **6**:436
DCCT (Diabetes Control and
Complications Trial)   **6**:436
UGDP (University Group Diabetes
Program)   **6**:436
UKPDS (United Kingdom
Prospective Diabetes
Study)   **6**:436
diagnostic criteria   **6**:424, **6**:425*T*
fasting plasma glucose   **6**:424
disease basis   **6**:426
epidemiology   **6**:419
etiological classification   **6**:421,
6:422*T*
associated genetic
syndromes   **6**:424
chemically-induced   **6**:423
drug-induced   **6**:423
endocrinopathies   **6**:423
exocrine pancreas diseases   **6**:423
genetic β-cell defects   **6**:423
infections   **6**:424
insulin action defects   **6**:423
type 1 *see* Diabetes mellitus type 1
type 2 *see* Diabetes mellitus type 2
uncommon-immune-
mediated   **6**:424
financial impact   **6**:420
gestational *see* Gestational diabetes
mellitus
management   **6**:702
prevalence   **6**:419, **6**:702
increase in   **4**:636
prevention trials   **6**:421
Diabetes Prevention Program
(DPP)   **6**:421
STOP-NIDDM study   **6**:421
TRIPOD study   **6**:421
protein structures on PSD   **3**:382
renal dysfunction *see* Renal
dysfunction
risk factors, obesity   **6**:419, **6**:576–
577
RNAi   **3**:180–181
Diabetes mellitus management
see also Insulin treatment
amylin analogs (pramlintide)   **6**:452
artificial pancreas   **6**:455
β-cell replacement therapy   **6**:455
gene therapy   **6**:455
islet cell transplantation   **6**:455
pancreas transplantation   **6**:455
stem cell transplantation   **6**:455

GLP-1 agonists (exanatide)   **6**:452
glycemic control benefits   **6**:436
goals   **6**:437, **6**:438*T*
imatinib mesylate   **3**:3
lifestyle modifications   **6**:437
UKPDS   **6**:437–438
weight management   **6**:438
new areas   **6**:453
dipeptidyl peptidase-IV   **6**:453–
454
glucagon-like peptide   **6**:453
glucose-dependent insulinotropic
polypeptide (GIP)   **6**:454
glycogen synthetase 3   **6**:454
incretin augmentation   **6**:453
protein tyrosine phosphatase-
1B   **6**:454
oral hypoglycemic agents   **6**:438
see also specific types
potential therapeutic targets   **6**:454
Diabetes mellitus type 1   **6**:421, **6**:702
animal models   **6**:433, **6**:434*T*
autoimmunity   **6**:434
chemical destruction of
β-cells   **6**:433
gastroesophageal reflux
disease   **6**:673
idiopathic   **6**:422
immune-mediated   **6**:421
incidence   **6**:418
insulin treatment   **6**:451
pathogenesis   **6**:430
autoantibodies   **6**:430
autoimmunity   **6**:430
congenital rubella
infection   **6**:430–431
environmental factors   **6**:430
genetic predisposition   **6**:430
hyperglycemia   **6**:430
obesity   **6**:430
preclinical stage   **6**:430
renal dysfunction   **6**:578
Diabetes mellitus type 2   **6**:423, **6**:702
animal models   **6**:433, **6**:434*T*, **6**:435
genetic alteration   **6**:435
insulin-resistant mutants   **6**:435
overnutrition-evoked   **6**:435
spontaneous models   **6**:435
development, obesity   **6**:702–703
endothelial dysfunction   **6**:580
fibrates   **2**:1021
genetic predisposition   **6**:431
glucocorticoid receptor
ligands   **2**:1017
glucose homeostasis   **6**:431
impaired insulin secretion   **6**:432
β-cell activity   **6**:432
incidence   **6**:420
insulin resistance   **6**:432
abdominal obesity   **6**:432
cytokines   **6**:432
dose–response curves   **6**:432
genetic factors   **6**:432

insulin treatment   **6**:451
metabolic abnormalities   **6**:432
metabolic syndrome *vs.*   **6**:383, **6**:577
metabonomics   **5**:1001–1002
mitochondria   **2**:1054–1055
pathogenesis   **6**:431, **6**:431*F*
renal dysfunction   **6**:578
unmet medical needs   **6**:453
Diabetes Prevention Program (DPP),
diabetes mellitus prevention
trials   **6**:421
Diabetic nephropathy   **6**:420
incidence   **6**:420
unmet medical needs   **6**:453
Diabetic neuropathy   **6**:420
unmet medical needs   **6**:453
Diabetic retinopathy   **6**:316, **6**:420
glycemic control effects   **6**:436–437
incidence   **6**:420
pathogenesis   **6**:316
prevalence   **6**:316
treatments   **6**:316
photocoagulation   **6**:316
unmet medical needs   **6**:453
vascular endothelial growth
factor   **3**:202
Diacylglycerol *O*-acyltransferase,
coenzyme A conjugation   **5**:157
Diagnostics
drug discovery, decline in   **1**:619
pharmaceutical companies   **2**:100
problems, traditional medicine
1:423
stability testing   **5**:495
DIALIGN, sequence alignment,
bioinformatics   **3**:320
Dialkoxyindoles, solid-phase pool
libraries   **3**:705*F*
3,4-Dialkoxy-phenyl,
bioisosterism   **2**:660*T*
N,N-Dialkyldipeptidylamines,
PASP   **3**:808–809, **3**:810*F*
4,5-Dialkylpyrazoles, triphenylethylene
anticancer agents   **7**:158,
7:158*F*
Dialysis
buffer exchange   **3**:423
provision, elderly patients   **1**:628–629
Diamine *N*-acetyltransferase,
acetylation   **5**:153
Diamines
solid-phase pool libraries   **3**:705*F*
structure   **5**:359
β-Diamino acids, peptide/protein
agonists   **2**:582
1,3-Diaminobenzene, structure   **5**:359
Diaminocyclitols,
aminoglycosides   **7**:632
Diamino-cyclobutene-diones, carboxylic
acid bioisosterism   **2**:663,
2:666*F*
Diamino-nitroethene, Ganellin, C R
and   **8**:44

Diaminopyrimidines, solid-phase pool libraries 3:705F
2,4-Diaminoquinazoline 2:1048
structure 2:1048F; 6:610F
Diaminoquinazolinones, solid-phase discrete libraries 3:709T
3,4-Diamino thiadiazole, bioisosterism 2:659, 2:670
Diamond, Robert, molecular graphs/graphics 4:68
Diarrheal diseases
see also specific diseases
antimicrobial agents 7:474
less developed countries 1:338
mortality/morbidity, developing countries 1:642T
Diaryl diamides, solid-phase pool libraries 3:705F
Diaryl ethers, solid-phase discrete libraries 3:709T
2,3-Diaryl piperazine series, endocrine-modulating anticancer agents 7:156, 7:156F
Diarylpropionitrile, estrogen receptor-β-selective ligands 7:167, 7:168F
Diarylquinolines, as antimycobacterial agents 7:723
comparisons within class 7:724
future work 7:724
limitations 7:724
mechanism of action 7:724
SAR 7:724
Diastabol see Miglitol
Diastereomeric diols, HIV protease inhibitors 8:189, 8:189F
Diasteromer resolution, stereoisomer synthesis 2:715, 2:718F
Diazaborinine, structure 2:699F
Diazepam
alcohol addiction 6:180T
analog design 2:652F
scaffold hopping 2:652F, 2:682F
anesthetic adjuvants 6:362
bioavailability 5:712T
defensive test battery 6:94
discovery 6:98
historical perspective 6:287
6-hydroxydopamine lesions 6:127
marmoset human threat test 6:95
metabolism, CYP 2C19 5:546T
oral absorption 5:468T, 5:680T
primary insomnia treatment 6:158
structure 2:682F; 6:88F, 6:181F, 6:363F
1,2-Diazepane-2-ones, solid-phase discrete libraries 3:709T
Diazepine derivatives, solid-phase discrete libraries 3:709T
Diazepinoindole, phosphodiesterase-4 inhibitors 2:936
Diazonamide A 1:363
structure 1:362F
Diazoxide, bioavailability 5:712T

Dibekacin 7:634
Dibenz (b,f)azepine analogs, mitoprotective drugs 2:1062
Dibenz[b,g]1,5-oxazocines, solid-phase libraries 3:706, 3:707F
Dibenzosuberene, hydrolysis 5:142, 5:144F
Dibenzylfluorescein (DBF), in vitro inhibition studies 5:247T
(S)-1,4-Dibromo-2-butanol, structure 2:720F
2,7-Dibromocryptolepine, malaria treatment 7:800T
1,2-Dibromoethane, oxidation 5:100–101
Dicarboxylic acids, ionization constants (pK_a) 5:368
Dice coefficients
similarity searching 4:176T, 4:178
3D molecular alignment 4:112
DICER see RNAi (RNA interference)
Dichloroacetamide, anti-NS5b nonnucleosides 7:403F, 7:404
Dichloroacetate, Huntington's disease treatment 6:222T
1,2-Dichlorobenzene, microwave heating 3:840T
Dichloroisoprotenerol, hypertension treatment 6:711
Dichloromethane, microwave heating 3:840T
3,4-Dichloro substituted benzyl piperidines, CCR3 inhibitors 7:888
Diclizumab 1:433T
Diclofenac 6:338
bioavailability 5:712T
metabolism, CYP 2C9 5:545T
O-methylation 5:146, 5:146F
oral absorption 5:468T, 5:680T
reaction phenotyping studies 5:243T
receptor binding, pH dependency 5:382
structure 2:9F; 4:787F; 5:369F
target information 4:750T
Dicloxacillin
atopic dermatitis treatment 7:950
bioavailability 5:712T
oral absorption 5:468T
Dictionary-based fingerprints, two-dimensional descriptors, similarity searching 4:170–171
Dictyostelium discoidae, as model system 3:647
Dicyclomine
GPCR targeting 2:773T
irritable bowel syndrome treatment 6:657
Didanosine 7:279–282
development 1:69, 1:528–529
drug compatibility 7:342
drug regimens 7:342T
exclusivity period 1:532T

medicinal chemistry 7:338–340
structure 1:70F; 7:338F
Didemnin B
discovery/development 1:395
structure 1:394F, 1:395
Dielectric constant, force-field molecular modeling 4:94
Dielectric heating, microwaves see Microwaves
Diels–Alder reaction, dynamic combinatorial library generation 3:962, 3:962F
Diepoxins
discovery/development 1:391–392
structure 1:393F
Diet
cancer risk factors 7:3
metabonomics 5:1002
metabotyping 5:1000
tetracycline pharmacokinetics 7:619
Dietary Supplement Health and Education Act (1994) 1:417
dietary supplement definition 1:408, 1:417
nutraceuticals 1:458–459
Dietary supplements 1:417
see also specific substances
adulteration 1:417
cholesterol reduction 6:472
definition 1:408, 1:417
Dietary Supplement Health and Education Act (1994) 1:408, 1:417
irritable bowel syndrome treatment 6:658
orphan drugs 1:658
Diethylcarbamazine, nematode infection treatment 7:760
Diethyldithiocarbamate (DDC), reaction phenotyping studies 5:243T
Diethyl ether 6:358
structure 6:352F
Diethylpropion
metabolism 6:397, 6:397F
obesity treatment 6:397
contraindications 6:397
indications 6:397
mechanism of action 6:397
side effects 6:397
structure 6:395T
Diethylstilbesterol
breast cancer treatment 8:97F
estrogen receptor, molecular dynamics simulations 4:660
structure 8:105F
Difference gel electrophoresis (DIGE) 2:334
Differential gene expression, microarrays see Microarrays

Differentiation
  best in class *vs.* first in class **2**:119, **2**:120*T*
  cell-based assays **3**:628
Differentiation markers, immunosuppression **5**:223–224
Diffraction data resolution, x-ray–crystal interaction **3**:456
Diffraction methods, GPCR structure determination **4**:675
Diffraction patterns
  indexing **3**:469
  x-ray–crystal interaction **3**:452
Diffraction symmetry, protein crystallography **3**:441–442
Diffraction theory, protein crystallography **3**:451
Diffuse gel electrophoresis (DIGE), 2D-PAGE protein quantitation **3**:35, **3**:35*T*
Diffusion cells, dermal/transdermal administration **5**:292
Diffusion coefficient, transmembrane permeation, solubility-diffusion model **5**:460–461
Diffusion-encoded spectroscopy (DECODES), mixture analysis **2**:527–528
Diffusion filtering, NMR, drug discovery **3**:908*T*, **3**:913
Diffusion-ordered spectroscopy (DOSY), mixture analysis **2**:527
Diffusion rate limited distribution, physiologically based pharmacokinetics **5**:868–870
Diffusivity
  affecting factors **5**:726
  definition **5**:725–726
  VolSurf descriptors **5**:540
Diflomotecan **7**:115
Difloxacin, development **7**:575
Diflucan *see* Fluconazole
Diflunisal, bioavailability **5**:712*T*
3,5-Difluoro-4-hydroxyphenyl, bioisosterism **2**:662*T*
Difunctional molecule synthesis, PASP **3**:795
DigiFab *see* Antidigoxin polyclonal antibody
Digitalis, discovery/development **1**:322
Digital micromirror devices, surface micromachining **2**:352
Digital Object Identifier (DOI), databases **3**:300–301
Digital signal processing, NMR **2**:529
Digoxin
  oral absorption **5**:680*T*
  permeability assays **5**:270–271
Dihydralazine, immunotoxicity **5**:221*T*
Dihydroartemisinin
  discovery/development **1**:370
  structure **1**:371*F*

Dihydrobenzoxanthin scaffold, estrogen receptor-α-selective ligands **7**:165, **7**:166*F*
Dihydro-β-erythroidine, structure **2**:887*F*
Dihydrocodeine, oral absorption **5**:680*T*
Dihydroeponemycin
  discovery/development **1**:384
  structure **1**:383*F*
Dihydrofolate reductase (DHFR)
  Molecular Interaction Fields **4**:250
  selection markers, mammalian protein expression systems **3**:117, **3**:118
  structure-based drug design **4**:72
  thymidine biosynthesis **7**:56–57, **7**:57*F*
Dihydrofolate reductase (DHFR) inhibitors
  see also specific drugs
  parallel library design **4**:327
Dihydroimidazolium salts, solid-phase pool libraries **3**:705*F*
Dihydroisoquinolines, phosphodiesterase-10 inhibitors **2**:946
Dihydrokacmpferol **1**:452*T*
Dihydrolipoic acid, structure **6**:260*F*
Dihydromotuporamine C, mechanism of action **3**:7
Dihydropyrancarboxamides, solid-phase pool libraries **3**:707*F*
1,4-Dihydropyridine derivatives, quantum mechanical calculations **4**:390–391, **4**:390*F*
Dihydropyridine neuropeptide Y₁ receptor antagonists, bioisosterism **2**:669–670, **2**:671*F*
Dihydropyridines
  as calcium channel blocker **2**:849–850
  SAR **6**:716*T*
Dihydroquinazolinethiones, non-nucleoside reverse transcriptase inhibitors (NNRTIs) **7**:344–345
Dihydrostilbene, solid-phase pool libraries **3**:705*F*
Dihydrotestosterone, topical, pharmacokinetics **6**:533*T*
5α-Dihydrotestosterone, structure **6**:522*T*
4-4′-Dihydroxybenzil (DHB), orthogonal ligand–receptor pairs **3**:227
1,25-Dihydroxyvitamin D (calcitriol) *see* Calcitriol
Diisocyanoadocaine, malaria treatment **7**:800*T*
Dilated cardiomyopathy, phosphodiesterase-3 inhibitors **2**:947–948

Diltiazem **6**:747
  bioavailability **5**:712*T*; **6**:715*T*
  characteristics **2**:829*T*
  dose **6**:715*T*
  half-life **6**:715*T*
  mechanism of action **6**:715–718
  oral absorption **5**:468*T*
  P-glycoprotein interaction **5**:781*T*
  structure **2**:75*F*, **2**:829*T*; **6**:715*F*
dim-5 gene **1**:264
Dimensional classification, 4D-QSAR **5**:579
Dimerization, oxidation **5**:492, **5**:493*F*
Dimers, x-ray structures, problems **3**:540
2,3-Dimethoxy-9-nitropaullone, structure **4**:409*F*
4-Dimethylaminoanhydrotetracycline
  structure **7**:600*F*
  tetracyclines biosynthesis **7**:600–601
6-(4-Dimethylaminobenzoyl)-aminocaproic acid hydroxamide
  discovery/development **1**:386
  structure **1**:385*F*
17-Dimethylamino-ethylaminogeldanamycin
  discovery/development **1**:386
  structure **1**:387*F*
Dimethylamino group epimerization, tetracyclines *see* Tetracyclines
6-Dimethylamino-purine (6-DMAP)
  discovery/development **1**:389–390
  structure **1**:388*F*
Dimethylaminosulfonamide, phenol bioisosterism **2**:670–671
Dimethylbenzopyrans, solid-phase discrete libraries **3**:707*F*
2,2-Dimethyl benzopyran scaffolds, solid-phase libraries **3**:745*F*, **3**:746
2′,4′-Dihydroxy-6′methoxy-3′,5′-Dimethylchalcone, mitochondria, drug targeting **2**:1068, **2**:1068*F*
N,N-Dimethylformamide, microwave heating **3**:840*T*
Dimethyl sulfoxide (DMSO)
  automation **5**:352, **5**:353*T*
  compound storage **3**:566, **3**:568
    measured concentrations **3**:568
    stock solutions **3**:862
  microwave heating **3**:840*T*
Dimethylthiourea
  central nervous system injury **6**:268
  structure **6**:268*F*
N,N-Dimethyltriazines **5**:1030
  structure **5**:1029*F*
Dimorphic systemic mycoses *see* Fungal disease
2,4-Dinitrochlorobenzene (DNCB), immunotoxicity **5**:222
2,4-Dinitrofluorobenzene (DNFB), immunotoxicity **5**:222

2'-O-2,4-Dinitrophenyl derivatives, siRNAs modification **3**:182–183
DI-NMR *see* Nuclear magnetic resonance (NMR)
Dinoprostone, vaginal administration **5**:281*T*
Diode array detectors, ionization **5**:350
Diols, carbon-silicon bioisosterism **2**:695, **2**:696*F*
Dioncophylline C, malaria treatment **7**:800*T*
Dioscorides, natural products **1**:356
Diotul *see* Dosmalfate
Diovan *see* Valsartan
Dioxins, immunosuppression **5**:223
3,5-Dioxo-1,2,4-oxadiazolidine, carboxylic acid bioisosterism **2**:661
DIP database, regulatory networks **3**:330
Dipeptide analogs, solid-phase pool libraries **3**:705*F*
Dipeptide surrogates, peptidomimetics **2**:606–608
Dipeptidyl boronic acids, proteasome inhibitors **7**:236
Dipeptidyl peptidase-IV
    diabetes mellitus management **6**:453–454
    peptide/protein drugs **2**:577
Diphenhydramine
    as antiemetic agent **6**:683
    bioavailability **5**:712*T*
    primary insomnia treatment **6**:157–158
    structure **6**:100*F*
Diphenoxylate-atropine, irritable bowel syndrome treatment **6**:656
1,2-Diphenylenediamine platinum compounds, endocrine-modulating anticancer agents **7**:155–156, **7**:155*F*
Diphenylpiperidine analogs, carbon-silicon bioisosterism **2**:696, **2**:697*F*
Diphenyl quinoline series, triphenylethylene anticancer agents **7**:161
2,3-Diphosphoglycerate (2,3-DPG) **4**:71
Diphtheria vaccine, development **1**:531
Diphyllobothrium latum see Cestode (tapeworm) infections
Dipolar coupling *see* Nuclear magnetic resonance (NMR), structure determination
Dipolar polarization, dielectric heating **3**:839
Dipolar relaxation, liquid-state NMR structure determination **3**:481
DIPPR **3**:278*T*

Diprafenone, bioavailability **5**:712*T*
Diprivan *see* Propofol
Dipropionate, pulmonary administration **5**:281*T*
Diprotic weak acids, solubility **5**:403, **5**:403*F*, **5**:404*T*
Diprotic weak base, solubility **5**:403*F*
Dipyridamole
    oral absorption **5**:468*T*
    as phosphodiesterase inhibitor **2**:931, **2**:947*T*
    structure **6**:776*F*
    thrombolytic drugs **6**:775
Direct (real-time) dimension, two dimensional NMR **2**:507
Directed enzyme-prodrug therapy (DEPT) **5**:1033
    see also Antibody-directed enzyme prodrug therapy (ADEPT); Gene-directed enzyme prodrug treatment (GDEPT)
    over-expressed enzyme targeting **5**:1033
Directed evolution, protein crystallization **3**:888
Direct enantioselective chromatography **2**:721
Directional atomic contact analysis, protein 3D structure validation **3**:517–518
Directly bonded nuclei, NMR structure determination **3**:485
Directly observed treatment, short course (DOTS), antimycobacterial agents **7**:704
Direct measurement, blood–breast milk transfer modeling **5**:738
Direct nerve trauma, neuropathic pain models **6**:334
Direct physiological scaling, physiologically based pharmacokinetics **5**:877
Direct public funding, governments, in drug discovery **1**:728–729
Direct reactions, toxicity, computational models **5**:914
Direct structure interpretation, electrotopological indices (E-state index) **5**:561
Direct-to-consumer (DTC) marketing **1**:549
    see also Marketing
Direct ultraviolet assay, high-throughput methods **5**:400–401, **5**:407
Dirithromycin **7**:540
    pharmacokinetics **7**:534*T*
Disallowed Ramachandran plot regions, x-ray structures, problems **3**:537
'Disappearing polymorphs,' solid-state physicochemistry **5**:521, **5**:522*F*
DISC1, schizophrenia **6**:24

Discarded research, innovation **2**:259
DISCO
    3D-QSAR **4**:582–583
    hERG blocker modeling **5**:940–941, **5**:941*F*
    pharmacophore searching **4**:129*T*
    substructure searching **4**:467
Discodermindol
    discovery/development **1**:391
    structure **1**:390*F*
Discodermolide **1**:363; **7**:101
    discovery/development **1**:380–381; **7**:104
    SAR **7**:101–102, **7**:102*F*
    structure **1**:362*F*, **1**:381*F*, **2**:28*F*; **7**:102*F*
(+)-Discodermolide, structure **2**:725*F*
DISCON, ionization prediction, in silico tools **5**:616*T*, **5**:620, **5**:622
Discovering, innovation *vs.* **2**:255
Discovery-1 **3**:686*T*
Discovery libraries, drug discovery **1**:50*T*
Discovery Partners International **2**:760*T*
Discrete compound synthesis, combinatorial chemistry *see* Combinatorial chemistry
Discretes, solid-phase libraries *see* Solid-phase libraries
Discriminant analysis, protein kinase library design **4**:370
Discriminating power, ligand-based library design **4**:316
Disease
    complexity **2**:274
    drug development **2**:175
    etiology, target *vs.* disease **2**:112
    expression profiling **2**:311
    global, epidemiological transition **1**:692
    heterogeneity
        nervous system autoimmune diseases, treatment **6**:250
        pharmacogenetics **3**:61
    history, clinical trials **2**:181
    mechanisms, proteomics **3**:42–43
    models
        observation-led drug discovery **1**:341
        preclinical trials **2**:177
        zebrafish (*Danio rerio*) **3**:652
    molecular mechanisms, genetically engineered animals **3**:159–160
    pathology, drug discovery, decline in **1**:617
    pharmacogenomics **1**:344
    social construction **1**:703
    susceptibility, pharmacogenomics **1**:344–345
Disease, PDB **3**:377
Disease activity assessment, IBD clinical trials **6**:619

Disease-modifying anti-inflammatory
    drugs
    see also specific drugs
    arthritis treatment **7**:856
Disease-modifying antirheumatoid
    drugs (DMARDs)
    see also specific drugs
    rheumatoid arthritis
        treatment **1**:440–441
Disease-modifying therapies
    see also specific drugs
    Alzheimer's disease **6**:203
    drug discovery, decline in **1**:617–618
Disease severity, IBD clinical
    trials **6**:619
'Diseases of affluence' **1**:691
Disease-specific proteins,
    proteomics **2**:386
DisEMBL, domain mapping **3**:879
Disk intrinsic dissolution rate
    (DIDR) **5**:400
    self buffering effects **5**:418, **5**:418*F*,
        5:419*F*
    weak acid **5**:417
        indomethacin **5**:417, **5**:417*T*,
            5:418*T*
    weak base **5**:417, **5**:418*T*
        phenazopyridine **5**:417
Disopain *see* Mofezolac
DISOPRED2, domain mapping **3**:879
Disopyramide **6**:742
    bioavailability **5**:712*T*
    characteristics **2**:829*T*
    effectiveness **6**:742
    mechanism **6**:742
    structure **2**:829*T*
    toxicity **6**:742
Dispensing
    compound storage/
        management **3**:561
    Ultra-High-Throughput Screening
        System (Aurora
        Biosciences) **2**:442–443,
        2:443*F*
Dispersion model, hepatic metabolic
    clearance models **5**:871
Displacement, plasma protein
    binding **5**:325
Displacement assays, radioactive
    assays **3**:585–586
DisProt, domain mapping **3**:879
Dissimilarity, reagent-based library
    design **4**:345
Dissimilarity-based compound selection
    (DCBS) **3**:253, **3**:254*F*,
    4:186
    algorithm **4**:186
    MaxMin **4**:186–187
    MaxSum **4**:186–187
    OptiSim program **4**:187
    similarity searching **4**:178
    sphere exclusion algorithms **3**:253,
        3:254*F*, **4**:187

Dissociation constant (K_D)
    biological activity databases **3**:302
    determination, fluorescence screening
        assays **3**:601
    ligand–receptor interactions **4**:194
Dissolution **5**:401
    see also Disk intrinsic dissolution rate
        (DIDR); Solubility
    absorption **5**:702
    aqueous boundary layer
        effects **5**:401
    buffer effects **5**:419, **5**:419*F*
        quantity **5**:419
    chemical reactions **5**:414, **5**:414*F*
        concentration–time profiles **5**:414,
            5:414*F*
    concentration–time profiles **5**:401
    convective diffusion mass
        transport **5**:413
    convective diffusion with
        simultaneous chemical reaction
        model (CDR) **5**:401
    electrolyte diffusion **5**:413
        charge separation **5**:413
        electrical neutrality **5**:413
    importance **5**:401
        scale-up synthesis **5**:402
        storage effects **5**:401–402
    rotating disk apparatus **5**:412, **5**:414,
        5:415*F*
        buffered media **5**:415
        drug purity **5**:415
        Noyes–Whitney equation **5**:412
        rotational speeds **5**:415
        thermostated baths **5**:415
    in situ fiber optic dip probe UV
        spectrophotometry **5**:415
    Rainbow Dynamic Dissolution
        Monitor **5**:415–416
    small volume dissolution apparatus
        (µDISS) **5**:415*F*, **5**:416, **5**:416*F*
    solubility *vs.* **5**:401
    spherical powder dissolution **5**:400,
        5:402, **5**:420
        model used by Wang–
            Flanagan **5**:416*F*, **5**:420
        nonlinear concentration
            gradients **5**:420
    Vinograd–McBain treatment **5**:413
Dissolution interface, dissolution
    rate **5**:517, **5**:517*F*
Dissolution rates
    BCS **5**:975, **5**:975*F*, **5**:976*F*,
        5:983–984
    solid-state physicochemistry *see* Solid-
        state physicochemistry
Dissolution titration template (DTT)
    method, solubility **5**:400–401
Distamycin, mechanism of
    action **2**:1040
Distance-based descriptors, similarity
    searching,
        representations **4**:172

Distance-based geometry, 3D structure
    generation **4**:461–462
Distance-based methods
    library design **4**:363
    ligand-based library design **4**:311
    phylogeny construction,
        bioinformatics **3**:321
Distance bounds, pharmacophore
    detection **4**:127
Distance coefficients, similarity
    searching **4**:175
Distance estimation, conformational
    analysis/searching **4**:108
Distance geometry
    conformational analysis/
        searching **4**:105
    QSAR substituent selection
        models **4**:52
    3D QSAR descriptors **4**:155
Distance matrices, pharmacophore
    detection **4**:127
Distance minimization,
    pharmacophores **4**:15, **4**:16*F*
Distance-scale finite ideal-gas reference
    (DFIRE), docking, scoring
    functions **4**:485–486
DistComp, pharmacophore
    searching **4**:129*T*
Distribution **1**:52
    affecting factors **5**:532, **5**:532*F*
    ampholyte lipophilicity
        measurement **5**:446, **5**:446*F*
    compound storage/
        management **3**:561
    ionization constants (pK_a)
        effects **5**:384
    pharmacokinetics *see*
        Pharmacokinetics
    pharmacokinetics *vs.* **5**:847
    prodrugs **1**:53, **1**:54*F*
Distribution coefficients
    log D, lipophilicity prediction **5**:653
    log*D*pH **5**:447
    ionization effects **5**:427
Distribution models, physiologically
    based pharmacokinetics *see*
    Physiologically based
    pharmacokinetics (PBPK)
Distribution studies
    intraspecies differences **5**:42
        muscle binding **5**:43, **5**:44*F*
        tissue binding **5**:43, **5**:45*T*
    plasma protein binding **5**:43
    tissue storage models **5**:43, **5**:44*F*
    volume of distribution at a steady
        state (V_{d-ss}) **5**:42
Distribution volume (V_D),
    pharmacokinetics **5**:20
3,4-Disubstituted isoxazoles, solid-
    phase discrete libraries **3**:709*T*
3,5-Disubstituted isoxazoles, solid-
    phase discrete libraries **3**:709*T*
N,N-Disubstituted triazines **5**:1029

Disulfide bridges
  cell-free expression systems **3**:123
  dynamic combinatorial
    libraries **3**:969, **3**:970*F*
  dynamic combinatorial library
    generation **3**:962*F*
  Escherichia coli protein expression
    systems *see Escherichia coli* protein
    expression systems
  protein crystallization **3**:436, **3**:437,
    3:437*F*
Disulfiram
  alcohol addiction **6**:179–182, **6**:180*T*,
    6:183
  aldehyde dehydrogenase
    inhibition **5**:120
  cocaine abuse **6**:181*T*, **6**:184*T*
  oral absorption **5**:680*T*
  reaction phenotyping studies
    5:243*T*
  stimulant addiction **6**:185
  structure **6**:181*F*
Dithranol
  psoriasis treatment **7**:977*T*
  structure **7**:977*T*
Diuretics **6**:585, **6**:586*F*, **6**:699–700,
  6:708
  see also Loop diuretics; Thiazide
    diuretics
  congestive heart failure
    treatment **6**:698
  hypertension treatment **1**:532*T*;
    6:586*F*, **6**:699–700
  mechanism of action **6**:587, **6**:708,
    6:711*F*
  potassium-sparing
    bioavailability **6**:709*T*
    dose **6**:709*T*
    half-life **6**:709*T*
    structure **6**:709*T*
  thiazide-like
    bioavailability **6**:709*T*
    dose **6**:709*T*
    half-life **6**:709*T*
    structure **6**:709*T*
Divergent side chains, peptide/protein
  agonists **2**:582
Diverse Property-Derived codes,
  multiproperty fingerprint/
  dataprint descriptors
  4:471
DiverseSolutions (DVS)
  ligand-based library design **4**:311
  screening library selection **4**:504
Diversified model companies **2**:102
Diversity
  libraries **3**:697
  library design **4**:308
Diversity analysis **3**:252; **4**:183
  aim **4**:184
  clustering *see* Clustering
  compound selection methods **3**:253
  definition **4**:184

dissimilarity-based compound
  selection *see* Dissimilarity-based
  compound selection (DCBS)
  HTS **4**:183–184
  optimization techniques **3**:256;
    4:187
    maximum diversity **4**:187
    Monte Carlo method **4**:187
    MoSELECT **4**:187–188
    sum of pairwise
      dissimilarities **4**:187
  partitioning **3**:254*F*, **3**:255; **4**:185
    BCUT (Burden, CAS and
      University of Texas) **3**:255;
      4:186
    definition **4**:185
    method **4**:186
    pharmacophore fingerprints **3**:255
  screening test design **3**:252–253
Diversity issues
  functional discipline
    organization **2**:108
  leadership *see* Leadership
  project teams **2**:145
Diversity-oriented combinatorial
  libraries **3**:133, **3**:134*F*
Diversity-oriented libraries, solution
  phase parallel chemistry **3**:767
Divide-and-conquer drug design **4**:292
  fragment linking methods **4**:292,
    4:292*F*
    advantages **4**:292
    chemistry problems **4**:293
    disadvantages **4**:292–293
    ligand structure **4**:293
  ligand growing methods **4**:293
    definition **4**:293
    disadvantages **4**:293
  mixed methods **4**:293
  structure-based library design **4**:321
'Divide–couple–recombine,'
  combinatorial chemistry **2**:461–
  462
Divinylbenzene
  polystyrene resin, PASP **3**:792
  resin beads, solid support
    synthesis **2**:493–494
Divinylethermaleic anhydride/acid,
  drug–polymer
    conjugates **5**:1045
DIVSEL procedure, reagent-based
  library design **4**:345
DK-507k **7**:589–590
δlog *P*, computational oral absorption
  prediction **5**:684
DLS
  light scattering, promiscuous
    ligands **2**:743
  promiscuous ligands,
    identification **2**:743
DLX5 gene imprinting, Rett
  syndrome **1**:269
DMDs, DNA microarrays **2**:357

DMP323, structure **2**:302*F*
DMP450, structure **2**:302*F*
DMP 696, structure **6**:103*F*
DMPK/efficacy
  oseltamivir **7**:384
  peramivir **7**:384
  zanamivir **7**:384
DN-1417, Alzheimer's disease
  treatment **6**:207
DNA **2**:1038
  see also specific enzymes
  duplication, cell cycle **7**:11–12
  functions **2**:1039
  HCS cell cycle **3**:691–692
  hybridization, nanotechnology **2**:361
  N-hydroxy-arylamines, reaction
    with **5**:918
  interactions, quantum mechanical
    calculations *see* Quantum
    mechanical calculations, enzyme
    mechanisms
  intercalation, DNA topoisomerase
    inhibitors **7**:112
  methylation *see* DNA methylation
  microinjection, transgenic
    animals **1**:436
  nomenclature **2**:1038
    International Union of Pure and
      Applied Chemistry
      (IUPAC) **2**:1039
  organization **2**:425–426
  pharmacology **2**:1039
    singe base change **2**:1039
  repair
    cell cycle **7**:11–12, **7**:11*F*
    nitrogen mustard resistance **7**:132
    Saccharomyces cerevisiae model
      system **3**:6
    triazene resistance **7**:138
  RNA duplexes **2**:1040
  sequencing
    nanotechnology **2**:361
    pharmaceutical productivity
      effects **2**:132
  shuffling
    glyphosate *N*-acetyltransferase
      crystallography **3**:884–885
    protein crystallization **3**:884–885,
      3:885*F*
  small molecule interactions, quantum
    mechanics/molecular mechanics
    (QM/MM) **4**:410
  structure **2**:1037, **2**:1038, **2**:1038*F*,
    2:1039*F*
    G-quartets **2**:1040
    helix **2**:1038–1039
    major groove **2**:1038–1039,
      2:1039*F*
    minor groove **2**:1038–1039,
      2:1039*F*
    residues **2**:1038–1039
  synthesis, antibacterial agent
    inhibition **7**:733

DCN **2**:154
definition **5**:240
erythromycin **7**:534
FDA **5**:241
ideal drug characteristics **2**:564*T*
importance **5**:4
K⁺ ion channels *see* K⁺ ion channels
metabolism studies **5**:38
omeprazole **1**:440
pharmacodynamic effects **1**:286–287
plasma protein binding *see* Plasma protein binding
side effects **5**:240
toxicity, computational models **5**:915
traditional medicine *see* Traditional medicine
transporters **2**:61
interference
cytochromes P450 *see* Cytochrome P450 family
fluorescent cell-based assays **3**:625
introductions *see* Drug introductions
low-density lipoproteins (LDLs) binding **5**:849–850
marketing *see* Market (drug/products)
mechanism of action
target *vs.* disease **2**:112
toxicity, computational models **5**:915
membrane binding **5**:382
molecular weight characteristics **2**:52
nomenclature **1**:535
common technical document (CTD) **2**:161–162
physicochemical properties
transmembrane permeation, solubility-diffusion model **5**:463
transmembrane permeation, two-step model (flip-flop model) **5**:464
plasma protein binding *see* Plasma protein binding
polymer conjugates *see* Drug–polymer conjugates
potency, lipophilicity *vs.* **5**:535
pricing
developing countries *see* Developing countries
ethical issues *see* Bioethics
profitability *see* Bioethics
purity, Woods' rotating disk apparatus **5**:415
receptor binding, ionizable drugs **5**:382
resistance
ABCC1 (MRP1) transporter **5**:60–61
antimicrobial agents **7**:473

glucan synthase inhibitors **7**:462
malaria **7**:754
responses
pharmacogenomics **1**:344–345
single nucleotide polymorphisms (SNPs) **2**:362
safety
ethical issues *see* Bioethics
monitoring, postmarketing **1**:720
orphan drugs evaluation **1**:657
saturation, pharmacokinetics **5**:22
specificity **3**:3
structure-based fragment screening **4**:793
synthesis, toxicity **5**:910
targets *see* Target(s)
top-selling **2**:714*T*
toxicity markers *see* Biomarkers
uptake **5**:726*T*
Drug ADME Associated Protein Database, BIDD Databases **3**:305
Drug Adverse Reaction Target database, BIDD Databases **3**:305
Drug approval **2**:161
see also Development chemistry
country variation **1**:746
FDA **2**:159–160
International Conference on Harmonization **2**:159–160
national rules **1**:746
physical/chemical properties **2**:161
particle size **2**:162–163
polymorphic forms **2**:161–162
time reductions **2**:8
Drug-based assays, mechanistic pharmacokinetic studies **5**:38
Drug-binding sites
albumin **5**:849, **5**:853*T*
P-glycoprotein (P-gp) *see* P-glycoprotein (P-gp)
Drug combinations
clinical trials **2**:187
drug discovery **2**:277
Drug Database **3**:309
Drug delivery systems
aminoglycosides **7**:644
drug–polymer conjugates *see* Drug–polymer conjugates
peptide/protein drugs *see* Peptide/protein drugs
transporters **2**:66, **2**:66*F*
wet age-related macular degeneration **6**:306–307, **6**:312
Drug design
attrition risk **4**:380
combinatorial chemistry **3**:763
computer-aided, justification **4**:380
de novo *see* Drug design, de novo
GPCR *see* G protein-coupled receptors (GPCRs)
late stage failures **4**:380

ligand-based *see* Ligand-based drug design
multitasking **1**:48, **1**:49*F*, **1**:50*T*
ADMET profiling **1**:48–49
HTS efficacy **1**:48
SAR **1**:48–49
target structure-based *see* Target-based drug design/discovery
toxicity **5**:910
Drug design, de novo **4**:4, **4**:283–305, **4**:285*F*, **4**:297
active ligand copying **4**:284
advantages **4**:284
atom-based methods **4**:288
construction algorithm **4**:288
GenStar method **4**:288
GroupBuild method **4**:288
GrowMol method **4**:288
Legend method **4**:288
ligand connectivity **4**:288
Optimus method **4**:288
binding energy estimation **4**:284
building blocks **4**:288
see also specific methods
CADD **4**:4
combinatorial library design **4**:289
combinatories **4**:284–285
divide-and-conquer methods *see* Divide-and-conquer drug design
exemplification **4**:297
carbonic anhydrase **4**:297
CAS registry **4**:297
estrogen receptor-alpha **4**:297
MARPAT **4**:297
MARPATPREV **4**:297
Skelgen **4**:297
synopsis **4**:297
flexibility **4**:297
branch and bound schemes **4**:298–299
co-crystal structure **4**:297
consensus protein models **4**:299
dead-end-elimination algorithm **4**:298–299
dynamic **4**:298, **4**:299*F*
ensembles **4**:297, **4**:298*F*
MCSS method **4**:299
soft scoring **4**:298
virtual target screening **4**:297
fragment/ligand selection *see* Fragment-based drug discovery
future work **4**:297, **4**:300
flexible residues **4**:300
ligand affinity **4**:300
selectivity **4**:300
GPCR library screening **4**:693
human genome sequencing **4**:284
hydration **4**:299
crystallographic water **4**:299–300
ligand assembly/design *see* Ligand assembly drug design
ligand building **4**:285
multiobjective optimization **4**:769

Drug design, de novo (*continued*)
    personalized medicine **4**:284
    presentation factors **4**:285
    reaction-based fragments **4**:289
    restriction **4**:285
    ring/acyclic fragments **4**:288
        databases **4**:288–289
        features of interest **4**:289
        fragments **4**:289
        recombination rules **4**:289
    scoring/binding energy estimation
        see Scoring/binding energy
        estimation drug design
    search factors **4**:285
    site analysis *see* Site analysis, drug
        design
    specificity **4**:300
        common motifs **4**:300
        electrostatic **4**:300
    storage factors **4**:285
    structure-based **4**:284
    targets
        common substrate binding
            sites **4**:286
        coordinate data **4**:285
        structure **4**:286
    target site identification **4**:285
    target-specific fragments **4**:289
Drug development **2**:26, **2**:**173–202**;
    5:6*F*
    see also Clinical trials; Pharmaceutical
        research and development
        (R&D)
    animal toxicity studies **2**:28–29,
        2:29*T*
    antisense **1**:601, **1**:602
    approval rates **1**:602*T*, **1**:603*T*,
        1:605*F*
    attrition *see* Attrition, drug discovery/
        development
    BCS **5**:978
    biotechnology *see* Biotechnology
    as bottleneck **1**:38*T*
    clinical trials *see* Clinical trials
    components **2**:304
    costs **2**:2, **2**:199, **2**:726
        attrition **2**:727
        average investment **2**:726–727
        clinical trials **2**:727
        cost of goods (COGS) **2**:727–728,
            2:728*F*
        failure rate **2**:199
        generic competition **2**:727
        marketing **2**:727
        pricing **2**:727, **2**:727*F*
        process improvements
            2:728
        safety **2**:727
        time in development **2**:199
    critical path **1**:600, **1**:600*T*
    DCN *see* Development candidate
        nomination (DCN)
    desired areas **2**:173

early
    cell-based assays **3**:617–618
    screening library selection **4**:506
formulations **1**:537
functional discipline
    organization **2**:108–109
gene expression studies **1**:600–601
general process **2**:174*F*
gene therapy **1**:601, **1**:602
Good Laboratory Practice
    (GLP) **2**:27
Good Manufacturing Practice
    (GMP) **2**:27
high throughput organic
    synthesis **2**:396, **2**:396*F*
history **1**:530
    anesthetics **1**:530
    vaccines **1**:530
hits/leads **2**:175, **2**:175*F*
    disease areas **2**:175
    'druggability' assessment **2**:176
    good manufacturing process **2**:177
    HTS **2**:177
    microarrays **2**:177
    pharmacophore **2**:176
    QSAR **2**:177
    strategies **2**:176, **2**:176*F*
    target assessment **2**:176
imaging technology **2**:192
    see also specific techniques
important properties **2**:75*T*, **2**:76
innovative improvement **2**:174
innovative research **2**:174
late stages, stability testing **5**:495
manufacturing methods **1**:537
'me-too' projects **2**:174
mitochondria **2**:1079
    complex I *see* Mitochondria
        complex I
monoclonal antibodies
    (MAbs) **1**:594
multitasking **1**:50, **1**:51*F*
    analytical standards **1**:50–51
    intellectual property **1**:50–51
pharmacogenetics **3**:62
philosophy *see* Philosophy (of drug
    discovery/development)
preclinical trials *see* Preclinical trials
process **2**:721, **2**:722*F*
    best in class *vs.* first in class **2**:119
    complex molecules **2**:724–725,
        2:725*F*
    element integration **2**:721
    manufacturing methods **2**:721–722
    medicinal chemistry
        overlap **2**:723, **2**:724*F*
    organization **1**:759*F*
    regulatory documentation **2**:721–
        722
    safety, health and environment
        information **2**:722–723, **2**:723*F*
    structural features **2**:723–724
    synthesis effects **2**:723–724

productivity **2**:126
    acute *vs.* chronic treatment **2**:128
    clinical trial organization **2**:130
    commercial fits **2**:130
    failure visibility **2**:126
    pharmacokinetics **2**:130
    scale-up **2**:130
    success rates **2**:126–127
    target types **2**:128
    therapeutic area **2**:128, **2**:129*T*
    toxicology **2**:130
proteomics **3**:42, **3**:42*F*
purity **2**:27
regulatory guidelines **2**:27, **2**:200,
    2:725
    approval by property **2**:726
    guidelines **2**:726
    industrialization **2**:725
    medical utility **2**:725
    molecular configuration **2**:726
    safety assessment **2**:725
required decisions **2**:174
salt forms **2**:27, **2**:27*T*
stages of **2**:27, **2**:29*F*
strategies **5**:342*F*
success *vs.* failure **2**:730
    attrition rate **2**:730
    'druglikeness' **2**:730–731
    molecular weight effects **2**:730
    'rule-of-five' **2**:730
surface-enhanced laser desorption
    ionization (SELDI) **2**:378,
    2:379
time lines **2**:396, **2**:396*F*
timetable **1**:536*T*
transition from research **2**:54
Drug discovery **1**:**29–96**, **1**:**321–353**,
    1:688; **2**:31; **5**:6*F*
    *see also specific research areas; specific
        technologies*
academe **1**:60
ADMET *see* ADMET
animal models **2**:277, **2**:284
    human disease *vs.* **2**:277
    pharmacogenomics *vs.* **3**:56
    productivity **2**:125
    side effects **2**:277
approaches **1**:45*T*
attrition *see* Attrition, drug discovery/
    development
BCS *see* Biopharmaceutical
    Classification Scheme (BCS)
bioisosterism *see* Bioisosterism
biological macromolecules *see*
    Biological macromolecules
biotechnology *see* Biotechnology
bottlenecks **1**:38*T*
changes in **4**:775
chemical genetics **3**:923, **3**:923*F*,
    3:923*T*
chemical space **2**:49, **2**:51*F*
chiral drugs *see* Chiral drug discovery
clinical orientation **1**:33–35, **1**:35*F*

Dual-labeling, enzyme-linked immunosorbent assays (ELISAs) 3:73–74
Dual pore system, paracellular transport 5:263F, 5:265
Dual promoters, gene expression vectors 3:108
Ductal carcinoma in situ (DCIS), breast cancer 7:150
Duloxetine 8:123–136
  clinical effects, bladder activity inhibition 8:126–127, 8:126F
  clinical trials 8:128
    American Urology Association 8:128
    considerations 8:128
    dose-dependence improvements 8:132
    FDA Guidance for Industry for the Development of Incontinence Trials 8:128
    improvement measures 8:128
    Incontinence Quality of Life Questionnaire (IQOL) 8:128, 8:129–130
    Incontinence Severity Index (ISI) 8:130, 8:131F
    Japanese trial 8:129
    neurogenic bladder 8:129
    overactive bladder 8:129, 8:133
    SAAA 8:128, 8:129F
    SAAB 8:129, 8:130F
    stress incontinence 8:131F, 8:132
    urinary urge incontinence 8:133
  development milestones 8:132T
  discovery 8:125
  incontinence effects 8:125
  regulatory approval 8:133
  structure 6:336F, 8:125
Duncan, Kennedy, industry/academia collaborations 1:584
Duncan, William, management strategies 2:260–261
Dundee Kinase Consortium, industry/academia collaborations 1:587
Duodenal ulcers, *Helicobacter pylori* infection 6:607
DuP-105
  clinical trial discontinuation 7:678
  SAR 7:679
DuP-697, arthritis treatment 7:852–853
DuP-721
  clinical trial discontinuization 7:678
  oxazolidinone development 8:159–160
  SAR 7:679
  tricyclic-fused analogs 7:682
Dura mater, migraine 6:370
Duration of action
  hydrophilic depoting 2:587–588
  prodrugs 5:1014

Dutasteride 1:151
  benign prostatic hyperplasia treatment 6:565
  structure 6:566F
Duteplase 1:169
DVS-233 6:74
DW-224a 7:590–591
DW-286 7:589–590
DX-619 7:590–591
DYLOMMS, 3D-QSAR 4:576–577
Dynamic combinatorial chemistry (DCC) 3:959–976
  see also Dynamic combinatorial libraries (DCLs)
  applications 3:966, 3:967T
  building block selection 3:960, 3:961
    functional groups 3:961, 3:961F
    interactional units 3:961
    organizational units 3:961
    recognition groups 3:961
  definition 3:960
  development 3:966
    boronic ester-based trypsin inhibitors 3:966
    metal ion assisted imine macrocycle synthesis 3:966
  solution phase library synthesis *see* Solution phase synthesis, library preparation
  types of 3:963
Dynamic combinatorial libraries (DCLs)
  see also Dynamic combinatorial chemistry (DCC)
  adaptive 3:963, 3:963F
    selection target presence 3:963
    system adaptability 3:963
  amino acid-based 3:966
    acyl hydrazones 3:968
    beta-endorphin ligand generation 3:967–968, 3:968F
    compartmentalization 3:967–968
    pseudodynamic systems 3:968
  carbohydrate-based 3:969
    acyl hydrazone exchange 3:969
    concanavalin A library 3:969
    difficulties 3:969
    disulfide interchange 3:969, 3:970F
    lysozyme inhibition library 3:969, 3:971F
    metal coordination 3:969, 3:970F
  catalytic screening 3:963, 3:964, 3:965F
    acetylcholinesterase inhibitors 3:973
  definition 3:960
  dynamic deconvolution 3:963
  fixation 3:963
  generation 3:960, 3:961
    aldol formation 3:962F
    boronic ester formation 3:962F

  carbonyl groups 3:961–962, 3:962F
    cycloaddition reactions 3:962
    Diels–Alder reaction 3:962, 3:962F
    disulfide formation 3:962F
    hemiacetal formation 3:962F
    imine formation 3:961–962, 3:962F
    metal coordination 3:962
    Metathesis reaction 3:962F
    Michael reaction 3:961–962, 3:962F
    multiple reactions 3:963
    reversible covalent bonds 3:961, 3:962F
    see also specific reactions
    transacylation 3:962F
  iterated 3:963
  nonnatural components 3:969
    acetylcholinesterase inhibitors 3:970–971, 3:972F
    acid hydrazone exchange 3:970–971
    carbonic anhydrase inhibitors 3:969–970
    cell surface ligands 3:973
    cyclin-dependent kinase 2 inhibitors 3:973
    metal coordination 3:973, 3:973F
    neuraminidase inhibitors 3:970, 3:971F
    nonnucleotide building blocks 3:973
    psammaplin A derivatives 3:973
    transamination 3:969–970, 3:971F
    whole cells 3:973
  nucleic acid-based 3:968
    DNA-templated oligonucleotide synthesis 3:966
    dynamic hydrogels 3:968–969
    nonnucleic acid residues 3:969
    nonnucleotide building blocks 3:973
  pre-equilibrated 3:963
    amplification 3:964, 3:965F
    biological targets 3:963–964
    definition 3:963–964
    screening 3:963–964, 3:964F
    separation 3:964
  pseudodynamic (deletion) screening 3:965, 3:968
  selection conditions 3:960
    see also specific methods
  tethering screening 3:965
  virtual 3:961
  x-ray crystallography 3:963, 3:965
    cyclin-dependent kinase 2 inhibitors 3:973
Dynamic contrast enhanced magnetic resonance imaging (DCE-MRI) 3:77

Dynamic deconvolution, dynamic combinatorial libraries **3**:963

Dynamic dimerization, GPCRs **2**:799

Dynamic hydrogels, dynamic combinatorial libraries **3**:968–969

Dynamic in vitro BBB model (DIV-BBB)
  blood–brain barrier permeation **5**:314
  brain drug uptake **5**:311

Dynamic ligand assembly **3**:*959–976*
  see also Dynamic combinatorial chemistry (DCC); Dynamic combinatorial libraries (DCLs)

Dynamic light scattering (DLS)
  protein analysis **3**:426
  protein purity determination **3**:881

Dynamic population shift, ligand–receptor interaction models **4**:205

Dynamic programming paradigm, sequence alignment, bioinformatics **3**:320

Dynamics, NMR structure determination *see* Nuclear magnetic resonance (NMR), structure determination

Dynamic structure–activity relationships, Ganellin, C R **8**:43

Dynamin I, GPCR interaction **2**:786*T*

Dynastat *see* Parecoxib sodium

Dynorphin A, protein conjugation **2**:589

Dysbindin-1, schizophrenia **6**:23

Dyshydrotic eczema, atopic dermatitis **7**:949

Dysidiolide **1**:363
  analogs, solid-phase discrete libraries **3**:707*F*
  structure **1**:362*F*

Dyslipidemia
  coronary artery disease **8**:65
  coronary heart disease **6**:462

Dyspareunia, sexual dysfunction **6**:547

Dysthymia
  disease state/diagnosis **6**:46
  DSM-IV classification **6**:47*T*
  ICD-10 classification **6**:47*T*

Dysuria, drug introductions **1**:246

DZ-2640, structure **7**:509*F*

E

E-64 **6**:272
  structure **6**:272*F*

E1010 **7**:507–508
  activity **7**:506*T*
  structure **7**:504*F*

E-2007 **2**:903; **6**:291

E5564/Eritoran, IBD treatment **6**:624–625

EA-Inventor program, ligand-based library design **4**:318

Earlier versus Later Levodopa Therapy in Parkinson's Disease (ELLDOPA) **6**:215

Early decision making, drug discovery, decline in **1**:619

Early drug attrition, toxicity **5**:910

Early menses, breast cancer **7**:151

Eastern Cooperative Oncology group (ECOG), tamoxifen clinical trials **8**:90

EB 1089
  structure **2**:1018*F*
  as vitamin D receptor ligand **2**:1019

EBCTCG, tamoxifen clinical trials **8**:93

Ebers papyrus **1**:356, **1**:409

Ebipenem, activity **7**:509*T*

Ebola virus **7**:278

Ebrocit *see* Ebrotidine

Ebrotidine **1**:198

Ebselen, structure **2**:702*F*

EC₅₀, biological activity databases **3**:302

E-cadherin, downregulation in cancer **7**:15

Ecdysteroid–ecdysone receptor, orthogonal ligand–receptor pairs **3**:229

E-Cell project **3**:330

Ecenofloxacin **7**:589

ECEPP/3, molecular dynamics simulations **4**:652

ECFP atoms, topological fingerprint/dataprint descriptors **4**:469–470

Echinacea (*Echinacea*) **1**:424

Echinocandin B **7**:457*T*
  clinical profiles **7**:459*T*

Echinocandins **7**:433, **7**:457*T*
  discovery **7**:457–458

Echinococcosis **7**:751*T*
  drug treatment **7**:753*T*

Echinococcus see Echinococcosis

Echocardiograms, heart disease **6**:461

ECLiPS libraries, bioavailability models **5**:708*T*

EcoCys database
  genomics, antibacterial target identification **7**:742
  metabolic networks **3**:328–329

E-commerce, orphan drugs marketing **1**:659

Econazole **7**:430*T*
  clinical profiles **7**:450*T*, **7**:455
  medicinal chemistry **7**:449
  pharmacokinetics **7**:454

Economic development *see* Governments, in drug discovery

Economic development funds, venture capital provision **1**:564

Economics
  see also Market (drug/products); Marketing
  bioethics **1**:688
  development chemistry **2**:160–161
  importance of **2**:240
  less developed countries **1**:338
  orphan drugs R&D, USA **1**:658
  parasitic infections **7**:750
  patenting issues **1**:805
  peptide/protein drugs **2**:596
  pharmaceutical R&D **2**:100, **2**:101*T*
  traditional medicine *see* Traditional medicine

Ecteinascidin 743 **1**:393
  clinical trials **1**:394
  discovery/development **1**:393
  mechanism of action **1**:393
    transcription-coupled nucleotide excision repair process **1**:393
  structure **1**:394*F*

Eczema, epinastine hydrochloride **1**:204

Eczema herpeticum, atopic dermatitis **7**:949

ED₅₀, biological activity databases **3**:302

ED-71, as vitamin D receptor ligand **2**:1019

Edaravone **1**:**188**
  introduction of **1**:107

EDETOX project, dermal penetration models **5**:735

EDG-3 receptor, activation **2**:785*T*

EDG-3 receptor antagonists, pharmacophores **4**:530*F*

EDG-4 kinase inhibitors, pharmacophores **4**:520*F*

EDG-5 receptor, activation **2**:785*T*

EDIC (Epidemiology of Diabetes Interventions and Complications) study, glycemic control effects **6**:437

Edirel *see* Erdosteine

Edman sequencing **2**:345
  protein identification **2**:336–337

Edotecarin **7**:115

Edronax *see* Reboxetine

EDS **3**:547*T*

Education
  American Association of Colleges of Pharmacy **2**:85
  curriculum recommendations **2**:85, **2**:86*T*, **2**:87*T*
  demography **2**:86
  foreign national applicants **2**:85–86
  government support **1**:729
  income differences **2**:85–86
  industry/academia collaborations **1**:574
  nature of **2**:85
  personalized medicine **1**:290–291
  range of fields **2**:87

science integration **2**:88–89
summer schools **2**:89, **2**:90*T*
traditional medicine **1**:407
EET mimetics, renal dysfunction
management **6**:592
Efalizumab **1**:**236**, **1**:334*T*, **1**:433*T*,
1:499
introduction of **1**:108
psoriasis treatment **7**:946*T*, **7**:947,
7:972
Efavirenz **1**:**122**; **7**:279–282
combination treatment **7**:342, **7**:347
discontinuation **7**:347
discovery/development **1**:71, **1**:71*F*;
7:344–345
dosing regimen **7**:346, **7**:346*T*
extent of use **7**:345
resistance **7**:345–346
structure **7**:345*F*, **7**:346*F*
target information **4**:750*T*
tolerability **7**:346
Effective communication, project
teams **2**:145
Effective fragment potentials,
ionization **5**:607
Effective intestinal membrane
permeability *see* Permeation
theory
Effective leaders *see* Leadership
Effectiveness measures, similarity
measures, comparison of **4**:180,
4:180*T*
Effective permeability (P_{eff}), human
perfusion studies **5**:672, **5**:673*T*
Effective prediction domain, QSAR
applicability domain **4**:159
Effective rate of absorption,
computational oral absorption
prediction **5**:686
Effexor *see* Venlafaxine
Efficacy
antimetabolites **7**:61
biological macromolecules **1**:444
comparisons, clinical trials **2**:184–
185
drug discovery **2**:11
estimates, adenosine A_1 receptor
agonists **5**:892
ideal drug characteristics **2**:564*T*
natural products **1**:58
project failure **1**:342–343
testing, future work **1**:81, **1**:81*F*
traditional medicine
classification system **1**:418
presumption **1**:414–415
Efficiency
discreditation, innovation **2**:263
outsourcing **2**:228, **2**:229*F*
Efflux mechanisms
inhibition, P-glycoprotein
models **5**:776
tetracycline bacterial
resistance **7**:616

Efflux transporters
active transport **5**:266
drug activity **2**:62
macrolide antibiotic
resistance **7**:529, **7**:530–531
quinolone antibiotic resistance **7**:586
Eflornithine
mechanism of action **7**:820, **7**:821*F*
structure **7**:820*F*
trypanosomiasis treatment **7**:757,
7:820
Efonidipine, SAR **6**:716*T*
Efonidipine hydrochloride
ethanol **1**:**155**
EGFP, HCS cell cycle **3**:691–692
Egg lecithin, PAMPA **5**:472, **5**:473*F*
EGIS-7229 **6**:755
EGSITE, pharmacophore
searching **4**:135*T*
EGSITE2 models, pharmacophore
detection **4**:134
Egualen sodium **1**:**199**
EHITS program
docking, scoring functions **4**:485–
486
ligand flexibility **4**:482
Ehrlich, Paul
antimicrobial agents **7**:469
structure-based drug design
(SBDD) **4**:66
Ehrlichiosis, tetracyclines **7**:612
Eicosapentaenoic acid (EPA)
as antiarrhythmic **6**:756
epilepsy treatment **6**:294
Eidogen-Sertanty, commercially
available databases **3**:308
EiRX Therapeutics **1**:512
EKB-569 **7**:195
characteristics **2**:984*T*
EGFR family kinase inhibitors **2**:968
structure **2**:968*F*; **7**:195*F*
Elacridar
ABCG2 transporter, inhibition
of **5**:63
structure **5**:770*F*
Elaspol *see* Sivelestat
Elastase inhibitors
see also specific drugs
peptidomimetics **2**:613, **2**:614*F*
quantum mechanical
calculations **4**:401–402
Elastases
quantum mechanical calculations *see*
Quantum mechanical
calculations, enzyme mechanisms
three-dimensional structures **2**:626*T*
ELB 139 **6**:106
structure **6**:105*F*
Elderly, pharmacokinetics **5**:27
Electrical neutrality, electrolyte
diffusion **5**:413
Electric correction factor, lipophilicity
measurement **5**:436–437

Electric field presence, microwave-
assisted chemistry **3**:842
Electriptan **1**:**182**
Electrocardiography (ECG), heart
disease **6**:461
Electrochemical cells, stability
testing **5**:499
Electrochemiluminescence, optical
assays **3**:592
Electroencephalography (EEG)
anesthetic, animal models **6**:358
psychotropic drug
development **6**:51*T*
Electrolyte diffusion, dissolution *see*
Dissolution
Electromagnetic field homogeneity,
microwave-assisted
chemistry **3**:855
Electromyography, somnolence
measures/sleep testing
6:142
Electron beam computed tomography
(EBCT), coronary heart disease
diagnosis **6**:696
Electron-conformational methods,
pharmacophores **4**:127,
4:135*T*
Electron count coding *see*
Electrotopological indices
(E-state index)
Electron density, x-ray structures,
problems **3**:537
Electron density maps
atomic model derivation **4**:763
autocorrelation **3**:457
isomorphous replacement **3**:457,
3:458*F*
protein crystallography model
generation **3**:464, **3**:464*F*
protein 3D structure **3**:519,
3:547
scaling **3**:464, **3**:464*F*
Electron Density Server **3**:510
Electron diffraction, structure-based
drug design (SBDD) **4**:69
Electron distribution, structure
information representation
4:541
Electron donation, ionization constants
(pK_a) **5**:381
Electronically enhanced hybridization,
DNA microarrays *see* DNA
microarrays
Electronic charge transfer, quantum
mechanical calculations
4:390–391
Electronic circular dichroism, chiral
drug analysis **2**:720
Electronic commerce, materials
management **2**:536
Electronic parameters, bioisosterism
modification analysis **2**:687,
2:688*F*

Electron impact ionization, gas chromatography-mass spectrometry (GC-MS) **5**:993–994, **5**:993*F*

Electron microscopy
  promiscuous ligands, identification **2**:745
  protein structure problems *see* Protein 3D structure, problems

Electron oscillation, x-ray–crystal interaction **3**:452, **3**:455*F*

Electron paramagnetic resonance studies, mitochondria diversity **2**:1065

Electron shuttle, ubiquinone (coenzyme Q) **2**:1073

Electron transfer system (ETS), mitochondria *see* Mitochondria

Electron withdrawal, ionization constants (pK_a) **5**:381

Electrooculography, somnolence measures/sleep testing **6**:142

Electrophilic aromatic substitutions **7**:607, **7**:607*F*

Electrophilic groups
  hydrolases **5**:135
  steroid-derived modulators **7**:172, **7**:172*F*

Electrophoresis *see* SDS-PAGE; 2D-PAGE
  single nucleotide polymorphisms *see* Single nucleotide polymorphisms (SNPs)

Electrophoretic mobility shift assay (EMSA) **3**:631

Electrophysiology
  arrhythmias **6**:739
  cell-based assays *see* Cell-based assays
  epilepsy, models **6**:286
  genetically engineered animals **3**:155–156

Electroporation, siRNAs delivery **3**:177

Electrospray, mass spectrometry proteomics **3**:32

Electrostatic fields
  GRID force field **4**:240
  ionization **5**:605
  three-dimensional descriptors, similarity searching **4**:173
  values, CoMFA **5**:589

Electrostatic properties
  force-field molecular modeling *see* Force-field molecular modeling
  ligand–receptor interactions *see* Ligand–receptor interactions
  lipophilicity **5**:429

Electrostatic repulsions, quantum mechanics **4**:411

Electrotopological indices (E-state index) **4**:541; **5**:555–576
  see also Molecular connectivity; Topological descriptors

aqueous solubility **5**:563
  cross-validation techniques **5**:563
  cSLogWS model **5**:563
  external validation **5**:563, **5**:564*T*
atoms as distinct entities **5**:557
atom sigma orbital electronegativity **5**:557
atom topology **5**:557
atom-type **4**:552, **4**:553*T*, **4**:569*T*, **4**:571*T*
  library similarity searches **4**:562
  molecular structure space **4**:558, **4**:559*F*
blood–brain partitioning **5**:568, **5**:569*F*
  molecular structure effects **5**:568
  nonlinear term removal **5**:568
bond E-state formalism **4**:554, **4**:554*F*
  bond type classification **4**:555, **4**:556*T*
  perturbation **4**:555
characterization **4**:549
electron count coding **4**:541, **4**:542*T*
  simple *vs.* valence delta values **4**:542, **4**:542*T*
features **5**:557
free valence **4**:551
group-type index **4**:553
  hydrogen bond acceptors **4**:554, **4**:569*T*, **4**:571*T*
  hydrogen bond donors **4**:554, **4**:569*T*, **4**:571*T*
hERG inhibitors **4**:565, **4**:566*F*, **4**:567*T*
human metabolic stability **5**:570
  artificial neural networks (ANN) **5**:570
  clearance (CL$_{int}$) **5**:570
  liver microsome data **5**:570
  phase I reactions **5**:570
information network models **5**:556
  atom–atom influences **5**:556
  atom–atom relationships **5**:556
  atomic states **5**:556
intestinal absorption **5**:563, **5**:565*F*
  topological polar surface area (TPSA) **5**:564
  validation tests **5**:564, **5**:565*F*
intrinsic state **4**:544
  first hypothesis **4**:544
  higher-row quantum states **4**:545*T*, **4**:546, **4**:546*F*
  second hypothesis **4**:545*T*
  third hypothesis **4**:546, **4**:547*F*
intrinsic state (*I*) **5**:557–558
log P calculation **5**:660
molecular descriptors **5**:584
molecular interactions **5**:557
molecular polarity index **4**:555
NMR *vs.* **4**:551

onium group parameterization **4**:548, **4**:549*T*
  amine groups **4**:548
  ammonium groups **4**:548
  quaternary nitrogen atoms **4**:548
pi/lone pair electrons **5**:557
procedures **5**:561
protein binding **5**:564
  artificial neural networks (ANN) **5**:566–567, **5**:567*F*
  cephalosporin example **5**:566
  data sets **5**:567–568
  hierarchical clustering model **5**:567, **5**:567*F*
  penicillin example **5**:566, **5**:566*F*
  QSAR **4**:568, **4**:568*T*, **4**:569*T*, **4**:571*T*
recent studies **5**:561
  direct structure interpretation **5**:561
  validated prediction **5**:561
significance **4**:550, **4**:550*T*
software **5**:573
special character of **4**:549
specific consideration of hydrogen **4**:548
  atomic electronegativity **4**:548
  independent considerations **4**:548
  Kier–Hall electronegativity **4**:548
statistical methods **5**:561
  acceptable criteria measures **5**:563
  artificial neural networks (ANN) **5**:562
  clustering techniques **5**:562
  external validation sets **5**:562
  k-nearest neighbor analysis (kNN) **5**:562
  modeling methods **5**:562
  multiple linear regression (MLR) models **5**:562
  partial least squares **5**:562
  randomization tests **5**:562
  terms and symbols **5**:562, **5**:562*T*
  test sets **5**:562
  training sets **5**:562
structure descriptors **5**:558, **5**:558*F*
  biological activity *vs.* **5**:558–559
  information encoded **5**:559
  related structures **5**:558
  valence electron structure **5**:559
structure information representation **5**:556
structure *vs.* **4**:551
toxicity **5**:570
  Ames test **5**:571, **5**:571*F*
  carcinogenicity **5**:572
  environmental toxicity/ pollution **5**:571
  fish toxicity **5**:570
  hepatotoxicity **5**:572

Epimaritidine
  PASP **3**:815, **3**:816*F*
  solution phase synthesis **2**:491–492, **2**:492*F*
Epinastine hydrochloride **1:204**
  allergic rhinitis **1**:204
  asthma **1**:204
  dermatitis **1**:204
  eczema **1**:204
  psoriasis vulgaris **1**:204
  urticaria **1**:204
Epinephrine
  discovery **1**:328–329
  side effects **6**:311
  structure **1**:374*F*; **2**:57*F*
EpiOral tissue model, buccal administration **5**:283
Epipodophyllotoxins **7**:122
Epiroprim, physiologically based pharmacokinetics **5**:876
Epirubicin, toxicity **7**:46*T*
Episkin
  non-animal skin corrosivity testing **1**:467*T*, **1**:472
  non-animal skin irritation testing **1**:467*T*, **1**:473
Episodic ataxia, calcium channels **2**:852
Epithelial cell cultures, vaginal administration **5**:290
Epithelial cells
  asthma **7**:878
  differentiation, retinoic acid receptor (RAR) **2**:1008
  intestinal absorption **5**:260, **5**:261*F*
  potassium channels **2**:860–861
Epithelial damage, asthma **7**:878
Epithienamycin, structure **7**:511*F*
EpiVaginal, vaginal administration **5**:290
Epivir *see* Lamivudine
Eplerenone **1:156**; **2**:1017–1018; **6**:711
  bioavailability **6**:709*T*
  characteristics **2**:1012*T*
  dose **6**:709*T*
  half-life **6**:709*T*
  structure **2**:1018*F*; **6**:587*F*, **6**:709*T*
  target information **4**:750*T*
Eplivanserin, primary insomnia treatment **6**:159*T*, **6**:161
EP$^{-/-}$ mutants, atopic dermatitis animal models **7**:972
Epoetin alpha **1**:334*T*; **1**:500; **2**:762–763
  clinical use **1**:606–607
  labeling changes **1**:758*T*
Epoetin beta **1**:334*T*
Eponemycin **7**:237–238
  discovery/development **1**:384
  structure **1**:383*F*
Epothilone A **1**:363
  structure **1**:362*F*; **2**:28*F*; **3**:700*F*

Epothilone D **1**:361–363
  structure **1**:362*F*
Epothilones **7**:100
  discovery/development **1**:380–381
  NMR structure determination **3**:494, **3**:496*F*
  PASP **3**:821, **3**:822*F*, **3**:823*F*, **3**:824*F*
  SAR **7**:100, **7**:101*F*
  solid support synthesis **2**:482, **2**:484*F*
  structure **1**:381*F*; **7**:101*F*
Epoxide hydrolase inhibitors, renal dysfunction management **6**:592
Epoxide hydrolases **5**:135, **5**:136*T*, **5**:140, **5**:179
  alkene oxides **5**:142, **5**:144*F*
    cycloalkenes **5**:142, **5**:144*F*
    species differences **5**:144
  arene oxides **5**:141, **5**:143*F*, **5**:144*F*
  characteristics **5**:169*T*
  drug metabolism **5**:203
  human liver microsomes **5**:234
  mechanism of action **5**:141, **5**:141*F*
  microsomal (mEH) **5**:141
  multiplicity **5**:180
    microsomal (EPHX1) **5**:180
    soluble (EPHX2) **5**:180
  ontogenic development **5**:174*T*, **5**:180
  specificity **5**:180
  structure **5**:180
  tissue distribution **5**:180
Epoxomicin **7**:237–238
  discovery/development **1**:384
  structure **1**:383*F*
25-Epoxycholesterol **2**:1023
  structure **2**:1024*F*
Epoxyeicosatrienoic acid, renal dysfunction **6**:582
Epoxyketones, proteasome inhibitors **7**:237
Epoxypironetin, structure **7**:86*F*
Epratuzumab **6**:240*T*
  clinical indication **6**:245
  systemic lupus erythematosus treatment **6**:248
Epristeride, acne treatment **7**:965
Eprosartan **1:156**
  bioavailability **6**:722*T*
  half-life **6**:722*T*
  structure **6**:722*T*
Epstein–Barr virus (EBV) **7**:295, **7**:296
  antivirals **7**:261
    acyclovir **7**:297
    adefovir dipivoxil **7**:314
  cancers **7**:4
  cell immortalization **3**:620
  Guillain–Barré syndrome **6**:234
  microRNA coding **3**:175
Eptifibatide **1:166**, **1**:502; **2**:579*T*; **6**:775
  structure **6**:776*F*

Epworth Sleepiness Scale (ESS), somnolence measures/sleep testing **6**:141–142
Equality in healthcare, personalized medicine **1**:291
Equilenin benzothiophene analogs, steroid-derived modulators **7**:173–174, **7**:174*F*
Equilibrium constants, benzoic acid **5**:404*T*
Equilibrium dialysis
  plasma protein binding *see* Plasma protein binding
  tissue homogenate binding **5**:328
'Equilibrium' solubility
  automation *see* Automation
  solid-state physicochemistry **5**:516
Equipment failure, HTS data management **2**:453
Equity, motivation **2**:246
ER-804191 **7**:237–238
ER-807446 **7**:237–238
ERB-041, structure **2**:1013*F*
ErbB2 **7**:195
  in disease **2**:963*T*
  mechanism of action **7**:196
  structure **7**:196
ErbB2 inhibitors **7**:195
  see also specific drugs
  future work **7**:197
  structure **7**:192*F*
ErbB3/HER3, in disease **2**:963*T*
Erb family tyrosine kinases **7**:193
Erbitux *see* Cetuximab
Erbstatin **7**:193–194
  structure **7**:193*F*
ERCC1 gene, cisplatin resistance **7**:142
Erdosteine **1:210**
ER drug delivery, BCS *see* Biopharmaceutical Classification Scheme (BCS)
Erectile dysfunction **6**:545
  definition **6**:545–546
  dopamine receptors and **6**:546
  drug targets **6**:546
  lower urinary tract symptoms (LUTS) **6**:562, **6**:563*F*
  phosphodiesterase-5 inhibitors **2**:948
  prevalence **6**:546
  treatment
    bioethics **1**:701–702
    centrally-acting **6**:546
    dopamine D2 agonists **6**:546
    PDE5 inhibitors **6**:546
    peripherally-acting **6**:546
    prostaglandin E$_1$ **6**:546
    sildenafil **6**:546
Ergocalciferol
  characteristics **2**:1012*T*
  structure **6**:499*F*
Ergosterol-affecting antifungal drugs *see* Antifungal drugs

Ergotamine, migraine
treatment **6**:373*T*, **6**:374
Eril *see* Fasudil hydrochloride
Eriodictycol **1**:452*T*
ERK1/2, GPCR MAP kinase
activation **2**:784–787, **2**:787*F*
ERK pathway, *Drosophila
melanogaster* **3**:650, **3**:650*F*
Erlotinib *see* OSI-774 (erlotinib)
Erm methylase induction, macrolide
antibiotic resistance **7**:529–530,
7:530*F*
ERP systems, materials
management **2**:536
ERRAT software, protein 3D structure
validation **3**:517–518
Error sources, HTS data management
see High-throughput screening
(HTS), data management
ERRs
ligand-independent transcriptional
activation **4**:737
structure **4**:737
Ertapenem sodium **1**:112; **7**:507
activity **7**:506*T*
structure **7**:504*F*
Erythema nodosum,
thalidomide **2**:759
Erythritol, oral absorption **5**:468*T*
Erythrocyte-binding antigen 175
(EBA-175), malaria **7**:767
Erythrocyte culture, malaria **7**:771
Erythromycin **7**:531
activity **7**:532
biosynthesis **7**:531
clinical indications
acne rosacea treatment **7**:963
acne treatment **7**:961, **7**:962–963
atopic dermatitis treatment **7**:950
malaria treatment **7**:795–796
discovery **7**:520
dual-action agents,
oxazolidinones **7**:685
IK$_r$ blockade **6**:746
mechanism of action, SAR **7**:528
oral administration **7**:533
pharmacokinetics **7**:533, **7**:534*T*
as prokinetic agent **6**:681
resistance, acne treatment **7**:939
safety properties **7**:534
drug-drug interactions **7**:534
hepatotoxicity **7**:534
SAR **7**:534
SAR **7**:532
macrolactone ring **7**:532, **7**:533*F*
solubility **7**:531
stability **7**:531
structure **2**:46*F*, **2**:1043*F*; **3**:700*F*;
7:531
Erythromycin A, taste **5**:1013
Erythropoietin, STAT5 signaling **3**:206
α-Erythropoietin, brands, global
sales **1**:605*T*

β-Erythropoietin, brands, global
sales **1**:605*T*
Erythropoietin receptor, phage
display **2**:576–577
Escherichia coli
genome sequencing **3**:317*T*
proteome complexity **3**:29*T*
Escherichia coli MsbA, P-glycoprotein
homology modeling **5**:785
Escherichia coli protein expression
systems **3**:111, **3**:880
chaperone co-expression **3**:112
bicistronic approach **3**:113
compatible dual-vector
approach **3**:113
DnaK/DnaJ/GrpE system
3:112–113
GroEL/GroES system **3**:112–113
cost **3**:111–112
disulfide bond expression **3**:116
BL21 Origami strains **3**:116
fermentation temperatures
3:116
expression stability **3**:113
temperature **3**:114
inclusion body formation **3**:112
insolubility problems **3**:880
membrane protein expression **3**:114,
3:115*F*
strains **3**:114
toxicity **3**:114
other systems *vs.* **3**:113*T*
posttranslational modification
3:880
protein degradation **3**:114
protease-free systems **3**:114–116
protein precipitation **3**:112
protein targeting **3**:114
cytoplasm **3**:114
periplasm **3**:114
secretion **3**:114
signal/leader sequences **3**:114
requirements **3**:112*T*
solubility tags **3**:113
glutathione-*S*-transferase
3:113–114
mannose-binding protein
3:113–114
protein D **3**:113–114
thioredoxin **3**:113–114
strains **3**:111–112
structural genomics **3**:553*T*,
3:554
Escitalopram oxalate **1**:176; **6**:69–70,
6:74
side effects **6**:60*T*
structure **2**:729*F*
transporter inhibition **6**:66*T*
E-selectin, NF-κB pathway
expression **3**:197
Eserine, structure **5**:359
ESI, proteomics **2**:347
Esmeron *see* Rocuronium bromide

Esmolol
bioavailability **6**:712*T*
dose **6**:712*T*
half-life **6**:712*T*
structure **6**:712*T*
Esmolol, as 'soft drug' **5**:1011,
5:1011*F*
Esmolol stat **1**:67*F*
development **1**:67
electronic physicochemical
properties **1**:67
negative SAR **1**:67
SAR **1**:67, **1**:68*F*
steric properties **1**:67
ESOL, aqueous solubility
models **5**:644–645
Esomeprazole magnesium **1**:199,
1:688–689; **8**:218
characteristics **2**:653*T*
gastroesophageal reflux
disease **1**:199
me-too prescribing **1**:713–714
pharmacokinetics **8**:218
structure **2**:653*T*, **2**:732*F*; **6**:606*F*
synthesis **2**:732*F*
Esophageal cancer
adenocarcinoma, prinomastat
7:243
gastroesophageal reflux
disease **6**:603–604
nedaplatin **1**:147
E$_s$ parameter, QSAR steric
effects **4**:46
Esp@cenet **3**:276*T*
Essential drugs, access in developing
countries **1**:643
Essential genes, genomics, antibacterial
target identification **7**:733
Essential targets, genomics,
antibacterial target
identification **7**:733
Essential thrombocytopenia, drug
introductions **1**:107
EST, pharmacogenomics **1**:345
Established protocols, outsourcing
scale-up **2**:211
E-state indices *see* Electrotopological
indices (E-state index)
Esterases **5**:181
human liver microsomes **5**:234
inhibitors **5**:181
multiplicity **5**:181
nomenclature **5**:181
ontogenic expression **5**:181
selectivity **5**:181
structure **5**:181, **5**:182*F*
tissue distribution **5**:181
Ester hydrolysis, quantum mechanical
calculations **4**:401
Ester oxidative cleavage, cytochrome
P450s **5**:109
Ester prodrugs, carboxylic
acids **5**:1018, **5**:1019*F*

Esters
  hydrolysis **5**:135
  intramuscular testosterone
    replacement therapy **6**:535
Ester sulfamate formation,
  sulfotransferases **5**:147, **5**:148*F*
Ester sulfate formation,
  sulfotransferases **5**:147
Estimated human dose, DCN **2**:154
Estradiol
  bioavailability **5**:712*T*
  derivatives, solid-phase discrete
    libraries **3**:709*T*
  dermal penetration models **5**:735
  orthogonal ligand–receptor
    pairs **3**:224*F*, **3**:225–227, **3**:225*F*,
    3:226*F*
  receptors, benign prostatic hyperplasia
    treatment **6**:564
  SAR **7**:94–95
  solid-phase libraries **3**:708, **3**:718*F*
  structure **6**:513*F*, **7**:95*F*
  transdermal administration **5**:281*T*
17α-Estradiol, structure **2**:1063*F*
17β-Estradiol, structure **2**:1063*F*
Estramustine
  toxicity **7**:46*T*
  transporters **7**:45*T*
Estriol, P-glycoprotein
  interaction **5**:781*T*
Estrogen
  bone disease treatment **6**:512
  bone resorption/remodelling **6**:500
  neuroprotective effects **2**:1005
  physiological effects **2**:1004–1005
  resistance, male
    hypogonadism **6**:525
  synthesis inhibition, endocrine-
    modulating anticancer
    agents **7**:175
  urinary incontinence
    treatment **6**:544
Estrogen receptors (ERs) **2**:994,
  2:1004; **7**:152
  analog binding **4**:735, **4**:736*F*
  arthritis treatment **7**:856
  cardiovascular homeostasis **2**:1005
  characteristics **4**:659
  crystal structure **4**:660
  drug mechanisms of action **4**:735–
    736
  genetic disease
    breast cancer **2**:1026
    endometrial cancer **2**:1026–1027
    psychiatric disease **2**:1026–1027
    splice variants **2**:1026
    systemic lupus
      erythematosus **2**:1026–1027
  genistein **4**:736–737
  homology modeling, retinoid X
    receptor **4**:739
  knockout animal models **2**:1005
  ligand affinities **4**:660

  ligand binding site topology **4**:660
  ligand-induced conformational
    changes **8**:109
  molecular dynamics *see* Estrogen
    receptors (ERs), molecular
    dynamics simulations
  nomenclature **2**:996*T*
  orthogonal ligand–receptor
    pairs **3**:224, **3**:224*F*, **3**:225*F*,
    3:226*F*
  plasma membrane-associated **2**:1005
  polyaromatic hydrocarbons **4**:660
  structure **4**:659, **4**:735
  structure-based drug design **4**:735
    comparative molecular field analysis
      (CoMFA) **4**:741
    ligand design/docking **4**:742
  subtypes **8**:108
  α*vs.*β **7**:165, **7**:166*F*
Estrogen receptors (ERs), molecular
  dynamics simulations **4**:659,
  4:740
  binding free energy
    calculation **4**:661
    thermodynamic integration **4**:661,
      4:661*F*, **4**:662*F*, **4**:663*F*
  drug design **4**:661*F*, **4**:662*F*, **4**:663
    coumestrol **4**:663
    free energy prediction **4**:664
  known ligands **4**:660
    diethylstilbesterol **4**:660
    pharmacophoric model **4**:660–661,
      4:661*F*
    simulation validation **4**:660
    water molecule inclusion **4**:661
  structural interpretation **4**:662,
    4:664*F*
  one-step perturbation
    method **4**:662–663
  rational drug design **4**:663
Estrogen receptor ligands **2**:1013,
  2:1013*F*
  see also specific drugs
  chemogenomics **3**:933, **3**:933*F*
  hormone replacement therapy
    (HRT) **2**:1013–1014
  phytoestrogens **2**:1014
  scaffold hopping **2**:683, **2**:684*F*,
    2:685*F*
  selective estrogen receptor
    downregulators
    (SERDs) **2**:1014
  selective estrogen receptor modulators
    (SERMs) **2**:1014
Estrogen receptor-α
  β*vs.* **7**:165, **7**:166*F*
  de novo drug design **4**:297
  selective ligands *see* Selective estrogen
    receptor modulators (SERMs)
Estrogen receptor-β
  selective ligands *see* Selective estrogen
    receptor modulators (SERMs)
  *vs.*α **7**:165, **7**:166*F*

Estrogen receptor-related receptors
  (ERRs)
  nomenclature **2**:996*T*
  structure-based drug design **4**:737
The Estrogen Replacement and
  Atherosclerosis Study **1**:686
Estrogen replacement therapy (ERT)
  breast cancer **7**:151
  endometrial cancer risk **8**:104
  high-density lipoprotein
    cholesterol **8**:103
  osteoporosis **8**:103
Estrogen sulfotransferase **5**:147
Estrone, Hirschmann, R F **1**:2–3
Eszopiclone, primary insomnia
  treatment **6**:158, **6**:159*T*
ET-743 **7**:144
  structure **7**:144*F*
Etanercept **1**:334*T*, **1**:500
  brands, global sales **1**:605*T*
  protein conjugation **2**:590
  psoriasis treatment **7**:946, **7**:946*T*,
    7:975
  skin inflammation treatment **7**:975
  tumor necrosis factor-α
    binding **1**:441
Etch stopping, lithography/
  photolithography **2**:352
Ethacrynic acid **6**:708
  bioavailability **6**:709*T*
  dose **6**:709*T*
  half-life **6**:709*T*
  nucleophilic additions **5**:160,
    5:160*F*
  structure **6**:709*T*
Ethambutol
  as antimycobacterial agent **7**:718
    limitations **7**:718
    mechanism of action **7**:718
  bioavailability **5**:712*T*
  mycobacterium treatment **7**:705*T*
  oral absorption **5**:468*T*, **5**:680*T*
  structure **7**:718*F*
Ethamoxytriphetol
  discovery **8**:83
  as postcoital contraceptive **8**:84
  raloxifene development **8**:106
  structure **8**:84*F*
Ethanol
  bioavailability **5**:712*T*
  microwave heating **3**:840*T*
Ethanol addiction **6**:173*T*
ether-a-go-go (HERG) gene, potassium
  channels **6**:736
Ether *O*-dealkylation, cytochrome
  P450 **5**:98
Ether–oxygen/methylene groups,
  bioisosterism **2**:692, **2**:693*F*
Ethical evaluation, clinical trials **2**:187
Ethics
  see also Bioethics
  corporate issues and challenges *see*
    Corporate ethics

Ethics (*continued*)
  health personnel migration
    issues 1:648
  human research *see* Human research
    ethics
  patenting refusals 1:810
  personalized medicine *see* Personalized
    medicine
  pharmacogenetics *see*
    Pharmacogenetics
  principles 1:710
  traditional medicine *see* Traditional
    medicine
  venture capital risk factors/
    problems 1:557, 1:559
  virtue approach 1:710
Ethics Officers Association 1:721
Ethinylestradiol
  bioavailability 5:712T
  labeling changes 1:758T
  oral absorption 5:680T
  structure 2:1013F
Ethionamide
  mycobacterium treatment 7:705T
    mechanism of action 7:709,
      7:710F
  oral absorption 5:468T
  oxidation 5:114, 5:115F
Ethiopia, health personnel 1:648
Ethnicity
  cardiovascular disease 6:694
  hypertension 6:699, 6:707
  metabolic syndrome 6:382
  obesity prevalence 6:391
  renal dysfunction 6:576
Ethosuximide 2:850; 6:288
  bioavailability 5:712T
  characteristics 2:829T
  historical perspective 6:287
  structure 2:829T; 6:281F
7-Ethoxycoumarin (EFC), in vitro
    inhibition studies 5:247T
Ethoxypiperidine side chains,
    triphenylethylene anticancer
    agents 7:160–161, 7:161F
Ethyl acetate, microwave
    heating 3:840T
Ethyl (1-methylbutyl) barbituric acid *see*
    Pentobarbital
6-Ethyl-CDCA, structure 2:1025F
Ethylenediamine, structure 5:359
Ethylenediamino piperidines, solid-
    phase pool libraries 3:705F
Ethylene glycol, microwave
    heating 3:840T
Ethylestrenole, structure 6:522T
Ethylmethane sulfonate (EMS),
    chemical mutagenesis
    3:13–14
N-Ethyl-*N*-nitrosourea (ENU), chemical
    mutagenesis 3:13–14
Ethynylbiphenyl derivatives,
    oxidation 5:1031

Ethyol *see* Amifostine
S(-)-Eticlopride, as antiemetic
    agent 6:687–688
Etidocaine 6:361
  structure 6:354F
Etidronate
  mechanism of action 8:205–206
  as prodrugs 5:1021
  structure 5:1021F; 6:510F
  toxicology 8:207
Etilefrine, bioavailability 5:712T
Etomidate 6:353, 6:360
  development 1:531
  structure 6:353F
Etonogestrel, vaginal
    administration 5:281T
Etoposide 7:119
  discovery/development 1:381–382
  oral absorption 5:468T
  structure 1:382F
  toxicity 7:46T
  transporters 7:45T
  xenograft models 7:39T
Etoricoxib 1:213
  arthritis treatment 7:853
  introduction of 1:108
  structure 2:658F, 2:681F
Etravirine
  discovery/development 1:74
  structure 1:74F
Etretinate
  psoriasis treatment 7:978T
  withdrawal 1:757T
Eubananin, SARS 7:272
EU Biotechnology Directive
  informed consent 1:789
  patent exemptions 1:788, 1:811
Euchromatin, structure 1:252
Euclidian distance, similarity
    searching 4:176T, 4:178, 4:180–
      181
EU Cosmetics Directive, toxicity
    testing 1:464
EUK-8, amyotrophic lateral sclerosis
    treatment 6:225
EUK 134, amyotrophic lateral sclerosis
    treatment 6:225
Eukaryotes, RNA polymerases 2:1044
Eukaryotic cell membranes, tetracycline
    activity *see* Tetracyclines
Europe
  see also specific countries
  orphan drugs legislation 1:662
  patient advocacy
    organizations 1:666–667
  SMEs 1:497
European Agency for the Evaluation of
    Medicinal Products (EMEA)
  chemogenomics 2:567
  regulation role 1:546
  safety testing importance 5:3
European Alliance of Genetic Support
    Groups (EAGS) 1:666–667

European Association of Urology (EAU)
  benign prostatic hyperplasia
    diagnosis 6:560
  benign prostatic hyperplasia
    treatment 6:566, 6:567T
European Bioinformatics Institute
    (EBI) 3:304–305
European Centre for the Validation of
    Alternative Methods (ECVAM),
    animal experimentation
    alternatives 1:465
European Collection of Cell
    Cultures 3:622
European College of
    Neuropsychopharmacology
    (ECNP) 6:96
European Medicines Agency,
    personalized medicine 1:289–
      290
European Parliament, Council on
    Orphan Medicinal
    Products 1:662
European Patent Convention
    (EPC) 1:782, 1:809
  nonobviousness requirement 1:785
  patent exclusions 1:787
  usefulness requirement 1:786
European Rare Diseases Therapeutic
    Initiative (ERDITI) 1:662
European Rare Disorders Organization
    (EURORDIS) 1:666–667
European Union (EU)
  drug discovery expenditure 1:738
  regulation role 1:546
  researcher numbers 1:573T
  research spending 1:572T,
      1:573T
European Venture Capital
    Association 1:561
EVA 5:581T
EVA/CoSA, 3D-QSAR
  alignment 4:585, 4:585F
Evaluation, innovation 2:260
Evaluation and Predictions of Dermal
    Absorption of Toxic Chemicals,
    dermal penetration
    models 5:730–731
Evanescent wave, surface plasmon
    resonance 2:371
Evans-type oxazolidinone condensation,
    ezetimibe synthesis 8:77
Evaporative light-scattering detection
    (ELSD), high throughput
    purification 3:863–864,
      3:864F
Evernimicin 7:667
  in vitro anti-aerobic Gram-positive
    activity 7:668T
Everolimus, transplant
    rejection 7:919–920
Evidence tags, Genome
    Reviews 3:356, 3:357F
Evista *see* Raloxifene hydrochloride

Evolutionary programing, ligand-based library design *see* Library design, ligand-based

Evotec Pharmaceuticals **1**:510

Evoxac *see* Cevimeline hydrochloride

Ewald summation, force-field molecular modeling cutoffs **4**:98

Exact-match searches, similarity searching **4**:167–168

Exatecan **7**:114, **7**:123

Excessive daytime sleepiness (EDS) **6**:148

Exchange rates, NMR, drug discovery **3**:917

Excipients
adducts/complexation/ formulation **5**:526
solubility effects *see* Solubility

Excised skin, non-animal skin irritation testing **1**:467*T*, **1**:473

Excitability
arrhythmias **6**:733
potassium channel patterns **2**:857

Excitable cells, sodium channels **2**:835

Excitatory amino acid transporter antagonists, pain **6**:323, **6**:341

Excitatory amino acid transporters **2**:60*T*

Exclusion, gender issues **2**:245

Exclusion volume methods, ligand-based library design **4**:317

Excretion **1**:56
affecting factors **5**:532, **5**:532*F*
cell-based assays **3**:629
OAT (*SLC22*) transporters **5**:71
pharmacokinetics *see* Pharmacokinetics
structure–excretion relationships (SER) **1**:56

Executive summation, DCN **2**:152

Exelixis **1**:504

Exelon *see* Rivastigmine tartrate

Exemestane **1**:138; **7**:175
breast cancer **1**:138

Exenatide **2**:579*T*; **6**:452, **6**:453
clinical trials **6**:452–453
GPCR targeting **2**:772

Exercise stress tests, heart disease **6**:461

Exifone, malaria treatment **7**:800*T*

Exocrine pancreas diseases, diabetes mellitus **6**:423

Exocyclic amino nucleosides, solid-phase discrete libraries **3**:709*T*

Exophilic acid **1**:374
structure **1**:373*F*

Exo-ring double bonds, oxidation, cytochrome P450 **5**:101

EXP3174, structure **1**:378*F*

Expectorants
see also specific drugs
introductions **1**:210

Expenditure *see* Costs
governments *see* Governments, in drug discovery

Experimental allergic encephalitis (EAE) **6**:239; **8**:174, **8**:175
glatiramer acetate **8**:175
immunological specificity **8**:175

Experimental animal use, skin sensitization, computational models **5**:923

Experimental artifacts, false positives **2**:737

Experimental autoimmune encephalomyelitis, multiple sclerosis **6**:239; **7**:924

Experimental error detection, HTS data management **2**:452

Experimental groups, microarrays **3**:100–101

Experimental models
antimetabolites *see* Antimetabolites
breast cancer **7**:151
depression *see* Depression

Experimental records, career guidelines **2**:18

Experimental value (EVA) adjusted approach, log P calculation **5**:655

Explicit conformer generation, 3D structure generation *see* 3D structure generation

Exposure, toxicity data **5**:911

Expressed proteins, posttranslational modifications **3**:880

Expression data, bioinformatics *see* Bioinformatics

Expression levels, messenger RNA (mRNA) **2**:1044–1045

Expression libraries, RNAi (RNA interference) **3**:180

Expression-modified proteins, peptide/ protein drugs *see* Peptide/protein drugs

Expression monitoring, functional genomics **2**:329

Expression profiling
antimetabolites **7**:75
proteomics **3**:42–43
target identification *see* Target identification, genomics

Expression systems, gene construct design **3**:880

Expression/trafficking regulation, GPCRs **2**:799

Expression vectors
Semliki Forest Virus expression systems **3**:119
yeast protein expression systems **3**:117

Extended Connectivity Fingerprints (ECFPs)
2D similarity searching **3**:246

two-dimensional descriptors, similarity searching **4**:171

Extended enterprises **2**:229
definition **2**:229
virtual companies **2**:766

Extended Huckel theory (EHT), quantum chemical indices **4**:56

Extended Scale for Dementia (ESD) **6**:201

Extended ternary complex model, GPCR activation models **2**:792–794, **2**:794*F*

External cross-references, UniProtKB **3**:359

External event monitoring, project team leaders **2**:144

External identifiers, database integration **3**:303

Externalities, economic market failure arguments **1**:727, **1**:728

External resource linking, databases **3**:301*F*

External suppliers
compound acquisition/ processing **3**:563–564
screening library selection **4**:498

External validation
aqueous solubility, E-state index **5**:563, **5**:564*T*
statistical methods, E-state index **5**:562

Extracellular matrix (ECM)
asthma **7**:878
cell culture **3**:622–623
deposition, glaucoma **6**:300
remodeling, cancer, invasive growth **7**:15–16

Extracellular signal-regulated kinase (ERK), in cancer **7**:23

Extracellular space
preservation, brain slices **5**:311
tissue slices **5**:329

Extracellular stimuli, phosphodiesterase(s) regulation **2**:923

Extracts, traditional medicine **1**:419

Extrahepatic metabolism routes, metabolism studies **5**:33–34

Extrapolation of data
microarrays **3**:103
QSAR applicability domain **4**:159
quantum mechanical calculations **4**:394

Extravascular administration, pharmacokinetics **5**:19

Extremophiles
antifungals **1**:361
natural products **1**:360

Extrinsic innervation, gastrointestinal tract **6**:645–646

Extrinsic pathway (death receptors), apoptosis *see* Apoptosis

Ex-vivo models, in vitro oral absorption
    prediction   **5**:677
Eye
    anatomy   **6**:298, **6**:298*F*
    drug introductions *see* Drug
        introductions
    irritation, animal experimentation
        alternatives *see* Animal
        experimentation alternatives
    phosphodiesterase-6   **2**:928
    segment culture, glaucoma
        models   **6**:303
Eye diseases *see* Ophthalmic diseases
Ezetimibe   **1**:172; **6**:485; **8**:**65–82**
    biological activity   **8**:22, **8**:23*T*, **8**:24*F*
        in vivo models   **8**:77, **8**:77*F*
    clinical trials   **8**:23, **8**:24*F*, **8**:77,
        8:79*T*
        statin combination   **8**:79*F*
    design   **8**:21, **8**:22*F*
    development   **8**:80
    discovery/development   **1**:19; **8**:18,
        8:72
        intestinally cannulated bile duct-
            diverted rat model   **8**:72–73,
            8:74*F*
        lead synthesis   **8**:19, **8**:20*T*
        SAR   **8**:20*F*, **8**:22*F*
        SCH 48461   **8**:19, **8**:20, **8**:20*F*,
            8:21*F*
    half-life   **6**:485–486
    mechanism of action studies   **8**:23,
        8:80
        Niemann–pick C1 like protein
            (NPC1L1)   **8**:80
    metabolite formation   **8**:75, **8**:75*F*,
        8:76*F*
        alcohols   **8**:75–76
        HPLC   **8**:75, **8**:75*F*
        intestinally cannulated bile duct-
            diverted rat model   **8**:74*F*
        ketones   **8**:75–76
        phenols   **8**:75–76
    second generation development
        8:76
    statin combination   **6**:477; **8**:77,
        8:78*F*
        clinical trials   **8**:79*F*
    structure   **6**:485*F*
    synthesis   **8**:22, **8**:23*F*, **8**:77, **8**:78*F*
        Corey oxazaborolidine
            reduction   **8**:77
        Evans-type oxazolidinone
            condensation   **8**:77
Ezetrol *see* Ezetimibe
Ezlopitant
    as antiemetic agent   **6**:682
    oxidation   **5**:109, **5**:109*F*
^{18}F
    PET   **3**:661–662, **3**:664*T*
    pharmacokinetic
        measurements   **3**:672
    proliferation probes   **3**:665

F

F_1-inhibitor protein (IF-1),
    mitochondria complex
    V   **2**:1078–1079
Fa2N-4, drug metabolism, in vitro
    studies   **5**:233
Fabry's disease, drug
    introductions   **1**:106–107,
    1:247
Face validity, schizophrenia
    models   **6**:25
Facilities, clinical trials   **2**:186–187
Factor V   **6**:764
Factor VII
    brands, global sales   **1**:605*T*
    clotting cascade   **6**:764
Factor VIIa   **1**:175
Factor VIIa inhibitors
    focused library design, structure-based
        methods   **4**:329
    PASP   **3**:809
Factor VIII   **6**:764
Factor IX   **6**:764
Factor X   **6**:764
Factor Xa, three-dimensional
    structures   **2**:626*T*
Factor Xa inhibitors
    CADD   **4**:9–10, **4**:9*F*
    focused library design, structure-based
        methods   **4**:329
    virtual screening   **4**:10
Factor XI, clotting cascade   **6**:764
Factor XII   **6**:764
Factor XIII   **6**:764
Factory acceptance tests, customized
    automation   **5**:348
Factual databases *see* Databases
Fadrazole hydrochloride   **1**:145
    breast cancer   **1**:145
Fail fast *vs*. 'follow the science' *see*
    Philosophy (of drug discovery/
    development)
Failure rates
    drug development   **2**:199
    drug discovery   **1**:340; **2**:124–125
Failure visibility, drug development
    productivity   **2**:126
Fairness, motivation   **2**:246
Falcilysin inhibitors, antimalarial
    drugs   **7**:784
Falecalcitriol   **1**:222
Falipamil   **6**:747
[^{18}F]Fallypride, central nervous system
    imaging   **3**:667
False discovery rate (FDR),
    microarrays   **3**:98–99, **3**:101–103
False-negatives
    clinical trials   **2**:30–31
    HTS   **3**:684
    protein interaction networks,
        bioinformatics   **3**:335
    toxicity, computational models   **5**:915

False-positives
    clinical trials   **2**:30–31
    fluorescent cell-based assays   **3**:625
    HTS   **3**:684
    protein interaction networks,
        bioinformatics   **3**:335
    proteomics   **3**:41–42
    toxicity, computational models   **5**:915
Famciclovir   **1**:123; **7**:299
    bioavailability   **7**:300–301
    herpesvirus infections   **7**:257–258
    oral absorption   **5**:680*T*
    oxidation   **5**:119–120, **5**:120*F*; **7**:301
    as prodrug   **5**:1024; **7**:299–300
    SAR studies   **7**:300–301
    varicella zoster virus   **7**:257–258
Familial amyotrophic lateral
    sclerosis   **2**:1071
Familial childhood epilepsy   **2**:840–841
Familial febrile convulsions,
    epilepsy   **6**:285
Familial hemiplegic migraine
    (FHM)   **2**:852
Familial hypocholesterolemia, gene
    therapy   **1**:303
Familial paragangliomas   **2**:1073
Familial partial lipodystrophy
    (FPLD)   **6**:384
    LMNA gene   **6**:384
    nondiabetic subjects   **6**:384
    PPARG gene   **6**:384
    subtypes   **6**:384
Family history
    breast cancer   **7**:151
    cancer risk factors   **7**:3–4
Family studies, schizophrenia
    6:22–23
Famotidine
    bioavailability   **5**:712*T*
    discovery/development   **1**:528;
        6:605
    GPCR targeting   **2**:773*T*
    ionization constant (pK_a)   **5**:366*T*
    oral absorption   **5**:468*T*, **5**:680*T*
    paracellular transport   **5**:265
    renal elimination   **5**:46
    structure   **5**:359; **6**:605*F*
Fampridine *see* 4-aminopyridine
Famvir *see* Famciclovir
Fandofloxacin   **7**:589–590
Fa% *see* Fraction of dose absorbed (Fa%)
Farglitazar
    as PPARγ ligand   **2**:1021–1022
    structure   **2**:1023*F*
Farnesoid X receptor (FXR)   **2**:1010
    nomenclature   **2**:996*T*
    pharmacology   **2**:1023, **2**:1025*F*
        see also specific drugs
    three dimensional structure
        2:999–1000
Farnesyl diphosphate synthetase
    inhibition, alendronate   **8**:202,
    8:203*F*

Farnesyl transferase inhibitors 7:233
  antimalarial drugs 7:791
  mechanism of action 7:233–234
  peptidomimetics 7:234
  pharmacophores 4:528*F*
  quinolinones 7:234
  sequence importance 7:234
  xenograft model predictive
    ability 7:43
Farnesyl transferases, as targets 1:369–
    370
Farom *see* Fropenam
Faropenem daloxate 7:503
  structure 7:503*F*
Fasciola hepatica see Liver flukes
Fasciolopsis see Intestinal flukes
FasL, apoptosis evasion 7:18
Faslodex *see* Fulvestrant; ICI182780
FASTA, comparative genomics 7:738
Fast acquisition schemes, NMR
    structure determination 3:491
'Fast follower' products 1:531–532,
    1:532*T*
Fast growth, free energy
    calculation 4:657*F*, 4:658
Fastic *see* Nateglinide
Fasting plasma glucose, diabetes
    mellitus 6:424
Fast-tracked drugs, postmarketing
    surveillance 1:760, 1:760*T*
Fasudil hydrochloride 1:188
  structure 7:203*F*
FATCAT, protein function prediction,
    bioinformatics 3:336–337
Fatigue, modafinil 8:152
Fatty acid amide hydrolase (FAAH)
    inhibitors, IBD
    treatment 6:629
Fatty acid-binding protein (FABP)
    inhibitors, ligand-based
    NMR 4:791
Fatty acids
  albumin binding 5:849
  as antiarrhythmic drugs 6:756
  biosynthesis
    antibacterial target
      identification 7:735
    antimalarial drugs 7:792, 7:792*F*,
      7:793*T*
  catabolism, coenzyme A
    conjugation 5:158
  oxidation, atherosclerotic plaque
    formation 6:467–468
^{123}I Fatty acids, as metabolic
    probe 3:665
Faulty genomic imprinting, epigenetics
    see Epigenetics
Fc domains, peptide/protein drug
    conjugation 2:590
FCFP atoms, topological fingerprint/
    dataprint descriptors 4:469–470
[^{18}F]FdUrd, radiotreatment
    imaging 3:670

Fear potentiated startle models, anxiety
    disorders, animal models 6:93
Feature-based alignment, virtual
    screening 4:474
Feature identifiers, UniProtKB 3:359
Feature table, DDBJ/EMBL/
    GenBank 3:352
Feature Tree descriptor, fingerprint/
    dataprint descriptors 4:472
Febrile seizures, epilepsy 6:284
Feces
  excretion 5:16–17
  metabolism studies *see* Metabolism
    studies
Federal Food, Drug, and Cosmetic Act,
    USA (1938), Kefauver-Harris
    Amendments 1:657
Federal sources, compound collections
    (libraries) 3:132
Feedback, communication 2:243
Felbamate 6:288
  oral absorption 5:680*T*
  structure 6:282*F*
Felodipine
  bioavailability 5:712*T*; 6:715*T*
  characteristics 2:829*T*
  dose 6:715*T*
  half-life 6:715*T*
  oral absorption 5:680*T*
  SAR 6:716*T*
Female reproductive tract disorders,
    vascular endothelial growth
    factor *see* Vascular endothelial
    growth factor (VEGF)
Femara *see* Letrozole
Fencanfamine, narcolepsy
    treatment 6:156
Fenclofenac, oral absorption 5:680*T*
Fenfluramine 6:413
  metabolism 6:400, 6:400*F*
  obesity treatment 6:399
  phentermine combination 6:400
  safety 1:686
  structure 2:65*F*; 6:400, 6:400*F*
  transporter interactions 2:63
  withdrawal 6:400
Fenobam, structure 6:107*F*
Fenofibrate 2:1021
  characteristics 2:1012*T*
  structure 2:1022*F*; 6:479*F*
  target information 4:750*T*
Fenoldopam 1:156
Fenoterol, oral absorption 5:468*T*,
    5:680*T*
Fenoxaprop, malaria treatment
    7:793*T*
Fenozan 7:782*T*
Fen-Phen, bioethics, safety
    issues 1:686
Fenpyroximate
  mitochondria complex I
    2:1068–1069
  structure 2:1069*F*

Fentanyl 6:353
  as analgesic 6:338
  anesthetic adjuvants 6:362
  buccal administration 5:283–284
  GPCR targeting 2:773*T*
  structure 6:336*F*; 6:354*F*
  transdermal administration 5:281*T*
Fermenters/fermentation 3:419
  disulfide bond expression 3:116
  traditional medicine 1:410
Ferrets
  emesis animal models 6:674
  influenza virus models 7:376
Ferroportin 1 identification, *Danio rerio*
    model system 3:10
Ferroquine 7:774*T*
Fertility control research program, ICI
    Pharmaceuticals *see* ICI
    Pharmaceuticals
Fertility enhancers
  see also specific drugs
  introductions 1:232
Fertility treatment, priority-
    setting 1:631–632
Ferulic acid, pharmacokinetics 1:454*T*
[^{18}F]FET, infection imaging 3:670
Fexofenadine 1:204
  GPCR targeting 2:773*T*
  structure 2:66*F*
FFDCA (Federal Food, Drug, and
    Cosmetic Act) *see* Federal Food,
    Drug, and Cosmetic Act, USA
    (1938)
[^{18}F]FDG imaging
  cellular metabolism 3:668
    androgen ablation 3:669
    cytotoxic agents 3:668
    infection 3:670
    photodynamic treatment 3:669
    viability 3:668
  cellular proliferation imaging 3:671
  5-fluorouracil 3:669
  infections 3:670
  as metabolic probe 3:665
  photodynamic treatment 3:669
  prostate cancer model 3:670
  radiotreatment imaging 3:670
  steroid hormone imaging 3:667
  toremifene 3:669
1,3-[^{18}F]fluoro-αmethyl-tyrosine
    ([^{18}F]FMT) 3:665
O-(2-[^{18}F]fluoroethyl)-l-tyrosine
    ([^{18}F]FET) 3:665
16-[^{18}F]fluoro-l-DOPA
    ([^{18}F]DOPA) 3:665
2-[^{18}F]fluoro-l-tyrosine
    ([^{18}F]Tyr) 3:665
FG-8205 6:104
  structure 6:105*F*
Fibrates 2:1021; 6:479, 6:479*F*
  see also specific drugs
  diabetes mellitus type 2 2:1021
  Helsinki Heart Trial 6:479

Fibrinogen receptor (GPIIa/IIIb)
    antagonists,
    peptidomimetics   2:608–609,
    2:611F
Fibrinolytic drugs
    see also specific drugs
    introductions   1:175
Fibrinolytic Therapy Trialists (FTT)
    meta-analysis   6:773
Fibroblast collagenase (MMP-1),
    peptidomimetics   2:632, 2:635F
Fibroblast growth factor (FGF)
    bone resorption/remodelling   6:502,
    6:516
    cancer angiogenesis   7:21
    receptor interaction   3:583
Fibroblast growth factor receptor(s)
    (FGFRs)   7:202
    in cancer   7:202
Fibroblast growth factor receptor 1
    (FGFR1), PDB entries
    2:965T
Fibroblast growth factor receptor 2
    (FGFR2), PDB entries   2:965T
Fibroblast growth factor receptor
    inhibitors   7:203
    see also specific drugs
    structure   7:203F
Fibrous bone dysplasia, Gα₃ subfamily
    mutations   2:780–782
Fick's law, pharmacokinetics   5:13
Field-based alignment, molecular
    modeling see Core molecular
    modeling, 3D molecular
    alignment
Field-based similarity searching (FBSS)
    three-dimensional
    descriptors   4:173–174
Field block anesthetics   6:355
FieldPrint   5:581T
Field representations, similarity
    searching   4:173
FieldScreen   5:581T
Field value hardness, CoMFA   5:590
Field visualization, molecular
    fields   5:580
50S subunit, bacterial
    ribosomes   7:523, 7:526
Figure of merit (FoM), isomorphous
    replacement   3:459
Filamin-A, GPCR interaction   2:797T
Filariasis   7:751T
    clinical manifestations   7:760
    drug treatment   7:753T, 7:760
File drawer effect, affective disorders,
    clinical trial issues   6:63
FileMaker Pro, desktop
    databases   3:294
Filgrastim   1:334T, 1:500;
    2:762–763
    direct-to-consumer
    advertising   1:659
Film, x-ray detection   3:468

Filoviruses
    antiviral drugs   7:254T, 7:278
    phosphorodiamidate morpholino
        oligomers   7:278
    ribavirin   7:278
    S-adenosylhomocysteine
        inhibitors   7:278
    inosine monophosphate
        dehydrogenase   7:278
Filter elements, NMR   3:909
Filter flexibility, KcsA   4:711
Filtering, fragment/ligand selection drug
    design   4:296
Filter probe extraction method,
    partition coefficient
    systems   4:48
Filtration assay see Radioactive assays
Final compound specification, solution
    phase library synthesis   3:771
Final drug characteristics,
    ADMET   2:563
Finances
    genetically engineered
        animals   3:159
    outsourcing   2:207
    venture capital see Venture capital
Finasteride
    acne treatment   7:940, 7:965
    benign prostatic hyperplasia
        treatment   6:565, 6:566
    bioavailability   5:712T
    structure   6:566F
Fingerprint/dataprint
    descriptors   4:467, 4:469F
    see also Biological fingerprints
    geometric   4:470
        atom pair descriptors   4:470,
        4:470F
        SQUID program   4:471
    'hashed fingerprints'   4:468
    'keyed fingerprints'   4:468
    molecular see Molecular fingerprints
    multiproperty   4:471
        BCUT (Burden, CAS and
            University of Texas)   4:471
        Diverse Property-Derived
            codes   4:471
    reduced graph
        representations   4:471, 4:472F
        DAYLIGHT program   4:471–472
        Feature Tree descriptor   4:472
        FTREES software   4:472
    structural key transformation   4:468
    topological   4:469
        BCI fingerprints   4:469
        DAYLIGHT software   4:469
        ECFP atoms   4:469–470
        FCFP atoms   4:469–470
        MACCS   4:469
        UNITY software   4:469
Fingerprint Models   3:298
Fipamazole   1:515
Firebird SQL   3:294

Firstcin see Cefozopran hydrochloride
First neighbor relationships, molecular
    connectivity   5:557
First order index (¹χ), molecular
    connectivity   5:559, 5:560F
First principal component, VolSurf
    descriptors   5:540
First-principle scoring, docking, scoring
    functions   4:260
FIRST program, ligand–receptor
    interaction models   4:208
Fischer, Emil, 'lock-and-key'
    metaphor   3:959–960, 3:960F
Fisher, Ronald, process variable
    statistical evaluation   2:164
Fish gills, modeling   5:739
    percentage uptake efficiency   5:740
    rate of absorption   5:740
Fish oils, as antiarrhythmic
    drugs   6:756
Fish toxicity, electrotopological indices
    (E-state index)   5:570
Fitness evaluations, multiobjective
    optimization   4:770
5′ cap structure, messenger RNA
    (mRNA) structure   2:1044
5D (social) molecular fields   5:578,
    5:578F
Five-choice serial reaction time test
    (5-CSRTT), ADHD
    models   6:118, 6:121, 6:122–
    123, 6:122F
Five-trial inhibitory avoidance (5-TIA),
    ADHD models   6:121
Fixation, dynamic combinatorial
    libraries   3:963
Fixed-interval/extinction schedule,
    ADHD models   6:121, 6:122
Fixed-temperature water baths, acute
    pain models   6:332
FK041
    activity   7:502T
    structure   7:500F
FK-288 (depsipeptide)   7:228
    heat shock proteins, action on   7:233
FK506 see Tacrolimus
FK-778, transplant rejection   7:920
FK-888, structure   6:110F
FKBP-12, three-dimensional
    structures   2:636T
FKBP-12 inhibitors
    quantum mechanics   4:414
    quantum mechanics/molecular
        mechanics (QM/MM)   4:414
FLAP (five lipoxygenase activating
    protein) inhibitors see
    Leukotriene biosynthesis
    inhibition
Flash Plate assays   3:585, 3:588
    charge coupled devices   3:589
    DNA helicase   3:589
    DNA polymerase   3:588
    DNA primase   3:588

GPCRs **3**:585–586
HTS **3**:587
kinases
LANCE technology *vs.* **3**:587
serine/threonine kinases **3**:587
small molecule inhibitors **3**:587
tyrosine kinases **3**:587
PARP-1 inhibitors **3**:589
peptide–protein binding
assays **3**:589
surface coatings **3**:585
washing steps **3**:585
FLAS program, feature-based
alignment **4**:474
Flatfiles
DDBJ/EMBL/GenBank **3**:352,
3:353F
Third Party Annotation
(TPA) **3**:355, **3**:355F
Flat microarrays, single nucleotide
polymorphisms *see* Single
nucleotide polymorphisms
(SNPs)
Flavanols, nutraceuticals **1**:452T
Flavin monooxygenases (FMO) **5**:112,
5:173
catalytic cycle **5**:112
active oxygen species **5**:112,
5:112F
characteristics **5**:169T
drug metabolism **5**:200
genes **5**:173
genetic deficiency **5**:200
human liver microsomes **5**:234
multiplicity **5**:112, **5**:173
occurrence **5**:112
ontogenic development **5**:174T,
5:175
oxidation at nitrogen atom **5**:113,
5:113F
oxidation at sulfur atom **5**:114,
5:115F
oxygen activation **5**:112
selectivity **5**:112, **5**:173
structure **5**:175
substrate specificity **5**:200,
5:200T
tissue expression **5**:175
Flavinoids, microtubule targeting
agents **7**:93
Flaviviruses, antiviral drugs **7**:254T,
7:267
ribavirin **7**:267
Flavone scaffold, estrogen receptor-α-
selective ligands **7**:165, **7**:166F
Flavonoids, mitochondria, drug
targeting **2**:1068
Flavonols, nutraceuticals **1**:452T
Flavopiridol **1**:389; **7**:205
CDK family kinase inhibitors **2**:979–
980
characteristics **2**:984T
clinical trials **7**:207

discovery/development **1**:389
structure **1**:388F; **2**:979F
Flecainide
bioavailability **5**:712T
characteristics **2**:829T
oral absorption **5**:468T, **5**:680T
structure **2**:829T
Fleroxacin **7**:569T
development **7**:574
structure **7**:570F
FlexE program **4**:256
ligand–receptor interaction
models **4**:207
protein flexibility **4**:482
Flexibility
drug design, de novo *see* Drug design,
de novo
Molecular Interaction Fields **4**:238,
4:248
oral absorption prediction, molecular
descriptors **5**:683
project teams **2**:145
reduction, hERG blockers *see* hERG
blockers
Flexible linkers, hERG blockers
5:946
Flexible receptors (cross-docking) *see*
Docking
Flexible regions, ligand–receptor
interaction models **4**:208
Flexible residues, de novo drug
design **4**:300
FlexPharm program, structure-based
virtual screening **3**:252
FLEXS system
feature-based alignment **4**:474
3D molecular alignment **4**:113
FlexX program **4**:256
carbonic anhydrase inhibitors
4:276
checkpoint kinase-1 (Chk1)
inhibitors **4**:276
clique detection **4**:479
consensus scoring **4**:486
docking **4**:257
flexible receptors (cross-
docking) **4**:269–270
library design **4**:359
ligand flexibility **4**:481–482
ligand–receptor interaction
models **4**:206–207
MM-PBSA **4**:264
multiple active site corrections
(MASC) **4**:271
protein–ligand docking
programs **3**:406
quantum mechanics **4**:411
3D structure generation **4**:460–461
water molecules **4**:270–271
FLIPR instrumentation, GPCR
functional assay
screening **2**:811
Flivas *see* Naftopidil

FLOG program
docking, scoring functions **4**:483
ligand flexibility **4**:480–481
Flomox *see* Cefcapene pivoxil
Flosulide, structure **2**:658F, **2**:681F
Flow-cell measurement, dissolution
rate **5**:517, **5**:518F
Flow cytometry
biomarkers **3**:74
cell-based assays *see* Cell-based assays
microfluidic devices *see* Microfluidic
devices
RNAi detection **3**:178
single nucleotide
polymorphisms **2**:368
Flower plots, drug discovery **2**:31–32,
2:33F
Flow injection, ionization constants
(pK_a) measurement **5**:393
Flow nuclear magnetic
resonance **2**:521
Flow processes, PASP *see* Polymer-
assisted solution phase synthesis
(PASP)
Flow-through system, CEM Discover
platform microwave-assisted
chemistry **3**:846, **3**:846F
Floxuridine, toxicity **7**:46T
[^{18}F]FLT
cellular proliferation imaging **3**:671
infection imaging **3**:670
Flt3 family kinases **2**:969
acute myeloid leukemia **2**:971
in disease **2**:963T
Flucam *see* Ampiroxicam
Fluconazole **7**:430T, **7**:432; **8**:25
bioavailability **5**:712T
candidosis **7**:426, **7**:432
clinical profiles **7**:452, **7**:453T,
7:455
candidosis **7**:455
exclusivity period **1**:532T
mechanism of action **8**:25
medicinal chemistry **7**:452,
7:453–454
oral absorption **5**:680T
pharmacokinetics **7**:454
target information **4**:750T
Flucytosine **7**:430T
bioavailability **5**:712T
oral absorption **5**:468T
Fludarabine **7**:62T
approval **7**:62T
clinical indication **7**:62T
toxicity **7**:46T
Fluid extracts, traditional
medicine **1**:419
Fluidics/electronics integration,
microtechnologies **2**:355
Fluidization, solubility **5**:628
Fluid shifts, equilibrium dialysis **5**:327
Flumadine *see* Rimantadine
Flumazenil, oral absorption **5**:680T

Flumequine
  ionization constant (pK_a) **5**:368
  SAR **7**:580
  structure **5**:369*F*; **7**:577*F*, **7**:580*F*
FluMist *see* Influenza vaccine, live
    vaccine
Flunarizine, as antiemetic agent
    6:683
Flunitrazepam, bioavailability
    5:712*T*
Fluocinolone
  structure **6**:307*F*
  wet age-related macular degeneration
    treatment **6**:306*T*
Fluocinolone acetonide, as fast-tracked
    drug **1**:760*T*
Fluocinonide, psoriasis
    treatment **7**:944
Fluorescein leakage assay (FLL), non-
    animal eye irritation
    testing **1**:466, **1**:467*T*
Fluorescence
  HCS labeling techniques **3**:688
  small animal imaging **3**:660*T*, **3**:661
Fluorescence anisotropy, fluorescence
    polarization **3**:603
Fluorescence assays, GPCR functional
    assay screening **2**:809
Fluorescence-based assays
  see also specific assays
  hERG blocker screening **5**:937
Fluorescence correlation spectroscopy
    (FCS) **2**:404
  fluorescence fluctuation
    techniques **3**:608, **3**:608*F*
Fluorescence correlation spectroscopy-
    related confocal fluorimetric
    techniques (FCS$^+$) **2**:404
  advantages **2**:406
  background elimination **2**:407
  case studies **2**:407
  cell assays **2**:411*F*
  enzyme assays **2**:409, **2**:409*F*
  GPCR assays **2**:410, **2**:410*F*
  homogeneous assays **2**:405*T*, **2**:406
  miniaturization **2**:406
    case studies **2**:407, **2**:408*F*
  multiple read-out modes **2**:407
    flexibility **2**:408
  multiple read-out parameters **2**:407
    case studies **2**:407, **2**:408,
      2:409*F*
  read-out **2**:405, **2**:427*T*, **2**:456*T*
  reagent safety **2**:407
  single component labeling **2**:407
  throughput **2**:407
Fluorescence detection systems
  HTS *see* High-throughput screening
    (HTS)
  microplate-based assays **2**:367
  Ultra-High-Throughput Screening
    System (Aurora
    Biosciences) **2**:445, **2**:446*F*

Fluorescence fluctuation
    techniques **3**:606
  data acquisition **3**:608
    multichannel scaler (MCS)
      trace **3**:608, **3**:608*F*
  fluorescence correlation
    spectroscopy **3**:608, **3**:608*F*
  fluorescence intensity distribution
    analysis **3**:609, **3**:610
    fluorescence intensity and lifetime
      distribution analysis
      (FILDA) **3**:610
    fluorescence intensity multiple
      distribution analysis
      (FIMDA) **3**:610
    quenching **3**:609–610, **3**:610*F*
    technique **3**:609
  instrumentation **3**:607
    confocal optics **3**:607
    laser excitation **3**:607–608
    signal **3**:607
  two-dimensional fluorescence
    intensity distribution analysis
    (2D-FIDA) **3**:610
    technique **3**:610
Fluorescence imaging
  biomarkers *see* Biomarkers
  drug development **2**:193
Fluorescence Imaging Plate Reader
    (FLIPR) **3**:683
Fluorescence intensity and lifetime
    distribution analysis
    (FILDA) **3**:610
Fluorescence intensity distribution
    analysis *see* Fluorescence
    fluctuation techniques
Fluorescence intensity multiple
    distribution analysis
    (FIMDA) **3**:610
Fluorescence lifetime assays **3**:605,
    3:611*F*
  frequency domain acquisition
    3:605–606
  future developments **3**:610
    confocal FLA **3**:610, **3**:611*F*
    confocal TLA **3**:611, **3**:612*F*
  time domain acquisition **3**:605–606
Fluorescence measurements, P-
    glycoprotein models **5**:777
Fluorescence-mediated tomography
    (FMT), drug
    development **2**:193
Fluorescence polarization (FP) **3**:602
  autofluorescence **3**:603
  fluorescence anisotropy **3**:603
  GPCR functional assay
    screening **2**:809
  molecular mass range **3**:603, **3**:604*F*
  polarization definition **3**:602
  quenching **3**:603, **3**:604*F*
  scintillation proximity assay *vs.*
    3:590
  target-oriented screening **3**:142–143

Fluorescence reflectance imaging
    (FRI), drug development **2**:193
Fluorescence resonance energy transfer
    (FRET) **3**:604
  acceptor–donor distances **3**:605
  autocalibration **3**:605
  autofluorescence **3**:605
  biomarker imaging **3**:77
  cell-based assays **3**:632
  fluorescent cell-based assays **3**:618,
    3:625
  GPCRs **2**:798–799
    functional assay screening **2**:813
    structure determination **4**:679–
      680
  homogeneous time-resolved
    fluorescence **2**:413–414
  HTS **2**:401
  single nucleotide
    polymorphisms **2**:366
  technique **3**:604–605
  Ultra-High-Throughput Screening
    System (Aurora
    Biosciences) **2**:437
Fluorescence screening assays **3**:599–
    616, **3**:601
  artifacts **3**:601
    autofluorescence **3**:601, **3**:602*F*
    inner-filter effects **3**:602
    instrument **3**:601
    optics **3**:602
    quenching **3**:601
  bulk fluorescence techniques **3**:602
    fluorescence lifetime *see*
      Fluorescence lifetime assays
    fluorescence polarization *see*
      Fluorescence polarization
    fluorescence resonance energy
      transfer (FRET) *see* Fluorescence
      resonance energy transfer
      (FRET)
    homogenous time-resolved
      fluorescence *see* Homogenous
      time-resolved fluorescence
      energy transfer (HTR-FRET)
    time-resolved anisotropy analysis *see*
      Time-resolved anisotropy analysis
    total fluorescence intensity
      (FLINT) **3**:602
  compound artifact correction **3**:611,
    3:612*F*
    artifact correction **3**:613
    autofluorescence **3**:611–612,
      3:614
    hit identification **3**:613
    identification step **3**:612–613,
      3:613*F*
    quenching **3**:611–612
    read out parameters **3**:614
    statistical accuracy **3**:612–613
  fluctuation techniques *see*
    Fluorescence fluctuation
    techniques

single molecule detection 3:600
statistics 3:600
   accuracy 3:600–601
   compound activity rating 3:601
   IC$_{50}$ determination 3:601, 3:609F,
      3:610F, 3:611F, 3:612F
   K_D determination 3:601
   suitability 3:600
Fluorescence spectroscopy, plasma
      protein binding
      measurement 5:328
Fluorescent dyes
   cell-based assays see Cell-based assays
   2D-PAGE protein quantitation 3:35,
      3:35T
Fluorescent labeled samples, 2D-
      PAGE 3:30–31
Fluorescent labeling, GPCR structure
      determination 4:678–679
Fluorescent proteins, HCS labeling
      techniques 3:690
Fluorescent reagents, high-content
      screening (HCS) see High-
      content screening (HCS)
Fluorescent stains, two-dimensional gel
      electrophoresis 2:334
Fluoride, bone disease
      treatment 6:514
Fluorinated hydrocarbons, solution
      phase synthesis 2:485
Fluorine–hydrogen bioisosterism see
      Bioisosterism, fluorine–hydrogen
3'-Fluoro-2',3'-dideoxyguanosine
      (FDG), hepatitis B 7:319
2-[^{18}F]Fluoro-2-deoxy-D-glucose
      (FDG), positron emission
      tomography 3:77
16 alpha-[^{18}F]Fluoroestradiol, steroid
      hormone imaging 3:667
Fluorogenic methods, CYP
      inhibition 5:352
Fluorogenic probe substrates
   in vitro expression studies 5:246,
      5:247T
   in vitro inhibition studies 5:247T
Fluorogenic substrates, target-oriented
      screening 3:142
(S)-3-Fluoropyrrolidine,
      structure 2:720F
Fluoroquinolone antibiotics,
      mycobacterium
      treatment 7:715
Fluorosome 5:477, 5:478F
   Fluorosome-trans-ent 5:477
   in vivo relevance 5:477
   liposomes 5:477
   method 5:477
Fluorosome-trans-ent 5:477
5-Fluorouracil 7:61–62, 7:62T, 7:64
   see also CMF regimen
   [^{18}F]FDG imaging 3:669
   adverse events 7:61T
   approval 7:62T

bioavailability 5:712T
clinical indication 7:62T
   glaucoma treatment 6:308
formulations 7:68–69
   microspheres 7:68–69
   polysaccharide matrix 7:69
intracellular transformations 7:59–60
mechanism of action 3:7; 7:64T
pharmacokinetics 7:64
prodrug of 5:1015
structure 5:1016F; 6:310F
toxicity 7:46T, 7:64
xenograft models 7:39T
Fluorous chemistry, solution phase
      library synthesis see Solution
      phase synthesis, library
      preparation
Fluorous split–mix library synthesis,
      high throughput
      purification 3:868
(S)-5-Fluorowillardine 2:900
   structure 2:901F
Fluoxetine
   bioavailability 5:712T
   bioethics 1:699–700
   cataplexy treatment 6:154
   chemogenomics 3:934F, 3:935
   CYP 2C9 metabolism 5:545T
   defensive test battery 6:94
   development 1:695; 6:99
   dopamine transporter knockout
      mouse 6:124
   major depressive disorder
      treatment 6:63
   obstructive sleep apnea (OSA)
      treatment 6:162
   off-label prescribing 1:718
   side effects 6:60T
   structure 2:57F, 2:62F; 5:359; 6:100F
   transporter inhibition 6:66T
Fluoxymesterone
   pharmacokinetics 6:533T
   structure 6:522T
Flupenthixol 6:29–31
   bioavailability 5:712T
Fluphenazine 6:29–31
Flupirtine, structure 6:315F
Flurazepam, structure 6:99F
Flurbiprofen, bioavailability 5:712T
S-Flurbiprofen, structure 4:787F
Flurithromycin ethylsuccinate 1:113;
      7:540
   pharmacokinetics 7:534T
   respiratory infections 1:113
9-[(1S;2R)-2-Fluro-1-methylpropyl]-2-
      methoxy-6(1-piperazinyl)purine,
      glucuronic acid
      conjugation 5:151, 5:153F
Flusilazole, structure 2:695F
Flutamide
   characteristics 2:1012T
   orthomyxoviruses 7:276
   structure 2:1081F

Fluticasone
   atopic dermatitis treatment 7:974T
   structure 7:974T
   target information 4:750T
Flutrimazole 1:131
Fluvastatin sodium 1:172, 1:532;
      6:473–474, 6:482
   absorption 5:680T; 6:483
   discovery/development 1:375
   metabolism, CYP 2C9 5:545T
   potency 6:483
   SAR 6:482, 6:482F, 6:483F
   structure 1:376F; 6:469F
   target information 4:750T
Fluvoxamine 6:69, 6:99–100
   bioavailability 5:712T
   side effects 6:60T
   structure 6:100F
   transporter inhibition 6:66T
Flux balance analysis, metabolic
      networks 3:329
Flux prediction, Caco-2 cell models
      see Caco-2 cell models
Flynn database, dermal penetration
      models 5:727, 5:728T,
      5:732–733
FMLP receptor, activation 2:785T
Fmoc peptide synthesis
   peptide synthesis 2:475–476
   sheets 2:503, 2:503F
FMR1 gene, fragile X
      syndrome 1:269–270, 1:274–
      275
Focal adhesion kinase (FAK)
   cancer, invasive growth 7:15–16
   in disease 2:963T
Focalin see Dexmedetomidine
      hydrochloride;
      Dexmethylphenidate
Focal ischemic stroke
   experimental models 6:256,
      6:257T
   unilateral middle cerebral artery
      occlusion (MCAO) model 6:256
   permanent 6:256
   temporary 6:256
Focused individuals, innovation 2:262
Focused libraries, solution phase library
      synthesis 3:769
Focused library design, structure-based
      methods 4:328
   docking 4:329
      cytochrome P450 inhibitors 4:329
      factor VIIa inhibitors 4:329
      factor Xa inhibitors 4:329
      HOX-PBX/DNA transcription
         factor 4:329
      PRO_SELECT 4:329
      scaffold–linker–functional
         group 4:329
   genetic algorithms 4:329
      thymidine monophosphate kinase
         inhibitors 4:329

Focused library design, structure-based
  methods (*continued*)
  homology models 4:328
    S-adenosyl homocysteine/
      methylthioadenosine
      nucleosidase inhibitors 4:329
    Cdk4 inhibitors 4:328, 4:328*F*
    LEGEND program 4:328
    phospholipase A₂
      inhibitors 4:328–329
    P-selectin inhibitors 4:329
    vitamin D receptor
      inhibitors 4:328
  pharmacophore models 4:329
Focusing effects, protein
  crystallography 3:440
FoldIndex, domain mapping 3:879
Fold reversal, P-glycoprotein
  models 5:776
Folic acid metabolism analogs, skin
  inflammation treatment 7:976
Folic acid metabolism antagonists,
  psoriasis treatment 7:976,
  7:978*T*
Follicle-stimulating hormone (FSH)
  brands, global sales 1:605*T*
  male hypogonadism 6:523, 6:538
Follicle-stimulating hormone receptors
  constitutively active 2:790*T*
  in human disease 2:775*T*
Follicular hyperkeratinization,
  acne 7:936, 7:936*T*, 7:958,
  7:959*F*
Follitropin alpha 1:233
Follitropin beta 1:233
Follow-up screening
  materials management 2:535
  screening library selection 4:500
Folylpolyglutamate synthetase,
  antimetabolites 7:59
Fomepizole 1:244
Fondaparinux sodium 1:166
  introduction of 1:107
Fontolizumab, IBD treatment 6:623–
  624
Food allergies, irritable bowel syndrome
  see Irritable bowel syndrome
  (IBS)
Food and Drug Administration (FDA)
  approval 2:159–160
    5-azacytidine 1:273
    biological macromolecules 1:434
    Caco-2 cell permeability
      assays 5:269
    cilomilast 2:948
    stereoisomer definitions 1:51–52
  approval time 2:200
  Center for Drug Evaluation and
    Research (CDER) 1:546–547
  chemogenomics 2:567
  clinical trials phase I, regulatory
    requirements 2:155
  drug discovery, decline in 1:622–623

drug interactions 5:241
fast-track approval 2:200
foundation, postmarketing
  surveillance 1:756
grants for clinical trial
  expenses 1:660–661
Guidance for Industry for the
  Development of Incontinence
  Trials 8:128
guidelines
  inhibition, in vitro studies 5:245
  light 5:497–498
  obesity clinical trials 6:394,
    6:394*T*
  osteoporosis clinical trials 6:509
  stability testing 5:493–494,
    5:494*T*
  personalized medicine 1:280,
    1:289
  safety testing importance 5:3
  tamoxifen approval 8:87
    breast cancer prevention 8:96
Food and Drug Administration
  Modernization Act (1997)
  (FDAMA) 1:603
  postmarketing surveillance 1:760
  productivity effects 2:123–124
Food–drug interactions, BCS *see*
  Biopharmaceutical Classification
  Scheme (BCS)
Food intake, oral absorption 5:672
Forbes coefficient, similarity
  searching 4:179, 4:181
Forced expiratory lung volume (FEV₁),
  asthma 7:875
Force-field molecular modeling 4:88
  angle bending 4:90
    Hooke's law 4:90
  applications 4:98
  'atom type' 4:89
  bond stretching/compressing 4:89,
    4:90*F*
    Morse curve 4:89, 4:90*F*
  class I 4:92
  class II 4:92
  class III 4:92
  comparisons 4:98
    derivation 4:99
    missing parameters 4:99
    parameterization 4:99
  cross terms 4:91
  cutoffs 4:98
    Ewald summation 4:98
  definition 4:88, 4:89*F*
  dielectric constant 4:94
  docking, scoring functions *see* Docking,
    scoring functions
  electrostatic interactions 4:92
    charge distribution 4:92
    Gastiger–Marsili approach 4:93
    interaction energies/
      geometries 4:93
    large molecules 4:93

partial atomic charge
  calculation 4:93–94
  quantum mechanical theory 4:93
  empirical approach 4:89
  generalized Born model 4:95
    docking, scoring functions 4:483
    surface area combination 4:96
  GRID force field *see* Molecular
    Interaction Fields (MIFs)
  implicit solvation models 4:94
    Poisson–Boltzmann equation
      4:95
  molecular descriptors 5:584
  nonbonded interactions 4:92
  out-of-plane bending terms 4:91
    'cross terms' 4:92
    improper torsion 4:91–92, 4:91*F*
  Poisson–Boltzmann equation 4:96
  polarization 4:94
    condensed phases 4:94
    definition 4:94
    methods 4:94
  quantum mechanical
    calculations 4:88
  scoring/binding energy estimation
    drug design 4:295
  structure-based virtual
    screening 3:251
  torsion potentials 4:90
  'transferability' 4:89
  van der Waals' interactions 4:96,
    4:97*F*
    Buckingham potential 4:98
    Hill potential 4:98
    Lennard–Jones function 4:97
    London forces 4:97
    Lorentz–Berthelot mixing
      rules 4:98
Force fields, computer-assisted drug
  design (CADD) 4:21
Forcing, conflict management 2:248
Foreign national applicants,
  education 2:85–86
Forkhead-associated domains,
  phosphoserine/
  phosphothreonine-interacting
  domains 2:985
Formal project reviews, project
  management teams 2:148
Formestane 7:175
  breast cancer treatment 8:97
Form factor (*f*), anomalous
  dispersion 3:455*F*, 3:459
Form/formulation enhancement,
  biotechnology companies 2:232
Form–function–fluctuation, molecular
  fields *see* Molecular fields
Formic acid, microwave
  heating 3:840*T*
Formivirsen sodium 1:123; 2:761,
  2:1045–1046
  clinical trials 2:1045
  cytomegalovirus 7:260

cytomegalovirus retinitis   1:123
structure   2:1046*F*
Formononetin   1:452*T*
Formoterol
asthma treatment   7:883
labeling changes   1:758*T*
Formulation excipients, BCS   5:975
Formulation patents, raloxifene   8:108
Formulations
absorption effects   5:703
aminoglycosides *see* Aminoglycosides
DCN   2:155
drug metabolism and
pharmacokinetics *see* Drug
metabolism and
pharmacokinetics (DMPK)
predevelopment   2:141
tobramycin   7:644
vaginal administration   5:290
N-Formyl-N-Met-L-Leu-L-Phe-OH,
NMR structure
determination   3:490
Förster resonance energy transfer
(FRET) *see* Fluorescence
resonance energy transfer
(FRET)
Fortimicin A   7:632
structure   7:633*F*
Fortovase *see* Saquinavir
Forward genetics
chemical biology   3:130, 3:131*F*
Danio rerio model system *see* Zebrafish
(*Danio rerio*)
Drosophila melanogaster model
system   3:9
human cells   3:14
Saccharomyces cerevisiae model
system   3:5–6
Fosamax *see* Alendronate
Fosamprenavir   1:123; 7:279–282
clinical uses   7:358
discovery   7:352
dosing regimens   7:357*T*
with ritonavir   7:358*T*
structure   1:76*F*
Foscan *see* Temoporphin
Foscarnet   7:305
cytomegalovirus   7:260, 7:305
intravenous   7:305
mechanism of action   7:305
medicinal chemistry   7:306
metabolism   7:305
oral absorption   5:468*T*, 5:680*T*
resistance   7:306
Fosfomycin
classification by target site   7:471*T*
oral absorption   5:680*T*
Fosinopril
bioavailability   6:719*T*
discovery/development   1:375
dose   6:719*T*
half-life   6:719*T*
in vitro potency   6:719*T*

structure   1:378*F*; 6:719*T*
target information   4:750*T*
Fosmidomycins
see also specific drugs
antimalarial activity   1:371–372
malaria treatment   7:794
mechanism of action   1:371–372
oral absorption   5:680*T*
SAR   7:794–795
structure   1:371*F*
Fosphenytoin sodium   1:**187**
characteristics   2:829*T*
structure   2:829*T*
4D descriptors, molecular
descriptors   5:533
4D (dynamic) molecular fields   5:578,
5:578*F*, 5:579
4D-QSAR   5:597
3D-QSAR alignment   4:581,
4:581*F*
ADME applications   5:598
bioavailability   5:598
blood–brain permeation
5:598
cytochrome P450s   5:599
conformations   5:597
dimensional classification   5:579
pharmacophore detection   4:133
pharmacophore searching   4:135*T*
property space   5:598
chemical libraries   5:598
definitions   5:598
single compound   5:598
receptor-to-ligand adaptation
5:597–598
4SC   1:517
4S trial   6:474*T*
Four-plate test, anxiety disorders,
animal models   6:93
Four-point pharmacophores, ligand-
based library design   4:314
FPR1 receptors, in human
disease   2:775*T*
FPT inhibition, lonafarnib   8:30–31
FR-233118, structure   6:408*F*
FR-252384, structure   6:408*F*
FR-901228
discovery/development   1:386
structure   1:385*F*
FR-901379   7:457*T*
Fractal structure, effective intestinal
membrane permeability   5:465
Fraction absorbed, absorption
studies   5:41, 5:42*F*
Fractional factorial design, QSAR
substituent selection
models   4:52
Fractionated precipitation, protein
purification   3:421
Fraction of dose absorbed (Fa%)
definition   5:869*T*, 5:974
immobilized liposome
chromatography   5:479

partition assessment tools   5:480,
5:481
permeation theory *see* Permeation
theory
Fracture incidence,
osteoporosis   6:502–503, 8:199
Fragile X syndrome
5-aza-2′deoxycytidine   1:274–275
epigenetics   1:270*T*
faulty genomic imprinting   1:269–
270
FMR1 gene   1:269–270, 1:274–275
Fragmental descriptors, log P
calculation *see* Log P calculation
'Fragmental values', lipophilicity
prediction, rule-based
systems   5:650
Fragmentation, genome
sequencing   3:317
Fragment-based drug discovery   2:49,
2:51*F*; 3:**939–957**; 4:289, 4:439
aqueous solubility models *see*
Solubility models, aqueous
Bcl-xL, fragment linking   3:950,
3:951*F*
caspase-3   3:950
chemical space coverage   3:943
c-Src   3:953, 3:954*F*
fragment linking   3:953, 3:954*F*
substrate activity screening
(SAS)   3:953–954
definition   3:940, 3:940*F*
examples   3:949
follow-up strategies   3:941, 3:942*F*
fragment elaboration   3:941
fragment evolution   4:451
example   4:451, 4:452*F*
GPCR   4:451
protein crystallography, rational
drug design   3:893, 3:893*F*
PTP1b inhibitors   4:451, 4:453*F*
structural information   4:451,
4:452*T*
fragment linking   4:454, 4:454*F*
fragment optimization   4:455, 4:455*F*
fragment self-assembly   4:455,
4:455*F*
GPCR library screening   4:691
historical perspective   3:941
hits to leads   4:450
bioactivity   4:451
HTS *vs.*   3:941
iterations   3:940–941, 3:941*F*
LFA-1, reverse screening   3:952,
3:953*F*
ligand efficiency   3:943
'rule of five'   3:943–944
'rule-of-three'   3:943–944
ligand selection   4:296
clustering   4:296
filtering   4:296
presentation   4:296
searching   4:296

Fragment-based drug discovery
  (*continued*)
    storage **4**:296
    synthetic analyses **4**:296
  mass spectrometry **3**:947
    advantages **3**:947
    disadvantages **3**:947
  NMR spectroscopy **3**:944
    advantages **3**:946
    chemical shifts **3**:945
    competition experiments **3**:946
    enzyme function **3**:946
    example **3**:952, **3**:953*F*
    fragment linking **3**:950, **3**:951*F*
    ligand observation **3**:945, **3**:945*F*,
      3:946
    nuclear Overhauser effects
      (NOEs) **3**:946
    protein observation **3**:945,
      3:945*F*
    relaxation rates **3**:946
    reporter ligand
      observation **3**:945*F*, **3**:946
    two-dimensional **3**:945
  p38α MAP kinase
    fragment elaboration **3**:949,
      3:949*F*
    fragment mapping **3**:949*F*, **3**:950
  SAR **3**:941
  screening set design **4**:447
    analog availability **4**:448
    availability **4**:448
    chemical complexity **4**:449
    computational techniques **4**:449
    framework definition **4**:449,
      4:450*F*
    high concentration screening
      (HCS) **4**:449
    large *vs.* small sets **4**:448, **4**:448*F*
    purity **4**:448
    reduced complexity screening
      (RCS) **4**:449
    retrosynthetic combinatorial
      analysis procedure
      (RECAP) **4**:449
    solubility **4**:448
    synthesis **4**:448–449
  screening technologies *see* Fragment
    screening
  SH2 domain inhibitors **4**:631,
    4:635*F*
  small libraries **3**:942, **3**:943*F*
  structure determination **3**:940
  surface plasmon resonance **3**:947
    advantages **3**:948
    technique **3**:947–948
  technologies **3**:944
    see also specific technologies
  tethering **3**:947
    definition **3**:947
    example **3**:950
    site-directed screening **3**:947
    technique **3**:947

2D similarity searching **3**:246
'undruggable' targets **3**:944
weak hits **3**:941
whole-molecule properties, similarity
  searching **4**:169–170
x-ray crystallography **3**:946; **4**:447
  advantages **3**:947
  example **3**:949, **3**:949*F*, **3**:950
  history **3**:946–947
  protein–ligand complexes **3**:946
Fragment bitstrings
  similarity searching **3**:246
  structure/substructure
    searching **3**:238, **3**:238*F*
Fragment construction, ligand
  flexibility **4**:481, **4**:481*F*
Fragment/correction factors, log P
  calculation **5**:655, **5**:656*T*
Fragment elaboration, p38α MAP
  kinase **3**:949, **3**:949*F*
Fragment evolution *see* Fragment-based
  drug discovery
Fragment libraries, SH2 domain
  inhibitors **4**:633, **4**:635*F*
Fragment linking
  see also Divide-and-conquer drug
    design
  fragment-based drug
    discovery **3**:953, **3**:954*F*; **4**:454,
    4:454*F*
  NMR spectroscopy **3**:950, **3**:951*F*
  protein crystallography, rational drug
    design **3**:893, **3**:893*F*
Fragment mapping, p38α MAP
  kinase **3**:949*F*, **3**:950
Fragment molecular orbital method
  (FMO), quantum mechanical
  calculations **4**:414
Fragment optimization
  fragment-based drug
    discovery **4**:455, **4**:455*F*
  protein crystallography, rational drug
    design **3**:893, **3**:893*F*
Fragment replacement, scaffold
  hopping **2**:681–683
Fragment screening **3**:940, **3**:944;
  4:446
  advantages **4**:439, **4**:439*T*
  binding affinities **4**:443–444
  biochemical assays **4**:446
  crystallography **4**:447
  functional group classification **4**:445
  historical perspective **4**:439,
    4:440*T*
    hit properties *vs.* lead
      properties **4**:440
    interaction models **4**:440–441,
      4:440*F*, **4**:441*F*
    'probability of a useful
      event' **4**:441
    terminology **4**:442
  HTS **4**:446
  NMR **4**:447

protein tyrosine phosphatase-1B
  inhibitor design **4**:643, **4**:643*F*
surface plasmon resonance **4**:447
tags **4**:447
Fragment self-assembly
  fragment-based drug
    discovery **4**:455, **4**:455*F*
  protein crystallography, rational drug
    design **3**:893, **3**:893*F*
Fragment stitching, optimization/
  evolutionary drug design **4**:294
Fragment/subfragment searches, QSAR
  validation **4**:161, **4**:162*F*
'Fragnomics,' protein crystallography,
  rational drug design **3**:892
Framework definition, fragment-based
  drug discovery **4**:449, **4**:450*F*
Frameworks analysis, ligand-based
  library design **4**:314
Framingham Heart Study **6**:470, **6**:471
France
  medicine marketing system **1**:632
  orphan drugs legislation **1**:662
  traditional medicine **1**:421
Francoeur, Eric, molecular
  modeling **4**:67–68
Frank, Ronald, combinatorial
  chemistry **2**:461
Franke's model, QSAR **4**:51
Franklin, Rosalind, protein
  crystallography **3**:450
Franz diffusion cell, buccal
  administration
  measurement **5**:280, **5**:282*F*
Frataxin mutation, Friedreich's
  ataxia **2**:1059–1060
FRED74 program, MM-PBSA **4**:264
FRED program
  multiple active site corrections
    (MASC) **4**:271
  postDock processing **4**:272
Free/bound fractions,
  pharmacokinetics **5**:14–15
Free consent, human research
  ethics **1**:715
Freedox *see* Tirilazad mesylate
Free energy
  definition **4**:652
  estrogen receptor, molecular dynamics
    simulations **4**:664
  perturbation **4**:655*F*, **4**:656
Free energy calculation **4**:653
  empirical free energy function **4**:659
  fast growth **4**:657*F*, **4**:658
  free energy perturbation **4**:655*F*,
    4:656
  Gibbs free enthalpy **4**:653
  Helmholtz free energy **4**:653–654
  λ-dynamics **4**:657
  linear free integration method **4**:659
  MM-PBSA method **4**:659
    Poisson–Boltzmann
      equation **4**:659

one-step perturbation **4**:655*F*, **4**:657*F*, **4**:658
  Jarzynski equation **4**:658
  overlapping ensembles **4**:654, **4**:655*F*
  slow growth method **4**:656
    drawback **4**:657
  thermodynamic cycles **4**:654, **4**:655*F*
    compound–receptor binding **4**:654–655
    mutation free energies **4**:654–655
  thermodynamic integration **4**:656
Free fatty acids, stratum corneum **5**:291, **5**:460
Free radicals
  Friedreich's ataxia pathogenesis **2**:1065
  mitochondria *see* Mitochondria
  mitochondria complex II **2**:1065
  mitochondrial permeability transition (MPT) **2**:1061
Free radical scavengers/antioxidants, clinical trials **6**:259, **6**:260*F*
Free R factor monitoring, protein crystallography model generation **3**:464
Free valence, electrotopological indices (E-state index) **4**:551
Freezer storage stability, drug reference standard **2**:167–168, **2**:168*T*
Frequency domain acquisition, fluorescence lifetime assays **3**:605–606
Frequency selective nuclear magnetic resonance **2**:509
'Frequent hitters' *see* Promiscuous ligands, computational prediction
Fresh skin, dermal/transdermal administration **5**:292–293
Friedel's law, isomorphous replacement **3**:459
Friedman's paradox, bioisosterism **2**:692
Friedreich's ataxia
  frataxin mutation **2**:1059–1060
  free radical production **2**:1065
  mitochondria **2**:1054–1055
  ubiquinone (coenzyme Q) treatment **2**:1074
Frog embryo teratogenesis assay (FETAX) **1**:467*T*, **1**:475
Fropenam **1**:113
  respiratory infections **1**:114
Frova *see* Frovatriptan
Frovatriptan **1**:182
  migraine treatment **6**:373*T*
Frozen skin, dermal/transdermal administration **5**:293
FRR1 receptor, constitutively active **2**:790*T*
Fructose-1,6-bisphosphate aldolase, malaria **7**:785–786

Frusemide, development **1**:530
^{19}F screening, NMR, drug discovery **3**:908*T*, **3**:910*F*, **3**:911
FT1, xenograft models **7**:39*T*
FTC *see* Emtricitabine
F test, microarrays **3**:92, **3**:101–103
FT (feature) lines, UniProtKB flatfile **3**:362, **3**:362*F*
FTREES software, fingerprint/dataprint descriptors **4**:472
FTY-210, multiple sclerosis treatment **7**:929
FTY-720
  IBD treatment **6**:636
  phosphate conjugation **5**:150, **5**:154*F*
  structure **7**:980*T*
  transplant rejection **7**:921
Fudosteine **1**:210
Fulcrum Pharmaceuticals **1**:504
FullDock algorithm, QXP (Quick Explore) program **4**:259
Full-factorial design, ligand-based library design **4**:310
Full file screen, screening library selection **4**:507
Full-text databases *see* Literature databases
Full-text searching, target databases **3**:300
Fully integrated pharmaceutical company (FIPCO) **2**:100
Fulstan *see* Falecalcitriol
Fulvestrant **1**:138
  breast cancer **1**:138
  as estrogen receptor ligand **2**:1014
  structure **2**:1013*F*
Fulvestrant, breast cancer treatment **8**:97*F*
Fumaric acid, structure **5**:369*F*
'Functional analogs,' bioisosterism *see* Bioisosterism
Functional analysis, protein sequence databases **3**:357
Functional areas
  project management **2**:149
  project team goals *see* Project teams
Functional Connectivity Fingerprints (FCFPs)
  2D similarity searching **3**:246
  two-dimensional descriptors, similarity searching **4**:171
Functional departments, project teams *vs.* **2**:145
Functional discipline organization *see* Pharmaceutical industry
Functional endpoints, genetically engineered animals **3**:155–156
Functional foods, definition **1**:408
Functional genomics **2**:319
  definition **1**:535; **2**:382*T*
  gene expression **2**:326, **2**:327*F*

autoregulation **2**:326
    techniques **2**:326
  molecular pathways **2**:324
    protein–protein interactions **2**:325
    signal transduction **2**:324
  signal transduction **2**:960, **2**:963*T*
  target identification **2**:308, **2**:329
  techniques **2**:326
    deletion detection **2**:329
    expression monitoring **2**:329
    HTS **2**:328
    hybridization differential display **2**:329
    microarrays **2**:328
Functional groups
  bioisosterism *see* Bioisosterism
  dynamic combinatorial chemistry **3**:961, **3**:961*F*
  fragment screening **4**:445
Functional imaging, biomarkers *see* Biomarkers
Functionalized polystyrene resins, solution phase synthesis **2**:485–486, **2**:487*F*
Functionalized pyrazole synthesis, PASP **3**:797, **3**:798*F*
Functional magnetic resonance imaging (fMRI)
  anxiety disorders **6**:89
  somnolence measures/sleep testing **6**:142
Functional proteomics
  definition **2**:381, **2**:382*T*
  gene-related **2**:389
Functional public health systems, developing countries **1**:641, **1**:642*T*
Function-based target validation **3**:3–4
Funding, SMEs *see* SMEs (small/medium-sized enterprises)
Fungal cell wall disruptors **7**:456
  see also specific drugs
Fungal cell walls **7**:433
  synthesis inhibitors *see* Antifungal drugs
Fungal disease **7**:419–443
  see also specific diseases/disorders
  classification **7**:420
    International Code of Biological Nomenclature **7**:420
  cutaneous mycoses **7**:420, **7**:423*T*
  diagnosis **7**:427
    culture **7**:428
    Fungitell assay **7**:428
    galactomannan detection **7**:428–429
    gene sequences **7**:429
    β-glucan detection **7**:428
    immune-compromised individuals **7**:427–428
    laboratory diagnosis **7**:427–428
    metabolite detection **7**:428
    microscopy **7**:428

Fungal disease (*continued*)
  dimorphic systemic mycoses   **7**:423, **7**:424*T*
    amphotericin B   **7**:423
    treatment   **7**:429
    voriconazole   **7**:423
  emerging   **7**:425
    antifungal resistance   **7**:425–426
  host defenses   **7**:427
    immunocompromised individuals   **7**:427
    skin barrier   **7**:427
    Toll-like receptor signaling   **7**:427
  opportunistic   **7**:424, **7**:425*F*, **7**:425*T*
  organ transplant recipients   **7**:425–426
  pathogenesis   **7**:425
  structure   **7**:420, **7**:421*F*
  subcutaneous mycoses   **7**:420, **7**:422*T*
    amphotericin B   **7**:420
  superficial mycoses   **7**:420, **7**:421*T*
    antifungal drugs   **7**:429
    terbinafine   **7**:429–431
    treatment   **7**:429
Fungi, mitochondria complex I   **2**:1070
Fungicidin, medicinal chemistry   **7**:456–457
Fungitell assay, fungal disease diagnosis   **7**:428
Funguard *see* Micafungin
Funiculosin, structure   **2**:1076*F*
Furafylline, reaction phenotyping studies   **5**:243*T*
Furan-4-ones
  solid-phase discrete libraries   **3**:709*T*
  solid-phase libraries   **3**:720–721, **3**:720*F*
Furans
  immunosuppression   **5**:223
  triphenylethylene anticancer agents   **7**:158–159
Furazano[3,4-*b*]pyrazines, solid-phase discrete libraries   **3**:709*T*
Furchgott, Robert, sildenafil citrate, invention of   **2**:257
Furo[3,2-b] pyridine, bioisosterism   **2**:660*T*
Furosemide   **6**:708
  BCS   **5**:973*T*
  bioavailability   **5**:712*T*; **6**:709*T*
  dose   **6**:709*T*
  effective permeability (P_{eff})   **5**:673*T*
  half-life   **6**:709*T*
  oral absorption   **5**:468*T*, **5**:680*T*
  paracellular transport   **5**:265
  structure   **5**:371*F*; **6**:586*F*, **6**:709*T*
Furoxane, bioisosterism   **2**:660*T*
Fused bicyclic lactams, solid-phase discrete libraries   **3**:709*T*
Fusicid acid, classification   **7**:471*T*
FusionPro, process variable statistical evaluation   **2**:164

Fusion proteins, protein crystallization   **3**:436–437
Fusion rules, similarity measure combination   **4**:182
Fuzeon *see* Enfuvirtide
Fuzzy adaptive least squares (FALS) analysis, bioavailability models   **5**:710
Fuzzy coordinates, bioinformatics   **1**:42–43
Fuzzy interference models, dermal penetration models   **5**:732
'Fuzzy pharmacophores,' biological fingerprints   **4**:812–813
FY-901739, medicinal chemistry   **7**:460–461

G

G1/S transition *see* Cell cycle
G2A receptor, activation   **2**:785*T*
G2/M transition *see* Cell cycle
$G_{12/13}$ subfamily, G proteins   **2**:780
G72, schizophrenia   **6**:24
^{67}Ga
  biomolecular probes   **3**:666
  SPECT   **3**:663*T*
^{68}Ga
  biomolecular probes   **3**:666
  PET   **3**:664*T*
$G\alpha_2$ subfamily, G proteins   **2**:782
$G\alpha_3$ subfamily *see* G proteins
$G\alpha_{12}$ subfamily, G proteins   **2**:783–784, **2**:783*F*, **2**:785*T*
GABA
  analgesia   **6**:339
  anxiety disorders *see* Anxiety disorders, genetic basis
  gabapentin mechanism of action   **8**:240
  gabapentin mimetic   **8**:228
  modafinil   **8**:152
  sleep   **6**:146
  sleep promotion   **6**:143
  structure   **2**:661*F*
GABA receptors
  see also specific receptors
  alphaxalone interactions   **6**:357
  anesthetics   **6**:357
  barbiturate interactions   **6**:357
  benzodiazepine interactions   **6**:357
  betaxalone interactions   **6**:357
  isoflurane interactions   **6**:357
  ketamine interactions   **6**:357
  pentobarbital interactions   **6**:357
GABA receptor antagonists
  see also specific drugs
  Alzheimer's disease treatment   **6**:206
GABA$_A$ receptor   **2**:891
  benzodiazepine binding site ligands   **4**:531*F*
  characteristics   **2**:881*T*

CNS disorders   **5**:896
  diseases   **2**:894
  as drug target   **5**:896
  heterodimerization   **2**:799
  mechanism of action   **5**:895–896
  molecular biology   **2**:891
  nomenclature   **2**:891
  pharmacology   **2**:892, **2**:893*F*
  physiological functions   **2**:892
    synaptic transmission   **2**:892
  regulation   **2**:892
  structure   **2**:891; **5**:895–896; **6**:98, **6**:98*F*
  therapeutics   **2**:893
    alphadolone   **2**:893
    benzodiazepines   **2**:893
    benzoxazines   **2**:893
    ganaxolone   **2**:893
    thiazoles   **2**:893
GABA$_A$ receptor agonists   **5**:895
  see also specific drugs
  benzodiazepines   **5**:896, **5**:897, **5**:897*F*
  comparative methods   **5**:896
  extrapolation to humans   **5**:898
  intrinsic efficacy   **5**:897, **5**:898*F*
  neuroactive steroids   **5**:897
  primary insomnia treatment   **6**:160
  quantitative EEG   **5**:896
  tiagabine   **5**:896
GABA$_A$ receptor antagonists
  see also specific drugs
  primary insomnia treatment   **6**:158, **6**:160
GABA$_A$ receptor ligands
  see also specific drugs
  scaffold hopping   **2**:681, **2**:682*F*
  subtype-specific
    anxiety treatment   **6**:103
    benzodiazepine site ligands   **6**:103
GABA$_B$ receptor
  allosteric modulators   **2**:790*T*
  constitutively active   **2**:790*T*
  in human disease   **2**:775*T*
  schizophrenia   **6**:24
  structural modeling   **4**:684
GABA$_B$ receptor agonists
  see also specific drugs
  antiemetic drugs   **6**:687
  prokinetic drugs   **6**:686
GABA$_B$ receptor antagonists
  see also specific drugs
  affective disorders   **6**:79
  carboxylic acid bioisosterism   **2**:661
GABA$_C$ receptor   **2**:891
  characteristics   **2**:881*T*
  diseases   **2**:894
  molecular biology   **2**:891
  nomenclature   **2**:891
  pharmacology   **2**:892, **2**:893*F*
  physiological functions   **2**:892
    synaptic transmission   **2**:892
  regulation   **2**:892

structure 2:891
therapeutics 2:893
benzodiazepines 2:893
benzoxazines 2:893
thiazoles 2:893
GABA hypothesis, schizophrenia 6:21
GABA transporters 2:60T
Gabapentin 6:108–6:109, 6:289;
8:227–246
see also Pregabalin
aliphatic side-chain
replacements 8:237
as analgesic 6:339
differential efficacy 6:337T
anticonvulsant properties 8:228–229
bipolar affective disorder
treatment 6:72
clinical development 8:238
clinical studies 8:238
cocaine abuse 6:184T
derivatives 8:231
aminobutyric acid motif 8:232–
233, 8:234T
binding conformations 8:231,
8:231F
cyclohexyl ring
substitutions 8:233, 8:233T
NMR 8:231–232, 8:232T
stereochemistry 8:232, 8:233T
stereoselectivity 8:231, 8:232T
discovery 8:228, 8:228T
early clinical work 8:227
epilepsy treatment 8:239
GABA as mimetic 8:228
mechanism of action 2:851–852;
8:240
$\alpha_2\delta$ binding site 8:240
$\alpha_2\delta$ binding site mutation
studies 8:241
calcium channel function 8:240
GABA concentration 8:240
neurotransmitter release 8:241
non-drug binding site
effects 8:241
neuropathic pain treatment 8:239
properties 8:229, 8:229T
ring system reorganization
(bioisosterism) 2:676, 2:678F
SAR 8:228, 8:229, 8:231, 8:231T
collective analysis 8:230, 8:230T
structure 6:108F, 6:181F, 6:282F,
6:336F
Gabapentinoids, anxiety
treatment 6:108
Gabitril see Tiagabine
Gaboxadol
primary insomnia treatment 6:159T,
6:160
structure 2:893F
GAD67 knockout models, anxiety
disorders, genetic basis 6:90
Gadoversetamide 1:247
GAERS, epilepsy, models 6:287

GAF domains, phosphodiesterase-5
2:927
'Gain of function' mutants
androgen receptor (AR) 2:1027
chemical mutagenesis 3:13–14
'Gain-of-function' screens, cDNA
functional screening 3:15,
3:15F
$G\alpha_{i/o}$ subfamily, G proteins 2:781F,
2:782
$G\alpha_i$ subfamily, G proteins 2:782
Galactomannan detection, fungal
disease diagnosis 7:428–429
β-Galactosidase, genetically engineered
animals 3:153–154
GALAHAD, pharmacophore
searching 4:129T
Galanin
affective disorders 6:77
anxiety disorders 6:90
sleep 6:147
sleep promotion 6:143
Galanin GA2 receptor,
activation 2:785T
Galantamine (galanthamine) 1:363
Alzheimer's disease
treatment 6:202, 6:203
cytochrome P450 metabolism 6:203
dementia with Lewy bodies
treatment 6:36
modification 1:366
nicotinic acetylcholine receptor
binding 2:887–888
solid-phase pool libraries 3:707F
structure 1:362F
Galanthamine analogs, solid-phase
libraries 3:745, 3:745F
Galapagos 1:516
Galen, natural products 1:356
Gallamine 6:748
Gallic acid, pharmacokinetics 1:454T
Gallopamil, oral absorption 5:680T
GALOPED, product-based library
design 4:350
GAMBIT, genomics, antibacterial target
identification 7:741
GAMBLER, consensus scoring 4:486
Gambling 6:9
Gametocytes, malaria 7:766
GAMMA, pharmacophore
searching 4:129T
Gamma amino acids, proteolysis
resistance 2:584
Gamma-aminobutyric acid (GABA) see
GABA
Gamma distribution, adverse drug event
frequency estimation 1:767
Gammaglobulins, plasma protein
binding 5:322
γ-scintigraphy, drug
development 2:192
γ-turn mimetics,
peptidomimetics 2:608

Gamma-herpesviruses, antiviral
drugs 7:254T, 7:261
Gamma interferon see Interferon-γ
Ganaxolone, GABA$_A$ receptors 2:893
Ganciclovir 7:301
[^{18}F]FDG imaging 3:669
cytomegalovirus 7:260, 7:301
mechanism of action 7:301–302
oral absorption 5:468T, 5:680T
pharmacokinetics 7:303
resistance 7:303
structure 7:302
Ganellin, C R 8:41–52
collaborators 8:44
Black, James 8:42–43
Grant, Ted 8:45
Jack, David 8:42
Prout, Keith 8:44
Richards, Graham 8:44
Schwartz, Jean-Charles 8:48
Spickett, Geoffrey 8:42
early life 8:41
Medicinal Chemistry Summer
School 8:45
research work 8:45
burimamide 8:42–43
cholecystokinin 8:48
diamino-nitroethene 8:44
dynamic SAR 8:43
histamine H$_2$ receptors 8:42–43
histamine H$_3$ receptors 8:49
histamine receptor
antagonists 8:45–46
icotidine 8:45–46
metiamide 8:42–43
potassium ion channels 8:47
SAR 8:43
temelastine 8:46
at SmithKline and French 8:42
University College London 8:47
Gangliosides, Guillain–Barré
syndrome 6:235
Ganguly, Ashit 8:17–40
see also Ezetimibe
on biotechnology 8:18
on new technology 8:18
on protein structure
determination 8:18
Ganirelix acetate 1:233; 2:579T
hydrophilic depoting 2:588
self-forming depots 2:593
$G\alpha_o$ subfamily, G proteins 2:782
Gap interpretation, homology-based
protein structure
prediction 3:326
Gap junction activators, antiarrhythmic
drugs new research areas 6:755
Gap junction intercellular
communication (GJIC) assay,
non-animal carcinogenicity
testing 1:478
$G\alpha_q$ subfamily, G proteins 2:782–783,
2:783F

GAR-936 *see* Tigecycline
Garenoxacin **7**:588–589
    antimicrobial characteristics **7**:716*T*
    structure **7**:716*F*
Garnier, Jean-Pierre, carvedilol **8**:138
Garrod, Archibald,
        pharmacogenetics **3**:58
Gas chromatography-mass spectrometry
        (GC-MS) **5**:993
    chemical ionization **5**:993–994,
        5:994*F*
    electron impact ionization **5**:993–
        994, **5**:993*F*
    time-of-flight (ToF)
        detectors **5**:993–994
    toxicity studies **5**:1001
GASP, pharmacophore
        searching **4**:129*T*
GASP-1/2, GPCR interaction **2**:797*T*
Gastiger–Marsili approach, force-field
        molecular modeling **4**:93
Gastric cancer
    Helicobacter pylori infection **6**:607
    marimastat **7**:243
Gastric H$^+$/K$^+$-ATPase
    proton pump inhibitor
        association **6**:607, **6**:609*F*
    selective inhibitors **6**:610–611,
        6:610*F*
    structure **6**:609*F*
Gastric/mucosal ulcers **6**:603–612
    see also Peptic ulcer disease (PUD)
    clinical trials **6**:604
    disease basis **6**:604
    disease state **6**:603
    Helicobacter pylori infection **6**:598,
        6:607
    models **6**:604
    treatment **6**:604
        see also specific drugs
        antacids **6**:604
        anticholinergic agents **6**:604
        atropine **6**:605
        new research areas **6**:610
        unmet medical needs **6**:609
Gastrin-releasing peptide (GRP), in
        obesity **6**:410
Gastrin-releasing peptide receptor
    activation **2**:785*T*
    constitutively active **2**:790*T*
Gastrin-releasing peptide receptor
        antagonists
    see also specific drugs
    nonpeptide drugs **2**:618–619, **2**:620*F*
Gastritis
    Helicobacter pylori infection **6**:607
    itopride hydrochloride **1**:203
Gastroesophageal reflux disease
        (GERD) **6**:603, **6**:672
    Barrett's esophagus **6**:603–604
    diabetes mellitus type 1 **6**:673
    disease basis **6**:673–674
    esophageal cancer **6**:603–604

etiology **1**:439
    knockout mouse model **3**:13
    prevalence **6**:598, **6**:603–604
    treatment **6**:604
        antacids **1**:439
        cisapride **6**:673–674
        COX-1 inhibitors **6**:603–604
        COX-2 inhibitors **6**:603–604
        esomeprazole magnesium **1**:199
        histamine H$_2$ receptor
            antagonists **6**:603–604
        histamine receptor
            antagonists **1**:439
        market size **6**:598
        pantoprazole sodium **1**:200
        prokinetic agents **6**:673–674
        see also Prokinetic drugs
        proton pump inhibitors **6**:603–604
Gastrointestinal diseases/
        disorders **6**:597–601
see also specific diseases/disorders
    etiology **6**:597
    historical overview **6**:597
    market size **6**:598, **6**:599*T*, **6**:600*T*
Gastrointestinal dysmotility, irritable
        bowel syndrome **6**:648
Gastrointestinal motility **6**:671–691
    see also Antiemetic drugs; Emesis/
        nausea; Gastroesophageal reflux
        disease (GERD); Prokinetic
        drugs
    clinical trials **6**:676
        magnetic resonance imaging **6**:676
        radioisotopes **6**:676
        ultrasound **6**:676
    disease basis **6**:672
    disease states **6**:672
    5-HT **6**:646, **6**:646*T*
        peristalsis **6**:646–647, **6**:647*F*
    future work **6**:688
    models **6**:675
        Caenorhabditis elegans **6**:675
        Danio rerio **6**:675
        dog **6**:675
        high-fat meals **6**:675
        octanoate metabolism **6**:675
    new research areas **6**:685
    treatment **6**:677
    unmet medical needs **6**:685
Gastrointestinal sensorimotor
        modulators, market size **6**:599*T*
Gastrointestinal stability, stability
        testing **5**:499
Gastrointestinal stromal tumors,
        KIT **2**:972
Gastrointestinal tract
    absorption *see* Intestinal absorption
    anatomy **6**:597
        musculature **6**:597
    cytochrome P450s **5**:171, **5**:172,
        5:172*T*
    diseases/disorders *see* Gastrointestinal
        diseases/disorders

drug introductions *see* Drug
        introductions
drug metabolism **5**:237, **5**:252–253
    pharmacokinetics and **5**:33
5-HT **6**:646
    secretion **6**:646
functions **6**:597
length, absorption studies **5**:40–41
metabolism, midazolam **5**:237
neurophysiology **6**:645
    acetylcholine **6**:645–646
    calcitonin gene-related
        peptide **6**:645–646
    enteric nervous system **6**:645–646
    extrinsic innervation **6**:645–646
pH **5**:458
physiology, absorption studies **5**:40
secretions, absorption studies **5**:41
tissue composition **5**:45*T*
transit time **5**:458
wall metabolism, physiologically based
        pharmacokinetics
        (PBPK) **5**:880
Gastrointestinal tract drugs
see also specific drugs
    introductions **1**:106*T*, **1**:107–108,
        1:198
        antiulcerative agents **1**:198
        gastroprokinetic agents **1**:202
        irritable bowel syndrome **1**:107–
            108, **1**:201
        ulcerative colitis **1**:203
GastroPlus **5**:679*T*
    compound prioritization **5**:880,
        5:880*F*
    metabolic drug–drug interaction
        software **5**:841
    oral absorption models **5**:872, **5**:873*F*
    physiologically based
        pharmacokinetics **5**:873–874,
        5:877, **5**:878*F*, **5**:879, **5**:879*F*
Gastroprokinetic drugs
see also specific drugs
    introductions **1**:202
    market **6**:600, **6**:600*T*
Gateway System, protein 3D structure,
        protein production **3**:414
Gatifloxacin **1**:114; **7**:567, **7**:568–569,
        7:571*T*
    antimicrobial characteristics **7**:716*T*
    clearance **7**:587
    discovery/development **7**:569,
        7:583–584, **7**:584*T*
    mycobacterium treatment **7**:715,
        7:717
    ocular infection treatment **6**:317
    resistance, efflux pumps **7**:586
    respiratory infections **1**:114
    structure **7**:572*F*, **7**:584*T*, **7**:715*F*,
        7:716*F*
    urinary tract infections **1**:114
Gating mechanism, inward rectifying
        K$^+$ ion channels **4**:715, **4**:716*F*

Goal awareness, career guidelines **2**:19
GOLD
  promiscuous ligands, computational prediction **2**:749
  protein–ligand docking programs **3**:406
Gold antirheumatics, immunotoxicity **5**:221*T*
Goldblatt technique, hypertension **6**:707
GOLD program **4**:256
  constraints **4**:271
  cytochromes P450 inhibitors **4**:784–785
  docking **4**:257, **4**:482–483
  genetic algorithms **4**:257
  homology models **4**:487–488
  ligand–receptor interaction models **4**:206–207
  multiple active site corrections (MASC) **4**:271
  quantum mechanics **4**:411
  regression-based **4**:484
  stochastic approaches **4**:480
  targeted libraries **4**:792
  water molecules **4**:270
GOLPE software **5**:581*T*
  molecular descriptors **5**:585
  monoamine oxidase **5**:593
Gombar–Polli rule, P-glycoprotein models **5**:780
Gonadotropin-releasing hormone (GnRH)
  analogs **2**:594–595
  male hypogonadism **6**:523
Gonadotropin-releasing hormone receptor agonists
  breast cancer treatment **8**:97*F*
  hydrophilic depoting **2**:587
Gonadotropin-releasing hormone receptor antagonist, nonpeptide drugs **2**:618–619, **2**:620*F*
Gonadotropin-releasing hormone (GnRH) receptor
  ligands **7**:175, **7**:176*F*
  imidazolopyrimidone series **7**:176
  pyrazolopyrimidone series **7**:176
  SAR **7**:175–176
  small molecule **7**:176, **7**:177*F*
  uracil series **7**:176
Gonadotropin-releasing hormone (GnRH) receptors, in human disease **2**:775*T*
Gonal-F *see* Follitropin alpha
Goodford, Peter, molecular modeling **4**:68
Good Laboratory Practice (GLP), drug development **2**:27
Good Manufacturing Practice (GMP)
  drug development **2**:27, **2**:177
  traditional medicine **1**:418
Gordon–Taylor equation, glass transitions **5**:512

Goserelin **2**:579*T*
Gossypol
  malaria treatment **7**:787
  SAR **7**:787
Gould–Jacobs synthesis, quinolone antibiotics **7**:574, **7**:574*F*
Governments
  developing countries **1**:548
  as purchasers **1**:548
  regulation **1**:527–528, **1**:546
  relationship with industry **1**:545
Governments, in drug discovery **1**:**725–753**; **2**:271
  economic development/ competitiveness outcomes **1**:750
    patents as measure **1**:750, **1**:751*F*
  economic market failure arguments **1**:727
    externalities **1**:727, **1**:728
    indivisibility **1**:727
    price setting **1**:728
    uncertainty **1**:727, **1**:728
  as end user **1**:746
    pricing structure **1**:747
    public healthcare provision **1**:747
  expenditure **1**:736
    actual levels **1**:738, **1**:739*F*
    by country **1**:732
    developed countries **1**:737–738, **1**:737*F*
    developing countries **1**:738, **1**:738*F*, **1**:740, **1**:741*F*
    EU **1**:738
    Japan **1**:738
    R&D trends **1**:736
    trends in **1**:738, **1**:740*F*
    US **1**:726, **1**:738, **1**:740*F*
  innovation role **1**:726, **1**:727
    basic science **1**:732
    biosafety role **1**:731–732
    generic drug markets **1**:731–732
    intellectual property protection **1**:731–732, **1**:732*F*
  market regulation **1**:729
    antitrust liability **1**:730
    intellectual property rights **1**:729–730
    open access **1**:730
  motivations **1**:731
  policies/programs **1**:733, **1**:742
    biotechnology **1**:742
    competitive grants **1**:744
    private sector incentives **1**:733
  R&D outcomes **1**:747, **1**:748*T*
  research enterprise/knowledge outcomes **1**:747, **1**:748*F*
    countries absorptive capacity **1**:748
  research priorities **1**:734
    competing disciplines **1**:734–735
    by country **1**:735

    new diseases **1**:736
    US **1**:735, **1**:735*F*
    roles **1**:732
  socially desirable outcomes **1**:727
  societal health outcomes **1**:748
    link **1**:749
    research success *vs.* **1**:749–750
  supply-side support **1**:728
    direct public funding **1**:728–729
    educational support **1**:729
    small innovative entrepreneurs **1**:729
    technology adoption **1**:730, **1**:735*T*
    venture capital **1**:729, **1**:730*F*
GP IIb/IIIa receptors, thrombus formation **6**:764, **6**:775
GP IIb/IIIa receptor antagonists, thrombolytic drugs **6**:775
GP-1650, IBD treatment **6**:628–629
GPC Biotech **1**:512
GPCR Annotator **3**:309
GPCRDB, biological information databases **3**:305
GPCR Ligand Database **3**:309
GPI-1046, Parkinson's disease treatment **6**:217–218
GPI-1485, Parkinson's disease treatment **6**:217–218
GPI-5693, schizophrenia treatment **6**:38–39
GPR3 receptor, constitutively active **2**:790*T*
GPR6 receptor, constitutively active **2**:790*T*
*GPR10 receptors, in human disease **2**:775*T*
GPR12 receptor, constitutively active **2**:790*T*
*GPR50 receptors, in human disease **2**:775*T*
*GPR54 receptors, in human disease **2**:775*T*
*GPR56 receptors, in human disease **2**:775*T*
*GPR154 receptors, in human disease **2**:775*T*
G protein-coupled receptor interacting proteins (GIPs) **2**:796, **2**:797*T*
G protein-coupled receptors (GPCRs) **2**:**771–826**
  see also G proteins; *specific receptors*
  allosteric modulation **2**:802, **2**:803*T*
    orthosteric site **2**:802
    signaling properties **2**:804–805
  amines, structure **2**:575*F*
  assays *see* G protein-coupled receptors (GPCRs), assays
  biological fingerprints **4**:813–815
  biological functions **3**:229–231
    calcium channel regulation **2**:848
  cell-free expression systems **3**:122–123
  chemogenomics **3**:923, **3**:924*F*

G protein-coupled receptors (GPCRs)
(*continued*)
class A (rhodopsin family) 2:575–
576, 2:777–778, 2:778*F*; 4:671
activation 2:796
see also Rhodopsin receptor
class B (secretin/glucagon
family) 2:575–576, 2:778; 4:671
large N-terminal GPCRs 2:779
modeling 4:683
class C 2:779; 4:671–672
modeling 4:684
transmembrane domains 4:684
classification 2:575–576; 4:670
see also specific classes
computer-assisted drug design
(CADD) 4:19
constitutively active 2:789, 2:790*T*
in human disease 2:789, 2:792*T*,
2:793*T*
inverse agonist
identification 2:789
definition 2:574–575
desensitization 2:800, 2:801*F*
agonist efficacy *vs.* 2:801
beta-arrestin 2:801
conformational changes 2:800
GRKs 2:802
internalization 2:800–801
phosphorylation 2:800, 2:801
recycling 2:802
discovery 1:329–330
disease relevance 2:772
multiple sclerosis treatment 7:929
drug design *see* G protein-coupled
receptors, drug design
drugs targeting 2:772, 2:773*T*
see also specific drugs
eicosanoids 2:575*F*
families 2:777
see also specific families
fragment-based drug discovery
4:451
genetic studies 2:772, 2:775*T*
gain-of-function mutations 2:772
glycoprotein hormones 2:575*F*
library design *see* Library design
library screening *see* G protein-coupled
receptors, library screening
ligands 2:576
chemogenomics 3:932
NMR structure
determination 3:492, 3:494*F*
physicochemical
properties 4:424*T*
MAP kinase activation 2:784
beta-arrestin 2:788
ERK1/2 2:784–787, 2:787*F*
human disease 2:784
protein kinase A activation 2:787
protein kinase C activation
2:787
Ras activation 2:787, 2:788*F*

mechanism of action 2:574–575
heterotrimeric G protein (GTP-
binding protein; Gαβγ)
2:574–575
members 3:932
models *see* G protein-coupled
receptors, models
neurotransmitters 2:575*F*
nucleosides 2:575*F*
oligomerization/
heterodimerization 2:798
bioluminescence energy transfer
(BRET) 2:798–799
drug discovery relevance 2:800
dynamic dimerization 2:799
expression/trafficking
regulation 2:799
fluorescence energy transfer
(FRET) 2:798–799
internalization 2:799
signaling 2:800
orthogonal ligand–receptor pairs *see*
Orthogonal ligand–receptor pairs
PDB 3:374*F*, 3:376
peptide hormones 2:575*F*
pharmacological target space
4:422*T*
promiscuous ligands, computational
prediction 2:748
protease activated receptor 2:575*F*
research, Timmerman, Henk 8:9
Saccharomyces cerevisiae expression
system 3:8
signaling molecules 2:796
GPCR interacting proteins
(GIPs) 2:796, 2:797*T*
signal transduction 2:574–575;
4:670
regulation 2:788
signal transduction, HCS receptor
internalization 3:690–691
SMEs 1:515
stimuli 4:670, 4:672*F*
structure 2:574–575, 2:575*F*, 2:774,
2:777*F*; 3:634, 3:636*F*
amino acid sequences 4:670
conserved domains 4:671
C-terminus 2:774–777
NpxxY motif 2:774–777
N-terminus 2:774–777
(D/E)RY motif 2:774–777
three-dimensional structural
models 2:623
transmembrane helices 2:774–777
structure determination 4:670,
4:671*F*, 4:675
diffraction methods 4:675
engineered metal binding
sites 4:680, 4:681*F*
fluorescence resonance energy
transfer 4:679–680
fluorescent labeling 4:678–679
NMR studies 4:675

nonsense stop codon suppression
techniques 4:679–680
SAR 4:676
site-directed mutagenesis 4:676
spin labeling 4:678
substituted cysteine accessibility
model (SCAM) 4:677, 4:677*F*
surface plasmon resonance 3:593
Xenopus laevis 3:649
G protein-coupled receptors, drug
design 2:70, 2:300, 2:757*T*,
2:758, 2:805; 3:3; 4:669–701
see also specific drugs
library design *see* Library design
library screening *see* G protein-coupled
receptors, library screening
ligand-based design 4:695
nonpeptide drugs 2:619–621, 2:623
peptide/protein drugs *vs.* 2:623
peptide/protein drugs 2:574, 2:580–
581
nonpeptide drugs *vs.* 2:623
subtype selectivity 2:575–576
peptidomimetics 2:595–596, 2:605,
2:623
structure-based 4:685
high-throughput docking 4:686
lack of structural
information 4:686
3D-QSAR 4:687
targeted libraries 4:315–316
target validation 4:814
therapeutic potential 2:772
virtual screening *see* G protein-coupled
receptors, models
G protein-coupled receptors, library
screening 4:690
one dimensional methods 4:690
property-based approach 4:690
three-dimensional methods 4:693
beta-turn peptidomimetics 4:693
cross-family screening 4:694
de novo techniques 4:693
microdomains 4:694
peptidomimetics 4:693
pharmacophores 4:693
receptor- based 4:694
two-dimensional methods 4:690
BCUT (Burden, CAS and
University of Texas) 4:692
fragment-based 4:691
pharmacophores 4:691
privileged structure
approach 4:690
recursive combinatorial analysis
procedure (RECAP)
algorithm 4:691
similarity/knowledge-based
approaches 4:691
support vector machine
technologies (SVMs) 4:692
target selectivity 4:691
topomers 4:692

G protein-coupled receptors,
   models **4**:680
   activation models **2**:789
      'cubic ternary complex
         model' **2**:792, **2**:794*F*
      extended ternary complex
         model **2**:792–794, **2**:794*F*
      multistate model *see below*
      ternary complex model **2**:789–
         792, **2**:794*F*
      'two-state model' **2**:792, **2**:794*F*
   agonist state modeling **4**:683
   automated models **4**:682
      homology models **4**:683
      Modeler software **4**:682
      PREDICT method **4**:682–683
   computational modeling **4**:672
      hydropathy plots **4**:673
      sequence identity **4**:672
      inward-facing residues **4**:685
   loop models **4**:680, **4**:682*F*
   microdomains **4**:685
   multistate activation model **2**:794
      deactivation conformations
         2:795
      definition **2**:794
      drug design implications
         2:795–796
      dual efficacy ligands **2**:795
      evidence **2**:794
      fluorescence labeling
         studies **2**:795
   virtual screening **4**:687
      homology models **4**:687–688
      MOBILE method **4**:688
      PREDICT method **4**:689
G protein-coupled receptors (GPCRs),
      assays **2**:805, **2**:806
   biosensors **2**:813
      bioluminescence energy transfer
         (BRET) **2**:813
      fluorescence energy transfer
         (FRET) **2**:813
   cAMP assays **2**:807
      AlphaScreen cAMP assay **2**:809
      chemiluminescent **2**:809
      fluorescence-based **2**:809
      fluorescence polarization **2**:809
      HitHunter cAMP assay **2**:809
      radiometric **2**:808
      receptor subtypes **2**:807–808
      time-resolved fluorescence-
         based **2**:809
   cell-based assays **2**:806; **3**:624*T*,
      3:634
      agonist/antagonist screening **3**:637
      cAMP EIA **3**:634–636
      Gi assays **3**:636, **3**:638*F*
      see also μ-Opiate receptor
      Gq assays **3**:636, **3**:637*F*
      Gs response **3**:634–636
   chimeric G proteins **2**:812
      melanophore platform **2**:813

constitutive activity **2**:806
filtration assay **3**:585–586
Flash Plate assays **3**:585–586
fluorescence correlation spectroscopy-
      related confocal fluorimetric
      techniques (FCS+) **2**:410,
      2:410*F*
fluorescent cell-based assays **3**:627
GTPγS binding assays **2**:778*F*, **2**:807
   time-resolved fluorescence **2**:807,
      2:807*T*
high-content imaging **2**:813
   assay throughput **2**:814
   CCD cameras **2**:813–814
   confocal microscopy **2**:813–814
   Transfluor technology **2**:813–814
inositol phosphate accumulation-IP$_3$
      assays **2**:809, **2**:810*F*
   homogeneous time-resolved
      fluorescence assays **2**:794*F*,
      2:810–811
intracellular calcium assays **2**:811
   AequoScreen **2**:811
   Calcium 3 assay **2**:811
   calcium-sensitive dyes **2**:811
   FLIPR instrumentation **2**:811
   HTS **2**:811
   subtypes **2**:811–812
membrane-based **2**:805
multiple signaling **2**:806
radioactive assays *see* Radioactive
      assays
receptor expression levels **2**:806
reporter gene assays **2**:812
   activator protein 1 (AP-1) **2**:812
   beta-lactamase **2**:812
   cAMP responsive element **2**:812
   luciferase **2**:812
scintillation proximity assay *see*
      Scintillation proximity assay
      (SPA)
signals measured **2**:806
universal screens **2**:806
G protein cycle **2**:779, **2**:779*F*
G proteins
   activation **2**:779
   α-subunit regulation **2**:780, **2**:781*T*
      adenylate cyclase **2**:780–782,
         2:781*F*
   βγ-regulation **2**:781*T*, **2**:784, **2**:786*T*
   families **2**:780
   G$_{12/13}$ subfamily **2**:780
   Gα$_2$ subfamily **2**:782
   Gα$_3$ subfamily **2**:780–782
      associated diseases **2**:780–782
   Gα$_{12}$ subfamily **2**:783–784, **2**:783*F*,
         2:785*T*
   Gα$_{i/o}$ subfamily **2**:781*F*, **2**:782
   Gα$_i$ subfamily **2**:782
   Gα$_o$ subfamily **2**:782
   Gα$_q$ subfamily **2**:782–783, **2**:783*F*
   G$_{i/o}$ subfamily **2**:780
   G$_q$ subfamily **2**:780

Gs subfamily **2**:780
   structure **2**:780
      α-subunits **2**:780
      βγ-subunits **2**:780
      palmitoylation **2**:780
Gq assays, GPCRs **3**:636, **3**:637*F*
G$_q$ subfamily, G proteins **2**:780
G-quartets, DNA structure **2**:1040
GR79236
   migraine treatment **6**:377
   obstructive sleep apnea (OSA)
      treatment **6**:162
   structure **6**:377
GR125743 **6**:75–76
GR135402, protein biosynthesis
      inhibitors **7**:439–440, **7**:440*F*
GR-205171 **6**:110
   structure **6**:110*F*
GR-705498, IBD treatment **6**:637
Graft polymers, drug–polymer
      conjugates **5**:1046*F*
Graft rejection, basiliximab **1**:433
Gram-negative bacteria
   aminoglycosides **7**:636
   multidrug resistant **7**:677
   tetracyclines **7**:610
      uptake **7**:613–614
Gram-positive bacteria,
      tetracyclines **7**:610
Granisetron
   as antiemetic agent **6**:681–682
   5-HT$_3$ receptor binding **2**:890;
      3:633–634
   nonulcer dyspepsia treatment **6**:600
   oral absorption **5**:680*T*
   structure **2**:890*F*
Grant, Ted, collaborators, Ganellin,
      C R **8**:45
Grants, SME funding **1**:495
Granulocyte colony-stimulating factor
      (G-CSF)
   brands, global sales **1**:605*T*
   STAT3 signaling **3**:206
Granulocyte-macrophage colony-
      stimulating factor (GM-CSF)
   asthma **7**:876–877
   STAT5 signaling **3**:206
Granulocytes, tetracyclines **7**:620
Graphite, microwave-assisted
      chemistry **3**:853
Graphs
   algorithms, ligand–receptor
      interaction models **4**:208
   HTS data management **2**:454
   lipophilicity measurement **5**:433
   screening library selection
      4:502–503
   structure/substructure
      searching **3**:236–237
   3D-similarity searching **4**:173
   2D similarity searching **3**:246
'Gray-baby' syndrome, cytochrome
      P450s **5**:172

Grb2 SH2, three-dimensional structures 2:636*T*

Greedy algorithms, product-based library design 4:352

'Green chemistry' 1:51–52

Green fluorescent protein (GFP)
  biomarker imaging 3:77
  genetically engineered animals 3:153–154
  HCS 3:690
  high-content screening (HCS) 2:427, 2:430*F*
  history 3:690
  protein crystallization 3:888
  Ultra-High-Throughput Screening System (Aurora Biosciences) 2:437
  variants 3:690

Green tea (*Camellia*) 1:425
  anticancer effects 1:327
  phase II clinical trials 1:455
  polyphenols 1:451

Grepafloxacin
  clearance 7:587
  development 7:579
  resistance, efflux pumps 7:586
  structure 5:934*F*; 7:572*F*, 7:579*F*
  withdrawal 1:757*T*

Grid cell occupancy descriptors (GCODs), 4D-QSAR 4:581

Gridding and partitioning (GaP), reagent-based library design 4:345–346

GRid-INdependent Descriptors (GRIND) 5:537, 5:541
  3D-QSAR 4:584, 4:584*F*
  active transport 5:541
  almond software 5:541, 5:542*F*
  library design 4:363
  metabolic stability 5:541
  relevant point extraction 5:541

GRID program 4:240, 5:581*T*
  see also Molecular Interaction Fields (MIFs)
  3D-QSAR 4:576–577
  computational oral absorption prediction 5:688
  cytochromes P450 inhibitor QSAR 4:784

GRID derived molecular interaction field 4:240
  historical perspective 4:68
  hydrophobic effect 4:201
  library design 4:359
  ligand-based library design 4:316–317
  ligand–receptor interaction models 4:208–209
  molecular descriptors 5:585
  site analysis, drug design 4:286

steric properties 4:196
transporter interaction modeling 5:775–776

Grid search, conformational analysis/searching 4:104

Grignard reaction, solid support synthesis 2:467–468, 2:469*F*

Griseofulvin 7:430*T*

GRKs, GPCR desensitization 2:802

GroEL, protein crystallization 3:887

GroEL/GroES system, *Escherichia coli* protein expression systems 3:112–113

GROMACS, K$^+$ ion channels 4:710

GROMOS, molecular dynamics simulations 4:652

Gross physiological properties, blood–placenta transfer modeling 5:737

Ground powders, traditional medicine 1:419

GroupBuild method, atom-based drug design 4:288

Group-type E-state index see Electrotopological indices (E-state index)

GrowMol method, atom-based drug design 4:288

Growth factor receptors
  Drosophila melanogaster 3:650
  STAT signal transduction/activation 3:204

Growth factors
  benign prostatic hyperplasia 6:556–557
  signal transduction, cancer see Cancer
  vascular endothelial growth factor expression 3:200

Growth hormone (GH)
  brands, global sales 1:605*T*
  development 1:556–557
  drug introductions 1:108, 1:231
  STAT3 signaling 3:206
  STAT5 signaling 3:206

Growth hormone releasing peptide (GHRP)
  nonpeptide drugs 2:618–619, 2:620*F*
  as peptide/protein drug 2:590
  peptidomimetics 2:608–609, 2:611*F*
  sleep 6:147

Growth hormone releasing peptide (GHRP) receptor agonists 2:620*F*, 2:621

Growth hormone secretagogue receptor agonists
  see also specific drugs
  prokinetic drugs 6:685

Growth hormone secretagogues (GHS), parallel library design 4:326, 4:326*F*

Growth rate slowing, drug discovery 1:340

Gruppo Italiano per lo Studio della Sopravvivenza nell'Infarto Miocardio see GISSI-1; GISSI-2

GSCORE program, regression-based 4:484

GSE, expression profiling 2:312

GSK3B, PDB entries 2:965*T*

GSK-189254 6:205–206
  narcolepsy treatment 6:156–157

GSK-766994
  asthma treatment 7:882
  clinical trials 7:882

GSK Imaging Centre, Imperial College London/Hammersmith Hospital collaborations 1:588

Gs subfamily
  GPCRs 3:634–636
  G proteins 2:780

GT-2331, narcolepsy treatment 6:156–157

GTPases
  biological functions 3:216–217
  orthogonal ligand–receptor pairs 3:216, 3:219*F*

GTPγS binding assays see G protein-coupled receptors (GPCRs), assays

GTP shift, adenosine A$_1$ receptor agonists 5:892

GTS-21
  Alzheimer's disease treatment 6:205
  nicotinic acetylcholine receptor binding 2:888
  structure 2:887*F*

Guanabenz
  oral absorption 5:468*T*, 5:680*T*
  oxidation 5:105–106

Guanethidine analogs, ring system reorganization (bioisosterism) 2:675–676, 2:678*F*

Guanfacine, ADHD treatment 6:135*T*

Guanidine
  hepatitis C virus inhibition 2:1047
  ionization constant (pK_a) 5:381
  structure 5:359
  trypanosomiasis treatment 7:829–830

Guanidine derivatives, solid-phase discrete libraries 3:709*T*

Guanidinium chloride, protein crystallization 3:881

Guanosine, structure 2:1038*F*

Guanoxon, oral absorption 5:680*T*

Guanylate cyclase activators, glaucoma treatment 6:313*T*

Guanylate kinase, drug metabolism 5:149

Guggulsterone, as nuclear receptor ligand 2:1010

Guidelines, regulatory problems 2:726

Guillain–Barré syndrome **6**:231*T*, **6**:234
 epidemiology **6**:234
 mortality **6**:234
 pathogenesis **6**:234
  autoantibodies **6**:230, **6**:235
  Campylobacter jejuni
   infections **6**:234, **6**:235
  cytomegalovirus infection **6**:234
  Epstein–Barr virus infection **6**:234
  gangliosides **6**:235
  genetic factors **6**:235
  Mycoplasma infection **6**:234
 subtypes **6**:234
 symptoms **6**:234
 treatment **6**:248
  4-aminopyridine **6**:248
  intravenous
   immunoglobulin **6**:248
  plasma exchange **6**:248
Guinea pig maximization test, skin
  sensitization, computational
  models **5**:923
Gusperimus trihydrochloride **1**:224
GUSTO-1, alteplase clinical
  trials **6**:772
GUSTO-3, reteplase **6**:773
GUSTO-5, reteplase **6**:773
Guttate psoriasis **7**:943
GV 196771A **6**:341–342
 structure **6**:342*F*
GVK Biosciences, commercially
  available databases **3**:309
GW-404, as farnesoid X receptor
  ligand **2**:1024
GW-2433
 as PPAR-δ ligand **2**:1022
 structure **2**:1023*F*
GW-3965 **2**:1023
 structure **2**:1024*F*
GW-4064, structure **2**:1025*F*
GW-5634
 discovery/development **1**:74
 structure **1**:74*F*
GW-5638 **8**:115
 structure **8**:116*F*
GW-9578, structure **2**:1022*F*
GW-353162 **6**:74
GW-405833 **6**:343
 structure **6**:343*F*
GW-468816 **2**:903
GW-501516
 as PPAR-δ ligand **2**:1022
 structure **2**:1023*F*
GW-572016 **7**:196
 characteristics **2**:984*T*
 EGFR family kinase inhibitors **2**:968
 structure **2**:968*F*, **7**:196*F*
GW-654652, structure **7**:202*F*
GW-742457 **6**:76
GW-873140 **7**:360
GYKI-16638 **6**:754
GYKI-53655, structure **2**:901*F*

Gymnema (*Gymnema sylvestre*) **1**:425
Gyration, size descriptors **5**:682–683
Gyros Microlabs, microfluidic
  devices **2**:350*F*, **2**:361

H

H2G **7**:306
 varicella zoster virus **7**:306–307
Habekacin **7**:634
 hydroxyl group replacement
  7:635
Haemophilus influenzae
 genome sequencing **3**:317*T*, **7**:732
 multidrug resistant **7**:677
 respiratory tract infections **7**:739
Haiti, health personnel **1**:648
Hajduk P. J., model developed by,
  albumin binding prediction
  models **5**:857
Half-life ($t_{1/2}$)
 alendronate **8**:201–202
 cytochromes P450 **4**:781
 definition **5**:869*T*
 PEGylation effects **2**:588, **2**:589*F*
 peptide/protein drugs **2**:583
 pharmacokinetics **5**:21
 volume of distribution (V_{ss}) *vs.* **5**:848
Halichondrin B **1**:363
 structure **1**:362*F*
Halichondrins **7**:84
 discovery/development **1**:380–381
 SAR **7**:84–85, **7**:84*F*
 structure **1**:381*F*, **7**:84*F*
Haloalkane dehalogenase substrates,
  quantum mechanics **4**:412
Haloalkenes, glutathione
  transferases **5**:162, **5**:162*F*
Halofantrine **7**:772, **7**:775–776
 IK$_r$ blockade **6**:746
Halogenated hydrocarbon toxification,
  glutathione *S*-transferases
  (GSTs) **5**:206
Halogens
 dealkylation, cytochrome P450 **5**:100
 intermolecular interactions **3**:404
Haloperidol **6**:29–31
 as antiemetic agent **6**:683
 bioavailability **5**:712*T*
 development **6**:15
 mechanism of action **6**:32
 nonulcer dyspepsia treatment **6**:600
 side effects, QT-prolongation **6**:35
 structure **5**:359
Haloperidol receptors, imaging **3**:667
Haloprogin **7**:430*T*
Halopyridol, reduction **5**:127
Halostilbene series,
  oxazolidinones **7**:684
Halothane **6**:353, **6**:359
 development **1**:531
 immunotoxicity **5**:221*T*

K$^+$ ion channel interactions **4**:720
 lipid interactions **6**:357
 mechanism of action **6**:359
 minimum alveolar
  concentration **6**:359
 oxidation **5**:100
 protein interactions **6**:356
 reduction **5**:122, **5**:123*F*
 side effects **6**:359
 structure **6**:352*F*
Hamilton anxiety scale (HAM-A)
 affective disorders **6**:59
 generalized anxiety disorder **6**:96
Hamilton Rating Scale for Depression
  (HAM-D)
 affective disorders, clinical trial
  issues **6**:63
 somnolence measures/sleep
  testing **6**:141–142
Hammerhead program **4**:256
Hammett equation
 ionization constants (pK_a)
  measurement **5**:394
 QSAR **4**:44, **4**:149
 quantum chemical indices **4**:56
Hammett–Taft constants, ionization
  prediction, in silico tools **5**:610
Hamming distance, similarity
  searching **4**:176*T*, **4**:178, **4**:180–181
Hanging drop method, protein
  crystallization **3**:438–439
Hanp *see* Caperitide
Hansch analysis, QSAR **4**:50, **4**:155
 parabolic model **4**:50
Hansch and Leo technique *see*
  Lipophilicity prediction, rule-
  based systems
Haplotype blocks, genetic variation
  analysis **3**:338
Haptens, immunotoxicology **5**:220, **5**:220*F*
Haptoglobin depletion,
  proteomics **3**:31–32
Ha-*ras* overexpression, genetically
  engineered animals **3**:161
Hardware developments, NMR **2**:515
Harker construction, isomorphous
  replacement **3**:457–459, **3**:458*F*
HARpick, product-based library
  design **4**:351–352
Harrison's Principles of Internal Medicine,
  beta-blocker
  contraindications **8**:141
Hartree–Fock (HF) approximation
 quantum mechanical
  calculations **4**:383–384
 Schrödinger equation **4**:383
Harvesting buffer, crystal
  handling **3**:465–467
Harvey–Bradshaw index, Crohn's
  disease **6**:620

Harzianopyridone
  mitochondria complex II   **2**:1072,
    2:1073*F*
  structure   **2**:1073*F*
'Hashed fingerprints'
  fingerprint/dataprint
    descriptors   **4**:468
  structure/substructure
    searching   **3**:239
  two-dimensional descriptors, similarity
    searching   **4**:171
HASL, 3D-QSAR alignment   **4**:579,
    4:579*F*
Hayfever *see* Allergic rhinitis
Hazard, definitions   **5**:910
Hazard Expert   **5**:926*T*, **5**:927*T*
  other systems *vs.*   **5**:926–927,
    5:927*T*
HBED, oral absorption   **5**:468*T*
HBMAP, site analysis, drug
    design   **4**:286
HCK, PDB entries   **2**:965*T*
HCN ion channels *see* K⁺ ion channels
HCR transgenic animals, psoriasis
    animal models   **7**:971
HCV-371, hepatitis C   **7**:270–271
HDL Atherosclerosis treatment
    (HATS) trial   **6**:478
Headache diary, migraine   **6**:369–370
Header, DDBJ/EMBL/GenBank   **3**:352
Head/neck cancer
  nedaplatin   **1**:147
  STATs   **3**:207–208
  temoporphin   **1**:148
Health Canada database, dermal
    penetration models   **5**:730
Healthcare costs, pain   **6**:321
Healthcare policy, developed
    countries   **1:627–635**
  community needs and desires   **1**:630
  definition   **1**:631
  demand categorization   **1**:633
  economic issues   **1**:634
    supply and demand   **1**:629, **1**:634
  government role   **1**:632
  individual and community
      demand   **1**:631
    priority setting   **1**:631
  individual *vs.* community
      needs   **1**:628
    medical necessity   **1**:629
    supply and demand   **1**:629, **1**:634
  life expectancy and   **1**:634
  media influences   **1**:629
  medicine market debate   **1**:632
Health claims, traditional
    medicine   **1**:417
Health insurance, expansion in
    developing countries   **1**:640
Health personnel, developing
    countries   **1**:647
Health status, determinants   **1**:628
Healthy People 2010   **1**:630

Hearing impairments, transient receptor
    potential ion channels   **2**:909–
    910
Heart
  cytochrome P450s   **5**:172*T*
  phosphodiesterase-4   **2**:926–927
  tissue composition   **5**:45*T*
Heart and Estrogen/Progestin
    Replacement Study
    (HERS)   **1**:686
Heart disease
  aspirin   **1**:323
  Danio rerio model system *see* Zebrafish
    (*Danio rerio*)
  imaging   **6**:461
  mortality   **6**:461
Heart failure
  adrenergic nervous system
      activation   **8**:141
  β-adrenergic receptor antagonists *see*
    β-Adrenergic receptor antagonists
  cardiac hypertrophy   **8**:141
  carvedilol *see* Carvedilol
  drug introductions   **1**:167
  milrinone   **2**:947–948
  remodeling   **8**:141
Heat dissipation, x-ray
    generation   **3**:465, **3**:466*F*
Heating effects, crystalline
    states   **5**:510
Heatmaps, microarrays   **3**:91
Heat shock protein 90 (hsp90)
    inhibitors   **7**:229
  see also specific drugs
  ansamycins   **7**:230
  biological functions   **7**:229–230
  C-terminal binders   **7**:232
  mechanism of action   **1**:387
  natural products   **1**:386
  N-terminal ATP pocket
      binders   **7**:230
  posttranslational modification
      inactivators   **7**:233
  purine-scaffold derivatives   **7**:232
    mechanism of action   **7**:232
  pyrazoles   **7**:232
  radicicol derivatives   **7**:231
  Saccharomyces cerevisiae model
      system   **3**:5–6
  stereochemistry   **1**:387
  structure   **1**:387*F*
Heat shock proteins
  cisplatin binding   **7**:233
  histone deacetylase (HDAC) inhibitor
      binding   **7**:233
  RNAi   **3**:181
Heat stability testing   **5**:498
  FDA guidelines   **5**:498
Heck reaction
  PASP   **3**:813
  solid support synthesis   **2**:467–468,
    2:470*F*
Hectorol *see* Doxercalciferol

HEK-293 cells
  active drug transport
      prediction   **5**:271–272
  cell-based assays   **3**:620–621
  mammalian cell expression
      systems   **3**:418
  protein expression systems   **3**:117
Helicobacter pylori eradication
  antibiotics   **6**:607–609
  gastric/mucosal ulcer
      treatment   **6**:607
  monotreatment   **6**:611
  optimal pH   **6**:607
  proton pump inhibitors   **6**:607–609
Helicobacter pylori infection
  acne rosacea   **7**:961
  adenocarcinoma   **6**:598
  B-cell gastric lymphoma   **6**:598
  discovery   **6**:598
  disease associations   **6**:598
  duodenal ulcers   **6**:607
  gastric cancer   **6**:607
  gastric ulcers   **6**:598, **6**:607
  gastritis   **6**:607
  genetics   **6**:607
  market size   **6**:598–599
  peptic ulcer disease (PUD)
      6:603
  rabeprazole sodium   **1**:201
  ranitidine bismuth citrate   **1**:201
  urea/carbonic anhydrase
      mechanism   **6**:607, **6**:610*F*
Helix, DNA structure   **2**:1038–1039
Helmholtz free energy, free energy
    calculation   **4**:653–654
Helminth administration, IBD
    treatment   **6**:624
Helminth infections   **7**:759
  see also specific infections
  cestode infections *see* Cestode
    (tapeworm) infections
  intestinal nematode infections
      7:759
  tapeworm infections *see* Cestode
    (tapeworm) infections
  tissue nematode infections   **7**:760
  trematode infections *see* Trematode
    infections
Helper T cells
  see also CD4+ T cells; T helper cells,
    type 1; T helper cells, type 2
  atopic dermatitis   **7**:947
  myasthenia gravis   **6**:236
Helsinki declaration, clinical
    trials   **2**:180
Helsinki Heart Trial, fibrates
    6:479
Hemagglutinin (HA)
  influenza   **4**:614
  influenza viruses   **7**:374
Hematological drugs
  see also specific drugs
  introductions   **1**:174

minocycline **7**:620–621
tetracyclines *see* Tetracyclines
Hemiacetal formation, dynamic
  combinatorial library
  generation **3**:962*F*
Hemiasterlins **7**:87
  discovery/development **1**:380–381
  SAR **7**:87, **7**:87*F*
  structure **1**:381*F*, **7**:87*F*
11a-Cl-6,12-Hemiketal
  oxytetracycline **7**:605
  methacycline synthesis **7**:605*F*
Hemodynamic effects
  metabolic syndrome **6**:381
  renal dysfunction **6**:579
Hemoglobin
  allosteric effectors **4**:71
  as model receptor **4**:71
  polyethylene glycol
    conjugates **5**:1059
  structure-based drug design **4**:70
Hemoglobin A1c, diabetes mellitus
  biomarker **3**:70
Hemophilia
  drug introductions **1**:175
  as monogenetic disease **3**:338
Hemorrhagic stroke **6**:253, **6**:697
  experimental models **6**:257*T*,
    6:258
  intracerebral hemorrhage
    (ICH) **6**:253
  subarachnoid hemorrhage
    (SAH) **6**:253
Hemostasis, biological
  macromolecules **1**:432–433
Henderson–Hasselbalch equation,
  solubility **5**:402
Hen's egg test-chorioallantoic
  membrane (HET-CAM), non-
  animal eye irritation
  testing **1**:466, **1**:467*T*
Hepaciviruses, antiviral drugs **7**:254*T*,
  7:267
Hepadnaviruses, antiviral
  drugs **7**:254*T*, **7**:263
Heparan sulfate proteoglycans,
  malaria **7**:766
HepaRG, drug metabolism, in vitro
  studies **5**:233
Heparin, thrombolytic drugs **6**:775
Hepatic circumvention, vaginal
  administration **5**:289
Hepatic clearance (CL$_H$)
  definition **5**:869*T*
  elimination studies **5**:44–45
  optimum pharmacokinetics
    5:26
  pharmacokinetics **5**:16, **5**:19
Hepatic intrinsic clearance (CL$_{INT}$)
  models **5**:871
  pharmacokinetics **5**:19
Hepatic metabolic clearance, models *see*
  Clearance models

Hepatic portal vein-cannulated animals,
  mechanistic pharmacokinetic
  studies **5**:38–39
Hepatitis B
  HIV coinfection **7**:265
  treatment **8**:59
  see also specific drugs
Hepatitis B antiviral drugs **7**:254*T*,
  7:263, **7**:312
  see also specific drugs
  acyclic nucleosides **7**:318
  acyclovir triphosphate **7**:318
  adefovir dipivoxil **1**:119; **7**:263, **7**:314
  azidothymidine **7**:319
  in development **7**:318
  emtricitabine **7**:264–265
  entecavir **7**:263, **7**:264
  future work **7**:321
  interferon-2α **7**:263
  introductions **1**:107
  lamivudine **7**:263
  2′,3′-modified nucleosides **7**:319
  non-nucleoside inhibitors **7**:320
  L-nucleosides **7**:264, **7**:319
  penciclovir **7**:318
  phenylpropenamide
    compounds **7**:321
  propagermanium **1**:128
  2,5-pyridineicarboxylic acid
    derivatives **7**:320
  resistance **7**:317
  tenofovir **7**:264–265
Hepatitis B virus **7**:311
  cancers **7**:4
  genome **7**:312
  incidence **7**:311
  interferon treatment **7**:312
  lamivudine **7**:338–340
  replication cycle **7**:311
Hepatitis C
  antiviral drugs **7**:254*T*, **7**:267
    A-782759 **7**:270–271
    albuferon-α **7**:267, **7**:390
    BILN 2061 (Ciluprevir) **7**:267
    HCV-371 **7**:270–271
    interferon-2α **7**:267
    interferon-α **7**:389
    interferon-α2a **7**:267
    interferon-α2b **7**:267
    interferon-α/ribavirin **7**:389, **7**:390
    merimepodib **7**:390
    resistance mutations **7**:271
    valopicitabine **7**:269
    viramidine **7**:267, **7**:390
    VX-950 **7**:268
  clinical trials **7**:388
    interferon **7**:389
  epidemiology **7**:388
  HIV coinfections **7**:267
  interferon alfacon-1 **1**:125
  models **7**:388
    animal model **7**:389
    cell models **7**:388–389

nonviral targets **7**:404
    isatoribine **7**:403*F*, **7**:405
    thymosin α1 **7**:403*F*, **7**:404, **7**:405
  prevalence **7**:386
  treatment **7**:389
Hepatitis C inhibitors
  see also specific drugs
  benzimidazoles **2**:1047, **2**:1048*F*
  carboxylic acid bioisosterism **2**:659
  guanidine **2**:1047
  phenazine **2**:1047
Hepatitis C virus **7**:386
  internal ribosomal entry site
    (IRES) **2**:1047
  isolates **7**:388
  NS3 protease, as drug target **7**:267–
    268
  NS5B RNA replicase, as drug
    target **7**:267–270
  protease
    as drug target *see* Hepatitis C virus,
      protease inhibitors
    mechanism of action **7**:390–391,
      7:392*F*
    structure **7**:391*F*
  replicative cycle **7**:386, **7**:387*F*
    attachment **7**:387
    envelop assembly **7**:388
    internal ribosome entry site **7**:387
    mutational frequency **7**:388
    nucleocapsid assembly **7**:388
    protein replication **7**:387–388
    translation initiation **7**:387
  RNA polymerase **7**:396–398
    see also Hepatitis C virus, RNA
      polymerase inhibitors
    structure **7**:396–398, **7**:397*F*
  therapeutic targets **7**:390
    see also specific targets
  future work **7**:405
  transmission **7**:388
Hepatitis C virus, protease
  inhibitors **7**:390
  see also specific drugs
  BILN 2061 **7**:391, **7**:393*F*, **7**:394*F*
    in vitro activity **7**:395
    peptidomimetics **7**:395
    pharmacodynamics **7**:395
  VX-950 **7**:395, **7**:396*F*
    pharmacokinetics **7**:396
    pyrrolidine-*trans*-lactams **7**:396,
      7:397*F*
Hepatitis C virus, RNA polymerase
  inhibitors **7**:390, **7**:396
  see also specific drugs
  anti-NS5b nonnucleosides **7**:400,
    7:406*F*
  benzimidazole compounds **7**:401,
    7:402, **7**:406*F*
  benzothiadiazine **7**:403*F*, **7**:404
  dichloroacetamide **7**:403*F*, **7**:404
  hydroxypyridimines **7**:403*F*, **7**:404
  indole derivatives **7**:402

Hepatitis C virus, RNA polymerase
    inhibitors (*continued*)
    pyranoindole compounds **7:**402
    SAR **7:**397*F,* **7:**401
    anti-NS5b nucleosides **7:**398,
        7:399*F*
        anti-HIV nucleosides *vs.* **7:**398
        2'-methyl nucleosides **7:**400
        2'-methyl pyrimidine
            nucleosides **7:**400
        structure **7:**398–399, **7:**399*F*
    kinetics **7:**398*T*
Hepatocellular insufficiency,
    pharmacokinetics **5:**27
Hepatocellularity definition, metabolic
    drug–drug interactions **5:**832
Hepatocyte intrinsic clearance (CL$_{INT}$)
    scaling, metabolic drug–drug
    interactions **5:**832
Hepatocyte nuclear factors (HNFs)
    diabetes mellitus **6:**423
    nuclear receptor interactions **2:**1000
Hepatocytes
    cell-based assays **3:**629
    cultures
        CYP induction, in vitro
            studies **5:**250
        metabolic drug–drug
            interactions **5:**830
    drug metabolism, in vitro studies *see*
        Drug metabolism, in vitro studies
    intestinal bacteria interactions, drug
        metabolism, in vitro
        studies **5:**233
    reaction phenotyping **5:**242*T*
    tissue binding *see* Tissue binding
Hepatoma, zinostatin
    stimalamer **1:**149
Hepatoma cell lines
    CYP induction, in vitro studies **5:**251
    drug metabolism, in vitro studies *see*
        Drug metabolism, in vitro studies
Hepatoprotectants, drug
    introductions **1:**245
Hepatotoxicity
    computational models **5:**925
    electrotopological indices (E-state
        index) **5:**572
    erythromycin **7:**534
    SLC transporters **5:**78
Hepatotoxicity testing
    Informatics and computational Safety
        Analysis Staff (ICAS) **5:**925
    Lhasa Limited **5:**925
    MCASE **5:**925
HepG2
    CYP induction studies **5:**251
    drug metabolism, in vitro
        studies **5:**233
Hepsera *see* Adefovir dipivoxil
1-Heptanol **6:**755
Heptoxdiazine Y, origins **6:**97–98
HER2 *see* ErbB2

Her2/neu expression, cancer
    biomarkers **7:**48
'Herbal-directed small libraries,' natural
    products **1:**59, **1:**59*F*
Herbal medicine *see* Traditional
    medicine
Herbimycin **7:**230
Herbs, definition **1:**407
Herceptin *see* Trastuzumab
hERG
    'binding promiscuity' **5:**940
    blockers *see* hERG blockers
    homology models **5:**940
    related cardiotoxicity **5:**935, **5:**948
        see also Long QT interval
        withdrawn drugs **5:**934, **5:**934*F*
    structure
        aromatic residues **5:**938–939,
            5:939*F*
        polar residues **5:**939, **5:**939*F*
        transmembrane domains **5:**938,
            5:938*F*
hERG blocker modeling **5:**938
    classification methods **5:**943
        atomic distance method
            5:943–944, **5:**944*F*
        combined 2D/3D approach **5:**943,
            5:943*F*
        decision-tree based
            approach **5:**943
        genetic programming model
            5:944
    in drug discovery **5:**945, **5:**946*F*
    homology modeling **5:**938
        bacterial potassium
            channels **5:**938, **5:**938*F*
    QSAR **5:**940, **5:**942*F*
        antiarrhythmic study **5:**942
        blockade prediction **5:**941
        calculated descriptor model
            5:942
        CoMSiA **5:**941–942
        DISCO interface **5:**940–941,
            5:941*F*
        ligand-based approaches **5:**940
        two-component model **5:**942
hERG blockers **5:**945
    basic center **5:**952, **5:**952*F*
    electrotopological indices (E-state
        index) **4:**565, **4:**566*F,*
        4:567*T*
    flexibility reduction **5:**950
        dopamine D$_4$ receptor
            antagonists **5:**951, **5:**951*F*
        5-HT$_{2A}$ receptor
            antagonists **5:**950–951, **5:**951*F*
    homology modeling **5:**940
    ligand-binding **5:**939*F,* **5:**943*F,*
        5:945
    lipophilicity/basicity
        reduction **5:**948, **5:**950*F*
        5-HT$_{2A}$ receptor
            antagonists **5:**950, **5:**950*F*

    lipophilicity reduction **5:**948,
        5:948*F*
        5-HT$_{2A}$ receptor
            antagonists **5:**948, **5:**949*F*
    modeling *see* hERG blocker modeling
    screening *see* hERG blocker screening
    structure **5:**945
        basic solubilizing groups **5:**946
        flexible linkers **5:**946
        physicochemical property
            profiling **5:**945, **5:**947*F*
        ring-linker topology **5:**947,
            5:947*F*
        zwitterions **5:**952, **5:**952*F*
hERG blocker screening **5:**937
    fluorescent cell-based assays **3:**625–
        626, **3:**626*F*
    high-throughput assays **5:**937
        fluorescence-based assays **5:**937
        radioligand binding assays **5:**937
    in vitro telemetry **5:**937
    patch clamp techniques **5:**937
        CHO cells **5:**937
        transfected cell lines **5:**937
    planar patch clamp **5:**937
    whole organism **5:**937
Heroin, structure **6:**173*F*
Heroin addiction **6:**173*T*
    comorbidities **6:**184*T*
    treatment **6:**9, **6:**179, **6:**180*T,*
        6:183
        buprenorphine **6:**9, **6:**180*T,*
            6:183
        levo-α-acetate-methadol
            (LAAM) **6:**180*T,* **6:**183
        methadone **6:**9, **6:**180*T,* **6:**183
        naltrexone **6:**180*T,* **6:**183–185
Herpes simplex viruses (HSV)
    acyclovir **7:**297
    adefovir dipivoxil **7:**314
    antiviral drugs **7:**257
        BAY 57-1293 **7:**259
        BILS 179BS **7:**259
    atopic dermatitis **7:**949
    gene therapy **1:**304
    thymidine kinase, quantum
        mechanics **4:**412
    transgenic mouse model
        promoters **3:**11
Herpes simplex virus type 1 (HSV-1)
    7:295
    antiviral drugs **7:**257
        A-5021 **7:**259
        acyclovir **7:**257–258, **7:**299
        brivudine **7:**257–258
        cidofovir **7:**303
        cyclohexenylguanine **7:**259
        famciclovir **7:**257–258
        idoxuridine **7:**257–258
        penciclovir **7:**257–258
        synguanol **7:**259
        trifluridine **7:**257–258
        valaciclovir **7:**257–258

Herpes simplex virus type 2 (HSV-2) **7**:295
  antiviral drugs **7**:257
    A-5021 **7**:259
    acyclovir **7**:257–258, **7**:299
    brivudine **7**:257–258
    cidofovir **7**:303
    cyclohexenylguanine **7**:259
    famciclovir **7**:257–258
    idoxuridine **7**:257–258
    penciclovir **7**:257–258
    synguanol **7**:259
    trifluridine **7**:257–258
    valaciclovir **7**:257–258
Herpesviruses **7**:295
  see also specific viruses
  antiviral drugs **7**:257, **7**:295
    see also specific drugs
    in development **7**:306
    future work **7**:311
    helicase/primase complex
      inhibitors **7**:309
    mechanism of action **7**:296–297
    recurrent infections **7**:311
    resistance **7**:311
    ZSM-I-62 **7**:259
  DNA polymerase **7**:296–297
  genome **7**:296
  protease inhibitors **7**:310
  pyrrolidine-5, 5′*trans*-
    lactams **7**:311
Herpes zoster infections,
  famciclovir **1**:123
Hervé F., model for α_1-acid glycoprotein
  (AGP) binding prediction
  models **5**:859, **5**:859*F*
Hesperetin **1**:452*T*
Hesperidin **1**:452*T*; **7**:207
  pharmacokinetics **1**:454*T*
  structure **7**:207*F*
Hessian matrix, core molecular
  modeling **4**:103
Heteroactivation, non Michaelis–
  Menten kinetics **5**:236
Heteroaromatic bases, ionization
  constants (pK_a) **5**:378
Heteroatom oxidation, cytochrome
  P450 *see* Cytochrome P450
  family
Heterobifunctional PEG, polyethylene
  glycol conjugates **5**:1053
Heterochromatin, structure **1**:252
Heterocompounds, protein 3D
  structure validation *see* Protein
  3D structure validation
Heterocycle/pharmacophores, solid
  support synthesis *see* Solid
  support synthesis
Heterocycles
  mitoprotective drugs **2**:1062,
    2:1062*F*
  ring equivalent bioisosterism *see*
    Bioisosterism, ring equivalents

Heterocyclic amines, genotoxicity,
  computational models **5**:918
Heterocyclic libraries
  solid-phase libraries **3**:719–720
  solution phase library synthesis *see*
    Solution phase synthesis, library
    preparation
Heterocyclic scaffolds, solid-phase
  libraries **3**:703–706, **3**:707*F*
Heterocyclic thiourea library
  preparation, PASP **3**:830
Heterodimerization, cytokine
  receptors **2**:576
Heterodimers, nuclear
  receptors **4**:730, **4**:731*F*
Heterogeneity, database
  integration **3**:303
Heterogeneous proteins
  crystallization *see* Protein
    crystallization
  protein crystallization *see* Protein
    crystallization
Heterogenous approach, microplate-
  based assays **2**:367
Heterogenous data sets, bioavailability
  models **5**:707–708
Heterostilbene series,
  oxazolidinones **7**:684
Heterotrimeric G protein (GTP-binding
  protein; G$\alpha\beta\gamma$), GPCRs **2**:574–
  575
20-HETE synthesis inhibitors, renal
  dysfunction management **6**:591
Hexachlorophene
  ionization constant (pK_a) **5**:372–373
  structure **5**:370*F*
Hexadecane membrane,
  PAMPA **5**:473, **5**:474*F*
Hexadecane sulfonic acid, hydrophobic
  patches **4**:244, **4**:245*F*
Hexafluroethane (HFE)
  lipid interactions **6**:357
  structure **6**:352*F*
6-(*N*,*N*-disubstituted)-3,4,5,6,7,8-
  Hexahydro-2*H*-naphthalen-1-one
  derivatives, oxidation **5**:1032
Hexamethylene bisacetamide
  discovery/development **1**:386
  structure **1**:385*F*
Hexamethylmelamine, tamoxifen
  discovery **8**:86
Hexamine
  as prodrug **5**:1010, **5**:1010*F*
  structure **5**:1010*F*
Hexane, microwave heating **3**:840*T*
Hexobarbital, bioavailability **5**:712*T*
[^{99m}Tc]Hexylmethylpropylene
  amineoxone, androgen ablation
  imaging **3**:670
Hi-Chip system, DNA
  microarrays **2**:357
HIC-Up database, protein 3D structure
  validation **3**:514

Hidden Markov model (HMM)
  protein function prediction,
    bioinformatics **3**:335
  sequence alignment,
    bioinformatics **3**:320
Hierarchical clustering model
  protein binding, E-state index **5**:567,
    5:567*F*
  screening library selection **4**:503
Hierarchical databases *see* Databases
Hierarchical shotgun sequencing
  genome sequencing **3**:318
  Human Genome Project **3**:318
HIFA (hydropathic intermolecular field
  analysis), 3D-QSAR
  alignment **4**:578
High-affinity binding site, P-
  glycoprotein **5**:775
High concentrations, PASP **3**:795
High-content imaging *see* G protein-
  coupled receptors (GPCRs),
  assays
High-content screening (HCS)
  3:685
  advantages **3**:693
  applications **3**:693
  ArrayScan II system **2**:427–428
  assay parameters **3**:685
  autofocusing **3**:687
  automation **2**:428
  CellChip System **2**:431, **2**:432*F*,
    2:433*F*, **2**:434*F*
    advantages **2**:432–433
  cell cycle **3**:691
    BrdU incorporation **3**:691–692
    DNA content information
      3:691–692
    EGFP **3**:691–692
    mitotic index **3**:691–692
  Cellomics Knowledgebase **2**:433,
    2:435*F*
    biochemical pathway
      diagrams **2**:435
    data mining **2**:436
  cellular toxicity **3**:691
  colony-formation assays **3**:692
  confocal systems **3**:687
    Nipkow disc **3**:687, **3**:688*F*
    resolution **3**:687
    signal-to-noise ratio **3**:687
  data acquisition **3**:685
  data analysis **3**:687, **3**:689*F*,
    3:692
    lack of standardization **3**:687
  data storage **3**:692
  drug-induced microtubule
    reorganization **2**:428, **2**:429*F*
  fluorescent reagents **2**:428
    green fluorescent protein
      (GFP) **2**:427, **2**:430*F*
    JC-1 **2**:429, **2**:430*F*, **2**:431*F*
  fragment-based drug discovery **4**:449
  future work **2**:436

High-content screening (HCS)
(*continued*)
  hardware **3**:685
  history **3**:685
  image detection **3**:686*T*, **3**:687
    CCD **3**:685, **3**:687
    laser scanning cytometer
      3:687
    photomultiplier tubes **3**:687
  labeling techniques **3**:688
    antibody staining **3**:689
    bright-field microscopy **3**:686*T*,
      3:688
    cellular stains **3**:689
    DNA dyes **3**:689
    fluorescence **3**:688
    fluorescent proteins **3**:690
    quantum dots **3**:690
    subcellular stains **3**:689
  living cells **2**:426
  micronucleus formation **3**:692
  multidimensional cytological
    profiling **3**:693
  neurite overgrowth **3**:692
  nuclear translocation assays **3**:691
    NF-κB **3**:691
  phenotype-oriented screening
    3:138
  receptor internalization **3**:690–691
    GPCR **3**:690–691
  translocation events **3**:690
High-density head group region, lipid
  bilayers **5**:456
High-density lipoprotein
  cholesterol **6**:464, **6**:701–702
  estrogen replacement therapy
    (ERT) **8**:103
High-density lipoprotein cholesterol-
  elevating drugs **6**:478, **6**:487
  see also specific drugs
  atherosclerosis **6**:486
  CETP inhibitors **6**:488
  PPAR agonists **6**:489*F*, **6**:490
  statin combination **6**:479
High-density lipoproteins
  (HDLs) **6**:462, **6**:464
  atheroprotective effects **6**:470
  as biomarker, cardiovascular
    disease **3**:70
  characteristics **6**:463*T*
  composition **6**:464*T*
  coronary heart disease *see* Coronary
    heart disease (CHD)
  drug binding **5**:849–850
  particle assembly **6**:465, **6**:467*F*
  plasma concentration **6**:465*T*
  plasma protein binding **5**:322
  reverse cholesterol transport **6**:465–
    466, **6**:467*F*
  size **6**:464–465
  structure **6**:466*F*
High-density tail region, lipid
  bilayers **5**:456

Higher education institutions (HELs),
  R&D investment **1**:572
Higher-row quantum states,
  electrotopological indices
  (E-state index) **4**:545*T*, **4**:546,
  4:546*F*
Highest occupied molecular orbital
  (HOMO)
  LUMO ratio, toxicity, computational
    models **5**:913
  molecular descriptors **5**:547, **5**:584
  QSAR *vs.*, genotoxicity **5**:919
High-fat diets
  gastrointestinal motility
    models **6**:675
  obesity models **6**:393
High gravity methods, protein
  crystallization **3**:439
High-intensity focused ultrasound,
  benign prostatic
  hyperplasia **6**:543
Highly active antiretroviral therapy
  (HAART) **8**:187
  AIDS **8**:53
  genetic variation analysis **3**:340
  mitochondria, effects on **2**:1082
  protease inhibitors **7**:356
High-performance liquid
  chromatography (HPLC)
  brain drug uptake **5**:310
  combination, light scattering,
    promiscuous ligands **2**:743
  dermal/transdermal
    administration **5**:292–293
  epigenetics **1**:266, **1**:267*T*
  equilibrium solubility **5**:349–350
  ezetimibe, metabolite
    formation **8**:75, **8**:75*F*
  high-throughput methods **5**:407
  high throughput purification *see* High
    throughput purification
  metabolism studies **5**:34
  partition assessment tools **5**:478
High-performance liquid
  chromatography-mass
  spectrometry (HPLC-
  MS) **3**:862; **5**:994, **5**:995*F*
  advantages **3**:863; **5**:994–996
  detection methods **3**:863
  historical perspective **3**:862–863
  ionization methods **5**:994–996
  metabolite identification
    5:996–998
  metabotyping **5**:1000
  parallel synthesis **3**:863
  protein identification *see* Protein
    identification
  proteomics **3**:37
  reverse-phase **5**:994–996, **5**:997*F*
  technology **3**:863
  toxicity studies **5**:1001
  unknown isolation/
    identification **5**:998–999

High-performance liquid
  chromatography-nuclear
  magnetic resonance
  (HPLC-NMR)
  correlation spectroscopy
    (COS) **5**:999
  marker identification **5**:999
  peak collection **5**:999
  sensitivity **5**:999
  unknown isolation/
    identification **5**:999
High-resolution magic angle spinning
  (HRMAS)
  drug development **2**:194
  nanoprobes **2**:519
High-resolution screening (HRS),
  compound libraries **1**:331
High-throughput chemistry
  (HTC) **4**:438
  lead identification/
    characterization **4**:436
High-throughput determination,
  log*D*pH **5**:447
High-throughput docking, GPCR drug
  design **4**:686
High throughput organic synthesis
  (HTOS) *see* High-throughput
  screening (HTS)
High throughput purification **3**:861–
  874
  combinatorial chemistry **3**:861
  fluorous split–mix library
    synthesis **3**:868
  hit generation support **3**:861
  HPLC/MS *see* High-performance
    liquid chromatography-mass
    spectrometry (HPLC-MS)
  intelligent method
    development **3**:867
  lead generation support **3**:861
  library purity assessment **3**:863
    chemiluminescence nitrogen
      detection (CLND) **3**:863
    evaporative light-scattering
      detection (ELSD) **3**:863–864,
      3:864*F*
    target purity **3**:863
    UV chromatograms **3**:863
  mass-directed **3**:865
    preparative LC/MS **3**:865–866,
      3:866*F*, **3**:867*F*
  monoliths **3**:867
  parallel analysis/purification
    3:868
    advances **3**:871
    advantages **3**:869
    commercial systems **3**:870
    limitations **3**:871
    rapid column switching **3**:868,
      3:870*F*
    technique **3**:869
  preparative liquid chromatography/MS
    demixing **3**:868

streamlining **3**:871
  efficiency **3**:871
  robustness **3**:871–872
  techniques **3**:864
    see also specific techniques
  turbulent flow media **3**:867
  UV-directed **3**:864
    automation **3**:865
    intelligent systems **3**:865
    MS coupling **3**:865
    multiple fractions **3**:865
High-throughput screening
    (HTS) **3**:679–696
  ADMET **1**:33; **2**:393–394, **2**:401
    candidate parallel
      development **2**:401, **2**:402*F*
  anticancer agents *see* Anticancer drugs
  assay development **2**:399
    cell-based **2**:400
    genomics, antibacterial target
      identification **7**:741
    lead validation **2**:400
    relevance *vs.* cost **2**:399, **2**:399*F*
    target selection **2**:399–400
  assay prerequisites **3**:682
  assay quality evaluation **3**:683
    false negatives **3**:684
    false positives **3**:684
    hit confirmation **3**:684
    nonspecific plate blank
      (NSB) **3**:683–684
    parallel screens **3**:684
    screening window
      coefficient **3**:684
    signal-to-background **3**:684
    signal-to-noise **3**:684
    target-unrelated effects **3**:684
  automation **3**:681
    advantages **3**:682
    assay dependency **3**:682
    combinatorial synthesis **3**:681
    large robotic system **3**:682
    operations needed **3**:681
    parallel synthesis **3**:681
    pipetting station/stacker
      dispensing **3**:681
    quality standards **3**:682
    small robotic system **3**:682
    workstations **3**:681–682
  bioinformatics **2**:402
    assay variation **2**:403
    data generation **2**:402
  brain drug uptake **5**:309
  cell-based assays **3**:618
  changes in **4**:775
  characteristics **3**:599
  chemical biology *see* Chemical biology
  comparative functional
    genomics **2**:401
  compound libraries **1**:331
  compound numbers **3**:680
  compound storage/
    management **3**:569

Danio rerio model system **3**:10
data acquisition **2**:449
  automation integration **2**:449
  information system
    integration **2**:450
  plate layouts **2**:451
  well types **2**:451
data analysis *see* High-throughput
  screening (HTS), data analysis
definition **3**:679
development **3**:130
  combinatorial chemistry **2**:39,
    2:42*F*
distribution, materials
  management **2**:535
diversity analysis **4**:183–184
drug development **2**:177
drug discovery **1**:36, **1**:331;
  2:139
  efficacy **1**:48
  productivity **2**:124–125
expression profiling **2**:310,
  2:311
false positives **2**:737
  see also Promiscuous ligands
  assay interference **2**:737
  compound impurity **2**:737
  experimental artifact **2**:737
Flash Plate assays **3**:587
fluorescence detection
  systems **2**:400
fluorescence resonance energy
  transfer **2**:401
fragment-based drug
  discovery **3**:941; **4**:446
functional genomics **2**:328
future work **1**:81; **2**:403, **2**:459
  microfluidics **2**:403
  plate densities **2**:459
  ultraHTS **2**:459
genetically engineered
  animals **3**:159
genomics **3**:5
  antibacterial target
    identification **7**:734
GPCR functional assay
  screening **2**:811
hERG blocker screening *see* hERG
  blocker screening
high throughput organic
  synthesis **2**:393, **2**:395
  drug development impact **2**:396,
    2:396*F*
  history **2**:395
  peptide-based **2**:390*F*, **2**:395
  solid supports **2**:396
  support industries **2**:395
historical perspective **3**:680
  assay development **2**:392
  Association of Laboratory
    Automation **3**:680
  perception of **2**:392
  'random screening' **2**:392

Society for Bimolecular
  Screening **3**:680
  standardization **2**:392
  support industries **2**:392–393
  throughput changes **2**:392
hit characterization **3**:684
  see also High-content screening
hit–to–lead process **3**:684
  see also High-content screening
hit validation **3**:902
HIV protease inhibitors **4**:610,
  4:611*F*
homogeneous soluble biological
  assays **3**:600
homogeneous time-resolved
  fluorescence **2**:416
  importance **3**:577
incorporation, ADMET **1**:33
knockout animal models **3**:159
lead identification **3**:940
library composition **3**:680
  compound stability **3**:681
  'drug-like' properties **3**:680
  natural lead analogs **3**:680
library design **4**:339
light **5**:497
logistics **2**:458
management/personnel **2**:403
metalloproteomics **3**:556–557
microplate formats **3**:682–683
  definition **3**:682–683
  high well densities **3**:682–683
  volumes **3**:682–683
N-myristoyltransferase
  inhibitors **7**:465
natural products **1**:363; **2**:397, **2**:397*F*
  marine organisms **2**:397
  microbials **2**:397
  plants **2**:397, **2**:398
  sensitivity issues **2**:398
optical biosensors **3**:594
phosphatase assays **3**:582
pH stability testing **5**:496
plasma, stability testing **5**:500
plate handling **2**:458
  consolidation **2**:459
  format changes **2**:459
  pooling/deconvolution **2**:459
present day **2**:393
  ADMET **2**:393–394
  de novo HTS **2**:393
  technology integration **2**:393
problems **3**:1
promiscuous ligands *see* Promiscuous
  ligands
protein crystallography, rational drug
  design **3**:890
protein tyrosine phosphatase-1B
  inhibitor design **4**:641, **4**:642*F*
purification *see* High throughput
  purification
RNA polymerase III
  inhibitors **7**:436–439

High-throughput screening (HTS)
    (*continued*)
    scaffold-based design *see*
        Scaffold-based design
    scintillation proximity assay   3:587
    single-parametric biochemical
        assays   3:683
        homogeneous *vs.* heterogeneous
            assays   3:683
        protein purity   3:683
        simplification   3:683
    single-parametric cellular
        assays   3:683
        Fluorescence Imaging Plate Reader
            (FLIPR)   3:683
        reporter gene assays   3:683
    stability testing   5:495
    structure-based drug design   2:397*F*,
        2:398
        computational design
            software   2:398, 2:398*F*
        success of   2:52
    surface plasmon resonance   3:594
    target selection   3:681
        'druggability'   3:681
    target *vs.* disease   2:114–115
    test substance supply   2:394
        see also Libraries
        sources   2:394
            see also specific sources
    turbidimetric assays   3:595
High-throughput screening (HTS), data
        analysis   2:451, 2:455
    cross-assay reports   2:455
    dose response tests   2:452
    kinetics experiments   2:452
    percent inhibition   2:451
        variation   2:451
    percent of control   2:451
        variation   2:451
    pivoted data   2:455
        classical structure activity   2:455
    ratios   2:452
    visualization   2:455
High-throughput screening (HTS), data
        management   2:448
    automation   2:449
    data volume   2:449
    decision support   2:453
        controls   2:453
        data integration   2:454
        hit location   2:454
        pattern searches   2:453
        references   2:453
        single run evaluation   2:454
        systematic error detection   2:453
    error sources   2:453
        environmental issues   2:453
        equipment failure   2:453
        reagent stability   2:453
    graphical display   2:454
    inherent variability   2:449
    reporting   2:455

result validation   2:452
    experimental error
        detection   2:452
    false positives *vs.* false
        negatives   2:453
    tabular display   2:454, 2:456*T*,
        2:458*T*
High-throughput synthesis, CEM
    Discover platform microwave-
    assisted chemistry   3:846,
    3:847*F*
High-throughput systems, lipophilicity
    measurement *see* Lipophilicity
    measurement
High-throughput x-ray crystallography,
    phosphodiesterase-4
    inhibitors   2:938
High well densities, HTS   3:682–683
Hill potential, van der Waals'
    interactions   4:98
HinCyc database, metabolic
    networks   3:328–329
HINT   5:581*T*
    log P calculation program   5:661*T*
    Molecular Interaction Fields
        4:242
Hippocampus, corticotrophin-releasing
    factor overexpression   3:99
Hippocratic Oath   1:682
Hirschmann, Ralph, medicinal
    chemistry, definition of   2:92
Hirschmann, R F   1:1–28
    at Merck   1:3
        exploratory research   1:8
        fundamental research   1:7
        process research   1:3
        transfer to West Point   1:12
        vice President of basic
            research   1:16
    research areas
        antibiotics   1:20
        β-adrenergic receptor
            antagonists   1:13
        N-carboxyanhydrides   1:8
        cortisol   1:5
        cortisone   1:2–3
        enzyme synthesis   1:8, 1:9
        estrone   1:2–3
        hypercholesterolemia   1:18
        Merrifield 'solid phase'
            synthesis   1:11
        methyldopa progenitors   1:13
        ophthalmic applications   1:6
        preg-4-eno [3,2-]pyrazole
            discovery   1:7
        prodrug design   1:7
        protease inhibitors   1:14
        somatostatin   1:14
        'stereoelectronic control'   1:6
        unnatural product synthesis   1:2
        uricosuric diuretics   1:13
    training   1:2
        graduate school   1:2

His₆ tag
    affinity chromatography   3:436–437
    problems in crystallization   3:436–
        437
His/Asn/Gln side chain flips, x-ray
    structures, problems   3:537
Histamine
    N-methylation   5:146–147, 5:146*F*
    sleep   6:143, 6:146
    structure   2:57*F*; 6:605*F*
Histamine H₁ receptor antagonists
    see also specific drugs
    antiemetic drugs   6:683
    atopic dermatitis treatment   7:950
    introductions   1:189
    as 'old' target   2:274
    parallel library design   4:325
    primary insomnia treatment   6:157
    Timmerman, Henk and   8:10, 8:11
Histamine H₁ receptors
    activation   2:785*T*
    constitutively active   2:790*T*
    in human disease   2:775*T*
Histamine H₂ receptor antagonists
    see also specific drugs
    analog design   2:651
    discovery/development   6:597,
        6:605
    gastric/mucosal ulcer
        treatment   6:605
    gastroesophageal reflux disease
        treatment   6:603–604
    as 'old' target   2:274
    structure   6:603
        quantum mechanical
            calculations   4:389–390, 4:389*F*
Histamine H₂ receptors
    constitutively active   2:793*T*
    Ganellin, C R and   8:42–43
    Timmerman, Henk and   8:10, 8:13
Histamine H₃ receptor antagonists
    see also specific drugs
    carboxamide bioisosterism   2:668
    narcolepsy treatment   6:156
Histamine H₃ receptors
    constitutively active   2:793*T*
    Ganellin, C R and   8:49
    Timmerman, Henk and   8:12
Histamine *N*-methyltransferase
    (HNMT)   5:819–820
Histamine receptor antagonists
    see also specific drugs
    Alzheimer's disease treatment   6:205
    Ganellin, C R and   8:45–46
    gastroesophageal reflux disease
        treatment   1:439
    narcolepsy   6:164
Histamine receptors
    see also specific types
    radioactive assays   3:586
    scintillation proximity assay   3:586
    Timmerman, Henk and   8:9
Histamine release, modafinil   8:152

Histochemistry, genetically engineered animal characterization 3:154
Histogram matching, protein crystallography model generation 3:462
Histological markers, cancer biomarkers 7:44–47
Histomycosis, amphotericin 7:457
Histone deacetylase (HDAC) inhibitors 1:273; 7:223
see also specific drugs
  acute promyelocytic leukemia 1:273–274
  benzamides 7:227
  carboxylic acid bioisosterism 2:659
  cyclic peptides 7:228
  in development 7:224
  heat shock proteins, action on 7:233
  hydroxamic acids 7:224
    aromatic dicarboxylic acid derivatives 7:226
    aroyl-pyrrole-hydroxy-amides (APHAs) 7:225–226
    succinimide derivatives 7:226
  isoform specific 7:228
    (aryloxopropenyl)pyrrolyl hydroxamides 7:229
  mechanism of action 5:225
    structural basis 7:223
  natural products 1:384
  nonhydroxamates 7:227
  PASP 3:811, 3:812F
  short-chain fatty acids 7:224
  solid-phase libraries 3:741
Histone deacetylases (HDACs) 1:263; 7:223
  acute myeloid leukemia (AML) 7:224
  gene silencing 1:263
  structure 7:224
  subclasses 7:223
  transcription suppression 1:265F
Histone deacetylation 1:274
Histones
  chromatin *see* Chromatin
  modifications
    cancer 1:272
    DNA methylation 1:264–265
  proteomics 3:41
Histoplasmosis 7:424T
  ketoconazole 7:455
Historical control groups, clinical trials 2:182
Histrelin 2:579T
  controlled release formulations 2:592
  hydrophobic depoting 2:586
Hit(s)
  confirmation 3:684
  follow up 2:205
  generation support 3:861

identification 1:519
  drug discovery 2:31, 2:32F
  fluorescence screening assays 3:613
  HTS 3:684
  location, HTS data management 2:454
  optimization 1:519
  validation 3:591, 3:592F
HitHunter cAMP assay, GPCR functional assay screening 2:809
Hit-to-lead *see* Drug discovery
HIV
  genome 7:331
    polypurine tract (PPT) 7:331
    primer binding site (PBS) 7:331
    sequencing in drug development 8:187
  integrase, as target 1:69
  life cycle 7:331, 7:332F
    CCR5R4 receptor 7:360
    CCR5 receptor 7:331, 7:358–359
    CD4 receptors 7:331, 7:358–359
    CXCR4 receptor 7:331, 7:358–359, 7:360
    host cell genome 7:331–333, 7:335F
    reverse transcriptase 7:331, 7:333F
    strand transfer 7:331
    transcription 7:331, 7:334F
    viral mRNA production 7:333
  protein structures on PSD 3:382
  reverse transcriptase, as target 1:69
Hivid *see* Zalcitabine
HIV infection 8:187
  see also AIDS
  antimicrobial agents 7:474
    aminoglycosides 7:641
      neomycin 7:641
      streptomycin 7:641
  antiviral drugs 7:254T, 7:279, 7:329–371
    see also specific drugs
    adefovir dipivoxil 7:314
    in clinical trials 7:282–285
    combination chemotherapy *see below*
    compliance 7:336, 7:362
    dosing regimens 7:336
    emtricitabine 7:314
    entry inhibitors *see below*
    integrase inhibitors 7:362, 7:363F, 7:364F
    long-term adverse effects 7:362
    maturation inhibitors 7:364, 7:364F
    naive *vs.* infected patients 7:337
    nonenzymic targets 7:365
    polyoxometallates 7:360

    preclinical development 7:285–288
    resistance 7:336, 7:362
    RNaseH inhibitors 7:365
    salvage treatment 7:337
    side effects 7:336
    structured treatment interruptions 7:337
  clade geographical distribution 7:335
  combination chemotherapy 7:288, 7:337
    cross-resistance 7:337
    drug interactions 7:337
    treatment sequencing 7:337
  developing countries 1:338; 7:361
    drug pricing 1:644–645
    impact 1:648
    mortality/morbidity 1:638, 1:642T, 1:648
  disease severity 7:335
    plasma viremia 7:335
    T cell numbers 7:335
  drug development 1:528–529
  entry inhibitors 7:279–282, 7:358
    clinical limitations 7:359
    compliance 7:359
    dosing regimens 7:359
    drug compatibility 7:359
    extent of use 7:359
    future work 7:359
    resistance 7:359
    tolerability 7:359
  experimental models 7:336
    cell culture 7:336
    simian models 7:336
  hepatitis B coinfection 7:265
  hepatitis C coinfection 7:267
  history 7:330
  infectious agent identification 7:331
  pharmacogenetics 3:61
  prevalence 7:330
  suboptimal treatment 7:333
  symptomatic classification 7:330
    CD4+ T cell numbers 7:330
  symptoms 7:330
  transmission 7:333
    drug-resistant 7:333
  unmet medical needs 7:360
HIV integrase inhibitors
  see also specific drugs
  discovery/development 1:76–79, 1:79F, 1:80F; 4:519
    pharmacophores 4:522F, 4:523F, 4:524F
  natural products 1:374
HIV protease
  crystal structure 4:600, 4:600F
    with inhibitor 4:600, 4:601F
  Molecular Interaction Fields 4:242, 4:243F
  Rous sarcoma virus *vs.* 3:508–509

HIV protease (*continued*)
  structure 3:508–509
    ritonavir binding 1:76, 1:78*F*
    three-dimensional
      structures 2:626*T*
HIV protease inhibitors 7:279–282,
  7:347
  see also specific drugs
  carboxamide bioisosterism 2:668
  clinical limitations 7:356
  compliance 7:356
  computer-assisted drug design
    (CADD) 4:22, 4:23*F*
  discovery/development 1:45*T*, 1:74–
    76, 1:75*F*, 1:76*F*, 7:347; 8:187
    aspartic acid proteases 7:347–349
    compound library
      screening 2:301–302
    diastereomeric diols 8:189,
      8:189*F*
    oral bioavailability problems 8:188
    pharmacophores 4:517*F*
    SAR 8:188
    surface plasmon resonance 3:593–
      594
    symmetry core structure 8:187–
      188, 8:188*F*
  dosing regimens 7:357*T*, 7:358*T*
  drug compatibility 7:357
  drug resistance, pharmacokinetic-
    pharmacodynamic
    models 8:191–192
  extent of use 7:356
  future work 7:358
  mechanism of action 7:191–192
  natural products 1:372
  nonpeptide drugs 2:618–619, 2:620*F*
  peptide/protein drugs 2:577
    structure 2:578*F*
  peptidomimetics 2:613, 2:614*F*,
    2:615*F*, 2:616*F*
  P-glycoprotein interactions 5:769*T*
  resistance 7:356
    mutations 7:356, 7:356*T*
    pharmacokinetic-pharmacodynamic
      models 8:192*F*
  scaffold-based design 4:28, 4:28*F*
  structure 7:347–349, 7:348*F*
    quantum mechanical
      calculations 4:391
    x-ray crystallography 2:625,
      2:630*F*
  structure-based design 4:599
    bioavailability 4:604
    HTS 4:610, 4:611*F*
    nonpeptide scaffolds 4:607,
      4:608*F*, 4:610, 4:612*F*
    substrate-based inhibitors 4:608,
      4:609*F*
  tolerability 7:356
HIV TAR 2:1044*F*, 2:1047
HKI-272 7:196
  structure 7:196*F*

HLA *see* Human leukocyte antigen
  (HLA)
³H-labeled compounds, metabolism
  studies 5:34
HLA-DR2/DQ1, narcolepsy 6:149
HLE, drug metabolism, in vitro
  studies 5:233
HlogP method, log P
  calculation 5:656
HMG-CoA reductase
  atherosclerosis 1:374–375
  inhibitors *see* Statins
  mechanism of action 6:469*F*
HMR1275 *see* Flavopiridol
HOBACGEN database, comparative
  genomics 7:739
Hobbs DeWitt, Sheila
  combinatorial chemistry, invention
    of 2:260
  management difficulties 2:260
Hodgkin, Dorothy Crowfoot, protein
  crystallography 3:450
HOE-694 6:756
Hoechst 33342, structure 5:778*F*
Hoffman-La Roche drugs,
  mitoprotective drugs 2:1062,
    2:1062*F*
Hofmeister Series, protein
  crystallization 3:881
Hohenberg–Kohn paradigm, quantum
  mechanical calculations
    4:384
Hollow-fiber assay, anticancer
  agents 7:34
Hologram QSAR, 3D-QSAR
  alignment 4:585
Homeopathy 1:419
  definition 1:408
Homeostatic control mechanisms,
  serotonin receptor
    agonists 5:899
Homeostatic feedback, mechanism-
  based pharmacokinetic/
  pharmacodynamic (PK/PD)
  modeling *see* Mechanism-based
  pharmacokinetic/
  pharmacodynamic (PK/PD)
  modeling
Homodimers
  nuclear receptors 4:730, 4:731*F*
  STAT (signal transducer and activator
    of transcription)
    association 3:204
Homogeneous approach, microplate-
  based assays 2:367
Homogeneous assays, fluorescence
  correlation spectroscopy-related
  confocal fluorimetric techniques
  (FCS⁺) 2:405*T*, 2:406
Homogeneous membranes,
  transmembrane permeation,
  solubility-diffusion
  model 5:460, 5:461*F*

Homogeneous time-resolved
  fluorescence (HTRF) 2:411
  enzyme assays 2:418, 2:419*F*
    sensitivity 2:419, 2:420*F*
  GPCR functional assay
    screening 2:794*F*, 2:810–811
  HTS applications 2:416
  immunoassays 2:417, 2:418*F*
    ELISA *vs.* 2:417
  lanthanide cryptates 2:412, 2:413*F*
    DELFIA 2:412–413
    fluorescence resonance energy
      transfer 2:413–414
    signal amplification 2:413
    XL665 2:413, 2:414*F*
  media, resistance to 2:412
  nucleic acid hybridizations 2:422
    DNA nuclease assays 2:422,
      2:422*F*, 2:423*F*
    reverse transcriptase assays 2:422,
      2:423*F*, 2:424*F*
    optimization 2:423
    buffers 2:423, 2:424*F*
    controls 2:425
    label choice 2:424
  protein–protein interactions 2:421,
    2:421*F*
    dose response curves 2:421,
      2:422*F*
  receptor-binding assays 2:419
    nuclear receptors 2:420
    peroxisome proliferator-activated
      receptor-γ 2:420, 2:420*F*,
      2:421*F*
  signal measurement 2:414
    data reduction 2:416
    dual-wavelength 2:415, 2:416*F*,
      2:417*F*
    time-resolved 2:415, 2:415*F*
  signal modulation 2:412
  unique tracers 2:412
Homogenous samples,
  proteomics 3:30–31
Homogenous time-resolved
  fluorescence energy transfer
  (HTR-FRET) 3:605
  scintillation proximity assay *vs.*
    3:590
Homologous genes, protein function
  prediction,
    bioinformatics 3:334
Homologous recombination, knockout
  mouse model generation
    3:11–13
Homolog screening
  protein crystallization 3:884
  protein structure prediction 3:325
Homology-based techniques, RNA
  structure prediction,
    bioinformatics 3:327
Homology models 3:544
  androgen receptor 4:739–740
  databases 3:544

focused library design *see* Focused
library design, structure-based
methods
generation, cytochrome P450s
5:813
glutathione *S*-transferases
(GSTs) **5**:818–819, **5**:819*F*
GPCRs **4**:683, **4**:687–688
hERG* *see* hERG blockers
5-HT$_{1A}$ receptor structure **4**:686
K$^+$ ion channels **4**:710
drug interactions **4**:720
Live Bench website **3**:545
methyltransferase molecular
modeling **5**:820
mineralocorticoid receptor **4**:739–
740
model databases
ModBase **3**:545
molecular dynamics (MD) **3**:544–
545
nuclear receptors, drug design **4**:739
P-glycoprotein models *see* P-
glycoprotein (P-gp), in silico
models
problems **3**:544
reliability **3**:544
CASP (Critical Assessment of tp
Structural Prediction) **3**:544
initial sequence alignment **3**:544
structure-based library
design **4**:322–323
SWISS-MODEL **3**:544
vitamin D receptors **4**:740
Homoplasty, mitochondria
genome **2**:1059–1060
Homo sapiens
genome sequencing **3**:317*T*
proteome complexity **3**:29*T*
Honesty, project teams **2**:145
Honolulu Heart Study, coronary heart
disease **6**:471
Hooke's law, force-field molecular
modeling **4**:90
Hookworm infections **7**:751*T*
biology **7**:759
clinical manifestations **7**:759
drug treatment **7**:753*T*
HOPE, number of patients needed to
treat **8**:144*T*
Horizontal structure, R&D
management **2**:249
Hormonal drugs, acne treatment *see*
Acne treatment
Hormone replacement therapy (HRT)
Alzheimer's disease therapy **6**:207
bioethics, safety issues **1**:686
bone diseases/disorders **6**:512
breast cancer risk **8**:104
estrogen receptor ligands **2**:1013–
1014
Million Women Study **8**:104
Women's Health Initiative **8**:104

Hormones, peptide/protein
drugs **2**:580
Hormone therapies, vaginal
administration **5**:289
Hornel *see* Falecalcitriol
Horner–Emmons reaction, solid support
synthesis **2**:469*F*
Horrabin, David, drug discovery **2**:269
Horse chestnut (*Aesculus*) **1**:425
Host cell genome, HIV life
cycle **7**:331–333, **7**:335*F*
Host cell preference, gene expression
vectors **3**:109
Host defenses, fungal disease *see* Fungal
disease
Host genetic factors, malaria
epidemiology **7**:770
Hot-plate assay, acute pain
models **6**:332
Hot spot/anchor residues, ligand–
receptor molecular
flexibility **4**:203–204, **4**:204*F*
Houghten, Richard, combinatorial
chemistry **2**:461
HOX-PBX/DNA transcription factor,
focused library design, structure-
based methods **4**:329
hPEPT1 inhibitors, development,
pharmacophores **4**:532*F*
HPLC *see* High-performance liquid
chromatography (HPLC)
HPMA, drug–polymer
conjugates **5**:1048
HPMPO-DAPy, poxviruses **7**:261–262
HSDB, ionization prediction, in silico
tools **5**:623*T*
HSITE, site analysis, drug
design **4**:286
5-HT (serotonin)
affective disorders **6**:74
anxiety disorders **6**:90
bioisosterism **2**:675, **2**:677*F*
gastrointestinal motility *see*
Gastrointestinal motility
gastrointestinal tract *see*
Gastrointestinal tract
hypothesis, schizophrenia **6**:20,
6:26–27
ionization constant (pK_a) **5**:366*T*
monoamine oxidase **5**:175
signaling abnormalities, irritable bowel
syndrome *see* Irritable bowel
syndrome (IBS)
sleep **6**:143, **6**:146
structure **2**:57*F*
5-HT receptor(s) **6**:647
D2 receptor interactions,
antipsychotic drugs **6**:33
molecular probes **3**:664*T*
polymorphisms **1**:288
regional distribution **5**:898
lower urinary tract **8**:124
spinal cord localization **8**:125

schizophrenia **6**:24
structure-based drug design **4**:692
3D QSAR **5**:594
5-HT receptor agonists
see also specific drugs
irritable bowel syndrome
treatment **6**:652, **6**:654, **6**:655*T*
mechanism-based PK/PD
modeling **5**:898
advantages **5**:900–901
homeostatic control
mechanisms **5**:899
hypothermic response **5**:898–899
In vivo activity prediction **5**:900
In vivo effects **5**:898–899
population approach **5**:900,
5:900*T*
response variation **5**:900–901,
5:901*F*
nonulcer dyspepsia **6**:600
5-HT receptor antagonists
see also specific drugs
Alzheimer's disease treatment **6**:207
carboxamide bioisosterism **2**:668
carboxylic ester bioisosterism **2**:663
discovery/development **3**:932,
3:932*F*
irritable bowel syndrome
treatment **6**:652, **6**:654, **6**:655*T*
irritable bowel syndrome (IBS)
treatment **6**:599
repositioning **2**:276
schizophrenia treatment **6**:37
subtypes **8**:125
urinary incontinence **8**:124
5-HT$_{1A}$ receptor
anxiety disorders **6**:91
in human disease **2**:775*T*
knockout animal models **3**:156
knockout animals, depression **6**:58
structure
CATALYST **4**:686–687
homology models **4**:686
structure-based drug design, virtual
screening **4**:689
5-HT$_{1A}$ receptor agonists **6**:75–76
see also specific drugs
anxiety disorder treatment **6**:101
defensive test battery **6**:94
glaucoma treatment **6**:313*T*
mechanism of action **6**:101–102
5-HT$_{1A}$ receptor antagonists,
development **4**:516*F*
5-HT$_{1B/D}$ receptor, allosteric
modulators **2**:790*T*
5-HT$_{1B}$ receptor
anxiety disorders **6**:91
in human disease **2**:775*T*
virtual screening **4**:689
5-HT$_{1C}$ receptor, anxiety
disorders **6**:91
5-HT$_{1C}$ receptor antagonists **6**:76
see also specific drugs

5-HT$_{1D}$ receptor, in human
  disease **2**:775*T*
5-HT$_{1D}$ receptor agonists,
  development **4**:519*F*
5-HT$_{1D}$ receptor antagonists
  see also specific drugs
  migraine treatment **6**:376
5-HT$_{1F}$ receptor antagonists
  see also specific drugs
  migraine treatment **6**:376
5-HT$_{2A/2C}$ receptor antagonists
  see also specific drugs
  primary insomnia treatment **6**:161
5-HT$_{2A}$ receptor, in human
  disease **2**:775*T*
5-HT$_{2A}$ receptor antagonists
  basicity reduction **5**:950, **5**:950*F*
  flexibility reduction **5**:950–951,
    5:951*F*
  lipophilicity reduction **5**:948, **5**:949*F*,
    5:950, **5**:950*F*
5-HT$_{2C}$ receptor
  activation **2**:785*T*
  allosteric modulators **2**:790*T*
  constitutively active **2**:793*T*
  homology models **4**:686
  in human disease **2**:775*T*
  polymorphisms **1**:286
    pharmacodynamics **1**:286, **1**:287*T*
5-HT$_{2C}$ receptor antagonists, obesity
  treatment **6**:413, **6**:414*F*
5-HT$_2$ receptor agonists
  see also specific drugs
  glaucoma treatment **6**:313*T*
5-HT$_3$ receptor **2**:888; **3**:624*T*, **3**:633,
    3:635*F*
  alosetron **6**:653
  antiemetic treatment **3**:633–634
  cell choice **3**:633–634
    receptor affinity **3**:635*T*
  diseases **2**:891
    bipolar disorders **2**:891
    neuropsychiatric disorders **2**:891
    Tourette's syndrome **2**:891
  distribution **3**:633–634
  molecular biology **2**:888
  nomenclature **2**:888
  pharmacology **2**:889, **2**:890
    see also specific drugs
    agonists **2**:890
    picrotoxin **2**:890
    'setrons' **2**:889–890
  physiological functions **2**:889
    neurotransmitter release
      2:889
  regulation **2**:889
  structure **2**:889
5-HT$_3$ receptor agonists
  see also specific drugs
  prokinetic drugs **6**:686
5-HT$_3$ receptor antagonists
  see also specific drugs
  antiemetic drugs **6**:681

discovery/development **3**:932,
    3:932*F*
irritable bowel syndrome
  treatment **6**:661*T*
5-HT$_{4A}$ receptor, activation **2**:785*T*
5-HT$_4$ receptor antagonists
  see also specific drugs
  discovery/development **3**:932,
    3:932*F*
  prokinetic drugs **6**:677
5-HT$_4$ receptor partial agonists
  see also specific drugs
  irritable bowel syndrome
    treatment **6**:661*T*
5-HT$_{5A}$ receptor, in human
  disease **2**:775*T*
5-HT$_6$ receptor, in human
  disease **2**:775*T*
5-HT$_7$ receptor
  allosteric modulators **2**:790*T*
  structure-based drug design **4**:693
5-HT$_7$ receptor antagonists **6**:76
  see also specific drugs
  development **4**:520*F*
HT29-MTK (mucus-producing) cell
  line, permeability assays **5**:270
HT-0712 **6**:80
HTML, database integration **3**:303
HU-210, IBD treatment **6**:629
HU-211
  as antiemetic agent **6**:684
  structure **6**:315*F*
HU-308 **6**:342–343
  structure **6**:343*F*
Huber, Robert, protein
  crystallography **3**:451
Humalog *see* Insulin Lispro
Human 3-D skin model phototoxicity
  test (H3D-PT), non-animal
  photoirritancy testing **1**:471–
  472
Human absorption studies
  dogs *vs.* **5**:41, **5**:42
  monkey *vs.* **5**:41–42, **5**:42*F*
  rats *vs.* **5**:40–41, **5**:42*F*, **5**:43*F*
Human antibody fragments, monoclonal
  antibodies **1**:336
Human cell use, animal
  experimentation
  alternatives **1**:479
Human chorionic gonadotropin (hCG),
  oral testosterone replacement
  therapy **6**:532, **6**:534*T*
Human disease, GPCR MAP kinase
  activation **2**:784
Human drug metabolism database, drug
  research, current status **1**:60
Human embryonic kidney cells *see*
  HEK-293 cells
Human ether-a-go-go related gene *see*
  hERG
Human evaluation step, development
  chemistry timeline **2**:170

Human Fertilization and Embryology
  Act (1990) **1**:559
Human Fertilization and Embryology
  Authority (HFEA) **1**:559
Human genome, mouse genome
  vs. **3**:10–11
Human Genome Project
  clinical practice effects **1**:604
  drug design, de novo **4**:284
  drug discovery **3**:146
  drug discovery, decline in **1**:618
  drug discovery effects **1**:595
  funding **1**:734
  genetic mapping objective **2**:306
  government role **1**:734
  hierarchical shotgun
    sequencing **3**:318
  industry/academia
    collaborations **1**:573
  target selection **2**:754–755
Human herpesvirus 6 (HHV-6) **7**:295,
    7:296
  antiviral drugs **7**:260
  cytomegalovirus 423 **7**:261
Human herpesvirus 7 (HHV-7) **7**:295,
    7:296
  antiviral drugs **7**:260
  cytomegalovirus 423 **7**:261
Human herpesvirus 8 (HHV-8) **7**:295,
    7:296
  antiviral drugs **7**:261
Human insulin
  brands, global sales **1**:605*T*
  development **1**:556–557
  drug discovery **1**:594
'Humanized' monoclonal
  antibodies **1**:336
Human kinome **7**:186*F*, **7**:187
Human leukocyte antigen (HLA)
  see also Major histocompatibility
    complex (MHC)
  class I, malaria epidemiology
    7:770
  IBD **6**:617*T*, **6**:618
  narcolepsy **6**:7–8
Human liver cytosolic fraction
  N-acetyltransferases (NAT) **5**:234
  drug metabolism, in vitro
    studies **5**:232*T*, **5**:234
  glutathione transferases **5**:234
  sulfotransferases **5**:234
Human liver S9 fraction, drug
  metabolism, in vitro
  studies **5**:232*T*, **5**:234
Human lymphotrophic virus type I
  (HTLV1) **7**:4
Human medulla cells, drug metabolism,
  in vitro studies **5**:234
Human metabolic stability,
  electrotopological indices
  (E-state index) *see*
  Electrotopological indices
  (E-state index)

Human nuclear factor 4 receptor, nomenclature 2:996*T*
Human oral absorption
  dog *vs.* 5:674
  rat *vs.* 5:673–674
Human organic cation transporter (hOCT1) 5:46
Human papilloma virus-16 (HPV16), cancers 7:4
Human papilloma virus-18 (HPV18), cancers 7:4
Human perfusion studies, oral absorption *see* Oral absorption
Human Pharmaco DNP Consortium 2:316
Human research ethics 1:714
  children 1:715, 1:718
  data access 1:718
  free consent 1:715
  informed consent 1:715, 1:716
  injury compensation 1:717
  monitoring 1:717
  placebo controls 1:715
  post-study responsibilities to subjects 1:717
  privacy 1:715
  protocol adherence 1:716
  risk *vs.* benefit 1:714
Human serum albumin (HSA) 4:776
  see also Albumin
  definition 4:776–777
  structure 4:777, 4:777*F*
Human serum albumin (HSA) binding, pharmacokinetics 5:14
Human skin–water partition coefficients (log K_{SC}), dermal penetration models 5:731
Human umbilical cord blood (HUCB), animal experimentation alternatives 1:480
Human whole-blood test (WBT), non-animal pyrogenicity testing 1:466, 1:467*T*
Humidity, stability testing 5:498
Humira *see* Adalimumab
Humoral immune response 7:846
Humulin *see* Human insulin
Hünefeld, Friedrich, protein crystallization 3:881
Hunter-Bolton reagent, direct iodination 3:666
Huntingtin protein, Huntington's disease 6:218
Huntington's disease 6:11, 6:218
  characteristics 6:199*T*
  clinical trial issues 6:219
  disease basis 6:218
    huntingtin protein 6:218
    mTOR sequestration 6:218–219
  disease state/diagnosis 6:218
  Drosophila melanogaster model system 3:9
  inheritance 6:195

mitochondria 2:1054–1055
  models 6:219
    genetic-based 6:219
    knockout mouse model 3:11–13
    mitochondria complex II *see* Mitochondria complex II
    neurotoxin-based 6:219
  motor function 6:195
  new research areas 6:221, 6:222*T*
  phosphodiesterase-10 2:930–931
  treatment 6:221
  unmet medical needs 6:221
Huperzine A 6:203–204
Huuskonen model, aqueous solubility models 5:638
Hyalohyphomycosis, opportunistic infections 7:425*T*
HYBOT 5:581*T*
  computational oral absorption prediction 5:687
Hybridization
  differential display, functional genomics 2:329
  single nucleotide polymorphisms 2:364, 2:365*F*
Hybridization probe uniqueness, repetitive elements, bioinformatics 3:324
Hybrid oligomers, solid support synthesis *see* Solid support synthesis
Hybridoma technology 1:436; 3:118
  see also Monoclonal antibodies
  chimeric antibodies 1:436
  disadvantages 1:436
    immunogenicity 1:436
    time 1:436
  discovery/development 1:432
  methods 1:436
Hybrid organization *see* Pharmaceutical industry
Hybrid pH-metric/ultraviolet methods, ionization constants (pK_a) measurement 5:392, 5:393*F*
Hybrid triacylglycerides, coenzyme A conjugation *see* Coenzyme A conjugation
Hybrid vendors, definition 2:206
Hycamtin *see* Topotecan
Hydantoin derivatives, solid-phase discrete libraries 3:709*T*
Hydantoin synthesis
  PASP 3:799–800, 3:800*F*
  solution phase library synthesis 3:786, 3:786*F*
Hydatid form moles, epigenetics 1:270*T*
Hydralazine
  acetylation 5:154, 5:154*F*
  bioavailability 5:712*T*
  drug action 1:532*T*
Hydrates, phase diagrams 5:512

Hydration
  chemical instability 5:492
  ionization 5:607
Hydrazides, acetylation 5:154
Hydrazines, acetylation 5:154
Hydrochlorothiazide
  BCS 5:973*T*
  benzoic acid 5:404*T*
  bioavailability 5:712*T*; 6:709*T*
  dose 6:709*T*
  effective permeability (P_{eff}) 5:673*T*
  half-life 6:709*T*
  oral absorption 5:680*T*
  structure 2:66*F*, 5:371*F*; 6:709*T*
Hydrocodone, GPCR targeting 2:773*T*
Hydrocortisone
  dermal penetration models 5:735
  oral absorption 5:468*T*, 5:680*T*
  P-glycoprotein interaction 5:781*T*
  psoriasis treatment 7:944
  structure 2:1017*F*
Hydrogen atoms, protein 3D structure validation 3:513
Hydrogen atom transfer, cytochrome P450 5:95, 5:95*F*, 5:96*F*
Hydrogen bond acceptor (HBA) groups
  group-type E-state index 4:554, 4:569*T*, 4:571*T*
  hydrogen bonding 5:587
  lipophilicity measurement 5:433
  logP' 5:440
Hydrogen bond acceptor (O) probes *see* VolSurf
Hydrogen bond descriptors
  QSAR 4:55
  VolSurf descriptors 5:539
Hydrogen bond donor (HBD) groups
  group-type E-state index 4:554, 4:569*T*, 4:571*T*
  hydrogen bonding 5:587
  lipophilicity measurement 5:433
  log$P$ 5:440
Hydrogen-bond entities, Molecular Interaction Fields 4:249–250
Hydrogen bond *J* coupling, NMR 2:514
Hydrogen bond potential, GRID force field 4:240
Hydrogen bonds
  absorption 5:704
  corneal permeability modeling 5:736
  equilibrium constants, molecular descriptors 5:536
  intermolecular interactions, CSD 3:401–402
  ligand–receptor interactions *see* Ligand–receptor interactions
  Mercury Visualizer 3:395
  molecular descriptors *see* Molecular descriptors
  oral absorption prediction *see* Oral absorption prediction, molecular descriptors

Hydrogen bonds (*continued*)
  polydimethylsiloxane **5**:741
  protein 3D structure validation *see*
      Protein 3D structure validation
  quantum mechanical
      calculations **4**:397
  site analysis, drug design **4**:287
  solvent effect calculations **4**:397
  strong, CSD **3**:402, **3**:403*F*, **3**:404*F*
  weak, CSD **3**:404
Hydrogen extension vectors (HEVs),
    3D-QSAR alignment **4**:579–580
Hydrogen gas, PASP **3**:800*F*, **3**:827,
    3:827*F*
Hydrogen peroxide, stability
    testing **5**:498
Hydrogen/potassium ATPase, knockout
    mouse model **3**:13
Hydrolases **5**:134, **5**:136*T*, **5**:179
  see also Epoxide hydrolases; *specific
      enzymes*
  almotriptan **5**:135
  carboxylesterases **5**:135
  carboxylic derivatives **5**:135
  catalytic features **5**:135, **5**:138*F*
  cholinesterases **5**:135
  drug metabolism **5**:203
  electrophilic component **5**:135
  International Union of Biochemistry
      and Molecular biology (IUBMB)
      classification **5**:134–135
  nucleophilic component **5**:135
  ontogenic development **5**:174*T*
  paraoxonases **5**:135
  peptidases **5**:135
  pharmacological target space **4**:422*T*
  protein data bank **5**:134–135
  proton donor **5**:135
  substrates **5**:135
Hydrolysis
  carbon-silicon bioisosterism **2**:694
  chemical instability *see* Chemical
      instability
  epoxide hydrolases *see* Epoxide
      hydrolases
  quantum mechanical
      calculations **4**:400
Hydrolysis, drug metabolism **5**:134
  amides **5**:138, **5**:139*F*
  definition **5**:134
  esters **5**:135
    cocaine *see* Cocaine, hydrolysis
  hydrolases *see* Hydrolases
  peptides **5**:139
    peptidases **5**:136*T*, **5**:139,
        5:140*F*
Hydrolytic enzymes
  see also specific enzymes
  plasma **5**:499–500
Hydropathic analysis, transmembrane
    helices **4**:674
Hydropathic intermolecular field
    analysis **4**:578

Hydropathy plots, GPCRs **4**:673
Hydroperoxidases, atherosclerotic
    plaque formation **6**:467–468
Hydrophilic depoting, peptide/protein
    drugs *see* Peptide/protein drugs
Hydrophilic drugs, paracellular
    transport **5**:265
Hydrophilic folding,
    lipophilicity **5**:431, **5**:432*F*
Hydrophilic–lipophilic balance (HLB)
  absorption **5**:704
  VolSurf descriptors **5**:539
Hydrophilic polymers, drug–polymer
    conjugates **5**:1046
Hydrophilic regions (W1-W8), VolSurf
    descriptors **5**:538, **5**:539*F*
Hydrophobic collapse,
    lipophilicity **5**:431, **5**:432*F*
Hydrophobic contour maps, Molecular
    Interaction Fields **4**:248
Hydrophobic depoting, peptide/protein
    drugs *see* Peptide/protein drugs
Hydrophobic field models, P-
    glycoprotein models
    5:782–783
Hydrophobic fields, similarity
    searching **4**:173
Hydrophobic forces,
    lipophilicity **5**:429
Hydrophobic integy moments
    (ID1-ID8), VolSurf descriptors
    5:539
Hydrophobic interaction
    chromatography, protein
    purification **3**:423
Hydrophobic interaction field (HINT),
    lipophilicity **5**:585
Hydrophobic interactions
  lipophilicity **5**:429–431
  molecular descriptors **5**:547–548
Hydrophobic ion pairs, lipophilicity
    measurement, ions/
    zwitterions **5**:442, **5**:442*F*
Hydrophobicity
  chemogenomics **4**:422
  ligand–receptor interactions *see*
      Ligand–receptor interactions
  lipophilicity *see* Lipophilicity
  molecular descriptors **5**:534
  phosphodiesterases **4**:760
  prediction, chemoinformatics,
      ADME property
      prediction **3**:259
  QSAR **4**:47
  transmembrane helices **4**:674,
      4:674*F*
Hydrophobicity vectors (HPV), 3D-
    QSAR alignment **4**:579–580
Hydrophobic patches, Molecular
    Interaction Fields *see* Molecular
    Interaction Fields (MIFs)
Hydrophobic peptides, P-glycoprotein
    interactions **5**:769*T*

Hydrophobic (DRY) probes *see*
    VolSurf
Hydrophobic regions
  GRID force field **4**:241, **4**:242*F*
  VolSurf descriptors **5**:539
6-Hydroxalkyl penams, β-lactamase
    inhibitors **7**:511
Hydroxamic acids
  see also Histone deacetylase (HDAC)
      inhibitors
  carboxylic acid bioisosterism *see*
      Bioisosterism, carboxylic acids
  derivatives, PASP **3**:811, **3**:811*F*
  structure **5**:371*F*
(*R*)-9-(4-Hydroxy-2-
    hydroxymethylbutyl)guanine *see*
    H2G
4-Hydroxy-12-deoxyanhydrotetracycline
  structure **7**:600*F*
  tetracyclines biosynthesis
      7:600–601
N-Hydroxy-arylamines, DNA, reaction
    with **5**:918
p-Hydroxybenzoate hydroxylase,
    quantum mechanics
    4:411–412
22(*R*)-Hydroxycholesterol **2**:1023
  structure **2**:1024*F*
Hydroxychromones,
    bioisosterism **2**:662*T*
6-Hydroxy-5a-dehydrotetracycline
  structure **7**:600*F*
  tetracyclines biosynthesis **7**:600–601
6-Hydroxydopamine lesions
  ADHD *see* Attention deficit
      hyperactivity disorder models
  Parkinson's models **6**:212–213
20-Hydroxyeicosatetraenoic acid, renal
    dysfunction **6**:582
Hydroxyethylureas, carboxamide
    bioisosterism **2**:668, **2**:669*F*
Hydroxyflutamide, structure **2**:1016*F*
3-Hydroxy-γ-pyrones, carboxylic acid
    bioisosterism **2**:661
2-Hydroxy groups, azetidinone
    cholesterol absorption
    inhibitors **8**:68–69
5-Hydroxy indoles, structure **2**:887*F*
Hydroxyisopropyl, phenol
    bioisosterism **2**:670–671
Hydroxylamides
  UDP-glucuronyltransferases **5**:151
  sulfotransferases **5**:148*F*, **5**:149
Hydroxylamines, UDP-
    glucuronyltransferases **5**:151
Hydroxylation, tamoxifen
    metabolism **8**:89
N-Hydroxylation, cytochrome
    P450s **5**:105
Hydroxyl group replacement,
    aminoglycosides **7**:635
Hydroxymethyl, phenol
    bioisosterism **2**:670–671

Inactive compounds, pharmacophore detection 4:122–123
Inactive metabolites, 'soft drug's 5:1010
Inamrinone, congestive heart failure treatment 6:698
Inborn errors of metabolism, metabonomics 5:1002
Incadronate, toxicology 8:207
Incandronic acid 1:219
IN Cell 1000 3:686T
IN Cell 3000 3:686T, 3:687
Incentive sensitization hypotheses, addiction 6:175
InChI, chemical information databases 3:298
Inclusion bodies
    formation, Escherichia coli protein expression systems 3:112
    purification, bacterial expression systems 3:415
Inclusion/exclusion criteria, clinical trials 2:186–187
Income differences, education 2:85–86
Incomplete combinatorials, library design 4:308
Incontinence Quality of Life Questionnaire (IQOL) 8:128, 8:129–130
Incontinence Severity Index (ISI) 8:130, 8:131F
Incretin
    analogs 2:583
    augmentation, diabetes mellitus management 6:453
    hydrophobic depoting 2:587
Incubational binding see Tissue binding
Indapamide
    bioavailability 6:709T
    dose 6:709T
    half-life 6:709T
    structure 6:709T
Indavir, structure 2:578F
Indazole, bioisosterism 2:660T
5-Indazole, bioisosterism 2:673T
Indels, sequence alignment, bioinformatics 3:318–319
Indenopyrazoles, cyclin-dependent kinase inhibition 7:205–206
Indenopyrrolocarbazoles, solid-phase libraries 3:703, 3:705F
Independence/initiative
    career guidelines 2:18
    drug discovery 2:272–273
Inderal see Propranolol
Index reaction
    IC50 determination 5:247
    reaction phenotyping 5:245
India
    healthcare systems reform 1:641
    IPA law 1:560
    market dominance 2:94

'Indicator organisms,' antimicrobial agent screening 7:472
Indigenous population worries, traditional medicine 1:410
Indinavir sulfate 1:124; 2:579T; 7:279–282
    approval 7:349
    discovery/development 1:76, 1:78F; 7:350
    dosing 4:607; 7:357T
        with ritonavir 7:358T
    structure 1:78F, 7:348F, 7:349F, 7:351F, 7:357T
    structure-based design 4:604, 4:607F
    target information 4:750T
Indiplon, primary insomnia treatment 6:158, 6:159T
Indirect competition enzyme-linked immunosorbent assay 3:579–580
Indirect enantioselective chromatography 2:721
Indirubins 1:365–366, 1:390
    see also specific drugs
    Alzheimer's disease treatment 6:210–211
    discovery/development 1:390
    structure 1:364F, 1:390F
Indisetron, 5-HT3 receptor binding 2:890
Individual combinatorial libraries, screening library selection 4:505
Individual dedicated workstations, automation 5:344
Individual evaluation, project teams 2:146
Individuality, drug discovery 2:269, 2:272, 2:282
Individualized medicine see Personalized medicine
Indivisibility, economic market failure arguments 1:727
Indocarbazole derivatives, UL97 protein kinase inhibitors 7:310
Indolactam V, solid support synthesis 2:480–482, 2:483F
Indole-2-carboxylate, bioisosterism 2:660T
Indole derivatives, solid-phase discrete libraries 3:709T
Indoles 1:449–450
    bioisosterism 2:660T
    derivatives 1:365–366
        anti-NS5b nonnucleosides 7:402
        structure 1:364F
    structure 5:359
    triphenylethylene anticancer agents 7:161–162
5-Indoles, bioisosterism 2:673T
Indolidan, structure 2:934

Indolizino[8,7-b]-5-carboxylates, solid-phase discrete libraries 3:709T
Indolocarbazoles 7:115
5-Indolone, bioisosterism 2:673T
Indomethacin
    aggregation reactions 5:407, 5:408F, 5:409F
    analogs, CRTH2/DP-1 receptor antagonists 7:890, 7:892
    benzoic acid 5:404T
    bioavailability 5:712T
    disk intrinsic dissolution rate (DIDR) 5:417, 5:417T, 5:418T
    oral absorption 5:468T, 5:680T
    structure 2:9F, 2:658F, 2:681F, 6:336F
Indomethacin analogs, solid-phase discrete libraries 3:707F
Indoramine, bioavailability 5:712T
Indoxyl 1:365–366
    structure 1:364F
IND timeline and resources, DCN 2:155
[111In]-DTPA-PEG annexin, apoptosis imaging 3:668
Induced-fit docking, docking 4:269
Induced-fit models, ligand–receptor interaction models 4:195F, 4:205
Inducible nitric oxide synthase (iNOS) 2:1078
    expression, NF-κB pathway 3:197
Inducible promoters
    gene expression vectors 3:108
    transgenic mouse model 3:11
Inductive effects, ionization 5:605
Industrial countries see Developed countries
Industrialization, regulatory problems 2:725
Industry, role of, venture capital see Venture capital
Industry interchange program, industry/academia collaborations 1:581
Ineffective leaders see Leadership
Inequality, motivation 2:246
Inert solvents, log P calculation 5:652
Infection imaging
    [18F]FDG 3:670
    [18F]FET 3:670
    [18F]FLT 3:670
Infections
    adalimumab 1:442
    age-related macular degeneration 6:302
    ATP-gated ion channels 2:905–907
    diabetes mellitus 6:424
    [18F]FDG 3:670
    multiple sclerosis 7:923
    small animal imaging 3:670

Infectious diseases
  see also specific diseases/disorders
  gene therapy **1**:303
  monoclonal antibodies **1**:434
  SMEs **1**:507
  target selection **2**:76
Infectious Diseases Society of
  America **1**:690
Infergen *see* Interferon alfacon-1
Infiltrating ductal carcinoma, breast
  cancer **7**:150
Infiltrating lobular carcinoma, breast
  cancer **7**:150
Infiltration anesthetics **6**:355
'Infinite-cylinder' approximation,
  nanoprobes **2**:517
Inflammation
  acne **7**:960
  age-related macular
    degeneration **6**:302
  angiotensin II **6**:584
  asthma **7**:875, **7**:881–882
  atherosclerotic plaque
    formation **6**:468–469
  benign prostatic hyperplasia **6**:559
  clinical trials **7**:881–882
  COPD **7**:875, **7**:881–882
  glucocorticoid receptor
    ligands **2**:1016–1017
  metabolism studies **5**:36, **5**:37*F*
  neuropathic pain models **6**:334
  NF-κB pathway **3**:192, **3**:197
  nuclear receptors **2**:1025
  phosphodiesterase-4 **2**:925–926
  Propionibacterium acnes **7**:938
  spinal cord injury **6**:273
  transient receptor potential ion
    channels **2**:908–910
  traumatic brain injury **6**:273
Inflammatory bowel disease
  (IBD) **6:613–642**
  areas of involvement **6**:614*T*
  characteristics **6**:614*T*
  clinical features **6**:614*T*
  clinical trials **6**:618
    biomarkers **6**:620
    concurrent medication **6**:621
    disease activity assessment **6**:619
    disease severity **6**:619
    endpoint selection **6**:619
    IBD questionnaire **6**:620
    patient selection **6**:619
    quality of life **6**:620
    ulcerative colitis *vs.* Crohn's
      disease **6**:619
  complications **6**:614*T*
  diagnosis **6**:614
  disease basis **6**:615
    abnormal immune response
      6:615
    genetic linkage analysis **6**:615
    human leukocyte antigens **6**:617*T*,
      6:618

    regulatory T cells **6**:618
    Toll-like receptors **6**:616
  disease models **6**:618
    Mdr1a-/- mice **6**:618
    transgenic animals **6**:617*T*
  epidemiology **6**:614*T*
  histology **6**:614*T*
  new research areas **6**:624
  NF-κB pathway **3**:197
  prevalence **6**:614–615, **6**:623
  transient receptor potential ion
    channels **2**:909–910
  unmet medical need **6**:623
Inflammatory bowel disease
  questionnaire, IBD clinical
  trials **6**:620
Inflammatory bowel disease
  treatment **6**:621, **6**:622*T*
  see also specific drugs
  ABT-874 **6**:623–624
  antiinflammatory treatment **6**:625
    cannabinoid receptor
      agonists **6**:629
    fatty acid amide hydrolase (FAAH)
      inhibitors **6**:629
    IκB kinase inhibitors **6**:626
    interleukin-1β converting enzyme
      inhibitors **6**:628
    p38 kinase inhibitors **6**:627
    peroxisome proliferator activated
      receptor agonists **6**:630
    phosphodiesterase-4
      inhibitors **6**:625
    poly(ADP ribose)polymerase
      inhibitors **6**:628
    retinoid X receptor agonists **6**:630
  anti-tumor necrosis-α agents
    6:623–624
  balsalazide **6**:621
  CDP-571 **6**:620–621
  CNTO 1275 **6**:623–624
  corticosteroids **6**:621–622
  cyclosporin **6**:622–623
  daclizumab **6**:623–624
  Enbrel **6**:622
  fontolizumab **6**:623–624
  immunosuppressants **6**:622–623
    azathioprine **6**:622–623
  infliximab **6**:620–621, **6**:622
  lymphocyte extravasation
    targeting **6**:631
    chemokine receptor
      antagonists **6**:633
    integrin receptor
      antagonists **6**:632
    sphingosine 1-phosphate receptor
      modulators **6**:636
  macrophage migration inhibitory
    factor antagonists **6**:637
  MDX-1100 **6**:623–624
  mesalazine **6**:621
  methotrexate **6**:622–623
  MLN-02 **6**:623–624

  mucosal tolerance re-establishment
    6:624
    E5564/Eritoran **6**:624–625
    helminth administration **6**:624
    probiotics **6**:624
    Toll-like receptor-4
      antagonists **6**:624–625
    Toll-like receptor-9
      antagonists **6**:624–625
  mycophenolate mofetil **6**:622–623
  natalizumab **6**:620–621
  olsalazine **6**:621
  remicade **6**:621, **6**:622
  sargramostim **6**:623–624
  sulfasalazine **6**:621
  tacrolimus **6**:622–623
  TRPV1 antagonists **6**:637
Inflammatory breast cancer **7**:150
Inflammatory cascade hypothesis,
  Alzheimer's disease **6**:198
Inflammatory cytokines
  hypertension **6**:581
  obesity **6**:581
  renal dysfunction **6**:581
Inflammatory diseases/
  disorders **7:845–872**
  see also specific diseases/disorders
  pathophysiology **7**:846
  respiratory diseases **7**:848
Inflammatory mediators
  atherosclerosis **6**:466
  cell-based assays, Toll
    receptors **3**:639
Infliximab **1**:334*T*, **1**:433*T*
  brands, global sales **1**:605*T*
  colorectal cancer treatment **1**:444
  as fast-tracked drug **1**:760*T*
  IBD treatment **6**:620–621, **6**:622
  irritable bowel syndrome
    treatment **6**:600
  psoriasis treatment **7**:946*T*, **7**:947,
    7:975
  rheumatoid arthritis treatment **7**:858
  skin inflammation treatment **7**:975
  tumor necrosis factor-α
    binding **1**:441
Influenza
  drug introductions **1**:107
  hemagglutinin (HA) **4**:614
  neuraminidase *see* Neuraminidases
  oseltamivir phosphate **1**:127
  pandemic strains **4**:614
  zanamivir **1**:131
Influenza endonuclease inhibitors,
  pharmacophores **4**:520*F*
Influenza vaccine **7**:376
  development **1**:531
  live vaccine **1:227**
  introduction of **1**:108
Influenza viruses **7**:373
  antiviral drugs **7**:274, **7**:376, **7**:377
  see also specific drugs
  antifusion drugs **7**:385, **7**:386*F*

clinical trials 7:376
endonuclease capping 7:385
future work 7:386
influenza polymerase complex as
    target 7:386
envelope 7:374
epidemiology 7:374
    antigenic drift 7:374–376
    antigenic shift 7:374–376
    epidemics 7:373, 7:374
    mortality 7:374
genome 7:374
hemagglutinin 7:374
ion channels 7:374
models 7:376
    ferret models 7:376
    Mardin–Darby canine kidney
        (MDCK) cells 7:376
    mouse models 7:376
neuraminidase 7:374
    structure 7:378, 7:378F
outbreaks 7:373
replicative cycle 7:374, 7:375F
    cell infection 7:377–378
    nuclear export protein 7:374
    vRNP complex 7:374
transmission 7:374
Informatics and computational Safety
    Analysis Staff (ICAS) 5:925
Information content, CSD see
    Cambridge Structural Database
    (CSD)
Information extraction, protein function
    prediction,
    bioinformatics 3:337
Information integration, protein
    function prediction,
    bioinformatics 3:338
Information network models,
    electrotopological indices
    (E-state index) see
    Electrotopological indices
    (E-state index)
Information retrieval, protein function
    prediction,
    bioinformatics 3:337
Information sources
    databases 3:285
    postmarketing surveillance 1:761
Information systems 3:265–291
    see also Databases
    definition 3:265
    integration, HTS data
        acquisition 2:450
    pharmaceutical R&D 2:131
    project management 2:156
    R&D budget amount 2:131
    R&D management 2:250
Informative design, product-based
    library design 4:353
Informed consent
    clinical trials 2:187
    EU Biotechnology Directive 1:789

human research ethics 1:715, 1:716
    mental illness, bioethics 1:696
    pharmacogenetics 3:65–66
Infrastructure, pharmaceutical
    R&D 2:130
ING-1, antibody-directed enzyme
    prodrug therapy 5:1034–1035
Ingres 3:294
Inhalation, peptide/protein
    drugs 2:595
Inhaled corticosteroids, asthma
    treatment 7:883
Inherent variability, HTS data
    management 2:449
Inhibition, in vitro studies 5:245
    FDA guidelines 5:245
    IC_{50} determination 5:246
        index reaction
            measurement 5:247
        major CYPs 5:246
        preliminary screening 5:247
        probe selection 5:247
    initial rate conditions 5:246
    interaction risk prediction 5:245
    In vivo risk prediction 5:235T, 5:247,
        5:248F
        absorption rate constant 5:248
        accumulation risk 5:248
        inhibitor concentrations 5:248
        maximum concentration reaching
            the liver (C_{in}) 5:248
        mean steady-state values 5:248
        therapeutic dose estimation
            5:248
    irreversible 5:249
        inactivation constant
            measurement 5:249
        inhibition constant
            measurement 5:249
        kinetics 5:249
        mechanism-based 5:249
        preincubation step 5:249
        time-dependent inhibition 5:249
    kinetic studies 5:245
    reversible 5:246
        competitive 5:246
        IC_{50} determination see above
        In vivo risk prediction see above
        K_i determination 5:246
        mixed 5:246
        noncompetitive 5:246
        types 5:235T, 5:246
        uncompetitive 5:246
    types of 5:245
Inhibition constant (K_i)
    irreversible inhibition 5:249
    metabolic drug–drug
        interactions 5:834, 5:835
Inhibition curves, promiscuous ligands,
    identification 2:743
Inhibitor concentration available to the
    enzyme ($[I]$), metabolic drug–
    drug interactions 5:834

Inhibitor concentrations, inhibition, in
    vitro studies 5:248
Inhibitor of NF-κB (IκB), ubiquitin –
    protein targeting 3:192
Inhibitors of apoptosis (IAPs)
    defects in cancer 7:14
    regulation, NF-κB pathway 3:196–
        197
In-house databases 3:284
In-house synthesis, compound
    acquisition/processing 3:563–
        564
In-house value, drug discovery
    2:279
Initial rate conditions, inhibition, in
    vitro studies 5:246
Initial sequence alignment, homology
    models reliability 3:544
Initiation factors (IFs), scintillation
    proximity assay 3:588
Initiator reactor, Biotage AB microwave-
    assisted chemistry 3:844,
    3:845F
Injection, peptide/protein drugs see
    Peptide/protein drugs, injection
INJECT trial, reteplase 6:773
Injury compensation, human research
    ethics 1:717
In-licensing activities
    drug discovery 2:278–279
    increase in 2:255
Innate immune response 5:217; 7:846
    psoriasis see Psoriasis
    toll-like receptors 7:846
Inner-filter effects, fluorescence
    screening assays 3:602
Inner membrane
    mitochondria see Mitochondria
    permeability, mitochondria 2:1059
Innovation 2:3, 2:20, 2:253–264
    basic research 2:253
    characteristics of 2:253, 2:259,
        2:259T
        management difficulties 2:259
    corporate strategies 2:261, 2:262T
        basic research 2:261
        efficiency discreditation 2:263
        Eli Lilly 2:262
        focused individuals 2:262
        multi-field individuals 2:262
        recruitment 2:262
        'skunk works' 2:262
    decline of 2:254–255
        biotechnology companies 2:231
        Lindberg, P (personal
            opinion) 8:222
    definition 2:255
        unplanned 2:255
        discovering vs. 2:255
    drug development 2:174
    fostering of 2:257
    generic competition, effects of 1:745
    importance of 2:240

Innovation (*continued*)
improvement, drug
  development **2**:174
individual members **2**:22
industry failure to **2**:254
  corporate goals **2**:254
  corporate mergers **2**:254–255
  ICI **2**:254
  marketing control **2**:254
  marketing efficiency **2**:254–255
intellectual property, effects
  of **1**:745
managerial strategies **2**:260,
  2:261*T*
  evaluation **2**:260
  isolation **2**:260–261
merger effects **1**:746
models of **1**:726
problem generation **2**:255
  failure readdressing **2**:256–257
  problem recognition **2**:256
regulation, effects of
  biosafety **1**:746
  competition **1**:746
research strategies **2**:22, **2**:257,
  2:258*T*
  basic science **2**:257–258
  discarded research **2**:259
  new approaches **2**:258
  questioning daily practice **2**:258
  quick/cheap/simple **2**:258
technology **2**:21, **2**:22*T*, **2**:23*T*
Innovator targets, target
  identification **1**:333, **1**:333*T*
Inorganic metal ions, calcium channel
  pharmacology **2**:849
Inositol phosphate accumulation-IP₃
  assays *see* G protein-coupled
  receptors (GPCRs), assays
Inositol phosphoryl ceramide synthase
  inhibitors **7**:463
  see also specific drugs
Inotropic drugs, congestive heart failure
  treatment **6**:698
INPADOC (International Patent
  Documentation Center
  Database) **3**:275, **3**:276*T*
INPHARMA, NMR **3**:948
Inpharmatica **1**:517
In-process controls (IPCs), common
  technical document (CTD)
  2:163
Input, Molecular Interaction
  Fields **4**:238
INS-50589 **6**:777
  structure **6**:776*F*
Insect cells
  cell-based assays **3**:619
  expression systems *see* Baculovirus
    expression systems
Insecticides
  malaria **7**:771
  mitochondria complex I **2**:1068–1069

INSERM (National Institute of Health
  and Medical Research) **1**:662
Insertional mutagenesis, genetically
  engineered animals **3**:156
In silico methods
  ADME *see* ADME
  brain drug uptake **5**:306*T*
  peptidomimetics **2**:622–623
  screening *see* Virtual screening
In situ fiber optic dip probe UV
  spectrophotometry *see*
  Dissolution
In situ hybridization, biomarkers **3**:75
In situ models, in vitro oral absorption
  prediction **5**:677
In situ perfusion, mechanistic
  pharmacokinetic studies **5**:38
Insolubility problems, *Escherichia coli* as
  expression system **3**:880
Insoluble/soluble polymers, solid
  support synthesis **2**:477
Insomnia
  prevalence **6**:7
  prognosis **6**:8
  treatment **6**:8
Insomnia, primary **6**:148
  clinical trials **6**:153
  disease basis **6**:150
  models **6**:152
  new research areas **6**:164
  treatment **6**:157, **6**:159*T*
    agomelatine **6**:160
    amitriptyline **6**:157
    antidepressants **6**:157
    antihistamines (H₁
      antagonists) **6**:157
    APD-125 **6**:159*T*
    benzodiazepines **6**:158
    BMS-214778 **6**:160
    chlorpheniramine **6**:157–158
    diazepam **6**:158
    diphenhydramine **6**:157–158
    doxepin **6**:157, **6**:161
    doxylamine **6**:157–158
    EMD-28104 **6**:159*T*, **6**:161
    eplivanserin **6**:159*T*, **6**:161
    eszopiclone **6**:158, **6**:159*T*
    5-HT₂ₐ/₂c receptor
      antagonists **6**:161
    GABA_A benzodiazepine receptor
      antagonists **6**:158
    GABA_A receptor agonists/
      antagonists **6**:160
    gaboxadol **6**:159*T*, **6**:160
    hydroxyzine **6**:157–158
    indiplon **6**:158, **6**:159*T*
    M-100907 **6**:159*T*
    melatonin receptor agonists **6**:158
    mirtazapine **6**:157, **6**:159*T*
    S-(+)-mirtazapine **6**:161
    NDG-96-3 **6**:159*T*
    nefazodone **6**:157
    ORG-50081 **6**:159*T*

PD-6735 **6**:159*T*, **6**:160
    ramelteon **6**:158, **6**:159*T*
    selective GABA reuptake
      inhibitors **6**:161
    TH-9507 **6**:161
    tiagabine **6**:159*T*, **6**:161
    trazodone **6**:157
    triazolam **6**:158
    tricyclic antidepressants **6**:157
    trimipramine **6**:157
    zaleplon **6**:158, **6**:159*T*
    zolpidem **6**:158, **6**:159*T*
    zopiclone **6**:158, **6**:159*T*
  unmet medical needs **6**:163
Inspra *see* Eplerenone
INSR, PDB entries **2**:965*T*
Institute of Tropical Diseases **1**:695
Instrumental variables, postmarketing
  surveillance **1**:769
Instruments, fluorescence screening
  assays **3**:601
Insulin **6**:448
  biological action **6**:429
    carbohydrate metabolism **6**:429–
      430
    lipid metabolism **6**:429–430
    protein metabolism **6**:429–430
  biosynthesis **6**:426, **6**:427*F*
  crystal packing **3**:442
  degradation **6**:450
  discovery/development, University of
    Toronto–Eli Lilly
    collaboration **1**:585
  genetically engineered, invention
    of **2**:262
  hydrophobic depoting **2**:586
  inhalation administration **2**:595
  obesity **6**:392
  pharmacodynamics **6**:450, **6**:450*T*
  pharmacokinetics **6**:450
  pulmonary administration **5**:281*T*
  refinement/development **1**:437
  SAR **6**:449
  secretion **6**:426, **6**:428*F*
    glucose-6-phosphate
      metabolism **6**:427–428
    phosphodiesterase-1 **2**:923
    phosphodiesterase-3 **2**:924
    plasma glucose levels **6**:428
    stages **6**:429
  signaling **6**:429, **6**:429*F*
  sources **6**:449
  structure **3**:382*F*
  subcutaneous administration **6**:450
  treatment *see* Insulin treatment
Insulin aspart **6**:450
  discovery/development **1**:437
  structure **6**:449*F*
Insulin-dependent diabetes mellitus *see*
  Diabetes mellitus type 1
Insulin detemir **6**:451
Insulin glargine **6**:451
  structure **6**:449*F*

In vitro tools
  automation *see* Automation
  metabonomics *see* Metabonomics
In vivo activity, DCN **2**:154
In vivo ADME studies **5**:31–50
  see also Absorption; Distribution
    studies; Drug metabolism and
    pharmacokinetics (DMPK);
    Elimination studies
  mechanistic pharmacokinetics *see*
    Mechanistic pharmacokinetic
    studies
  metabolism studies *see* Metabolism
    studies
  prediction to humans **5**:47
    allometric scaling **5**:47
    physical transport processes
      5:47
In vivo assays
  carcinogenicity, computational
    models **5**:920
  genomics, antibacterial target
    identification **7**:733
  hepatic metabolic clearance
    models **5**:871, **5**:871*F*
  In vitro testing *vs.*, Lindberg, P
    (personal opinion) **8**:223
  rectal administration **5**:290
  regulations, non-animal
    carcinogenicity testing **1**:477
  risk assessment, animal
    experimentation
    alternatives **1**:481
  stability testing **5**:494
In vivo blood–brain barrier correlation,
    surface activity
    parameters **5**:481, **5**:482*F*,
    5:483*F*
In vivo cancer cell inhibition,
    tamoxifen **8**:90
In vivo clearance (CL), plasma protein
    binding **5**:322
In vivo DBMA studies, tamoxifen
    8:90
In vivo expression technology (IVET),
    genomics **7**:743
In vivo gene silencing
  RNAi (RNA interference) **3**:182
  siRNAs **3**:182
In vivo gene targeting, anticancer
    agents, tumor models **7**:37
In vivo–in vitro correlations,
    distribution models **5**:870
In vivo knock-down approach,
    mechanism-based PK/PD
    modeling **5**:894–895
In vivo micronucleus assay,
    chromosomal damage **5**:919
In vivo models
  antimetabolites **7**:60
  azetidinone cholesterol absorption
    inhibitor development **8**:66,
    8:66*F*

brain drug uptake *see* Brain drug
    uptake
  ezetimibe activity **8**:77, **8**:77*F*
In vivo mutagenicity testing, transgenic
    animals **3**:161
In vivo profiling, DCN **2**:154
In vivo screening, costs, in vitro
    screening *vs.* **5**:4–5
Inward-facing residues, GPCRs
    4:685
Inwardly-rectifying potassium channels
    see K$^+$ ion channels, inward
    rectifying
Iodine, isotopes
  see specific isotopes (under I)
Iodobananin, SARS **7**:272
(*S*-)5-Iodowillardine, structure
    2:901*F*
Ioflupane **1**:196
Ion beam milling, lithography/
    photolithography **2**:354
Ion-binding sites, KcsA **4**:706–707
Ion channel-directed research,
    SMEs **1**:516
Ion channels
  acid sensitive *see* Acid-sensitive ion
    channels (ASICs)
  anesthetics **6**:357
  arrhythmias **6**:733
  cardiac *see* Cardiac ion channels
  influenza viruses **7**:374
  pain **6**:321
  see also specific types
  pharmacological target space
    4:422*T*
  surface plasmon resonance **3**:593
  as targets **2**:757*T*, **2**:758
    natural products **2**:758
  Xenopus laevis **3**:649
Ion channels, ligand-gated **2**:877–918
  see also specific ion channels
  ATP (purinergic) gated family
    2:879
    expression **2**:879
    structure **2**:880*F*
  characteristics **2**:881*T*
  Cys-loop superfamily **2**:879
    anion-conducting **2**:879
    cation-conducting **2**:879
    expression **2**:879
    structure **2**:879, **2**:880*F*
  definition **2**:878
  future work **2**:912
    HTS, changes in **2**:912
    transmission modulation **2**:912
  ligands, physicochemical
    properties **4**:424*T*
Ion channels, voltage-gated
    2:827–875
  see also specific types
  cell function regulation **2**:828–831
  definition **2**:827
  drugs acting on **2**:829*T*

future work **2**:865, **2**:865*F*
  automated planar chip-based
    electrophysiology **2**:866
  subunit expression **2**:865–866
gating properties **2**:828
structure **2**:827, **2**:828*F*
transmembrane segments **2**:828
Ion exchange chromatography, protein
    purification **3**:422
Ion fluxes, cell-based assays **3**:624,
    3:624*T*
Ionic conduction mechanism, dielectric
    heating **3**:839
Ionic exchange chromatography,
    polyethylene glycol
    conjugates **5**:1063
Ionic liquids, solvent reactions,
    microwave-assisted
    chemistry **3**:854
Ionic product (K_W), water pH **5**:359
Ionic strength, promiscuous ligands,
    identification **2**:740
Ionix Pharmaceuticals **1**:508
Ionizable drugs **5**:368, **5**:382
  absorption **5**:383
  dissolution *see* Dissolution
  drug–membrane binding **5**:382
  drug–receptor binding **5**:382
    brain membrane receptors **5**:382
  importance **5**:603
  solubility *see* Solubility
Ionization **5**:357–397
  see also Ionization constants (pK_a)
  acid/base reaction equilibria **5**:607,
    5:609*F*
    cetirizine **5**:608, **5**:609*F*
    iteration **5**:608–609
    macroconstant derivation **5**:608–
      609
    microconstant
      measurement **5**:608–609
    proton count **5**:608
    sequential ionization **5**:608
  automation *see* Automation
  blood–brain distribution **5**:753–754
  corneal permeability modeling **5**:736
  importance **5**:603
  lipophilicity effects **5**:385, **5**:385*F*
    ampholytes **5**:386
    aqueous distribution **5**:386
    monoprotic acids **5**:386
    monoprotic bases **5**:386
  macromolecules **5**:384
    RNA–RNA duplexes **5**:384
    serine proteases **5**:384
  molecular factors **5**:605
    conformational effects **5**:606
    effective fragment
      potentials **5**:607
    electrostatic field effects **5**:605
    hydration effects **5**:607
    inductive effects **5**:605
    mesomeric effects **5**:605, **5**:606

Ionization (*continued*)
  solvent effects 5:606–607
  steric effects 5:605, 5:606
  tautomerism 5:606
  NMR 5:384
    advantages 5:384
  permeation *vs.* 5:603
  physicochemical properties 5:385
  in silico prediction *see* Ionization
    prediction, in silico tools
  solubility effects 5:387, 5:387F
    ampholytes 5:387
    monoprotic acid 5:387
    monoprotic base 5:387
  theoretical basis 5:604
  thermodynamic forces 5:604
    aqueous free energy of
      dissociation 5:605
    proton exchange 5:604–605
    thermodynamic cycle 5:604,
      5:604F
  water pH 5:359
Ionization class, hepatocyte
  binding 5:334–335
Ionization constants (pK_a) **5:357–397**
  see also Ionization; pH
  alcohols 5:372
  aliphatic bases 5:377
  amides 5:381
  ampholytes 5:365F
    microspecies ionization 5:365,
      5:367F
  aromatic bases 5:378
  carboxylic acids 5:368, 5:369F
    affecting factors 5:368, 5:381T
    dicarboxylic acids 5:368
  definition 5:358, 5:604
  distribution effects 5:384
  drug analysis 5:384
  drug published values 5:385
  guanidines 5:381
  heteroaromatic bases 5:378
  ionization equations 5:362,
    5:364T
  phenols 5:372
  software prediction 5:395
  species distribution 5:362
  structural affecting factors 5:381,
    5:381T
    electron donation 5:381
    electron withdrawal 5:381
    mesomeric effects 5:381
  weak acids 5:361
  weak bases 5:361
  zwitterions 5:363, 5:365F
    microspecies ionization 5:365,
      5:367F
Ionization constants (pK_a)
  measurement 5:388
  capillary electrophoresis 5:392
  chemical structure
    calculations 5:394
  conductivity 5:386

co-solvent effects 5:388, 5:389F
  extrapolations 5:388
  Yasuda–Shedlovsky plots 5:388
flow injection 5:393
Hammett equation 5:394
hybrid pH-metric/ultraviolet
  methods 5:392, 5:393F
ionization prediction, in silico
  tools 5:617
pH/fluorometry 5:393
pH gradient titration 5:392
pH-metric titration 5:389
  automation 5:389–390
  lipophilicity measurement 5:436
  weak acids/bases 5:390
solubility measurements 5:387F,
  5:393
Taft equation 5:394
Ionization effects, lipophilicity *see*
  Lipophilicity
Ionization equations, ionization
  constants (pK_a) 5:362,
    5:364T
Ionization methods, high-performance
  liquid chromatography-mass
  spectrometry (HPLC-MS)
    5:994–996
Ionization partitioning,
  lipophilicity 5:427
Ionization prediction, in silico
  tools **5:603–626**
  accuracy 5:619
    chemical space
      interpolation 5:619–620
  algorithms 5:610
  biopolymers 5:618
    permanent charge 5:619
    Poisson–Boltzmann
      equation 5:619
    small molecules *vs.* 5:618
    titration curve 5:619
  database lookup methods 5:616
  data sources 5:622, 5:623T
  empirical methods 5:610, 5:616T
    see also specific methods
    artificial neural networks
      5:616
  future work 5:622
  linear free energy
    relationships 5:610
    assessment of 5:612–613
    Hammett–Taft constants 5:610
    multi-case methodology 5:612
    perturbed molecular orbital
      theory 5:611
    SPARC method 5:611
  quantitative structure–property
    relationships 5:610, 5:613
    atomic descriptors 5:613
    commercial models 5:615
    molecular tree structure
      fingerprints 5:614
    Xing method 5:614, 5:614F

quantum chemical methods 5:617
  atoms in molecule (AIM)
    energy 5:617
  comparative molecular field analysis
    (CoMFA) 5:617
  energy shifts 5:617
  geometrical properties 5:617
  ionization constant
    calculation 5:617
  quantum mechanical
    descriptors 5:617
  quantum mechanical calculations *see*
    Quantum mechanical calculations
  property prediction
  software packages 5:620
    ACD/pK_a 5:616T, 5:620
    ADME Boxes 5:616T, 5:620,
      5:621
    ADMET Predictor 5:616T, 5:620,
      5:621, 5:622F
    CSpKa 5:615, 5:616T, 5:620,
      5:621
    DISCON 5:616T, 5:620, 5:622
    Marvin 5:615, 5:616T, 5:620,
      5:621
    Pallas 5:616T, 5:620
    pKalc module 5:620
    SPARC 5:620
Ionization profiles **5:357–397**
IONIZER program, 2D- to 3D
  structure 4:464
Ionizing radiation, cancer risk
  factors 7:4
Ionophores, tetracyclines 7:618
Ionotropic glutamate receptors 6:322,
  6:323
Ion pair structure, lipophilicity 5:429
Ion permeation mechanisms,
  KcsA 4:711
Ions, lipophilicity measurement *see*
  Lipophilicity measurement, ions/
  zwitterions
Iontophoresis, dermal/transdermal
  administration 5:292
IPA (International Pharmaceutical
  Abstracts) 3:271T
IPD *see* Suplatast tosylate
IPG gels, isoelectric focusing 2:332
IPI (International Protein
  Index) 3:364
  building of 3:365
  Ensembl 3:365–366
  genomes 3:365–366
  Integr8 3:367
  nonredundant complete UniProt
    proteome sets 3:365
  predicted coding sequences 3:366
Ipratropium bromide
  GPCR targeting 2:773T
  pulmonary administration 5:281T
Iproniazid
  development 6:15
  structure 2:68F

Iproplatin
  as prodrug **5**:1033
  structure **5**:1032*F*
Ipso substitution, cytochrome
      P450s **5**:110, **5**:111*F*
Irbesartan **1**:**157**
  bioavailability **6**:722*T*
  drug action **1**:532*T*
  GPCR targeting **2**:773*T*
  half-life **6**:722*T*
  hypertension treatment **6**:721–723
  structure **6**:722*T*
Iressa *see* Gefitinib
Irinotecan hydrochloride **1**:**146**;
      7:113–114
  cervical cancer **1**:147
  lung cancer **1**:147
  ovarian cancer **1**:147
  as prodrug **5**:1024
  structure **5**:1024*F*
  xenograft models **7**:39*T*, **7**:43
IRMA-2 **6**:589*T*
Iron, tetracyclines **7**:618
Iron chelation, drug
      introductions **1**:173
Iron-responsive element (IRE) **2**:1045
Irregular sleep–wake pattern
      (ISWP) **6**:148–149
  treatment **6**:162
Irreversible inhibition, in vitro studies
      see Inhibition, in vitro studies
Irritable bowel syndrome (IBS) **6**:598–
      599, **6**:**643–670**
  brain–gut axis communication **6**:649
  clinical observations **6**:648
  clinical trials **6**:653
    patient mix **6**:652
    Rome criteria **6**:652
  disease basis **6**:645
  disease state **6**:644
  epidemiology **6**:644
    gender **6**:644
  food allergies **6**:651
    immunoglobulin E levels **6**:651
  gastrointestinal dysmotility **6**:648
  genetics **6**:651
  gut flora **6**:651
  5-HT signaling abnormalities **6**:649
    postprandial measurements **6**:650
    serotonin reuptake transporters
        (SRTs) **6**:649
  immune system alterations **6**:651
  infections **6**:650
    Campylobacter **6**:650
    Salmonella **6**:650
    Shigella **6**:650
  microarrays *see* Microarrays
  models **6**:651
    maternal separation model **6**:652
    neonatal irritation model **6**:652
    PI-IBS model **6**:652
    SERT knockout mouse **6**:652
    Trichinella spiralis model **6**:652

  socioeconomic burden **6**:644
  stress/psychological factors **6**:650
    corticotropin-releasing
        factor **6**:650
    interleukin-1 **6**:650
  symptoms **6**:644, **6**:645*T*
  visceral sensitivity **6**:649
  neuroimaging **6**:649
Irritable bowel syndrome (IBS)
      treatment **6**:599, **6**:599*T*, **6**:600,
      6:653, **6**:655*T*
  alosetron **6**:657
  antidepressants **6**:655*T*, **6**:659
  antidiarrheal agents **6**:655*T*, **6**:658
  antispasmodic agents **6**:659
  benzodiazepines **6**:655*T*
  bulking agents **6**:655*T*, **6**:658
  calcium-channel blockers **6**:657
  cilansetron **6**:599
  dietary aids **6**:658
  drug introductions **1**:201
  5-HT receptor agonists **6**:652, **6**:654,
      6:655*T*
  5-HT receptor antagonists **6**:652,
      6:654, **6**:655*T*
  laxatives **6**:655*T*, **6**:658
    osmotic laxatives **6**:658
    stimulant laxatives **6**:658
  market **6**:644
  new research areas **1**:107–108; **6**:660
    α₂-adrenergic agonists **6**:661*T*,
        6:664
    cannabinoid receptor
        modulators **6**:661*T*, **6**:665
    chloride channel openers **6**:661*T*,
        6:665
    cholecystokinin receptor
        antagonists **6**:661*T*, **6**:664
    cilansetron **6**:661*T*, **6**:662
    corticotropin-releasing factor-1
        antagonists **6**:661*T*, **6**:664
    5-HT₃ receptor
        antagonists **6**:661*T*
    5-HT₄ receptor partial
        agonists **6**:661*T*
    neurokinin receptor
        modulators **6**:661*T*, **6**:662
    opiate receptor
        antagonists **6**:661*T*, **6**:663
    renzapride **6**:661*T*, **6**:662
  selective serotonin reuptake
      inhibitors **6**:655*T*, **6**:657, **6**:660
  serotonin receptor antagonists **6**:599
  stool softeners **6**:658
  symptom management **6**:652
  tegaserod **6**:599, **6**:654
  tricyclic antidepressants **6**:655*T*,
      6:657, **6**:659
  unmet medical needs **6**:660
Isatin **1**:365–366
  structure **1**:364*F*
  thiosemicarbazone
      derivatives **1**:365–366

Isatoribine, hepatitis C **7**:403*F*,
      7:405
ISAtx-247
  psoriasis treatment **7**:973
  transplant rejection **7**:919
Ischemia
  acid-sensitive ion channels
      (ASICs) **2**:911
  uncoupling proteins **2**:1080
Ischemia reperfusion induced lung
      apoptosis, RNAi (RNA
      interference) **3**:182
Ischemia–reperfusion injury,
      mitochondria **2**:1054–1055
Ischemic heart disease **6**:696
  mortality/morbidity, developing
      countries **1**:642*T*
Ischemic stroke **6**:697
  pathogenesis **6**:256
  prevalence **6**:253
Iscover *see* Clopidogrel hydrogensulfate
Isepamicin **7**:634
ISI **3**:282
ISIS
  foundation **2**:761
  library design **4**:360–361
ISIS-2 trial, streptokinase **6**:772
ISIS-3, alteplase clinical trials **6**:772
ISIS/3D program, 3D structure
      generation **4**:465
Isis Pharmaceuticals **1**:516
Islet cell transplantation, diabetes
      mellitus management **6**:455
Isocarboxazid, side effects **6**:60*T*
Isocephalosporins **7**:497
Isocombretastatin **7**:91
  structure **7**:92*F*
Isocyanate-epoxide cyclization,
      oxazolidinones **7**:686
Isocyanates, glutathione
      transferases **5**:160–161,
      5:160*F*
Isoelectric focusing (IEF)
  protein analysis **3**:425, **3**:425*F*
  2D-PAGE **3**:34
  two-dimensional gel electrophoresis
      see 2D-PAGE
Isoelectric heterogeneity, heterogeneous
      protein crystallization **3**:882
Isoenzymes
  'classical' pharmacogenetics **3**:58
  cytochrome P450s **5**:810, **5**:811*T*
Isoetharine, library similarity
      searches **4**:564*T*
Isoflavones, nutraceuticals **1**:452*T*
Isoflurane **6**:359
  GABA receptor interactions **6**:357
  malignant hyperthermia **6**:359
  mechanism of action **6**:359
  minimum alveolar
      concentration **6**:359
  oxidation **5**:100
  protein interactions **6**:356

Isoflurane (*continued*)
structure **6**:352*F*
voltage-gated potassium channel interaction **6**:357
Isoforms
cytochrome P450s **5**:170
glutathione *S*-transferases (GSTs) **5**:818
polymorphisms, cytochromes P450 **4**:781
Isoguvacine, structure **2**:661*F*
Isolated cerebral microvessels, brain drug uptake **5**:307, **5**:311
Isolated chicken eye (ICE) assay, non-animal eye irritation testing **1**:470
Isolated double bond oxidation, cytochrome P450 **5**:101
Isolated rabbit eye (IRE) assay, non-animal eye irritation testing **1**:470
Isolation, innovation **2**:260–261
Isolaulimalide, structure **1**:364*F*
Isoleucyl-tRNA synthetase inhibitors (ITRS), antifungal drugs **7**:433
'Isometric ballast,' chiral drug discovery **2**:714–715
Isomorphous replacement *see* X-ray crystallography, phase problem solutions
Isoniazid
acetylation **5**:154, **5**:154*F*
bioavailability **5**:712*T*
mycobacterium treatment **7**:704, **7**:705*T*, **7**:709
comparisons within class **7**:711
future work **7**:711
limitations **7**:711
mechanism of action **7**:709, **7**:710*F*
resistance **7**:711
SAR **7**:709, **7**:710*F*, **7**:711*F*
oral absorption **5**:468*T*, **5**:680*T*
Isooxacephalosporins **7**:497
Isopentenyladenine
discovery/development **1**:389–390
structure **1**:388*F*
Isophosphoylceramide synthase inhibitors, antifungal drugs **7**:434, **7**:435*F*
Isoprenaline
bioavailability **5**:712*T*
discovery/development **1**:328–329, **1**:374
structure **1**:374*F*
Isoprenoid biosynthesis inhibitors, antimalarial drugs **7**:794, **7**:795*F*
Isopropyl-β-D-1-thiogalactopyranoside (IPTG), gene expression vectors **3**:108
Isoprostanes, neuroprotective drug biomarker **6**:262–263

Isoquine **7**:775
Isoquinolines
phosphodiesterase-7 inhibitors **2**:944
triphenylethylene anticancer agents **7**:161
Isosalsoline, *N*-methylation **5**:146–147, **5**:146*F*
Isosorbide 2-nitrate, bioavailability **5**:712*T*
Isosorbide 5-nitrate, bioavailability **5**:712*T*
Isospora belli see Isosporiasis
Isosporiasis **7**:751*T*
biology **7**:758
clinical manifestations **7**:758
drug treatment **7**:753*T*
IsoStar (intermolecular interactions) *see* Cambridge Structural Database (CSD)
Isotaxel
as prodrug **5**:1012
structure **5**:1013*F*
Isothiazolones, asthma treatment **7**:903
Isothiocyanates
glutathione transferases **5**:160–161, **5**:160*F*
nutraceuticals **1**:449–450
Isotope-coded affinity tags (ICAT)
LC-MS (peptide-based) proteomics **3**:37–38
protein expression mapping **2**:385
Isotope labeling
cell-free expression systems **3**:120–121
LC-MS (peptide-based) proteomics **3**:38
Isotopic analog peak ratios, LC-MS (peptide-based) proteomics **3**:38
Isotopic field orientation (IFO), bioavailability models **5**:711–718, **5**:719*T*
Isotopic labels, NMR **2**:512
structure determination **3**:475, **3**:477*F*
Isotretinoin
acne treatment **7**:939, **7**:963
bioavailability **5**:712*T*
characteristics **2**:1012*T*
Isotropic *B*-factors, x-ray structures, problems **3**:538
Isotropic etching *see* Lithography/photolithography
Isotropic systems, lipophilicity measurement *see* Lipophilicity measurement
Isoxazoles
bioisosterism **2**:662*T*
ring cleavage, cytochromes P450 **5**:799

Isoxicam, oral absorption **5**:680*T*
Ispronicline
nicotinic acetylcholine receptor binding **2**:888
structure **2**:887*F*
Isradipine
bioavailability **5**:712*T*
cocaine abuse **6**:184*T*
oral absorption **5**:680*T*
structure **6**:181*F*
Israel, orphan drugs legislation **1**:662
IST-662 **7**:120
Istamycin **7**:632
resistance **7**:642
Istradefylline, Parkinson's disease treatment **6**:11, **6**:217
Italy
medicine marketing system **1**:632
stem cell research programs **1**:558–559
Iterated dynamic combinatorial libraries **3**:963
Iteration, acid/base reaction equilibria **5**:608–609
Iterative follow-up, screening library selection *see* Screening library selection
I_{to} potassium channel **6**:736
Itopride hydrochloride **1**:202
gastritis **1**:203
as prokinetic agent **6**:679
Itraconazole **7**:430*T*; **8**:25
absorption, pH dependency **5**:383
chromoblastomycosis **7**:420
clinical profiles **7**:451, **7**:451*T*, **7**:455
candidosis **7**:455
dermatophytosis **7**:455
onychomycosis **7**:455
exclusivity period **1**:532*T*
mechanism of action **8**:25
pharmacokinetics **7**:454
reaction phenotyping studies **5**:243*T*
sporotrichosis **7**:420
structure **5**:360
target information **4**:750*T*
Ituximab **1**:499
[^{131}I]IUDR, androgen ablation imaging **3**:670
IUPHAR-approved scheme, calcium channel nomenclature **2**:842, **2**:843*T*
IUPred, domain mapping **3**:879
Ivabradine **6**:753
Ivermectin
discovery/development **1**:21, **1**:22
MSD/Kitasato Institute collaboration **1**:22
human formulation **1**:23
nematode infection treatment **7**:760
Ixel *see* Milnacipran

Labor cost reduction,
  outsourcing **2**:208
Lac derived promoters, gene expression
  vectors **3**:108
Lacidipine, SAR **6**:716*T*
Lack of knowledge, personalized
  medicine **1**:290
Lacosamide **6**:291
  structure **6**:290*F*
Lactacystin **7**:237
  discovery/development **1**:384
  structure **1**:383*F*
β-Lactam antibiotics *see* β-lactam
  antibiotics (*under beta*)
β-Lactamase *see* β-lactamase (*under beta*)
β-Lactam synthesis, PASP **3**:829,
  3:829*F*
Lactate dehydrogenase,
  malaria **7**:785–787
Lactic acid, acne treatment **7**:966
Lactic acid bacteria, protein expression
  systems **3**:116
Lactic acidosis
  biguanides **6**:445
  metformin **6**:445
  pyruvate dehydrogenase kinase **4**:755
Lactivicin, structure **7**:511*F*
Lactobacillus lactis, protein expression
  systems **3**:116
Lactone ring hydrolysis
  lovastatin activation **5**:1030
  prodrug activation **5**:1030
β-Lactones, proteasome
  inhibitors **7**:237
S-D-Lactoylglutathione hydrolysis,
  glyoxalase assays **3**:581
Lactulose, oral absorption **5**:468*T*
Lafutidine **1**:199
Lag time (*t*₁), pharmacokinetics **5**:17
λ-dynamics, free energy
  calculation **4**:657
Lambert–Eaton syndrome **6**:11
Lambert review **1**:582, **1**:583, **1**:583*T*
Lamivudine **1**:125; **7**:279–282, **7**:312;
  8:59
  CNS penetration **7**:314
  drug compatibility **7**:342
  drug regimens **7**:342*T*
  as fast-tracked drug **1**:760*T*
  hepatitis B virus **7**:263, **7**:338–340
  mechanism of action **7**:312–313
  medicinal chemistry **7**:338–340
  oral absorption **5**:680*T*
  pharmacokinetics **7**:314
  resistance **7**:263, **7**:317, **7**:340–341
  safety **7**:312–313
  SAR **7**:313–314
  structure **1**:70*F*; **7**:312–313, **7**:338*F*
Lamotrigine **6**:289
  as analgesic **6**:339
    differential efficacy **6**:337*T*
  bipolar affective disorder
    treatment **6**:72

characteristics **2**:829*T*
  immunotoxicity **5**:221*T*
  oral absorption **5**:680*T*
  structure **2**:829*T*; **6**:282*F*, **6**:336*F*
LANCE technology, Flash Plate assays
  vs. **3**:587
Landel *see* Efonidipine hydrochloride
  ethanol
Landiolol **1**:163
Langerhans cells, prohaptens **5**:221
Langmuir, Irving, management
  strategies **2**:261
Langmuir binding isotherm **3**:903
Language problems, protein function
  prediction,
  bioinformatics **3**:337
Laniquidar, structure **5**:770*F*
Lanoconazole **1**:132
Lanosterol-14α-demethylase
  inhibitors **7**:449
  see also specific drugs
  clinical profiles **7**:455
  drug resistance **7**:455
    cytochrome P450 51 affinity **7**:455
    MDR transporter **7**:455
  medicinal chemistry **7**:449
  pharmacokinetics **7**:454
Lanoteplase **6**:771
  clinical trials **6**:774
    InTIME-1 trial **6**:774
    InTIME-2 trial **6**:774
Lanreotide acetate **1**:231; **2**:579*T*
  introduction of **1**:108
  self-forming depots **2**:593
Lansoprazole
  characteristics **2**:653*T*
  CYP 2C19 metabolism **5**:546*T*
  oral absorption **5**:468*T*, **5**:680*T*
  structure **2**:653*T*; **6**:606*F*
Lanthanide cryptates *see* Homogeneous
  time-resolved fluorescence
  (HTRF)
β-Lapachol
  discovery/development **1**:382
  structure **1**:383*F*
β-Lapachone
  discovery/development **1**:382
  structure **1**:383*F*
Lapatinib *see* GW-572016
LAQ824, heat shock proteins, action
  on **7**:233
Large, diverse libraries, solution phase
  library synthesis **3**:769
Large databases, 3D molecular
  alignment **4**:113
Large molecules
  conformational analysis/
    searching **4**:104
  force-field molecular modeling **4**:93
Large ribosomal subunit (*Deinococcus
  radiodurans*) **3**:535
Large robotic system, HTS
  automation **3**:682

Large-scale domain motion (proteins),
  ligand–receptor molecular
  flexibility **4**:204, **4**:204*F*
Large-scale genomic screens, *Drosophila
  melanogaster* model system **3**:9
Large scale simulations, CADD **4**:4
Large scale synthesis, stereoisomer
  synthesis **2**:718
Large-scale virus vector libraries *see*
  Genomics
Laronidase **1**:211
Laser ablation, lithography/
  photolithography **2**:354
Laser angioplasty **6**:696
Laser capture microdissection (LCM)
  brain slices **5**:311
  proteomics **3**:30–31
Laser excitation, fluorescence
  fluctuation techniques
  3:607–608
Laser flow cytometry, kinetic
  solubility **5**:349
Laser-induced CNV model **6**:304
Laser scanning confocal
  microscopy **5**:294
Laser scanning cytometer **3**:687
Laser treatment
  acne treatment **7**:941
  psoriasis treatment **7**:945
Lasofoxifene **7**:154; **8**:115
  cardiovascular effects **2**:1014
  as estrogen receptor ligand **2**:1014
  structure **2**:1013*F*; **7**:153*F*; **8**:116*F*
Lassa fever **7**:278
Last Observation Carried Forward
  (LOCF) protocols, affective
  disorders, clinical trial
  issues **6**:62
Latanoprost **1**:240
  GPCR targeting **2**:773*T*
  introduction of **1**:108
  structure **6**:309*F*
Late menopause, breast cancer **7**:151
Latent inhibition, schizophrenia
  models **6**:25
Laue symmetry, molecular
  replacement **3**:461–462
Laufberger, Vilem, protein
  crystallization **3**:885
Laulimalide **1**:366; **7**:102–103
  analogs **1**:367
  structure **1**:364*F*; **7**:103*F*
  synthesis **1**:367
Laxatives, irritable bowel syndrome
  treatment *see* Irritable bowel
  syndrome (IBS) treatment
LB11058 **7**:497
  activity **7**:496*T*, **7**:497*T*
  structure **7**:495*F*
Lck kinase
  PDB entries **2**:965*T*
  SH2, three-dimensional
    structures **2**:636*T*

Lck kinase (*continued*)
  structure   **2**:976
  transplant rejection   **7**:921–922
LCM technology, cellular
    proteomics   **2**:385–386
LD$_{50}$, toxicity, computational
    models   **5**:913
LE135, IBD treatment   **6**:636
Lead(s)
  differentiation, biological
      fingerprints   **4**:811, **4**:811*F*
  druglikeness *see* 'Druglikeness'
  identification/characterization
    ADMET as consequence   **4**:435–
        436
    biomarkers   **3**:69
    as bottleneck   **1**:38*T*; **4**:440
    brain drug uptake   **5**:317
    comparative genomics   **7**:739
    definition   **3**:939
    fragment-based drug discovery *see*
        Fragment-based drug discovery
    high-throughput chemistry   **4**:436
    high throughput
        purification   **3**:861
    historical perspective   **4**:435,
        4:436*F*
    HTS   **3**:940
    new paradigms   **1**:33
    NMR, drug discovery   **3**:901
    protein crystallography *see* X-ray
        crystallography, rational drug
        design
    screening library selection *see*
        Screening library selection
    solid-phase libraries   **3**:741,
        3:742*T*
    surface plasmon resonance   **3**:593–
        594
    virtual screening *see* Virtual
        screening
  immunosuppression   **5**:223
  likeness prediction, computational
      filters   **3**:258
  optimization   **2**:140, **2**:205
    activity interdependency   **2**:214
    ADME   **2**:139
    animal models   **2**:139
    attrition   **2**:205, **2**:206*F*
    candidate seeking   **2**:205
    genetically engineered
        animals   **3**:159
    hit follow up   **2**:205
    HTS   **2**:400
    development   **2**:205
    NMR, drug discovery   **3**:901,
        3:902
    pharmacokinetics   **2**:139
    pharmacology   **2**:139
    physiologically based
        pharmacokinetics
        (PBPK)   **5**:880
    process outcome   **2**:213

protein crystallography, rational
    drug design   **3**:890
  SMEs   **1**:519
  subactivities   **2**:213
  toxicology   **2**:139
quality   **2**:560
  genomics, antibacterial target
      identification   **7**:739–740
screening   **2**:425
shortage of   **7**:734
structure-based drug design
    (SBDD)   **4**:776
throughput rate   **2**:560
Leadership   **2**:239–252, **2**:271–272
  becoming a leader   **2**:242,
      2:243
    administrative
        responsibilities   **2**:243
    delegation   **2**:243
    interpersonal skills   **2**:243
  behavioral aspects   **2**:240
  collaborative research   **2**:240
  communication   **2**:243, **2**:247
    conflict resolution *see below*
    criticism   **2**:243
    definition   **2**:247
    feedback   **2**:243
    nonjudgemental   **2**:247
    open-mindedness   **2**:247
  conflict management   **2**:242, **2**:243,
      2:248
    avoidance   **2**:242, **2**:248
    compromise   **2**:248
    confrontation   **2**:249
    forcing   **2**:248
    smoothing   **2**:248
    sources of conflict   **2**:248
    superordinate goals   **2**:249
  diversity issues   **2**:244
    communication   **2**:244
    nationalities   **2**:244
  effective leaders   **2**:241
    intellectuality   **2**:241
    management skills   **2**:241
    role models   **2**:241
    working climate   **2**:242
  gender issues   **2**:244
    being ignored   **2**:245
    career obstacles   **2**:244
    exclusion   **2**:245
    judgment of   **2**:244–245
    male preconceptions   **2**:245
    male stereotyping   **2**:245
  importance of   **2**:239
    economic effects   **2**:240
    innovation   **2**:240
    issues   **2**:240
    social effects   **2**:240
  ineffective leaders   **2**:240,
      2:241
    impact of   **2**:241
    long-term effects   **2**:240
    working climate   **2**:242

laboratory climate   **2**:242
    positive feedback   **2**:242
Lindberg, P (personal
    opinion)   **8**:221
motivation   **2**:243, **2**:245
  effort *vs.* outcome   **2**:246
  equity   **2**:246
  fairness   **2**:246
  inequality   **2**:246
  micromanagement dangers   **2**:247
  reasonable working
      situation   **2**:246
  recruitment   **2**:246
  target setting   **2**:246
  technical competencies   **2**:246
narrow training   **2**:240
project team leader
    assessment   **2**:146
R&D management   **2**:249
  approval systems   **2**:250
  decision making systems   **2**:250
  horizontal structure   **2**:249
  information systems   **2**:250
  performance appraisal   **2**:250
  problem definitions   **2**:249
  recruitment   **2**:250
  reward systems   **2**:250
  size   **2**:250
  structure   **2**:249
  systems   **2**:250
  vertical structure   **2**:249
scientific training   **2**:240
  behavioral aspects   **2**:240
  collaborative research   **2**:240
  narrow training   **2**:240
  'soft' science perception   **2**:240
'soft' science perception   **2**:240
'Lead-like' libraries, library
    design   **4**:308
Leaky expression, gene expression
    vectors   **3**:108
Learned helplessness model (LHM),
    affective disorders   **6**:54*T*
Learning/memory, phosphodiesterase-
    2   **2**:923–924
'Least developed countries' (LDCs) *see*
    Developing countries
Least-squares regression, solubility
    models   **5**:630
Leave-one-out (LOO)
  3D-QSAR   **4**:587
  QSAR validation   **4**:157
'Leave-out-some' methods, 3D-
    QSAR   **4**:587
Leber's hereditary optic neuropathy
    (LHON), mitochondria complex
    I   **2**:1071
Lectins, buccal administration
    5:283–284
Leflunomide   **1**:214
  arthritis treatment   **7**:857
  introduction of   **1**:108
  transplant rejection   **7**:920

Liposome electrokinetic chromatography **5**:479
Liposomes
 aminoglycosides **7**:644
 Fluorosome **5**:477
 lipophilicity measurement **5**:434, **5**:435–436
5-Lipoxygenase, atherosclerotic plaque formation **6**:467
15-Lipoxygenase, atherosclerotic plaque formation **6**:467
5-Lipoxygenase activating protein inhibitors **6**:491
Lipoxygenase expression, NF-κB pathway **3**:197
Lipoxygenase inhibitors
 see also specific drugs
 arthritis therapy **7**:849
5-Lipoxygenase inhibitors *see* Leukotriene biosynthesis inhibition
Lipoxygen Technologies **1**:492*T*
LIPS **6**:474*T*
Lipscomb, William, protein crystallography, rational drug design **3**:876
Liquid assays, cell-based assays **3**:631
Liquid chromatography-mass spectrometry (LC-MS)
 brain drug uptake **5**:310
 CYP inhibition **5**:352
 metabolic clearance **5**:352
 product structure determination **5**:493*F*, **5**:501–502
Liquid chromatography-NMR *see* Nuclear magnetic resonance (NMR)
Liquid crystals, solid-state physicochemistry **5**:513
Liquid/hematologic models, anticancer agents, tumor models **7**:36
Liquid–liquid extraction, solution phase library synthesis **3**:776
Liquid–liquid phases, microwave-assisted chemistry **3**:853
Liquid sample collection, Pfizer *see* Compound storage/management (Pfizer)
Liraglutide, hydrophobic depoting **2**:587
Liranaftate **1:132**
Lisinopril
 ACE complex structure **4**:17, **4**:18*F*
 ATP-gated ion channels **2**:905–907
 BCS **5**:973*T*
 bioavailability **6**:719*T*
 dose **6**:719*T*
 effective permeability (*P*eff) **5**:673*T*
 half-life **6**:719*T*
 in vitro potency **6**:719*T*
 structure **6**:587*F*, **6**:719*T*
 target information **4**:750*T*

List-based nearest searching, ligand-based library design **4**:312–313, **4**:313*F*
Literature databases **3**:269, **3**:270, **3**:271*T*, **3**:273
 bibliographic **3**:273
 specialist **3**:274
 full-text **3**:275
 searching **3**:275
 patent **3**:275, **3**:276*T*
 primary **3**:270
 searching **3**:270
 secondary **3**:270
 specific **3**:275
 tertiary **3**:273
Literature models, plasma protein binding *see* Plasma protein binding, prediction
Literature work, career guidelines **2**:18
Lithium
 bipolar affective disorder treatment **6**:71
 development, history **1**:695
 tetracyclines **7**:618
Lithocholic acid, structure **2**:1025*F*
Lithography/photolithography **2**:349
 anisotropic etching **2**:352
 solutions **2**:352, **2**:352*T*
 bulk micromatching **2**:349, **2**:350*F*
 deep reactive ion etching (DRIE) **2**:353
 DNA microarrays **2**:357
 etch stopping **2**:352
 ion beam milling **2**:354
 isotropic etching **2**:350, **2**:351*F*
 solutions **2**:351, **2**:351*T*
 laser ablation **2**:354
 metal electrodeposition **2**:353
 electroless process **2**:353
 plasma etching **2**:353
 polymer replication **2**:354
 porous silicon **2**:355
 surface micromachining **2**:352, **2**:353*F*
 digital micromirror devices **2**:352
 ultraprecision machining **2**:354
 ultraviolet LIGA **2**:354
 x-ray LIGA **2**:353, **2**:354*F*
 drawback **2**:305
 polymethyl methacrylate (PMMA) **2**:305
Livalo *see* Pitavastatin
Live Bench website, homology models **3**:545
Liver
 cytochrome P450s **5**:171–172, **5**:172*T*
 first-pass studies, mechanistic pharmacokinetic studies *see* Mechanistic pharmacokinetic studies
 OATP (*SLC21/SLCO*) transporters **5**:78
 phosphodiesterase-3 **2**:924

Liver cancer, tamoxifen **8**:92
Liver disease, pharmacodynamic effects **1**:286–287
Liver flukes **7**:751*T*
 biology **7**:760
 clinical manifestations **7**:761
 drug treatment **7**:753*T*, **7**:761
Liver/hepatic clearance, membrane transport protein pharmacokinetics **5**:77, **5**:77*F*
Liver microsome data, electrotopological indices (E-state index) **5**:570
Liver receptor homologous protein 1, nomenclature **2**:996*T*
Liver slices
 drug metabolism, in vitro studies **5**:232, **5**:232*T*
 tissue composition **5**:45*T*
Liver X receptor (LXR) **2**:1010
 anti-inflammatory properties **2**:1010
 biological functions **4**:738
 ligand binding pocket flexibility **4**:738–739
 nomenclature **2**:996*T*
 oxysterol binding **4**:738
 peroxisome proliferator-activated receptor-γ (PPAR-γ) effects **2**:1009
 pharmacology **2**:1023, **2**:1024*F*
 see also specific drugs
 structure-based drug design **4**:738
 pharmacophore models **4**:742
 three dimensional structure, ligand-binding pocket **2**:999–1000
Lividomycin B **7**:630–631
LL3858
 mycobacterium treatment **7**:725
 structure **7**:725*F*
LLC-PK1 cells, active drug transport prediction **5**:271
LMNA gene, familial partial lipodystrophy (FPLD) **6**:384
Loa loa see Loiasis
Lobeline
 cocaine abuse **6**:184*T*
 structure **6**:181*F*
Local anesthesia
 see also specific drugs
 introductions **1**:193
 sodium channels **2**:839
Local correlation tests, protein 3D structure validation **3**:510
Local interaction energy minima (Emin1-3), VolSurf descriptors **5**:539
Local interaction energy minima distances (d12/13/23), VolSurf descriptors **5**:539
Local lymph node assay (LLNA) **5**:923
Local sample handing, compound storage/management **3**:562

Local treatment, vaginal
  administration **5**:290
Loc-I-Gut technique, permeability
  assays **5**:269*F*, **5**:270
'Lock-and-key' metaphor
  Fischer, Emil **3**:959–960, **3**:960*F*
  ligand–receptor interactions **4**:194
Locked nucleic acids, siRNAs
  modification **3**:183
Locomotor activity, schizophrenia
  models **6**:25
Locus control regions (LCRs), gene
  therapy **1**:313
Locus de novo design method, site
  analysis, drug design **4**:287
Lofepramine
  bioavailability **5**:712*T*
  side effects **6**:60*T*
  transporter inhibition **6**:66*T*
Log D calculation **5**:660
Log*D*pH, lipophilicity measurement,
  ions/zwitterions *see* Lipophilicity
  measurement, ions/zwitterions
Logistics, HTS **2**:458
LOGKOW, ionization prediction, in
  silico tools **5**:623*T*
Log *P*
  aqueous solubility models **5**:633*T*,
    5:640, **5**:641*F*
  calculation *see* Log *P* calculation
Log *P*, ions/zwitterions **5**:439
  chemical class **5**:440*T*, **5**:441
  hydrogen bond donors/
    acceptors **5**:440
  p*K*a effects **5**:440*T*, **5**:441
  steric factors **5**:439
Log *P*, oral absorption
  prediction **5**:693
Log *P* calculation **5**:650, **5**:654,
    5:660
  fragmental descriptors **5**:654
    AlogP method **5**:655
    experimental value (EVA) adjusted
      approach **5**:655
    fragment/correction factors **5**:655,
      5:656*T*
    Ghose/Crippen approach **5**:655
    group correction methods **5**:655
    HlogP method **5**:656
    KlogP program **5**:656
    KOWWIN program **5**:655, **5**:661*T*
    missing fragment problem **5**:655
    Rekker method **5**:655
    substructural molecular fragment
      (SFM) method **5**:657
  linear models **5**:654
  nonfragmental descriptors **5**:657
    BlogP method **5**:658
    COSMOtherm program **5**:658
    MlogP method **5**:658, **5**:661*T*
    mobile order and disorder (MOD)
      theory **5**:657
    QikProp method **5**:658

quantum-chemical
  calculations **5**:658
  SLIPPER method **5**:659,
    5:661*T*
nonlinear methods **5**:659
  artificial neural networks **5**:659
  AUTOLOGP program **5**:659–660
  AutoQSAR models **5**:660
  E-state indices **5**:660
  nearest neighbor analog **5**:659
  XLOGP program **5**:660, **5**:661*T*
non-octanol systems **5**:652, **5**:653*F*
  amphiprotic solvents **5**:652
  Collander equation **5**:652
  inert solvents **5**:652
  linear solvation energy
    relationships **5**:652–653
  propyleneglycol dipelargonate
    (PGPD) **5**:652
  pure proton acceptors **5**:652
  pure proton donors **5**:652
  on-line systems **5**:650, **5**:652*T*
  programs **5**:661*T*
  substituted derivatives **5**:649
Log *P*/log *D*, computational oral
  absorption prediction **5**:684
Log *P*oct values, brain uptake
  prediction **5**:759–760
Loiasis **7**:751*T*
  biology **7**:760
  clinical manifestations **7**:760
  drug treatment **7**:753*T*, **7**:760
  epidemiology **7**:760
Lomefloxacin **7**:569*T*
  structure **7**:570*F*
Lomerizine hydrochloride **1**:183
  introduction of **1**:107
Lometrexol **4**:74–75
  SAR **4**:75, **4**:75*F*
  structure **4**:75*F*
Lomustine **7**:138–139
  toxicity **7**:46*T*
Lonafarnib **7**:234; **8**:30
  biological activity **8**:34, **8**:36*T*
  FPT inhibition **8**:30–31
  library screening **8**:31
  SAR **8**:31, **8**:32*T*
  synthesis **8**:35, **8**:37*F*
  x-ray crystallography studies **8**:33,
    8:33*F*, **8**:34*F*, **8**:34*T*, **8**:35*F*,
    8:36*F*
London forces, van der Waals'
  interactions **4**:97
London Pharmacopoeia **1**:530
  natural products **1**:357
Lone pair vectors (LPVs), 3D-QSAR
  alignment **4**:579–580
Long-acting β-adrenergic receptor
  agonists, asthma
  treatment **7**:883
Long-acting insulin treatment **6**:451
Long-chain acyl-CoA synthetase,
  coenzyme A conjugation **5**:155

Long-chain fatty acyl coenzyme A
  synthase
  radioactive assays **3**:588
  scintillation proximity assay **3**:588
Longitudinal relaxation rate, NMR,
  drug discovery **3**:909, **3**:910*F*
Long QT interval **5**:935, **5**:936*F*
  see also Drug-induced QT interval
    prolongation
  as adverse event **1**:772
  drugs withdrawn **1**:773, **1**:773*T*
  haloperidol **6**:35
  models **5**:933–955
    see also hERG; hERG blocker
      modeling; hERG blocker
      screening
    hERG potassium channel
      blockage **5**:924
    molecular basis **5**:935
  pimozide **6**:35
  schizophrenia treatment side
    effects **6**:33, **6**:35
  sertindole **6**:35
Long QT syndrome (LQTS)
  see also QT interval
  acquired **5**:935
  congenital **5**:935
  potassium channels **2**:864
  sodium channels **2**:841; **6**:735
Long-term adverse effects,
  antivirals **7**:362
Long-term memory, phosphodiesterase-
  4 **2**:926–927
Long-term oxygen therapy, COPD
  treatment **7**:883–884
Loop diuretics **6**:585–587, **6**:708
  bioavailability **6**:709*T*
  dose **6**:709*T*
  half-life **6**:709*T*
  hypertension treatment **6**:585–587
  side effects **6**:587
  structure **6**:709*T*
Loop models, GPCRs **4**:680, **4**:682*F*
Loop motion (proteins), ligand–
  receptor molecular
  flexibility **4**:204, **4**:204*F*
Loperamide, irritable bowel syndrome
  treatment **6**:656
Lopinavir **1**:125; **2**:579*T*; **7**:279–282
  discovery **7**:353; **8**:193
  metabolism **8**:194
  with ritonavir *see* Kaletra
  structure **1**:77*F*; **2**:578*F*; **7**:353*F*;
    8:189*F*, **8**:193–194, **8**:194*F*
  structure-based design **4**:602,
    4:602*F*
    potency improvement **4**:602–604
    resistance problems **4**:602–604
  target information **4**:750*T*
Lopressor **1**:532*T*
Loprinone hydrochloride **1**:169
Loracarbef **7**:500
  structure **7**:500*F*

regulation *vs.* pathology   2:1078
   MELAS   2:1078
   MERRF   2:1078
Mitochondria complex V   2:1078
   catalysis   2:1078
   composition   2:1078
   inhibitors   2:1079
      Bz-423   2:1079
      oligomycin   2:1079
      oligomycin sensitivity-conferring
         protein (OSCP)   2:1079
   pathology   2:1079
      Leigh syndrome   2:1079
      neurogenic muscle weakness, ataxia
         and retinitis pigmentosa
         (NARP)   2:1079
   regulation   2:1078
      ADP availability   2:1078–1079
      F_1-inhibitor protein (IF-1)
         2:1078–1079
Mitochondrial encephalopathy, lactic
      acidosis and stroke-like episodes
      syndrome (MELAS)   1:658;
      2:1060
   mitochondria complex I   2:1071
   mitochondria complex IV
      2:1078
Mitochondrial membrane
      permeabilization (MMP),
      apoptosis   7:13
Mitochondrial permeability transition
      (MPT)
   adenine nucleotide translocator
      conformation shifts   2:1060
   ATP production as protection
      2:1061
   calcium exposure   2:1060
   free radical production   2:1061
   mitoprotective drugs *see*
      Mitoprotective drugs
Mitogen-activated protein kinase
      (MAPK) *see* MAP kinase
      (MAPK)
Mitogen-activated protein kinase kinase
      (MEK), in cancer   7:23
Mitogenesis assays, non-animal
      endocrine toxicity
      testing   1:467*T*
Mitomycin C   7:140
   activation   7:141*F*
   clinical use   7:140
   glaucoma treatment   6:308
   as prodrug   5:1032
   resistance   7:140
   side effects   7:140
   structure   5:1032*F*; 6:310*F*; 7:140,
      7:141*F*
   toxicity   7:46*T*
Mitoprotective drugs   2:1061
   see also specific drugs
   caffeic acid phenethyl ester
      (CAPE)   2:1064, 2:1064*F*
   celastrol   2:1064, 2:1064*F*

mitochondrial permeability transition
      as target   2:1061
   5-(benzylsulfonyl)-4-bromo-2-
      methyl(2H)-pyridazione
      (BBMP)   2:1061, 2:1061*F*
   cytotoxicity   2:1063
   dibenz (b,f)azepine
      analogs   2:1062
   heterocycles   2:1062, 2:1062*F*
   Hoffman-La Roche drugs   2:1062,
      2:1062*F*
   N-substituted
      phenothiazines   2:1062
   novel steroids   2:1063, 2:1063*F*
   SAR   2:1063
Mitotic index, HCS cell cycle
      3:691–692
Mitotic spindle defects, RNA
      interference   3:18
Mitoxantrone   6:240*T*; 7:117
   clinical indication   6:245
   discovery/development   1:380
   multiple sclerosis treatment   6:248;
      7:926
   structure   1:381*F*; 6:238*F*
   toxicity   7:46*T*
   transporters   7:45*T*
Mitoxantrone-resistance protein *see*
      ABCG2 transporter
Mitozolomide   7:136
Mitsunobu reaction, PASP   3:793,
      3:798–799
Mivacurium
   anesthetic adjuvants   6:362
   structure   6:363*F*
Mivotilate   1:245
Mixed inhibition, in vitro
      studies   5:246
Mixed-lineage kinase inhibitors,
      Parkinson's disease
      treatment   6:218
Mixed Model Repeated Measures
      (MMRM), affective disorders,
      clinical trial issues   6:62
Mixture analysis, NMR *see* Nuclear
      magnetic resonance (NMR)
Mizolastine   1:190
Mizollen *see* Mizolastine
MK-0767, structure   5:799*F*
MK-771
   Alzheimer's disease treatment   6:207
   structure   6:274*F*
MK-801
   central nervous system injury
      6:266
   NMDA-sensitive receptors   2:902–
      903
MK-0869   6:77
MKC-733, as prokinetic agent   6:686
MLAB, metabolic drug–drug interaction
      software   5:841
MLN-02, IBD treatment   6:623–624,
      6:632

MLN-518   7:199
   acute myeloid leukemia
      2:971
   characteristics   2:984*T*
   KIT inhibitor   2:972
   structure   2:971*F*; 7:199*F*
MLN-944   7:122
MlogP method, log P
      calculation   5:658, 5:661*T*
MLP (molecular lipophilicity
      potential)   5:581*T*
   3D-QSAR alignment   4:578
   lipophilicity   5:585
MM, scoring/binding energy estimation
      drug design   4:295
MMI-270   7:243
MM-PBSA method
   free energy calculation *see* Free energy
      calculation
   scoring/binding energy estimation
      drug design   4:295
MM-PSA, docking, scoring
      functions   4:263
MN-001   1:505
MNNG   7:138–139
   structure   7:139*F*
MNU   7:138–139
   structure   7:139*F*
MOAT-B *see* ABCC4 (MRP4: MOAT-B)
      transporter
MOAT-C *see* ABCC5 (MRP5: MOAT-C)
      transporter
MOAT-D (MRP3:aBCC3) transporter
      see ABCC3 (MRP3: MOAT-D)
      transporter
Mobec *see* Meloxicam
MOBILE method
   GPCRs   4:688
   homology models   4:487–488
   structure-based virtual
      screening   4:487–488
Mobile order and disorder (MOD)
      theory
   log P calculation   5:657
   solubility   5:629, 5:629*T*
Mobile regions, protein crystallization
      see Protein crystallization
Moclobemide   6:67–68
   bioavailability   5:712*T*
   metabolism, CYP 2C19   5:546*T*
   side effects   6:60*T*
   structure   2:68*F*
Modafinil   1:194; 8:149–156
   see also Adrafinil
   ADHD treatment   6:132
   adverse effects   6:8
   animal studies   8:150
   clinical indications   8:152
      ADHD   6:118, 6:134,
         6:135*T*; 8:149, 8:152
      antidepressant drug
         enhancement   8:153
      cocaine abuse   6:184*T*; 8:153

Modafinil (*continued*)
  cognitive defects in
      schizophrenia **8**:152–153
  fatigue **8**:152
  narcolepsy **6**:154, **6**:155*T*, **6**:156;
      8:151
  psychogeriatrics **8**:151
  sleep apnea **8**:152
  somnolence **8**:152
  wake-promoting properties **8**:151
  discovery **8**:150
  drug discovery process **8**:149
  mechanism of action **8**:151
      α-adrenoceptor antagonists **8**:151
      amphetamines *vs.* **8**:151
      central dopaminergic
          systems **8**:151
      dopamine **8**:151–152
      gamma-aminobutyric acid
          (GABA) **8**:152
      histamine release **8**:152
  preclinical data **8**:150
  sleep disorders treatment **6**:7, **6**:8
  stimulant addiction **6**:185
  structure **8**:150*F*
MODBAS, metalloproteomics **3**:556–
      557
ModBase, homology models **3**:545
MODDE, ligand-based library
      design **4**:310–311
Modeler software, GPCRs **4**:682
ModelMaker, metabolic drug–drug
      interaction software **5**:841
Models/modeling
  see also specific models
  bias, protein crystallography model
      generation **3**:463–464
  comparisons, oral absorption
      prediction **5**:692, **5**:692*T*
  failure, ADMET in silico tools **5**:959
  fitting, solubility models **5**:630
  polydimethylsiloxane **5**:741
  selection, solubility models **5**:630
  size, adaptive in combo
      strategy **5**:966
  statistical methods, E-state
      index **5**:562
Modification analysis, bioisosterism *see*
      Bioisosterism
Modification-sensitive restriction
      enzymes (MSREs),
      epigenetics **1**:266–268, **1**:267*T*
Modified active analogs, pharmacophore
      searching **4**:129*T*
Modified natural products, historical
      perspective **1**:325
2′,3′-Modified nucleosides, hepatitis B
      antivirals **7**:319
Modiodal *see* Modafinil
Modulated receptor hypothesis, sodium
      channels **2**:839
MOE descriptors, 3D QSAR
      descriptors **4**:155

Moersch–Woltman syndrome **6**:11
Moexipril hydrochloride **1**:159
  bioavailability **6**:719*T*
  dose **6**:719*T*
  half-life **6**:719*T*
  in vitro potency **6**:719*T*
  structure **6**:719*T*
Mofezolac **1**:**223**
Mogul (intramolecular geometry) *see*
      Cambridge Structural Database
      (CSD)
Molar refraction, QSAR steric
      effects **4**:47
MOLCAD
  computational oral absorption
      prediction **5**:689–690
  P-glycoprotein models **5**:785
MolConnZ software
  QSAR **4**:153, **4**:154*F*
  similarity searching,
      representations **4**:170
Molecular accessibility, nervous system
      autoimmune diseases,
      treatment **6**:243
Molecular biology
  definition **2**:87
  paradigms **2**:87
Molecular comparisons, ligand-based
      library design **4**:317
Molecular composition, ligand assembly
      drug design **4**:291
Molecular configuration, regulatory
      problems **2**:726
Molecular conformational energies,
      quantum mechanical
      calculations **4**:392
  density functional theory **4**:392
  structure *see* Structure, quantum
      mechanical calculations
Molecular conformation assessment,
      chemoinformatics **1**:40
Molecular connectivity indices
      (MCIs) **4**:541, **4**:556; **5**:559
  see also Electrotopological indices
      (E-state index); Topological
      descriptors
  derivation **4**:557, **4**:557*F*
  first neighbor relationships **5**:557
  first order index ($^1\chi$) **5**:559,
      5:560*F*
  library screening **4**:563
      antivirals **4**:563
  ligand assembly drug design **4**:291
  molecular structure space **4**:560,
      4:560*F*
  objective **5**:559
  QSAR **4**:153
  topological information **5**:557
Molecular descriptors **5**:531–**554**,
      5:581, **5**:583
  see also specific descriptors
  Abraham descriptors **5**:544
  adaptive in combo strategy **5**:963

ADME correlation **5**:583, **5**:587
  blood–brain permeation **5**:587
  'chameleon effect' **5**:588, **5**:588*F*
  lipophilicity **5**:588
apolar surface area (ASA) **5**:582
core molecular modeling **4**:88
Dragon software **5**:544
drug potency, lipophilicity *vs.*
      5:535
electronic properties **5**:583
  charge densities **5**:583–584,
      5:584*F*
  electrotopological state
      (E-state) **5**:584
  force fields **5**:584
  GOLPE software **5**:585
  GRID **5**:585
  highest occupied molecular orbital
      (HOMO) **5**:584
  lowest unoccupied molecular orbital
      (LUMO) **5**:584
  Molecular Interaction Fields
      (MIFs) **5**:584
  Poisson–Boltzmann
      equation **5**:584
4D descriptors **5**:533
hydrogen bonding **5**:536, **5**:586
  atomic increment
      projection **5**:586–587
  combination with other
      properties **5**:536–537
  hydrogen bond acceptor (HBA)
      groups **5**:587
  hydrogen bond donor (HBD)
      groups **5**:587
  hydrogen bonding equilibrium
      constants **5**:536
  lipophilicity *vs.* **5**:535, **5**:536
  measurement **5**:536
  MHBPs **5**:587
  Molecular Interaction Fields
      (MIFs) **5**:586–587
  molecular recognition **5**:536
  permeation **5**:536
  polar surface area (PSA)
      calculation **5**:536, **5**:537*F*
  quantum mechanics **5**:586–587
  thermodynamics **5**:536
hydrophobicity **5**:534
lipophilicity **5**:534, **5**:585, **5**:588
  advantages **5**:585
  assessment **5**:535
  atomic fragmental system **5**:586,
      5:586*F*
  drug potency *vs.* **5**:535
  DRY **5**:586
  hydrogen bonding **5**:536
  hydrogen bonding *vs.* **5**:535
  hydrophobic interaction field
      (HINT) **5**:585
  linear free energy relationship
      (LFER) **5**:535
  MLP fields **5**:585

Molecular Interaction Fields
(MIFs) 5:586
molecular size *vs.* 5:535
n-octanol as gold standard 5:535–536
pharmacokinetics 5:535
pH-independent partition
coefficient 5:535–536
molecular fingerprints *see* Molecular
fingerprints
molecular interaction field (MIF)
derived *see* Molecular Interaction
Fields (MIFs)
molecular size/shape 5:533
lipophilicity *vs.* 5:535
molecular surface 5:533–534
molecular volume 5:533–534
molecular weight 5:533–534
permeability effects 5:533
shape descriptors 5:534
solvent-accessible surface (Connolly
surface) 5:533–534, 5:534F
topological indices 5:534
1D descriptors 5:533
physicochemical descriptors 5:533
'rule of five' 5:533
polar surface area 5:582
projection to latent structure (PLS)
method 5:548–549
properties 5:533
quantum chemical descriptors
5:546
highest occupied molecular orbitals
(HOMO) 5:547
lowest unoccupied molecular
orbitals (LUMO) 5:547
MolSurf program 5:546
special surface-related descriptor
sets 5:547
charged partial surface area
(CPSAs) 5:547
hydrophobic interactions 5:547–548
molecular hashkeys 5:548
MS-WHIM descriptors 5:548
partitioned total surface
areas 5:548
weighted holistic invariant
molecular approach 5:548
steric fields 5:582
comparative molecules similarity
indices analysis
(CoMSIA) 5:582
Lennard–Jones potential
energy 5:582
structure information
representation 4:540
surfaces 5:581
molecule alignment 5:582
point collection 5:581, 5:586F
solid rendering 5:581, 5:584F
solvent accessible surface
(SAS) 5:581, 5:582F

solvent excluded surface
(SES) 5:581, 5:582F
van der Waals surface 5:581,
5:582F
3D descriptors 5:533
2D descriptors 5:533
Molecular diagnostics,
proteomics 3:45
Molecular diversity, protein display
chips 2:370
Molecular dynamics (MD)
3D-QSAR alignment 4:581–582
androgen receptor 4:740–741
conformational analysis/
searching 4:105
cytochrome P450s *see* Cytochrome
P450 family
estrogen receptors 4:740
homology models 3:544–545
impact of 4:4
ligand–receptor interaction
models 4:208
nuclear receptors, drug design *see*
Nuclear receptors, drug design
protein crystallography model
generation 3:463–464
simulations *see* Molecular dynamics
simulations
Molecular dynamics
simulations **4:651–668**
equations in motion 4:652
AMBER 4:652
Beeman algorithms 4:652
CHARMM 4:652
ECEPP/3 4:652
ENCAD 4:652
GROMOS 4:652
OPLS 4:652
Verlet algorithms 4:652
estrogen receptor *see* Estrogen
receptors (ERs), molecular
dynamics simulations
free energy calculation *see* Free energy
calculation
KcsA 4:711
KirBac1.1 4:713
statistical mechanics 4:653
partition function 4:653
phase-space probability
4:653
utility 4:651
Molecular electrostatic potentials
(MEP) *see* Quantum mechanical
calculations property prediction
Molecular evolution, proteins *see*
Protein(s)
Molecular factors, ionization *see*
Ionization
Molecular features, pharmacophore
detection *see* Pharmacophores,
detection
Molecular fields **5:577–602**
descriptors *see* Molecular descriptors

dimensionality 5:578, 5:578F
environmental effects 5:579
5D (social) 5:578, 5:578F
4D (dynamic) 5:578, 5:578F,
5:579
*n*D (multidimensional) 5:578
1D (elementary) 5:578, 5:578F
3D (spatial) 5:578, 5:578F
2D (symbolic) 5:578, 5:578F
form–function–fluctuation 5:579
see also Molecular descriptors
definitions of 5:579–580
physical representations
5:580
4D-QSAR *see* 4D-QSAR
theoretical basis 5:580
field visualization 5:580
molecular interaction fields
(MIFs) 5:580, 5:581T
molecular surfaces 5:580, 5:584F,
5:586F
QSAR tools 5:581T
scalar fields 5:580
3D-QSAR *see* 3D quantitative
structure–activity relationship
(3D-QSAR)
Molecular fingerprints 5:548
2D *vs.* 3D 5:548
chemical information
databases 3:298
definition 5:548
Fingerprint Models 3:298
Mesa Analytics 3:298
metabolic fingerprints 5:548
similarity assessment 5:548
Molecular fractionation with
conjugation capes (MFCC),
quantum mechanics
4:414
Molecular fragment characterization,
structure information
representation 4:561
Molecular Function, PDB 3:377
Molecular globularity (*G*), VolSurf
descriptors 5:538
Molecular graphs/graphics
historical perspective 4:68
Bilder 4:68
Diamond, Robert 4:68
Johnson, Carroll 4:68
Massachusetts Institute of
Technology 4:68
ORTEP 4:68
raster graphs 4:68
ligand-based virtual screening 4:460,
4:460F
QSAR 4:52, 4:155–156
structure/substructure
searching 3:236–237, 3:237F
Molecular hashkeys
molecular descriptors 5:548
oral absorption prediction, molecular
descriptors 5:683

Molecular hologram quantitative
    structure–activity relationship
    (HQSAR) 4:54
Molecular hydrogen-bonding potential
    (MHBP), hydrogen
    bonding 5:682
Molecular Initiatives Roadmap
  chemical biology 3:129
  industry/academia
    collaborations 1:574,
    1:580–581
Molecular Interaction Fields
    (MIFs) 4:237–253
  applications 4:241
    see also specific applications
  calculation 4:238
  definition 4:238
  derived molecular descriptors see
    GRid-INdependent Descriptors
    (GRIND); Molecular descriptors;
    VolSurf
  flexible on protein target 4:247,
    4:248F
  GRID force field 4:240
    electrostatic contribution 4:240
    hydrogen bond potential 4:240
    hydrophobic regions 4:241,
    4:242F
    Lennard–Jones potential 4:240
    structure reconstruction 4:241,
    4:241F
  grid-type for ligands 4:247
    dopamine 4:247, 4:248F
    flexibility 4:248
    hydrophobic contour maps
    4:248
  hydrogen bonding 5:586–587
  hydrophobic patches 4:244
    comparison 4:244
    cytochrome P450 2C
    enzymes 4:244–245
    hexadecane sulfonic acid 4:244,
    4:245F
  lipophilicity 5:586
  low-molecular weight
    molecules 4:238
  macromolecules 4:238
  molecular descriptors 5:537, 5:537F,
    5:548–549, 5:585
    see also GRid-INdependent
    Descriptors (GRIND); VolSurf
    comparative molecular field analysis
    (CoMFA) 5:537
    principal component analysis
    (PCA) 5:537
    projection to latent structures
    (PLS) 5:537
    selective site of
    metabolism 5:543, 5:545T,
    5:546T, 5:547T
    site of metabolism
    prediction 5:543
  molecular fields 5:580, 5:581T

  probe 4:239
    aromatic sp^2 hydroxyl
    probe 4:239–240, 4:239F
    carbonyl oxygen probe 4:239,
    4:239F
    chemical environment 4:239
    chemical properties 4:239
    DRY probe 4:241
    O:: carboxylate probe 4:239F
  probe-water competition 4:248
    dihydrofolate reductase 4:250
    hydrogen-bond entities 4:249–250
    ligand selectivity 4:250
    methotrexate 4:250, 4:250F
  protein binding site
    identification 4:245
    aliphatic amidine multiatom
    probe 4:246, 4:246F
    β-lactamase enzymes 4:247
    N3+ probe 4:246
    neuraminidase as target 4:245,
    4:246F
    O:: carboxylate probe 4:246
    O1 aliphatic hydroxyl probe
    4:246
    Relenza 4:245
  target 4:238
    aqueous environment 4:238
    flexibility 4:238
    input 4:238
    structural water molecules 4:238
  water position calculation 4:242
    cytochrome P450 thymidine
    kinases 4:242–243
    docking programs 4:242–243
    glycogen phosphorylase
    inhibitors 4:243, 4:244F
    HINT 4:242
    HIV protease 4:242, 4:243F
    QSAR studies 4:243
Molecular interactions,
    electrotopological indices
    (E-state index) 5:557
Molecular Libraries Initiative (MLI)
  PubChem 3:307
  QSAR 4:150
Molecular Libraries screening Center
    Network, compound collections
    (libraries) 3:132
Molecular markers
  cancer 7:3
  genetic variation analysis 3:338–339
Molecular mass based methods, NMR,
    drug discovery 3:908
Molecular mass range, fluorescence
    polarization 3:603, 3:604F
Molecular mechanics CADD 4:21
Molecular mechanics (force field
    methods), quantum mechanical
    calculations see Quantum
    mechanical calculations
Molecular mimicry, multiple
    sclerosis 7:923

Molecular modeling
  N-acetyltransferases 5:820
  cytochrome P450s 5:810
  drug metabolism 5:809–825
  historical perspective 4:67, 4:68
    Francoeur, Eric 4:67–68
    Goodford, Peter 4:68
    Martz, Eric 4:67–68
    Richards boxes 4:67
    Richardson, Jane 4:67–68
    Rubin, Byron 4:67–68
  K^+ ion channels 4:710
  sulfotransferases 5:820
Molecular Modeling Database
    (MMDB), blueprint
    databases 3:305–306
Molecular networks,
    bioinformatics 3:327
  metabolic networks 3:328, 3:328F,
    3:329F
    BRENDA database 3:328–329
    databases 3:328–329
    definition 3:328
    EcoCyc database 3:328–329
    ENZYME database 3:328–329
    flux balance analysis 3:329
    GEPASI 3:329
    HinCyc database 3:328–329
    IntEnz database 3:328–329
    KEGG database 3:328–329
    MetaCyc database 3:328–329
    METATOOL 3:329
    MPW database 3:328–329
    PseudoCyc database 3:328–329
  regulatory networks 3:330
    BIND database 3:330
    DIP database 3:330
    DOQCS database 3:330
    nonstationary analysis 3:330
    protein interaction
    networks 3:330
    signal transduction
    pathways 3:330
    TRANSPATH database 3:330
  virtual cells 3:330
    E-Cell project 3:330
    Genomic Object Net 3:330
Molecular orbital methods, cytochrome
    P450 molecular dynamics
    5:817
Molecular pathology, pharmacogenetics–
    environmental effects 3:57
Molecular pathways, functional
    genomics see Functional
    genomics
Molecular polarity index,
    electrotopological indices
    (E-state index) 4:555
Molecular probes, opiate
    receptors 3:664, 3:664T
Molecular profiling, cancer
    biomarkers 7:48
Molecular properties, prodrugs 5:1037

drug introductions **1**:106–107, **1**:193
FTY210 **7**:929
gene therapy **6**:247
glatiramer acetate **6**:246, **6**:248; **7**:924
glucocorticoid treatment **6**:12
GPCR targeting **7**:929
immunomodulation **6**:12
immunomodulatory therapies **8**:173–174
interferon-β **6**:12, **6**:244, **6**:248; **7**:924
intravenous immunoglobulin **6**:248; **7**:928
methotrexate **6**:12; **7**:925
methylprednisolone **7**:924–925
mitoxantrone **6**:248; **7**:926
natalizumab **6**:12, **6**:232, **6**:243, **6**:248
peptide copolymers **7**:929
plasma exchange **6**:248
statins **6**:246; **7**:928
steroids **7**:928
tetracyclines **7**:620
vaccines **7**:929
Multiple sequences, sequence alignment, bioinformatics **3**:320
Multiple sites, organization, drug discovery *see* Organization, drug discovery
Multiple Sleep Latency Test (MSLT)
narcolepsy **6**:148
somnolence measures/sleep testing **6**:142
Multiple structure ensembles, ligand–receptor interaction models *see* Ligand–receptor interactions, models
Multiple targets, biological activity databases **3**:302
Multiple wavelength anomalous dispersion (MAD)
anomalous dispersion **3**:459–460
protein crystallography, rational drug design **3**:889–890
Multiple well dispensing, Ultra-High-Throughput Screening System (Aurora Biosciences) **2**:444, **2**:444*F*
Multiplexing
enzyme-linked immunosorbent assays (ELISAs) **3**:73–74
single nucleotide polymorphisms **2**:363
Multiprotein thyroid-associated protein (TRAP), nuclear receptor coactivators **2**:1002
Multistate activation model, GPCRs *see* G protein-coupled receptors (GPCRs)

Multistep synthesis, PASP *see* Polymer-assisted solution phase synthesis (PASP)
Multisystem expression vectors, gene expression vectors **3**:108
Multiterm gamma Poisson shrinker (MGPS), adverse drug event frequency estimation **1**:768
Multitherapeutic areas, SMEs **1**:503
Multi-use manufacturing equipment, cleaning **2**:165
Multivalent polymers, drug–polymer conjugates **5**:1046*F*
Multivariate library design
PCA combination **4**:330
statistical molecular design combination **4**:330
Multivariate models, computational oral absorption prediction *see* Oral absorption prediction, computational
Mulundocandin **7**:457*T*
medicinal chemistry **7**:460
MUMmer, repetitive elements, bioinformatics **3**:323
Mumps, antiviral drugs **7**:276
Mupirocin **7**:658
administration **7**:658–660
analog stability **7**:662*T*
clinical indication **7**:658
mechanism of action **7**:658
MUPP1, GPCR interaction **2**:797*T*
Muqubilin **1**:371
structure **1**:371*F*
Murad, Ferid, sildenafil citrate, invention of **2**:257
Muraglitazar **6**:490
as PPARγ ligand **2**:1021–1022
structure **2**:1023*F*; **6**:489*F*
Muramyl dipeptide analogs, solid-phase discrete libraries **3**:709*T*
Murine erythroleukemia cells *see* MEL cells
Murko fragment analysis, screening library selection **4**:498
Muromonab **1**:334*T*, **1**:433*T*
clinical use **1**:606–607
Muscarinic M$_1$ receptors
activation **2**:785*T*
allosteric modulators **2**:790*T*
drug discovery **2**:302
in human disease **2**:775*T*
structure **2**:798
structure-based drug design
homology models **4**:687–688
virtual screening **4**:688
structure-based library design **4**:322–323
Muscarinic M$_2$ receptors
allosteric modulators **2**:790*T*
constitutively active **2**:793*T*
in human disease **2**:775*T*

Muscarinic M$_3$ receptors
activation **2**:785*T*
allosteric modulators **2**:790*T*
in human disease **2**:775*T*
inhibitor development **4**:529*F*
Muscarinic M$_4$ receptors, allosteric modulators **2**:790*T*
Muscarinic receptor(s)
agonists
glaucoma treatment **6**:308
schizophrenia treatment *see* Schizophrenia treatment
Alzheimer's disease treatment **6**:204
clozapine interactions **6**:37
constitutively active **2**:793*T*
ligands, carboxylic ester bioisosterism **2**:663, **2**:667*F*
Muscimol, structure **2**:893*F*
MUSCLE, sequence alignment, bioinformatics **3**:320
Muscle
binding, distribution studies **5**:43, **5**:44*F*
pain/weakness, statins **6**:476–477
tissue composition **5**:457*T*
Muscle cells, calcium channels **2**:844
Muscle relaxants
see also specific drugs
introductions **1**:190
'Muscle-type' nicotinic acetylcholine receptors **2**:879–880
Muscular dystrophy model, *Caenorhabditis elegans* model system **3**:8
Musculature, gastrointestinal tract **6**:597
Musculoskeletal disease, SMEs **1**:512
MUSK, PDB entries **2**:965*T*
Muskelin, GPCR interaction **2**:797*T*
Mutagenesis
models, proteomics **2**:389
phosphodiesterase-5 **2**:922
protein crystallization **3**:882
Mutagenicity testing
adenine phosphoribosyltransferase (Aprt) knockouts **3**:161
hypoxanthine phosphoribosyltransferase (Hprt) knockouts **3**:161
Mutamycin *see* Mitomycin C
Mutation free energies **4**:654–655
Mutations
cytochrome c **2**:1076
frequency, hepatitis C virus **7**:388
ion channels, antiarrhythmic drugs, rational design **6**:757
mitochondria complex II **2**:1073
mitochondria genome **2**:1059–1060
PDB **3**:381
protein 3D structure validation **3**:521
spontaneous, obesity models **6**:393

Mutual induced-fit mechanism, nuclear receptors 4:732
Myasthenia gravis 6:11, 6:231T, 6:236
 diagnosis 6:11–12
 epidemiology 6:236
  sexual dimorphism 6:230–232
 generalized 6:236
 ocular 6:236
 pathogenesis 6:236
  autoantibodies 6:230, 6:236
  B cells 6:236
  helper T cells 6:236
  nicotinic acetylcholine receptors 2:888
  thymus 6:236
 pathology 6:11
 symptoms 6:236
 treatment 6:12, 6:249
  acetylcholinesterase inhibitors 6:12, 6:249
  corticosteroids 6:249
  intravenous immunoglobulin 6:247
  mycophenolate mofetil 6:249
  thymectomy 6:249
Mycamine see Micafungin
Mycobacterium avium intracellulare complex (MAC) 7:700
Mycobacterium tuberculosis
 see also Antimycobacterial drugs
 amikacin 7:637
 infection see Tuberculosis
 kanamycin 7:637
 multidrug resistant 7:676, 7:677, 7:690
 spectinomycin 7:637
 Structural Proteomic Project (XMTB) 3:552T
Mycophenolate mofetil 1:225; 6:240T
 clinical indication 6:245
  arthritis treatment 7:856–857
  Coxsackie B virus infections 7:266
  IBD treatment 6:622–623
  myasthenia gravis treatment 6:249
  systemic lupus erythematosus treatment 6:248
  transplant rejection 7:920
 as prodrug 5:1019–1020
 structure 5:1019F; 6:238F
 target information 4:750T
Mycophenolate sodium 1:225
Mycoplasma, tetracyclines 7:612
Mycoplasma genitalium, genome sequencing 7:732
Mycoplasma infection, Guillain–Barré syndrome 6:234
Mycoses
 cutaneous 7:420, 7:423T
  Trichophyton 7:420–423
 subcutaneous see Fungal disease
 superficial, terbinafine 7:429–431
 systemic, amphotericin 7:432
Mycoses fungoides, OCT-43 1:140

Mycotic mycetoma 7:422T
Myelin damage, multiple sclerosis 7:923
Myelodysplastic syndromes
 5-azacytidine 1:273
 demethylating agents 1:273
Myeloperoxidase, atherosclerotic plaque formation 6:467
Myelosuppression, oxazolidinones 7:691
Myfortic see Mycophenolate sodium
Mylotarg see Gemtuzumab ozogamicin
Myocardial infarction (MI) 6:696–697
 carvedilol 8:139
 mitochondria 2:1054–1055
 mitochondria dysfunction 2:1060
 prevalence 6:695T
 risk factors, hypertension 6:706
Myoclonus epilepsy with ragged-red fibers (MERRF) 2:1060
Myogen 1:508
Myoglobin, protein crystallization 3:434
Myoinositol 1,4,5-triphosphate, structure 5:360
Myosin–actin complex, structure 3:543
Myriad 1:506
Myricetin 1:452T
N-Myristoyltransferase inhibitors see Antifungal drugs
MySQL 3:294
 server databases 3:294
Myxothiazol, mitochondria complex III 2:1075

N

^{13}N, PET 3:661–662, 3:664T
N3+ probe, Molecular Interaction Fields 4:246
Nabilone, as antiemetic agent 6:684
Nabumetone, bioavailability 5:712T
Nacyl see Ampiroxicam
NADH-CoQ oxidoreductase complex see Mitochondria complex I
NADH peroxidase, crystal packing 3:442, 3:444F
NAD(P)H quinone oxidoreductase
 drug metabolism 5:202–203
 mechanism of action 5:203
NADH:ubiquinone oxidoreductase inhibitors, solid-phase libraries 3:751F, 3:752–753
Nadifloxacin 7:569T
 acne treatment 7:961
 structure 7:570F, 7:577F
Nadolol
 absorption studies 5:42
 bioavailability 5:712T; 6:712T
 dose 6:712T
 half-life 6:712T

hypertension treatment 6:711
 oral absorption 5:468T, 5:680T
 structure 6:712T
NADP$^+$ reduction, oxidoreductase assays 3:581
NAD$^+$ reduction, oxidoreductase assays 3:581
Nafarelin 2:579T
 hydrophobic depoting 2:585
 intranasal administration 2:593–594
 self-forming depots 2:593
Nafarelin acetate, nasal administration 5:281T, 5:284
Nafcillin, bioavailability 5:712T
Nafion-H resin, PASP 3:821
Nafoxidine 8:85
 clinical trials 8:87T, 8:106–107
 kidney cancer 8:85
 raloxifene development 8:106–107
 structure 8:85F, 8:107F
Naftifine 7:430T
 clinical profiles 7:447
  aspergillosis 7:447
  Candida 7:447T
  candidosis 7:447
 clinical trials 7:449
 pharmacokinetics 7:448
Naftopidil 1:246
Nafuridin, structure 2:1067F
Naked DNA, gene therapy 1:312
Nakijiquinone C
 derivatives 1:392F
 discovery/development 1:391
 structure 1:392F
Nalbuphine, bioavailability 5:712T
Nalidixic acid 7:567
 discovery 7:567–568
 mycobacterium treatment 7:715
 structure 7:568F, 7:715F
Nalmefene hydrochloride 1:194
 alcohol addiction 6:184T
 structure 6:181F
Naloxone
 bioavailability 5:712T
 oral absorption 5:680T
 reduction 5:126–127, 5:126F
 structure 6:181F, 6:412F
Naltrexone
 alcohol abuse treatment 6:9, 6:180T, 6:183
 bioavailability 5:712T
 heroin addiction 6:180T, 6:183–185
 oral absorption 5:468T
 prodrugs of, carbonate/carbamate esters 5:1023–1024
 structure 6:181F
Naltrindole derivatives, solid-phase discrete libraries 3:709T
NAMD, K$^+$ ion channels 4:710
Nandrolone decanoate
 intramuscular testosterone replacement therapy 6:528
 structure 2:1016F; 6:522T

Nandrolone phenylpropionate
intramuscular testosterone
replacement therapy 6:528
structure 6:522*T*
NanoBio Design 1:492*T*
Nanoprobes, NMR *see* Nuclear
magnetic resonance (NMR)
Nanotechnology 2:46–47, 2:361
antigen-antibody reactions 2:361
definition 2:349
DNA hybridization 2:361
DNA sequencing 2:361
NAPH analogs, mitochondria complex I
2:1069, 2:1070*F*
Naphthalene ring, tetracyclines 7:602
Naphthalene-sulfone derivatives,
selective estrogen receptor
modulators (SERMs) 7:164–
165, 7:165*F*
2-Naphthoic acid, benzoic acid
5:404*T*
Naphthol, structure 5:370*F*
Naproxen
Alzheimer's disease treatment 6:211
BCS 5:973*T*
bioavailability 5:712*T*
effective permeability (P_{eff}) 5:673*T*
metabolism, CYP 2C9 5:545*T*
migraine treatment 6:373–374,
6:373*T*
oral absorption 5:468*T*, 5:680*T*
safety 1:684
structure 6:373–374
2-Napthoic acid, aggregation
reactions 5:407, 5:408*F*,
5:409*F*
Naramig *see* Naratriptan hydrochloride
Naratriptan hydrochloride 1:183
migraine treatment 6:373*T*
Narcolepsy 6:7, 6:141
animal models 6:149
daschund 6:149
orexin knockout mice 6:149,
6:152
clinical trials 6:153
Multiple Sleep Latency
Test 6:148
diagnosis 6:148
Multiple Sleep Latency
Test 6:148
polysomnography 6:148
disease basis 6:149
clock genes 6:150
COMT polymorphisms 6:149
HLA-DR2/DQ1 6:149
HLA markers 6:7–8
models 6:152
new research areas 6:164
histamine receptor
antagonists 6:164
prokineticin 6:164
transforming growth factor-α
6:164

prognosis 6:8
treatment 6:8, 6:153, 6:155*T*
ABT-239 6:156–157
ampakanes 6:156
amphetamine 6:153–154, 6:155*T*
amphetamine-like
substances 6:155
armodafinil 6:155*T*, 6:156
bupropion 6:156
caffeine 6:154
CX-516 6:155*T*
CX-717 6:155*T*
dexamphetamine 6:155*T*
dexmethylphenidate 6:155,
6:155*T*
drug introductions 1:194
fencanfamine 6:156
GSK-189254 6:156–157
GT-2331 6:156–157
histamine (H_3) receptor
antagonists 6:156
JNJ-5207852 6:156–157
JNJ-17216498 6:155*T*, 6:156–157
mazindol 6:156
methamphetamine 6:153–154,
6:155
methylphenidate 6:153–154,
6:155, 6:155*T*
modafinil 6:154, 6:155*T*, 6:156;
8:151
monoamine oxidase
inhibitors 6:154
pemoline 6:153–154, 6:155
selegiline 6:153–154, 6:155
sodium oxybate 6:154, 6:155*T*,
6:156
tricyclic antidepressants 6:154
unmet medical needs 6:163
selective orexin antagonists 6:163
Naringenin 1:452*T*
Naringin, pharmacokinetics 1:454*T*
Narirutin 1:452*T*
Naropin *see* Ropivacaine
Nartograstim 1:152
NAS-91, malaria treatment 7:793*T*
Nasal administration 5:284
absorption enhancers 5:285*F*, 5:286
animal nasal mucosa 5:285, 5:286
interspecies differences 5:286
bioadhesive drug delivery 5:286
CNS delivery 5:286
in vitro studies 5:284
see also specific studies
mucociliary clearance 5:286
peptide drugs 5:284
permeation studies 5:285, 5:285*F*
drug metabolism 5:285
permeation coefficient (Pc) 5:285
transport rates 5:285
primary human cell cultures 5:284
Calu-3 lines 5:284–285
disadvantages 5:284–285
RPMI 2650 5:284–285

products 5:281*T*
rat models 5:286
Nasal mucosa, cytochrome
P450s 5:172*T*
Nasea *see* Ramosetron
Natalizumab 1:334*T*; 6:240*T*
adverse reactions 1:772
progressive multifocal leukoenceph-
alopathy 1:772
clinical trial suspension 6:243–244
IBD treatment 6:620–621, 6:632
irritable bowel syndrome
treatment 6:600
multiple sclerosis treatment 6:12,
6:232, 6:243, 6:248
withdrawal 1:757*T*, 1:772
Nateglinide 1:228; 6:439*T*, 6:443
advantages 6:444
clinical use 6:444
disadvantages 6:444
mechanisms of action 6:443
pharmacokinetics 6:443
side effects/contraindications 6:444
structure 6:440*F*
National Acute Spinal Cord Injury
Study (NASCIS II) 6:264–265
National Cancer Institute
(NCI) 3:282*T*
biological fingerprints 4:813
chemogenomics 4:816
compound collections
(libraries) 1:327; 3:132
database 3:308–309
Diversity Set 2:749
structure databases 3:281
3D database pharmacophore
searching 4:138
National Center for Biotechnology
Information (NCBI)
messenger RNA (mRNA) 2:1043
nucleotide sequence
databases 3:350
National Center for Computational
Biology, cDNA
microarrays 3:331
National Center for
Toxicogenomics 2:569
National Cholesterol Education
Program 8:65
Adult Treatment Panel 6:472
coronary heart disease 6:471
National Economic Research
Associates 1:646
National Health and Nutrition
Examination Survey 1:683
National Institute for Aging (NIA),
Alzheimer's disease 6:197
National Institute for Environmental
Health Sciences 1:57*T*
National Institute for Health
(NIH) 1:736
budget allocations 1:736
composition 1:736

National Institute for Health (NIH) (*continued*)
  funding **1**:736
  research topics **1**:736
National Institute of General Medical Science (NIGMS) **3**:551
National Institute of Health and Medical Research (INSERM) **1**:662
National Institute of Neurological Disorders and Stroke (NINDS) **3**:132
Nationalities, diversity issues **2**:244
National Obesity Forum **1**:693
National Organization for Rare Disorders (NORD) **1**:657
National Research Council, industry/academia collaborations **1**:580
National Research Fund, industry/academia collaborations **1**:580
National Surgical Adjuvant Breast and Bowel Project (NSABP)
  breast cancer prevention **8**:112–113
  tamoxifen clinical trials **8**:91, **8**:95
National University of Singapore Bioinformatics Drug Design Group, BIDD Databases **3**:305
Native peptides, peptide/protein drugs **2**:580
Natrecor *see* Neseritide
Natural analogs, peptide/protein drugs **2**:580
Natural killer cells, psoriasis **7**:942
Natural Law **1**:805
Natural lead analogs, HTS library composition **3**:680
Naturally proliferating cells, cell-based assays **3**:620–621
Natural membranes, lipophilicity measurement **5**:434
Natural products **1**:**355–403**
  see also specific drugs
  anti-Gram-positive *see* Anti-Gram-positive natural products
  antimicrobial agents **7**:469, **7**:470–471
  chemical genetics **1**:366
  combinatorial biosynthesis **1**:361
    nonribosomal peptides **1**:361
    nonribosomal peptide synthesis **1**:361
    polyketide synthase enzymes **1**:361
  combinatorial chemistry **1**:363
  combinatorial chemistry *vs. see* Combinatorial chemistry
  compound collections (libraries) **1**:327; **3**:133
  disadvantages **3**:133
  disease treatment **1**:367
    see also specific drug classes
  diversity-oriented synthesis **1**:363
  drug design **1**:45*T*

drug discovery **1**:357; **2**:296, **2**:296*T*, **2**:297
  proportions **1**:357, **1**:358*F*
  efficacy assessment **1**:58
  future work **7**:475
  'herbal-directed small libraries' **1**:59, **1**:59*F*
  history **1**:356
    Ancient Egypt **1**:356
    Ancient Greece **1**:356
    Arabic culture **1**:356–357
    Avicenna **1**:356–357
    Canon Medicinae **1**:356–357
    China **1**:356
    Corpus of Simples **1**:356–357
    Dioscorides **1**:356
    Galen **1**:356
    London Pharmacopoeia **1**:357
    Mesopotamia **1**:356
    middle ages **1**:356–357
  HTS *see* High-throughput screening (HTS)
  importance **1**:358
  mast cell tryptase/chymase inhibitors **7**:898
  materials management **2**:531
  microbes **1**:359
    culture conditions **1**:359
    endophytes **1**:361
    environmental sample extraction **1**:360, **1**:360*F*
    extremophiles **1**:360
    marine sediments **1**:361
    'nutrient-sparse' media **1**:359–360
    symbionts **1**:361
  multicomponent systems **1**:58–59
  original functions **1**:358
    acquisition through diet **1**:358–359
    antimicrobial agents **1**:359
    as chemical defense **1**:358, **1**:358*F*
    defense **1**:359
    predation **1**:359
  outsourcing **2**:205
  privileged structures **1**:363
  protein–protein interaction disruption **1**:327
  purification **3**:133
  skeleton analyses **1**:363
  solid-phase libraries **3**:718, **3**:726, **3**:728*T*, **3**:740, **3**:740*F*
  synthesis, PASP **3**:814
  target identification **3**:133
  total synthesis **1**:363
  use of, WHO estimates **1**:357
Naturopathy, definition **1**:409
Nausea *see* Emesis/nausea
Nazasetron hydrochloride **1**:**150**
NBI 30775, structure **6**:103*F*
NBI-74330, IBD treatment **6**:635
N-cadherin, upregulation in cancer **7**:15

NCBI Reference Sequences *see* RefSeq (NCBI Reference Sequences)
ND1251 **6**:80
NDG-96-3, primary insomnia treatment **6**:159*T*
*n*D (multidimensional) molecular fields **5**:578
Neamine derivatives **7**:636
  O-glycosylation **7**:635
  glycosyl mimetics **7**:636
Nearest neighbor analog, log P calculation **5**:659
Nebilet *see* Nebivolol
Nebivolol **1**:**159**
  bioavailability **6**:712*T*
  dose **6**:712*T*
  half-life **6**:712*T*
  hypertension treatment **6**:711
  structure **6**:712*T*
Neboglamine **2**:903
Nebramine, derivatives **7**:636
Necator americanus see Hookworm infections
Necrosis, cell-based assays **3**:629
Nedaplatin **1**:**147**
  cancer treatment **1**:147
'Need to know' *vs.* 'nice to know,' culture **2**:121–122
Neem oil, patenting issues **1**:802
Nefazodone hydrochloride **1**:**177**; **6**:101
  oral absorption **5**:680*T*
  primary insomnia treatment **6**:157
  side effects **6**:60*T*
  structure **6**:101*F*
  transporter inhibition **6**:66*T*
Nefiracetam **2**:903
Negamycin analogs, solid-phase discrete libraries **3**:709*T*
Negative phenotype interpretation, genetically engineered animals *see* Genetically engineered animals
Negative symptoms, schizophrenia **6**:18
Neglected diseases, orphan drugs **1**:666
'Neighborhood behavior' principle, definition **4**:807
Neighbor mutation operator, product-based library design **4**:349–350
Neighbors, adaptive in combo strategy **5**:963
Neisseria gonorrhoeae, spectinomycin **7**:637
Nelarabine **7**:62*T*
  approval **7**:62*T*
  clinical indication **7**:62*T*
  clinical trials **7**:67
  mechanism of action **7**:67
Nelfinavir mesylate **1**:**126**; **2**:579*T*; **7**:279–282
  clinical uses **7**:358
  discovery **7**:351

discovery/development **1**:76*F*
dosing regimens **7**:357*T*
mitochondria, effects on
    2:1082
SARS **7**:273–274
structure **1**:76*F*; **7**:352*F*
structure-based design **4**:612,
    4:613*F*
target information **4**:750*T*
Nematode infections *see* Parasitic
    infections; *see specific infections*
Nembutal *see* Pentobarbital
Nemorubicin **7**:116, **7**:123
Neocarcinostatin, drug–polymer
    conjugates **5**:1051
Neolignans, PASP **3**:818
Neomycin
    dual-action agents,
        oxazolidinones **7**:685
    non-microbial activities **7**:641
        HIV infection **7**:641
    oral absorption **5**:680*T*
    structure **2**:1041–1042
Neomycin B **7**:630
    structure **7**:631*F*
Neonatal epilepsy syndromes,
    epilepsy **6**:285
Neonatal hypoxia, ADHD
    models **6**:128
Neonatal irritation model, irritable
    bowel syndrome **6**:652
Neonatal respiratory distress syndrome,
    vascular endothelial growth
    factor treatment **3**:203
Neonates, pharmacokinetics **5**:27
Neoorthoforms, bioisosterism **2**:674,
    2:677*F*
Neostigmine
    autoimmune neuromuscular
        disorders **6**:12
    as prokinetic agent **6**:680
Neovastat **3**:202–203
    psoriasis treatment **3**:202
Nepafenac
    structure **6**:307*F*
    wet age-related macular
        degeneration **6**:306–307
Nephelometry, kinetic solubility
    5:349
Nephrotoxicology,
    aminoglycosides **7**:644, **7**:715
Neprilysin, oxytocin hydrolysis
    5:140*F*
Neridronic acid **1**:224
Nerixia *see* Neridronic acid
Nernst equation, mitochondria *see*
    Mitochondria
Nerve block **6**:355
Nerve compression, neuropathic pain
    models **6**:3346:334*F*
Nerve growth factor
    in analgesia **6**:345
    psoriasis **7**:982

Nervous system
    ATP-gated ion channels **2**:904–905
    drug introductions *see* Drug
        introductions
    phosphodiesterase-4 **2**:926–927
Nervous system autoimmune
    diseases **6**:229–251, **6**:231*T*
see also specific diseases/disorders
    animal models **6**:237
        disease-specific **6**:239
        Drosophila **6**:237
        primates **6**:237–239
        rodents **6**:237
        symptom replication **6**:237
    autoantibody-mediated **6**:230
    classification systems **6**:230
    clinical trials **6**:242
        design flaws **6**:242
        measures **6**:242
        recruitment issues **6**:242
    CNS **6**:230, **6**:231*T*, **6**:232
    CNS/PNS **6**:230, **6**:231*T*, **6**:233
    PNS **6**:230, **6**:231*T*, **6**:234
    presentation **6**:230
    sexual dimorphism **6**:230–232
    T-cell mediated **6**:230
Nervous system autoimmune diseases,
    treatment **6**:231*T*, **6**:238*F*,
    6:240*T*, **6**:242
see also specific drugs
    anti-inflammatory cytokines
        6:244
    B cell /T cell depletion
        mechanisms **6**:245
    corticosteroids **6**:246
    co-stimulatory molecule
        inhibitors **6**:245
    dynamic changes **6**:242–243
    future work **6**:250
        disease heterogeneity
            problems **6**:250
        DNA microarray screening
            6:250
        magnetic resonance imaging
            techniques **6**:250
    gene therapy **6**:247
    intravenous immunoglobulin
        6:247
    molecular accessibility **6**:243
    plasma exchange **6**:247
    proinflammatory cytokines **6**:244
    selective adhesion molecule
        inhibitors **6**:243
    statins **6**:246
    stem cell treatment **6**:247
    therapeutic vaccines **6**:246
    tolerogens **6**:246
Nervous system diseases/disorders
see also specific diseases/disorders
    autoimmune disease *see* Nervous
        system autoimmune diseases
    potassium channels **2**:864–865
Nervous system trauma **6**:12

Neseritide **1**:168; **2**:579*T*
    congestive heart failure
        treatment **6**:698
    introduction of **1**:107
Nested case-control study,
    postmarketing
    surveillance **1**:773, **1**:774*F*
Netherlands
    medicine marketing system **1**:632
    orphan drugs legislation **1**:662
Netilmicin **7**:634
Netivudine, oral absorption **5**:680*T*
Netoglitazone
    as PPARγ ligand **2**:1021–1022
    structure **2**:1023*F*
Net present value (NPV) analysis,
    definition **1**:567
Network models, databases **3**:268,
    3:268*F*
Neupogen *see* Filgrastim
Neural networks
    3D-QSAR **4**:586, **4**:586*F*
    blood–brain distribution **5**:751
    computational oral absorption
        prediction *see* Oral absorption
        prediction, computational
    corneal permeability modeling **5**:736
    protein function prediction,
        bioinformatics **3**:335
Neural specificity, irritable bowel
    syndrome **3**:100
Neuraminidase inhibitors **4**:614,
    4:615*F*; **7**:377
see also specific drugs
    benzoic acid-based **7**:382*F*, **7**:383
    clinical studies **7**:384
    DMPK/efficacy **7**:384
    dynamic combinatorial
        libraries **3**:970, **3**:971*F*
    introductions **1**:107
    in vitro activity **7**:383
    oral absorption **5**:1014
    orthomyxoviruses **7**:275
    pyrrolidine-based **7**:382, **7**:382*F*
    resistance **7**:385
    structure-based design **4**:614
        cyclopentane scaffolds **4**:620,
            4:622*F*
        docking experiments **4**:622–623
        new apolar pocket
            identification **4**:619, **4**:619*F*
        oral available drugs **4**:620,
            4:621*F*
        resistant strains **4**:623, **4**:624*F*
        SAR **4**:622–623
        transition state-like analogs **4**:617,
            4:618*F*
Neuraminidases
    influenza viruses *see* Influenza viruses
    quantum mechanical
        calculations **4**:405
    quantum mechanics/molecular
        mechanics (QM/MM) **4**:405

Neuraminidases (*continued*)
  structure 4:614, 4:615*F*
    sialic acid complex 4:615,
      4:616*F*
    transition-state-like analog 4:617,
      4:617*F*
  as target, Molecular Interaction Fields
    (MIFs) 4:245, 4:246*F*
Neuregulin, schizophrenia 6:23
Neurex *see* Fenoldopam
Neurite overgrowth, HCS 3:692
Neuritis, neuropathic pain
    models 6:334
Neuroactive steroids, mechanism-based
    PK/PD modeling 5:897
Neuroadaptations, addiction 6:173–
    175
Neuroanatomy, migraine 6:370
Neurochemistry, anxiety disorders *see*
    Anxiety disorders
Neurocircuitry, anxiety disorders *see*
    Anxiety disorders
Neurocrine Biosciences 1:510
Neurodegenerative diseases 6:1, 6:2,
    6:9, **6:193–228**
  see also specific diseases
  AMPA-sensitive receptors 2:903
  animal models 6:195, 6:196
    Danio rerio model system 3:10
    genetic models 6:196
  autoimmune disorders 6:11
  cognitive function 6:195
  future work 6:195, 6:225
  incidence 6:196
  inheritance 6:195
  kainate-sensitive receptors 2:903
  mediators 6:195
  mitochondria dysfunction 2:1060
  motor function 6:195
  neuromuscular disorders 6:11
  progressive nature 6:194
  selective neuronal
    vulnerability 6:195
  treatment 6:195
    see also specific drugs
    curative 6:196
    protective 6:196
    symptomatic 6:196
Neuroendocrine factors, acne *see* Acne
Neuroethics, definition 1:699
Neurogen 1:509
Neurogenesis
  Caenorhabditis elegans model
    system 3:8
  depression 6:57
Neurogenic bladder, duloxetine clinical
    trials 8:129
Neurogenic inflammation, skin
    inflammation treatment 7:981
Neurogenic muscle weakness, ataxia
    and retinitis pigmentosa
    (NARP), mitochondria complex
    V 2:1079

Neuroimaging
  see also specific methods
  anxiety disorders *see* Anxiety disorders
  generalized anxiety disorder 6:89
  irritable bowel syndrome 6:649
  obsessive compulsive disorder
    (OCD) 6:90
  posttraumatic stress disorder
    (PTSD) 6:90
  social anxiety disorder (SAD)
    6:89
  somnolence measures/sleep
    testing 6:142
Neuroimmunophilins, Parkinson's
    disease treatment 6:217
Neurokinin 1 receptor
    antagonists 6:14
  see also specific drugs
  parallel library design 4:325
Neurokinin 1 receptor
  irritable bowel syndrome 6:658
  virtual screening 4:689
NK1 receptor antagonists
  see also specific drugs
  affective disorders 6:76
  antiemetic drugs 6:682
Neurokinin 2 receptor
  irritable bowel syndrome 6:658
  structure determination 4:680,
    4:681*F*
NK2 receptor antagonists, structure-
    based drug design 4:687
Neurokinin 3 receptor antagonists
  see also specific drugs
  schizophrenia treatment 6:40
Neurokinin A receptor antagonist,
    peptidomimetics 2:608–609,
    2:611*F*
Neurokinin B receptor antagonist,
    peptidomimetics 2:608–609,
    2:611*F*
Neurokinin receptor antagonists
  see also specific drugs
  anxiety treatment 6:109
    clinical trials 6:111
    preclinical studies 6:110
Neurokinin receptor modulators,
    irritable bowel syndrome
    treatment 6:661*T*, 6:662
Neuroleptic drugs
  see also specific drugs
  calcium-channel blockers 2:849–850
  introductions 1:184
  side effects 1:696
Neurology, potassium channels 2:864
Neuromedins, in obesity 6:410
Neuromedin U (NmU), obesity
    treatment 6:409
Neuromodulators
  see also specific drugs
  peptide/protein drugs 2:580
Neuromuscular ataxia, potassium
    channels 2:864

Neuromuscular blocking drugs
  see also specific drugs
  anesthetic adjuvants 6:362
Neuromuscular disorders 6:11
Neuronal distribution, acid-sensitive
    ion channels (ASICs)
    2:911
Neuronal hyperexcitability, calcium-
    channel blockers 2:851
α7-Neuronal nicotinic receptor agonist,
    clozapine 6:39–40
Neuronal nicotinic receptor agonists
  see also specific drugs
  schizophrenia treatment 6:39
Neuronal nicotinic receptor antagonists
  see also specific drugs
  pain 6:340
α7-Neuronal nicotinic receptor gene
    polymorphisms,
    schizophrenia 6:39
Neuronal nicotinic receptors 2:879–
    880
  anesthetics 6:358
Neuronal nitric oxide synthase
    (nNOS) 2:1078
Neuron-derived orphan receptor 1,
    nomenclature 2:996*T*
Neurons, calcium channels 2:844–845,
    2:847–848
Neuropathic pain 6:328–329
  definition 6:328–329
  models 6:334, 6:334*F*
    chemotherapy-induced 6:335
    chronic constriction injury
      (CCI) 6:334, 6:334*F*
    diabetes 6:335
    direct nerve trauma 6:334
    inflammation 6:334
    nerve compression 6:334
    neuritis 6:334
    partial nerve ligation model 6:334,
      6:334*F*
    SIN model 6:334–335
    spinal nerve ligation 6:334,
      6:334*F*
  pathophysiology 6:329
    tissue injury 6:330
  treatment
    analgesics, differential
      efficacy 6:337*T*
    gabapentin 8:239
    unmet medical needs 6:340
Neuropathic Pain Scale 6:335
Neuropathic Pain Symptoms
    Inventory 6:335
Neuropathophysiology, pain *see* Pain
Neuropathy
  cisplatin 7:142
  knock-in animal models 3:160
Neuropeptides
  acid-sensitive ion channels
    (ASICs) 2:911
  affective disorders 6:74

Neuropeptide Y
affective disorders **6**:77
anxiety disorders **6**:90
in obesity **6**:406–407
Neuropeptide Y agonists **6**:406
Neuropeptide Y antagonists **6**:406
structure **6**:407*F*, **6**:408*F*
Neurophysiology, gastrointestinal tract
see Gastrointestinal tract
Neuroprostanes, neuroprotective drug
biomarker **6**:262–263
Neuroprotective drugs **6**:254
see also specific drugs
Alzheimer's disease treatment
6:211
amyotrophic lateral sclerosis
treatment **6**:224
biomarkers **6**:262
calcium-channel blockers **2**:849–850
glaucoma **6**:312–314, **6**:315*F*
introductions **1**:188
oxidative phosphorylation
uncoupling **2**:1079–1080
Neuropsychiatric disorders, 5-HT₃
receptors **2**:891
Neurorestorative drugs **6**:275
see also specific drugs
Neurospora crassa, as model
system **3**:647
Neurotensin, sleep **6**:147
Neurotensin receptor antagonists
see also specific drugs
binding **2**:623–624, **2**:624*F*
nonpeptide drugs **2**:618–619,
2:620*F*
Neurotensin receptor downregulation,
corticotrophin-releasing factor
overexpression **3**:100
Neurotoxicity
animal experimentation
alternatives **1**:467*T*, **1**:474
artemisinin **7**:778
cell-based assays **3**:639, **3**:642*F*,
3:643*F*
measures used **3**:641
Multicenter Evaluation of in Vitro
Cytotoxicity **3**:641
Neurotoxin receptor site 1, sodium
channels **2**:837
Neurotransmitters
chronic pain **6**:331
release
gabapentin mechanism of
action **8**:241
5-HT₃ receptors **2**:889
sleep *see* Sleep
sodium channels **2**:836
Neurotransmitter transporters **2**:59,
2:60*T*
Neurotrophic drugs
see also specific drugs
age-related macular degeneration
treatment **6**:315*T*

amyotrophic lateral sclerosis
treatment **6**:224
anxiety disorders **6**:90
Neutral compounds
PAMPA **5**:471
permeability **5**:672
NEUTRALIZER program, 2D- to 3D
structure **4**:464
Neutral protamine Hagedorn
(NPH) **6**:451
Neutral species
aggregation, solubility **5**:410
lipophilicity measurement **5**:439
NeuTrexin *see* Trimetrexate glucuronate
Neutron diffraction studies, crystal
size **3**:440
Neutrophil elastase, COPD **7**:880
Neutrophil elastase inhibitors
see also specific drugs
clinical trials **7**:882
COPD treatment **7**:882
Neutrophils
COPD **7**:879
psoriasis **7**:942
tetracyclines **7**:620
NeutroSpec (Technetium (99mTc)
Fanolesomab)
adverse reactions **1**:770
withdrawal **1**:770
Neu-Up *see* Nartograstim
Nevirapine **1**:**126**; **7**:279–282
combination treatment **7**:347
discontinuation **7**:347
discovery/development **1**:69, **1**:71*F*;
7:343
dosing regimen **7**:346, **7**:346*T*
extent of use **7**:345
structure **7**:343*F*
reverse transcriptase-
binding **1**:72–73, **1**:73*F*
target information **4**:750*T*
tolerability **7**:346
New Clinical Drug Evaluation Unit
Program (NCDEU), affective
disorders, clinical trial
issues **6**:63
New compound registration, materials
management **2**:534
New diseases
governments, in drug
discovery **1**:736
target *vs.* disease **2**:113, **2**:114*T*
New drug applications (NDAs) **2**:561,
2:561*F*
pharmaceutical R&D **2**:104–105
procedures **1**:546–547
New molecular entities
(NMEs) **1**:688–689
see also 'Me-to drugs'
Newron Pharmaceuticals **1**:511
New technology
intellectual property **1**:745
outsourcing **2**:208

Newton–Raphson method, core
molecular modeling **4**:102–103
New York Structural GenomiX Research
(NYSGRC) **3**:551–552, **3**:552*T*
metalloproteomics **3**:556–557
progress **3**:553*T*
New Zealand hypertensive rats **6**:707–
708
New Zealand obese (NZO)
mouse **6**:435
Nexium *see* Esomeprazole;
Esomeprazole magnesium
NF-κB *see* Nuclear factor kappa B
(NF-κB)
NGB 2904, structure **6**:187*F*
NGD-91-3 **6**:106
NGD-96-3 **1**:510
NGF-induced factor B,
nomenclature **2**:996*T*
NGX-424 **2**:903
NH-3
structure **2**:1020*F*
as thyroid hormone receptor
ligand **2**:1019
NHERF, GPCR interaction **2**:797*T*
Niacin **6**:479
structure **6**:479*F*
Nicardipine
aneurysmal subarachnoid hemorrhage
treatment **6**:263
bioavailability **5**:712*T*; **6**:715*T*
characteristics **2**:829*T*
dose **6**:715*T*
half-life **6**:715*T*
P-glycoprotein interaction **5**:781*T*
SAR **6**:716*T*
structure **2**:829*T*
Niched Pareto genetic algorithm
(NPGA), multiobjective
optimization **4**:768
Niche pharmaceutical market,
value **1**:659
Niclosamide
cestode infection treatment **7**:761
SARS **7**:273–274
Nicotine
bioavailability **5**:712*T*
glucuronic acid conjugation **5**:152–
153
nicotine addiction **6**:180*T*
oral absorption **5**:680*T*
PASP **3**:817
structure **2**:655*F*, **2**:887*F*; **5**:359;
6:173*F*
Nicotine addiction **6**:173*T*
see also Tobacco use
comorbidities **6**:184*T*
medications for **6**:179, **6**:180*T*, **6**:183,
6:184*T*
bupropion **6**:180*T*, **6**:183
methoxsalen **6**:184*T*
nicotine **6**:180*T*
replacement therapy **6**:183

Nicotine addiction (*continued*)
  rimonabant 6:184*T*
  varenicline 6:184*T*
Nicotinic acetylcholine receptors
    (nAchRs) 2:879
  characteristics 2:881*T*
  diseases 2:888
    autosomal dominant nocturnal
      frontal lobe epilepsy
      (ADNFLE) 2:888
    myasthenia gravis 2:888
  distribution 2:884
  ligand stereoisomerism, quantum
      mechanical calculations 4:405,
      4:406*F*
  molecular biology 2:884
    autonomic synaptic
      transmission 2:884
    CNS receptors 2:884
  nomenclature 2:879
    'muscle-type' 2:879–880
    neuronal 2:879–880
  pharmacology 2:887, 2:887*F*
  physiological functions 2:886
  regulation 2:886
    cAMP response element binding
      protein (CREB) 2:886–887
    SNARE 2:886–887
  structure 2:885
    accessory proteins 2:886
    ACh binding site 2:885
    muscle-type 2:885
    neuronal 2:886
    pentameric 2:885
    subunits 2:880–884
    TM3-TM4 intracellular
      loop 2:885–886
  therapeutics 2:887
    see also specific drugs
    Alzheimer's disease
      treatment 6:205
Nicotinic acid
  ionization constant (pK_a)
    5:366*T*
  oral absorption 5:468*T*
Nicotinic acid analogs, solid-phase
    discrete libraries 3:709*T*
Niemann–Pick C1 like protein
    (NPC1L1), ezetimibe
    8:80
Nifedipine
  bioavailability 5:712*T*, 6:715*T*
  characteristics 2:829*T*
  dose 6:715*T*
  drug action 1:532*T*
  half-life 6:715*T*
  mechanism of action 6:715–718
  SAR 6:716*T*
  structure 2:75*F*, 6:587*F*
Nifekalant hydrochloride 1:164
Niflumic acid, ionization constant
    (pK_a) 5:366*T*
Nifuroxime, structure 5:371*F*

Nifurtimox
  bioavailability 5:712*T*
  structure 7:823*F*
  trypanosomiasis treatment 7:758,
    7:822–823
Niger, life expectancy 1:638
Nigeria
  functional public health
    system 1:642
  health insurance coverage 1:641
Niguldipine
  sila analog 2:697*F*
  structure 2:697*F*
NIH 3T3 cells, protein expression
    systems 3:117
NIH Chemical Genomics Center
    (NCGC), compound collections
    (libraries) 3:132
Nilsson, J Lars G, collaborators,
    Lindberg, P 8:214
Nilutamide, structure 2:1016*F*
Nimbex *see* Cisatracurium besylate
Nimble Gen, microarrays 3:89
Nimesulide, structure 2:658*F*, 2:681*F*
Nimodipine
  aneurysmal subarachnoid hemorrhage
    treatment 6:263
  bioavailability 5:712*T*
  characteristics 2:829*T*
  nervous system trauma
    treatment 6:13
  SAR 6:716*T*
  stroke treatment 6:259
  structure 2:829*T*, 6:315*F*
  traumatic brain injury
    treatment 6:259
NINDS screening program,
    anticonvulsant drugs 6:287
NIP-142 6:752–753
Nipecotic acid, structure 2:895*F*
Nipkow disc, HCS confocal
    systems 3:687, 3:688*F*
Nisoldipine
  bioavailability 5:712*T*, 6:715*T*
  characteristics 2:829*T*
  dose 6:715*T*
  half-life 6:715*T*
  oral absorption 5:680*T*
  SAR 6:716*T*
Nisoxetine
  chemogenomics 3:934*F*, 3:935
  structure 6:100*F*
NIST Databases 3:278*T*
NIST webBook 3:278*T*
Nitazoxanide, protozoal infection
    treatment 7:758
Nitisinone 1:247
Nitrazepam, ionization constant
    (pK_a) 5:366*T*
Nitrendipine
  bioavailability 5:712*T*
  characteristics 2:829*T*
  oral absorption 5:680*T*

  SAR 6:716*T*
  structure 2:829*T*
Nitric oxide
  enhancement, calcium-channel
    blockers 6:588
  function, diabetes mellitus 6:581
  mitochondria complex IV 2:1077–
    1078
  production, central nervous system
    injury 6:254–255
  production inhibition, selenium
    bioisosterism 2:703
  renal dysfunction 6:580
  vasodilation 6:580–581
Nitric oxide inhibitors
  see also specific drugs
  glaucoma treatment 6:312–314
Nitric oxide synthase (NOS), knockout
    animal models 3:157–158
Nitric oxide synthase inhibitors
  see also specific drugs
  migraine treatment 6:376
Nitrilase and Nitrile Hydratase
    Knowledgebase 3:309
Nitrile oxidative transformation,
    cytochrome P450s 5:109
Nitrobenzene
  microwave heating 3:840*T*
  reduction 5:123–124, 5:123*F*
4-Nitrocatechol, structure 5:370*F*
Nitroflurbiprofen, Alzheimer's disease
    treatment 6:211
Nitrofurantoin
  bioavailability 5:712*T*
  structure 5:371*F*
[13]N, PET 3:661–662, 3:664*T*
Nitrogen atoms
  cytochrome P450s 5:105
  oxidation, flavin monooxygenases
    (FMO) 5:113, 5:113*F*
Nitrogen-containing functional group
    reduction, cytochrome
    P450s 5:123*F*, 5:124*F*
Nitrogen mustards 7:130
  see also specific drugs
  carcinogenicity 7:132
  clinical use 7:131
  history 7:129
  mechanism of action 7:130, 7:131*F*
  resistance 7:132
    cellular transport 7:132, 7:133
    DNA repair enzymes 7:132
    glutathione-*S*-transferases 7:132
  structure 7:130
[13]N, PET 3:661–662, 3:664*T*
Nitrogen/oxygen atom numbers, brain
    uptake prediction 5:760
Nitroglycerin, buccal
    administration 5:281*T*
Nitroheterocycle
  structure 7:834*F*
  trypanosomiasis treatment
    7:834–835

HTS hit validation   **3**:902
intermolecular NOE   **3**:908*T*, **3**:912,
    3:913*F*, **3**:914*F*
intramolecular NOE   **3**:908*T*, **3**:911
lead optimization   **3**:901, **3**:902
lead search   **3**:901
ligand-based methods
    labeling   **3**:907
    molecular mass based
        methods   **3**:908
    speed   **3**:907
    target-based techniques *vs.*   **3**:907
NOE pumping   **3**:908*T*, **3**:913
paramagnetic spin labels   **3**:909
    second-site screening   **3**:910,
        3:911*F*
    SLAPSTIC   **3**:910
relaxation filtering   **3**:908*T*, **3**:913
relaxation rate measurement   **3**:909,
    3:910*F*
    diffusion rate *vs.*   **3**:909
    filter elements   **3**:909
    longitudinal relaxation rate   **3**:909,
        3:910*F*
    subtraction artifacts   **3**:909
saturation transfer difference (STD)
    experiments   **3**:908*T*, **3**:912,
        3:912*F*, **3**:913*F*, **3**:914*F*
target-based techniques   **3**:915
    CRIPT   **3**:915–916
    ligand-based methods   **3**:907
    pH effects   **3**:907
    receptor signal changes   **3**:915–916
    specific *vs.* non-specific
        binding   **3**:907
    structural information   **3**:907
    TROSY   **3**:915–916
techniques   **3**:908*T*
technology improvements   **3**:901
trNOESY   **3**:908*T*
WaterLOGSY   **3**:908*T*, **3**:913, **3**:914*F*
weak binding detection   **3**:901–902
Nuclear magnetic resonance (NMR),
    structure determination   **2**:512;
    3:473–505, **3**:540, **3**:902, **3**:948
assignment   **3**:476
    amino acid side chains   **3**:476,
        3:478*F*
    dipolar couplings   **3**:476, **3**:478*F*
    liquid *vs.* solid-state   **3**:476
    scalar couplings   **3**:476, **3**:478*F*
chemical shifts   **3**:486, **3**:489*F*
    solid-state   **3**:487
COSY   **3**:474
dipolar coupling   **2**:514; **3**:480*T*,
    3:481, **3**:483, **3**:489
    alignment   **3**:484–485
    cross-relaxation rate   **3**:481
    directly bonded nuclei   **3**:485
    internuclear vectors   **3**:484, **3**:485*F*
    membrane proteins   **3**:484
    NOESY   **3**:482, **3**:483*F*
    orientation information   **3**:484

recoupling   **3**:482
    as restraints   **3**:484
dipolar relaxation, liquid-state   **3**:481
dynamics   **3**:495, **3**:496*F*
    µs to ms dynamics   **3**:497, **3**:498*F*
    ns to µs dynamics   **3**:499, **3**:500*F*
    relaxation dispersion   **3**:498
    solid-state NMR   **3**:500*F*, **3**:501,
        3:501*F*
    sub-correlation dynamics   **3**:496
    examples   **3**:490
    see also specific proteins/molecules
fast acquisition schemes   **3**:491
fragment screening   **4**:790, **4**:792
global quality measures   **3**:546
    restraints per residue   **3**:546
hardware developments   **2**:515
hydrogen bond *J* coupling   **2**:514
interaction cross-correlation   **3**:486
    intervector angle   **3**:486, **3**:488*F*
    side chain angles   **3**:486
    torsional angles   **3**:486, **3**:488*F*
isotopic labels   **2**:512; **3**:475, **3**:477*F*
library design   **4**:323
liquid-state   **3**:474, **3**:474*F*
    CSA   **3**:479–481
    dipolar relaxation   **3**:481
    examples   **3**:490
    proton detection   **3**:479–481
    relaxation optimized
        sequences   **3**:479–481
    samples   **3**:476*F*
    scalar coupling   **3**:486
    solid-state *vs.*   **3**:476
magic angle spinning (MAS)   **3**:475
model ensembles   **3**:541
    Olderado (On Line Database of
        Ensemble Representatives and
        Domains)   **3**:541–542
    representative member   **3**:541
molecular weights   **2**:512; **3**:479
    interactions   **3**:479, **3**:480*F*, **3**:480*T*
    liquid *vs.* solid   **3**:476–477
    solid-state   **3**:479
multidomain proteins   **3**:492
    binding affinities   **3**:492
    domain motion   **3**:500
    preparation   **3**:492
    tight binding   **3**:492
    weak binding   **3**:493
multiple entries   **3**:543
multiple RF channel   **2**:512,
    2:513*F*
NOE assignments   **3**:542
NOESY   **3**:474
    numbers of   **3**:540
paramagnetic relaxation   **3**:482
parameters   **3**:481
protein–ligand complexes   **3**:492
    binding affinities   **3**:492
    preparation   **3**:492
    tight binding   **3**:492
    weak binding   **3**:493

protein–protein interactions   **3**:492
    binding affinities   **3**:492
    preparation   **3**:492
    tight binding   **3**:492
    weak binding   **3**:493
quality problems   **3**:542
    accuracy   **3**:542
    precision   **3**:542
    x-ray crystallography *vs.*   **3**:542
re-refined structures   **3**:542
    DRESS (Database of Refined
        Solution NMR
        Structures)   **3**:542
    RECOORD   **3**:542
samples   **3**:475
    concentrations   **3**:475
    solid *vs.* liquid state   **3**:475,
        3:476*F*
scalar coupling   **3**:485, **3**:489
    assignment   **3**:485–486
    bonds   **3**:485–486, **3**:487*F*
    liquid-state   **3**:486
    TOBSY spectra   **3**:486
solid-state
    calculations   **3**:489
    chemical shift anisotropy   **3**:479
    dynamics   **3**:500*F*, **3**:501, **3**:501*F*
    examples   **3**:490
    INEPT   **3**:478
    liquid *vs.*   **3**:476
    magic angle spinning   **3**:473–474,
        3:482
    molecular tumbling   **3**:479
    molecular weights   **3**:479
    restrained molecular
        dynamics   **3**:489
    samples   **3**:476*F*, **3**:477
    simulated annealing
        protocol   **3**:489
    TOCSY   **3**:478
    transfer schemes   **3**:477, **3**:479*F*,
        3:480*F*
structural genomics   **3**:555
TALOS routine   **3**:489
techniques   **2**:512
transverse relaxation-optimized
    spectroscopy (TROSY)   **2**:514
water (solvent) suppression
    2:514
x-ray structure *vs.*   **3**:540
Nuclear magnetic resonance
    spectroscopy
    drug development   **2**:194
    fragment-based drug discovery *see*
        Fragment-based drug discovery
    nuclear receptors   **2**:997
    peptidomimetics   **2**:622–623
Nuclear Overhauser effects (NOEs),
    NMR spectroscopy   **3**:946
Nuclear protein kinase(s)   **7**:185
Nuclear receptor ligands,
    chemogenomics   **3**:932, **3**:933*F*,
        3:934*F*

Nuclear receptors 2:757*T*, 2:993–1036; 4:739; 4:725
see also specific receptors; *specific types*
acetylation 2:1003
activator assays, CYP induction studies 5:251
class I 2:994, 2:995*F*
  physiological functions 2:1004, 2:1012*T*
  regulation 2:1002
class II 2:994, 2:995*F*
  physiological functions 2:1007, 2:1012*T*
  regulation 2:1002
class III 2:994, 2:995*F*
classification 2:994
coactivators 2:1002
  ATP-dependent chromatin remodelers 2:1002
  multiprotein thyroid-associated protein (TRAP) 2:1002
  p160/steroid receptor coactivator 2:1002
  protein arginine methyltransferase 1 (PRMT1) 2:1002
cofactor recruitment 4:726
constitutive receptors 4:726
coregulators 2:1002
corepressors 2:1003
diseases 2:1025
  cancer 2:1025
  genetic disease 2:1026
  inflammation 2:1025
  metabolic syndrome 2:1025–1026
  osteoporosis 2:1025
DNA-binding region 4:725
drug design *see* Nuclear receptors, drug design
future work 2:1028
homogeneous time-resolved fluorescence 2:420
ligand-binding domain (LBD) 2:998, 2:998*T*; 4:726
  conformational change 4:726
  dimerization sites 2:1001
  domain fold 4:726, 4:727*F*
  hypervariable region 4:726
  ligand absence 2:999
  ligand as structural cofactor 2:1000
  ligation states 4:726
  structure 4:728*T*
ligand-binding mechanisms 4:730
  conformational change 4:731
  ligand adaptability 4:731
  mutual induced-fit mechanism 4:732
  protein plasticity 4:731
  selectivity 4:731–732
  transcriptional activation 4:731
ligand dependence 4:726, 4:730
  activity *vs.* 4:730, 4:732*F*
  heterodimers 4:730, 4:731*F*

homodimers 4:730, 4:731*F*
  loosely-defined ligands 4:730
  well-defined ligands 4:730
ligands 4:728*T*, 4:732
  physicochemical properties 4:424*T*
mechanism of action 4:726, 4:727*F*
methylation 2:1003
nomenclature 2:995, 2:996*T*
orphan 2:994, 2:995*F*, 2:996*T*
  as drug targets 2:994
pharmacological target space 4:422*T*
pharmacology 2:1011, 2:1012*T*
  see also specific drugs
phosphorylation 2:1004
physiological functions 2:1004
structure 2:995, 2:997*F*
  AF-1 domain 2:1001
  central domain 2:995–997
  C-terminal domain 2:995–997
  DNA-binding domain 2:1002
  N-terminal domain 2:995–997
  three-dimensional structure 2:997, 2:998*T*
    coregulator binding site 2:998*T*, 2:999*F*, 2:1000
    NMR spectroscopy 2:997
  transcriptional regulation 4:726
  ubiquitination 2:1003
Nuclear receptors, drug design 4:739, 4:725–747
see also specific receptors
CADD 4:739
current drugs 4:732, 4:734*T*
  side effects 4:733
homology modeling 4:739
ligand design/docking 4:742
molecular dynamics 4:740
  ligand entry/exit 4:741
pharmacophore models 4:741
  definition 4:741–742
QSAR 4:741
  comparative molecular field analysis (CoMFA) 4:741
structure-based 4:732
virtual screening 4:742
Nuclear transfer, transgenic animals 1:436
Nuclear translocation assays, HCS *see* High-content screening (HCS)
Nuclease-based assays, single nucleotide polymorphisms *see* Single nucleotide polymorphisms (SNPs)
Nucleic acid-based dynamic combinatorial libraries *see* Dynamic combinatorial libraries (DCLs)
Nucleic acid constructs, gene therapy *see* Gene therapy
Nucleic acids 2:1037–1052
see also DNA; RNA
  cell-based assays 3:628, 3:631
  drug discovery 2:34
  as drug targets 2:1037

homogeneous time-resolved fluorescence *see* Homogeneous time-resolved fluorescence (HTRF)
structure 2:1037, 2:1038*F*
Nucleocapsid assembly, hepatitis C virus 7:388
Nucleophilic additions, glutathione transferases 5:160, 5:160*F*
Nucleophilic component, hydrolases 5:135
Nucleophilic substitution, conjugation, drug metabolism 5:161, 5:162*F*
Nucleoside analogs
see also specific drugs
  development 1:54*F*, 1:69
  natural products 1:372
  solid-phase pool libraries 3:705*F*
Nucleoside mimics *see* Antimetabolites
Nucleoside-phosphate kinase, drug metabolism 5:149
(deoxy)Nucleoside-phosphate kinase, drug metabolism 5:149
Nucleoside phosphotransferase, drug metabolism 5:149
Nucleoside reverse transcriptase inhibitors (NRTIs) 7:279–282, 7:338
see also specific drugs
  clinical limitations 7:340
  comparisons 7:342
  compliance 7:341
  discovery 7:338
  dosing regimens 7:342, 7:342*T*
  drug compatibility 7:342
  extent of use 7:340
    combination chemotherapy 7:340
  future work 7:342
  mitochondria, effects on 2:1082
  resistance 7:340
  tolerability 7:341
L-Nucleosides
  hepatitis B 7:264, 7:319
Nucleosomes, structure 1:252
Nucleotide binding domains, ABC transporters 5:57–58, 5:57*F*
Nucleotide identification, single nucleotide polymorphisms 2:363
Nucleotide-like libraries, solid-phase libraries 3:751–752
Nucleotides, solid-phase libraries 3:698
Nucleotide sequence databases 3:351
see also specific databases
  history 3:350
  RefSeq (NCBI Reference Sequences) 3:355
Nucleotide synthesis inhibitors, transplant rejection 7:920
Nucleotide transporters, trypanosomiasis treatment 7:828, 7:830*F*, 7:831*F*

Nucleus accumbens, corticotrophin-releasing factor overexpression **3**:99
Nucleus of the solitary tract, emesis **6**:672–673
Number of patients needed to treat, carvedilol **8**:143, **8**:144*T*
Numerical databases **3**:277
Numerical descriptors, similarity searching **3**:245
Numerical encoding of structure, ADMET in silico tools **5**:556
Nur related factor 1, nomenclature **2**:996*T*
NusA, gene expression vectors **3**:110*T*
Nutlin-1 **7**:239
Nutlin-2 **7**:239
  crystallization **7**:239
Nutlin-3 **7**:239
Nutlins **7**:239
Nutraceuticals **1**:449–461, **1**:451*T*, **1**:452*T*
  see also Traditional medicine
  allylic sulfur compounds **1**:449–450
  assays **1**:456
  bioavailability **1**:455
  carotenoids **1**:449–450
  chemical classes **1**:450
    phenolic compounds *see* Phenolic nutraceuticals
  classification **1**:449–450, **1**:450*F*
  definition **1**:449–450
  indoles **1**:449–450
  intelligent product delivery **1**:458
    quality management **1**:458*T*
    species names **1**:458
    traceability **1**:458, **1**:458*T*
  isolation **1**:456
  isothiocyanates **1**:449–450
  metabolic principles **1**:455
  monoterpenes **1**:449–450
  nutragenomics **1**:457
  regulatory considerations **1**:458
    Dietary Supplement Health and Education Act **1**:458–459
    market forces **1**:458–459
  screening techniques **1**:456, **1**:457*T*
Nutragenomics, nutraceuticals **1**:457
Nutrient content claims, traditional medicine **1**:417
'Nutrient-sparse' media, natural products **1**:359–360
Nutropin *see* Somatotrophin
NVP-LAK974 **7**:225
NVP-LAQ824 **7**:225
NVP-LAQ-824
  discovery/development **1**:356
  structure **1**:385*F*
NXL103 **7**:662
  clinical indication **7**:662
  in vitro anti-aerobic Gram-positive activity **7**:663*T*

NXY-059 **6**:266
  structure **6**:267*F*
Nymphomania **6**:9
Nystatin **7**:430*T*
  clinical profile **7**:457
  mechanism of action **7**:456

O

O:: carboxylate probe, Molecular Interaction Fields **4**:239*F*, **4**:246
^{15}O, PET **3**:661–662, **3**:664*T*
O1 aliphatic hydroxyl probe, Molecular Interaction Fields **4**:246
O6-methylguanine DNA methylase methylation, pharmacogenetics **3**:61
O-1057 **6**:342–343
  structure **6**:343*F*
O-acetylation, acetylation **5**:154
Oak Ridge National Laboratories system, microfluidic devices **2**:359, **2**:360*T*
OATP (*SLC21/SLCO*)
  transporters **5**:64, **5**:65*T*
  blood–brain barrier **5**:67, **5**:79
  characterization **5**:66–67
  hepatotoxicity **5**:78
  in kidney **5**:67
  kidney/renal clearance **5**:79–80
  in liver **5**:78
  modeling **5**:788
  nomenclature **5**:64–66
  structure **5**:64–66, **5**:70*F*
OAT (*SLC22*) transporters **2**:59; **5**:70
  cloning/structure elucidation **5**:67*T*, **5**:71
  detoxification **5**:71
  energy requirements **5**:70–71
  excretion role **5**:71
  kidney/renal clearance **5**:79–80
  members **5**:67*T*, **5**:70–71
  renal clearance **2**:61
  structure **2**:59, **2**:61*F*, **5**:70–71, **5**:70*F*
    bacterial mimetics **2**:59–60, **2**:61*F*
Obese mouse strain, QTL analysis **3**:14
Obesity **6**:389–416, **6**:702
  see also Hypertension; Renal dysfunction
  abdominal *see* Abdominal obesity
  adiponectin **6**:577
  associated medical conditions **6**:390, **6**:390*T*
    see also specific conditions
  breast cancer **7**:151
  Caenorhabditis elegans model system **3**:9
  cancer risk factors **7**:3
  cardiovascular disease **6**:420
  clinical trial issues **6**:394
    Committee for Proprietary Medical Products (CPMP) **6**:394, **6**:394*T*
    FDA guidelines **6**:394, **6**:394*T*

  definition **6**:576–577
  diabetes mellitus **6**:419, **6**:430, **6**:576–577, **6**:702–703
  diagnosis **6**:390–391
    body mass index (BMI) **6**:390–391, **6**:390*T*, **6**:394
  disease basis **6**:392
    energy expenditure *vs.* energy intake **6**:392
    energy metabolism defects **6**:392
    insulin **6**:392
    leptin **6**:392
    weight gain genetics **6**:392
  disease state **6**:390
  endothelial dysfunction **6**:580
  endothelin **6**:591
  epidemiology **6**:419
  experimental models **6**:392
    dual energy x-ray absorptiometry **6**:393
    high-fat diets **6**:393
    humans *vs.* **6**:392–393
    nuclear magnetic resonance imaging **6**:393
    spontaneous mutations **6**:393
  genetics **6**:391
    environmental influences **6**:391
    genes **6**:391
  incidence **6**:419, **6**:576–577
  inflammatory cytokines **6**:581
  interleukin-6 (IL-6) **6**:581
  metabolic syndrome relationship **6**:391, **6**:577
  monocyte chemoattractant protein-1 (MCP-1) **6**:581
  mortality **6**:390–391
  prevalence **6**:390, **6**:391, **6**:419, **6**:702–703
    age **6**:391
    ethnicity **6**:391
  renal dysfunction **6**:576, **6**:580
  tumor necrosis factor-α (TNF-α) **6**:581
  unmet medical needs **6**:453
Obesity treatment **6**:394
  see also specific compounds
  antioxidants **6**:590–591
  CNS mechanisms **6**:395
    marketed compounds **6**:395, **6**:395*T*
  efficacy **6**:403
  history **6**:394
  multiple mechanisms **6**:403
  new research **6**:404
    bombesin **6**:410
    cholecystokinin agonists **6**:411
    cocaine and amphetamine-related transcript (CART) **6**:409
    corticotropin-releasing factor (CRF) **6**:409
    enterostatin **6**:410
    ghrelin antagonists **6**:408
    glucagon-like peptide 1 **6**:413
    5-HT$_{2C}$ receptor antagonists **6**:413, **6**:414*F*

Obesity treatment (*continued*)
MC4 receptor agonists   6:409
melanin-concentrating hormone
antagonists   6:408, 6:409*F*
neuromedin U (NmU)   6:409
opioid antagonists   6:412, 6:412*F*
orexins   6:410, 6:411*F*
oxyntomodulin   6:413
uncoupling proteins   6:412
urocortins   6:409
patient selection   6:404
peripheral mechanisms   6:402
safety/tolerability   6:403
unmet medical needs   6:403
Object-based models, databases *see*
Databases
Objectiveness, culture   2:122–123
Obligate intracellular parasites,
tetracyclines   7:612
Obliquine, PASP   3:819
Observation-led drug discovery *see* Drug
discovery
Obsessive compulsive disorder
(OCD)   6:6, 6:87
clinical trials   6:96
Y-BOCS (Yale–Brown obsessive
compulsive scale)   6:96
human genetics   6:92
neuroimaging   6:90
positron emission tomography   6:6
prevalence   6:87
prognosis   6:6
SPECT   6:6
treatment   6:6
selective serotonin reuptake
inhibitors   6:99–100
Obstruction, benign prostatic
hyperplasia diagnosis   6:561
Obstructive sleep apnea (OSA)   6:7,
6:149
disease basis   6:151
prognosis   6:8
treatment   6:162
continuous positive airway pressure
(CPAP)   6:162
fluoxetine   6:162
GR79236   6:162
mirtazapine   6:162
OCA-983, structure   7:509*F*
Occupational Safety and Health
Administration, development
chemistry timeline   2:170
Occupations, cancer risk factors   7:3
Ocinaplon   6:104
structure   6:105*F*
OC (organism classification) lines,
UniProtKB flatfile   3:360
OCT-43   1:139
mycoses fungoides   1:140
Octadedecyl-poly(vinyl alcohol)
columns, lipophilicity
measurement   5:435
Octanoate metabolism, gastrointestinal
motility models   6:675

Octanol–water partitioning   5:649
see also Log D calculation; Log P
calculation; Partition coefficient
(*P*)
historical perspective   4:48
in vitro oral absorption
prediction   5:676
lipophilicity   5:682
lipophilicity measurement   5:433,
5:703–704
Octin *see* OCT-43
Octopamine, affective disorder
treatment   6:80
Octreotide   2:579*T*
development   2:581–582
structure   2:581–582
OCT (*SLC22*) transporters   2:59;
5:66*T*, 5:68
distribution   5:68–69
drug pharmacokinetics   5:69–70
mechanism of action   5:70
structure   2:59, 2:61*F*; 5:68, 5:70*F*
bacterial mimetics   2:59–60, 2:61*F*
substrate specificities   5:68
Ocular neovascularization, RNAi (RNA
interference)   3:181–182
Oenin   1:452*T*
Office of Orphan Product Development
(OOPD)   1:660
'Off-label' drug use
children, regulations   1:547
definition   1:712
postmarket monitoring   1:720–721
see also Safety
Prozac   1:718
Offshore outsourcing
definition   2:205, 2:206
determination of   2:215
communication   2:215
existence of   2:216
import/export
considerations   2:215
IP risk   2:215
suitability   2:215
Off-target activity
biological fingerprints   4:809–810,
4:810*F*
RNAi (RNA interference)   3:178
Off-the-shelf workstations, automation
see Automation
Ofloxacin   7:569*T*
activity   7:577
mycobacterium treatment   7:705*T*,
7:715, 7:717
ocular infection treatment   6:317
structure   7:570*F*, 7:577*F*, 7:578*T*
OG (organelle) line, UniProtKB
flatfile   3:360
Okamoto–Aoki rats,
hypertension   6:707–708
Olamufloxacin   7:589–590
Olanzapine   1:185; 6:32
characteristics   2:653*T*
cognition effects   6:36

mechanism of action   6:32
GPCR targeting   2:772,
2:773*T*
side effects   6:33, 6:35
metabolic syndrome   6:35–36
structure   2:653*T*
Olcegepant, structure   6:373–374
Olderado (On Line Database of
Ensemble Representatives and
Domains), NMR
structures   3:541–542
Oleandomycin, discovery   7:520–521
Olefin metathesis, solid support
synthesis   2:469, 2:470*F*
Olefin polymerization, alumoxane-
activated metallocene
catalysts   1:791
Ole–Tozer equation
distribution models   5:870
volume of distribution (V_{ss}) *see* Volume
of distribution (V_{ss})
Olfactory bulbectomy, affective
disorders   6:54*T*
Olfactory signaling, phosphodiesterase-
2   2:923–924
2'-5' Oligoadenylate synthetase, RNAi
(RNA interference)   3:173
Oligodeoxynucleotides, antisense-based
therapeutics   2:1045
Oligomers, solid support
synthesis   2:475
Oligomycin, mitochondria complex V
2:1079
Oligomycin sensitivity-conferring
protein (OSCP), mitochondria
complex V   2:1079
Oligonucleotide arrays *see* Microarrays
Oligonucleotide drugs
see also specific drugs
chemical *vs.* biological   2:117–118
Oligonucleotides
length, arrays   3:89
ligation, single nucleotide
polymorphisms (SNPs)   2:365*F*,
2:366
solid support synthesis   2:476,
2:477*F*
Oligosaccharide-binding fold, x-ray
structures, problems   3:537
Oligosaccharides, solid support
synthesis *see* Solid support
synthesis
Olmesartan medoxomil   1:159
bioavailability   6:722*T*
GPCR targeting   2:773*T*
half-life   6:722*T*
hypertension treatment   6:721–723
structure   6:722*T*
Olomucine   7:205
discovery/development   1:389–390
structure   1:388*F*
quantum mechanical
calculations   4:392, 4:392*F*,
4:393*F*, 4:394*F*

protein kinases **3:**218
CDK1 substrates **3:**218
inhibitors **3:**218–220, **3:**221*F*
(4-amino-3-phenyl-1-*t*-
butyl)pyrazolo[3,4-*d*]pyrimidine
(PPI) derivatives **3:**218–220,
3:221*F*
v-Src **3:**218, **3:**220*F*
technique **3:**215
Orthomyxoviruses, antiviral
drugs **7:**254*T,* **7:**274
A-192558 **7:**275–276
A-315675 **7:**275–276
amantadine **7:**274–275
flutamide **7:**276
neuraminidase inhibitors **7:**275
oseltamivir **7:**275
ribavirin **7:**276
rimantadine **7:**274–275
RWJ-270201 **7:**275–276
T-705 **7:**276
zanamivir **7:**275
Orthosomycins **7:**667
Orthosteric site, GPCR allosteric
modulation **2:**802
Orthostrapper, protein function
prediction,
bioinformatics **3:**334
Orthotopic/metastatic models,
anticancer agents, tumor
models **7:**36
Osanetant, schizophrenia
treatment **6:**40
Oscient Therapeutics **1:**507
Oseltamivir phosphate **1:127,** **1:**501;
7:379
approval **7:**378
clinical studies **7:**384–385
DMPK/efficacy **7:**384
influenza **1:**127
oral absorption **5:**1014
orthomyxoviruses **7:**275
resistance **7:**385
transmission efficacy **7:**385
SAR **7:**379, **7:**381*T*
structure **5:**1015*F,* **7:**377*F*
structure-based design **4:**619,
4:619*F*
cell culture evaluation **4:**620
target information **4:**750*T*
zanamivir *vs.* **7:**381
OSI-774 (erlotinib) **1:**500; **7:**194, **7:**195
characteristics **2:**984*T*
EGFR family kinase inhibitors **2:**968
mechanism of action **7:**185–187
structure **2:**71*F,* **2:**968*F;* **7:**193*F,*
7:194
target information **4:**750*T*
toxicity **7:**46*T*
OSI-7904L **7:**72
development **7:**72
Osiris, log P calculation
program **5:**661*T*

OS (organism species) line, UniProtKB
flatfile **3:**360
OSM, STAT3 signaling **3:**206
Osmotic laxatives
irritable bowel syndrome
treatment **6:**658
side effects **6:**654–656
Osmotic volume shifts, equilibrium
dialysis **5:**327
OSPREYS (Oriented Substituent
Pharmacophore PRopErtY
Space), reagent-based library
design **4:**344
Osteoarthritis (OA) **6:**506
animal models **7:**866
clinical markers **7:**847–848
magnetic resonance imaging
7:847–848
pain, nociceptive pain models
6:333
pathogenesis **7:**846
treatment **7:**847–848
x-ray **7:**847–848
Osteoblasts **6:**497
morphology **6:**497*F*
regulation **6:**501
Osteoclasts **6:**497
apoptosis, bone resorption/
remodelling **8:**205
clear zone **6:**497
morphology **6:**497*F*
numbers, bone resorption/
remodelling **8:**205
osteopetrosis **6:**507
regulation **6:**500, **6:**501*F*
CSF-1 **6:**500
OPG **6:**501
RANKL **6:**501
receptor activator of NFκB
(RANK) **6:**501
vacuolar pump inhibitors, parallel
library design **4:**325
Osteocytes **6:**497
morphology **6:**497*F*
Osteogenesis, cDNA functional
screening **3:**15–16
Osteogenesis imperfecta **6:**507
bisphosphonates **6:**512
pamidronate **6:**507
Osteogenic sarcoma, bone
tumors **6:**505
Osteoinducers
see also specific drugs
introductions **1:**224
Osteomalacia, treatment, vitamin
D **6:**514
Osteopetrosis **6:**507
carbonic anhydrase II
mutations **6:**507
knockout animal models **3:**156,
3:157*F*
osteoclasts **6:**507
transgenic animals **3:**156, **3:**157*F*

Osteoporosis **6:**502; **8:**199
animal models **6:**507
genetically modified mice **6:**508
knockout animal models **3:**156,
3:157*F*
large animal models **6:**507
menopausal bone loss **6:**507
rat aged oophorectomy
models **6:**507
rat oophorectomy models **6:**507
bone mineral density **6:**503
definitions **6:**502; **8:**199
fractures **6:**502–503
incidence **8:**199
genetics **6:**503, **6:**506
glucocorticoid induced **6:**503
male hypogonadism **6:**536
menopause **8:**103
nuclear receptors **2:**1025
pathogenesis **6:**503
postmenopausal **6:**503
prevalence **6:**502
reduction, tamoxifen **8:**95
risk factors **6:**503
Osteoporosis treatment **6:**509
see also Bone diseases/disorders,
treatment
bisphosphonates **6:**512
clinical trials **6:**509
FDA guidelines **6:**509
drug introductions **1:**219
estrogen replacement therapy
(ERT) **8:**103
phase I clinical trials **6:**509
phase II clinical trials **6:**509
phase III clinical trials **6:**509
fracture incidence
endpoint **6:**509–510
secondary endpoints **6:**510
vertebral fractures **6:**510
testosterone replacement
therapy **6:**536
vitamin D **6:**513
Osteosclerosis **6:**507
OSU 6162 (PNU-9391A), schizophrenia
treatment **6:**35
Otitis media, macrolide
antibiotics **7:**522
Ototoxicity, aminoglycosides **7:**644–
645, **7:**715
Otsuka Long–Evans Tokushima Fatty
(OLETF) rats, diabetes mellitus
type 2 animal models **6:**435
Ouabain
oral absorption **5:**680*T*
structure **6:**309*F*
Outer membrane permeability, β-lactam
antibiotic resistance **7:**487
Outlier labeling, solubility
models **5:**632
Out-of-plane bending terms, force-field
molecular modeling *see* Force-
field molecular modeling

Outsourcing   **2**:20, **2**:**203–223**
  see also Contract research organizations
    (CROs); Service companies;
    specific types
  advantages   **2**:207, **2**:227, **2**:228*T*
    capability improvement   **2**:207
    commercial considerations   **2**:209
    efficiency   **2**:228, **2**:229*F*
    financial fluctuations   **2**:207
    intellectual property (IP)
      generation   **2**:208
    internal staff activity value   **2**:208
    labor cost reduction   **2**:208
    multidisciplinary   **2**:228, **2**:229
    new technology acquisition   **2**:208
    performance improvement   **2**:208
    resource lack   **2**:227–228
    risk minimization   **2**:208
    staffing flexibility   **2**:207
    talent pool diversity   **2**:207
    time saving   **2**:228, **2**:229*F*
  analog production (singletons)   **2**:211
    compounds prepared per unit
      time   **2**:218
    IP protection   **2**:212
    performance assessment   **2**:218
    problem-solving   **2**:212
    reactions per unit time   **2**:218
  basic research   **2**:20
  chemistry *vs.* biotechnology   **2**:204–
    205
  compound acquisition/
    processing   **3**:564
  compound resynthesis   **2**:211,
    2:213
  context   **2**:204
    natural products   **2**:205
  disadvantages   **2**:227, **2**:228*T*
    project management   **2**:228
  history   **2**:205
  internal performance *vs.*   **2**:219
  parallel chemistry-enabled
    synthesis   **2**:205, **2**:212
    market commoditization   **2**:212
    performance assessment   **2**:218
    value chain migration   **2**:212
  performance assessment   **2**:217
    metric drawbacks   **2**:218
  risks   **2**:219, **2**:220
    communication   **2**:220
    cost differential   **2**:220
    IP loss   **2**:219
    organizational complexity   **2**:220
    publicity   **2**:220
    quality   **2**:219
    staff morale   **2**:219
  scale-up   **2**:211
    cost   **2**:211
    established protocols   **2**:211
    IP   **2**:211
    monomers and templates
      (M&Ts)   **2**:211
    performance assessment   **2**:218

  precandidates   **2**:211
  time   **2**:211
Ovarian cancer
  biomarkers   **7**:49
  docetaxel   **1**:145
  irinotecan hydrochloride   **1**:147
  metabonomics   **5**:1004
  nedaplatin   **1**:147
  Src kinases   **7**:191
Ovaries
  cytochrome P450s   **5**:172*T*
  removal, breast cancer
    treatment   **8**:83
Overactive bladder
  definition   **8**:124
  duloxetine clinical trials   **8**:129, **8**:133
Overeating   **6**:9
Overexpression models, small animal
    test systems   **3**:649
Overexpression pools, cDNA functional
    screening   **3**:15
Overfitting
  solubility models   **5**:631
  toxicity, computational
    models   **5**:913, **5**:916
Overlapping ensembles, free energy
    calculation   **4**:654,
    4:655*F*
Overlapping fragments, aqueous
    solubility models   **5**:637
Overnutrition-evoked, diabetes mellitus
    type 2 animal models   **6**:435
Oversaturation number (O_{lumen}), in
    vitro oral absorption
    prediction   **5**:676
Over-the-counter medication,
    ephedrine   **6**:401
Ovulation induction
  clomiphene   **8**:84–85, **8**:87
  ICI Pharmaceuticals fertility control
    research program   **8**:86
Ownership, culture   **2**:122
Oxacillin
  bioavailability   **5**:712*T*
  oral absorption   **5**:468*T*
1,2,4-Oxadiazole-5(4-*H*)-ones,
    bioisosterism   **2**:662*T*
1,2,4-Oxadiazole-5(4-*H*)-thiones,
    bioisosterism   **2**:662*T*
1,2,4-Oxadiazoles
  carboxamide bioisosterism   **2**:668,
    2:670*T*
  carboxylic ester bioisosterism   **2**:663
1,3,4-Oxadiazoles, carboxamide
    bioisosterism   **2**:668, **2**:670*T*
Oxagen (company)   **1**:506
Oxalic acid, structure   **5**:369*F*
Oxalic acid arylamides, solid-phase
    discrete libraries   **3**:709*T*
Oxaliplatin   **1**:**140**; **7**:142
  activation   **7**:142–143
  bioavailability   **7**:143
  structure   **7**:143*F*

  toxicity   **7**:46*T*
  xenograft models   **7**:39*T*
Oxametacin, carboxylic acid
    bioisosterism   **2**:659
Oxamflatin   **7**:224–225
Oxandrolone, structure   **2**:1016*F*; **6**:522*T*
Oxapenems   **7**:503
  β-lactamase inhibitors   **7**:512
Oxapiperazines, solid support
    synthesis   **2**:482*F*
Oxaprozin, bioavailability   **5**:712*T*
Oxarol *see* Maxacalcitol
Oxatomide, oral absorption   **5**:680*T*
Oxazaborolidin, structure   **2**:699*F*
Oxazaphosphorines   **7**:133
  clinical use   **7**:134
  resistance   **7**:135
Oxazepam
  bioavailability   **5**:712*T*
  development   **6**:98
  glucuronic acid conjugation   **5**:151
  oral absorption   **5**:680*T*
  structure   **6**:99*F*
Oxazole rings, carboxamide
    bioisosterism   **2**:668, **2**:670*T*
Oxazolidin-2-ones
  see also specific drugs
  drug introductions   **1**:107
Oxazolidines, solid-phase discrete
    libraries   **3**:709*T*
Oxazolidinones   **7**:**673–698**
  see also specific drugs
  5-acetamidomethyl
    derivatives   **7**:683
  carboxamide side chain   **7**:683–684
  halostilbene series   **7**:684
  heterostilbene series   **7**:684
  thiocarbonyl substitution   **7**:683–
    684
  advantages   **7**:689
  adverse effects   **7**:687
    anemia   **7**:691
    monoamine oxidase
      inhibition   **7**:691
    myelosuppression   **7**:691
    thrombocytopenia   **7**:691
  as antimycobacterial agents   **7**:725
  bacteriostatic activity   **7**:688
  classification, by target site   **7**:471*T*
  clinical trials   **7**:678
  core replacements   **7**:685
  development   **8**:159
  dual-action agents   **7**:685
    azithromycin   **7**:685
    erythromycin   **7**:685
    lincosamide   **7**:685
    neomycin   **7**:685
    streptogramins   **7**:685
  enriched oxazolidinone
    synthesis   **7**:686
  isocyanate-epoxide
    cyclization   **7**:686
  Manninen reaction   **7**:686

extent of use **7**:686
5′-indolinyl oxazolidinones **7**:680
future work **7**:692
   adverse effect reduction **7**:692
mechanism of action **7**:687; **8**:166
   protein synthesis initiation **7**:687
   ribosome interactions **2**:1042–1043
pharmacokinetics **7**:689
   clearance profile **7**:689
   metabolism **7**:689
   pharmacokinetic/pharmacodynamics **7**:689
3-(4-piperazinyl-3-fluorophenyl)-2-oxazolidinones **7**:681
relative potency **7**:688
resistance **7**:690
rigid ring-fused derivatives **8**:161
SAR **7**:679, **7**:683; **8**:165
site of action **7**:687–688
spectrum **7**:687, **7**:688
structure–toxicity relationship **8**:160
synthesis development **8**:163, **8**:163*F*, **8**:164*F*
troponyl oxazolidinones **7**:681
virulence factor suppression **7**:688
5′-indolinyl Oxazolidinones **7**:680
Oxazolidinone Working Group **8**:164
Oxazolines, solid-phase discrete libraries **3**:709*T*
Oxcarbazepine **6**:288
   characteristics **2**:829*T*
   structure **2**:829*T*, **6**:282*F*
Oxetane, development of **1**:62
Oxford Biomedica **1**:516
Oxford index, Crohn's disease **6**:620
Oxford Instruments, Oxford University collaboration **1**:588
Oxford Protein Production Facility (OPPF) **3**:552*T*
Oxford University, Oxford Instruments collaboration **1**:588
Oxiconazole **7**:430*T*
Oxidases **5**:168
   ontogenic development **5**:174*T*
Oxidation **5**:88
   see also specific enzymes
   chemical instability *see* Chemical instability
   ethynylbiphenyl derivatives **5**:1031
   famciclovir **7**:301
   6-(*N*,*N*-disubstituted)-3,4,5,6,7,8-hexahydro-2*H*-naphthalen-1-one derivatives **5**:1032
   proguanil **5**:1031
   stability testing *see* Stability testing
Oxidative cross-linking, P-glycoprotein homology modeling **5**:786–787, **5**:787*F*
Oxidative dehydrogenation, cytochrome P450s **5**:107, **5**:108*F*, **5**:109*F*
Oxidative phosphorylation
   uncoupling **2**:1079

pharmacology **2**:1079
   cerebral infarct **2**:1079–1080
   neuroprotection **2**:1079–1080
   weight-loss drugs **2**:1079
uncoupling proteins **2**:1080
   brain expression **2**:1080
   glucose-stimulated insulin secretion **2**:1080
   ischemia **2**:1080
   Parkinson's disease treatment **2**:1080
   uncoupling protein-1 (thermogenin) **2**:1080
Oxidative stress
   age-related macular degeneration **6**:302
   phenolic nutraceuticals **1**:451
   reactive oxygen species (ROS) **6**:581
   renal dysfunction **6**:581
Oxidoreductases
   assays
      NADP$^+$ reduction **3**:581
      NAD$^+$ reduction **3**:581
      direct assay format, optical assays **3**:581
   drug metabolism **5**:195
   pharmacological target space **4**:422*T*
   3D-QSAR **5**:591
Oxindole series, rolipram **7**:894
OX (organism taxonomy) line, UniProtKB flatfile **3**:360
Oxolinic acid **7**:567
   structure **7**:568*F*
Oxomaritidine
   PASP **3**:815, **3**:816*F*
   synthesis, solution phase synthesis **2**:491–492, **2**:492*F*
Oxprenolol
   bioavailability **5**:712*T*
   oral absorption **5**:468*T*, **5**:680*T*
Oxxon Therapeutics **1**:516
Oxybutynin
   mechanism of action, GPCR targeting **2**:773*T*
   oral absorption **5**:468*T*
   transdermal administration **5**:281*T*
   urinary incontinence treatment **6**:543–544
Oxybutyric acid tamoxifen derivatives, triphenylethylene anticancer agents **7**:163, **7**:163*F*
Oxycodone
   as analgesic **6**:338
   mechanism of action, GPCR targeting **2**:773*T*
   structure **6**:336*F*
Oxyfedrine, oral absorption **5**:680*T*
Oxygen activation
   cytochrome P450s **5**:88
   flavin monooxygenases (FMO) **5**:112
   monoamine oxidase *see* Monoamine oxidase (MAO)

Oxygen free radical inhibition, carvedilol **8**:139
Oxygen-induced retinopathy (OIR), wet age-related macular degeneration models **6**:304
Oxygen removal, chemical stability **5**:490
Oxygen species, flavin monooxygenases (FMO) **5**:112*F*
Oxymetholone, structure **6**:522*T*
Oxyntomodulin, obesity treatment **6**:413
Oxysterol, liver X receptor binding **4**:738
Oxytetracycline **7**:599
   antimicrobial activity **7**:610
      acne **7**:611–612
      efficacy **7**:610
      Onchocerca infections **7**:612
      ophthalmic infections **7**:610
      urinary tract infections **7**:610
      Wolbachia infections **7**:612
   antioxidant activity **7**:619
   biosynthesis **7**:600–601, **7**:600*F*, **7**:602
   C6-hydroxyl group **7**:604–605, **7**:605*F*
   discovery **7**:599
   mechanism of action, ribosome interactions **2**:1042–1043
   methacycline synthesis **7**:605*F*
   oral absorption **5**:468*T*
   patent protection **1**:534
   structure **7**:599*F*, **7**:600*F*, **7**:601*F*, **7**:611*F*
Oxytetracycline, discovery, HTS **3**:680
Oxytocin
   hydrolysis **5**:140, **5**:140*F*
   as peptide/protein drugs **2**:579*T*, **2**:580
Oxytocin receptors
   antagonists, nonpeptide drugs **2**:618–619, **2**:620*F*
   desensitization, GRKs **2**:802
   in human disease **2**:775*T*
   as targets **2**:300

P

P1 receptor agonists
   see also specific drugs
   pain **6**:344
P2 receptor agonists **6**:322, **6**:344
P13 kinases, radioactive assays **3**:587
P23 rat, dry age-related macular degeneration models **6**:304
p38 kinase
   arthritis treatment **7**:861
   fragment-based drug discovery *see* Fragment-based drug discovery
p38 kinase inhibitors
   chemical proteomics **3**:44

p38 kinase inhibitors (*continued*)
  IBD treatment   6:627
  structure-based fragment
        screening   4:799, 4:801*F*
  structure-based library design   4:323
p53
  alpha helix scaffold-based
        design   4:30–31
  as biomarker   3:72
  cancer   1:271–272
  cell cycle   7:12
  deletion, genetically engineered
        animals   3:161, 3:163*F*
  large-scale virus vector
        libraries   3:18–19
  node network theory   3:4
  protein degradation   3:193, 3:194*F*
  regulator identification   3:15–16
  RNAi   3:181
  ubiquitin -proteasome pathway
        inhibition   3:194
p85 SH2, three-dimensional
        structures   2:636*T*
p160/steroid receptor coactivator,
        nuclear receptor
        coactivators   2:1002
P450 *see* Cytochrome P450 family
p561ck inhibitors, solution phase library
        synthesis   3:779–780, 3:781*F*
PA-824   7:722
  mycobacterium treatment   7:721
    mechanism of action   7:722
  structure   7:721*F*
Pacemakers
  anesthetics   6:358
  arrhythmias   6:733, 6:749
Pacing currents, cardiac ion
        channels   6:734
Packaging
  development chemistry   2:168
  materials, common technical
        document (CTD)   2:163
Packing arrangements, protein
        crystallography *see* X-ray
        crystallography
Packing densities, protein
        crystallography   3:442
Paclitaxel   7:99
  adverse events   7:61*T*
  analogs, solid-phase pool
        libraries   3:705*F*
  cytotoxicity   1:63–64, 1:64*F*
  decomposition   5:493*F*, 5:501–502,
        5:502*F*
  discovery/development   1:325–326
  mechanism of action   7:99
  prodrug of   5:1012
  proteomics, stathmin/OP-18   3:41
  SAR   1:64–65, 1:64*F*, 1:65*F*, 7:99
  strategic drug design   1:66*F*
  structure   2:28*F*, 2:1026*F*, 3:700*F*,
        5:1013*F*
  modifications   1:48*F*, 1:62, 1:63*F*

synthesis   7:98*F*
toxicity   7:46*T*
transporters   7:45*T*
water solubility   5:1012
xenograft models   7:39*T*
Pactimibe   6:487
Paget, Edward, management
        strategies   2:260–261
Paget's disease   6:504
  bisphosphonate treatment   6:504,
        6:512
  characteristics   6:504
  diagnosis   6:504
  genetics   6:504, 6:506
  prevalence   6:504
  risedronate sodium   1:220
  tiludronate disodium   1:221
Pagoclone   6:106
  structure   6:105*F*
PagP, NMR structure
        determination   3:490, 3:491*F*
Pain   6:321–326, 6:327–349
  acute *see* Acute pain
  chronic *see* Chronic pain
  clinical trials   6:335
    design   6:335
    McGill Pain Questionnaire   6:335
    Neuropathic Pain Scale   6:335
    Neuropathic Pain Symptoms
        Inventory   6:335
    nociceptive tests   6:335–336
    visual analog scale   6:335
  definition   6:328
  future work   6:324
  healthcare costs   6:321
  ion channels/receptors   6:321
    see also specific types
    acid-sensitive ion channels
        (ASICs)   2:911
    ATP-gated ion channels   2:905–907
    glutamate-gated ion
        channels   2:900
    nociceptors   6:321
  models   6:331
    analgesics, differential
        efficacy   6:335, 6:337*T*
  neuropathic *see* Neuropathic pain
  neuropathophysiology   6:328,
        6:328*F*
    psychological parameters   6:328–329
  new research areas   6:321, 6:340
    acid-sensitive ion channels
        (ASICs)   2:911; 6:322
    calcium channel modulators   6:342
    cannabinoids   6:342
    excitatory amino acid
        antagonists   6:323, 6:341
    neuronal nicotinic receptor
        antagonists   6:340
    P1 receptor agonists   6:344
    P2 receptor agonists   6:322, 6:344
    purines   6:344
    sodium channel modulators   6:343

    transient receptor potential ion
        channels   2:908–909; 6:322
    vanilloid receptor
        modulators   6:340
    voltage-gated calcium
        channels   6:323
    voltage-sensitive ion
        channels   6:323
  nociceptive *see* Nociceptive pain
  translational medicine   6:337
  treatment   6:337
    see also Analgesics
    analgesic adjuvants   6:339
  unmet medical needs   6:340
PAK1, PDB entries   2:965*T*
PAL-287, structure   6:187*F*
Palivizumab   1:334*T*, 1:433*T*
  brands, global sales   1:605*T*
  respiratory syncytial virus
        infections   1:434, 1:444
Palladium-mediated *N*-arylation, solid
        support synthesis   2:470,
        2:471*F*
Pallas, ionization prediction   5:616*T*,
        5:620, 5:623*T*
Palliative drugs
  future work   1:82
  Parkinson's disease treatment   6:11
  pharmacogenetics   3:54, 3:58
Palmer, John, oxazolidinone
        development   8:160
^{11}C Palmitate, as metabolic
        probe   3:665
Palmitoylation, G protein
        structure   2:780
Palmoplantar psoriasis   7:943
Palonosetron   1:150
  as antiemetic agent   6:682
  5-HT$_3$ receptor binding   2:890
PAMAM dendrimers, gene
        therapy   1:309–310, 1:310*F*
Pamaquine   7:776
Pamelor *see* Nortriptyline
Pamidronate
  oral absorption   5:1026–1027
  osteogenesis imperfecta   6:507
  structure   5:1027*F*; 6:510*F*
  toxicology   8:207
Pamidronate sodium   1:501
para-aminobenzoic acid,
        structure   7:719*F*
para-aminosalicylic acid (PAS)
  as antimycobacterial agent   7:719
  mechanism of action   7:719
  structure   7:719*F*
PAM matrices, sequence alignment,
        bioinformatics   3:319–320
PAMPA *see* Parallel artificial membrane
        permeation assay (PAMPA)
Pancreas
  islet cells   6:426
  transplantation, diabetes mellitus
        management   6:455

Phase-space probability, molecular
  dynamics simulations **4**:653
Phase switch concept, PASP **3**:799–
  800
Phase-transfer catalysis, microwave-
  assisted chemistry *see*
  Microwave-assisted chemistry
Phasing information, x-ray structures,
  problems **3**:533–534
Phasing methods, crystallography *see*
  X-ray crystallography, rational
  drug design
pH-dependent K$^+$ ion channels **4**:705
PhDock program, constraints **4**:271
Phenacyl-CoA synthetase, coenzyme A
  conjugation **5**:155
Phenazine, hepatitis C virus
  inhibition **2**:1047
Phenazopyridine
  aggregation reactions **5**:409*F*, **5**:410
  benzoic acid **5**:404*T*
  disk intrinsic dissolution rate
    (DIDR) **5**:417
Phenbenazamine, structure **2**:654*F*
Phendimetrazine
  metabolism **6**:398*F*
  obesity treatment **6**:398
    contraindications **6**:398
    indications **6**:398
    mechanism of action **6**:398,
      6:398*T*
    side effects **6**:398
  structure **6**:395*T*
Phenelzine **6**:67
  defensive test battery **6**:94
  side effects **6**:60*T*
  structure **2**:68*F*
1,4-Phenenebis-
  (methylene)selenocyanate,
  structure **2**:701*F*
Phenglutamide, oral absorption **5**:680*T*
Phenobarbital **6**:288
  α$_1$-acid glycoprotein (AGP)
    binding **5**:849
  bioavailability **5**:712*T*
  historical perspective **6**:287
  immunotoxicity **5**:221*T*
  oral absorption **5**:468*T*
  structure **5**:371*F*; **6**:281*F*
4-Phenol, bioisosterism **2**:673*T*
Phenolic nutraceuticals **1**:450, **1**:451*T*,
  1:452*T*
  see also specific types
  bioavailability **1**:450
  classification **1**:451
  mechanism of action **1**:451
  pharmacokinetics **1**:450–451, **1**:454*T*
Phenolphthalein,
  bioavailability **5**:712*T*
Phenols
  ezetimibe metabolite
    formation **8**:75–76
  ionization constants (pK_a) **5**:372

structure **5**:370*F*
sulfotransferases **5**:148, **5**:148*F*
Phenothiazines **6**:29–31
  analogs **5**:782
  P-glycoprotein models **5**:779, **5**:782,
    5:783
N-Phenothiazines, mitoprotective
  drugs **2**:1062
Phenotype
  definition **2**:382*T*
  genetically engineered animals *see*
    Genetically engineered animals
Phenotype-oriented screening **3**:135
  detection methods **3**:138
    high-content screening **3**:138
    uniform readout **3**:138
  goals **3**:135
    biological HTS **3**:135
    target identification **3**:135
  model organisms **3**:135
    cell-based assays **3**:136
    cell differentiation **3**:136
    human material **3**:136
    multicellular organisms **3**:137
  specificity **3**:140
    targetless mutant methods **3**:140–
      141, **3**:141*F*
    techniques **3**:140
  target identification **1**:332; **3**:138
    affinity-based approaches **3**:138
    affinity chromatography **3**:138
    drug-induced
      haploinsufficiency **3**:139,
      3:140*F*
    genetic approaches **3**:139
    protein microarrays **3**:139
    tagged libraries **3**:139
    three-hybrid reporter gene
      systems **3**:140
Phenoxymethylpenicillin, effective
  permeability (P$_{eff}$) **5**:673*T*
Phenoxypropanolamines, solid-phase
  pool libraries **3**:705*F*
Phenprocoumon, discovery **7**:355–356
Phentermine
  fenfluramine combination **6**:400
  obesity treatment **6**:396
    contraindications **6**:396
    indications **6**:396
    mechanism of action **6**:396
    side effects **6**:396
  safety **1**:686
  structure **6**:395*T*
Phentolamine, discovery/
  development **1**:328–329
1-Phenyethylimidazoles, medicinal
  chemistry **7**:449
Phenylalanine, effective permeability
  (P$_{eff}$) **5**:673*T*
Phenylalkylamides
  solid-phase libraries **3**:737–738,
    3:737*F*
  solid-phase pool libraries **3**:705*F*

Phenylalkylamines, as calcium channel
  blocker **2**:849–850
Phenylbutazone
  bioavailability **5**:712*T*
  structure **5**:371*F*
4-Phenylbutylamine
  receptor binding, pH
    dependency **5**:382
  structure **5**:360
4-Phenylbutyric acid
  amino acid-coenzyme A
    conjugation **5**:156–157, **5**:156*F*
  β-oxidation/chiral inversion **5**:156*F*,
    5:158
p-Phenylenediamine,
  immunotoxicity **5**:221*T*
Phenylenediamine sulfonamide,
  cytomegalovirus
    antivirals **7**:309
Phenylephrine
  bioavailability **5**:712*T*
  discovery/development **1**:328–329
Phenylethanolamine *N*-
  methyltransferase
    (PNMT) **5**:819–820
Phenylethylamine, structure **6**:395*T*
β-Phenylethylamine, affective disorder
  treatment **6**:80
Phenylethylamines, biological
  activities **3**:925*T*
Phenylethylmalonamide,
  bioavailability **5**:712*T*
Phenyl group, quinolone
  antibiotics **7**:575, **7**:576*T*
Phenylhydroxylamine,
  reduction **5**:123–124, **5**:124*F*
Phenylimidazole, carboxamide
  bioisosterism **2**:668, **2**:670*T*
2-Phenylnaphthalenes, estrogen
  receptor-β-selective
    ligands **7**:169, **7**:171*F*
α-Phenyl-*N*-tert-butylnitrone
  (PBN) **6**:266
  structure **6**:267*F*
Phenyl *O*-methyltransferase
  5:819–820
Phenylpropenamide compounds,
  hepatitis B antivirals **7**:321
Phenyltetrazolyl acetophenone,
  structure **6**:107*F*
5-Phenylvaleric acid
  receptor binding, pH
    dependency **5**:382
  structure **5**:360
Phenytoin **6**:288
  bioavailability **5**:712*T*
  characteristics **2**:829*T*
  glucuronic acid conjugation **5**:151–
    152, **5**:153*F*
  historical perspective **6**:287
  hydrolysis **5**:142, **5**:144*F*
  metabolism, CYP 2C9 **5**:545*T*
  oral absorption **5**:468*T*, **5**:680*T*

Phenytoin (*continued*)
  prodrugs of **5**:1027, **5**:1027*F*
    Mannich base **5**:1028
  structure **2**:829*T*; **6**:281*F*
Pheochromocytomas
  diabetes mellitus **6**:423
  mitochondria complex II **2**:1073
φ-angle, protein 3D structure
  validation **3**:516–517
Philadelphia chromosome, Bcr-Abl
  kinases **2**:972
Philantotoxins
  solid-phase discrete libraries **3**:709*T*
  solid-phase libraries **3**:718–719,
    3:719*F*
Philosophy (of drug discovery/
  development) **2**:111
  best in class *vs.* first in class **2**:118,
    2:118*F*, **2**:119*F*
    assessment responsibilities **2**:119–
      120
    development process
      effects **2**:119
    differentiation **2**:119, **2**:120*T*
    merits **2**:118–119
  chemical *vs.* biological **2**:115
    antibody responses **2**:117
    biological production **2**:117
    gene therapy **2**:117–118
    investment **2**:117
    multidimensional profiling **2**:117
    oligonucleotide drugs **2**:117–118
    parallel development **2**:117,
      2:117*F*
    route of administration **2**:117
    sulfanilamides **2**:115, **2**:116*F*
  culture **2**:121
    collaborative *vs.* insular **2**:121–122
    competitive collaborations **2**:122
    entrepreneurism **2**:123
    market research **2**:122–123
    'need to know' *vs.* 'nice to
      know' **2**:121–122
    objectiveness **2**:122–123
    ownership **2**:122
    transparency **2**:122–123
  decision making *vs.* organizational
    alignment **2**:120
    small *vs.* large organizations **2**:120
    speed definitions **2**:120
    transparency **2**:120
  fail fast *vs.* 'follow the science' **2**:121
    ADME **2**:121
    budget constraints **2**:121
    predictive screens **2**:121
  NME *vs.* life cycle
    management **2**:115
    investments **2**:115
  partnering **2**:123
  target *vs.* disease **2**:112
    disease etiology **2**:112
    'druggable genome' **2**:113,
      2:113*F*

'druggable target'
    definition **2**:112, **2**:112*F*
    drug mechanism of action **2**:112
    HTS **2**:114–115
    molecular target classes **2**:112,
      2:112*F*
    new diseases **2**:113, **2**:114*T*
    structure-based drug design
      (SBDD) **2**:114–115
    targets by therapeutic area **2**:114–
      115, **2**:115*T*
pH-independent partition coefficient,
  lipophilicity **5**:535–536
PHKG1, PDB entries **2**:965*T*
Phlorizin analogs, solid-phase pool
  libraries **3**:705*F*
pH-metric titration *see* Ionization
  constants (pK_a) measurement
Phobias **6**:5
  treatment **6**:6
Phosphatase assays
  high-throughput screens
    (HTS) **3**:582
  optical assays **3**:580–581, **3**:582,
    3:582*F*, **3**:582*T*, **3**:583*T*
Phosphatase inhibitors
  see also specific drugs
  glaucoma treatment **6**:313*T*
Phosphate conjugates, conjugation, drug
  metabolism **5**:149
Phosphates, prodrugs **5**:1025
Phosphatidic acid, structure **5**:455*F*
Phosphatidylcholine (PC)
  biological membranes **5**:455–456
  intestinal membrane **5**:458
  mitochondria membranes **2**:1057
  structure **5**:455*F*
Phosphatidylethanolamine (PE)
  biological membranes **5**:455–456
  intestinal membrane **5**:458
  mitochondria membranes **2**:1057
  structure **5**:455*F*
Phosphatidylglycerol (PG)
  biological membranes **5**:455–456
  structure **5**:455*F*
Phosphatidylinositol (PI)
  biological membranes **5**:455–456
  intestinal membrane **5**:458
  structure **5**:455*F*
Phosphatidylinositol 4,5-bisphosphonate
  (PIP$_2$), inward rectifying K$^+$ ion
  channels **4**:713–714, **4**:715*F*
Phosphatidylinositol-3-kinase (PI3K)
  activation, vascular endothelial growth
    factor receptor-2 (VEGF-R2)
    3:200
  GPCR interaction **2**:786*T*
  ubiquitin-protein targeting **3**:192
Phosphatidylserine (PS)
  biological membranes **5**:455–456
  intestinal membrane **5**:458
  structure **5**:455*F*
Phosphinates, bioisosterism **2**:662*T*

3′-Phosphoadenosine 5′phosphosulfate
  (PAPS),
    sulfotransferases **5**:147, **5**:148*F*
Phosphodiesterase(s) **2**:919–957
  see also specific enzymes
  active site **4**:758, **4**:758*F*
  binding site promiscuity **4**:760,
    4:760*F*
    complex structures **4**:760
    hydrophobicity **4**:760
  biological activity **2**:920, **2**:923;
    4:756–757
  cAMP levels **2**:920
  cGMP levels **2**:920
  classification **2**:920, **2**:921*T*
    class I **2**:920
    class II **2**:920
  disease states **4**:757
  as drug targets **3**:3
  future work **2**:950
  genes/molecular biology **2**:920,
    2:921*T*
  genetic diseases **2**:950
  inhibitors **4**:756
    see also specific drugs
    atopic dermatitis treatment *see*
      Atopic dermatitis treatment
    chemogenomics **3**:930, **3**:931*F*
    multiple idiotype inhibition **2**:950
    selectivity **2**:950
  ligands, physicochemical
    properties **4**:424*T*
  nomenclature **2**:920
  optical assays **3**:582
  pharmacological target space **4**:422*T*
  pharmacology **2**:931, **2**:947, **2**:947*T*
  reaction **2**:920
  regulation **2**:923
    competition **2**:923
    extracellular stimuli **2**:923
    phosphorylation **2**:923
    second messengers **2**:931
  selectivity **4**:758
  splice variants **2**:920
  structure **2**:920, **2**:922*F*; **4**:759*F*
    catalytic domain **4**:757, **4**:757*F*
    C-terminal catalytic
      domain **2**:920–922
    N-terminal regulatory
      domain **2**:920, **2**:922–923
    substrate specificity **2**:922
  subcellular localization **2**:922–923
  substrate recognition **4**:758, **4**:759*F*
    cAMP **4**:758–759
    cGMP **4**:758–759
    stabilizing residues **4**:759
  substrate specificity **2**:920
  as targets **2**:757*T*
  target selection **2**:73
Phosphodiesterase-1 **2**:923
  genetics **2**:921*T*, **2**:923
  insulin secretion **2**:923
  regulation, calcium by **2**:923

splice variants **2**:923
structure **2**:922*F*
   C-terminal catalytic
      domain **2**:920–922
   substrate specificity **2**:921*T*, **2**:923
   as therapeutic target **2**:950–951
Phosphodiesterase-1 inhibitors **2**:931
  see also specific drugs
  milrinone derivatives **2**:932
  tetracyclic guanine
    derivatives **2**:931–932
  therapeutic applications **2**:947
  zaprinast derivatives **2**:932
Phosphodiesterase-2 **2**:923
  atrial naturietic peptide
    interaction **2**:923–924
  conformational changes **2**:923–924
  genetics **2**:921*T*, **2**:923–924
  learning/memory **2**:923–924
  olfactory signaling **2**:923–924
  platelet aggregation **2**:923–924
  splice variants **2**:923–924
  structure **2**:922*F*
    crystal structure **2**:933
    C-terminal catalytic
      domain **2**:920–922
  subcellular localization **2**:923–924
  substrate specificity **2**:921*T*
  as therapeutic target **2**:950–951
  tissue expression **2**:923–924
Phosphodiesterase-2 inhibitors **2**:933
  see also specific drugs
  benzodiazepines **2**:933
  morpholinochromes **2**:933
  pyrido[2,3-*d*]pyrimidine
    derivatives **2**:933
  therapeutic applications **2**:947
  tricyclic oxindoles **2**:933
Phosphodiesterase-3 **2**:924
  genetics **2**:921*T*, **2**:924
  insulin secretion **2**:924
  leptin interaction **2**:924
  phosphorylation **2**:924
  platelet function **2**:924
  structure **2**:922*F*
    crystal structure **2**:934–935
    C-terminal catalytic
      domain **2**:920–922
  substrate specificity **2**:921*T*, **2**:924
  as therapeutic target **2**:950–951
  tissue distribution **2**:924
    adipocytes **2**:924
    cardiac tissue **2**:924
    liver **2**:924
    platelets **2**:924
    smooth muscle **2**:924
Phosphodiesterase-3 inhibitors **2**:933
  see also specific drugs
  structural requirements **2**:934–935
  therapeutic applications **2**:947
    dilated cardiomyopathy **2**:947–948
Phosphodiesterase-4 **2**:924
  β-arrestin association **2**:926

β$_2$-adrenergic receptors **2**:926
CNS disorders **2**:925–926
conformations **2**:925
emesis/nausea **2**:925
genetics **2**:921*T*, **2**:924–925
inflammation **2**:925–926
intracellular distribution **2**:926
long-term memory **2**:926–927
membrane-bound **2**:925
phosphorylation **2**:925
point mutation **2**:925–926
rolipram affinity **2**:925–926
soluble form **2**:925
splice variants **2**:924–925
structure **2**:922*F*
substrate specificity **2**:921*T*,
  2:924–925
subtypes **7**:893–894
as therapeutic target **2**:950
tissue distribution **2**:926
  CNS **2**:926–927
  heart **2**:926–927
  nervous system **2**:926–927
  smooth muscle **2**:926–927
upstream conserved regions
  (UCRs) **2**:924–925
Phosphodiesterase-4 inhibitors **2**:935
  see also specific drugs
  amino diazepinoindole **2**:937–938
  catechol derivatives **2**:936
  clinical trials **7**:882
  high-throughput x-ray
    crystallography **2**:938
  phthalazinone derivatives **2**:936
  pyridylisoquinolines **2**:936
  scaffold-based design **2**:938
  therapeutic applications **2**:948
    anticancer agents **2**:948
    asthma treatment **2**:935–936,
      2:948; **7**:893
    COPD **2**:935–936, **2**:948;
      7:882
    emesis **2**:936
    IBD treatment **6**:625
    memory impairment **2**:935–936
    psoriasis **2**:935–936
    skin inflammation
      treatment **7**:979
    ulcerative colitis **2**:935–936
Phosphodiesterase-5 **2**:927
  atrial natriuretic peptide **2**:928
  conformational chains **2**:927
  genetics **2**:921*T*
  mutagenesis **2**:922
  phosphorylation **2**:927
  point mutations **2**:927
  regulatory domain **2**:927
    GAF domains **2**:927
  splice variants **2**:927
  structure **2**:922*F*
    crystal structure **2**:942
  substrate specificity **2**:921*T*, **2**:922,
    2:927

as therapeutic target **2**:950
vascular tone regulator **2**:928
Phosphodiesterase-5 inhibitors **2**:938
  see also specific drugs
  carboxamide **2**:940
  pyrazolopyridines **2**:939–940
  pyrazolopyridopyridazine **2**:941
  pyrazolopyrimidines **2**:940
  pyrroloquinolines **2**:941–942
  quinazoline derivatives **2**:939
  sildenafil citrate, invention of **2**:257
  therapeutic applications **2**:948
    aspirin-induced gastric
      damage **2**:949
    benign prostatic
      hypertrophy **2**:948
    chronic lymphocytic
      leukemia **2**:949
    cognitive disorders **2**:949
    diabetes insipidus **2**:949
    erectile dysfunction **2**:948
    smoking-induced pulmonary
      endothelial dysfunction **2**:948–
      949
Phosphodiesterase-6 **2**:928
  in eye **2**:928
  genetics **2**:921*T*, **2**:928
  in light transduction **2**:928
  posttranslational modification **2**:928
  retinitis pigmentosum **2**:928, **2**:950
  structure **2**:922*F*
    cGMP allosteric binding
      sites **2**:928
    regulatory domain **2**:928
  substrate specificity **2**:921*T*
Phosphodiesterase-6 inhibitors **2**:942
  see also specific drugs
  sildenafil **2**:942–943
  therapeutic applications **2**:949
Phosphodiesterase-7 **2**:929
  B cells **2**:929
  chronic lymphocytic leukemia **2**:929
  COPD **2**:929
  genetics **2**:921*T*, **2**:929
  splice variants **2**:929
  structure **2**:922*F*
  substrate specificity **2**:921*T*
  T-cell proliferation **2**:929
  as therapeutic target **2**:950–951
  tissue distribution **2**:929
Phosphodiesterase-7 inhibitors **2**:943
  see also specific drugs
  benzothienol[3,2-*a*]thiadiazine 2,2-
    dioxides **2**:943
  isoquinolines **2**:944
  phthalazinones **2**:944
  pyrazolopyrimidinones **2**:944
  spiroquinazolinones **2**:943–944
  therapeutic applications **2**:949
  thiadiazoles **2**:944
Phosphodiesterase-8 **2**:929
  genetics **2**:921*T*, **2**:929–930
  splice variants **2**:929–930

Phosphodiesterase-8 (*continued*)
structure **2:**922*F*
substrate specificity **2:**921*T*
thyroid gland **2:**930
tissue expression **2:**929–930
Phosphodiesterase-8 inhibitors **2:**945
see also specific drugs
therapeutic applications **2:**949
Alzheimer's disease **2:**949
Phosphodiesterase-9 **2:**930
genetics **2:**921*T*
memory acquisition **2:**930
splice variants **2:**930
structure **2:**922*F*
crystal structure **2:**945–946
substrate specificity **2:**921*T*
as therapeutic target **2:**950–951
tissue expression **2:**930
Phosphodiesterase-9 inhibitors **2:**945
see also specific drugs
therapeutic applications **2:**950
glucose homeostasis **2:**950
Phosphodiesterase-10 **2:**930
CNS diseases **2:**930–931
genetics **2:**921*T*, **2:**930
Huntington's disease **2:**930–931
inhibitor sensitivity **2:**930
schizophrenia **2:**930–931
splice variants **2:**930
structure **2:**922*F*
substrate specificity **2:**921*T*
as therapeutic target **2:**950–951
tissue expression **2:**930
Phosphodiesterase-10 inhibitors
2:946
see also specific drugs
dihydroisoquinolines **2:**946
imidazotriazines **2:**946
papaverine **2:**946
therapeutic applications **2:**950
glucose homeostasis **2:**950
Phosphodiesterase-11 **2:**931
genetics **2:**921*T*
isoenzymes **2:**931
sperm characteristics **2:**931
structure **2:**922*F*
substrate specificity **2:**921*T*
tissue expression **2:**931
Phosphodiesterase-11 inhibitors
2:946
see also specific drugs
dipyridamole **2:**931
therapeutic applications **2:**950
zaprinast **2:**931
Phosphodiesters, structure **2:**35*F*
Phospholipase A$_2$ inhibitors
see also specific drugs
arthritis treatment *see* Arthritis
treatment
focused library design, structure-based
methods **4:**328–329
Phospholipid-binding proteins,
radioactive assays **3:**587

Phospholipidosis
amiodarone **5:**924
toxicity, computational models **5:**924
Phospholipids
metabolism inhibitors
antimalarial drugs *see* Antimalarial
drugs
PAMPA **5:**470–471
Phosphoprotein-interacting
domains **2:**961*T*
genomic relationships **2:**963*T*
Phosphoric acid, structure **5:**371*F*
Phosphoric acid prodrugs **5:**1021
Phosphorodiamidate morpholino
oligomers, filoviruses **7:**278
Phosphorothioate, structure **2:**35*F*
Phosphorus, primary
hyperparathyroidism **6:**504
Phosphorus-derived acidic functions,
carboxylic acid bioisosterism *see*
Bioisosterism, carboxylic acids
Phosphorylation
GPCR desensitization **2:**800, **2:**801
heterogeneous protein
crystallization **3:**882
nuclear receptors **2:**1004
phosphodiesterase-3 **2:**924
phosphodiesterase-4 **2:**925
phosphodiesterase-5 **2:**927
phosphodiesterase regulation **2:**923
proteomics **3:**41
sodium channels **2:**836
Phosphoryl transfer reaction, quantum
mechanical calculations **4:**404
Phosphoserine/phosphothreonine-
interacting domains **2:**985
forkhead-associated domains **2:**985
Phosphotransferases, aminoglycoside
resistance **7:**642–643
Phosphotyrosine-interacting
domains **2:**983
SH2 domain inhibitors **2:**983,
2:983*F*
Photoactive yellow protein
(*Ectothiorhodospira halophila*)
structure
correct **3:**532, **3:**533*F*
incorrect **3:**532, **3:**533*F*
Photoaffinity labeling, P-glycoprotein
structure **5:**775
Photo-Ames test, non-animal
photogenotoxicity
testing **1:**467*T*, **1:**472
Photo-chromosome aberration test, non-
animal photogenotoxicity
testing **1:**467*T*, **1:**472
Photocoagulation, diabetic retinopathy
treatments **6:**316
Photodynamic treatment (PDT)
[^{18}F]FDG imaging **3:**669
acne treatment **7:**940
cancer treatment **7:**7
[^{18}F]FDG **3:**669

limitations **6:**310
psoriasis treatment **7:**945, **7:**978*T*
selenium bioisosterism **2:**702*T*,
2:703
small animal imaging **3:**669
wet age-related macular degeneration
treatment **6:**309–310
Photogenotoxicity, animal
experimentation alternatives *see*
Animal experimentation
alternatives
Photolithographic masks
DNA microarrays *see* DNA microarrays
solid supports **2:**503, **2:**504*F*
Photoluminescence, optical
assays **3:**578, **3:**590
Photolysis, oxidation **5:**492
Photo-micronucleus test (PMNT), non-
animal photogenotoxicity
testing **1:**467*T*, **1:**472
Photomultiplier tubes
HCS image detection **3:**687
scintillation proximity assay **3:**584
Photon counters, x-ray
detection **3:**468
Photoreceptor function, age-related
macular degeneration **6:**300
Photoreceptor-specific nuclear receptor,
nomenclature **2:**996*T*
Photosensitization
animal experimentation
alternatives **1:**467*T*, **1:**471
drug introductions **1:**237
Photostability chambers, light **5:**497–
498
Photo-thymidine kinase test, non-
animal photogenotoxicity
testing **1:**467*T*, **1:**472
Phototoxicity
animal experimentation
alternatives **1:**467*T*, **1:**471
tetracyclines *see* Tetracyclines
Phototreatment
atopic dermatitis treatment **7:**950
psoriasis treatment *see* Psoriasis
treatment
pH-partition hypothesis, passive
transcellular transport **5:**264–
265
pH sensitive linkers, drug–polymer
conjugates **5:**1055
Phthalazine, oxidation **5:**119–120
Phthalazinones
phosphodiesterase-4 inhibitors **2:**936
phosphodiesterase-7 inhibitors **2:**944
Phthalic acid, structure **5:**369*F*
Phthalides
solid-phase discrete libraries **3:**709*T*
solid-phase libraries **3:**720*F*, **3:**721
Phthalimido amides, solid-phase
discrete libraries **3:**709*T*
Phylogenetic profiles, genomic context
methods, bioinformatics **3:**336

PEG-succinimidyl
derivatives 5:1052
PEG-vinylsulfone (VS-
PEG) 5:1053
'Y'-shaped branched PEG 5:1052,
5:1053*F*
peptide linkers 5:1054
peptide/protein drugs *see* Peptide/
protein drugs
polymerization process 5:1051–1052
optimization 5:1052
polymer purity 5:1052
purification/characterization 5:1063
ionic exchange
chromatography 5:1063
labile linker incorporation 5:1063–
1064, 5:1064*F*
PEGylation site
identification 5:1063–1064
site identification 5:1063–1064
site-selective conjugation 5:1060
amino group accessibility 5:1060–
1061
enzymes 5:1053*F*, 5:1061, 5:1061*F*
reactive group blocking 5:1061
solvation properties 5:1052
thiol PEGylation 5:1059
cysteine muteins 5:1059
hemoglobin 5:1059
Poly(ethylene glycol)-containing resins
see Resin beads, solid support
synthesis
Polyexponential functions,
pharmacokinetics 5:17
Polyfunctional 1,3-dioxanes, solid-phase
pool libraries 3:707*F*
Polygenic disease
pharmacogenetics 3:63
psoriasis 7:941
Polyglutamine disease, *Saccharomyces
cerevisiae* model system 3:6
Poly-His tag, as affinity tag 3:421
Polyisoprenyl benzophenones,
microtubule targeting
agents 7:103
Polyketide synthase enzymes, natural
products 1:361
Polymerase chain reaction (PCR)
candidosis diagnosis 7:429
epigenetics 1:258
invention of 2:261
Cetus Corporation 2:256
Mullis, Kary 2:256
protein cloning 3:554
protein drug discovery 1:333
protein production 3:414
single nucleotide
polymorphisms 2:363, 2:369
slab gel electrophoresis combination,
epigenetics 1:266–268
Polymer-assisted solution phase
synthesis (PASP) 3:**791–836**,
3:798–799

advantages 3:793
environmental advantages
3:793
Mitsunobu reaction 3:793
safety 3:793
separation simplicity 3:793
spent reagent removal 3:792,
3:793
Staudinger reaction 3:793
time saving 3:792
Wittig reaction 3:793, 3:793*F*
catch-and-release strategies 3:795,
3:796*F*
chemical library synthesis 3:804
2-alkylthiobenzimidazoles 3:805
amides 3:806, 3:807*F*
amine derivatization 3:806,
3:806*F*
2-aminothiazole scaffolds 3:805–
806, 3:805*F*
antirhinoviral candidates 3:804,
3:804*F*
5-arylidine 4-
thiazolidinones 3:806–807,
3:807*F*
benzimidazole derivatives 3:804
benzimidazolin-2-ones 3:805,
3:805*F*
1,5-biaryl pyrazoles 3:808, 3:809*F*
cathepsin D inhibitors 3:808
cyanoborohydride 3:806, 3:806*F*
N,N-dialkyldipeptidyl-
amines 3:808–809, 3:810*F*
factor VIIa inhibitors 3:809
histone deacetylase
inhibitors 3:811, 3:812*F*
hydroxamic acid derivatives 3:811,
3:811*F*
α-ketothiazole peptidyl protease
inhibitors 3:809, 3:810*F*
matrix metalloproteinase
inhibitors 3:811
perruthenate 3:806, 3:806*F*
scaffold decoration 3:811
substituted pyrazoles 3:807,
3:808*F*
3-thioalkyl-1,2,4-triazoles 3:804,
3:804*F*
thiomorphine analogs 3:811,
3:813*F*, 3:814*F*
triazole analogs 3:804, 3:804*F*
disadvantages, reaction
kinetics 3:793–794
flow processes 3:826
advantages 3:826, 3:827–828
byproduct scavenging 3:826
catch-and-release protocols 3:826
chromatographic separation 3:826
condition optimization 3:827–828
heterocyclic thiourea library
preparation 3:830
hydrogen gas 3:800*F*, 3:827,
3:827*F*

imine hydrogenation 3:827, 3:827*F*
β-lactam synthesis 3:829, 3:829*F*
micro-reactors 3:826, 3:826*F*
miniaturization 3:826
organic synthesis 3:828
penicillin sulfoxide
synthesis 3:828
O-silylated steroid
derivatization 3:828–829,
3:829*F*
thioester library 3:830, 3:830*F*
historical perspective 3:792
compound library
construction 3:792
ligand bound synthesis *vs.* 3:792
ocadec-9-enoic butyl ester
synthesis 3:792
immobilized scavengers 3:795,
3:795*F*
library generation 3:797–798, 3:799*F*
microencapsulation 3:812
aryl boronic acid synthesis 3:813
aryl bromide synthesis 3:813
definition 3:812
Heck reaction 3:813
polyurea microcapsules 3:813
preparation 3:812
Stille couplings 3:813
supercritical carbon dioxide
(seCO₂) 3:813
Suzuki couplings 3:813
Suzuki–Miyaura coupling
procedure 3:812
multistep synthesis 3:792, 3:800
Amaryllidaceae alkaloids 3:815,
3:816*F*, 3:819
carpanone 3:818, 3:819*F*
chemical library synthesis *see above*
commercially-available
drugs 3:801
see also specific drugs
epibatidine 3:816, 3:817*F*, 3:818*F*
epimaritidine 3:815, 3:816*F*
epothilones 3:821, 3:822*F*, 3:823*F*,
3:824*F*
natural product synthesis 3:814
see also specific compounds
neolignans 3:818
nicotine 3:817
nornicotine 3:817
obliquine 3:819
oxomaritidine 3:815, 3:816*F*
plicamine 3:819, 3:820*F*
plicane 3:819
polysyphorin 3:818
rhaphidecursinol B 3:818
rosiglitazone 3:802, 3:802*F*
salmeterol 3:802–804, 3:803*F*
sildenafil 3:801, 3:801*F*
one-pot reactions 3:794*F*, 3:795,
3:796
alkane functionalization 3:796,
3:796*F*

Polymer-assisted solution phase
    synthesis (PASP) (*continued*)
    amino alcohol synthesis **3**:797,
        3:797*F*
    functionalized pyrazole
        synthesis **3**:797, **3**:798*F*
    propranolol synthesis **3**:797,
        3:798*F*
    trihydroxy nucleoside
        synthesis **3**:797
    unsaturated nitrile
        synthesis **3**:796–797
    Wadsworth–Horner–Emmons
        alkenylation **3**:796–797
    'wolf-and-lamb' approach **3**:796
    reaction quenching **3**:797
    scavenging techniques **3**:797
        amide library construction **3**:798,
            3:799*F*
        benzodiazepine synthesis **3**:799–
            800, **3**:800*F*
        catch-and-release principle **3**:799,
            3:800*F*
        chemical library generation **3**:797–
            798, **3**:799*F*
        hydantoin synthesis **3**:799–800,
            3:800*F*
        Mitsunobu reagents **3**:798–799
        phase switch concept **3**:799–800
        solution phase library
            construction **3**:798–799
        sulfonamide library
            construction **3**:798, **3**:799*F*
        thiourea library
            construction **3**:798, **3**:799*F*
        urea library construction **3**:798,
            3:799–800, **3**:799*F*, **3**:800*F*
    site isolation effects **3**:794, **3**:794*F*
        difunctional molecule
            synthesis **3**:795
        high concentrations **3**:795
        macrocyclization **3**:795
        one-pot transformations **3**:794*F*,
            3:795
        three-dimensional
            environment **3**:794–795
    solution phase synthesis **2**:477*F*,
        2:489
    supports **3**:792
        acrylamides **3**:792–793
        Amberlyst 15 ion exchange
            resin **3**:821
        cellulose **3**:793
        controlled pore glass **3**:793
        cross-linking **3**:792–793
        divinyl benzene polystyrene
            resin **3**:792
        Jandagel **3**:792–793
        microreticular resins **3**:792–793
        monoliths **3**:793
        Nafion-H resin **3**:821
        PEGylated polystyrenes **3**:792–
            793

polyureas **3**:792–793
silica **3**:793
zeolites **3**:793
Polymer-bound ^1H-benzotriazole,
    solution phase synthesis **2**:490,
    2:490*F*
Polymeric resins, cholesterol
    absorption **6**:472
Polymer purity, polyethylene glycol
    conjugates **5**:1052
Polymer replication, lithography/
    photolithography **2**:354
Polymer-supported synthesis, solution
    phase synthesis *see* Solution
    phase synthesis
Polymethyl methacrylate (PMMA), x-
    ray LIGA **2**:305
Polymorph form, solid-state
    physicochemistry *see* Solid-state
    physicochemistry
Polymorphic forms, drug
    approval **2**:161–162
Polymyxin B, target site
    classification **7**:471*T*
Polyomaviruses, antiviral drugs **7**:254*T*,
    7:256
    cidofovir **7**:256
Polyoxometallates, antivirals **7**:360
Polypeptides, as drug
    candidates **3**:374–375
Polyphenols, green tea **1**:451
Poly(ADP ribose)polymerase inhibitors,
    IBD treatment **6**:628
Polypurine tract (PPT), HIV
    genome **7**:331
Polysaccharide matrix,
    5-fluorouracil **7**:69
Polysaccharides, drug–polymer
    conjugates *see* Drug–polymer
    conjugates
Polysomnography
    narcolepsy **6**:148
    somnolence measures/sleep
        testing **6**:142
Polyspecific transport, membrane
    transport protein
    expression **5**:74
Polystyrene-PEG, solid support
    synthesis **2**:496, **2**:496*F*,
    2:497*T*
Polysyphorin, PASP **3**:818
Polytherics **1**:492*T*
Polyureas, PASP **3**:792–793
    microcapsules **3**:813
Polyvinylidene difluoride (PVDF)
    filters, PAMPA **5**:472
Polyvinylpyrrolidone, drug–polymer
    conjugates **5**:1045
Poly(His)$_x$ tag, gene expression
    vectors **3**:110, **3**:110*T*
PONDR, domain mapping **3**:879
Pooled phenotypes, reaction
    phenotyping **5**:241

Pool libraries, solid-phase libraries *see*
    Solid-phase libraries
Poorly soluble substances, dissolution
    rate **5**:517, **5**:519*T*
Pople, John A, quantum mechanical
    calculations **4**:381
Population approach, serotonin receptor
    agonists **5**:900, **5**:900*T*
Populations, affective disorders, clinical
    trial issues **6**:62
Population studies, cancer
    biomarkers **7**:47–48
Porcine buccal mucosa, buccal
    administration **5**:280–282
Porcine models, blood–brain barrier
    permeation **5**:313
Pore-blocking mechanism, potassium
    channels **2**:862
Porin-like proteins, protein
    crystallization **3**:434
Porin mutations, quinolone antibiotic
    resistance **7**:586–587
Porous silicon, lithography/
    photolithography **2**:355
Portal hypertension **6**:725
    definition **6**:706
    treatment **6**:725–726
Portfolio management, organization,
    drug discovery **2**:25
'Portioning–mixing,' combinatorial
    chemistry **2**:461–462
Posaconazole **7**:430*T*; **8**:24
    antifungal activity **8**:30
    design **8**:29
    initial lead synthesis **8**:25
    medicinal chemistry **7**:451
    SCH 45012
        isomers **8**:26, **8**:26*F*, **8**:27*F*
        synthesis **8**:26
    SCH 51048 **8**:28*F*, **8**:29
    synthesis **8**:29, **8**:29*F*
Posatirelin, Alzheimer's disease
    treatment **6**:207
'Pose,' docking **4**:255
Pose prediction *see* Docking assessment
Posicor *see* Mebefradil hydrochloride
Positional (flat) assays, cell-based
    assays **3**:631
Positional cloning
    Danio rerio model system **3**:10
    pharmacogenomics **1**:344, **1**:345
    target identification *see* Target
        identification, genomics
Position C13 bacterial efflux protein
    inhibitors, tetracyclines **7**:605
Positive and Negative Syndrome Scale
    (PANSS), schizophrenia
    treatment **6**:29
Positive controls, CYP induction, in
    vitro studies **5**:250, **5**:251*T*
Positive phenotype interpretation,
    genetically engineered animals
    see Genetically engineered animals

Positive symptoms, schizophrenia
  6:18
Positron emission tomography (PET)
  applications 3:659
    affective disorders 6:50
    anxiety disorders 6:89
    cancer research 3:77
    coronary heart disease
      diagnosis 6:696
    dopamine transporter knockout
      mouse 6:124–125
    drug development 2:192
    functional small animal
      imaging 3:661
    migraine 6:371
    obsessive compulsive disorder
      (OCD) 6:6
    psychotropic drug
      development 6:51T
    small animal imaging 3:660T,
      3:661, 3:663–664
    somnolence measures/sleep
      testing 6:142
    transporters 2:65
  biomarkers 3:77
  combination with other imaging
    technology 2:193
  mechanism of action 2:192–193
  microdose studies 2:193
  pharmacokinetic studies 2:193
  pulmonary administration 5:287–288
  radionuclides 3:663T
    see also specific nuclides
    ^{75}Br 3:664T
    ^{76}Br 3:664T
    ^{11}C 3:664T
    ^{55}Co 3:664T
    ^{62}Cu 3:664T
    ^{64}Cu 3:664T
    ^{18}F 3:661–662, 3:664T
    2-[^{18}F]fluoro-2-deoxy-D-glucose
      (FDG) 3:77
    ^{68}Ga 3:664T
    ^{124}I 3:664T
    ^{13}N 3:661–662, 3:664T
    ^{15}O 3:661–662, 3:664T
    ^{82}Rb 3:661–662
    ^{94m}Tc 3:664T
    ^{86}Y 3:664T
Postabsorptive pharmacokinetics,
  membrane transport
  proteins 5:55
Postantibiotic effect
  aminoglycosides 7:638
  linezolid 7:689
Postcoital contraceptive,
  ethamoxytriphetol 8:84
Postcrystallization treatments see
  Protein crystallization
PostDock processing see Docking
Poste, George, carvedilol 8:138
Posterior hypothalamic gray,
  migraine 6:371

PostgreSQL 3:294
  server databases 3:294
Postmarketing surveillance 1:755–777
  see also Adverse drug events (ADEs)
  causality assessment 1:764
    algorithms 1:764, 1:764T
    Bradford–Hill criteria 1:764,
      1:764T
    covariate adjustment 1:769
      aggregate data use 1:770
      instrumental variables 1:769
      propensity scores 1:769
      sensitivity analysis 1:769
  data acquisition 1:761
    information sources 1:761
    spontaneous reporting
      systems 1:761, 1:763T
    Weber effect 1:763
  decision making 1:764, 1:766F
    acceptable risk 1:764
    average vs. individual risk/
      benefit 1:765
    'benefit-harm assessment' 1:765
    quality-adjusted life years 1:765
    risk–benefit ratio 1:764
  definition 1:758
    adverse reaction alerts 1:759
    labeling changes 1:758, 1:758T
  drug withdrawals see Drug withdrawals
  examples 1:770
    see also specific drugs
  historical perspective 1:756, 1:756T
    Biologics Control Act
      (1904) 1:756
    Committee on Safety of
      Medicines 1:756
    FDA foundation 1:756
    Medicines Act (1962) 1:756
  international/multinational
    collaborations 1:757
    Council for International
      Organization of Medical Sciences
      (CIOMS) 1:757
    International Conference on
      Harmonization (ICH) 1:757
  rationale 1:759
    clinical trial design 1:761, 1:763T
    clinical trial sizes 1:761
    drug discovery process 1:759
    fast-tracked drugs 1:760, 1:760T
    Food and Drug Administration
      Modernization Act
      (FDAMA)(1997) 1:760
    orphan diseases/drugs 1:761
  regulatory bodies 1:772
  study limitations 1:772
    General Practice Research
      database 1:773, 1:774F
    nested case-control study 1:773,
      1:774F
    open prospective cohort
      study 1:773
    pharmacogenetics 1:774

Prescription Event Monitoring
  Study 1:773
  retrospective cohort study 1:773,
    1:774, 1:774F
Postmenopausal osteoporosis 6:503
  alendronate 8:200
Post-mortem diagnosis, Alzheimer's
  disease 6:197
Postprandial measurements, irritable
  bowel syndrome 6:650
Postpubertal male
  hypogonadism 6:525T
Post-reaction cooling, Milestone s.r.l.
  microwave-assisted
  chemistry 3:849
Post-source decay (PSD), mass
  spectrometry proteomics
  3:33–34
Posttranscriptional gene silencing,
  RNAi (RNA
  interference) 3:173
Posttranslational modifications
  definition 2:381
  as drug targets 2:383
  epigenetics 1:258
  Escherichia coli as expression
    system 3:880
  expressed proteins 3:880
  heterogeneous protein
    crystallization 3:881
  pathogenesis associated 2:381
  phosphodiesterase-6 2:928
  protein display chips 2:370
  proteome 3:29
  proteomics 3:41
  2D-PAGE 3:41
  UDP-glucuronyltransferases 5:183
  UniProtKB 3:359
Posttraumatic stress disorder
  (PTSD) 6:5, 6:87
  clinical trials 6:96
    clinician-administered PTSD scale
      (CAPS-2) 6:96
    Davidson trauma scale
      (DTS) 6:96
    treatment outcome PTSD
      rating 6:96
  DSM-IV-R classification 6:87
  human genetics
    association studies 6:92
    dopamine receptors 6:92
    neuroimaging 6:90
  prevalence 6:87
  treatment, benzodiazepines 6:99
Postvoid residue, benign prostatic
  hyperplasia diagnosis 6:561
Potassium, tetracyclines 7:618
Potassium channels see K$^+$ ion channels
Potassium-competitive acid blockers
  (p-CABs) 8:219
Potassium cyanide, reactive metabolite
  studies 5:240
Potassium-sparing diuretics see Diuretics

Potency
  aminoglycosides **7**:636
  improvement, lopinavir **4**:602–604
Potential energy surfaces, core
    molecular modeling *see* Core
    molecular modeling
Potential of mean force (PMF),
    docking, scoring
    functions **4**:485
Potentiometric method, solubility *see*
    Solubility
Potentiometric titration, equilibrium
    solubility **5**:350
Potts and Guy equation, dermal
    penetration models **5**:729,
    5:730
Potts and Guy notation, dermal
    penetration models **5**:733
Powderject, mergers and
    acquisitions **1**:541–542
Powders, dissolution rate **5**:517,
    5:518*F*
Powell–Tuck index, ulcerative colitis
    (UC) **6**:619
Power outputs, CEM Corporation
    microwave-assisted
    chemistry **3**:851
Poxviruses, antiviral drugs **7**:254*T*,
    7:261
  cidofovir **7**:261–262
  HPMPO-DAPy **7**:261–262
  idoxuridine **7**:261–262
  imatinib mesylate **7**:261–262
  8-methyladenoisine **7**:261–262
  S2242 **7**:261–262
  ST-246 **7**:262–263
PP-1 **7**:191–192
  Bcr-Abl kinase inhibitor **2**:973
  Src kinase inhibitor **2**:976–977
  structure **2**:974*F*, **2**:978*F*,
    7:191*F*
  three-dimensional structures **2**:636*T*
PP2, structure **7**:191*F*
PP12 **7**:191–192
PPARG gene, familial partial
    lipodystrophy (FPLD) **6**:384
PPI-0903, structure **7**:495*F*
3-PPP, schizophrenia treatment **6**:34
P/Q voltage-gated calcium channel,
    migraine **6**:370
Practolol
  immunotoxicity **5**:221*T*
  oral absorption **5**:468*T*, **5**:680*T*
  structure **2**:677*F*
Pradamicins, antifungal drugs **7**:433
Prader–Lahart–Willi syndrome, male
    hypogonadism **6**:523
Prader–Willi syndrome (PWS)
  epigenetics **1**:270*T*
  faulty genomic imprinting **1**:271
  SNPRN gene **1**:271
Pragmatic innovation, project
    teams **2**:145

Pralatrexate **7**:72
  approval **7**:62*T*
  clinical indication **7**:62*T*
Pralnacasan, rheumatoid arthritis
    treatment **7**:860
Pramipexole hydrochloride **1**:180
  Parkinson's disease treatment **6**:11,
    6:215
  quantum mechanical
    calculations **4**:388–389
  structure **4**:388*F*
Pramlintide **2**:579*T*
  amylin analog **2**:580
Prandin *see* Repaglinide
Pranlukast **1**:208; **7**:907, **7**:908
  arthritis treatment **7**:854
Pravachol *see* Pravastatin
Pravastatin **6**:240*T*, **6**:473–474, **6**:481
  blood to plasma ratios **5**:325*T*
  blood volume of distribution
    (V$_{SS,b}$) **5**:325*T*
  discovery/development **1**:374–375
  oral absorption **5**:468*T*
  permeability assays **5**:272
  plasma clearance (CL$_p$) **5**:325*T*
  structure **1**:376*F*, **6**:238*F*, **6**:469*F*
  target information **4**:750*T*
  transplant rejection **7**:920–921
Pravastatin or Atorvastatin Evaluation
    and Infection Thrombolysis in
    Myocardial Infarction (PROVE-
    IT TIMI-22) **6**:475
Pravastatin sodium, structure **2**:66*F*
Praziquantel
  cestode infection treatment **7**:761
  oral absorption **5**:680*T*
  trematode infection treatment **7**:761
Prazosin
  benign prostatic hyperplasia
    treatment **6**:556*T*
  bioavailability **5**:712*T*
  discovery/development **1**:328–329
  drug action **1**:532*T*
  hypertension treatment **6**:714
  mechanism of action **6**:145
  oral absorption **5**:468*T*
  P-glycoprotein interaction **5**:781*T*
  structure **6**:714*F*
Pre-approval costs, drug discovery **2**:4,
    2:6*F*
Precandidates, outsourcing scale-
    up **2**:211
Precedex *see* Dexmedetomidine
    hydrochloride
Precipitants, protein
    crystallization **3**:438
Precision
  adaptive in combo strategy **5**:964
  NMR structures, problems **3**:542
Preclamol, schizophrenia
    treatment **6**:34
Preclinical Alliances Group, Lindberg,
    P **8**:220

Preclinical efficacy testing, central
    nervous system injury **6**:261
Preclinical research **2**:52
  antifungal drugs **7**:434
  BCS *see* Biopharmaceutical
    Classification Scheme (BCS)
  bottlenecks **2**:3
  central nervous system disease **2**:52
  decision making **2**:70
  modafinil **8**:150
  neurokinin receptor
    antagonists **6**:110
  pharmacokinetics **5**:12–13
  physiologically based
    pharmacokinetics **5**:877
  safety evaluation, drug discovery
    productivity **2**:125
  solid-phase chemistry **2**:52
Preclinical trials **2**:177
  adverse event prediction **2**:179–180,
    2:180*F*
  amyotrophic lateral sclerosis **6**:223
  as bottleneck **1**:38*T*
  definition **2**:177
  documentation **2**:178–179
  dosage **2**:178, **2**:179*F*
  models **2**:177
  pharmacokinetics **2**:178
  pharmacological studies **2**:177,
    2:178*T*
  proteomics **3**:42, **3**:42*F*
  'rule-of-five' **2**:178
  toxicology studies **2**:178, **2**:179*T*
Precompetitive research, industry/
    academia collaborations **1**:575
Precursor cells, cell-based assays **3**:643
Predevelopment, drug discovery *see*
    Drug discovery
Predicted coding sequences, IPI
    (International Protein
    Index) **3**:366
Predicted performance measures,
    screening library
    selection **4**:506
Predicted residual error sum of squares
    (PRESS), solubility
    models **5**:631
Prediction coefficients, metabolism
    prediction **5**:804
Prediction models, volume of
    distribution (V$_{ss}$) *see* Volume of
    distribution (V$_{ss}$)
Predictive abilities, QSAR **4**:58
Predictive Array Design (PAD),
    noncombinatorial library
    design **4**:354
Predictive genotyping,
    pharmacogenetics *vs.* **3**:66
Predictive models, CADD **4**:3–4
Predictive performance, non-animal
    carcinogenicity testing **1**:477
Predictive purposes, solubility
    models **5**:630

S9 **5**:242*T*
specific antibodies **5**:241, **5**:242
limitations **5**:242
specific inhibitors **5**:241, **5**:242,
5:243*T*
CYP dependence **5**:242
sulfaphenazole **5**:243*T*
ThioTEPA **5**:243*T*
tranylcypromine **5**:243*T*
Reaction quenching, PASP **3**:797
Reaction times, microwave-assisted
chemistry **3**:857
Reaction types, combinatorial
chemistry **3**:762
Reaction volumes, microwave-assisted
chemistry scale-up **3**:856
Reactive azetidinone group, azetidinone
cholesterol absorption
inhibitors **8**:70, **8**:72*F*
Reactive groups
blocking, polyethylene glycol (PEG)
conjugates **5**:1061
Reactive metabolites
drug metabolism, in vitro
studies **5**:240
formation, CYP induction, in vitro
studies **5**:250
glutathione (GSH) **5**:240
metabolism prediction *see* Metabolism,
prediction of
potassium cyanide **5**:240
prediction **5**:796
Reactive oxygen species (ROS)
atherosclerosis **6**:466
COPD **7**:879–880
enzymatic sources **6**:581
formation, tetracycline
phototoxicity **7**:618
oxidative stress **6**:581
REACT++ programs, structure-based
library design **4**:321
Read out parameters, fluorescence
screening assays **3**:614
Reagent-based selection methods,
library design **4**:309
Real-space correlation coefficient,
protein 3D structure
validation **3**:520
Real Space R-factor (RSR), protein 3D
structure **3**:510, **3**:547
Real-time measurements, cell-based
assays **3**:618
Reasonable working situation,
motivation **2**:246
Reboxetine **1**:**178**
side effects **6**:60*T*
structure **2**:62*F*
transporter inhibition **6**:66*T*
Recainam, oral absorption **5**:680*T*
RECEPTOR, pharmacophore
detection **4**:127*T*, **4**:128, **4**:133
Receptor activator of NFκB (RANK),
osteoclast regulation **6**:501

Receptor averaging, docking **4**:270
Receptor-binding assays
cell-based assays *see* Cell-based assays
homogeneous time-resolved
fluorescence *see* Homogeneous
time-resolved fluorescence
(HTRF)
Receptor-mediated transcytosis (RMT),
blood–brain barrier
permeation **5**:302–303, **5**:304*F*
Receptor protein tyrosine kinases
(RPTKs)
ligand binding **2**:576
mechanism of action **2**:576
oncogene carcinogenesis **2**:576
peptide/protein drugs **2**:576
protein hormone binding **2**:576
structure **2**:576
Receptors
affinity, 5-HT₃ ligand gated ion
channel assays **3**:635*T*
crystal structure, ligand-based library
design **4**:316–317
genes, CYP induction, in vitro
studies **5**:252
homology modeling, steroid-derived
modulators **7**:173, **7**:173*F*
internalization, HCS *see* High-content
screening (HCS)
mechanism-based pharmacokinetic/
pharmacodynamic (PKPD)
modeling *see* Mechanism-based
pharmacokinetic/
pharmacodynamic (PKPD)
modeling
peptidomimetics *see* Peptidomimetics
promiscuous ligands **2**:750
signal changes, NMR, drug
discovery **3**:915–916
target design **4**:314
unknown, CADD **4**:15
Receptor surface models (RSM), 3D-
QSAR alignment **4**:579–580
Receptor-to-ligand adaptation, 4D-
QSAR **5**:597–598
Receptor tyrosine kinase FLT3
identification, target search
genomics **3**:5
Receptor tyrosine kinases
(RTKs) **7**:185, **7**:192
ligand upregulation **7**:185
structure **7**:185
Receptor tyrosine kinases (RTKs), type
III **7**:197
see also specific types
in cancer **7**:197
Receptor tyrosine kinases (RTKs), type
III inhibitors **7**:198
in clinic **7**:199
future work **7**:199
structure **7**:199*F*
Reciprocal lattice, x-ray–crystal
interaction **3**:456–457

Recognition
dynamic combinatorial
chemistry **3**:961
ligand–receptor interactions **4**:194
pharmacophore detection **4**:124,
4:125*F*
Recognition 'hot spots,' scaffold-based
design *see* Scaffold-based design
Recombinant cytochrome P450
3A4, clearance (CL₍int₎) in vitro
screens **5**:958–959
metabolic drug–drug
interactions **5**:829
Recombinant enzymes, reaction
phenotyping *see* Reaction
phenotyping
Recombinant parathyroid hormone
(rPTH) *see* Teriparatide
Recombinant Plasminogen Activator
Phase II International Dose-
Finding Study *see* RAPID-I;
RAPID-II
Recombinant proteins
protein crystallization **3**:436
tagging, protein display chips
2:370
Recombinant tissue plasminogen
activator, stroke
treatment **6**:263
Recombinant tumor necrosis factor,
introduction of **1**:107
Recombination rules, de novo drug
design **4**:289
Recommendation, DCN **2**:155
Reconstituted Human Vaginal
Epithelium, vaginal
administration **5**:290
RECOORD, NMR structures,
problems **3**:542
Recording/reporting process, clinical
trials **2**:187
Recoupling, NMR structure
determination **3**:482
Recruitment
see also Organization, drug discovery
innovation **2**:262
motivation **2**:246
R&D management **2**:250
Rectal administration **5**:279–280,
5:290
in vitro cell cultures **5**:290
in vivo experiments **5**:290
Recursive combinatorial analysis
procedure (RECAP) algorithm,
GPCR library screening **4**:691
Recursive feature elimination (RFE),
CNS uptake **5**:758–759
Recursive partitioning (RP)
computational oral absorption
prediction **5**:689
genetic algorithm
combination **4**:327–328
machine learning methods **3**:250

Recursive partitioning (RP) (*continued*)
  promiscuous ligands, computational
      prediction 2:746, 2:747*F*
  virtual screening, large data
      sets 4:476, 4:477*F*
Recycling, GPCR
    desensitization 2:802
Red blood cell (RBC) assay, non-animal
    eye irritation testing 1:467*T*
Red cell membrane binding,
    pharmacokinetics 5:14
Redox capacity, glutathione
    (GSH) 5:158
Redox reactions 5:87–132
  see also Oxidation; *specific enzymes*
  as defense system 5:87
Redox targets 2:1053–1087
  see also Mitochondria; Mitoprotective
      drugs
Redrugs
  materials management 2:531
  receiving/tracking 2:534
  sourcing 2:533
  stability, HTS data
      management 2:453
Reduced complexity screening (RCS),
    fragment-based drug
    discovery 4:449
Reduced folate carrier (RFC) 7:58–59
  defective 7:59
Reduced graphs
  fingerprint/dataprint descriptors *see*
      Fingerprint/dataprint descriptors
  library design 4:361
  ligand–receptor interaction
      models 4:206
  2D similarity searching 3:247
5α-Reductase inhibitors
  see also specific drugs
  benign prostatic hyperplasia
      treatment 6:564, 6:566
  discovery/development 1:17
  introductions 1:151
  side effects 6:566
5α-Reductase isozymes, benign
    prostatic hyperplasia 6:558
Reductases 5:168, 5:178
  drug metabolism 5:202
  reversible reactions 5:178
Reductil *see* Sibutramine
Reductionism
  drug discovery 2:272
  structure information
      representation 4:540
Reduction-of-function (hypomorphic),
    chemical mutagenesis 3:13–14
Reductions 5:121
  see also specific enzymes
Reductive dehalogenation, cytochrome
    P450s 5:119*F*, 5:121, 5:122*F*,
    5:123*F*
Redux *see* Dexfenfluramine
Reference linking, databases 3:284

Reference panel, biological
    fingerprints 4:808
References
  DCN 2:155
  HTS data management 2:453
Refinement
  biological macromolecules 1:443
  protein crystallography model
      generation 3:463
Reflection
  core molecular modeling 4:101,
      4:101*F*
  dielectric heating 3:840
  measurement, x-ray–crystal
      interaction 3:454
Reflection intensity strength, protein
    crystallography 3:470
Reflectometric interference
    spectroscopy (RIfS), optical
    biosensors 3:594–595
Refludan *see* Lepirudin
Reflux/uptake transporters, metabolic
    drug–drug interactions 5:830
Refolding, protein purification 3:420,
    3:421
Reformulation, salt form effects 5:524
Refractoriness, arrhythmias 6:733
Refrigeration temperature stability, drug
    reference standard 2:167–168,
    2:168*T*
RefSeq (NCBI Reference Sequences)
  curation 3:355–356
  data acquisition, DDBJ/EMBL/
      GenBank 3:354
  features 3:355
  nucleotide sequences 3:355
  protein sequences 3:358
Regadenoson 1:508
Registration, compound acquisition/
    processing 3:564
REGISTRY 3:282*T*
Regression analysis
  CoMFA 5:589
  distribution models 5:870
Regression-based scoring *see* Docking,
    scoring functions
Regression-extension PCA,
    metabonomics 5:996
Regression models, bioavailability
    models 5:718
Regulator of K$^+$ conductance (RCK)
    domain, MthK 4:707
Regulatory aspects 2:131
  BCS *see* Biopharmaceutical
      Classification Scheme (BCS)
  chemogenomics 2:567
  clinical trails, phase I 2:155
  documentation 2:721–722
  drug development 2:200
  drug discovery, decline in 1:620,
      1:621–622
  guidelines, drug development 2:27
  innovation, effects on 1:744

  biosafety 1:746
  competition 1:746
  nutraceuticals *see* Nutraceuticals
  personalized medicine *see* Personalized
      medicine
  pharmacogenetics *see*
      Pharmacogenetics
  traditional medicine *see* Traditional
      medicine
Regulatory authorities
  see also specific regulatory bodies
  metabolic drug–drug
      interactions 5:828
  phase IV clinical trials 2:190
  postmarketing surveillance 1:772
Regulatory DNA sequences, gene
    expression vectors *see* Gene
    expression vectors
Regulatory elements
  genetically engineered
      animals 3:156
  phosphodiesterase-6 2:928
  pyruvate dehydrogenase
      kinase 4:755–756, 4:756*F*
Regulatory networks, bioinformatics *see*
    Molecular networks,
    bioinformatics
Regulatory pathways, cDNA
    microarrays 3:332
Regulatory regions, bioinformatics *see*
    Molecular sequence analysis,
    bioinformatics
Regulatory systems, cell-based
    assays 3:617–618
Regulatory T cells 5:218–219
  IBD 6:618
Reimbursement policies,
    pharmaceutical productivity
    effects 2:133
Reinvention patent, raloxifene 8:108
Rekker method
  lipophilicity prediction, rule-based
      systems 5:650, 5:651*F*
  log P calculation 5:655
  partition coefficient systems 4:49
Relapse, addiction 6:175
Relapsing-remitting multiple sclerosis
    (RRMS) 7:923
Related chi index, similarity searching,
    representations 4:170
Relational database management system
    (RDBMS)
  databases 3:294
  object-based databases 3:268
Relational databases 3:268, 3:269*F*
  advantages 3:268
  SQL (structured query
      language) 3:268
Relational networks, topological
    descriptors 5:561
Relative activity factor (RAF), reaction
    phenotyping 5:243, 5:245
Relative potency, oxazolidinones 7:688

Relaxation dispersion, NMR structure determination **3**:498

Relaxation-edited spectroscopy, mixture analysis **2**:528

Relaxation filtering, NMR, drug discovery **3**:908*T*, **3**:913

Relaxation optimized sequences, liquid-state NMR structure determination **3**:479–481

Relaxation rates, NMR spectroscopy **3**:946

Relenza *see* Zanamivir

Relevant point extraction, GRIND descriptors **5**:541

Rel homology domain (RHD), NF-κB **3**:195

Reliability, project teams **2**:145

Reliability factor *see* R-factor

Relpax *see* Electriptan

Rel protein family **3**:195

Remacemide **6**:291
    Huntington's disease treatment **6**:222*T*
    structure **6**:290*F*

Remeron *see* Mirtazapine

Remicade *see* Infliximab

Remifentanil
    anesthetic adjuvants **6**:362
    characteristics **6**:364
    mechanism-based PK/PD modeling **5**:894–895
    structure **6**:363*F*

Remifentanil hydrochloride **1**:223

Remodeling, benign prostatic hyperplasia **6**:557

Remodulin *see* Treprostinil sodium

REMOTEDISC, 3D-QSAR **4**:576

Remoxipride
    structure **2**:717*F*
    synthesis **2**:717*F*

Remune, data publication issues **1**:711–712

RENAAL **6**:589*T*

Renal clearance (CL$_r$)
    ABCC transporters **5**:79–80
    ABC transporters **5**:80
    active transport **5**:46
    elimination studies **5**:46
    mechanisms **5**:52
    membrane transport protein pharmacokinetics **5**:79
    models *see* Clearance models
    OATP (*SLC21/SLCO*) transporters **5**:79–80
    OAT (*SLC22*) transporters **5**:79–80
    plasma protein binding **5**:323–324, **5**:324*F*

Renal disease
    benign prostatic hyperplasia **6**:563
    risk factors, hypertension **6**:706

Renal dysfunction **6**:545, **6**:575–595
    see also Hypertension; Obesity
    chronic kidney disease **6**:579

clinical trials **6**:585
    ACE inhibitors **6**:585
    angiotensin receptor blockers **6**:585
    antihypertensives **6**:585
    β-adrenergic receptor antagonists **6**:585
    calcium-channel blockers **6**:585
    Candesartan and Lisinopril Microalbuminuria (CALM) study **6**:585
    losartan **6**:585
definition **6**:575
diabetes mellitus **6**:578
    type 1 **6**:578
    type 2 **6**:578
endothelial dysfunction **6**:580
epoxyeicosatrienoic acid **6**:582
experimental models **6**:584
20-hydroxyeicosatetraenoic acid **6**:582
hypertension **6**:576
inflammatory cytokines **6**:581
management **6**:585, **6**:586*F*
    20-HETE synthesis inhibitors **6**:591
    antioxidants **6**:590
    EET mimetics **6**:592
    emerging approaches **6**:590
    endothelin receptor antagonists **6**:591
    epoxide hydrolase inhibitors **6**:592
mechanisms **6**:579, **6**:579*F*
    hemodynamic changes **6**:579
microalbuminuria **6**:578, **6**:579
nitric oxide **6**:580
obesity **6**:576
oxidative stress **6**:581
prevalence **6**:575
    ethnicity **6**:576
prognosis **6**:579
renin–angiotensin system **6**:583, **6**:583*F*
    angiotensin II **6**:583–584
    sodium resorption **6**:584
    water resorption **6**:584
risk factors **6**:576*F*
stages **6**:579
sympathetic nervous system **6**:579
    hypertension correlation **6**:579–580
    leptin **6**:580
    obesity **6**:580
    unmet medical needs **6**:590

Renal excretion
    modeling **5**:738
        Manga method **5**:738
    optimum pharmacokinetics **5**:25–26
    pharmacokinetics *see* Pharmacokinetics

Renal metabolism (CL$_{rm}$), plasma protein binding **5**:323–324

Renal toxicity, cisplatin **7**:142

Renin, biological functions **3**:929

Renin–angiotensin system
    hypertension *see* Hypertension
    renal dysfunction *see* Renal dysfunction

Renin inhibitors
    see also specific drugs
    active site binding **2**:628*F*
    carboxamide bioisosterism **2**:668
    hypertension treatment **6**:723
    nonpeptide drugs **2**:625
    peptidomimetics **2**:613, **2**:614*F*, **2**:615*F*, **2**:625
    three-dimensional structures **2**:626*T*

Renormax *see* Spirapril hydrochloride

Renpress *see* Spirapril hydrochloride

Renzapride
    irritable bowel syndrome treatment **6**:661*T*, **6**:662
    as prokinetic agent **6**:677–678

ReoPro *see* Abciximab

Reoviruses, antiviral drugs **7**:279

ReOx **1**:506

Repaglinide **1**:229; **6**:439*T*, **6**:443
    advantages **6**:444
    clinical use **6**:444
    disadvantages **6**:444
    mechanisms of action **6**:443
    pharmacokinetics **6**:443
    side effects/contraindications **6**:444
    structure **6**:440*F*

RepeatMasker, repetitive elements, bioinformatics **3**:322–323

Repeats, genome sequencing **3**:317

Repetitive elements, bioinformatics *see* Molecular sequence analysis, bioinformatics

Replacement therapy, nicotine addiction **6**:183

Replagal *see* Agalsidase alfa

Replication interference, DNA therapeutic agents **2**:1039

Repolarizing currents, cardiac ion channels **6**:734

Reporter genes/assays
    cell-based assays *see* Cell-based assays
    constructs, genetically engineered cells **1**:479
    GPCR functional assay screening *see* G protein-coupled receptors (GPCRs), assays
    single-parametric cellular assays **3**:683
    small animal imaging **3**:665

Reporter ligand observation, NMR spectroscopy **3**:945*F*, **3**:946

Reporting, HTS data management **2**:455

Representative conformation generation, conformational analysis/searching *see* Conformational analysis/searching

Representative subsets, screening library selection **4**:502

Representivity of Target Families in the Protein data Bank (PDBRTF) **3**:304–305

Reprocessing, development chemistry **2**:169

Reproductive/developmental toxicity animal experimentation alternatives **1**:467*T*, **1**:475

MCASE **5**:925–926

TOPKAT models **5**:925

toxicity, computational models **5**:925

Reproductive medicine **6**:548

cancer risk factors **7**:4

drug introductions *see* Drug introductions

SMEs **1**:512

Reprofiling, drug discovery *see* Drug discovery

Reproterol, oral absorption **5**:680*T*

REPuter, repetitive elements, bioinformatics **3**:323

ReQuip *see* Ropinirole hydrochloride

Required permeability, in vitro oral absorption prediction **5**:676

Required solubility, in vitro oral absorption prediction **5**:676, **5**:677*T*

Rescoring *see* Docking, scoring functions

Rescriptor *see* Delaviridine mesylate

Rescula *see* Unoprostone isopropyl ester

Research and development (R&D) *see* Pharmaceutical research and development (R&D)

Research Collaboratory for Structural Bioinformatics (RCSB)

data nomenclature **3**:543

ligand nomenclature standardization **3**:539

macromolecular Crystallographic Information File (mmCIF) **3**:543

PDB **3**:373

see also Protein Data Bank (PDB)

Research Council restructuring, industry/academia collaborations **1**:581

Research databases, industry/academia collaborations *see* Collaborations, industry/academia

Research project management *see* Project management

Research strategies, innovation **2**:22

Reserpine, structure **6**:88*F*

Residual standard deviation (RSD)

QSAR applicability domain **4**:159

solubility models **5**:632

Residues, DNA structure **2**:1038–1039

Resin beads, solid support synthesis **2**:493

macroporous resins **2**:495

definition **2**:495

Merrifield resins *vs.* **2**:495, **2**:495*T*

poly(ethylene glycol)-containing resins **2**:495, **2**:496*T*

CLEAR resins **2**:496, **2**:497*F*

dendrimers **2**:499

polyethylene glycol-poly (*N*,*N*-)dimethylacrylamide (PEGA) **2**:497, **2**:499*F*

polystyrene-PEG **2**:496, **2**:496*F*, **2**:497*T*

tetraethylene glycol diacrylate cross-linked polystyrene support (TTEGDA-PS) **2**:497, **2**:498*F*, **2**:498*T*

polystyrene-based resins **2**:493

divinylbenzene **2**:493–494

functional groups **2**:493–494, **2**:494*F*

Merrifield resins **2**:493–494

Resin-bound scavengers/redrugs, solution phase synthesis *see* Solution phase synthesis

Resin capture/release, solution phase library purification **3**:777, **3**:777*F*

Resiniferatoxin, structure **2**:909*F*

Resin scavenging, solution phase library purification **3**:776, **3**:777*F*

Resin-to-resin transfer reactions, solid support synthesis **2**:474, **2**:475*F*

Resistance

adefovir dipivoxil **7**:318

aminoglycosides *see* Aminoglycosides

antimicrobial agents **7**:475

antivirals **7**:336, **7**:362

azidothymidine **7**:336

emtricitabine **7**:317

entecavir **7**:317

gentamicin **7**:642

hepatitis B antivirals **7**:317

HIV protease inhibitors *see* HIV protease inhibitors

Kaletra **8**:195, **8**:195*F*

lamivudine **7**:317

linezolid **7**:691

neuraminidase inhibitor design **4**:623, **4**:624*F*

non-nucleoside reverse transcriptase inhibitors (NNRTIs) **1**:74

oseltamivir *see* Oseltamivir

oxazolidinones **7**:690

ritonavir **8**:191–192

streptomycin **7**:642

zanamivir *see* Zanamivir

Resolution

HCS confocal systems **3**:687

imaging mass spectrometry **3**:39–40

protein crystallization **3**:438

protein crystallography model generation **3**:462–463, **3**:464*F*

protein 3D structure **3**:545

2D-PAGE **3**:34–35

x-ray crystallography **4**:763

Resorption suppression, bone resorption/remodelling **8**:205–206

Resorufin benzyl ether, in vitro inhibition studies **5**:247*T*

Resource lack, outsourcing **2**:227–228

Respiratory infections

see also specific infections

Haemophilus influenzae infections **7**:739

Moraxella catarrhalis infections **7**:739

mortality/morbidity, developing countries **1**:642*T*

SMEs **1**:514

Streptococcus pneumoniae infections **7**:739

Streptococcus pyogenes infections **7**:739

treatment

flurithromycin ethylsuccinate **1**:113

fropenam **1**:114

gatifloxacin **1**:114

moxifloxacin hydrochloride **1**:116

pathogen range **7**:739, **7**:740*F*

telithromycin **1**:118

trovafloxacin mesylate **1**:119

Respiratory sensitization, toxicity, computational models **5**:924

Respiratory syncytial virus (RSV) **7**:405

antiviral drugs **7**:276

palivizumab **1**:434, **1**:444

epidemiology **7**:406

reinfection **7**:406

fusion protein **7**:407, **7**:408*F*

replicative cycle **7**:405

structure **7**:405, **7**:406*F*

transmission **7**:406

treatment **7**:407

fusion inhibitors *see* Respiratory syncytial virus (RSV), fusion inhibitors

future work **7**:411

passive immunization **7**:407

ribavirin **7**:407

vaccines **7**:406–407

Respiratory syncytial virus (RSV), fusion inhibitors **7**:407

see also specific drugs

benzazapine carboxamides **7**:409*F*, **7**:411

BMS-433771 **7**:408, **7**:408*F*, **7**:409*F*, **7**:410*F*

peptides **7**:407

Fuzeon **7**:407–408

R170591 **7**:409*F*, **7**:410

RFI-641 **7**:410

small-molecules **7**:408

VP-14637 **7**:409*F*, **7**:410

Respiratory tract, cytochrome P450s **5**:171

Respiratory zones, pulmonary administration **5**:287

'Respiratosomes,'
mitochondria **2**:1057–1058
Response element sequence, cell-based
assays **3**:633
Response variation
mechanism-based PK/PD
modeling **5**:900–901
serotonin receptor agonists **5**:901*F*
Responsibilities
project management teams **2**:147
project teams **2**:145
Restenosis, gene therapy **1**:303
Resting state, sodium channels **2**:835
Restraints per residue, NMR
structures **3**:546
Restriction, drug design, de
novo **4**:285
Restriction enzymes
definition **1**:259
DNA methylation **1**:259
Restriction fragment length
polymorphisms (RFLPs),
discovery/development **1**:259
Restriction proteases, gene expression
vectors **3**:111, **3**:111*T*
REST/RSF, identification, large-scale
virus vector libraries **3**:18–19
Result scale, clinical trials **2**:184
RET, in disease **2**:963*T*
Retaane *see* Anecortave acetate
Retention times, partition assessment
tools **5**:478
Reteplase **1**:175; **6**:770
clinical trials **6**:773
GUSTO-3 **6**:773
GUSTO-5 trial **6**:773
INJECT trial **6**:773
RAPID-I **6**:773
RAPID-II **6**:773
Retevase *see* Reteplase
Reticulocyte binding protein,
malaria **7**:767
Reticulocytes, cell-free expression
systems **3**:122
Retigabine **6**:291
structure **6**:290*F*
Retinal
acid-sensitive ion channels
(ASICs) **2**:911
oxidation **5**:103–104
Retinaldehyde, age-related macular
degeneration **6**:300
Retinitis pigmentosum
mitochondria **2**:1054–1055
phosphodiesterase-6 **2**:928,
2:950
Retinoblastoma protein (Rb)
apoptosis induction **7**:11
as biomarker **3**:72
cell cycle **7**:9
Retinoic acid, metabolism, CYP
2C8 **5**:547*T*
9-*cis*-Retinoic acid, structure **2**:1020*F*

Retinoic acid receptor-α (RAR-α)
acute myeloid leukemia **2**:1027
three dimensional structure **2**:999
Retinoic acid receptor ligands
see specific drugs
Retinoic acid receptor-related orphan
receptor, nomenclature **2**:996*T*
Retinoic acid receptors (RARs) **2**:1007
in cancer **2**:1020–1021
embryogenesis **2**:1007
epithelial cell differentiation **2**:1008
ligands **2**:1020, **2**:1020*F*
see also specific drugs
isotype-selective ligands **4**:733
nomenclature **2**:996*T*
retinoid X receptors *vs.* **4**:733
reverse cholesterol transport **2**:1008
structure-based drug design **4**:733
subtype sequence alignments **4**:733
Retinoid receptors **7**:964
Retinoids
acne treatment *see* Acne treatment
clinical uses **4**:733
mechanism of action **7**:963–964
psoriasis treatment **7**:946
SARs **7**:964
cyclohexyl ring **7**:964
Retinoid X receptors (RXRs)
agonists, IBD treatment **6**:630
heterodimers **2**:1001
homology modeling, estrogen
receptors **4**:739
nomenclature **2**:996*T*
orthogonal ligand–receptor
pairs **3**:221–222, **3**:222*F*, **3**:223*F*
retinoic acid receptors (RARs)
vs. **4**:733
structure-based drug design **4**:733
ligand design/docking **4**:742
three dimensional structure, ligand-
binding pocket **2**:1000, **2**:1000*F*
vitamin D receptor
heterodimerization **2**:1007
Retinopathy of prematurity, vascular
endothelial growth factor **3**:202
Retrospective cohort study,
postmarketing
surveillance **1**:773, **1**:774,
1:774*F*
Retrospective follow-up, screening
library selection **4**:503, **4**:507,
4:509*F*
Retrosynthetic combinatorial analysis
procedure (RECAP), fragment-
based drug discovery **4**:449
Retrovir *see* Zidovudine
(azidothymidine: AZT)
Retroviruses
see also specific viruses
antiviral drugs **7**:254*T*, **7**:279
large-scale virus vector
libraries **3**:18–19
vectors

gene therapy **1**:301, **1**:304
RNA interference **3**:17
siRNAs **3**:177
transgenic animals **1**:436
Rett syndrome
DLX5 gene imprinting **1**:269
epigenetics **1**:270*T*
germline mutations **1**:269
methyl-CpG-binding protein
mutation **1**:269
Reuptake blockers,
characteristics **2**:647*T*
Revenue
biological macromolecules **1**:434
SMEs **1**:496*T*
funding **1**:494
Reversal, genome rearrangements,
bioinformatics **3**:324
Reversal of Atherosclerosis with
Aggressive Lipid Lowering
(REVERSAL) **6**:474*T*, **6**:475
Reverse chemical genetics,
definition **7**:734
Reverse cholesterol transport
high-density lipoproteins
(HDLs) **6**:465–466, **6**:467*F*
retinoic acid receptor **2**:1008
Reverse engineering, biomarkers **3**:71–
72
Reverse erbA receptor,
nomenclature **2**:996*T*
Reverse genetics
chemical biology **3**:130, **3**:131*F*
*Drosophila melanogaster see Drosophila
melanogaster*
human cells **3**:14
Saccharomyces cerevisiae model system
see Saccharomyces cerevisiae model
system
Reverse mapping, target
identification **7**:734
Reverse-phase arrays, proteomics **3**:40
Reverse-phase high-performance liquid
chromatography, partition
coefficient systems **4**:48
Reverse screening, fragment-based drug
discovery **3**:952, **3**:953*F*
Reverse transcriptase
epigenetics **1**:258
HIV life cycle **7**:331, **7**:333*F*
homogeneous time-resolved
fluorescence **2**:422, **2**:423*F*,
2:424*F*
Reverse transcriptase polymerase chain
reaction (RT-PCR)
genetically engineered animal
characterization **3**:153–154
RNAi detection **3**:178
Reverse transcription, ribosome
display **1**:435
Reversible covalent bonds, dynamic
combinatorial library
generation **3**:961, **3**:962*F*

Reversible H⁺, K⁺ -ATPase inhibitors **8**:219

Reversible inhibition, in vivo kinetic studies *see* Inhibition, in vitro studies

Reversible inhibitors of monoamine oxidase type A (RIMAs) **6**:69
see also specific drugs
molecular action **6**:69

Reversible interactions, time-dependent metabolic drug-drug interactions **5**:838

Reversine, discovery/development **3**:922, **3**:923*F*

Revex *see* Nalmefene hydrochloride

Revised proposals, customized automation **5**:348

Reward, definition **6**:170*T*

Reward systems, R&D management **2**:250

Reworking, development chemistry **2**:169

Reyataz *see* Atazanavir

Rezulin *see* Troglitazone

R-factor
atomic model derivation **4**:763
chloromuconate cycloisomerase structure **3**:533–534
definition **3**:546
protein 3D structure **3**:510, **3**:545
testing, x-ray structures, problems **3**:535

RF homogeneity, NMR **2**:515

RFI-641, respiratory syncytial virus (RSV), fusion inhibitors **7**:410

*R*ᵣₑₑ, testing of **3**:535

RG (reference group), UniProtKB flatfile **3**:360

RG2077 **6**:240*T*

RG-13022 **7**:193–194
structure **7**:193*F*

R-group selection, ligand-based library design **4**:310

RGS4, schizophrenia **6**:23

Rhabdoviruses, antiviral drugs **7**:254*T*, **7**:278

Rhaphidecursinol B, PASP **3**:818

Rheophoresis, dry age-related macular degeneration **6**:305

Rheumatic disease, bone diseases/disorders **6**:506

Rheumatic heart disease **6**:694

Rheumatoid arthritis (RA) **1**:440; **6**:506
clinical markers **7**:847–848
etiology, tumor necrosis factor-α **1**:441
NF-κB pathway **3**:197
pathogenesis **7**:846
STATs **3**:208
therapeutic targets **7**:847
treatment **1**:440–441
adalimumab **1**:433–434

anti-TNF-α therapies **6**:244
belimumab **6**:245
disease-modifying antirheumatoid drugs (DMARDs) **1**:440–441
methotrexate **1**:440–441
monoclonal antibodies **1**:433–434
nonsteroidal anti-inflammatories **1**:440–441
structure-modifying anti-inflammatory drugs **7**:857–858

Rheumatrex *see* Methotrexate

Rhinosporidiosis **7**:422*T*

Rhinoviruses
antiviral drugs **7**:265
ruprintrivir **7**:266
protein structures on PSD **3**:384
structures **3**:382, **3**:383*F*

Rhizopuspepsin, three-dimensional structures **2**:626*T*

Rhizoxin **7**:82
SAR **7**:82–84, **7**:84*F*
structure **7**:84*F*

Rhodacyanine dye MKT-077, mitochondria, drug targeting **2**:1067, **2**:1068*F*

Rhodamine 123, structure **5**:778*F*

Rhodopsin, orthogonal ligand–receptor pairs **3**:230*F*, **3**:231–232

Rhodopsin receptor **2**:575–576, **2**:777–778
activation **2**:796
TM3 residue displacement **2**:798
in human disease **2**:775*T*, **2**:792*T*
light-induced conformational changes **2**:796
structure **2**:796; **4**:671
structure-based drug design, virtual screening **4**:689

RhoGDI, protein crystallization **3**:887

Rhop complex, malaria **7**:768

Ribavirin
Crimean–Congo hemorrhagic fever virus **7**:278
filoviruses **7**:278
flaviviruses **7**:267
oral absorption **5**:468*T*
orthomyxoviruses **7**:276
respiratory syncytial virus (RSV) **7**:407
side effects **7**:390
structure **7**:389*F*

Riboflavin deficiency, carbon-boron bioisosterism toxicity **2**:698

Ribonuclease
alpha helix scaffold-based design **4**:30, **4**:30*F*
protein crystallization **3**:434

Ribonuclease protection assay, genetically engineered animal characterization **3**:153–154

Ribonucleotide reductase **7**:58, **7**:58*F*

Ribosomal RNA (rRNA) **2**:1040
aminoglycosides **2**:1041, **2**:1042*F*

antimicrobial agents **2**:1041
bacterial **2**:1041
eukaryotic **2**:1041
macrolides **2**:1042
pharmacology **2**:1041
antibacterial chemotherapy *see specific antibacterial agents*
physiological function **2**:1041
protein synthesis **2**:1041
rRNA domains, bacterial ribosomes **7**:526
structure **2**:1037, **2**:1040

Ribosome binding site (RBS; Shine–Dalgarno sequence), gene expression vectors **3**:109

Ribosomes
alterations, aminoglycoside resistance **7**:642
assembly blocking
ketolide antibiotics **7**:528–529
macrolide antibiotics **7**:528–529
binding
aminoglycosides **7**:638–639
tetracyclines **7**:614, **7**:615*F*
display **1**:435
antibody libraries **1**:435
phage display *vs.* **1**:435
reverse transcription **1**:435
interactions
chloramphenicol **2**:1042–1043
clindamycin **2**:1042–1043
linezolid **2**:1043
ketolide antibiotic binding site **7**:528
modifications, macrolide antibiotic resistance **7**:529
oxazolidinones interactions **2**:1042–1043
oxytetracycline interactions **2**:1042–1043
protection, tetracycline bacterial resistance **7**:616
structure, problems with **3**:543
tetracycline interactions **2**:1042–1043

Ribostamycin **7**:630–631

Riboswitches, messenger RNA (mRNA) **2**:1048

Ribozymes, transgenic animal generation **3**:152

Rib-X **1**:507

Richards, Graham, Ganellin, C R collaboration **8**:44

Richards boxes, molecular modeling **4**:67

Richardson, Jane, molecular modeling **4**:67–68

Ricinoleic acid, irritable bowel syndrome treatment **6**:654

Rickettsias, tetracyclines **7**:612

Ridogrel
ring equivalent bioisosterism **2**:655
structure **2**:655*F*

Rifabutin
  development 7:707
  mycobacterium treatment 7:705,
      7:705T
    characteristics 7:709T
    resistance 7:708
    structure 7:708F
Rifalazil
  mycobacterium treatment 7:707
    characteristics 7:709T
    resistance 7:708
    structure 7:708F
Rifampicin
  malaria treatment 7:795–796
  mycobacterium treatment 7:704,
      7:705, 7:705T
    characteristics 7:709T
    in HIV infection 7:701
    resistance 7:708
  as nuclear receptor ligand 2:1010
    structure 2:1026F; 7:708F
Rifamycin B, structure 2:46F
Rifamycins
  see also specific drugs
  classification, by target site 7:471T
  mycobacterium treatment 7:705,
      7:706F
    comparisons within class 7:708,
        7:709T
    drug-drug interactions 7:708
    future work 7:708
    limitations 7:708
    mechanism of action 7:705
    resistance 7:705–706, 7:708
    SAR 7:706, 7:707F
    structure 7:706F, 7:708F
Rifapentine
  mycobacterium treatment 7:705,
      7:705T
    characteristics 7:709T
    pharmacokinetics 7:707
    structure 7:708F
Rigel 1:506
Rigid analogs, azetidinone cholesterol
      absorption inhibitors see
      Azetidinone
Rigid body alignments, pharmacophore
      identification 3:243, 3:244F
Rigid searching, chemoinformatics,
      three-dimensional
      structure 3:242
RIKEN Structural Genomics Initiative
      (RSGI) 3:552T
Rilpivirine 7:347
  structure 7:364F
Rilutek see Riluzole
Riluzole 1:189; 6:240T
  amyotrophic lateral sclerosis
      treatment 6:223
  central nervous system injury 6:266
  Huntington's disease
      treatment 6:222T
  structure 6:238F

Rimantadine
  licensing 7:377
  mechanism of action 7:377
  orthomyxoviruses 7:274–275
  structure 7:377F
Rimexolone 1:218
Rimiterol, oral absorption 5:680T
Rimonabant
  alcohol addiction 6:184T
  clinical trials 6:405
  nicotine addiction 6:184T
  obesity treatment 6:405
  structure 6:181F; 6:405F
Ring/acyclic fragments, drug design, de
      novo see Drug design, de novo
RING_CONF program, 2D- to 3D
      structure 4:464
Ring equivalents, bioisosterism see
      Bioisosterism, ring equivalents
Ring-linker topology, hERG
      blockers 5:947, 5:947F
Ringsmi programs, screening library
      selection 4:498
Ring system reorganization,
      bioisosterism see Bioisosterism,
      ring system reorganization
Riotech 1:492T
Riquent see Abetimus sodium
RISC (RNA-induced silencing complex)
      see RNAi (RNA interference)
Risedronate sodium 1:220; 6:512
  Paget's disease 1:220
  structure 6:510F
  target information 4:750T
Risk, definitions 5:910
Risk assessment
  blood–placenta transfer
      modeling 5:737
  corneal permeability modeling 5:735
  definitions 5:910
Risk–benefit ratio, postmarketing
      surveillance 1:764
Risk factors
  cardiovascular disease 6:700, 6:701T
  clinical trials 2:181
Risk minimization, outsourcing 2:208
Risperidone 6:32
  cognition effects 6:36
  decomposition 5:502
  invention of, Colpaert,
      Francis 2:257–258
  mechanism of action 6:32
    GPCR targeting 2:772, 2:773T
  side effects 1:696, 1:713; 6:33
Risperidone receptors, imaging 3:667
Ritalin see Methylphenidate
Ritipenem acoxil 7:503
  structure 7:503F
Ritodrine, bioavailability 5:712T
Ritonavir 1:128; 2:579T; 7:279–7:282;
      8:187–198
  approval 7:349
  clinical uses 7:358

  combination treatment 7:358T
    atazanavir 7:358T
    with lopinavir see Kaletra
  CYP inhibition/induction 5:840;
      8:192–193, 8:193F
  discovery/development 1:76, 1:77F;
      7:349; 8:190
  dosing regimens 7:357T
  exclusivity period 1:532T
  HIV protease inhibitors vs. 8:190
  mechanism of action 7:349–350,
      7:350F
  mitochondria, effects on 2:1082
  phase III clinical trials 8:190–191
  resistance development 8:191–192
  serum protein binding 8:192
  single-dose studies 8:190
  structure 1:77F; 2:578F; 8:189F
    HIV protease binding 1:76, 1:78F
  structure-based design 4:602
    core element 4:602, 4:602F
    pseudosymmetry 4:602
  synthesis 8:191F
  target information 4:750T
Ritro see Flurithromycin ethylsuccinate
Rituvan see Rituximab
Rituxan see Rituximab
Rituximab 1:334T; 1:433T; 1:501;
      6:240T
  brands, global sales 1:605T
  cancer treatment 1:434
  clinical indication 6:245
  non-Hodgkin's lymphoma
      treatment 1:434, 1:444
  systemic lupus erythematosus
      treatment 6:248
  toxicity 7:46T
Rivanicline, nicotinic acetylcholine
      receptor binding 2:888
Rivastigmine tartrate 1:192
  Alzheimer's disease treatment 6:10,
      6:202, 6:203
  dementia with Lewy bodies
      treatment 6:36
Rizatriptan benzoate 1:183
  migraine treatment 6:373T
RL (reference location), UniProtKB
      flatfile 3:360
RMS deviation, pose prediction 4:266
RN (reference number), UniProtKB
      flatfile 3:360
RNA
  see also Messenger RNA (mRNA);
      Ribosomal RNA (rRNA)
  sizes 2:1037
  structure 2:1037, 2:1038F
RNA-based strategies, target selection
      see Target selection
RNA-binding proteins, cell-based
      assays 3:631
RNA constructs, gene therapy see Gene
      therapy
RNA duplexes, DNA 2:1040

RNA editing, glutamate-gated ion
 channel regulation 2:899
RNAi (RNA interference) 2:47,
 2:761–762, 2:1045, 2:1046; 3:17
 see also Short interfering RNAs
  (siRNAs)
 adenovirus delivery 3:17
 applications 3:178
  biomarker identification/
   validation 3:172
  disease links 3:180
  disease treatment *see Below*
  drug action validation 3:181
  ischemia reperfusion induced lung
   apoptosis 3:182
  kinase inhibition 3:17–18
  mitotic spindle defects 3:18
  ocular neovascularization 3:181–
   182
  stress responses 3:181
  target identification 3:172
  target validation 3:172, 3:181
 arrayed DNA-based siRNA
  libraries 3:18
 biological function 3:171
  viral infections 3:173
 Caenorhabditis elegans model
  system 3:8, 3:179
  advantages 3:179
  delivery 3:8, 3:179
  growth 3:179
  history 3:172
  obesity research 3:9
 depression 6:58
 DICER 1:313; 3:173, 3:174F
  discovery 3:172
  PAZ domain 3:173
  species homologs 3:173
  structure 3:173
 DNA methylation *see* DNA
  methylation
 Drosophila melanogaster 3:9, 3:179,
  3:651
  cytokinesis 3:179
  delivery 3:179
 endocytosis 3:17–18
 experimental controls 3:177
 future work 3:183
  cancer cell targeting 3:184
  lethal gene identification 3:184
 gene product inhibition *vs.* 3:17
 gene silencing detection 3:178
  flow cytometry 3:178
  immunofluorescence 3:178
  mRNA 3:178
  Northern blot 3:178
  protein 3:178
  reverse transcriptase polymerase
   chain reaction (RT-PCR) 3:178
  Western blot analysis 3:178
 genome-wide screens 3:179
  see also specific models
  cytokinesis 3:180

 expression libraries 3:180
  humans 3:180
 history 3:172
  clinical trials 3:172
  technology development 3:172
 hypoxia response 3:17–18
 interferon response 3:172, 3:176
 in vivo gene silencing 3:182
 lentivirus delivery 3:17
 limitations 3:176, 3:182
  cellular uptake 3:182
  nonspecific immune
   response 3:182
  off-target effects 3:178
  stability 3:182–183
  tissue distribution 3:182, 3:183
 molecular mechanisms 3:173,
  3:174F
  dsRNA dependent protein
   kinase 3:173
  message amplification 3:174
  2′-5′ oligoadenylate
   synthetase 3:173
  posttranscriptional gene
   silencing 3:173
 pharmacogenomics 1:347
 promoters 3:17
 reagent synthesis 3:18
 retroviral delivery 3:17
 RISC (RNA-induced silencing
  complex) 3:173, 3:174F
  activated 3:173
  composition 3:174
  discovery 3:172
 RNA-induced silencing complex
  (RISC) 2:1046
 short hairpin RNA (shRNA) 3:17
 siRNA 3:17
 target search genomics 3:17
 TRAIL inhibition 3:17–18
 transgenic animal generation 3:152
RNAi consortium (TRC), large-scale
 virus vector libraries 3:19
RNA-induced silencing complex
 (RISC) 2:1046
RNA maturation, antisense-based
 therapeutics 2:1047
RNA polymerase, hepatitis C virus *see*
 Hepatitis C virus
RNA polymerase II promoters,
 siRNAs 3:177
RNA polymerase III inhibitors,
 antifungal drugs *see* Antifungal
 drugs
RNA polymerase III promoters,
 siRNAs 3:177
RNA polymerases, mRNA
 synthesis 2:1043
 eukaryotes 2:1044
 prokaryotes 2:1043
RNA–RNA duplexes, ionization 5:384
RNAse-deficient *Escherichia coli*, cell-
 free expression systems 3:122

RNase H based cleavage, antisense-
 based therapeutics 2:1045,
 2:1046F
RNase H inhibitors, antivirals 7:365
RNA splicing, messenger RNA (mRNA)
 structure 2:1044
RNA structure prediction,
 bioinformatics 3:323F, 3:327
 base-paring patterns 3:327
 comparative sequence analysis 3:327
 homology-based techniques 3:327
 secondary structure motifs 3:327
 thermodynamics 3:327
RNA synthesis inhibitors, antibacterial
 agents 7:733
RNA viruses **7:329–371**
Ro 04-2843, structure 2:1062F
Ro 06-9075, malaria treatment 7:800T
Ro 09-1428 7:492–493
 activity 7:493T
 structure 7:494F
Ro 09-2210 7:204
 structure 7:204F
Ro 12-9424, dual-action
 antibiotic 7:503, 7:510F
Ro 22-8014, malaria treatment 7:800T
Ro 31-8220 7:209–210
 structure 7:209F
Ro 41-0986 7:463–464
Ro 47-1317, structure 7:512F
Ro 48-1220 7:512
 structure 7:511F
Ro 60-0175 6:76
 structure 6:414F
Ro 60-0332 6:76
Ro 63-0563 6:76
Ro 64-0802
 oral absorption 5:1014
 structure 5:1015F
Ro 64-6198 6:412
 structure 6:412F
Ro 68-3400
 mitoprotective activity 2:1062
 structure 2:1062F
Ro 68-3406, structure 2:1062F
Robotic liquid-handling, materials
 management 2:536
Robotic weighing, materials
 management 2:536
Robustaflavone, hepatitis B
 antivirals 7:320
Roche 1:558
 R&D spending 2:101T
 sales/R&D expenditure 1:542T
 SME acquisition 1:496T
Rocky Mountain Spotted Fever
 (RMSF), tetracycline
 treatment 7:612
ROCS
 ligand-based library design 4:317
 pharmacophore 3D database
  searching 4:139T, 4:142
Rocuronium bromide **1:191**

anesthetic adjuvants 6:362
structure 6:363F
ROD-188, structure 2:893F
Rodent models
see also specific diseases/disorders; specific
models
antibacterial agents 7:677–678
asthma models 7:880–881
COPD models 7:880–881
nervous system autoimmune
diseases 6:237
Rofecoxib 1:215; 6:339
Alzheimer's disease treatment
6:211
arthritis treatment 7:853
hydrolysis 5:142
introduction of 1:108
prostatitis treatment 6:571
safety 1:684, 1:690, 1:712
structure 6:336F
target information 4:750T
withdrawal 1:106–107, 1:757T
Roferon-A 1:499
Roflumilast 7:893
analogs
fluorine-hydrogen
bioisosterism 2:692, 2:693F
structure 2:935
clinical trials 7:882
COPD treatment 7:882
IBD treatment 6:626
Rohitukine
discovery/development 1:389
structure 1:388F
Role models, effective leaders 2:241
Role/responsibility definitions,
collaborative project
management 2:151
Rolipram 2:935; 7:893
animal models, cognition
2:926–927
aspirin-induced mucosal injury 2:948
benzimidazole ring analogs 7:894
carbachol derivatives 7:894
clinical evaluation 2:935–936
clinical trials 7:882
COPD treatment 7:882
IBD treatment 6:625
oxindole series 7:894
phosphodiesterase-4 affinity 2:925–
926
Rolitetracycline see Pyrrolidinylmethyl
tetracycline (29)
Roller bottles, protein
fermentation 3:418
Rolliniastatin-1, mitochondria complex
I 2:1068–1069
Rome criteria, irritable bowel syndrome
clinical trials 6:652
Rondomycin see Methacycline
RONN, domain mapping 3:879
Röntgen, Wilhelm Conrad, protein
crystallography 3:449–450

Room temperature
compound storage 3:567
crystal handling 3:467
Room temperature ionic liquids,
microwave-assisted
chemistry 3:854
Root mean square distance (RMSD)
conformational analysis/
searching 4:108, 4:108F
x-ray structures, problems 3:538
Root mean square error (RMSE)
aqueous solubility models 5:633T,
5:643–644
blood–brain distribution 5:751–752
solubility models 5:631
Ropinirole hydrochloride 1:180
Parkinson's disease treatment 6:11
Ropivacaine 1:193; 6:361
structure 2:716F; 6:354F
synthesis 2:716F
Roscovitine see CYC202 (roscovitine)
ROSETTA, protein structure
prediction,
bioinformatics 3:327
Rosiglitazone maleate 1:229
characteristics 2:1012T
IBD treatment 6:630
labeling changes 1:758T
mechanism of action 2:1081
mitochondria complex I 2:1081–
1082
as PPARγ ligand 2:1021–1022
metabolism, CYP 2C8 5:547T
PASP 3:802, 3:802F
structure 2:1023F; 2:1081F; 5:799F;
6:440F
target information 4:750T
Rostaporfin
structure 6:307F
wet age-related macular degeneration
treatment 6:306T
Rosuvastatin 6:473–474, 6:476, 6:484
adverse effects 6:477
introduction of 1:107
SAR 6:484, 6:484F
structure 2:650F; 6:469F
target information 4:750T
Rotamer libraries, ligand–receptor
interaction models 4:207
Rotating disk apparatus, dissolution see
Dissolution
Rotational degrees of freedom, ligand–
receptor interaction models see
Ligand–receptor interactions,
models
Rotational–entropy, ligand–receptor
interactions 4:201–202
Rotational speeds, Woods' rotating disk
apparatus 5:415
Rotational treatment, psoriasis
treatment 7:978T, 7:982
Rotation component, molecular
replacement 3:462

Rotenone
mitochondria complex I
2:1068–1069
Parkinson's models 6:212–213
structure 2:1069F
Rotigotine 1:503–504
Parkinson's disease treatment 6:215
Rous sarcoma virus promoter,
mammalian protein expression
systems 3:117
Rous sarcoma virus protease, HIV-1
protease vs. 3:508–509
Route design, process outcome 2:213
Routes of administration
biological macromolecules 1:438
chemical vs. biological 2:117
Route validation, solution phase library
synthesis see Solution phase
synthesis, library preparation
Roxithromycin 7:539
pharmacokinetics 7:534T
Royal College of Surgeons rat, dry age-
related macular degeneration
models 6:304
Royal Marooned studies,
tamoxifen 8:95
RP (reference position), UniProtKB
flatfile 3:360
RP-67580 6:110
structure 6:110F
RPI-4610 (Angiozyme) 7:200
RPMI 2650, in vitro nasal
administration models
5:284–285
RPMI cells, protein expression
systems 3:117
RPR 101511A 7:198
structure 7:198F
RPR 119990 6:224
RPR196972 7:662–664
RPR202868 7:663T
RS-102221, structure 6:414F
RSCB PDB 3:547T
RSD 921 6:753–754
RSD 1000 6:753–754
RSD1019 6:753–754
RSD1030 6:753–754
RSD 1235 1:508; 6:753–754
development 6:743F, 6:758
SAR 6:759–760
RSR 13 4:71
RT (reference title), UniProtKB
flatfile 3:360
RTI 336, structure 6:187F
RTK1, xenograft models 7:39T
RU-486 6:548
RU-59863 7:492–493
activity 7:493T
structure 7:494F
RUBICON program, 3D structure
generation 4:464, 4:465
Rubidium-86, cell-based
electrophysiology assays 3:625

Rubin, Byron, molecular modeling **4**:67–68

Rubitecan **7**:114

Rubredoxin, structure, coordinate errors **3**:532

Ruffigallol, malaria treatment **7**:800*T*

Ruffled border formation, bone resorption/remodelling **8**:206

Rufin *see* Pantoprazole sodium

Rufinamide **6**:290
   structure **6**:290*F*

Rufloxacin **7**:569*T*
   structure **7**:570*F*, **7**:577*F*

Rugosity *(R)*, VolSurf descriptors **5**:538

Rule-based machine learning, constraint satisfaction programming (CSP) **4**:771

Rule-based methods
   conformational analysis/searching *see* Conformational analysis/searching
   P-glycoprotein models **5**:780
   toxicity, computational models *see* Toxicity, computational models

Rule Discovery System, aqueous solubility models **5**:643, **5**:644*T*

'Rule-of-five' **4**:438; **5**:671–672, **5**:700
   absorption/permeability/transport prediction **5**:260
   chemogenomics **4**:422
   computational filters **3**:258
   drug development **2**:730
   drug discovery **1**:332
      decline in **1**:620
      fragment-based **3**:943–944
   druggability **3**:2
   library design **4**:364
   lipophilicity **5**:426
   molecular descriptors **5**:533
   molecular weight **5**:684
   preclinical trials **2**:178
   screening library selection **4**:498–499

'Rule-of-three'
   adverse drug event frequency estimation **1**:766
   fragment-based drug discovery **3**:943–944
   structure-based fragment screening **4**:793

Rupafin *see* Rupatadine fumarate

Rupatadine fumarate **1**:206
   perennial rhinitis **1**:207
   seasonal rhinitis **1**:207

Ruprintrivir, rhinoviruses **7**:266

Russell–Rao coefficient, similarity searching **4**:179, **4**:181

Rutin **1**:452*T*
   pharmacokinetics **1**:454*T*

RV2002, protein crystallization **3**:888

R-value, protein 3D structure validation **3**:519

RWJ-51204 **6**:104–106
   structure **6**:105*F*

RWJ-54428 **7**:497
   activity **7**:496*T*, **7**:497*T*

RWJ-270201, orthomyxoviruses **7**:275–276

RWJ-442831, structure **7**:495*F*

RX (reference cross-reference), UniProtKB flatfile **3**:360

RX-77368, Alzheimer's disease treatment **3**:207

(D/E)RY motif, GPCR structure **2**:774–777

S

S4 domain, potassium channels **2**:858–859

S6 helix bending, voltage-dependent K^+ ion channels **4**:716–717, **4**:717*F*

S9, reaction phenotyping **5**:242*T*

S-11, structure **1**:393*F*

S-12, structure **1**:393*F*

S-2242, poxviruses **7**:261–262

S-3578
   activity **7**:496*T*, **7**:497*T*
   structure **7**:495*F*

S-16020-2 **7**:118, **7**:123–124

SA, scoring/binding energy estimation drug design **4**:295

SAAA, duloxetine clinical trials **8**:128, **8**:129*F*

SAAB, duloxetine clinical trials **8**:129, **8**:130*F*

Saccharin, oral absorption **5**:680*T*

Saccharomyces cerevisiae
   genome sequencing **3**:317*T*, **3**:647–649
   proteome complexity **3**:29*T*

Saccharomyces cerevisiae model system **3**:5, **3**:647
   cell cycle control **3**:5–6
   forward genetics **3**:5–6
   gene deletion combinations **3**:6–7
   gene function assessment **3**:6
   genome resources **3**:6, **3**:6*T*
   hsp90 identification **3**:5–6
   human genome functional study model **3**:8
      Bax repressor identification **3**:8
      GPCR expression **3**:8
      ligand identification **3**:8
   induced haploinsufficiency (chemical genomics) **3**:7
      limitations **3**:7
      target identification **3**:7
      target validation **3**:7
   limitations **3**:8
   mitogen activated protein kinase (MAPK) signaling **3**:5–6
   random mutation screening **3**:5–6

reverse genetics **3**:6, **3**:7*F*
   DNA repair mechanisms **3**:6
   mitochondrial function **3**:6
   polyglutamine disease **3**:6
   viral replication **3**:6
   'synthetic lethality' **3**:6–7
   'three-hybrid system' **3**:8
      cyclin-dependent kinase inhibitor identification **3**:8
      methotrexate target identification **3**:8
   transcriptional profiling studies **3**:6
   two-hybrid assay **3**:6

Saccharomyces Genome Deletion Project **3**:139, **3**:140*F*

Saccharomycin **7**:634

Safe laboratory conduct, career guidelines **2**:18

Safety
   antibacterial agent assays **7**:677–678
   biological macromolecules **1**:438, **1**:444
   definitions **5**:910
   drug development costs **2**:727
   drug discovery **2**:2, **2**:11
      decline in **1**:621–622, **1**:623
   drug metabolism, in vitro studies **5**:238
   microwave-assisted chemistry scale-up **3**:856–857
   monitoring **2**:8–9
   PASP **3**:793
   pharmacogenetics **3**:64
   prediction **5**:4
   regulatory problems **2**:725
   traditional medicine *see* Traditional medicine

Safety, health and environment information, drug development **2**:722–723, **2**:723*F*

'Safety-catch' linkers, solution phase synthesis **2**:489–490, **2**:490*F*

Safety ratio definition, drug metabolism, in vitro studies **5**:238

Safety studies
   early, importance of **5**:3
   metabolism studies **5**:34

Safinamide **1**:511; **6**:291
   structure **6**:290*F*

SAHA **1**:513

Salacia oblonga, blood sugar/insulin levels **1**:327–328

Salbutamol
   bioavailability **5**:712*T*
   discovery/development **1**:328–329

Salicylaldoxime-based ligands, triphenylethylene anticancer agents **7**:159–160, **7**:160*F*

Salicylic acid
   acne treatment **7**:966

amino acid-coenzyme A
      conjugation **5**:155, **5**:156*F*
   bioavailability **5**:712*T*
   oral absorption **5**:468*T*, **5**:680*T*
   psoriasis treatment **7**:944
Salinosporamide A **1**:361
   discovery/development **1**:384
   structure **1**:360*F*
Saliva, excretion **5**:16–17
Salmedix **2**:767
Salmeterol
   asthma treatment **7**:883
   labeling changes **1**:758*T*
   PASP **3**:802–804, **3**:803*F*
Salmeterol xinafoate, pulmonary
      administration **5**:281*T*
Salmon calcitonin *see* Calcitonin
Salmonella infections, irritable bowel
      syndrome **6**:650
Salsolidine, *N*-methylation **5**:146–147,
      5:146*F*
Salt forms
   drug development **2**:27, **2**:27*T*
   solid-state physicochemistry *see* Solid-
      state physicochemistry
Salt tolerance, NMR probes **2**:515–
      516
Salvage treatment, antivirals **7**:337
Salvarsan *see* Arsphenamine
SAM algorithm, microarrays **3**:98–99
Sample amounts, screening library
      selection **4**:500
Sample analysis
   preparation
      isoelectric focusing (IEF) (first
         dimension) **2**:332
      microarrays **3**:97
      single nucleotide
         polymorphisms **2**:363
Sample classification, microarrays **3**:93
Sample sizes, adverse drug event
      frequency estimation **1**:766
Sample volumes, NMR probes **2**:516,
      2:516*F*
Sancycline
   derivatives **7**:608
   D-ring chemistry, electrophilic
      aromatic substitutions **7**:607,
      7:607*F*
   minocycline synthesis **7**:607–608
   structure **7**:601*F*
   synthesis **7**:601
   transition-metal catalyzed
      reactions **7**:610
Sandoz, research reporting **1**:711–712
Sandwich enzyme immunoassay, cell-
      based assays **3**:631
Sandwich enzyme-linked
      immunosorbent assay **3**:579–
      580, **3**:579*F*
Sanfetrinem
   activity **7**:509*T*
   structure **7**:509*F*

Sangamo Biosciences **1**:516
Sanger, Fred, x-ray
      crystallography **4**:66
Sannamycin **7**:632
Sanofi-Aventis
   pipeline of products **1**:542
   R&D spending **2**:101*T*
   sales/R&D expenditure **1**:542*T*
   SME acquisition **1**:496*T*
Saperconazole
   mechanism of action **8**:25
   medicinal chemistry **7**:451
Saphenamycin analogs, solid-phase
      discrete libraries **3**:709*T*
Saponin, promiscuous ligands **2**:740
Sapresta *see* Aranidipine
Saquinavir mesylate **1**:128; **2**:579*T*;
      7:279–282
   design/development **7**:349
   peptidomimetics **2**:613
   dosing regimens **7**:357*T*
      with ritonavir **7**:358*T*
   exclusivity period **1**:532*T*
   mechanism of action **7**:349, **7**:350*F*
   physiologically based
      pharmacokinetics **5**:881
   structure **1**:75*F*; **2**:578*F*
   structure-based design **4**:604
      transition-state mimetics **4**:604,
         4:605*F*
   structure-based drug design **4**:77*T*
   target information **4**:750*T*
Sarcodictyins **1**:364–365; **7**:102
   structure **1**:364*F*; **7**:103*F*
Sarcoma, definition **7**:2, **7**:55
Sarcosine, schizophrenia
      treatment **6**:38
Sargramostim, IBD treatment **6**:623–
      624
SARs *see* Structure–activity relationships
      (SARs)
SARS (severe acute respiratory
      syndrome)
   antiviral drugs **7**:254*T*, **7**:271
      bananin **7**:272
      chloroquinine **7**:273–274
      eubananin **7**:272
      glycyrrhizin **7**:273–274
      interferons **7**:274
      iodobananin **7**:272
      MDL 28170 **7**:271
      nelfinavir **7**:273–274
      niclosamide **7**:273–274
      peptidomimetic compounds **7**:272
      rational drug design **7**:272
      valinomycin **7**:273–274
      vanillinbananin **7**:272
   NTPase/helicase **7**:272
   RNA-dependent polymerase **7**:272–
      273
   siRNAs **7**:274
Saruplase
   animal models **6**:767

clinical trials **6**:774
   COMPASS **6**:774
Satraplatin **7**:143
   clinical indication **7**:143–144
   dose **7**:143–144
   mechanism of action **7**:143*F*
   as prodrug **5**:1033
   structure **5**:1032*F*; **7**:143*F*
Saturated fat diet, breast cancer **7**:151
Saturation transfer difference (STD)
      experiments, NMR, drug
      discovery **3**:908*T*, **3**:912,
      3:912*F*, **3**:913*F*, **3**:914*F*
SAVE, number of patients needed to
      treat **8**:144*T*
Saw palmetto *(Serenoa repens)* **1**:426
SB-27701-A, structure **6**:187*F*
SB-42920 **7**:229
SB-171046 **6**:76
SB-203580
   chemical proteomics **3**:44
   structure **2**:979*F*
   TGFβR kinase inhibitors **2**:977–979
SB-224289 **6**:75–76
SB-238569, structure **7**:493*F*
SB-242084 **6**:413
   structure **6**:414*F*
SB-242235, IBD treatment **6**:627–628
SB-243213 **6**:76
SB-331750 **6**:515
SB-334867 **6**:410–411
   structure **6**:411*F*
SB-366791, structure **6**:329*F*
SB-379872 **7**:229
S*BIO **1**:512
Scaffold-based design **4**:22
   ADME prediction **4**:35
      cytochrome P450 enzymes **4**:35
   alpha helices **4**:28, **4**:29*F*
      Bak BH3 domain **4**:31, **4**:31*F*,
         4:32*F*
      Fuzeon (enfuvirtide) **4**:31, **4**:32*F*
      p53/Hdm interaction **4**:30–31
      preorganization **4**:30
      problems **4**:29–30, **4**:29*F*
      ribonucleases **4**:30, **4**:30*F*
      thyrotropin-releasing
         hormone **4**:30–31
   beta-sheets/beta-turns **4**:28
      HIV protease **4**:28, **4**:28*F*
   combinatorial chemistry **4**:31
      diversity **4**:34
      screening prioritization **4**:34
   entropy reduction **4**:26
      peptidomimetics **4**:28
      preorganization **4**:26–28
   failures **4**:35
   genome-wide modeling **4**:34
   genomics **4**:22–23
   HTS **4**:31, **4**:34
      solid phase chemistry **4**:34
   phosphodiesterase-4 inhibitors **2**:938
   privileged organic scaffolds **4**:32

Scaffold-based design (*continued*)
α-Amylase inhibitors 4:33–34
benzodiazepines 4:32–33,
4:33*F*
disadvantages 4:33
integrin inhibitors 4:33–34
substance P antagonist 4:33–34
protein crystallography, rational drug
design 3:892
protein engineering 4:24
bacterial receptors 4:25, 4:25*F*
cysteine knot proteins 4:25, 4:26*F*,
4:27*F*
enzyme, conversion to 4:25, 4:25*F*
examples 4:24
ligand activation 4:25
zinc finger flexible loops 4:25
proteomics 4:22–23
recognition 'hot spots' 4:24
charged groups 4:24
protein–protein interactions 4:24
similarity 4:34
solution phase library
synthesis 3:779, 3:779*F*,
3:780*F*
structure-based fragment
screening 4:792
Scaffold classes, GPCR drug
design 4:367
Scaffold decoration, PASP 3:811
Scaffold hopping
bioisosterism *see* Bioisosterism,
scaffold hopping
two-dimensional descriptors, similarity
searching 4:171
Scaffold–linker–functional group,
focused library design, structure-
based methods 4:329
Scaffold prioritization, protein kinase
library design 4:370
Scaffold set, library design 4:309
Scalable methodologies, chiral
switches 2:729–730
Scalar coupling, NMR structure
determination *see* Nuclear
magnetic resonance (NMR),
structure determination
Scalar fields, molecular fields 5:580
Scalar objectives, multiobjective
optimization *vs.* 4:767
SCALE record, protein 3D structure
validation 3:511
Scale-up
dissolution 5:402
drug development
productivity 2:130
microwave-assisted chemistry *see*
Microwave-assisted chemistry
outsourcing *see* Outsourcing
Scaling, protein crystallography 3:468–
470
SCAMPI, pharmacophore
searching 4:129*T*

Scattered light intensity, light
scattering, promiscuous
ligands 2:743
Scattering factor, x-ray–crystal
interaction 3:452–453,
3:455*F*
Scatterplots, IsoStar (intermolecular
interactions) 3:405
Scavenging resins, solution phase
synthesis *see* Solution phase
synthesis
Scavenging techniques, PASP *see*
Polymer-assisted solution phase
synthesis (PASP)
Scedosporium infection, as emerging
disease 7:425–426
SCH 6636 7:234
SCH 28080, structure 6:610*F*
SCH 33256
biosynthesis 7:602
structure 7:601*F*
SCH 34164
biosynthesis 7:602
structure 7:601*F*
SCH 37370 7:234
SCH 45012 *see* Posaconazole
SCH 48461 *see* Ezetimibe
SCH 417690
discovery/development 1:79,
1:80*F*
structure 1:80*F*
Schering-Plough
sales/R&D expenditure 1:542*T*
SME acquisition 1:496*T*
Schistosoma see Schistosomiasis
Schistosomiasis 7:751*T*
biology 7:760
clinical manifestations 7:761
drug treatment 7:753*T*,
7:761
epidemiology 7:761
Schizophrenia 6:2, 6:17–44
comorbid substance abuse 6:39
competence and consent
issues 1:698
disease basis 6:4, 6:19
clozapine hypothesis 6:22
dopamine hypothesis *see*
Schizophrenia, dopamine
hypothesis
epigenetics 6:19–20
GABA hypothesis 6:21
genetic factors 6:19–20
glutamate hypofunction hypothesis
see Schizophrenia, glutamate
hypofunction hypothesis
phosphodiesterase-10 2:930–931
serotonin hypothesis 6:20
subtypes 6:19–20
disease state/diagnosis 6:18
cognitive dysfunction 6:18
comorbid mood disorders 6:19
DSM-IV-TR 6:2–4, 6:18

Measurement and treatment
Research to Improve Cognition in
Schizophrenia
(MATRICS) 6:19
negative symptoms 6:18
positive symptoms 6:18
dopamine hypothesis 6:4, 6:20
dopamine D2 receptor 6:20
genetic models 6:28
knockout animal models 6:28
genetics 6:22
CAPON 6:24
CHRNA7 6:24
COMT 6:23
DISC1 6:24
dopamine D3 receptor 6:24
dysbindin-1 6:23
family studies 6:22–23
G72 6:24
GABA_B receptor 6:24
glutamate receptors 6:24
neuregulin 6:23
α7-neuronal nicotinic receptor gene
polymorphisms 6:39
RGS4 6:23
serotonin receptor 6:24
glutamate hypofunction
hypothesis 6:4, 6:20
induced psychosis evidence
6:20–21
incidence 6:2–4, 6:18
lesion models 6:27
symptom delay 6:27
linkage studies, proteomics 3:43
models 6:24
see also specific models
behavioral assays 6:25
conditioned avoidance 6:25
construct validity 6:25
face validity 6:25
latent inhibition 6:25
locomotor activity 6:25
positive symptoms *vs.* negative
symptoms 6:28
prepulse inhibition 6:26
personalized medicine *see* Personalized
medicine
pharmacological models 6:26
dopamine releasing agents 6:26
drug efficacy evidence 6:27
NMDA receptor
hypofunction 6:26–27
serotonin hypothesis 6:26–27
prognosis 6:4
social isolation models 6:27
construct validity 6:28
symptoms 6:2–4, 6:18
Schizophrenia treatment 6:4, 6:29
antipsychotic drugs *see* Antipsychotic
drugs
clinical trials 6:28
Abnormal Involuntary Movement
Scale 6:29

Brief Psychiatric Rating Scale
(BPRS) **6**:29
exclusions **6**:28
Positive and Negative Syndrome
Scale (PANSS) **6**:29
sample sizes **6**:28
Simpson–Angus Experimental
Scale **6**:29
washout period **6**:28–29
dopamine receptor modulators/partial
agonists **6**:34
receptor reserve system
variations **6**:34
future work **6**:40
glutamatergic agents **6**:37
N-acetyl-L-glutamate **6**:38
muscarinic agonists **6**:36
N-desmethylclozapine **6**:37
neurokinin₃ antagonists **6**:40
neuronal nicotinic receptor
agonists **6**:39
new research areas **6**:36
side effects
metabolic syndrome **6**:35
QT prolongation **6**:33, **6**:35
unmet medical needs **6**:35
broader efficacy **6**:36
reduced side effects **6**:35
Schlavo, mergers and
acquisitions **1**:541–542
Schrödinger equation, quantum
mechanical calculations *see*
Quantum mechanical
calculations
Schultz, Theodore, universities in
commerce **2**:93
Schwartz, Jean-Charles, collaborators,
Ganellin, C R **8**:48
SCI (Science Citation Index) **3**:271T,
3:274
searching **3**:274
SCID transplant models, psoriasis
animal models **7**:971
Science, marketing, replacement
by **2**:284
Science integration, education **2**:88–89
Scientific Patent Support,
Lindberg, P **8**:220
Scientific training, leadership *see*
Leadership
SciFinder Scholar, database
interfaces **3**:283
Scintillation proximity assay
(SPA) **3**:584
bead type **3**:584
fluorescence polarization (FP)
vs. **3**:590
GPCRs **3**:585–586
adenosine 2a receptor **3**:586
histamine receptors **3**:586
homogenous time-resolved
fluorescence energy transfer
(HTR-FRET) *vs.* **3**:590

human cytochrome P450 s **3**:589
iodine-125 **3**:584
kinases
HTS **3**:587
protein kinase A **3**:587
long-chain fatty acyl coenzyme A
synthase **3**:588
radioisotope characteristics **3**:584
signal measurement **3**:584
charge-coupled device
(CCD) **3**:584
photomultiplier tube **3**:584
target-oriented screening **3**:142–143
technique **3**:584, **3**:585F
translation factors **3**:588
initiation factors (IFs) **3**:588
tritium **3**:584
Scitegic circular fingerprints, two-
dimensional descriptors,
similarity searching **4**:171
SCN1A gene, migraine **6**:370
SCOP (Structural Classification of
Proteins) **3**:374
protein function prediction,
bioinformatics **3**:336–337
search capabilities **3**:376–377
Scopolamine
as antiemetic agent **6**:683
bioavailability **5**:712T
oral absorption **5**:680T
Scopus **3**:271T, **3**:274
SCORE system, sleep disorders **6**:152
Scoring/binding energy estimation drug
design **4**:295
empirical scoring **4**:295
definition **4**:295
energy landscapes **4**:296
force-field methods **4**:295
knowledge-based **4**:295
definition **4**:295
ligand-based design **4**:296
physical chemistry **4**:295
GB **4**:295
linear interaction energy (LIE)
method **4**:295
MM **4**:295
MM/PBSA **4**:295
PB **4**:295
SA **4**:295
Scoring functions, structure-based
virtual screening **3**:251
Scoring system, bioavailability
models **5**:721
Scotland, Archangel syndicate **1**:564
Screening
see also specific assays; *specific screens*
animal experimentation
alternatives **1**:464
cell-based *see* Cell-based assays
chemogenomics **4**:423,
4:427F
fluorescence screening assays *see*
Fluorescence screening assays

pre-equilibrated dynamic
combinatorial libraries **3**:963–
964, **3**:964F
protein crystallization **3**:439
small animal test systems *see* Small
animal test systems
target-oriented *see* Target-oriented
screening
targets *see* Target-oriented screening
Screening cascade
drug discovery **2**:565, **2**:565F
hit-to-lead **2**:139
Screening Information Data Set
(SIDS) **1**:57T
Screening library selection **4:495–513**
analysis **4**:506
assay validation effects **4**:506
early development **4**:506
full file screen **4**:507
retrospective follow-up **4**:503,
4:507, **4**:509F
assessment methods **4**:495
compound numbers **4**:442, **4**:443F
definition **4**:495
examples **4**:497
iterative follow-up **4**:506
definition **4**:507
goals **4**:507, **4**:508F
ligand efficiency **4**:503, **4**:505,
4:509F
methods **4**:501
algorithms **4**:501
clustering **4**:502–503
collection nature **4**:501
Diverse Solutions (DVS) **4**:504
graph-based analysis **4**:502–503
hierarchical clustering **4**:503
individual combinatorial
libraries **4**:505
maximally diverse selection
model **4**:504
maximum auto cross correlation
(MACCS) **4**:502–503
molecular weight clustering **4**:503
nonhierarchical clustering **4**:503
Optism **4**:504
Pipeline Pilot **4**:505
predicted performance
measures **4**:506
random selection **4**:501–502
representative subsets **4**:502
subset modification **4**:504
reasons for **4**:496
sampling, amount of **4**:443, **4**:444F
subset composition **4**:496, **4**:497F
availability **4**:497
capacity issues **4**:500
clustering methods **4**:498
'druglikeness' **4**:498–499
dry powder samples **4**:497
external sources **4**:498
follow-up effects **4**:500
Murko fragment analysis **4**:498

Screening library selection (*continued*)
  non-ADME compounds 4:499
  preparative work 4:496
  quality assessment 4:501
  ringsmi programs 4:498
  'rule-of-five' 4:498–499
  sample amounts 4:500
  screening technology
    effects 4:501
  single series 4:498
  structural filters/alerts 4:499–500
  triage 4:509
    exempts 4:509–510
    3σ level 4:510, 4:510*F*,
      4:511*F*
Screening set design, fragment-based
  drug discovery *see* Fragment-
  based drug discovery
Screening technology effects, screening
  library selection 4:501
Screening test design,
  chemoinformatics, diversity
  analysis 3:252–253
Screening window coefficient,
  HTS 3:684
Scripps Research Institute 1:565
  collaborations 1:711–712
Scriptaid 7:224–225
SD-2038580, IBD treatment 6:627–
  628
SDBS 3:280*T*
SDF-1 alpha receptor,
  activation 2:785*T*
SDR superfamily, carbonyl reductases
  (CBRs) 5:178
SDS-PAGE
  protein analysis 3:424, 3:425*F*
  protein purity determination 3:881
  two-dimensional gel
    electrophoresis 2:332–333
SEA0400 6:756
SEAL program, 3D-QSAR 4:582
Search engine design, structural
  databases 1:44
Search factors, drug design, de
  novo 4:285
Search-generated conformations, 3D
  database pharmacophore
  searching 4:141
Searching, fragment/ligand selection
  drug design 4:296
Search Ligands, PDB searches 3:377
Search Sequence, PDB searches 3:377
Search tree generation, CSD 3:396
Search Unreleased, PDB
  searches 3:377
Searle, mergers and acquisitions 1:541
Seasonal rhinitis, rupatadine
  fumarate 1:207
Sebum production, acne 7:936,
  7:936*T*, 7:958, 7:958*F*
Secloramine 6:67–68
Secobarbital, hydrolysis 5:142, 5:144*F*

Secondary carbons, cytochrome P450
  oxidation 5:90
Secondary databases *see* Databases
Secondary literature databases 3:270
Secondary metabolism, traditional
  medicine *see* Traditional
  medicine
Secondary progressive multiple sclerosis
  (SPMS) 7:923
Secondary structure, protein structure
  prediction *see* Protein structure
  prediction
Secondary structure mimetics,
  peptidomimetics 2:608, 2:610*F*
Secondary structure motifs, RNA
  structure prediction,
  bioinformatics 3:327
Second International Study of Infarct
  Survival *see* ISIS-2
Second messengers
  potassium channels 2:861
  sodium channels 2:836
Second principal component, VolSurf
  descriptors 5:540–541
Secramine 1:366
  structure 1:364*F*
β-Secretase, Alzheimer's disease
  6:209
γ-Secretase, Alzheimer's disease 6:209
Secretase inhibitors
  see also specific drugs
  Alzheimer's disease
    treatment 6:207, 6:208*F*
Secretin 2:575–576
  as peptide/protein drugs 2:579*T*
Secretin receptors 2:778
Security, collaborative project
  management 2:151
Sedatives
  see also specific drugs
  clonidine side effects 6:134
  introductions 1:196
Sediel *see* Tandospirone
Seed, Brian, cDNA functional
  screening 3:14
'Seed and soil' hypothesis,
  metastasis 7:19
SEED approach, site analysis, drug
  design 4:288
Seizures, epilepsy 6:13
SELDI *see* Surface-enhanced laser
  desorption ionization (SELDI)
Selection conditions, dynamic
  combinatorial libraries 3:960
Selection markers
  gene expression vectors *see* Gene
    expression vectors
  mammalian protein expression
    systems 3:117
  yeast protein expression
    systems 3:117
Selection target presence, dynamic
  combinatorial libraries 3:963

Selective absorption, intestinal
  absorption 5:263–264
Selective adhesion molecule inhibitors,
  nervous system autoimmune
  diseases, treatment 6:243
Selective androgen receptor modulators
  (SARMs) 2:1016
  male hypogonadism 6:538
Selective estrogen receptor
  downregulators
  (SERDs) 2:1014
Selective estrogen receptor modulators
  (SERMs) 2:1014; 7:163
  see also specific agents; specific drugs
  benzothiophenes 8:104–106
  bone disease treatment 6:512,
    6:513*F*
  definition 7:154
  development 8:104–106
  estrogen receptor-α-selective
    ligands 7:165
    chromane scaffold 7:165,
      7:166*F*
    dihydrobenzoxanthin
      scaffold 7:165, 7:166*F*
    flavone scaffold 7:165, 7:166*F*
  estrogen receptor-β-selective
    ligands 7:165, 7:166*F*
    aryl benzothiophenes 7:166–167,
      7:167*F*
    benzimidazole analogs 7:169,
      7:169*F*
    chromane derivatives 7:169,
      7:170*F*
    diarylpropionitrile 7:167, 7:168*F*
    genistein analogs 7:168–169,
      7:168*F*
    2-Phenylnaphthalenes 7:169,
      7:171*F*
    tetrahydrofluorenones 7:169,
      7:170*F*
  mechanism of action 7:174
  naphthalene-sulfone
    derivatives 7:164–165, 7:165*F*
  steroid-derived modulators 7:170
    16α-Carboxylic esters/
      acids 7:172–173, 7:172*F*
    electrophilic group
      attachment 7:172, 7:172*F*
    equilenin benzothiophene
      analogs 7:173–174, 7:174*F*
    receptor homology
      modeling 7:173, 7:173*F*
    7α-thioestratrienes 7:171,
      7:171*F*
  tamoxifen *see* Tamoxifen
  tetrahydrochrysene
    derivatives 7:164, 7:164*F*
  tissue selectivity 7:174–175
  triphenylethylenes 8:104–106
Selective GABA reuptake inhibitors,
  primary insomnia
  treatment 6:161

Selective gene expression repression, Tet-On-Tet-Off operons 7:622F
Selective heating, microwave-assisted chemistry 3:842
Selective norepinephrine reuptake inhibitors (SNRIs) 6:69
see also specific drugs
Selective optimization offside activities (SOSA), scaffold hopping 2:683
Selective orexin antagonists, narcolepsy unmet medical needs 6:163
Selective repression, Tet-On-Tet-Off operons 7:622
Selective serotonin receptor antagonists, pH dependency 5:383
Selective serotonin reuptake inhibitors (SSRIs) 2:62; 6:56–57, 6:59F, 6:69
see also specific drugs
analgesia 6:339–340
dopamine transporter knockout mouse 6:124
efficacy 6:70
irritable bowel syndrome treatment 6:655T, 6:657, 6:660
marketing issues 1:720
mechanism of action 2:67; 6:51, 6:76
cytochrome P450 inhibition 6:70
GPCR targeting 2:772
obsessive compulsive disorder (OCD) treatment 6:99–100
panic disorders treatment 6:99–100
potency 6:70
premature ejaculation 6:547–548
side effects 6:60T
social anxiety disorder (SAD) treatment 6:99–100
social phobia treatment 6:99–100
structure 6:69–70
suicidal behavior association 1:687, 1:713
Selective site of metabolism, molecular descriptors 5:543, 5:545T, 5:546T, 5:547T
Selectivity
de novo drug design 4:300
genomics, antibacterial target identification 7:732–733
Selectivity filter, KcsA 4:707
SELECT programs
library design 4:356
ligand-based library design 4:319–320
product-based library design 4:350–351
Selegiline
bioavailability 5:712T
cocaine abuse 6:184T
narcolepsy treatment 6:153–154, 6:155
Parkinson's disease treatment 6:217
structure 2:68F, 2:1071F; 6:181F

1,3-Selenazol-4-one
selenium bioisosterism 2:703
structure 2:701F
Selenazolidine prodrugs, structure 2:701F
Selenium, bioisosterism see Bioisosterism, selenium
Selenopyrylium derivatives, structure 2:701F
Seletracetam 6:289
structure 6:282F
Self administration/reinstatement model, addiction models 6:176
Self buffering effects, disk intrinsic dissolution rate (DIDR) 5:418, 5:418F, 5:419F
Self-complementary ssDNA, gene therapy 1:304
Self-consistent-field procedure (SCF), quantum mechanical calculations 4:383–384
Self-consistent reaction field (SCRF), solvent effect calculations 4:385–386
Self-contained projects, project management 2:12–13, 2:13F
Self-forming depots, peptide/protein drugs 2:593
Self-organizing map, bioavailability models 5:711
Selfotel 6:259
nervous system trauma treatment 6:13
Seliciclib, clinical trials 7:207
Selodenoson 1:503–504
Semaxanib see SU-5416
Semiempirical methods
aqueous solubility models 5:642
distribution models 5:870
quantum mechanical calculations see Quantum mechanical calculations
Semliki Forest Virus expression systems 3:119
cytotoxicity 3:119
expression vectors 3:119
membrane proteins 3:115F, 3:119
plasmid construction 3:119
safety 3:119
two-vector systems 3:119
Senescence, benign prostatic hyperplasia 6:557
Senju see Bromfenac sodium
Senna, irritable bowel syndrome treatment 6:654
Sensitivity
analysis, postmarketing surveillance 1:769
homogeneous time-resolved fluorescence 2:419, 2:420F
microarrays 3:88
NMR probes 2:515

Sensitization
definition 6:170T
testing, toxicogenomics 1:481
Sensorgram, surface plasmon resonance 2:371–372, 2:372F
Sensory information, anxiety disorders 6:87
Sensory receptors, transient receptor potential ion channels 2:908
SEP-22589 6:74
Separation
PASP 3:793
pre-equilibrated dynamic combinatorial libraries 3:964
Sepracor 1:499, 1:499T
characteristics 2:226T
SeqHound, blueprint databases 3:305–306
SEQRES record, protein 3D structure validation 3:513
Sequence
DDBJ/EMBL/GenBank 3:352, 3:353F
siRNAs 3:176
Sequence alignment/analysis
bioinformatics see Molecular sequence analysis, bioinformatics
protein 3D structure, protein production 3:413
Sequence clustering, biological fingerprints see Biological fingerprints
Sequence databases 3:281
Sequenced Treatment Alternatives to relieve Depression (STAR*D) project 6:62
Sequence identity
DDBJ/EMBL/GenBank 3:353–354
GPCRs 4:672
Sequence motifs, protein function prediction, bioinformatics 3:334–335
Sequence–register shift errors
large ribosomal subunit (Deinococcus radiodurans) structure 3:535
x-ray structure see X-ray crystallography, problems
Sequencing, single nucleotide polymorphisms 2:364
Sequential clinical trials 2:182, 2:183F
Sequential ionization, acid/base reaction equilibria 5:608
Sequential processing, microwave-assisted chemistry 3:857
Sequential tests, non-animal skin irritation testing 1:473
SEQUEST
protein expression mapping 2:385
raw MS/MS data searching 2:345
Seractil see Dexibuprofen
Seratrodast 1:208
introduction of 1:108

Serdolect *see* Sertindole
Serenoa repens (saw palmetto) **1**:426
Serine hydroxymethyl transferase
    (SHMT), thymidine
    biosynthesis **7**:56–57, **7**:57*F*
Serine protease inhibitors
    see also specific drugs
    asthma treatment **7**:882
    clinical trials **7**:882
    COPD treatment **7**:882
Serine proteases
    COPD **7**:880
    as drug targets **3**:3
    ionization **5**:384
    ligands **4**:424*T*
    pharmacological target space
        4:422*T*
Serine–threonine protein
    kinases **2**:977; **7**:203
    see also specific kinases
    in cancer **7**:203
    as drug targets **3**:3
    Flash Plate assays **3**:587
    as targets **2**:758, **2**:960–962
    three-dimensional structures
        2:636*T*
Serine–threonine protein
    phosphatases **2**:981, **2**:982
    see also Phosphoserine/
        phosphothreonine-interacting
        domains; *specific types*
    three-dimensional structures
        2:636*T*
Serinyl proteases
    nonpeptide drugs **2**:628
    peptidomimetics **2**:628
    three-dimensional structures
        2:626*T*
Seroquel *see* Quetiapine fumarate
Serosion *see* Propagermanium
Serotone *see* Nazasetron hydrochloride
Serotonin *see* 5-HT (serotonin)
Serotonin reuptake transporters (SRTs),
    irritable bowel syndrome **6**:649
Serotonin-selective reuptake inhibitors
    (SSRIs) *see* Selective serotonin
    reuptake inhibitors (SSRIs)
Serotonin transporter (SERT) **2**:59,
    2:60*T*
    affective disorders **6**:52
    antidepressant efficacy **6**:66*T*
    knockout animals **6**:91
        depression **6**:58
    polymorphisms
        depression **6**:52
        personalized medicine **1**:288
Seroxat *see* Paroxetine
Sertindole **1**:186; **6**:32
    side effects **6**:33
        QT-prolongation **6**:35
    structure **5**:934*F*, **5**:948*F*
SERT knockout mouse, irritable bowel
    syndrome **6**:652

Sertraline **6**:69
    characteristics **2**:653*T*
    cocaine abuse **6**:184*T*
    side effects **6**:60*T*
    structure **2**:653*T*; **6**:100*F*, **6**:181*F*
    transporter inhibition **6**:66*T*
Serum **5**:848
    composition **5**:848
Serum albumin, peptide/protein drug
    conjugation **2**:589
Serum markers, cancer
    biomarkers **7**:44–47
Server databases **3**:294
Service companies **2**:226
    see also Contract research organizations
        (CROs)
    history **2**:226
    preclinical trials **2**:226
    SMEs *see* SMEs (small/medium-sized
        enterprises)
Serzone *see* Nefazodone hydrochloride
Sesquiterpene derivatives, P-
    glycoprotein models **5**:783
7TM Pharma **1**:514
Seven-descriptor model, computational
    oral absorption
    prediction **5**:688
Severe acute respiratory syndrome *see*
    SARS (severe acute respiratory
    syndrome)
Severe Impairment Battery (SIB),
    Alzheimer's disease
    assessment **6**:201
Severe myoclonic epilepsy of infancy
    (SMEI), sodium
    channels **2**:840–841
Severity index, psoriasis
    treatment **7**:943–944,
    7:944*T*
Sevoflurane **6**:360
    minimum alveolar
        concentration **6**:360
    structure **6**:352*F*
Sex hormone-binding globulin (SHBG),
    androgens **6**:526
Sex hormones, systemic lupus
    erythematosus **6**:234
Sexual dimorphism
    genetically engineered
        animals **3**:154
    metabolism **5**:210
    multiple sclerosis **6**:230–232
    myasthenia gravis **6**:230–232
    nervous system autoimmune
        diseases **6**:230–232
    systemic lupus
        erythematosus **6**:230–232
Sexual dysfunction **6**:545
    see also specific diseases/disorders
    female **6**:547
        dyspareunia **6**:547
        hypoactive sexual disorder **6**:547
        vaginismus **6**:547

Sexual polymerase chain
    reaction **1**:437
SFCHECK, protein 3D structure
    validation **3**:519
SGD, Universal Protein
    Resource **3**:362–363
SGS-742 **6**:79
    ADHD treatment **6**:135*T*
    Alzheimer's disease
        treatment **6**:206–207
SGX Pharmaceuticals **2**:761*T*
SH2 domain
    in disease **2**:963*T*
    in signaling **4**:624–625, **4**:625*F*
    STAT (signal transducer and activator
        of transcription) **3**:203
    structure **4**:625, **4**:625*F*
        binding site **4**:627, **4**:628*F*
        high-affinity peptide
            binding **4**:626, **4**:626*F*
SH2 domain inhibitors
    see also specific drugs
    structure-based design **4**:624
        fragment approaches **4**:631,
            4:635*F*
        fragment libraries **4**:633, **4**:635*F*
        high-affinity **4**:627, **4**:629*F*
        NMR **4**:631–633
        nonpeptide inhibitors **4**:626,
            4:628*F*, **4**:633–636, **4**:635*F*
        peptidomimetics **4**:631, **4**:632*F*
Shake-flask method
    lipophilicity measurement **5**:436
    lipophilicity prediction **5**:650
    partition coefficient systems **4**:48
    solubility measurement **5**:406
Shape-based alignment, virtual
    screening **4**:474, **4**:475*F*
Shape-based screening,
    pharmacophores **4**:533*F*, **4**:535
Shape descriptors, molecular
    descriptors **5**:534
Shaped pulses, NMR *see* Nuclear
    magnetic resonance (NMR)
Shape features, pharmacophore
    detection **4**:124
Shape/geometric combination searching,
    3D database pharmacophore
    searching **4**:121
Shape matching, scaffold
    hopping **2**:681–683
Shape searching, 3D database
    pharmacophore searching
    4:142
SHAPES methodology
    ligand-based NMR **4**:791
    structure-based fragment
        screening **4**:792
Shared assay workstations,
    automation **5**:344
Sheets, solid supports *see* Solid supports
Shift laminar pulse (SLP),
    NMR **2**:510

Shift work sleep disorder 6:148–149
  disease basis 6:150
    chronobiological factors 6:150
  treatment 6:162
    melatonin 6:162
    zolpidem 6:162
    zopiclone 6:162
Shigella infections, irritable bowel
    syndrome 6:650
Shinbit *see* Nifekalant hydrochloride
Short-acting insulin treatment 6:451
Short-chain fatty acids, histone
    deacetylase (HDAC)
    inhibitors 7:224
Short-chain fatty acyl-CoA synthetase,
    coenzyme A conjugation 5:155
Shorter, Edward, innovation 2:258
Short hairpin RNA (shRNA)
  precursors (vectors) *see* Short
    interfering RNAs (siRNAs)
  RNA interference 3:17
Short interfering RNAs (siRNAs) 3:3–
    4, 3:17, 3:171–187
  see also RNAi (RNA interference)
  cDNA precursors 3:172
  definition 2:1046
  delivery 3:177
    aerosol 3:182
    electroporation 3:177
    liposomal transfection 3:177
    protein turnover rate 3:177
    stability problems 3:181–182
  design criteria 3:176
    base composition 3:176
    base length 3:176
    genetic redundancy 3:177
    internal thermodynamic
      stability 3:176
    on-line design tools 3:176
    polymorphisms 3:176
    sequence 3:176
  as drugs 1:347
  duplexes, gene therapy 1:313
  generation 3:172
  hepatoma cell lines 5:233–234
  history 3:172
  microRNA precursors 3:175, 3:175F
    mismatches 3:175
    transcription 3:175
    translational repression 3:175
    in viruses 3:175
  modification 3:182–183
    2'-O-2,4-dinitrophenyl
      derivatives 3:182–183
    locked nucleic acids 3:183
  shRNA precursors (vectors) 3:172,
    3:177
    adenoviral vectors 3:177
    in vivo gene silencing 3:182
    lentivirus vectors 3:177
    retroviral vectors 3:177
    RNA polymerase III
      promoters 3:177

RNA polymerase II
  promoters 3:177
SMEs 1:516
stability problems 3:182
synthetic 3:176
  history 3:172
systemic administration 3:181
target site choice 3:176
  specificity 3:176
  technical considerations 3:176
tissue distribution 3:183
  cell-surface receptors 3:183
  cell targeting 3:183
  cholesterol tagging 3:183
  topical administration 3:181
'Shotgun proteomics,' LC-MS (peptide-
    based) proteomics 3:37
Siah1A, GPCR interaction 2:797T
Sialic acid, neuraminidase complex,
    crystal structure 4:615, 4:616F
Siani, Mike, peptoid-based antibiotics,
    discovery of 2:259
SIB-1553A, Alzheimer's disease
    treatment 6:205
Sibutramine 1:243
  metabolism 6:398, 6:399F
  as monoamine reuptake
    inhibitor 6:398, 6:399T
  obesity treatment 6:398
    contraindications 6:399
    indications 6:399
    mechanism of action 6:398
    side effects 6:399
  side effects 6:402–403
  structure 6:395T
Sicilian gambit, antiarrhythmic drug
    classification 6:742
Sickle cell anemia
  5-aza-2'deoxycytidine 1:273
  demethylating agents 1:273
  etiology 4:70
  as monogenetic disease 3:338
  structure-based drug design 4:70
Sickle-cell trait, malaria
    epidemiology 7:770
Sick sinus syndrome 6:730
Side chain angles, NMR structure
    determination 3:486
Side chain planarity, protein 3D
    structure validation 3:516
Side-chain rotamer libraries, PDB
    errors 3:523
Side effects, drug interactions 5:240
σ, QSAR 4:44–45
Sigmoidal surface fit, computational oral
    absorption prediction 5:685
Sigmosceptrellin 1:371
  structure 1:371F
Signal amplification, homogeneous
    time-resolved
    fluorescence 2:413
Signaling pathways 3:189–214
  see also specific pathways

activation 3:190
cell-based assays 3:618
definition 3:190
GPCRs 2:800
μ-opiate receptor 5:893
Signaling proteins, cell-based
    assays 3:643–644
Signal/leader sequences, *Escherichia coli*
    protein expression
    systems 3:114
Signal measurement, radioactive
    assays 3:578
Signal recognition particle (SNP),
    messenger RNA
    (mRNA) 2:1049
Signal sensors, gene detection,
    bioinformatics 3:321–322
Signal-to-background ratio,
    HTS 3:684
Signal-to-noise ratio
  HCS confocal systems 3:687
  HTS 3:684
Signal transducer and activator of
    transcription *see* STAT (signal
    transducer and activator of
    transcription)
Signal transduction
  affective disorders 6:73–74
  biomarkers 3:73
  cDNA functional screening
    3:15–16
  in disease 2:960, 2:963T
    cancer 2:960–962
    EGFR family kinases 2:967
  functional genomics 2:324, 2:960,
    2:963T
  genetically engineered
    animals 3:160
  GPCRs 2:574–575; 4:670
  immunosuppression 5:223
  large-scale virus vector
    libraries 3:18–19
  modulators, peptidomimetics 2:615,
    2:617F
  regulatory networks 3:330
  structural genomics 2:964
    x-ray crystallography 2:964
  ubiquitin -proteasome
    pathway 3:190
Signature method 3:363
Signature-tagged mutagenesis,
    genomics 7:743
'Significant activity' definition,
    anticancer agents, tumor
    models 7:38
Sila-atipamezole, structure 2:695F
Silabolin, structure 2:695F
Sila-meprobamate, structure
    2:695F
Sila-pridinol, structure 2:695F
Sildenafil citrate 1:244; 2:927
  absorption, pH dependency 5:383
  derivatives 2:938–939

Sildenafil citrate (*continued*)
  discovery/development **2**:257;
      3:930–931, **3**:931*F*
    phosphodiesterase-5
      blockers **2**:257
  erectile dysfunction treatment **6**:546
  introduction of **1**:106–107
  ionization constant (pK_a) **5**:366*T*
  medical needs *vs.* individual
      desires **1**:628
  PASP **3**:801, **3**:801*F*
  phosphodiesterase inhibition **2**:942–
      943, **2**:947*T*
  priority-setting **1**:632
  quinazoline derivatives **2**:939
  sales revenue **1**:691
  structure **5**:360
  target information **4**:750*T*
Silencer regions, regulatory regions,
    bioinformatics **3**:322
Silica
  microwave-assisted chemistry **3**:853
  PASP **3**:793
Silicon diols, carbon-silicon
    bioisosterism **2**:696
Silicone *see* Artificial membranes
Silperisone
  clinical studies **2**:697
  structure **2**:697*F*
Silver stain, two-dimensional gel
    electrophoresis **2**:333; **3**:35,
    3:35*T*
Silybum marinanum (milk thistle) **1**:425
O-Silylated steroid derivatization,
    PASP **3**:828–829, **3**:829*F*
SIMCYP **5**:679*T*
  physiologically based
      pharmacokinetics **5**:874
Simdax *see* Levosimendan
Simian models, HIV infection **7**:336
Simian monkey fibroblasts(COS), cell-
    based assays **3**:620–621
Simian virus 40 (SV40)
  microRNA coding **3**:175
  promoter, mammalian protein
      expression systems **3**:117
  transgenic mouse model
      promoters **3**:11
Similarity coefficients *see* Similarity
    searching
Similarity distance, QSAR applicability
    domain **4**:159
Similarity profiles, library design **4**:342
Similarity searching **3**:245; **4**:**167–192**
  see also Diversity analysis
  best-match searches **4**:167–168
  coefficients **3**:245; **4**:169, **4**:174,
      4:174*T*, **4**:176*T*
    association *vs.* distance **4**:178
    continuous *vs.* binary **4**:179
    cosine coefficient **4**:176*T*, **4**:178,
        4:180–181
    dice coefficients **4**:176*T*, **4**:178

dissimilarity-based compound
    selection (DCBS) **4**:178
distance coefficients **4**:175
Euclidian distance **4**:176*T*, **4**:178,
    4:180–181
Forbes coefficient **4**:179, **4**:181
Hamming distance **4**:176*T*, **4**:178,
    4:180–181
Jaccard coefficient **4**:177–178
Kulczynski coefficient **4**:179
Minkowski distances **4**:178
monotonic **4**:178
nonmetric **4**:175
pseudometric **4**:175
quantification **4**:174
Russell–Rao coefficient **4**:179,
    4:181
Soergel distance **4**:176*T*
Tanimoto coefficient **4**:176*T*,
    4:177–178, **4**:180–181
zero-valued **4**:175
evaluation **3**:248
exact-match searches **4**:167–168
fragment bitstrings **3**:246
GPCR library screening **4**:691
numerical descriptors **3**:245
partial-match searches **4**:167–168
rationale **3**:245
representations *see* Similarity
    searching, representations
scaffold hopping **2**:681–683
similarity coefficients **3**:245
similarity measure
    combination **4**:182
  available reference
      structures **4**:183
  database searching **4**:182
  definition **4**:182
  fusion rules **4**:182
  process **4**:182
  turbo similarity searching **4**:183
similarity measures, comparison
    of **4**:179
  active molecules **4**:179–180
  effectiveness measures **4**:180,
      4:180*T*
  G–H score **4**:180
  molecule size effects **4**:181
  similar-property principle **4**:180
  single-component performance
      measures **4**:180
  two- *vs.* three-dimensional **4**:181
similar property principle *see* Similar
    property principle
structure searching **4**:168
substructure searching *see*
    Substructure searching
Tanimoto coefficient **3**:246
target (reference) structure **4**:168
  nearest neighbors **4**:168
Similarity searching,
    representations **4**:169
local regions **4**:169

quantum mechanics **4**:169
three-dimensional descriptors **4**:169,
    4:172
  atom types **4**:172–173
  conformational properties **4**:172
  distance-based **4**:172
  electrostatic fields **4**:173
  field-based similarity searching
      (FBSS) program **4**:173–174
  field representations **4**:173
  geometric atom pairs **4**:173
  geometric binding property
      pairs **4**:173
  graph-matching **4**:173
  hydrophobic fields **4**:173
  pharmacophore keys **4**:173
  steric fields **4**:173
  substructure searches **4**:172
  valance angle descriptors
      4:172–173
topological index **4**:169
  branching index **4**:170
  kappa shape index **4**:170
  Molconn-Z program **4**:170
  related chi index **4**:170
  Wiener index **4**:170
two-dimensional descriptors **4**:169,
    4:170
  atom pair descriptors **4**:171
  BCI fingerprints **4**:170–171
  chemically advanced template
      searches (CATS) **4**:171
  dictionary-based
      fingerprints **4**:170–171
  element type **4**:171
  extended connectivity fingerprints
      (ECFPs) **4**:171
  functional connectivity fingerprints
      (FCFPs) **4**:171
  hashed fingerprints **4**:171
  MACCS structural keys **4**:170–
      171
  maximum common
      subgraph **4**:171–172
  RASCAL **4**:171–172
  scaffold hopping **4**:171
  Scitegic circular fingerprints **4**:171
  Similog keys **4**:171
  vector representations **4**:171–172
whole-molecule properties **4**:169
  atom-based methods **4**:169–170
  ClogP program **4**:169–170
  fragment-based methods
      4:169–170
  physicochemical
      properties **4**:169–170
Similar property principle **4**:168, **4**:180
  QSAR **4**:168
Similog keys, two-dimensional
    descriptors, similarity
    searching **4**:171
Simon Foundation, urinary incontinence
    markets **8**:127–128

Simple endoscopic score for Crohn's disease (SES-CD) 6:620
Simplification, single-parametric biochemical assays 3:683
Simplified Molecular Input Line Entry System (SMILES)
  dermal penetration models 5:733
  ligand-based virtual screening 4:460, 4:461F
Simpson–Angus Experimental Scale, schizophrenia treatment 6:29
Simulated annealing
  conformational analysis/ searching 4:107
  ligand-based library design 4:318–319
  pharmacophore searching 4:129T
  product-based library design 4:349, 4:351–352
  protein crystallography model generation 3:463
Simulect see Basiliximab
Simvastatin 6:240T, 6:473–474, 6:481
  bioavailability 5:712T; 6:481
  characteristics 2:653T
  discovery/development 1:19, 1:374–375; 2:1074
  number of patients needed to treat 8:144T
  structure 1:376F, 2:653T, 6:238F, 6:469F
  target information 4:750T
Singapore, orphan drugs legislation 1:663
Single base extension (SBE), single nucleotide polymorphisms 2:365, 2:365F
Single cell systems, non-animal chronic toxicity testing 1:476–477
Single-component performance measures, similarity measures, comparison of 4:180
Single descriptors, computational oral absorption prediction see Oral absorption prediction, computational
Single-domain prokaryotic proteins, structural genomics 3:554
Single-dose studies, ritonavir 8:190
Single-electron transfer (SET) mechanism, cytochrome P450 oxidation 5:95, 5:95F
Single ion channels, arrhythmias 6:739
Single-isomorphous replacement with anomalous scattering (SIRAS), protein crystallography, rational drug design 3:889–890
Single-mode instruments, microwave-assisted chemistry 3:844
Single molecule detection, fluorescence screening assays 3:600
Single Nucleotide Polymorphism (SNP) Consortium 2:569

Single nucleotide polymorphisms (SNPs) 2:362
  allele-specific chain extension 2:364, 2:365F
  allele-specific PCR 2:365
  applications 2:362
    drug responses 2:362
    protein expression 2:362
    protein function 2:362
  automation 2:363
  bead-based fiberoptic arrays 2:357
  detection 2:328
  discovery 2:362
  drug development 2:195, 2:195T
  drug discovery 2:315–316
  electrophoresis 2:364, 2:367, 2:369T
    limitations 2:367
  flat microarrays 2:364, 2:368, 2:369T
    hybridization assays, with 2:368
  genetic variation 1:282–283; 3:338
  Human Pharmaco DNP Consortium 2:316
  hybridization 2:364, 2:365F
  mass spectrometry 2:367, 2:369T
    limitations 2:367–368
  microarrays 3:90
  microplate-based assays 2:363, 2:367, 2:369T
    fluorescence detection 2:367
    heterogenous approach 2:367
    homogeneous approach 2:367
  multiplexing 2:363
  nuclease-based assays 2:366
    fluorescence-based energy transfer (FRET) 2:366
    Invader 2:366
    Taq-Man 2:366
  oligonucleotide ligation 2:365F, 2:366
  PCR 2:369
  pharmacogenomics 1:344–345
  platforms 2:367, 2:369T
    see also specific platforms
  proteome 3:29
  QTL analysis 3:14
  sample analysis 2:363
  sample preparation 2:363
    nucleotide identification 2:363
    PCR 2:363
  sequencing 2:364
  single base extension (SBE) 2:365, 2:365F
  soluble arrays 2:364, 2:368, 2:369T
    advantages 2:368
    flow cytometry 2:368
Single-parametric biochemical assays, HTS see High-throughput screening (HTS)
Single-parametric cellular assays, HTS see High-throughput screening (HTS)
Single-pass perfusion technique, BCS 5:975F, 5:977

Single photon emission computed tomography (SPECT)
  affective disorders 6:50
  anxiety disorders 6:89
  applications 3:659
  coronary heart disease diagnosis 6:696
  functional small animal imaging 3:661
  obsessive compulsive disorder (OCD) 6:6
  psychotropic drug development 6:51T
  pulmonary administration 5:287–288
  radionuclides 3:662T, 3:663T
    see also specific nuclides
  small animal imaging 3:660T, 3:661, 3:663–664
  99mTc 3:661–662, 3:663T
Single run evaluation, HTS data management 2:454
Single series, screening library selection 4:498
Singlet oxygen formation, tetracycline phototoxicity 7:618, 7:619
Single wavelength anomalous dispersion (SAD), anomalous dispersion 3:460
Singulair see Montelukast; Montelukast sodium
SIN model, neuropathic pain models 6:334–335
Sinoatrial (SA) node
  definition 6:730
  tachycardia/bradycardia 6:730
Sirna 2:761
siRNA-027, wet age-related macular degeneration treatment 6:306T
siRNAs see Short interfering RNAs (siRNAs)
Sirolimus
  discovery/development 1:327
  transplant rejection 7:919–920
Siscard see Clinidipine
Sisomicin 7:631
Sitafloxacin
  clearance 7:587
  development 7:583
Sitamaquine
  leishmaniasis treatment 7:836–838
  structure 7:838F
Sitaxsentan
  pulmonary artery hypertension (PAH) treatment 6:725
  structure 6:726F
Site acceptance tests, customized automation 5:348
Site analysis, drug design 4:286
  docking/minimization 4:287
    docking 4:287F
  Locus de novo design method 4:287

Site analysis, drug design (*continued*)
  multiple copy simultaneous
    search 4:287–288
    SEED approach 4:288
  experimental data 4:286
    Cambridge Structural
      Database 4:286
    PDB 4:286
    SuperStar 4:286
  site derivation 4:286
    GRID 4:286
    HBMAP 4:286
    HSITE 4:286
    hydrogen bonds 4:287
    LUDI 4:286
Site-directed mutagenesis
  (SDM) 1:33, 1:37F
  GPCR structure
    determination 4:676
  historical perspective 1:33
  protein tyrosine phosphatases 4:753
Site-directed screening, fragment-based
  drug discovery 3:947
Site isolation effects, PASP *see* Polymer-
  assisted solution phase synthesis
  (PASP)
Site of metabolism prediction,
  molecular descriptors 5:543
SitePrint, ligand-based library
  design 4:317
Site-selective conjugation, polyethylene
  glycol conjugates *see*
  Polyethylene glycol (PEG)
  conjugates
Sitting drop method, protein
  crystallization 3:439
Sivelestat 1:218
16-membered ketolide
  antibiotics 7:556
Size
  pharmaceutical industry
    organization 2:110
  R&D management 2:250
Size descriptors, oral absorption *see* Oral
  absorption prediction, molecular
  descriptors
Size exclusion chromatography
  see also specific methods
  buffer exchange 3:424
  protein analysis 3:423F, 3:426
  protein purification 3:423,
    3:423F
SJA6017 6:272
  structure 6:272F
Sjögren's syndrome, dry eye 6:317
Skeletal dysplasia 6:507
Skeletal muscle
  calcium channels 2:847
  sodium channels 2:834
Skelgen, de novo drug design 4:297
Skelgen software, scaffold
  hopping 2:683
Skelid *see* Tiludronate disodium

SKF-38 393, ring system reorganization
  (bioisosterism) 2:678–679,
  2:679F
SKF-104365
  structure 2:979F
  TGFβR kinase inhibitors 2:977–979
SKI-166, characteristics 2:984T
SKI-606 7:192
  Src kinase inhibitor 2:976–977
  structure 2:974F, 2:978F, 7:192F
SKI-2053R 1:140
Skin
  corrosivity, animal experimentation
    alternatives *see* Animal
    experimentation alternatives
  cytochrome P450s 5:172T
  drug introductions *see* Drug
    introductions
  equivalents, dermal/transdermal
    administration 5:293
  fungal disease 7:427
  hydration, atopic dermatitis
    treatment 7:949
  inflammation *see* Skin inflammation
  integrity function tests, non-animal
    skin irritation testing 1:473
  irritation
    animal experimentation alternatives
      see Animal experimentation
      alternatives
    testosterone replacement
      therapy 6:531
  permeability
    animal experimentation
      alternatives 1:467T, 1:472
    skin sensitization, computational
      models 5:923
    transmembrane permeation,
      solubility-diffusion
      model 5:463–464
  sensitization
    animal experimentation alternatives
      see Animal experimentation
      alternatives
    computational models *see* Skin
      sensitization, computational
      models
  stratum corneum 5:291
    lipids 5:291
    structure 5:291, 5:291F
    tissue composition 5:45T
SkinEthic, non-animal skin irritation
  testing 1:473
Skin inflammation 7:969–985
  see also Atopic dermatitis; Psoriasis
  animal models 7:970
    classic model 7:970
    genetically engineered
      models 7:971
    immunodeficient models 7:971
  treatment 7:972
    adalimumab 7:975
    angiogenesis blocking 7:978

    antileukotriene drugs 7:979
    CCR3 inhibitors 7:981
    cytokine targeting 7:975
    etanercept 7:975
    folic acid metabolism
      analogs 7:976
    glucocorticoid receptor
      antagonists 7:976
    infliximab 7:975
    interleukin-15 7:975
    interleukin-18 7:975
    MAPK p38 inhibitors 7:979
    neurogenic inflammation 7:981
    phosphodiesterase 4
      inhibitors 7:979
    T-cell activation/signaling 7:972
    topical
      immunosuppressants 7:973
    vitamin D_3 analogs 7:976
Skin sensitization, computational
  models 5:922
  Beuhler test 5:923
  DEREK for windows (DfW) 5:923
  experimental animal use 5:923
  guinea pig maximization test 5:923
  local lymph node assay
    (LLNA) 5:923
  QSAR models 5:923
  random forest algorithm 5:923
  skin penetration 5:923
Skin stripping, dermal/transdermal
  administration 5:294
SklogP, log P calculation
  program 5:661T
'Skunk works,' innovation 2:262
SkyPharma, controlled release
  formulation technology 2:232
SL-65.0155, as prokinetic agent 6:678
SL-651498 6:106
  structure 6:105F
Slab gel electrophoresis, epigenetics *see*
  Epigenetics
SLAPSTIC, NMR, drug
  discovery 3:910
SLATE program, 3D-QSAR 4:582
Slater–Kirkwood equation, steric
  properties 4:195
SLC15A1 (PEPT1) transporters 5:68T,
  5:71
SLC15A2 (PEPT2) transporters 5:68T,
  5:71
SLC21 see OATP (*SLC21/SLCO*)
  transporters
SLC22 see OAT (*SLC22*) transporters;
  OCT (*SLC22*) transporters
SLC28 see CNT (*SLC28*) transporters
SLC29 see ENT (*SLC29*) transporters
SLCO see OATP (*SLC21/SLCO*)
  transporters
SLC transporters 2:58; 5:64, 5:65T,
  5:66T, 5:67T, 5:68T
  see also CNT (*SLC28*) transporters;
    ENT (*SLC29*) transporters;

OATP (*SLC21/SLCO*)
transporters; OAT (*SLC22*)
transporters; OCT (*SLC22*)
transporters
blood–brain barrier **5**:79
distribution **5**:54
genes **5**:64
intestinal absorption **5**:75
polymorphisms, hepatotoxicity **5**:78
structure **5**:55
transmembrane domains **5**:55
Sleep **6**:143
circadian rhythm
synchronization **6**:143
disorders *see* Sleep disorders
measures **6**:141
neurotransmitters **6**:143
acetylcholine **6**:146
adenosine **6**:145
corticotropin-releasing
hormone **6**:147
dopamine **6**:143, **6**:145
GABA **6**:146
galanin **6**:147
glutamate **6**:145
growth hormone-releasing
factor **6**:147
histamine **6**:143, **6**:146
melatonin **6**:147
neurotensin **6**:147
norepinephrine **6**:143, **6**:145
orexin/hypocretin **6**:146
serotonin **6**:143, **6**:146
substance P **6**:147
vasoactive intestinal peptide
(VIP) **6**:147
promotion **6**:143
GABA system **6**:143
galanin systems **6**:143
regulation **6**:143, **6**:144*F*
stages of **6**:142
non-rapid eye movement
(non-REM) **6**:142–143
rapid eye movement
(REM) **6**:142–143
superchiasmatic nuclei (SCN) **6**:143
wake promotion **6**:143
Sleep apnea, modafinil **8**:152
Sleep disorders **6**:7, **6:139–167**
see also specific diseases/disorders
breathing-related *see* Breathing-related
sleep disorders
categorization **6**:141
circadian rhythm-related *see* Circadian
rhythm sleep disorders
classification **6**:147
DSM-IV-R **6**:147, **6**:147*T*
International Classification of Sleep
Disorders **6**:147
clinical trials **6**:153
diagnosis **6**:7
disease basis **6**:149
incidence **6**:141

models **6**:152
SCORE system **6**:152
Tg2576 mice **6**:152–153
transgenic models **6**:152
new research areas **6**:164
prevalence **6**:141
prognosis **6**:8
socioeconomic impact **6**:141
somnolence measures/sleep
testing **6**:141
electromyography **6**:142
electrooculography **6**:142
encephalography **6**:142
Epworth Sleepiness Scale
(ESS) **6**:141–142
functional magnetic resonance
imaging **6**:142
Hamilton Rating Scale for
Depression (HAM-D) **6**:141–
142
Maintenance of Wakefulness
Test **6**:142
Multiple Sleep Latency Test
(MSLT) **6**:142
neuroimaging **6**:142
polysomnography **6**:142
positron emission
tomography **6**:142
Stanford Sleepiness Scale **6**:141–
142
Wake After Sleep Onset
(WASO) **6**:142
treatment **6**:8, **6**:153
modafinil **6**:7, **6**:8
unmet medical needs **6**:163
Sleeping sickness *see* Trypanosomiasis,
African
Sleep paralysis **6**:148
SLIPPER
log D calculation program **5**:662*T*
log P calculation program **5**:659,
5:661*T*
SLIPPER 2001 **5**:679*T*
S+logP, log P calculation
program **5**:661*T*
Slow growth method, free energy
calculation *see* Free energy
calculation
Slow release, prodrugs **5**:1014
SLV-308, Parkinson's disease
treatment **6**:215
SM-232724, structure **7**:507*F*
SM-232725, activity **7**:506*T*
Small animal imaging **3:659–677**
amino acid metabolism **3**:672
anatomical **3**:661
CT **3**:661
MRI **3**:661
ultrasound **3**:661
cell surface receptors **3**:667
radiolabeled annexin **3**:667
cellular metabolism **3**:668
androgen ablation **3**:669

cytotoxic agents **3**:668
infection **3**:670
photodynamic treatment **3**:669
radiotreatment **3**:670
viability **3**:668
cellular proliferation **3**:671
contrast agents **3**:660
functional **3**:661
CT **3**:661
MRI **3**:661
PET **3**:661
SPECT **3**:661
internal cell processes **3**:665
amino acid probes **3**:665
metabolic probes **3**:665,
3:666*F*
proliferation probes **3**:665
radiolabeled biomolecular
probes **3**:666
reporter genes **3**:665
methods **3**:660*T*
see also specific methods
microscopy **3**:660*T*
molecular **3**:661
angiogenesis probes **3**:665
bioluminescence **3**:661, **3**:668
cell receptor probes **3**:664
CT **3**:662
fluorescence **3**:661
NMR **3**:662
PET **3**:661, **3**:663–664
see also Positron emission
tomography (PET)
SPECT **3**:661, **3**:663–664
see also Single photon emission
computed tomography
(SPECT)
ultrasound **3**:662
MRS **3**:660*T*
pharmacokinetics **3**:660, **3**:672
direct measurements **3**:672
indirect measurements **3**:672
receptor occupancy **3**:667
central nervous system **3**:664*T*,
3:667
steroid hormones **3**:667
SPECT **3**:660*T*
transgenic animals **3**:660
ultrasound **3**:660*T*
Small animal test systems **3:647–657**
see also Animal experimentation
alternatives; *specific species*
history **3**:647
knockout models **3**:649
overexpression models **3**:649
protein conservation **3**:647–649
transgenic models **3**:649
Small cell lung carcinoma (SCLC)
amrubicin hydrochloride **1**:143
nedaplatin **1**:147
treatment **7**:7–8
Small GTPases, bone resorption/
remodelling **8**:205

Small innovative entrepreneurs, governments, in drug discovery **1:**729
Small interfering RNAs *see* Short interfering RNAs (siRNAs)
Small intestine, drug metabolism **5:**974
Small libraries
    fragment-based drug discovery **3:**942, **3:**943*F*
    solution phase parallel chemistry **3:**765
Small ligands, structure-based library design *see* Library design, structure-based
Small-medium enterprise (SME) *see* SMEs (small/medium-sized enterprises)
Small molecule drugs
    see also specific drugs
    biological macromolecules *vs. see* Biological macromolecules
    biotechnology companies **2:**102
    calcium channel pharmacology **2:**849–850
    future work **1:**442
        lead discovery changes **1:**442
    light scattering, promiscuous ligands **2:**744
    mechanisms of action **1:**431
    specificity **2:**299–300, **2:**300*T*
Small-molecule microarrays
    surface plasmon resonance **3:**594
    target-oriented screening **3:**142, **3:**143*F*
Small molecule models, glutathione *S*-transferases (GSTs) **5:**818
Small neutral molecules, transmembrane permeation, solubility-diffusion model **5:**462, **5:**462*F*
Small organic molecules
    combinatorial chemistry **2:**462
    solid-phase libraries **3:**698
Smallpox
    antiviral drugs **7:**262
    vaccine, development **1:**531
Small robotic system, HTS automation **3:**682
Small-scale motion, ligand–receptor molecular flexibility **4:**203
Small volume dissolution apparatus (μDISS), dissolution **5:**415*F*, **5:**416, **5:**416*F*
SMANCS *see* Zinostatin stimalamer
SMART
    InterPro, member of **3:**363
    protein function prediction, bioinformatics **3:**334
    signature method **3:**363
SMARTS program, 2D- to 3D structure **4:**464

SMEs (small/medium-sized enterprises) **1:489–525**
see also specific companies
acquisition
    by major companies **1:**496*T*, **1:**503
    by other SMEs **1:**503
antibodies **1:**516
definitions **1:**490; **2:**225
    Japan **1:**490, **1:**490*T*
drug discovery **1:**491, **1:**503
    cardiovascular disease **1:**508
    CNS disease **1:**509
    genitourinary tract disease **1:**512
    infectious diseases **1:**507
    metabolic disease **1:**512
    multitherapeutic areas **1:**503
    musculoskeletal disease **1:**512
    oncology **1:**512
    reproductive health **1:**512
    respiratory disease **1:**514
failures **1:**503
funding **1:**493
    business angels **1:**494
    grants **1:**495
    public offerings **1:**494
    revenue generation **1:**494, **1:**496*T*
    stock market listings **1:**494
    venture capitalists **1:**494
gene therapy **1:**516
locations **1:**495
    Europe **1:**497
    Japan **1:**498
    'Medicon Valley' **1:**498
    North America **1:**495
    UK **1:**497–498
'major players' **1:**498, **1:**499*T*
origins **1:**521*F*, **1:**521*T*
    academia **1:**491, **1:**521*T*
    big pharmaceutical companies **1:**493, **1:**521*T*
proteins **1:**516
service providers **1:**518
    computer-aided drug design **1:**518
    drug candidate validation **1:**520
    hit discovery **1:**519
    hit optimization **1:**519
    lead optimization **1:**519
    target validation **1:**518
siRNA **1:**516
target-led **1:**515
    chemokine receptor-directed research **1:**515
    GPCR-directed research **1:**515
    ion channel-directed research **1:**516
    kinase-directed research **1:**516
    zinc-finger directed research **1:**516
vaccine companies **1:**517
value **1:**520
SMID (Small Interaction Database), blueprint databases **3:**305–306

SMILES (simplified molecular input line entry specification)
    chemical information databases **3:**298
    chemoinformatics, molecule representation **3:**236
    THOR **3:**295
    two-dimensional representations **3:**298
SMIRKS, library design **4:**341–342
Smith, George, phage display discovery **1:**435
Smith Kline Beckman, history **2:**100–101
Smith Kline Beecham, history **2:**100–101
SmithKline Beecham Pharmaceuticals, carvedilol **8:**137–138
Smoking cessation
    COPD treatment **7:**883–884
    drugs, bioethics **1:**701–702
Smoking-induced pulmonary endothelial dysfunction, phosphodiesterase-5 inhibitors **2:**948–949
Smoothing, conflict management **2:**248
Smooth muscle
    asthma **7:**878
    ATP-gated ion channels **2:**904–905
    calcium channels **2:**847
    phosphodiesterase-3 **2:**924
    phosphodiesterase-4 **2:**926–927
SMP-707 **6:**487
SN-38, transporters **7:**45*T*
SNAP-7941 **6:**77–78, **6:**109, **6:**408
    structure **6:**108*F*, **6:**409*F*
SNARE, nicotinic acetylcholine receptor regulation **2:**886–887
SNP Consortium **1:**57*T*
SNPRN gene, Prader–Willi syndrome **1:**271
Sobuzoxane **1:148**
    adult T-cell leukemia **1:**148
    lymphoma **1:**148
Social anxiety disorder (SAD) **6:**86
    clinical trials **6:**96
        Liebowitz social anxiety scale (LSAS) **6:**96
    DSM-IV-R classification **6:**86
    human genetics **6:**92
    medicalization issue **1:**720
    neuroimaging **6:**89
    prevalence **6:**86
    treatment, selective serotonin reuptake inhibitors **6:**99–100
Social effects, importance of **2:**240
Social interaction test, anxiety disorders, animal models **6:**95
Social isolation models, schizophrenia *see* Schizophrenia

Socially desirable outcomes, governments, in drug discovery 1:727
Social phobia treatment, selective serotonin reuptake inhibitors 6:99–100
Societal health outcomes, governments, in drug discovery 1:748
Society of Biomolecular Screening
    HTS history 3:680
    microplate standardization 3:682–683
Sodium, tetracyclines 7:618
Sodium butyrate, Huntington's disease treatment 6:222T
Sodium–calcium exchanger blockers, antiarrhythmic drugs 6:756
Sodium channel blockers 6:742
    characteristics 2:829T
    new research areas 6:750
    vitamin E bifunctional agents 6:751
Sodium channel modulators, pain 6:343
Sodium channels 2:831; 6:734, 6:735
    action potentials 6:734F
    Brugada syndrome 6:735
    genes 6:735
    genetic diseases 2:840
        Brugada syndrome 2:841
        cardiac arrhythmias 2:840–841
        epilepsy 2:840–841
        familial childhood epilepsy 2:840–841
        hyperkalemic periodic paralysis 2:841
        long-QT syndrome 2:841
        paramyotonia congenita 2:841
        primary erythermalgia 2:841
        severe myoclonic epilepsy of infancy (SMEI) 2:840–841
    GPCR interaction 2:786T
    hypertension 6:707
    long QT syndrome 6:735
    molecular biology 2:833
        TTX_s subtypes 2:833–834
    mutations 6:735
    nomenclature 2:831
        International Union of Pharmacology (IUPHR) 2:831–832
        subunits 2:832, 2:833T
    overexpression, lower urinary tract symptoms (LUTS) 6:560
    pharmacology 2:837, 2:838T
        see also Sodium channels, therapeutics; specific drugs
        clinical useful drugs 2:839
        modulated receptor hypothesis 2:839
        neurotoxin receptor site 1 2:837
        tetrodotoxin (TTX) 2:837
        physiological functions 2:835
        β subunits 2:836

    cardiac action potentials 2:836
    cell membrane depolarization 2:836
    excitable cells 2:835
    resting state 2:835
    second messenger functions 2:836
    regulation 2:836
        atrial/ventricular contraction 2:837
        neurotransmitters 2:836
        phosphorylation 2:836
        prostaglandin E_2 2:836–837
        protein kinase A 2:836
        protein kinase C 2:836
    structure 2:832F, 2:834
        β subunits 2:835
        intracellular loop 2:834–835
        ion selectivity/permeation 2:834–835
    subcellular distribution 2:833
    subunit classification 2:833T
    therapeutics 2:839
        see also Sodium channels, pharmacology; specific drugs
        antiepileptic drugs 2:839–840
        chronic pain 2:840
        conduction anesthesia 2:839
        epilepsy 2:839–840
        local anesthetics 2:839
        tissue distribution 2:840
    tissue distribution 2:833
        auxiliary subunits 2:834
        cardiac myocytes 2:834
        CNS 2:833–834
        skeletal muscle cells 2:834
        voltage-sensitive 6:323
Sodium codependent bile acid transporter inhibitors 2:48F, 2:63
Sodium concentrations, cell-based assays 3:627
Sodium estrone sulfate, structure 2:66F
Sodium–hydrogen exchanger blockers, antiarrhythmic drugs 6:756
Sodium lauryl sulfate (SLS)
    dermal penetration models 5:734–735
    solubility effects 5:411, 5:412T
Sodium oxybate
    cataplexy treatment 6:154
    narcolepsy treatment 6:154, 6:155T, 6:156
Sodium phenyl butyrate (buphenyl)
    acute myelogenous leukemia 1:274
    prostate cancer 1:274
    β-thalassemia 1:274
Sodium/potassium ATPase inhibitors, glaucoma treatment 6:308, 6:313T
Sodium resorption, renal dysfunction 6:584

Sodium stibogluconate
    leishmaniasis treatment 7:756, 7:825
    structure 7:826F
Sodium sulfacetamide
    acne rosacea treatment 7:961
    acne treatment 7:961
Soergel distance, similarity searching 4:176T
'Soft drugs' 5:1010
    see also specific drugs
    bioisosteres 5:1010
    current status 1:67
    definition 1:67; 5:1009, 5:1010
    design limitations 5:1011
    inactive metabolites 5:1010
    labile bridges 5:1010
    metabolism 1:54F, 1:55
    prodrugs vs. 5:1010
Soft independent modeling of class analog (SIMCA), bioavailability models 5:709
Soft representations, ligand–receptor interaction models 4:206
Soft scoring, de novo drug design 4:298
Soft tissue sarcoma
    new drugs 1:107
    tasonermin 1:141
Software
    see also specific programs
    advanced compartmental absorption and transit (ACAT) models 5:678, 5:679T
    computational oral absorption prediction 5:690
    CSD 3:393
    development, metabolism prediction 5:805
    electrotopological indices (E-state index) 5:573
    ionization constants (pK_a) 5:395
    ionization prediction see Ionization prediction, in silico tools
    oral absorption prediction, molecular descriptors 5:682
    pharmacophore identification 3:242
    scaffold hopping 2:683
Solcot see Dutasteride
Solenoid-controlled liquid dispensing, Ultra-High-Throughput Screening System (Aurora Biosciences) 2:444, 2:444F
Solid dispersions, adducts/complexation/formulation 5:526
Solid–liquid phases, microwave-assisted chemistry 3:853
Solid-phase chemistry
    scaffold-based design 4:34
    triphenylethylene anticancer agents 7:157–158, 7:158F
Solid-phase extraction/chromatography (SPEC) nuclear magnetic

resonance (NMR), unknown
    isolation/identification 5:998
Solid-phase libraries 3:702
    chemical assessment 3:703, 3:704T,
        3:738, 3:739T
    alkyl-tethered diisopropylarylsilane
        linker 3:738F, 3:740
    anandamide 3:703
    benzodiazepine analogs 3:703,
        3:706F
    bicyclic amines 3:706–708, 3:708F
    clavulone analogs 3:703
    cyclopent-2-enones 3:708, 3:708F
    heterocyclic scaffolds 3:703–706,
        3:707F
    indenopyrrolocarbazoles 3:703,
        3:705F
    natural products 3:740, 3:740F
    dibenz[b,g]1,5-oxazocines 3:706,
        3:707F
    peptide nucleic acids
        (PNAs) 3:740–741
    6,6-spiroketals 3:740F, 3:741
    discretes 3:708, 3:709T, 3:746,
        3:746F, 3:747T
    arylsulfonamide
        hydroxamates 3:725–726,
        3:726F
    ceramides 3:746F, 3:751–752
    chiral 1,3-oxazolidines 3:722–724,
        3:723F
    estradiols 3:708, 3:718F
    furan-4-ones 3:720–721, 3:720F
    heterocyclic libraries 3:719–720
    lidocaine analogs 3:725, 3:725F
    NADH:ubiquinone oxidoreductase
        inhibitors 3:751F, 3:752–753
    natural products 3:718
    nucleotide-like libraries 3:751–
        752
    peptidomimetic scaffolds 3:719–
        720, 3:720F
    philantotoxins 3:718–719, 3:719F
    phthalides 3:720F, 3:721
    procainamide analogs 3:725,
        3:725F
    quinazolinones 3:722, 3:722F
    recurrent motifs 3:721–722,
        3:721F
    spiroimidazolidinones 3:722,
        3:723F
    tellimagrandin analogs 3:751F,
        3:752
    1,3,5-trisubstituted
        Triazines 3:724–725, 3:724F
    Wnt antagonists 3:751–752
    examples 3:738
    nucleotides 3:698
    peptides 3:698
    pool libraries 3:726, 3:741
    benzimidazolones 3:735, 3:736F
    benzodiazepines 3:736–737,
        3:736F

CCR5 antagonists 3:727–735,
    3:734F
2,2-dimethyl benzopyran
    scaffolds 3:745F, 3:746
galanthamine analogs 3:745, 3:745F
histone deacetylase
    inhibitors 3:741
ketoamides 3:741, 3:741F
lead discovery 3:741, 3:742T
macrolides 3:726, 3:727F
natural products 3:726, 3:728T
phenylalkylamides 3:737–738,
    3:737F
pyridines 3:735, 3:735F
tetrapeptide libraries 3:746
tyrphostins 3:727, 3:734F
ureido acids 3:737, 3:737F
size 3:698
small organic molecules 3:698
split-and-mix libraries 3:698
Solid phase parallel synthesis
    applicability 3:764
    intermediates 3:765
    purification concerns 3:765
    solution phase parallel chemistry
        vs. 3:763, 3:764T
Solid-phase peptide synthesis
    CEM Discover platform microwave-
        assisted chemistry 3:846–847,
        3:847F
    microwave-assisted
        chemistry 3:843F
Solid rendering, surfaces 5:581,
    5:584F
Solid samples, compound storage see
    Compound storage/management
    (Pfizer)
Solid–solvent interfaces, solid-state
    physicochemistry 5:513
Solid-state NMR structure
    determination see Nuclear
    magnetic resonance (NMR),
    structure determination
Solid-state physicochemistry 5:509–
    530
    adducts/complexation/
        formulation 5:526
    co-crystals 5:526
    cyclodextrins 5:526, 5:527T
    excipients 5:526
    micellar solubilization 5:526
    solid dispersions 5:526
    amorphous states 5:510, 5:514
    glass transitions 5:512
    characterization/detection modeling/
        screening 5:513
    temperature of transition 5:514,
        5:515T
    thermodynamic stable
        forms 5:514
    x-ray diffraction 5:514
    chemical stability 5:520, 5:521T
    light 5:521

crystalline states 5:510
    heating effects 5:510
    definitions 5:510
    dissolution rate 5:517
        dissolution interface 5:517,
            5:517F
        flow-cell measurement 5:517,
            5:518F
        per surface area 5:517
        pH dependency 5:517, 5:519F
        poorly soluble substances 5:517,
            5:519T
        powders 5:517, 5:518F
    drug formulation 5:515
    history 5:509–510
    HTS, effects on 5:514
    hygroscopicity 5:518
    intrinsic dissolution rate
        (IDR) 5:517
    kinetics 5:521, 5:522F
        anhydrous to monohydrate
            transition 5:523, 5:523F
        'disappearing polymorphs' 5:521,
            5:522F
        mechanical stress 5:523
    liquid crystals 5:513
    modulation of 5:523
    particle size effects 5:526, 5:527F
    phase diagrams 5:511, 5:511F
        hydrates 5:512
        solvates 5:512
    polymorph form 5:525
        definition 5:510
    salt form effects 5:523, 5:525T
        absorption 5:519F, 5:523
        bioavailability 5:523, 5:524
        counterion effects 5:524
        reformulation 5:524
    solubility 5:515
        'equilibrium' solubility 5:516
        measurement 5:515–516
        solubility curves 5:514F, 5:515
    sorption isotherms 5:518, 5:520F
    thermodynamics 5:511
        Gibbs energy 5:511
        Gibbs energy plots 5:511,
            5:512F
        Gibbs' phase rule 5:511
        G-isobar position 5:512
        solid–solvent interfaces 5:513
Solid-supported lipid membranes
    (SSLM) 5:435–436
Solid supports 2:493
    glass 2:503, 2:504F
        DNA arrays 2:504–505, 2:505F
        photolithographic masks 2:503,
            2:504F
    high throughput organic
        synthesis 2:396
    pins and crowns 2:499
        improvements 2:500, 2:501F
        monomer alternatives 2:501–502,
            2:502F

radiation grafting **2**:499–500, **2**:501*F*

removable crown **2**:501–502, **2**:501*F*

resin beads *see* Resin beads, solid support synthesis

sheets **2**:502

Fmoc peptide synthesis **2**:503, **2**:503*F*

Solid support synthesis **2**:464

see also Solid supports

N-acetylation **2**:466*F*, **2**:467, **2**:467*F*

O-acetylation **2**:466*F*, **2**:467, **2**:468*F*

advantages/disadvantages **2**:464

N-alkylation **2**:466*F*, **2**:467, **2**:467*F*

O-alkylation **2**:466*F*, **2**:467, **2**:468*F*

N-arylation **2**:466*F*, **2**:467, **2**:467*F*

O-arylation **2**:466*F*, **2**:467, **2**:468*F*

carbon–carbon bond-forming reactions **2**:467

Baylis–Hilman reaction **2**:469*F*

Grignard reaction **2**:467–468, **2**:469*F*

Heck reaction **2**:467–468, **2**:470*F*

Horner–Emmons reaction **2**:469*F*

Stille reaction **2**:467–468, **2**:470*F*

Suzuki reaction **2**:467–468, **2**:470*F*

Wittig reaction **2**:468–469, **2**:469*F*

carbon–heteroatom coupling reactions **2**:466, **2**:466*F*

chemical libraries **2**:760

complex multistep synthesis **2**:475

see also specific reactions

cycloadditions **2**:472, **2**:473*F*

heterocycle/pharmacophores **2**:479

1,4-benzodiazepines **2**:479, **2**:481*F*

benzopiperazinones **2**:479, **2**:481*F*

bicyclic guanidines **2**:481*F*

'complexity-generating' reactions **2**:479–480, **2**:483*F*

oxapiperazines **2**:482*F*

prostaglandins **2**:479, **2**:482*F*

quinolones **2**:481*F*

history **2**:464–465

Merrifield, Bruce **2**:464–465, **2**:465*F*

hybrid oligomers **2**:478

peptide nucleic acids **2**:478, **2**:479*F*

intramolecular cyclization-resin-cleavage **2**:470, **2**:471*F*, **2**:472*F*

library construction **2**:464

multiple-component reactions **2**:472, **2**:473*F*

Passerini reaction **2**:472–474, **2**:474*F*

natural products **2**:480, **2**:484*F*

epothilones **2**:482, **2**:484*F*

glyantrypine **2**:482

indolactam V **2**:480–482, **2**:483*F*

oligomers **2**:475

oligonucleotides **2**:476, **2**:477*F*

oligosaccharides **2**:477, **2**:477*F*, **2**:478*F*

glycopeptides **2**:477, **2**:478*F*

insoluble/soluble polymers **2**:477

peptides **2**:475, **2**:476*F*

Fmoc-amino acid approach **2**:475–476

library generation **2**:476

tBoc-protected approach **2**:476

resin-to-resin transfer reactions **2**:474, **2**:475*F*

solution phase synthesis *vs.* **2**:465*T*

transition metal-mediated reactions **2**:469, **2**:470*F*

copper-mediated *N*-arylation **2**:470, **2**:471*F*

olefin metathesis **2**:469, **2**:470*F*

palladium-mediated *N*-arylation **2**:470, **2**:471*F*

unnatural oligomers **2**:478, **2**:480*F*

SOLON, pharmacophore 3D database searching **4**:139*T*

Solubility **5**:400, **5**:402

see also Dissolution

absorption effects *see* Absorption

absorption *vs.* **5**:627

acid–base titration method **5**:401

aggregation reactions **5**:407

anion aggregation **5**:407

cation aggregation **5**:410

charged aggregates **5**:408–409

neutral species aggregation **5**:410

speciation anomalies **5**:407, **5**:408*F*

ampholytes **5**:403*F*, **5**:405

automation **5**:346*T*, **5**:348–349

bioisosterism modification analysis **2**:687, **2**:688*F*

chemogenomics **4**:422

definition **5**:628

aqueous solution **5**:628

pH-dependent solubility **5**:628

thermodynamic solubility **5**:628

diprotic weak acids **5**:403, **5**:403*F*, **5**:404*T*

diprotic weak base **5**:403*F*

dissolution titration template (DTT) method **5**:400–401

dissolution *vs.* **5**:401

errors **5**:629

experimental data errors **5**:629–630

excipient effects **5**:411

sodium lauryl sulfate (SLS) **5**:411, **5**:412*T*

experimental artifacts **5**:400

fragment-based drug discovery **4**:448

Gibbs pK_a **5**:403*F*, **5**:405, **5**:405*F*

'sdiff 3–4' effect **5**:405

Henderson–Hasselbalch equation **5**:402

high-throughput methods **5**:406

direct ultraviolet assay **5**:400–401, **5**:407

HPLC-based assays **5**:407

turbidimetric assays **5**:406

kinetic/pharmokinetic aspects **5**:628

apparent *vs.* equilibrium solubility **5**:628

fluidization **5**:628

mobile order and disorder thermodynamics (MOD) **5**:629, **5**:629*T*

solution phase effect model **5**:629

low-solubility compounds **5**:407

measurement **5**:406

see also specific methods

aggregation reactions *see above*

ionization constants (pK_a) measurement **5**:387*F*, **5**:393

models *see* Solubility models

monoprotic weak acids **5**:402

intrinsic solubility (S_0) **5**:402, **5**:403*F*

monoprotic weak bases **5**:403

oral absorption prediction, molecular descriptors **5**:682

permeability *vs.* **5**:627

pK_a shift method **5**:407

advantages **5**:411

disadvantages **5**:411

potentiometric method **5**:406

data collection **5**:406

sample preparation **5**:406

titrants **5**:406

prediction

chemoinformatics, ADME property prediction **3**:259

in vitro oral absorption prediction **5**:676, **5**:677*T*

prodrugs **5**:1012

quantum mechanical calculations **4**:397

relevance in drug discovery **5**:627

shake flask measurement **5**:406

solid-state physicochemistry *see* Solid-state physicochemistry

VolSurf descriptors **5**:540

Solubility curves **5**:514*F*, **5**:515

Solubility-diffusion model

see also Transmembrane permeation

passive transcellular transport **5**:264–265

Solubility models **5**:627–648

applicability domain **5**:632

Mahalanobis distance **5**:632

outlier labeling **5**:632

residual standard deviation (RSD) **5**:632

applications **5**:630

intellectual property **5**:630

predictive purposes **5**:630

underlying properties **5**:630

Solubility models (*continued*)
  aqueous *see* Solubility models, aqueous
  organic solvents **5**:645
    CODESSA PRO software **5**:645–646
    COSMO-RS method **5**:645
    QSPR studies **5**:645–646
  relevance **5**:630
  statistical methods **5**:630
    absolute average error (AAE) **5**:631
    average error (AE) **5**:631
    cross-validation **5**:631
    least-squares regression **5**:630
    model fitting **5**:630
    model selection **5**:630
    multiple linear regression **5**:630
    overfitting **5**:631
    predicted residual error sum of squares (PRESS) **5**:631
    randomization **5**:631
    root mean square (rms) error **5**:631
    squared correlation coefficient of test set (q^2) **5**:631
    squared correlation coefficient of training set (r^2) **5**:631
    standard deviation (s) **5**:631
    training set **5**:630
    validation **5**:630
Solubility models, aqueous **5**:632, **5**:633*T*
  accuracy **5**:638*F*, **5**:641*F*, **5**:643
    ESOL **5**:644–645
    IALOGS **5**:644–645, **5**:645*T*
    root mean square errors **5**:633*T*, **5**:643–644
  applicability **5**:643
  commercial software **5**:643, **5**:644*T*
    COSMO-RSol **5**:643, **5**:644*T*
    CSlogWS **5**:643, **5**:644*T*
  consensus (ensemble) modeling **5**:642
    KnowItAll program **5**:643
    Rule Discovery System **5**:643
    virtual library screening **5**:642–643
  data sets **5**:632
    AQUASOL database **5**:632
    origins **5**:633*T*, **5**:636
    PHYSPROP database **5**:632
    principal component analysis (PCA) **5**:632–635
  fragmental methods **5**:633*T*, **5**:636
    atom type *E*-state indices **5**:637–638
    model used by Huuskonen **5**:638
    overlapping fragments **5**:637
    partial least squares model **5**:637, **5**:637*F*
  property-based methods **5**:633*T*, **5**:638
    log *P* **5**:633*T*, **5**:640, **5**:641*F*

semiempirical methods **5**:642
solvation properties **5**:633*T*, **5**:639, **5**:639*F*
Solubility tags, *Escherichia coli* protein expression systems *see Escherichia coli* protein expression systems
Soluble arrays, single nucleotide polymorphisms *see* Single nucleotide polymorphisms (SNPs)
Soluble epoxide hydrolase (sEH)
  drug metabolism **5**:203–204
  expression **5**:204
  substrate specificity **5**:204
Soluble proteins, protein crystallization *see* Protein crystallization
Solute carriers *see* SLC transporters
Solute entropy, lipophilicity measurement **5**:434
Solution equilibria **5**:358
Solution phase effect model, solubility **5**:629
Solution phase parallel chemistry **3**:**761–790**
  see also specific drugs
  applicability **3**:765
    advantages **3**:767
    chemistry limitations **3**:767
    diversity-oriented libraries **3**:767
    intermediate purification **3**:767
    microwave-assisted chemistry **3**:767
    small/focused libraries **3**:765
    target-focused libraries **3**:767
    validation times **3**:767
  future trends **3**:786
  solid phase *vs.* **3**:763, **3**:764*T*
Solution phase synthesis **2**:482
  library preparation *see* Solution phase synthesis, library preparation
  polymer-supported synthesis **2**:482
    advantages **2**:482–485
    dendrimers **2**:485, **2**:485*F*
    examples **2**:482–485, **2**:485*F*
    fluorinated hydrocarbons **2**:485
  resin-bound reagents **2**:488
    cycloaddition reactions **2**:490, **2**:491*F*
    polymer-assisted solution phase PASP chemistry **2**:477*F*, **2**:489
    polymer-bound ^1H-benzotriazole **2**:490, **2**:490*F*
    'safety-catch' linkers **2**:489–490, **2**:490*F*
  resin-bound scavengers/reagents **2**:491, **2**:491*F*
    epibatidine synthesis **2**:492, **2**:493*F*
    epimaritidine synthesis **2**:491–492, **2**:492*F*
    oxomaritidine synthesis **2**:491–492, **2**:492*F*
  scavenging resins **2**:485

    Dess–Martin oxidization procedure **2**:487*F*, **2**:488
    disadvantages **2**:486
    functionalized polystyrene resins **2**:485–486, **2**:487*F*
    product trapping **2**:486, **2**:486*F*
    sulfonic acid resins **2**:488, **2**:488*F*
  solid-phase synthesis *vs.* **2**:465*T*
Solution phase synthesis, library preparation **3**:768
  analytical characterization **3**:778
  dynamic combinatorial chemistry **3**:785
    CDK2 inhibitor synthesis **3**:785, **3**:785*F*
    examples **3**:778
    fluorous chemistry **3**:785, **3**:785*F*
    hydantoin synthesis **3**:786, **3**:786*F*
    pyrimidine synthesis **3**:786, **3**:786*F*
  heterocyclic libraries **3**:779
    benzazepine-3-one synthesis **3**:782–783, **3**:784*F*
    benzimidazole synthesis **3**:780, **3**:782*F*
    p56lck inhibitors **3**:779–780, **3**:781*F*
    N-substituted piperidinone synthesis **3**:780, **3**:783*F*
    pyrimidine synthesis **3**:780, **3**:782*F*
    pyrrole synthesis **3**:780, **3**:783*F*
    1,2,4-triazole synthesis **3**:780, **3**:783*F*
  intermediate synthesis/purification **3**:771
    novel intermediates **3**:771, **3**:775*F*
  library design **3**:768
    compound numbers **3**:769–770
    computational chemistry **3**:770
    focused libraries **3**:769
    large, diverse libraries **3**:769
    library size **3**:769
    SAR **3**:769
    targeted libraries **3**:769
  multicomponent reactions **3**:784
    quinoline synthesis **3**:784, **3**:784*F*
  parallel work-up **3**:776
    intermediate purification **3**:776
    liquid–liquid extraction **3**:776
  PASP *see* Polymer-assisted solution phase synthesis (PASP)
  preproduction **3**:771
  purification **3**:776
    resin capture/release **3**:777, **3**:777*F*
    resin scavenging **3**:776, **3**:777*F*
  reaction monitoring **3**:775
  route validation **3**:770
    analytical techniques **3**:771
    final compound specification **3**:771
    library purification **3**:771, **3**:772*T*

reaction conditions **3**:770
step assessment **3**:770
scaffold derivatization **3**:779, **3**:779*F*,
   3:780*F*
Solvates, phase diagrams **5**:512
Solvation properties
   aqueous solubility models **5**:633*T*,
      5:639, **5**:639*F*
   polyethylene glycol
      conjugates **5**:1052
Solvatochromic approach, lipophilicity
   factorization **5**:432
Solvent-accessible surface (SAS;
   Connolly surface)
   molecular descriptors **5**:533–534,
      5:534*F*
   steric properties **4**:196
   surfaces **5**:581, **5**:582*F*
Solvent excluded surface (SES),
   surfaces **5**:581, **5**:582*F*
Solvent-free reactions, microwave-
   assisted chemistry *see*
   Microwave-assisted chemistry
Solvent resonance suppression
   liquid chromatography-NMR **2**:522
   NMR **2**:511
Solvents
   calculations, quantum mechanical
      calculations *see* Quantum
      mechanical calculations
   categorization **2**:162*T*
   development chemistry **2**:161,
      2:162*T*
   entropy **4**:202
   ionization **5**:606–607
   liquid chromatography-NMR **2**:522–
      523
   microwave-assisted chemistry *see*
      Microwave-assisted chemistry
Solvent suppression, NMR **2**:509
Solvent surroundings, ligand–receptor
   interactions **4**:194
Somatoliberin **2**:579*T*
Somatomedin-1 **1**:231
   introduction of **1**:108
Somatostatin
   Hirschmann, R F and **1**:14
   structure **2**:581–582
Somatostatin-2 receptor agonists, beta-
   turn peptidomimetics **4**:694
Somatostatin receptor partial agonist,
   peptidomimetics **2**:608–609,
      2:611–612, **2**:611*F*
Somatostatin receptors st5, in human
   disease **2**:775*T*
Somatotrophin **1**:231
   introduction of **1**:108
   recombinant **1**:499
Somatuline LP *see* Lanreotide acetate
Somavert *see* Pegvisomant
SOMFA (self-organizing molecular field
   analysis), 3D-QSAR
   alignment **4**:578

Somnolence, modafinil **8**:152
Sonata *see* Zaleplon
Soraphen A, acetyl-CoA carboxylase
   inhibitors **7**:439, **7**:439*F*
Sordarin
   discovery/development **1**:369
   structure **1**:370*F*
Sorivudine, oral absorption **5**:680*T*
Sormodren, oral absorption **5**:680*T*
Sorption isotherms, solid-state
   physicochemistry **5**:518, **5**:520*F*
Sosei **1**:506
Sotalol
   bioavailability **6**:712*T*
   discovery/development **1**:328–329
   dose **6**:712*T*
   half-life **6**:712*T*
   hypertension treatment **6**:711
   ionization constant (pK_a) **5**:366*T*
   oral absorption **5**:468*T*, **5**:680*T*
   structure **6**:712*T*
D-Sotalol **6**:745
   effectiveness **6**:745
   mechanism **6**:745
   toxicity **6**:746
Sourcing
   see also Outsourcing
   definitions **2**:205
Sourcing development **2**:209
   activity cluster analysis **2**:216,
      2:216*F*, **2**:217*F*
   activity determination **2**:212
      activity interdependency **2**:214
      company/industry practice **2**:214
      competitive advantage **2**:213
      existence of **2**:214
      IP generation **2**:214
      offshore outsourcing *see* Offshore
         outsourcing
      process maturity **2**:213
      process outcome control **2**:213
   discrete activities **2**:209
   primary chemistry activity
      definition **2**:209
      mapping processes **2**:209
      subactivities **2**:209, **2**:210*F*
      subcategories **2**:210, **2**:210*F*
South Africa
   drug pricing **1**:644–645
   health insurance coverage **1**:641
Southeast Collaboratory for Structural
   Genomics (SECSG) **3**:552*T*
   metalloproteomics **3**:556–557
Southern blots, discovery/
   development **1**:259
South Korea
   orphan drugs legislation **1**:662
   stem cell research programs **1**:558–
      559
Soya isoflavones **1**:426
sp^2 carbon–hydrogen bond oxidation,
   cytochrome P450 *see* Cytochrome
   P450 family

sp^3 carbon–hydrogen bond oxidation,
   cytochrome P450 *see* Cytochrome
   P450 family
Space-directed hydrophobicity, P-
   glycoprotein models **5**:778–779
Space groups *see* X-ray crystallography
Spanidin *see* Gusperimus
   trihydrochloride
SPARC method, ionization
   prediction **5**:611, **5**:620, **5**:623*T*
Sparfloxacin **7**:571*T*
   clearance **7**:587
   development **7**:579
   structure **7**:572*F*, **7**:579*F*
Sparse matrices, library design **4**:308
Spatial regions, pharmacophore
   detection **4**:127
SPC-839, arthritis treatment **7**:862
sp carbon–hydrogen bond oxidation,
   cytochrome P450s **5**:104,
      5:104*F*
Spearman rank-correlation coefficient,
   3D molecular alignment **4**:112
Special surface-related descriptor sets
   see Molecular descriptors
Specialty pharma companies **2**:102
Speciation anomalies, solubility **5**:407,
      5:408*F*
Species dependence, microsome
   binding **5**:332
Species differences
   drug metabolism, in vitro
      studies **5**:237
   malaria **7**:754
   metabolism prediction **5**:796, **5**:797
Species extrapolation, blood–placenta
   transfer modeling **5**:737
Species loss, traditional
   medicine **1**:410
Species names, nutraceuticals **1**:458
Species-specific enzymes, metabolism
   studies **5**:33
Species-specific metabolism,
   tamoxifen **8**:94
Species-specific parameters,
   physiologically based
   pharmacokinetics **5**:868,
      5:869*T*
Specific binding sites, biological
   fingerprints **4**:809
Specificity
   drug design, de novo *see* Drug design,
      de novo
   drug discovery *see* Drug discovery
   enzyme-linked immunosorbent assays
      (ELISAs) **3**:73–74
   monoclonal antibodies **1**:432–433
   'Specificity scale,' drug
      discovery **2**:298, **2**:299*T*
Specific polarization effects, ligand–
   receptor interactions **4**:197–198
Specific reversible binding, distribution
   models **5**:870

SpecInfo **3**:279, **3**:280*T*
SPECITOPE program, ligand–receptor
    interaction models **4**:206–207
Spectinomycin **7**:632
    mechanism of action **7**:639
        ribosome interactions **2**:1042–
        1043
    Neisseria gonorrhoeae infection **7**:637
    resistance **7**:642
    structure **7**:633*F*
Spectral maps, microarrays **3**:92,
    3:101, **3**:102*F*
Spectromycin, spectrum **7**:637
    Mycobacterium tuberculosis **7**:637
Spectroscopic databases **3**:279, **3**:280*T*
Spectrum Pharmaceuticals **1**:513
Speed
    microwave-assisted chemistry **3**:838
    NMR, drug discovery **3**:907
Speed definitions, decision making *vs.*
    organizational alignment **2**:120
Speedel, characteristics **2**:226*T*
Speed phenotyping, genetically
    engineered animals **3**:153
Spent reagent removal, PASP **3**:792,
    3:793
SPERM software, virtual
    screening **4**:472–473
Sphere exclusion algorithms **3**:253,
    3:254*F*; **4**:187
Spherical powder dissolution *see*
    Dissolution
SPHGEN, ligand-based library
    design **4**:316–317
Sphingolipid biosynthesis inhibitors,
    antifungal drugs **7**:434, **7**:435*F*
Sphingomyelin, intestinal
    membrane **5**:458
Sphingosine 1-phosphate receptor
    modulators, IBD
    treatment **6**:636
Spickett, Geoffrey, collaborators,
    Ganellin, C R **8**:42
Spinal anesthetics **6**:356
Spinal cord injury (SCI) **6**:12–13
    see also Central nervous system injury;
        Neuroprotective drugs
    experimental models **6**:257*T*, **6**:258
        inflammation **6**:273
    incidence **6**:13
    pathogenesis **6**:254–256
    prevalence **6**:254
    treatment **6**:263
        antioxidants **6**:260*F*, **6**:266
        dexamethasone **6**:263, **6**:264
        glucocorticoids **6**:263
        methylprednisolone **6**:263
        National Acute Spinal Cord Injury
            Study (NASCIS II) **6**:264–265
Spinal cord localization, 5-HT
    receptors **8**:125
Spinal nerve ligation, neuropathic pain
    models **6**:334, **6**:334*F*

Spin cycle *see* Start-up companies
Spin labeling, GPCR structure
    determination **4**:678
Spinner flasks, protein
    fermentation **3**:418
Spinocerebellar ataxia
    calcium channels **2**:852
    Drosophila melanogaster model
        system **3**:9
Spinophilin, GPCR interaction **2**:797*T*
Spirapril hydrochloride **1**:**160**
Spiriva *see* Tiotropium bromide
Spirocyclic series, azetidinone
    cholesterol absorption
    inhibitors **8**:72, **8**:73*F*
Spiro derivatives, ring system
    reorganization (bioisosterism) *see*
        Bioisosterism, ring system
        reorganization
Spirogermanyl, structure **2**:698*F*
Spirohydantoins
    bioisosterism **2**:660*T*
    solid-phase discrete libraries **3**:709*T*
Spiroimidazolidinones
    solid-phase discrete libraries **3**:709*T*
    solid-phase libraries **3**:722,
        3:723*F*
6,6-Spiroketals, solid-phase
    libraries **3**:740*F*, **3**:741
Spironolactone **6**:711
    acne treatment **7**:965
    bioavailability **5**:712*T*; **6**:709*T*
    dose **6**:709*T*
    half-life **6**:709*T*
    hypertension treatment **6**:585–587
    mechanism of action **6**:590
    structure **2**:1018*F*; **6**:587*F*, **6**:709*T*
Spirooxindoles, solid-phase pool
    libraries **3**:707*F*
Spiroquinazolinones, phosphodiesterase-
    7 inhibitors **2**:943–944
Spiruchostatin **7**:228
Spleen
    cytochrome P450s **5**:172*T*
    tissue composition **5**:45*T*
Splice variants
    AMPA-sensitive receptor
        regulation **2**:899
    estrogen receptors (ERs) **2**:1026
    phosphodiesterase-1 **2**:923
    phosphodiesterase-2 **2**:923–924
    phosphodiesterase-4 **2**:924–925
    phosphodiesterase-5 **2**:927
    phosphodiesterase-7 **2**:929
    phosphodiesterase-8 **2**:929–930
    phosphodiesterase-9 **2**:930
    phosphodiesterase-10 **2**:930
Splicing redirection, antisense-based
    therapeutics **2**:1047
Split-and-mix libraries, solid-phase
    libraries **3**:698
Split diagrams, phylogeny construction,
    bioinformatics **3**:321, **3**:321*F*

Spongistatins **7**:85
    structure **7**:85*F*
Spongothymidine **1**:372
    structure **1**:373*F*
Spongouridine **1**:372
    structure **1**:373*F*
Spontaneous animal models
    diabetes mellitus type 2 animal
        models **6**:435
    hypertension **6**:707
    spontaneously hypertensive rat *see*
        Attention deficit hyperactivity
        disorder models
Spontaneous reporting systems,
    postmarketing
    surveillance **1**:761,
    1:763*T*
Sporanox *see* Itraconazole
Sporaricin **7**:632
Sporiofungin A **7**:457*T*
Sporotrichosis **7**:420, **7**:422*T*
    itraconazole **7**:420
Sporozoite microneme protein essential
    for cell traversal (SPECT),
    malaria **7**:766
Sporozoites, malaria **7**:766
SPRESI **3**:281–282, **3**:282*T*
SPROUT, library design **4**:360
SPS resins, NMR *see* Nuclear magnetic
    resonance (NMR)
SQi inhibitors, mitochondria complex
    III **2**:1075
SQL (structured query language),
    relational databases **3**:268
SQ (sequence header) lines, UniProtKB
    flatfile **3**:362, **3**:362*F*
Squalamine lactate
    structure **6**:307*F*
    wet age-related macular degeneration
        treatment **6**:306*T*
Squalene epoxidase inhibitors **7**:447
    see also specific drugs
    clinical profiling **7**:449
    drug resistance **7**:449
    pharmacokinetics **7**:448
Squared correlation coefficient of test
    set (q^2), solubility
    models **5**:631
Squared correlation coefficient of
    training set (r^2), solubility
    models **5**:631
SQUID program, fingerprint/dataprint
    descriptors **4**:471
SR 5722A, structure **2**:890*F*
SR 16234
    as estrogen receptor ligand **2**:1014
    structure **2**:1013*F*
SR 58611, structure **6**:406*F*
Src homology domains
    identification **2**:633–634
    members **2**:633, **2**:636*T*
    nonpeptide drugs **2**:633
    peptidomimetics **2**:633

three-dimensional structures **2**:635–637, **2**:636*T*, **2**:639*F*
 x-ray crystallography **2**:634, **2**:637*F*, **2**:638*F*
Src kinase inhibitors **7**:191
 see also specific drugs
 potency **7**:191–192, **7**:192*F*
 structure **7**:191*F*, **7**:192*F*
 structure-based fragment screening, x-ray crystallography **4**:789–790, **4**:789*F*
Src kinases **2**:975
 angiogenesis **2**:975–976
 in cancer **7**:190
  breast cancer **7**:191
  colorectal cancer **7**:191
  ovarian cancer **7**:191
  prostate cancer **7**:191
 in disease **2**:963*T*
 GPCR interaction **2**:786*T*
 inhibitors **2**:976
 members **2**:975–976; **7**:191
 metastases **2**:975–976
 PDB entries **2**:965*T*
 SH$_2$
  domain inhibitors **2**:615–616, **2**:617, **2**:617*F*
  three-dimensional structures **2**:636*T*
 SH3, three-dimensional structures **2**:636*T*
 structure **2**:976, **2**:976*F*
 substrate complex, x-ray crystallography **2**:964, **2**:966*F*
 transduction pathways **2**:975–976, **2**:975*F*
 vascular endothelial growth factor receptor-2 (VEGF-R2) **3**:200
SRD5A2, benign prostatic hyperplasia **6**:556
SSAP, protein function prediction, bioinformatics **3**:336–337
SSR 125543, structure **6**:103*F*
St. John's Wort (*Hypericum perforatum*) **1**:426
 antidepressant effects **1**:327
 depression **1**:411
 P-450 metabolism induction **1**:422
ST-246, poxviruses **7**:262–263
Stability, RNAi (RNA interference) **3**:182–183
Stability testing **5**:493
 see also Chemical stability
 applications **5**:502
  biological buffers **5**:503–505
  compound selection **5**:502
  reaction kinetics **5**:503, **5**:504*F*
  structural optimization **5**:503, **5**:503*F*, **5**:504*F*
 biological matrix **5**:499
  half-life effects **5**:499
 buffers **5**:499
 development late stages **5**:495

diagnostic methods **5**:495
 early drug discovery **5**:495
 FDA guidelines **5**:493–494, **5**:494*T*
 gastrointestinal stability **5**:499
 heat **5**:498
  FDA guidelines **5**:498
  HTS **5**:495
 humidity **5**:498
  FDA guidelines **5**:498
 in vivo testing **5**:494
 light **5**:497
  FDA guidelines **5**:497–498
  HTS **5**:497
  photostability chambers **5**:497–498
 low-throughput methods **5**:501
 oxidation **5**:498
  electrochemical cells **5**:499
  hydrogen peroxide **5**:498
  plasma stability **5**:498, **5**:499*T*
 pH **5**:496, **5**:497*T*
  absorption prediction **5**:496
  HTS **5**:496
  microtiter plates **5**:496
  physiologically applicable **5**:496
  prodrug evaluation **5**:496
 plasma **5**:499, **5**:500*T*
  blood *vs.* **5**:500
  HTS **5**:500
  hydrolytic enzymes **5**:499–500
  species **5**:500
  throughput **5**:500
 product structure determination **5**:501
  LC/MS **5**:493*F*, **5**:501–502
  modification during development **5**:502
  NMR **5**:502
 qualitative data **5**:494
  toxicity testing **5**:494
 quantitative data **5**:494
  structure–property relationships (SPR) **5**:494
 as solid **5**:501
 in solution **5**:501
 stereochemistry **5**:501
Stabilizing residues, phosphodiesterase substrate recognition **4**:759
Stable isotope labeling by amino acids in cell culture (SILAC), LC-MS (peptide-based) proteomics **3**:38
Stable transfection, mammalian cell expression systems **3**:418
Staden Package, molecular sequence analysis, bioinformatics **3**:318
Staffing flexibility, outsourcing **2**:207
Staff morale, outsourcing risks **2**:219
Staging, cancer **7**:2
STAN **3**:547*T*
Standard conformation libraries, protein 3D structure errors **3**:509

Standard deviation (*s*), solubility models **5**:631
Standard error of mean, biological activity databases **3**:302
Standard free energy of hydration, blood–brain distribution **5**:753
Stanford Sleepiness Scale, somnolence measures/sleep testing **6**:141–142
Stanford Synchrotron Radiation Laboratory, metalloproteomics **3**:556–557
Stanozolol, structure **2**:1016*F*; **6**:522*T*
Staphylococcus aureus infections **7**:675
 aminoglycosides **7**:636
 atopic dermatitis **7**:948, **7**:949
 cefozopran hydrochloride **1**:111
 macrolide antibiotics **7**:522
 penicillin resistance **7**:653–654
Staphylokinase **6**:771
STAR **1**:577*T*
Stars, drug–polymer conjugates **5**:1046*F*
Starsis *see* Nateglinide
Starting material definition, common technical document (CTD) **2**:163
Start-up companies **2**:764
 chemistry *vs.* biology **2**:764–765
 collaborations **2**:765
 medicinal chemistry importance **2**:765
 spin cycle **2**:765
  technology licensing **2**:765
 virtual companies **2**:766
  advantages **2**:766
  contract research organizations **2**:766
  extended enterprises **2**:766
STAT (signal transducer and activator of transcription) **3**:203
 see also specific types
 activation pathway **3**:205*F*
  homodimer association **3**:204
 alternative RNA splicing **3**:204
 chromosomal localization **3**:203, **3**:204*T*
 in disease **3**:207
  allergic inflammation **3**:208
  autoimmune disease **3**:208
  tumorigenesis **3**:207
 gene isolation, cDNA functional screening **3**:14
 signal transduction/activation **3**:203
  cytokines **3**:205–206, **3**:206*T*
  growth factor receptors **3**:204
  Janus tyrosine kinases **3**:204
 structure **3**:203, **3**:204*F*
  C-terminal transcriptional activation domain **3**:203
  DNA-binding domain **3**:203

STAT (signal transducer and activator
of transcription) (*continued*)
  N-terminal domain (oligomerization
    domain) **3**:203
  SH2 domain **3**:203
  truncated isoforms **3**:204
STAT1 **3**:203
  alternative RNA splicing **3**:204
  asthma **3**:208
  chromosomal localization **3**:204*T*
  chromosomal location **3**:203
  heterodimer formation **3**:204
  rheumatoid arthritis **3**:208
  signaling **3**:205
    cytokines **3**:205–206, **3**:206*T*
  truncated isoforms **3**:204
STAT2 **3**:203
  chromosomal location **3**:203, **3**:204*T*
  heterodimer formation **3**:204
  signaling **3**:206
    cytokines **3**:206*T*
    interferons **3**:206
STAT3 **3**:203
  alternative RNA splicing **3**:204
  chromosomal location **3**:203, **3**:204*T*
  heterodimer formation **3**:204
  knockout animals, embryonic
    lethals **3**:206
  oncogenesis **3**:207
  rheumatoid arthritis **3**:208
  signaling **3**:206
    cytokines **3**:206, **3**:206*T*
  truncated isoforms **3**:204
STAT4 **3**:203
  chromosomal localization **3**:204*T*
  chromosomal location **3**:203
  signaling **3**:207
    cytokines **3**:206*T*
    IL-12 **3**:206
STAT5 **3**:203
  signaling **3**:207
    cytokines **3**:206
  truncated isoforms **3**:204
STAT5a **3**:203
  chromosomal location **3**:203, **3**:204*T*
  heterodimer formation **3**:204
  signaling, cytokines **3**:206*T*
STAT5b **3**:203
  chromosomal location **3**:203, **3**:204*T*
  heterodimer formation **3**:204
  natural killer cell development **3**:207
  signaling, cytokines **3**:206*T*
STAT6 **3**:203
  asthma **3**:208
  chromosomal location **3**:203, **3**:204*T*
  signaling **3**:207
    cytokines **3**:206*T*, **3**:207
Static blocking, computer-assisted drug
  design (CADD) **4**:22
Statins **1**:106–107; **6**:469*F*, **6**:480
  see also specific drugs
  adverse effects **6**:476
    muscle pain/weakness **6**:476–477

analog design **2**:651
bioisosterism **2**:650, **2**:650*F*
clinical trials **6**:473, **6**:474*T*
  see also specific clinical trials
  ezetimibe combination **8**:79*F*
clinical use **1**:606–607
  age-related macular degeneration
    treatment **6**:315*T*
  Alzheimer's disease
    treatment **6**:211
  multiple sclerosis
    treatment **6**:246; **7**:928
  nervous system autoimmune
    diseases, treatment **6**:246
  systemic lupus erythematosus
    treatment **6**:246
  transplant rejection **7**:920
combinations
  antioxidants **6**:478
  cholesterol absorption
    inhibitors **6**:477
  ezetimibe **8**:77, **8**:78*F*
discovery/development **1**:18, **1**:325–
  326, **1**:532; **2**:1074
glucuronic acid conjugation **5**:151
high-density lipoprotein cholesterol-
  elevating agent
  combination **6**:479
new drugs **1**:107
polymorphism effects **2**:67
subclasses **6**:480
Statin Therapies for Elevated Lipid
  Levels Compared Across Doses
  (STELLAR) **6**:474*T*, **6**:476
Statistical classification of activities of
  molecules for pharmacophore
  identification (SCAMPI) **4**:133
Statistical genetics, genetic variation
  analysis **3**:338–339
Statistical learning methods,
  cytochrome P450 molecular
  dynamics **5**:817
Statistical modeling, computer-assisted
  drug design (CADD) *see*
  Computer-assisted drug design
  (CADD)
Statistical molecular design,
  multivariate library design
  combination **4**:330
Statistical significance, clinical
  trials **2**:184
Statistics
  accuracy, fluorescence screening
    assays **3**:612–613
  affective disorders, clinical trial
    issues **6**:64
  clinical trials *see* Clinical trials
  CoMFA **5**:589
  electrotopological indices (E-state
    index) *see* Electrotopological
    indices (E-state index)
  fluorescence screening assays *see*
    Fluorescence screening assays

Genome Reviews **3**:357
microarrays **3**:90
solubility models *see* Solubility models
toxicity, computational models *see*
  Toxicity, computational models
two-dimensional gel
  electrophoresis **2**:335–336
Status epilepticus **6**:284
  models **6**:286
Staudinger reaction
  PASP **3**:793
  polyethylene glycol
    conjugates **5**:1060, **5**:1060*F*
Staurosporine
  discovery/development **1**:381–382
  structure **1**:382*F*
Stavudine **1**:129; **7**:279–282
  development **1**:69
  drug compatibility **7**:342
  drug regimens **7**:342*T*
  medicinal chemistry **7**:338–340
  oral absorption **5**:680*T*
  resistance **7**:340–341
  structure **1**:70*F*
  target information **4**:750*T*
Steady state, pharmacokinetics **5**:21,
  5:21*F*
Steatohepatitis, antivirals **2**:1082
Steepest descents algorithms, core
  molecular modeling **4**:101–102,
  4:102*F*
Steganacin **7**:91
  SAR **7**:91
  structure **7**:91*F*, **7**:92*F*
STELLA, metabolic drug–drug
  interaction software **5**:841
Stem cells
  ABCG2 transporter **5**:63–64
  animal experimentation alternatives
    see Animal experimentation
    alternatives
  bioethics **1**:559, **1**:682–683
  cell-based assays **3**:641–642, **3**:643
  consumer advocacy groups **2**:271
  genomics **2**:46
  Italy **1**:558–559
  Japan **1**:558–559
  South Korea **1**:558–559
  UK **1**:558–559
  venture capital risk factors/
    problems **1**:558
Stem cell transplantation
  age-related macular degeneration
    treatment **6**:315*T*
  bone disease treatment **6**:516
  diabetes mellitus management **6**:455
  nervous system autoimmune diseases,
    treatment **6**:247
Stem loops, messenger RNA (mRNA)
  structure **2**:1044, **2**:1044*F*
Stemmer, William, sexual PCR **1**:437
Step assessment, solution phase library
  synthesis **3**:770

Stephensen, John, propranolol, invention of 2:254
Stepwise elaboration, structure-based library design 4:321
Stepwise methods, P-glycoprotein models 5:782
Stereochemistry
  checks, protein 3D structure 3:547
  conversion, chemical instability 5:493, 5:493F
  gabapentin derivatives 8:232, 8:233T
  protein 3D structure 3:546
  stability testing 5:501
'Stereoelectronic control,' Hirschmann, R F 1:6
Stereoisomers
  ephedrine 6:401, 6:401F
  QSAR 4:154
Stereoisomer synthesis see Chiral drug discovery
STEREOIZER program, 2D- to 3D structure 4:464
STEREOPLEX program, 2D- to 3D structure 4:462
Stereoselectivity
  gabapentin derivatives 8:231, 8:232T
  metabolism 5:199F, 5:211
  stereoisomer synthesis 2:715–718, 2:719F
Steric factors
  ionization 5:605, 5:606
  lipophilicity measurement 5:434
  log P' 5:439
Steric fields
  CoMFA 5:589
  molecular descriptors see Molecular descriptors
  three-dimensional descriptors, similarity searching 4:173
Steric properties
  ligand–receptor interactions see Ligand–receptor interactions
  QSAR see Quantitative structure–activity relationship (QSAR)
STERIMOL descriptors
  QSAR steric effects 4:46
  QSAR substituent selection models 4:52
Sterix 1:492T, 1:513
Sternbach, Leo
  management difficulties 2:259
  Valium, invention of 2:256–257
Steroid-derived modulators see Selective estrogen receptor modulators (SERMs)
Steroid hormone imaging
  [^{18}F]FDG 3:667
  16 alpha-[^{18}F]fluoroestradiol 3:667
  small animal imaging 3:667
Steroid hormone receptors, molecular probes 3:664
Steroidogenic factor 1, nomenclature 2:996T

Steroids
  acne treatment see Acne treatment
  arthritis treatment see Arthritis treatment
  biological activities 3:926T
  COPD treatment 7:848
  metabolism, alcohol dehydrogenases (ALDs) 5:199F
  multiple sclerosis treatment 7:928
  P-glycoprotein interactions 5:769T
Steroid sulfotransferase 5:147–148
Sterol esters, stratum corneum 5:460
Sterol O-acyltransferase, coenzyme A conjugation 5:157
Sterol synthesis, alendronate 8:203
STI-571 2:972
  Bcr-Abl kinase inhibitor 2:972–973
  characteristics 2:984T
  structure 2:971F
Stiff-man syndrome 6:424
Stigmatellin, mitochondria complex III 2:1075
Stigmatization, psychiatry 1:288
Stilbazulenyl nitrone 6:266
  structure 6:267F
Stille couplings, PASP 3:813
Stille reaction, solid support synthesis 2:467–468, 2:470F
Stimulant laxatives
  see also specific drugs
  irritable bowel syndrome treatment 6:658
  side effects 6:654
Stimulants
  see also specific drugs
  addiction medication 6:181T, 6:185
    baclofen 6:185
    disulfiram 6:185
    modafinil 6:185
    tiagabine 6:185
    topiramate 6:185
  ADHD treatment 1:701
  introductions 1:197
STK6, PDB entries 2:965T
Stock–Einstein equation, light scattering, promiscuous ligands 2:743
Stock market listings, SME funding 1:494
Stogar see Lafutidine
Stool softeners, irritable bowel syndrome treatment 6:658
STOP-NIDDM study, diabetes mellitus prevention trials 6:421
Storage
  chemical stability see Chemical stability
  compound storage/management 3:567
  dissolution effects 5:401–402
  drug design, de novo 4:285
  fragment/ligand selection drug design 4:296

Strain specificity, gene expression vectors 3:109
Strand transfer, HIV life cycle 7:331
Strategic thinking, project teams 2:145
Strategy, protein crystallography 3:468–469
Strattera see Atomoxetine
Stratum corneum 5:459
  intercellular lipids 5:460
    ceramides 5:456, 5:456F, 5:460
    cholesterol 5:460
    free fatty acids 5:460
    sterol esters 5:460
  structure 5:459, 5:460F
Streisinger, George, Danio rerio as model system 3:10
Strength Pareto archived evolutionary algorithm (SPEA), multiobjective optimization 4:768
Strengths, Weaknesses, Opportunities and Threat analysis (SWOT), project management 2:14
StrepII tag, affinity chromatography 3:436–437
Strep-tag, as affinity tag 3:421
Streptavidin
  gene expression vectors 3:110, 3:110T
  structure, coordinate errors 3:532
Streptococci, dalbavancin 7:659T
Streptococcus pneumoniae
  infections 7:675
  linezolid 7:686
  macrolide antibiotic resistance 7:521
  macrolide antibiotics 7:522
  resistance mechanisms, efflux pumps 7:586
  respiratory tract infections 7:739
Streptococcus pyogenes infections, respiratory tract infections 7:739
Streptogramins 7:660
  see also specific drugs
  dual-action agents, oxazolidinones 7:685
  resistance 7:664
  structural classes 7:660–662
Streptokinase 6:767, 6:768
  clinical indication 6:768–769
  clinical trials 6:772
    GISSI-1 6:772
    ISIS-2 trial 6:772
  drug–polymer conjugates 5:1047
  history 6:767
  inactivation 6:769
  pharmacokinetics 6:769
Streptomyces aureofaciens, chlortetracycline production 7:599
Streptomycin 7:630
  clinical use

Streptomycin (*continued*)
   mycobacterium treatment 7:704–705, 7:705T, 7:713, 7:714–715
     tularemia 7:637
     Yersinia pestis 7:637
   discovery/development 1:325–326
   mechanism of action 2:1041–1042; 7:639
   non-microbial activities 7:641
     HIV infection 7:641
   oral absorption 5:468T, 5:680T
   resistance 7:642
   structure 7:630F, 7:713F
Streptozocin 7:138–139
   diabetes mellitus type 1 animal models 6:433
   structure 7:139F
   toxicity 7:46T
Stress-induced hyperthermia, anxiety disorders, animal models 6:95
Stress responses, RNAi (RNA interference) 3:181
Stress-responsive kinases, ubiquitin protein targeting 3:192
Stress signaling, large-scale virus vector libraries 3:18–19
Stress urinary incontinence
   definition 8:124
   duloxetine clinical trials 8:131F, 8:132
Stress urinary incontinence (SUI) 6:544
Stretch activated channel blockers, antiarrhythmic drugs 6:755
Strida Pharmaceuticals 1:513
STRIDE, protein structure prediction, bioinformatics 3:325
STRING database, protein interaction networks, bioinformatics 3:335–336
Stroke 6:12, 6:694, 6:697
   see also Central nervous system injury; Neuroprotective drugs
   definition 6:697
   diagnosis 6:13
   experimental models 6:257T
   hemorrhagic 6:697
   incidence 6:13
   ischemic 6:697
   management 6:697–698
     tissue plasminogen activator (tPA) 6:697–698
   mitochondria 2:1054–1055, 2:1060
   prevalence 6:253, 6:695T, 6:697
   risk factors, hypertension 6:706
   thrombolysis 6:763
   treatment 6:263
     nimodipine 6:259
     recombinant tissue plasminogen activator 6:263
     thrombolytic agents 6:263
Stroke-prone rats, hypertension 6:707–708

Stromelysin (MMP-3) inhibitors, peptidomimetics 2:613, 2:614F, 2:632
Strong acids
   pH 5:360
   pH calculations 5:360
Strong bases
   pH 5:360
   pH calculations 5:360
Strong hydrogen bonds
   CSD 3:402, 3:404F
   intermolecular interactions, CSD 3:403F
Strong proton acid/base role, brain perfusion 5:755
Strongyloides stercoralis see Strongyloidiasis
Strongyloidiasis 7:751T
   biology 7:759
   clinical manifestations 7:759
   drug treatment 7:753T
Strontium ranelate, bone diseases/disorders 6:516
Strontium salts, bone disease treatment 6:514
Structural abnormalities, genetically engineered animals 3:155
Structural alerts
   drug metabolism, in vitro studies 5:240
   genotoxicity, computational models 5:917
Structural biology and CADD
   partnership 4:6
   x-ray crystallography 4:6–8
Structural clustering, biological fingerprints *see* Biological fingerprints
Structural criteria, comparative computational modeling 3:557
Structural databases
   architectural importance 1:43–44
   drug design 1:43
   search engine design 1:44
   tiered 1:43, 1:44F
   two-dimensional 1:43
Structural data (SD) file, ligand-based virtual screening 4:460
Structural descriptors
   P-glycoprotein models 5:780
   QSAR *vs.* 5:919
Structural errors, x-ray structures, problems 3:538
Structural filters/alerts, screening library selection 4:499–500
Structural flexibility, computer-assisted drug design (CADD) *see* Computer-assisted drug design (CADD)
Structural genomics 3:551–560
   see also Protein Structural Initiative (PSI)
   auto-publishing 3:558

   comparative computational modeling 3:557
     accuracy 3:557
     energetic criteria 3:557
     PDB related structures 3:557
     structural criteria 3:557
   definition 2:382T
   future work 3:558
     technology improvement 3:558
   metalloproteomics 3:556
     HTS 3:556–557
     x-ray absorption spectroscopy (XAS) 3:556
   protein cloning 3:554
     PCR 3:554
   protein expression 3:554
     automation 3:554–555
     baculovirus 3:554
     Escherichia coli 3:553T, 3:554
     Pichia pastoris 3:554
     SUMO-based expression 3:555
   protein purification 3:555
     affinity tags 3:555
     automation 3:555
     characterization studies 3:555
   signal transduction 2:964
   structure determination 3:555
     NMR 3:555
     x-ray crystallography *see below*
   structure–function
     relationship 3:557
     DALI 3:557
     ligand presence 3:557–558
     PINT 3:557
     VAST 3:557
   target selection 3:553
     biomedical interest 3:554
     Pfam5000 strategy 3:554
     protein families 3:553–554
     protein orthologs 3:554
     single-domain prokaryotic proteins 3:554
     structure determination methods 3:554
   x-ray crystallography 3:555
     automation 3:556
     crystallization improvements 3:555
     limited proteolysis combined with mass spectrometry (LP/MS) 3:555
     methodology determination 3:556
     'phase problem' solution 3:556
     synchrotron radiation 3:556
Structural Genomics Consortium 1:574
Structural Genomics for Pathogenic protozoa Consortium (SGPP) 3:552T
Structural information
   Integr8 3:368
   structure-based fragment screening 4:796

Structural interaction fingerprints (SIFs), postDock processing 4:272

Structural motif replacement, peptide/protein drugs 2:590

Structural novelty, intellectual property 1:51–52

Structural optimization, stability testing 5:503, 5:503F, 5:504F

Structural parameters, bioisosterism modification analysis 2:686, 2:687F

Structural properties, blood–testis transfer modeling 5:739

Structural proteomics, definition 2:382T

Structural similarity
  lipophilicity prediction, rule-based systems 5:664
  sequence similarity vs. 3:462

Structural target of CADD 4:19

Structural water molecules, Molecular Interaction Fields 4:238

Structure
  bioavailability models 5:708
  R&D management 2:249

Structure, quantum mechanical calculations 4:386
  atom hybridization 4:387
    aniline 4:387, 4:387T
    histamine H$_2$ antagonists 4:389–390, 4:389F
    MH3 4:388–389, 4:389F
    pramipexole 4:388–389
  bioisosterism 4:386–387
  molecular conformational energies 4:391
    ABT-594 4:391, 4:391F
    analgesics 4:391
    bioactive confirmations 4:392
    HIV proteinase inhibitors 4:391
    ligand binding information 4:392–393
    olomucine 4:392, 4:392F, 4:393F, 4:394F
    3D-QSAR 4:392
  pharmacophore concept 4:386–387
  QSAR 4:390
    1,4-dihydropyridine derivatives 4:390–391, 4:390F
    electronic charge transfer 4:390–391

Structure–absorption relationships (SABRs), absorption 1:52

Structure–activity relationships (SARs) 4:14, 4:538
  A-80987 8:190
  carcinogenicity, computational models 5:921
  chemical proteomics 3:44
  Comer, William (views on) 8:4
  computer-assisted drug design (CADD) 4:14

data 4:538, 4:539T
development 1:33–35
drug design 1:48–49
drug discovery 2:276
  fragment-based 3:941
  productivity 2:124–125
flumequine 7:580
fluoroquinolone antibiotics see Fluoroquinolone antibiotics
Ganellin, C R and 8:43
glycinamide ribonucleotide formyltransferase (GARFT) 4:75, 4:75F, 4:76F
GPCR structure
  determination 4:676
  historical perspective 4:437–438
HIV protease inhibitors 8:188
ligand-based library design 4:310–311, 4:311F
model combination, parallel library design 4:327
NMR 2:525
norfloxacin 7:583
pharmacophores 4:120
promiscuous ligands 2:750
solution phase library synthesis 3:769
specific drugs
  acyclovir 7:298
  aminoglycosides 7:634–635
  AMPA-sensitive receptors 2:900
  anti-NS5b nonnucleosides 7:397F, 7:401
  azetidinone cholesterol absorption inhibitors see Azetidinone
  biological macromolecules see Biological macromolecules
  calcium-channel blockers 6:718
  CCR3 inhibitors 7:884
  cidofovir 7:304–305
  DuP-105 7:679
  DuP-721 7:679
  eperezolid 7:680
  famciclovir 7:300–301
  glucan synthase inhibitors 7:458
  gonadotropin-releasing hormone (GnRH) agonist/antagonists 7:175–176
  lamivudine 7:313–314
  linezolid 7:680
  lometrexol 4:75, 4:75F
  neuraminidase inhibitor design 4:622–623
  oseltamivir 7:379, 7:381T
  oxazolidinones 7:679, 7:683
  pneumocandins 7:460
  ravuconazole 7:452–453
  terbinafine 7:448
  tetracyclines 7:599, 7:617
  triphenylethylene anticancer agents 7:162
target development 2:139

toxicity prediction 5:910
toxicogenomics combination 2:568

Structure-based drug design (SBDD) 1:45T, 2:760; 4:65–86, 4:775–806
  A-77003 4:77T
  ADME properties 4:776
    see also Cytochrome P450 inhibitors; Human serum albumin (HSA); Plasma protein binding, prediction
  amprenavir 4:77T
  anti-AIDS drugs 1:68
  antifolates 4:72
  blood substitutes 4:71
  captopril 4:77T
  companies 2:761T
  crystallization problems 4:598–599
  de novo 4:284, 4:286
  dihydrofolate reductase (DHFR) 4:72
  electron diffraction 4:69
  enzymes see Enzyme(s)
  examples 4:70, 4:77T, 4:598
  fragment-based 2:760–761
    screening see Structure-based fragment screening
  future work 1:81; 4:786, 4:802
  glycinamide ribonucleotide formyltransferase (GARFT) 4:74
  hemoglobin 4:70
  historical perspectives 4:66, 4:69
    computing 4:67
    Ehrlich, Paul 4:66
    Levinthal, Cyrus 4:67
    molecular graphics 4:68
    molecular models 4:67
  HTS see High-throughput screening (HTS)
  in-silico design, drug discovery 2:301, 2:301T
  L-735,524 4:77T
  lead molecule binding 4:776
  methods 4:66
    x-ray crystal data collection 4:66
    x-ray crystallography 4:66
  peptidomimetics see Peptidomimetics
  protein production 4:598–599
  saquinavir mesylate 4:77T
  sickle cell anemia 4:70
  structure determination process 4:599
  target vs. disease 2:114–115
  theory 4:66
  thymidylate synthase see Thymidylate synthase
  Viracept 4:77T

Structure-based fragment screening 4:786
  bioassays 4:788
  cocktailing 4:795

Structure-based fragment screening
   (*continued*)
   crystallographic shape
      similarity 4:795
   procedure 4:795
definition 4:788
development 4:786–787
examples 4:798, 4:798*F*,
      4:800*F*
   conformational changes 4:799
   p38 kinase inhibitors 4:799,
      4:801*F*
   thrombin inhibitors 4:799,
      4:801*F*
general purpose libraries 4:792
   drug fragment set 4:793
   NMR 4:792
   'rule-of-three' 4:793
   scaffold-based approach 4:792
   SHAPES methodology 4:792
   version 1.0 4:793, 4:794*F*
   version 2.0 4:793
informatic requirements 4:795
   AstexViewer 4:796, 4:797*F*
   AutoSolve 4:796
   project pages 4:797, 4:797*F*
   pyramid database 4:796
   structural information 4:796
libraries 4:791
   targeted 4:792
noncrystallographic methods 4:790
   bioassays 4:789*F*, 4:791
   ligand-based NMR 4:791
   mass spectrometry 4:790
   NMR 4:790
   virtual screening 4:789*F*, 4:790
x-ray crystallography 4:789, 4:789*F*
Structure-based ligand design,
   definition 4:237
Structure-based QSAR, oral absorption
   prediction, molecular
   descriptors 5:682
Structure-based virtual
   screening 3:250–251,
   4:478
aim 3:251
docking program evaluation 3:251–
   252
DOCK program 3:251
examples 4:487, 4:488*T*
FlexPharm program 3:252
force-field scoring 3:251
homology models 4:487–488
   MOBILE program 4:487–488
knowledge-based scoring 3:251
ligand pose prediction 3:252
protein flexibility 3:251
regression-based scoring 3:251
scoring functions 3:251
ZINC database 3:252
Structure correlation principles,
   CSD 3:399
Structure databases *see* Databases

Structure descriptors, electrotopological
   indices (E-state index) *see*
   Electrotopological indices
   (E-state index)
Structure determination
   NMR 3:948
   structural genomics 3:554
   x-ray crystallography 3:948
Structured query language (SQL),
   databases *see* Databases
Structure elucidation, NMR *see* Nuclear
   magnetic resonance (NMR)
Structure–excretion relationships
   (SER), excretion 1:56
Structure–function relationship,
   structural genomics *see* Structural
   genomics
Structure generation programs 3:240
Structure information
   representation **4:537–574**
   see also Electrotopological indices
      (E-state index)
   advantages 4:540
   applications 4:558
      library/database screening 4:558
   group characterization 4:561
   as information network 4:540
      electron distribution 4:541
      reductionism 4:540
   library screening 4:540
      ligand identification 4:540
      molecular descriptors 4:540
   molecular connectivity indices *see*
      Molecular connectivity indices
      (MCIs)
   molecular fragment
      characterization 4:561
   pharmacophores 4:561
   QSAR 4:564
   topological nature 4:539–540
Structure–metabolism rule
   development 5:805
Structure-modifying anti-inflammatory
   drugs, arthritis treatment *see*
   Arthritis treatment
Structure optimization, Comer, William
   (views on) 8:3
Structure–permeation relations, dermal/
   transdermal administration *see*
   Dermal/transdermal
   administration
Structure prediction
   bioinformatics 3:324
   transmembrane helices 4:673
Structure–property relationships
   (SPR)
   proton pump inhibitors (PPIs) 6:607
   stability testing 5:494
Structure quality measures, 3D
   structure generation *see* 3D
   structure generation
Structure reconstruction, GRID force
   field 4:241, 4:241*F*

Structure searching, similarity
   searching 4:168
Structure/substructure searching *see*
   Chemoinformatics, molecule
   representation
Structure to Function (S2F) 3:551,
   3:552*T*
   Target DB 3:551
Structure–toxicity relationships,
   toxicity 1:56–58
Strychnine, structure 2:895*F*
ST-segment elevation myocardial
   infarction (STEMI),
   thrombolysis 6:763
Study of Tamoxifen and Raloxifene
   (STAR) 8:96, 8:104–106
   breast cancer prevention 8:112–113,
      8:113*F*
   endocrine-modulating anticancer
      agents 7:152
SU-101 7:198
   structure 7:198*F*
SU-668 7:200
SU-5402 7:203
   structure 7:203*F*
SU-5416 (semaxanib) 2:969–970;
   7:200
   clinical trials 7:201
   structure 2:970*F*; 7:201*F*
   VEGFR family kinase
      inhibitors 2:969
SU-6656
   Src kinase inhibitor 2:976–977
   structure 2:978*F*
SU-6668 2:969–970
   characteristics 2:984*T*
   clinical trials 7:201
   KIT inhibitor 2:972
   structure 2:970*F*; 7:201*F*
   VEGFR family kinase
      inhibitors 2:969
SU-9516 7:205–206
   structure 7:206*F*
SU-11248 1:513; 2:969–970; 7:199,
   7:200
   acute myeloid leukemia 2:971
   characteristics 2:984*T*
   clinical trials 7:201
   KIT inhibitor 2:972
   structure 2:970*F*; 7:201*F*
   VEGFR family kinase
      inhibitors 2:969
SU-11925 7:196
   structure 7:196*F*
Subactivities
   lead optimization 2:213
   primary chemistry activity
      definition 2:209,
      2:210*F*
Sub-ambient chemistry, CEM Discover
   platform microwave-assisted
   chemistry 3:847–848
Subarachnoid anesthetics 6:356

mesterolone **6**:532
oral testosterone **6**:532
testosterone decanoate **6**:532
osteoporosis effects **6**:536
precautions **6**:536
side effects
benign prostate
enlargement **6**:537
prostate cancer **6**:537
skin irritation management **6**:531
transdermal **6**:529
Androgel **6**:531, **6**:531*F*
intramuscular *vs.* **6**:531
percutaneous
dihydrotestosterone **6**:530
potential partner testosterone
transfer **6**:532
testosterone patch therapy
(Androderm) **6**:530, **6**:530*F*
transscrotal **6**:530
Testosterone undecanoate
intramuscular testosterone
replacement therapy **6**:528
oral testosterone replacement
therapy **6**:532
pharmacokinetics **6**:533*T*
structure **6**:522*T*
Test protocols, cell-based assays **3**:623
Test sets, statistical methods, E-state
index **5**:562
Tethering
drug discovery *see* Fragment-based
drug discovery
'scaffold-based' drug design **3**:892
screening, dynamic combinatorial
libraries (DCLs) **3**:965
Tetomilast
clinical trials **7**:882
COPD treatment **7**:882
structure **2**:935
Tet-On-Tet-Off operons **7**:622
molecular biology applications **7**:622
selective gene expression
repression **7**:622*F*
selective repression **7**:622
transgenic animals **7**:622
Tetracaine, structure **5**:359
2,3,7,8-Tetrachlorodibenzo-*p*-dioxin
(TCDD),
immunosuppression **5**:223
Tetracyclic guanine derivatives,
phosphodiesterase-1
inhibitors **2**:931–932
Tetracyclines **7**:**597–628**, **7**:599
see also specific types
acne treatment **7**:939
antioxidant activity **7**:619
A-ring chemistry **7**:603, **7**:603*F*
metal-binding **7**:617
biosynthesis **7**:600–601, **7**:600*F*
acetyl-CoA **7**:600–601
Actinomadura **7**:600
Actinomycetes **7**:601

4-aminotetracycline **7**:600–601
Dactylosporangium **7**:600
4-dimethylamino-
anhydrotetracycline **7**:600–601
6-hydroxy-5a-
dehydrotetracycline **7**:600–601
4-hydroxy-6-
methylpretetramide **7**:600–601
4-hydroxy-12-deoxyan-
hydrotetracycline **7**:600–601
intermediates **7**:600–601, **7**:602
malonylamyl-CoA **7**:600–601
6-methylpretetramide **7**:600–601
organisms **7**:600
B-ring chemistry **7**:604, **7**:605*F*
C2 modifications **7**:603
C4 derivatives **7**:603, **7**:604*F*
activity **7**:603
antimicrobial activity **7**:604
quaternary ammonium betaine
analogs **7**:604
chemical modification **7**:602
chemical properties **7**:617
see also specific properties
classification, by target site **7**:471*T*
C-ring chemistry **7**:604
second generation **7**:605, **7**:606*F*
definition **7**:598
dimethylamino group
epimerization **7**:603
kinetics **7**:603–604
discovery **7**:599
D-ring chemistry **7**:607, **7**:607*F*
electrophilic aromatic
substitutions **7**:607, **7**:607*F*
eukaryotic cell membranes **7**:620
apoptosis induction **7**:620
cytotoxicity **7**:620
formulations **7**:619
generation/origins **7**:599
hematological effects **7**:620–621
activated T-cells **7**:620
alveolar macrophages **7**:620
granulocytes **7**:620
neutrophils **7**:620
history **7**:598, **7**:598*F*
ionization constant (pK_a) **5**:366*T*,
5:373
mechanism of action, ribosome
interactions **2**:1042–1043
metal binding **7**:617
albumin binding **7**:618
A-ring chemistry **7**:617
calcium **7**:617–618
cobalt **7**:618
dedimethylamino
tetracyclines **7**:618
DNA binding **7**:618
ionophores **7**:618
iron **7**:618
lithium **7**:618
magnesium **7**:617–618
manganese **7**:618

matrix metalloproteinase
binding **7**:618
phototoxicity **7**:619
potassium **7**:618
sodium **7**:618
spectroscopy **7**:617–618
zinc **7**:618
non-antibiotic effects **7**:599, **7**:620
see also specific effects
Alzheimer's disease **7**:620
amyotrophic lateral sclerosis **7**:620
cerebral ischemia **7**:620
cytokines **7**:621
matrix metalloproteinases **7**:621,
7:621*T*
multiple sclerosis **7**:620
Parkinson's disease **7**:620
oral absorption **5**:468*T*
mechanisms **7**:620
pharmacokinetics **7**:619
diet **7**:619
phototoxicity **7**:618
metal binding effects **7**:619
reactive oxygen species
formation **7**:618
screening **7**:619
singlet oxygen formation **7**:618,
7:619
superoxide anion radical
formation **7**:618
physicochemical properties **7**:619
pH effects **7**:619
solubility **7**:619
position C13 bacterial efflux protein
inhibitors **7**:605
resistance, acne treatment **7**:939
SAR **7**:599, **7**:617
structure **2**:1043*F*; **5**:370*F*, **5**:444*F*;
7:598, **7**:598*F*, **7**:599*F*, **7**:600*F*,
7:602*F*, **7**:611*F*
C6-hydroxyl group **7**:604–605,
7:605*F*
deprotonations **7**:617
intramolecular H bonding **7**:617
keto-enolate group
formation **7**:617
naphthalene ring **7**:602
tautomerism **7**:617
synthesis
Nocardia **7**:600
Penicillium **7**:600
third-generation **7**:608
transition-metal catalyzed
reactions **7**:609*F*, **7**:610
2N prodrugs **7**:603
Tetracyclines, antimicrobial
activity **7**:610, **7**:611, **7**:611*F*,
7:613
acne **7**:611–612
acne rosacea treatment **7**:963
bacterial resistance **7**:615, **7**:616*T*
efflux mechanisms **7**:616
inactivation mechanisms **7**:617

Tetracyclines, antimicrobial activity (*continued*)
ribosomal protection **7**:616
Tet repressor **7**:605, **7**:614, **7**:615
bacterial uptake **7**:613
Gram-negative bacteria **7**:613–614
passive diffusion **7**:613
zwitterion *vs.* lipophilic **7**:613, **7**:613*F*
Chlamydia infections **7**:612
Cryptosporidium parvum infections **7**:613
efficacy **7**:610
ehrlichiosis **7**:612
Entamoeba histolytica infections **7**:613
Giardia lamblia infections **7**:613
Gram-negative **7**:610
Gram-positive **7**:610
malaria treatment **7**:610, **7**:612
mechanism of action **7**:614
membrane perturbation **7**:613, **7**:613*F*, **7**:614
protein synthesis inhibition **7**:614
ribosome binding **7**:614, **7**:615*F*
Mycoplasma **7**:612
obligate intracellular parasites **7**:612
Q-fever treatment **7**:612
QSAR **7**:613
Rickettsias **7**:612
Rocky Mountain Spotted Fever (RMSF) treatment **7**:612
Toxoplasma gondii infections **7**:613
Ureaplasma infections **7**:612
Wolbachia infections **7**:610, **7**:612
Tetracyn *see* Tetracyclines
Tetraethyl ammonium (TEA), K^+ ion channel interactions **4**:719
Tetraethylene glycol diacrylate cross-linked polystyrene support (TTEGDA-PS), solid support synthesis **2**:497, **2**:498*F*, **2**:498*T*
delta-9-Tetrahydrocannabidiol, structure **6**:405*F*
Tetrahydrocannabinol
bioavailability **5**:712*T*
structure **6**:173*F*
Tetrahydrochrysene derivatives, selective estrogen receptor modulators (SERMs) **7**:164, **7**:164*F*
Tetrahydrofluorenones, estrogen receptor-β-selective ligands **7**:169, **7**:170*F*
Tetrahydrofuranones, solid-phase pool libraries **3**:705*F*
Tetrahydrofurans
microwave heating **3**:840*T*
solid-phase pool libraries **3**:707*F*
Tetrahydroisoquinolines, *N*-methylation **5**:146–147, **5**:146*F*
Tetrahydrooxazepines, solid-phase discrete libraries **3**:709*T*

Tetrahydroquinolines, solid-phase pool libraries **3**:705*F*
Tetraiodophenolphthalein, β-lactamase, inhibition of **2**:740, **2**:741*F*
Tetrakistriphenylphosphine palladium reagent, PASP **3**:795
Tetramers, x-ray structures, problems **3**:540
4-hydroxy-2,2,6,6-Tetramethylpiperidin-*N*-oxide (TEMPOL), mitochondria, drug targeting **2**:1067
Tetramisole
bioisosterism **2**:678, **2**:679*F*
structure **2**:679*F*
Tetrapeptide libraries, solid-phase libraries **3**:746
Tetraplatin
as prodrug **5**:1033
structure **5**:1032*F*
meso-Tetra(3-hydroxylphenyl)porphine, structure **5**:360
Tetrazoles
bioisosterism **2**:662*T*
structure **5**:371*F*
Tet repressor
anhydrotetracycline binding **7**:615
13-cyclopentyl-5-hydroxytetracycline **7**:605, **7**:606*F*
tetracycline bacterial resistance **7**:605, **7**:614, **7**:615
Tetrodotoxin (TTX), sodium channels **2**:837
Tetrodotoxin-resistant voltage-gated sodium channels (Na$_V$) **6**:322, **6**:323, **6**:343, **6**:345
Teveten *see* Eprosartan
Textbook of Medicine, β-adrenergic receptor antagonist contraindications **8**:141
Text databases **3**:277
Text formats, two-dimensional representations **3**:298
Text mining, protein function prediction *see* Protein function prediction
TFV *see* Tenofovir
Tg2576 mice, sleep disorders **6**:152–153
TG-973980 **7**:590
TH-9507, primary insomnia treatment **6**:161
Thailand, healthcare systems reform **1**:641
β-Thalassemia
5-azacytidine **1**:273
demethylating agents **1**:273
sodium phenyl butyrate (buphenyl) **1**:274
Thalidomide **2**:759; **6**:240*T*
biological activity **2**:759

clinical activity **2**:759
amyotrophic lateral sclerosis treatment **6**:237
erythema nodosum treatment **2**:759
clinical trials **2**:759
hydrolysis **5**:139, **5**:139*F*
structure **2**:759*F*; **6**:238*F*
toxicity **7**:46*T*
Theaflavine **1**:452*T*
Theaflavine-3,3′-digallate **1**:452*T*
Theaflavine-3′-monogallate **1**:452*T*
Theaflavine-3-monogallate **1**:452*T*
T helper cells, type 1 **5**:222
Th2 cell balance, glatiramer acetate (GA) **8**:181, **8**:182
T helper cells, type 2 **5**:222
asthma **7**:877
Th1 cell balance, glatiramer acetate (GA) **8**:181, **8**:182
Theophylline
asthma treatment **7**:883
bioavailability **5**:712*T*
oral absorption **5**:468*T*, **5**:680*T*
as phosphodiesterase inhibitor **2**:947*T*
structure **5**:371*F*
Theoretical descriptors, BCS **5**:978
Therapeutically Relevant Multiple Pathways Database, BIDD Databases **3**:305
Therapeutic area
decisions, functional discipline organization **2**:109
drug development productivity **2**:128, **2**:129*T*
organization *see* Pharmaceutical industry
Therapeutic area teams, Comer, William **8**:3
Therapeutic dose estimation, inhibition, in vitro studies **5**:248
Therapeutic irradiation, breast cancer **7**:151
Therapeutic Target Database, BIDD Databases **3**:305
Therapeutic vaccines, nervous system autoimmune diseases, treatment **6**:246
Therapeutic window, central nervous system injury **6**:262
Theravance **1**:513
GlaxoSmithKline alliance **2**:234
Thermal vibration, x-ray crystallography **4**:763–764
Thermodynamic cycles
free energy calculation *see* Free energy calculation
ionization **5**:604, **5**:604*F*
Thermodynamic forces, ionization *see* Ionization

Thermodynamic integration
  binding free energy
    calculation **4**:661, **4**:661*F*,
      4:662*F*, **4**:663*F*
  free energy calculation **4**:656
Thermodynamics
  hydrogen bonding **5**:536
  ligand–receptor interactions **4**:194
  lipophilicity **5**:428–429
  protein structure prediction,
    bioinformatics **3**:326
  RNA structure prediction,
    bioinformatics **3**:327
  solid-state physicochemistry
    see Solid-state physicochemistry
Thermodynamic solubility,
  definition **5**:628
Thermodynamic stable forms, solid-
  state physicochemistry **5**:514
Thermogenin (uncoupling protein-
  1) **2**:1080
Thermolysin
  peptidomimetics **2**:631, **2**:634*F*
  three-dimensional structures **2**:626*T*
Thermostated baths, Woods' rotating
  disk apparatus **5**:415
Thiabendazole
  bioisosterism **2**:679*F*
  nematode infection treatment **7**:760
  structure **2**:679*F*
Thiadiazole ethers, solid-phase pool
  libraries **3**:705*F*
Thiadiazoles, phosphodiesterase-7
  inhibitors **2**:944
1,2,4-Thiadiazoles, carboxylic ester
  bioisosterism **2**:663
Thiakis **1**:492*T*
Thiamylal, structure **6**:352*F*
Thiazide diuretics **6**:585–587
  see also specific drugs
  α₁-adrenoceptor antagonist
    combination **6**:714
  β-adrenergic receptor antagonists
    with **6**:724
  bioavailability **6**:709*T*
  dose **6**:709*T*
  half-life **6**:709*T*
  hypertension treatment **6**:585–587
  mechanism of action **2**:63
  side effects **6**:587, **6**:708
  structure **6**:709*T*
Thiazide-like diuretics *see* Diuretics
Thiazol-2-ylamine,
  bioisosterism **2**:660*T*
Thiazole peptide analogs, solid-phase
  discrete libraries **3**:709*T*
Thiazoles, GABA receptors **2**:893
Thiazole urea derivatives, herpes
  viruses, antivirals **7**:309–310
Thiazolidinediones **2**:1021–1022;
  6:439*T*, **6**:446
  see also specific drugs
  advantages **6**:448

bioisosterism **2**:662*T*
  carboxylic acid bioisosterism **2**:661
  clinical use **6**:447
  disadvantages **6**:448
  mechanism of action **2**:1081;
    6:446
    peroxisome-proliferator activated
      receptors (PPARs) **6**:446,
      6:447*T*
  pharmacokinetics **6**:447
  side effects/contraindications **6**:448
  structure **6**:440*F*
Thienamycin
  discovery/development **1**:20; **7**:480
  structure **7**:480
6H-Thieno[2,3-b]pyrrole,
  bioisosterism **2**:660*T*
Thienylpyridine **7**:201
Thiethylperazine
  as antiemetic agent **6**:683
  nonulcer dyspepsia treatment **6**:600
Thin-layer chromatography (TLC)
  biomarkers **3**:76
  epigenetics **1**:266, **1**:267*T*
3-Thioalkyl-1,2,4-triazoles,
  PASP **3**:804, **3**:804*F*
Thiobarbital, development **1**:531
Thiocarbonyl substitution,
  oxazolidinones **7**:683–684
Thiocolchicoside, buccal
  administration **5**:283
Thioester library, PASP **3**:830,
  3:830*F*
Thioester methyltransferase
  5:819–820
7α-Thioestratrienes, steroid-derived
  modulators **7**:171, **7**:171*F*
Thioguanine, toxicity **7**:46*T*
6-Thioguanine, transporters **7**:45*T*
Thiolactomycin
  malaria treatment **7**:793*T*
  structure **7**:835*F*
  trypanosomiasis treatment **7**:835–
    836
Thiol group, glutathione (GSH) **5**:159
Thiol methyltransferase **5**:145
Thiol PEGylation *see* Polyethylene
  glycol (PEG) conjugates
Thiol reactivity index, promiscuous
  ligands, computational
  prediction **2**:749
Thiomorphine analogs, PASP **3**:811,
  3:813*F*, **3**:814*F*
Thiopental **6**:353
  propofol *vs.* **6**:360
  structure **6**:352*F*
Thiophene aspartyl ketones, solid-phase
  discrete libraries **3**:709*T*
Thiophenes
  toxicity potential **5**:800, **5**:801*F*
  triphenylethylene anticancer
    agents **7**:158–159
Thioproperazine **6**:29–31

Thiopurine methyltransferase **5**:145
  deficiency, adverse drug
    effects **1**:282*T*, **1**:284
Thioredoxin
  Escherichia coli protein expression
    systems **3**:113–114
  gene expression vectors **3**:110,
    3:110*T*
Thioredoxin reductase-1 (TRX-1),
  inhibition **1**:391–392, **1**:393*F*
Thioridazine **6**:29–31
  bioavailability **5**:712*T*
  mechanism of action **6**:32
Thiorphan
  inhibition values **2**:677*F*
  structure **2**:677*F*
Thiostrepton, malaria
  treatment **7**:795–796
Thiotepa
  reaction phenotyping studies **5**:243*T*
  toxicity **7**:46*T*
Thiothixene **6**:29–31
Thiourea
  'functional analogs' **2**:668
  PASP **3**:798, **3**:799*F*
  bis-substituted Thioureas,
    cytomegalovirus
    antivirals **7**:309
Thioxanthenes **6**:29–31
  P-glycoprotein models **5**:779
Third-generation calcium-channel
  blockers **6**:716*T*, **6**:718
Third-generation tetracyclines **7**:608
Third International Study of Infarct
  Survival *see* ISIS-3
'Third-party' informational materials,
  traditional medicine **1**:417
30S subunit, bacterial
  ribosomes **7**:523, **7**:526
THLE, drug metabolism, in vitro
  studies **5**:233
THOR
  database management systems
    3:295
  Daylight CIS **3**:295
  SMILES **3**:295
Thoughtful responsiveness, project
  teams **2**:145
Threat information, anxiety
  disorders **6**:87–89
3D database searching
  pharmacophores **4**:518
  substructure searching *see*
    Substructure searching
3D descriptors
  molecular descriptors **5**:533
  similarity searching *see* Similarity
    searching, representations
3D-environment, PASP **3**:794–795
3D geometrical parameters,
  CSD **3**:393
3D graphs, chemoinformatics **3**:241
3-Dimensional Pharma **2**:760*T*

3D (spatial) molecular fields **5**:578, **5**:578*F*
3D pharmacophore fingerprints, library design **4**:363
3D pharmacophore frequency counts **4**:316
3D quantitative structure–activity relationship (3D-QSAR) **4**:575–595; **5**:589
  ADME prediction **5**:591
    drug-metabolizing enzymes **5**:591
    see also specific enzymes
    permeation processes **5**:595
    transporters **5**:594
    see also specific transporters
  advantages/disadvantages **4**:156
  alignment-dependent methods **4**:577
    see also specific methods
    CoMISA (comparative molecular similarity indices analysis) **4**:578
    COMPA **4**:578
    CoRSA (comparative receptor surface analysis) **4**:580
    4D-QSAR **4**:581, **4**:581*F*
    genetic algorithms **4**:579–580
    GERM (genetically evolved receptor models) **4**:580
    HASL **4**:579, **4**:579*F*
    HIFA (hydropathic intermolecular field analysis) **4**:578
    hydrogen extension vectors (HEVs) **4**:579–580
    hydrophobicity vectors (HPV) **4**:579–580
    lone pair vectors (LPVs) **4**:579–580
    MLP (molecular lipophilicity potential) **4**:578
    MQSM (molecular quantum similarity methods) **4**:578
    pseudoreceptors **4**:579
    receptor surface models (RSM) **4**:579–580
    SOMFA (self-organizing molecular field analysis) **4**:578
    VFA (Voronoi Field Analysis) **4**:578
  alignment-independent methods **4**:576, **4**:583
    autocorrelation **4**:584
    CoMMA (comparative molecular moment analysis) **4**:585
    correlograms **4**:584, **4**:584*F*
    EVA/CoSA **4**:585, **4**:585*F*
    hologram QSAR **4**:585
    WHIM (weighted holistic invariant molecule) **4**:584
  alignment methods **4**:581
    atom-based **4**:582
    conformational analyzer with molecular dynamics and sampling (CAMDAS) **4**:581–582
  field-based **4**:582
  molecular dynamics **4**:581–582
  multiple coformers **4**:583
  pharmacophore-based **4**:582
  assumptions **4**:587; **5**:589
  Caco-2 cell models, flux prediction **5**:691
  CoMFA **4**:155, **4**:577, **4**:577*F*; **5**:589
    boundary element method (BEM) **4**:578
    descriptor space **4**:577
    generating optimal linear partial least squares estimation (GOLPE) **4**:577
    history **4**:576–577
    partial-least squares **4**:576–577
    q^2-guided region selection (q^2-GRS) **4**:577
  comparative molecular moment analysis (CoMMA) **5**:591
  compounds **4**:589
    activity values **4**:589–590
    selection of **4**:589
    training vs. test set **4**:590
  computer-assisted drug design (CADD) **4**:18
  CoMSIA *see* CoMISA (comparative molecular similarity indices analysis)
  database design **1**:43–44
  definition **4**:156
  descriptors **4**:155, **4**:590
    atom pair descriptors **4**:155
    CoMFA **4**:155
    distance geometry **4**:155
    MOE descriptors **4**:155
    molecular shape analysis **4**:155
    selection of **4**:589
    VolSurf descriptors **4**:155
    Voronoi techniques **4**:155
  dimensional classification **5**:579
  goals **4**:588
  GPCR drug design **4**:687
  guidelines **4**:588
  historical perspective **4**:576; **5**:589
    DYLOMMS **4**:576–577
    GRID **4**:576–577
    hypothetical active site lattice (HASL) **4**:576–577
    minimum topological difference **4**:576
    molecular shape analysis (MSA) **4**:576
    REMOTEDISC **4**:576
  inverse QSAR **4**:588
  mathematics **4**:586
    genetic algorithms **4**:587
    neural nets **4**:586, **4**:586*F*
    PCA/PLS **4**:586
    support vector machines **4**:587
  monoamine oxidase **5**:593
  observations **4**:589
  activity values **4**:589
  homogeneity **4**:589
  oxidoreductases **5**:591
  P-glycoprotein models *see* P-glycoprotein (P-gp), in silico models
  pharmacophores **4**:122; **5**:591
  prediction/interpretability **4**:588
    cross-validation **4**:588
    data overfitting **4**:588
  proton-coupled peptide transporters **5**:595, **5**:596*F*
  quantum mechanical calculations **4**:392
  transferases **5**:593
  transporter interaction modeling **5**:775–776
  UDP-glucuronyltransferases **5**:593, **5**:821
  validation **4**:587
    'leave-out-one' (LOO) **4**:587
    'leave-out-some' methods **4**:587
  weighted holistic invariant molecular descriptors (WHIM) **5**:591
3D representations, chemical information databases **3**:298
3D Search **4**:139*T*
3D similarity searching *see* Chemoinformatics, virtual screening
3D solubility parameters, in vitro oral absorption prediction **5**:676
3D-structure
  see also specific databases
  assignment **1**:40–41
  bioavailability models **5**:711–718
  chemoinformatics *see* Chemoinformatics, three-dimensional structure
  conformational relevance **1**:42
  databases **3**:283
  derivation **1**:40, **1**:41–42, **1**:41*F*
  messenger RNA (mRNA) **2**:1044
  'random walk' effects **1**:40–41, **1**:41*F*
3D structure generation **4**:460
  conformational analysis/searching *see* Conformational analysis/searching
  distance-based geometry **4**:461–462
  explicit conformer generation **4**:464
    CONFIRM program **4**:464
    CONFORT program **4**:464
    OMEGA program **4**:464
    RUBICON program **4**:464
  knowledge-based techniques **4**:461–462
  on-the-fly conformation generation **4**:465
    CATALYST program **4**:465
    CHEM-DBSD program **4**:465
    ISIS/3D program **4**:465
    torsion fitting **4**:465
    UNITY program **4**:465

structure quality measures **4**:465
  CONCORD program **4**:465
  CONFIRM program **4**:465, **4**:466*F*
  CONFORT 500 program **4**:464, **4**:465
  CONVERTER program **4**:465
  CORINA program **4**:465
  LIGPREP program **4**:465
  OMEGA program **4**:465
  RUBICON program **4**:465
2D-structure, conversion from **4**:462
  CONCORD program **4**:462
  CONVERTER program **4**:463
  CORINA program **4**:462–463, **4**:463*F*
  IONIZER program **4**:464
  LIGPREP program **4**:464
  MACROMODEL program **4**:464
  Merck Molecular Force Field (MMFF) program **4**:463
  NEUTRALIZER program **4**:464
  OMEGA program **4**:463
  PROTOPLEX program **4**:462
  RING_CONF program **4**:464
  SMARTS program **4**:464
  STEREOIZER program **4**:464
  STEREOPLEX program **4**:462
  TAUTOMERIZER program **4**:464
3D substructures, CSD **3**:393
3D superposition, protein kinase library design **4**:370
3D tissue constructs, non-animal eye irritation testing **1**:466, **1**:467*T*
3σ level, screening library triage **4**:510, **4**:510*F*, **4**:511*F*
3T3 neutral red uptake phototoxicity test (3T3-NRU-PT), non-animal photoirritancy testing **1**:471
3TC *see* Lamivudine
Three by Five program, WHO **1**:645–646
Three-dimensional topics *see entries beginning 3D*
'Three-hybrid system'
  see also Saccharomyces cerevisiae model system
  phenotype-oriented screening **3**:140
3-PPP, schizophrenia treatment **6**:34
'Threshold cutoffs,' ligand-based library design **4**:310
Thrombin
  peptide/protein drugs **2**:577
  quantum mechanics *see* Quantum mechanical calculations, protein–small molecules
  three-dimensional structures **2**:626*T*
Thrombin inhibitors **3**:929
  nonpeptide drugs **2**:628–629, **2**:631*F*, **2**:632*F*
  peptidomimetics **2**:613–614, **2**:614*F*, **2**:616*F*, **2**:628–629, **2**:631*F*, **2**:632*F*

quantum mechanical calculations **4**:412
quantum mechanics/molecular mechanics (QM/MM) **4**:412
structure-based fragment screening **4**:799, **4**:801*F*
Thrombin receptor, activation **2**:785*T*
Thrombocytopenia, oxazolidinones **7**:691
Thrombolysis **6**:765, **6**:766*F*
  medical management **6**:774
  pulmonary embolism **6**:763
  regulation **6**:766
  stroke **6**:763
  ST-segment elevation myocardial infarction (STEMI) **6**:763
  tissue-type plasminogen activator (tPA) **6**:765–766
  urokinase-type plasminogen activator (uPA) **6**:765–766
Thrombolysis in Myocardial Infarction *see* TIMI 10A; TIMI 10B
Thrombolytic drugs **6**:763–781
  see also specific drugs
  animal models **6**:766
    time of administration **6**:766
  clinical indication **6**:768*T*
  clinical trials **6**:772
  contraindications **6**:768*T*
  first-generation **6**:767
  fourth-generation **6**:771
  future work **6**:777
  new approaches/adjuvant treatment **6**:775
    aspirin **6**:775
    dipyridamole **6**:775
    GP IIb/IIIa receptor antagonists **6**:775
    heparin **6**:775
    low-molecular-weight heparins **6**:775
    P2Y$_{12}$ antagonists **6**:777
    warfarin **6**:775
  second-generation **6**:769
  stroke treatment **6**:263
  structure **6**:776*F*
  third-generation **6**:770
Thromboplastin, clotting cascade **6**:764
Thrombospondin-related anonymous protein, malaria **7**:766
Thromboxane A$_2$ receptors, in human disease **2**:775*T*
Thromboxane A$_2$ synthetase inhibitors, ring equivalent bioisosterism **2**:655
Thrombus formation **6**:764
  see also Clotting cascade
  affecting factors **6**:764*F*, **6**:765
  atherosclerotic plaques **6**:765
  GP IIb/IIIa receptors **6**:764, **6**:775
  vascular injury **6**:764
Throughput, NMR automation **2**:530

Thymectacin **7**:62*T*, **7**:68
  approval **7**:62*T*
  clinical indication **7**:62*T*
  mechanism of action **7**:64*T*
Thymectomy, myasthenia gravis treatment **6**:249
[^{3}H]Thymidine
  proliferation probes **3**:665
  radiotreatment imaging **3**:670
Thymidine biosynthesis **7**:56, **7**:57*F*
  dihydrofolate reductase **7**:56–57, **7**:57*F*
  serine hydroxymethyl transferase (SHMT) **7**:56–57, **7**:57*F*
  thymidylate synthase **7**:56–57, **7**:57*F*
Thymidine kinase, drug metabolism **5**:149
Thymidine monophosphate kinase inhibitors, focused library design, structure-based methods **4**:329
Thymidine phosphorylase, capecitabine activation **5**:1015
Thymidylate synthase
  pharmacodynamics **1**:287*T*
  structure-based drug design **4**:72
    de novo lead generation **4**:74, **4**:74*F*
    structure-guided optimization **4**:72, **4**:73*F*
    thymidine biosynthesis **7**:56–57, **7**:57*F*
Thymidylate synthetase inhibitors **7**:75
  see also specific drugs
  gemcitabine combination **7**:75
Thymine, structure **1**:256*F*
Thymol blue, structure **5**:371*F*
Thymosin α1, hepatitis C **7**:403*F*, **7**:404, **7**:405
Thymus
  atrophy, immunosuppression **5**:223
  cytochrome P450s **5**:172*T*
  myasthenia gravis **6**:236
Thyroid gland, phosphodiesterase-8 **2**:930
Thyroid hormone receptor (TR) **2**:1008
  ligands **2**:1019, **2**:1020*F*
  see also specific drugs
  nomenclature **2**:996*T*
  orthogonal ligand–receptor pairs **3**:227*F*, **3**:228
  subtypes **2**:1008
Thyroid hormone receptor inhibitors
  chemogenomics **3**:933, **3**:934*F*
  pharmacophores **4**:528*F*
Thyroid hormone
  sulfotransferase **5**:147
Thyroid-stimulating hormone (TSH) receptor
  activation **2**:785*T*, **2**:798
  human disease **2**:792*T*
Thyroliberin **2**:579*T*

Thyrotropin-releasing hormone (TRH)
  active-analog drug design 4:17,
    4:17*F*
  alpha helix scaffold-based
    design 4:30–31
  structure 6:274*F*
Thyrotropin-releasing hormone (TRH)
  receptor, activation 2:785*T*
Thyrotropin-releasing hormone receptor
    agonists
  see also specific drugs
  Alzheimer's disease treatment 6:207
  peptidomimetics 2:608–609, 2:611*F*
Thyrotropin-releasing hormone receptor
    analogs
  see also specific drugs
  central nervous system injury 6:275
Tiacrilast, oral absorption 5:468*T*
Tiagabine 1:187; 6:289
  cocaine abuse 6:181*T*, 6:184*T*
  mechanism-based PK/PD
    modeling 5:896
  mechanism of action 6:282*F*, 6:284
  primary insomnia treatment 6:159*T*,
    6:161
  stimulant addiction 6:185
  structure 6:181*F*, 6:282*F*
Tianeptine, structure 2:62*F*
Tibetan medicine, new drug
    source 1:456
Ticarcillin, structure 7:488*F*
Ticlopidine 6:777
  metabolism 5:546*T*
  structure 6:776*F*
Ticonazole 7:430*T*
  medicinal chemistry 7:449,
    7:451–452
TID
  lipid interactions 6:357
  structure 6:354*F*
Tie-2 inhibitors, structure 1:392*F*
Tienilic acid
  immunotoxicity 5:221*T*
  structure 5:801*F*
Tigecycline 7:599
  antimicrobial activity 7:610
  clinical indication 7:679
  structure 7:599*F*
  synthesis 7:608, 7:609*F*
Tight junctions
  blood–brain barrier
    permeation 5:313
  intranasal drug administration
    2:595
  paracellular transport 5:265
TIGR assembler, comparative
    genomics 7:738
TIGRFAMS
  InterPro, member of 3:363
  signature method 3:363
Tikosyn *see* Dofetilide
TILLING, *Danio rerio* model
    system 3:10

Tiludronate disodium 1:220
  Paget's disease 1:221
  toxicology 8:207
Timcodar, structure 5:770*F*
Time, outsourcing scale-up 2:211
Time course, central nervous system
    injury 6:261
Time-dependent inhibition, irreversible
    inhibition 5:249
Time domain acquisition, fluorescence
    lifetime assays 3:605–606
Timeliness, career guidelines 2:19
Time-of-flight (ToF) detectors
  gas chromatography-mass
    spectrometry (GC-MS)
    5:993–994
  mass spectrometry proteomics 3:32
  matrix-assisted laser desorption/
    ionization (MALDI)-MS 2:339
Time-of-flight mass spectrometry,
    protein display chips 2:374
Time-resolved anisotropy
    analysis 3:606
  mechanism of action 3:606
Time-resolved fluorescence-based
    assays
  GPCR functional assay
    screening 2:807, 2:807*T*, 2:809
Time saving
  outsourcing 2:228, 2:229*F*
  PASP 2:792
Time-specific gene deletions, knockout
    animal models 3:152
TIMI 10A, tenecteplase 6:773
TIMI 10B, tenecteplase 6:773, 6:774
Timing, microarrays 3:103
Timmerman, Henk 8:7–16
  collaborations, Ariëns, Eef 8:8
  on combinatorial synthesis 8:13
  education 8:7
  on individualized medicine 8:13
  research
    antihistamines 8:10, 8:11
    GPCRs 8:9
    histamine H$_2$ receptor 8:10, 8:13
    histamine H$_3$ receptor 8:12
    histamine receptors 8:9
    QSAR 8:8
  at Vrije Universiteit Amsterdam 8:9
Timolol
  bioavailability 5:712*T*, 6:712*T*
  dose 6:712*T*
  GPCR targeting 2:773*T*
  half-life 6:712*T*
  oral absorption 5:468*T*, 5:680*T*
  structure 6:309*F*, 6:712*T*
Timoprazole
  discovery/development 6:605–606
  structure 6:606*F*
Tinctures, traditional medicine 1:419
Tinea nigra 7:421*T*
Tinidazole, protozoal infection
    treatment 7:758

Tiotropium bromide 1:210
Tipifarnib 7:234–235
  structure 7:835*F*
  trypanosomiasis treatment
    7:835–836
Tipranavir 7:279–282, 7:355–356
  discovery 7:355
  structure 1:78*F*; 2:302*F*; 7:355*F*
Tirapazamine 5:1032
  as prodrug 5:1017
  structure 5:1018*F*
Tirilazad mesylate 1:174
  central nervous system injury
    6:259–260
  clinical trials 6:260–261
  introduction of 1:107
  nervous system trauma
    treatment 6:13
  structure 6:260*F*
Tirofiban hydrochloride 1:167;
    6:775
  structure 6:776*F*
Tissue binding 5:328, 5:850
  adipose tissue binding 5:331, 5:853
    adipose storage index 5:331
    prediction models 5:861
  cells 5:850–852
  distribution studies 5:43, 5:45*T*
  hepatocyte binding 5:334
    drug binding mechanisms 5:334
    ionization class 5:334–335
    lipophilicity effects 5:334–335
    physicochemical properties 5:334
    unbound intrinsic clearance
      measurement 5:334
  incubational binding 5:331
    in vitro *vs.* in vivo 5:331–332
  interstitial fluid 5:850–852
  in vitro measurements 5:328
    see also specific methods
    tissue-to-plasma
      concentrations 5:330
    volume of distribution 5:330
  in vivo methods 5:330
  microsome binding *see* Microsomes
  prediction models *see* Tissue binding,
      prediction models
  tissue homogenate binding 5:328
    equilibrium dialysis 5:328
  tissue slices 5:329, 5:329*T*
    advantages 5:329
    albumin diffusion 5:329
    extracellular space 5:329
    tissue-to-medium partition
      coefficient 5:330
    total water 5:329
  volume of distribution *see* Volume of
      distribution (V_{ss})
Tissue binding, prediction
    models 5:860
  tissue/blood distribution
    coefficients 5:860, 5:861
  Zhang, model used by 5:861–862

tissue/plasma distribution coefficients ($D_{T/P}$) 5:860
  adipose tissue 5:861
  Poulin, model used by 5:861
  Rodgers, model used by 5:861
  tissue/plasma water model 5:861
  volume of distribution (V_{ss}) 5:336
Tissue/blood distribution coefficients *see* Tissue binding, prediction models
Tissue composition models, distribution models 5:870
Tissue concentrations, physiologically based pharmacokinetics 5:868–869
Tissue destruction, COPD 7:880
Tissue distribution
  RNAi (RNA interference) 3:182, 3:183
  sodium channels 2:840
Tissue extraction, proteomics 3:30
Tissue factor pathway inhibitor, clotting cascade 6:764–765
Tissue homogenate binding *see* Tissue binding
Tissue inhibitors of metalloproteinases (TIMPs) 7:241–242
  cancer, invasive growth 7:17
Tissue injury, neuropathic pain 6:330
Tissue microarrays, biomarkers *see* Biomarkers
Tissue partitioning, distribution models 5:870
Tissue/plasma distribution coefficients ($D_{T/P}$) *see* Tissue binding, prediction models
Tissue/plasma water model, tissue binding, prediction models 5:861
Tissue plasminogen activator (tPA)
  see also Tissue-type plasminogen activator (tPA)
  stroke management 6:697–698
Tissue remodeling
  angiotensin II 6:584
  cancer, invasive growth 7:14
Tissue selectivity, selective estrogen receptor modulators (SERMs) 7:174–175
Tissue slices *see* Tissue binding
Tissue-specific gene deletions, knockout animal models 3:152
Tissue specificity, membrane transport protein expression 5:73
Tissue-specific promoters, transgenic animal generation 3:152
Tissue storage models, distribution studies 5:43, 5:44F
Tissue-to-medium partition coefficient, tissue slices 5:330
Tissue-to-plasma concentrations, tissue binding 5:330

Tissue-type plasminogen activator (tPA)
  cancer, invasive growth 7:17–18
  recombinant
    see also specific drugs
    animal models 6:766
    stroke management 6:697–698
    thrombolysis 6:765–766
Titin mutation, *Danio rerio* model system 3:10
Titration, lipophilicity measurement 5:436
Titration curve, biopolymers 5:619
Tivirapine
  discovery/development 1:71, 1:72F
  structure 1:72F
Tizanidine, GPCR targeting 2:773T
TKS-159, as prokinetic agent 6:678
TlogP, log P calculation program 5:661T
TM3-TM4 intracellular loop, nicotinic acetylcholine receptors 2:885–886
TMC-95-A
  discovery/development 1:384
  structure 1:383F
TMC-114, structure 1:76F
TMC-207
  mycobacterium treatment 7:723
    clinical trials 7:724
  structure 7:723F
TM domain, inward rectifying K$^+$ ion channels 4:713
TMI-005, arthritis treatment 7:859–860
TNP-ATP, structure 2:906F
Tobacco use 6:701
  see also Nicotine addiction
  age-related macular degeneration 6:302
  biological effects 6:701
  cancer risk factors 7:3, 7:4
  cessation *see* Smoking cessation
  COPD 7:848, 7:873–874, 7:875
  prevalence 6:701
  schizophrenia 6:39
Tobi (tobramycin inhalant formulation) 1:501
Tobramycin 7:631, 7:632
  applications 2:1042
  beta-lactam combination 7:637
  formulation 7:644
  O-glycosylation 7:635–636
  resistance 7:642
  structure 2:1041–1042; 7:632F
TOC-39 7:497
  activity 7:496T, 7:497T
  structure 7:495F
Tocainide 6:743
  bioavailability 5:712T
  effectiveness 6:743
  mechanism 6:743
  toxicity 6:743

α-Tocopherol, structure 6:260F
Tofranil *see* Imipramine
Tolazamide 6:442T
  structure 6:440F
Tolazoline, discovery/development 1:328–329
Tolboxane, carbon-boron bioisosterism 2:698
Tolbutamide 6:438, 6:442T
  advantages 6:443
  bioavailability 5:712T
  clinical use 6:443
  disadvantages 6:443
  mechanisms of action 6:440
  oral absorption 5:468T, 5:680T
  pharmacokinetics 6:440
  side effects/contraindications 6:443
  structure 6:440F
Tolcapone 1:181
  cocaine abuse 6:184T
  introduction of 1:107
  Parkinson's disease treatment 6:11, 6:217
  structure 6:181F
Tolerance, amphotericin 7:432
Tolerogens, nervous system autoimmune diseases, treatment 6:246
Tolfenamic acid, migraine treatment 6:373–374, 6:373T
Toll-like receptor-4 (TLR-4), IBD 6:616, 6:617T
Toll-like receptor-4 antagonists, IBD treatment 6:624–625
Toll-like receptor-5 (TLR-5), IBD 6:616
Toll-like receptor-9 antagonists, IBD treatment 6:624–625
Toll-like receptors
  cell-based assays 3:638, 3:640F
    cytokine production 3:639
    inflammatory mediators 3:639
    peripheral blood mononuclear cells 3:639
  IBD 6:616
  psoriasis 7:942
  signaling, fungal disease 7:427
Tolmesoxide, oral absorption 5:680T
Tolmetin, bioavailability 5:712T
Tolnaftate 7:430T
Toloxatone 6:67–68
Tolterodine
  GPCR targeting 2:773T
  urinary incontinence treatment 6:543–544
Toluene, microwave heating 3:840T
Tomudex *see* Raltitrexed
Tonerg *see* Pivagabine
Tonic-clonic (*grand-mal*; major motor seizures) 6:283
TOP-53 7:120
Topamax *see* Topiramate
Topical anesthetics 6:355

Topical antibiotics, acne
    treatment **7**:961
Topical applications, dermal/transdermal
    administration **5**:290–291, **5**:726
Topical corticosteroids, atopic
    dermatitis treatment **7**:949
Topical immunosuppressants
    atopic dermatitis treatment *see* Atopic
        dermatitis treatment
    psoriasis treatment **7**:973
    skin inflammation treatment **7**:973
Topical retinoids, acne
    treatment **7**:939
Topigen **1**:513
Topiramate **1**:**188**; **6**:289
    alcohol addiction **6**:184*T*
    characteristics **2**:829*T*
    cocaine abuse **6**:181*T*, **6**:184*T*
    introduction of **1**:107
    migraine **6**:372
    oral absorption **5**:680*T*
    stimulant addiction **6**:185
    structure **2**:829*T*; **6**:181*F*, **6**:282*F*
TOPKAT models **5**:926*T*, **5**:927*T*
    Ames assay *vs.* **5**:927
    other systems *vs.* **5**:926–927, **5**:927*T*
    performance characteristics **5**:928*T*
    reproductive/developmental
        toxicity **5**:925
Topliss decision tree, QSAR substituent
    selection models **4**:52
Topoisomerase I/II inhibitors **7**:120,
    7:124
Topoisomerase I inhibitors **7**:113
    see also specific drugs
    introduction **1**:107
    structure **1**:382*F*
Topoisomerase II inhibitors **7**:116
    see also specific drugs
    natural products **1**:381
    structure **1**:382*F*
Topoisomerase IV
    inhibition, quinolone
        antibiotics **7**:570–572
    mechanisms of action **7**:571–572
Topoisomerase inhibitors
    DNA therapeutic agents **2**:1040
    toxicity **7**:46*T*
Topological descriptors **5**:560
    see also Electrotopological indices
        (E-state index); Molecular
        connectivity
    assumed dimensionality **5**:560
    chromosomal damage **5**:920
    computational oral absorption
        prediction **5**:688
    oral absorption prediction, molecular
        descriptors **5**:683
    QSAR *see* Quantitative structure–
        activity relationship (QSAR)
    relational networks **5**:561
    two-dimensional descriptors
        vs. **5**:560

Topological fingerprint/dataprint
    descriptors *see* Fingerprint/
    dataprint descriptors
Topological index *see* Similarity
    searching, representations
Topological indices, molecular
    descriptors **5**:534
Topological polar surface area (TPSA)
    electrotopological indices (E-state
        index) **5**:564
    hydrogen bonding **5**:682, **5**:683*F*
Topological similarities
    parallel library design **4**:327
    scaffold hopping **2**:683–685, **2**:685*F*
TOPological Sub-structural MOlecular
    DEsign (TOPS-MODE), dermal
    penetration models **5**:731
Topology, x-ray structures,
    problems **3**:532, **3**:534*F*
Topomers
    GPCR library screening **4**:692
    reagent-based library design **4**:345
Topotarget **1**:513
Topotecan **1**:**141**; **7**:113–114
    DNA interactions, quantum
        mechanical calculations **4**:410
    structure **4**:410*F*
    transporters **7**:45*T*
Topotecin *see* Irinotecan hydrochloride
TOPS-MODE, blood–brain
    distribution **5**:752
TORC1, identification, cDNA
    functional screening **3**:15–16
Torcetrapib **6**:490, **6**:491–492
    clinical trials **6**:490
    structure **6**:489*F*
Toremifene **7**:154; **8**:104–106
    [^{18}F]FDG imaging **3**:669
    breast cancer treatment **8**:97*F*,
        8:98
    as estrogen receptor ligand **2**:1014
    oral absorption **5**:680*T*
    structure **2**:1013*F*; **7**:153*F*
Torsades de pointes **5**:935; **6**:731
    ECG **6**:731, **6**:732*F*
Torsemide **6**:708
    bioavailability **6**:709*T*
    bioisosterism **2**:668–669, **2**:670*F*
    dose **6**:709*T*
    half-life **6**:709*T*
    metabolism, CYP 2C9 **5**:545*T*
    oral absorption **5**:680*T*
    structure **6**:586*F*, **6**:709*T*
Torsion angles
    ligand structure **3**:538
    NMR structure
        determination **3**:486, **3**:488*F*
    peptides **2**:604
    protein 3D structure
        validation **3**:511, **3**:516
    rotations, ligand–receptor molecular
        flexibility *see* Ligand–receptor
        interactions, molecular flexibility

Torsion fitting, 3D structure
    generation **4**:465
Torsion potentials, force-field molecular
    modeling **4**:90
Tositumomab **1**:**142**, **1**:334*T*, **1**:433*T*
    non-Hodgkin's lymphoma **1**:142,
        1:444
    toxicity **7**:46*T*
Tosufloxacin **7**:569*T*
    development **7**:575, **7**:581
    structure **7**:570*F*, **7**:576*T*, **7**:582*F*
Total clearance (CL),
    definition **5**:869*T*
Total correlated spectroscopy (TOCSY),
    metabolite identification **5**:996
Total fluorescence intensity (FLINT),
    fluorescence screening
    assays **3**:602
Total plasma clearance,
    pharmacokinetics **5**:18
Total surface area, size
    descriptors **5**:682–683
Total volume cleared (V_{CL}),
    permeability assays **5**:310
Total water, tissue slices **5**:329
Totelle Sekvens *see* Trimegestone
Tourette's syndrome, 5-HT$_3$
    receptors **2**:891
ToxBox **5**:926*T*
TOXCENTER **3**:271*T*
Tox Express database **1**:57*T*
Toxicity **1**:56
    see also Toxicology; *specific classes*
    carbon–boron bioisosterism *see*
        Bioisosterism, carbon–boron
    Comer, William (views on) **8**:3
    computational models *see* Toxicity,
        computational models
    cytochromes P450 **4**:781
    data
        dose **5**:911
        exposure **5**:911
        metabolic fate **5**:911
        toxicity tests **5**:912
    databases **1**:56–57, **1**:57*T*
    DCN **2**:154
    definitions **5**:910
    drug metabolism, in vitro
        studies **5**:240
    electrotopological indices (E-state
        index) *see* Electrotopological
        indices (E-state index)
    emtricitabine **7**:314
    ideal drug characteristics **2**:564*T*
    importance **5**:910
        drug design **5**:910
        drug synthesis **5**:910
        early drug attrition **5**:910
    legislation **5**:912
    prediction **5**:910, **5**:911*F*
        see also Toxicity, computational
            models
    SAR **5**:910

QSAR **4**:58
  quantum mechanical calculations
    polycyclic aromatic
      compounds **4**:396–397
    quantum topological molecular
      similarity (QTMS) **4**:396–397
    triazines **4**:396–397
  selenium bioisosterism **2**:701
  structure–toxicity
    relationships **1**:56–58
  testing *see* Toxicity testing
Toxicity, computational models **5**:909–
    932
  advantages **5**:928–929
  carcinogenicity *see* Carcinogenicity,
    computational models
  commercial systems **5**:926, **5**:926*T*
    see also specific systems
    comparisons of **5**:926, **5**:927*T*
  current methods **5**:912
    see also specific methods
  data **5**:911, **5**:912
  developmental toxicity **5**:925
  genotoxicity *see* Genotoxicity,
    computational models
  hepatotoxicity **5**:925
  history **5**:912–913
  LD$_{50}$ **5**:913
  limitations **5**:929
    data **5**:929
  linear free energy approaches **5**:912–
    913
  maximum tolerated dose **5**:913
  phospholipidosis **5**:924
  QSARs **5**:912–913
  QT prolongation *see* QT interval
  reproductive toxicity **5**:925
  respiratory sensitization **5**:924
  rule-based approaches **5**:913, **5**:914
    advantages **5**:915
    chemical properties **5**:914
    data **5**:914
    direct reactions **5**:914
    disadvantages **5**:915
    false-negatives **5**:915
    false-positives **5**:915
  skin sensitization *see* Skin
    sensitization, computational
    models
  statistical methods **5**:913
    advantages **5**:913
    compound properties *vs.*
      activity **5**:913
    disadvantages **5**:913
    HOMO/LUMO ratio **5**:913
    overfitting **5**:913
  3D-approaches **5**:913, **5**:915
    advantages **5**:916
    application limits **5**:916
    Comparative Molecular Field
      Analysis (CoMFA) **5**:915
    data needed **5**:915
    disadvantages **5**:916

  drug–drug interactions **5**:915
  drug mechanisms **5**:915
  overfitting **5**:916
  protein structure **5**:916
  toxicity mechanisms **5**:911
Toxicity Databases **3**:309
Toxicity testing
  animal testing **5**:912
    alternatives *see* Animal
      experimentation alternatives
  cell-based assays *see* Cell-based assays
  drug discovery, decline in **1**:620
  early, importance of **5**:3
  gas chromatography-mass
    spectrometry (GC-MS) **5**:1001
  high-performance liquid
    chromatography-mass
    spectrometry (HPLC-MS)
    5:1001
  importance **5**:4
  metabonomics *see* Metabonomics
  NMR **5**:1000–1001
  pharmacokinetics **5**:12–13
  stability testing **5**:494
  toxicity data **5**:912
Toxicogenomics **1**:347, **1**:480; **2**:568;
    3:56; **5**:225
  animal experimentation **3**:56
  constraints **1**:480
  cost reduction **2**:198
  drug development **2**:197
  endocrine disrupter testing **1**:481
  gene expression profiles **5**:225
  goals **2**:568–569
  mouse models **3**:654
  principles **1**:480
  QSAR combination **2**:568
  SAR combination **2**:568
  sensitization testing **1**:481
Toxicokinetics, animal experimentation
    alternatives *see* Animal
      experimentation alternatives
Toxicology
  see also Toxicity
  carbon-silicon bioisosterism **2**:694
  drug development
    productivity **2**:130
  drug metabolism and
    pharmacokinetics **5**:32
  lead optimization **2**:139
  preclinical trials **2**:178, **2**:179*T*
  predevelopment **2**:141
  tamoxifen **8**:95
  target identification **1**:347
Toxification **5**:**193–214**
  see also Detoxification; Metabolism;
    specific enzymes
  detoxification *vs.* **5**:194
  drug metabolism **5**:194
    see also specific enzymes
Toxins, acid-sensitive ion channels
    (ASICs) **2**:911
TOXNET **3**:275, **3**:278*T*

Toxoplasma gondii infections *see*
    Toxoplasmosis
Toxoplasmosis **7**:751*T*, **7**:758
  biology **7**:758
  clinical manifestations **7**:759
  drug treatment **7**:753*T*, **7**:759
    see also specific drugs
    tetracyclines **7**:613
  epidemiology **7**:758
TOXSYS **1**:57*T*
TP53 gene, cell cycle **7**:12
TPMA, structure **2**:893*F*
TR-14035, IBD treatment **6**:632–633
Traceability, nutraceuticals **1**:458,
    1:458*T*
Trace amines, affective disorder
    treatment **6**:80
Trace metal ions, oxidation **5**:492
Tracer molecules, blood–brain barrier
    permeation **5**:313–314
Tracleer *see* Bosentan
Tractocile *see* Atosiban
Trade Act, USA (1974) **1**:798, **1**:803
Trade-Related Aspects of Intellectual
    Property (TRIPS) **2**:94, **2**:211
  establishment **1**:780–781
  impact on developing
    countries **1**:643, **1**:644, **1**:645,
    1:694
  principles **1**:534–535
Traditional Chinese medicine
    (TCM) **1**:415, **1**:416
  database **3**:308
    Daylight CIS **3**:308
  new drug source **1**:456
Traditional healers, developing
    countries **1**:638
Traditional medicine **1**:58, **1**:405–430,
    1:455
  see also specific preparations; *specific types*
  bioavailability **1**:422
  bioprospecting **1**:414
    intellectual property issues **1**:414
  chemo-cure **1**:422
  chemo-prevention **1**:422
  clinical trials, lack of **1**:415
  crossover **1**:411
  definition **1**:406
  developing countries **1**:410–411
  dosage forms **1**:419
    extracts **1**:419
    fluid extracts **1**:419
    ground powders **1**:419
    homeopathic solutions **1**:419
    tinctures **1**:419
  drug interactions **1**:413–414, **1**:422
    concealment of use **1**:422
  economic issues **1**:412
    clinical information lack **1**:413
    patents, lack of **1**:413
  education **1**:407
  efficacy presumption **1**:414–415
  ethical issues **1**:413

Traditional medicine (*continued*)
  dosage variability 1:413
  drug interactions 1:413–414
  identity 1:414
  purity 1:414
  safety 1:413
  evaluation 1:423
   clinical trial problems 1:424
   diagnosis problems 1:423
   imprecision in terminology 1:423
  future work 1:426
  historical aspects 1:406, 1:409
   ancient Greece 1:409
   Ebers papyrus 1:409
   fermentation 1:410
   indigenous population
    worries 1:410
   species loss 1:410
  lifestyle issues 1:423
  malaria 1:413
  'official' medicine *vs.* 1:414
  'organic' fallacy 1:406
  plant species 1:406–407
  processing 1:415
  regulatory status 1:417
   adulteration 1:417
   efficacy classification
    system 1:418
   good manufacturing
    practice 1:418
   health claims 1:417
   legal structure/function
    claims 1:417
   names 1:417–418
   nutrient content claims 1:417
   proof of safety 1:418
   'third-party' informational
    materials 1:417
  resurgence of interest 1:410
  safety 1:413, 1:421
   adulteration 1:421
   chronic toxicity 1:421
   important species 1:421
   metabolism issues 1:421
   presumption 1:414–415
  safety presumption 1:414–415
  sales 1:410
  scientific studies
   costs 1:407
   lack of 1:406–407
  secondary metabolism 1:411
   definition 1:411
   purpose 1:411
  single agents *vs.* combination
   treatment 1:412
  sources of material 1:418
   cultivation 1:418
   wild-grown 1:418
  standardization 1:419
   active ingredients 1:420
   botanical descriptions 1:419
   Canada 1:420
   France 1:421

   Germany 1:420
   labeling 1:420
   UK 1:420–421
  variation in preparations 1:412
TRAIL
  apoptosis evasion 7:18
  inhibition, RNA interference 3:17–18
Training (in medicinal
  chemistry) 1:83
  broad-base training 1:84
  physical organic chemistry 1:83–84
Training sets
  solubility models 5:630
  statistical methods, E-state
   index 5:562
Tramadol
  mechanism of action 2:773*T*
  oral absorption 5:680*T*
  structure 6:336*F*
Trandolapril
  bioavailability 6:719*T*
  dose 6:719*T*
  half-life 6:719*T*
  in vitro potency 6:719*T*
  structure 6:719*T*
Tranexamic acid, oral
  absorption 5:468*T*
Transactivation systems, non-animal
  endocrine toxicity
   testing 1:467*T*, 1:475
Transacylation, dynamic combinatorial
  library generation 3:962*F*
Transamination, dynamic combinatorial
  libraries 3:969–970, 3:971*F*
Transcellular route
  blood–brain barrier
   permeation 5:302–303, 5:304*F*
  buccal administration 5:280
Transcription
  DNA 2:1037
  HIV life cycle 7:331, 7:334*F*
  interference, DNA therapeutic
   agents 2:1039
  profiling, *Danio rerio* model
   system 3:10
  regulation
   large-scale virus vector
    libraries 3:18–19
   nuclear receptors 4:726
   ubiquitin -proteasome
    pathway 3:190
  repression
   DNA methylation 1:265*F*
   histone deacetylases
    (HDACs) 1:265*F*
   siRNAs 3:175
  termination, gene expression
   vectors 3:108
Transcriptional activation, nuclear
  receptors 4:731
Transcriptional profiling studies,
  Saccharomyces cerevisiae model
  system 3:6

Transcription-coupled nucleotide
  excision repair process,
  ecteinascidin 743 1:393
Transcription factors
  affective disorders 6:79
  cell-based assays 3:631
   production 3:632
  methyl-CpG-binding
   proteins 1:262–263
  modular design 2:301*T*, 2:323–324
Transcription–translation systems, cell-
  free expression systems 3:121
Transcriptomics
  definition 3:330–331
  drug development 2:196
Transdermal administration *see* Dermal/
  transdermal
Transdermal patches, dermal/
  transdermal
  administration 5:290–291
Transduction, mechanism-based
  pharmacokinetic/
  pharmacodynamic (PKPD)
  modeling *see* Mechanism-based
  pharmacokinetic/
  pharmacodynamic (PKPD)
  modeling
Transendothelial electrical resistance
  (TEER) 5:313–314
Transepidermal resistance (TER), non-
  animal skin corrosivity
  testing 1:467*T*, 1:472
Transfected cell lines
  brain drug uptake *see* Brain drug
   uptake
  hERG blocker screening 5:937
Transfection, hepatoma cell
  lines 5:233–234
'Transferability,' force-field molecular
  modeling 4:89
Transferases 5:181
  see also specific enzymes
  conjugation, drug metabolism 5:145
  drug metabolism 5:205
  ontogenic development 5:174*T*
  pharmacological target space 4:422*T*
  3D-QSAR 5:593
Transfer/permeation 5:453–487
  see also Permeation
  assessment tools *see* Partition
   assessment tools; Permeation
  biological membranes *see* Biological
   membranes
  surface activity *see* Surface activity
Transfer/permeation, in vivo
  prediction 5:466
  biological data complexity 5:467
  blood brain barrier (BBB) *see* Blood–
   brain barrier (BBB)
  oral absorption 5:467, 5:468*T*
  probe compound data set 5:466
  inconsistency 5:466
  structural diversity 5:466

Tubulin, GPCR interaction **2**:786*T*, **2**:797*T*
Tubulin interactive drugs (TIAs)
  see also specific drugs
  natural products **1**:380
  structure **1**:381*F*
Tubulozole **7**:96
Tularemia, streptomycin **7**:637
Tularik **1**:506
Tumor-activated prodrugs (TAPs) **5**:1032
Tumorigenesis
  NF-κB pathway **3**:192
  STAT (signal transducer and activator of transcription) **3**:207
Tumor mass, as surrogate marker **3**:80
Tumor necrosis factor-α (TNF-α)
  adalimumab binding **1**:441
  apoptosis induction, NF-κB pathway **3**:196–197
  bone resorption/remodelling **6**:501
  etanercept binding **1**:441
  hypertension **6**:581
  infliximab binding **1**:441
  NF-κB pathway activation **3**:197
  obesity **6**:581
  psoriasis **7**:942
  rheumatoid arthritis **1**:441
Tumor necrosis factor-α (TNF-α) antagonists
  see also specific drugs
  age-related macular degeneration treatment **6**:315*T*
  arthritis treatment **7**:857–858
Tumor necrosis factor-α converting enzyme inhibitors
  arthritis treatment **7**:858
  carboxylic acid bioisosterism **2**:659
Tumors
  see also Cancer
  cell-based assays **3**:619
  expression profiling **2**:311
  mitochondria, drug targeting **2**:1067
  NMR **3**:77
  positron emission tomography **3**:77
Tumor suppresser genes
  Drosophila melanogaster **3**:650–651
  removal **3**:161, **3**:163*F*
TUniRef, UniProtKB **3**:362
Turbidimetric assays **3**:595
  high-throughput methods **5**:406
  HTS **3**:595
  kinetic solubility **5**:349
  technique **3**:595
Turbo similarity searching, similarity measure combination **4**:183
Turbulent flow media, high throughput purification **3**:867
TV3326 **6**:69
Tween-20, promiscuous ligands **2**:740
23S methylation, macrolide antibiotic resistance **7**:529–530

Twinning, protein crystallography **3**:470
2/4/A1 cell line, permeability assays **5**:270
2D (symbolic) molecular fields **5**:578, **5**:578*F*
2D-PAGE **2**:331; **3**:29–30, **3**:34
  advantages/disadvantages **3**:36
    protein class **3**:37
    protein properties **3**:39
  biomarkers, diagnostic implications **3**:45–46
  denaturing conditions **3**:34
  detergents **3**:34
  fluorescent labeled samples **3**:30–31
  imaging/image analysis **2**:335
    autoradiography **2**:335
    CCD devices **2**:335
    data management **2**:335–336
    QUEST system **2**:335
    statistical analysis **2**:335–336
  isoelectric focusing (IEF) (first dimension) **2**:331–332; **3**:34
    ampholyte carriers *vs.* IPG gels **2**:332
    sample preparation **2**:332
  LC-MS (peptide-based) proteomics *vs.* **3**:39
  pH gradient **3**:34–35, **3**:34*F*
    narrow range **3**:34–35
  protein characterization, posttranslational modifications **3**:41
  protein expression mapping **2**:384–385
  protein extraction/prefractionation **3**:30–31
  protein identification **3**:36
    MALDI analysis **3**:36
    MALDI-TOFTOF analysis **3**:36
  protein quantification **3**:29–30, **3**:35
    comparison methods **3**:34*F*, **3**:35–36, **3**:36*F*
    Coomassie stain **3**:35, **3**:35*T*
    diffuse gel electrophoresis (DIGE) **3**:35, **3**:35*T*
    fluorescent dyes **3**:35, **3**:35*T*
    gel stains **3**:35
    pre-separation labeling **3**:35
    replicates **3**:35–36
    silver stain **3**:35, **3**:35*T*
    Sypro ruby **3**:35*T*
  protein spot excision **2**:336
  proteomics **2**:314
  resolution **3**:34–35
  SDS-PAGE (second dimension) **2**:332–333
    staining **2**:333
      Coomassie stains **2**:333
      fluorescent stains **2**:334
      silver staining **2**:333

2D similarity searching *see* Chemoinformatics, virtual screening
2D structural representations, CSD **3**:390–391
2D substructures, CSD **3**:393
2D topological descriptors, Caco-2 cell models, flux prediction **5**:691
2D topological networks, computational oral absorption prediction **5**:688
2D to 3D conversion, 3D database pharmacophore searching **4**:141
2N prodrugs, tetracyclines **7**:603
Two-component model, hERG blocker modeling **5**:942
Two-dimensional descriptors
  molecular descriptors **5**:533
  similarity searching *see* Similarity searching, representations
  topological descriptors *vs.* **5**:560
Two-dimensional fluorescence intensity distribution analysis (2D-FIDA) *see* Fluorescence fluctuation techniques
Two-dimensional NMR spectroscopy, fragment-based drug discovery **3**:945
Two dimensional nuclear magnetic resonance *see* Nuclear magnetic resonance (NMR)
Two-dimensional representations
  chemical information databases **3**:298
  SMILES **3**:298
  TDT (THOR data tree) **3**:298
  text formats **3**:298
Two-dimensional spectra, metabolite identification **5**:996
Two-hybrid assay, *Saccharomyces cerevisiae* model system **3**:6
264W94 **6**:487
Two-stage process, conformational analysis/searching **4**:108, **4**:109*F*
'Two-state model'
  'flip-flop *see* Transmembrane permeation
  GPCR activation models **2**:792, **2**:794*F*
Two-vector systems, Semliki Forest Virus expression systems **3**:119
TWS119, target identification **3**:138–139
Tylosin **7**:556–557, **7**:558
  discovery **7**:520–521
Tyramine, affective disorder treatment **6**:80
Tyrosine kinase inhibitors, transplant rejection **7**:920–921
Tyrosine kinase receptor ligands, surface plasmon resonance **3**:593

Tyrosine kinases **2**:967; **7**:188
  see also specific kinases
  as drug targets **3**:3
  Flash Plate assays **3**:587
  non-receptor **7**:188
  oncogenic **2**:960–962
  structure **2**:967*F*
  three-dimensional structures **2**:636*T*
Tyrosine phosphatases **2**:981
  see also Phosphotyrosine-interacting
    domains; *specific types*
  inhibitors **2**:981–982, **2**:982*F*
  nonpeptide drugs **2**:638, **2**:640*F*
  peptidomimetics **2**:638, **2**:640*F*
  three-dimensional structures
    2:636*T*
Tyrphostin analogs, solid-phase pool
    libraries **3**:705*F*
Tyrphostins, solid-phase
    libraries **3**:727, **3**:734*F*
Tysabri *see* Natalizumab
TZP-101, as prokinetic agent
    6:685–686

U

U-0126 **7**:204
  structure **7**:204*F*
U-72099E, structure **6**:260*F*
U-78517F
  central nervous system injury
    6:268–269
  structure **6**:260*F*
U-80493E **7**:344
  structure **7**:344*F*
U-85910, oxazolidinone
    development **8**:161
U-97456 **7**:680
  oxazolidinone development **8**:161
U-100480 **7**:682
  development **8**:167
  mycobacterium treatment **7**:725
  structure **7**:726*F*
U-100766 *see* Linezolid
U-101033E
  central nervous system injury **6**:268
  structure **6**:268*F*
U-7101033, structure **6**:260*F*
UbiQGel, R&D **1**:658
Ubiquinone (coenzyme Q) **2**:1073
  biology **2**:1073
    as antioxidant **2**:1073
    electron shuttle **2**:1073
  drug development **2**:1074
    Friedreich's ataxia
      treatment **2**:1074
    idebenone **2**:1074
    mevalonic acid **2**:1074
    Parkinson's disease
      treatment **2**:1074
    simvastatin **2**:1074
    statins **2**:1074

pathology **2**:1074
  encephalomyopathies **2**:1074
  structure **2**:1073*F*, **2**:1074*F*
Ubiquitin
  gene expression vectors **3**:110
  NMR structure
    determination **3**:476, **3**:478*F*,
      3:490–491, **3**:491*F*
  structure **3**:190–191
Ubiquitin-activating enzyme
    (E1) **3**:190
Ubiquitin-conjugating enzyme
    (E2) **3**:190
Ubiquitin-ligase enzyme (E3) **3**:190
Ubiquitin-proteasome pathway **3**:190,
    3:191*F*
  definition **3**:190
  functions **3**:190
    antigen processing **3**:190
    cell cycle **3**:190
    protein turnover regulation **3**:191–
      192
    signal transduction **3**:190
    transcription regulation **3**:190
  inhibition **3**:194
    see also Bortezomib
    Bax protein **3**:194
    p53 levels **3**:194
  JNK-activated pathway
    c-Jun-associated protein
      degradation **3**:192, **3**:194*F*
    p53-associated protein
      degradation **3**:193, **3**:194*F*
  NF-κB activation *see* Nuclear factor
    kappa B (NF-κB)
  protein targeting **3**:191–192
    dephosphorylation **3**:192
    stress-responsive kinases **3**:192
  ubiquitin -activating enzyme
    (E1) **3**:190
  ubiquitin -conjugating enzyme
    (E2) **3**:190
  ubiquitin -ligase enzyme (E3) **3**:190
Ubiquitous chromatin opening
    elements (UCOEs), gene
    therapy **1**:313
UC781, structure **1**:72*F*
UCN-01 **7**:205, **7**:209–210
  CDK family kinase inhibitors **2**:979–
    980
  characteristics **2**:984*T*
  structure **2**:979*F*; **7**:209*F*
UC spectrography, kinetic
    solubility **5**:349
Uganda
  health insurance coverage **1**:641
  health personnel **1**:648
UGDP (University Group Diabetes
    Program), diabetes mellitus
    clinical trials **6**:436
UDP-glucuronyltransferases
    see UDP-Glucuronyltransferases
      (*in 'G' section*)

UK-370.485, *N*-myristoyltransferase
    inhibitors **7**:436, **7**:437*T*
UK-47265, medicinal
    chemistry **7**:451–452
UK-118,005, RNA polymerase III
    inhibitors **7**:436–439, **7**:438*T*
UK-362,091, *N*-myristoyltransferase
    inhibitors **7**:436, **7**:437*T*
UK-427857, IBD treatment **6**:633–635
UKPDS (United Kingdom Prospective
    Diabetes Study) **6**:436
  glycemic control effects **6**:437
  management goals **6**:437–438
UK Science Parks Association, industry/
    academia collaborations **1**:584
UL97 protein kinase inhibitors *see*
    Cytomegalovirus (CMV)
Ulcerative colitis (UC) **6**:599–600
  see also Inflammatory bowel disease
    (IBD)
  areas of involvement **6**:614*T*
  Baron score **6**:619
  characteristics **6**:614*T*
  clinical features **6**:614*T*
  complications **6**:614*T*
  diagnosis **6**:615
  drug introductions **1**:203
  epidemiology **6**:614*T*
  histology **6**:614*T*
  location **6**:613
  NF-κB pathway **3**:197
  phosphodiesterase-4
    inhibitors **2**:935–936
  Powell–Tuck index **6**:619
  treatment **6**:622*T*
  Truelove and Witts index **6**:619
Ullmann algorithm, structure/
    substructure searching
    3:239–240
Ultair *see* Pranlukast
Ultiva *see* Remifentanil hydrochloride
Ultracentrifugation, plasma protein
    binding measurement **5**:328
Ultrafiltration
  buffer exchange **3**:424
  plasma protein binding
    measurement **5**:328
  protein concentration **3**:424
Ultrahigh-throughput screening, HTS
    data analysis **2**:459
Ultra-High-Throughput Screening
    System (Aurora
    Biosciences) **2**:436
  assay development **2**:437
    fluorescence resonance energy
      transfer **2**:437
    green fluorescent protein **2**:437
  assay examples **2**:447, **2**:448*F*
  assay loading **2**:438
  compound distribution **2**:438
  detection **2**:439
  fluorescence detection **2**:445,
    2:446*F*

incubation 2:439
key technologies 2:436
microfluidics 2:441, 2:442*F*
dispensing accuracy 2:442–443,
2:443*F*
multiple well dispensing 2:444,
2:444*F*
plate design/performance 2:440,
2:441*F*
cell-substrate adhesion 2:431,
2:441*F*
results database 2:440
solenoid-controlled liquid
dispensing 2:444, 2:444*F*
Ultraprecision machining, lithography/
photolithography 2:354
'Ultrarapid metabolizers,' cytochrome
P450 2D6 1:284
Ultrasonic vocalization, anxiety
disorders, animal models 6:95
Ultrasound
gastrointestinal motility 6:676
small animal imaging 3:660*T*, 3:661,
3:662
Ultraviolet (UV) radiation
chromatograms, high throughput
purification 3:863
LIGA, lithography/
photolithography 2:354
protein concentration
determination 3:426
psoriasis treatment 7:945
Ultraviolet spectroscopy, plasma protein
binding measurement 5:328
'Ulysses contracts' 1:697
Unbound intrinsic clearance
measurement
hepatocyte binding 5:334
microsome binding 5:332
Uncaria tomentosa (cat's claw) 1:424
Uncertainty, economic market failure
arguments 1:727, 1:728
Uncharged membranes, permeation
theory 5:466
Uncompetitive inhibition, in vitro
studies 5:246
Uncontrolled trials 2:182
β-adrenergic receptor antagonists in
heart failure 8:140
UN Convention on Biological
Diversity 1:789, 1:801
Uncoupling protein-1
(thermogenin) 2:1080
in obesity 6:412–413
Uncoupling proteins
see also Oxidative phosphorylation
uncoupling
obesity treatment 6:412
Undecanediamidine
structure 7:831*F*
trypanosomiasis treatment 7:829–
830
Undecylenate 7:430*T*

Undisturbed sinusoidal (parallel-tube)
model, hepatic metabolic
clearance models 5:871
'Undruggable' targets, fragment-based
drug discovery 3:944
Uniform Resource Locators
(URLs) 3:299
Unilateral inflammation, nociceptive
pain models 6:333
Unilateral middle cerebral artery
occlusion (MCAO) model *see*
Focal ischemic stroke
UniParc
see also Universal Protein Resource
DDBJ/EMBL/GenBank
checking 3:354
Integr8 3:367
UniProt *see* Universal Protein Resource
(Uni-Prot)
Unit cells 3:451, 3:452*F*
axis restrictions 3:451–452
United Kingdom (UK)
drug introductions 1:105
economic development funds 1:564
industry/academia collaborations *see*
Collaborations, industry/academia
medicine marketing system 1:632
regulations, prospective human
studies 1:479
SMEs 1:497–498
stem cell research programs
1:558–559
traditional medicine 1:420–421
United Nation's World Intellectual
Property Organization
(WIPO) 1:534
UNITY software
database management systems
3:295
homology models 4:487–488
pharmacophore 3D database
searching 4:139*T*
3D structure generation 4:465
topological fingerprint/dataprint
descriptors 4:469
Tripos 3:295
Univasc *see* Moexipril hydrochloride
Universal Protein Resource
(Uni-Prot) 3:358
data acquisition 3:354
DDBJ/EMBL/GenBank
checking 3:354
history 3:358
specialized protein databases 3:362
MEROPS 3:362–363
PDB 3:362–363
SGD 3:362–363
UniParc 3:358
design 3:358–359
Ensembl 3:358–359
primary sequences 3:358–359
sources 3:358–359
structure 3:358–359

UniProtKB 3:358, 3:359
design 3:359
external cross-references 3:359
feature identifiers 3:359
posttranslational modifications 3:359
TUniRef 3:362
UniProtKB flatfile 3:360, 3:361*F*
accession number 3:360
CC (comments) 3:360, 3:361*F*
DE (description) lines 3:360
DR (database crossreference) 3:360,
3:361*F*
DT (date) lines 3:360
FT (feature) lines 3:362, 3:362*F*
GN (gene name) line 3:360
ID lines 3:360
KW (keyword) 3:361*F*, 3:362
OC (organism classification)
lines 3:360
OG (organelle) line 3:360
OS (organism species) line 3:360
OX (organism taxonomy) line
3:360
RA (reference author) 3:360
RC (reference comment) 3:360
RG (reference group) 3:360
RL (reference location) 3:360
RN (reference number) 3:360
RP (reference position) 3:360
RT (reference title) 3:360
RX (reference cross-
reference) 3:360
SQ (sequence header) lines 3:362,
3:362*F*
UniProtKB/Swiss-Prot 3:359
annotation 3:359–360
curation 3:359–360
gene name standardization
3:359–360
UniProtKB/TremBL 3:359, 3:360
UniRef 3:358
Universities
academic environment 2:89–90
challenges 2:91
changes 2:91, 2:94
commercial activities 2:93
definition 2:89–90
history 2:90
industry collaborations *see*
Collaborations, industry/academia
science, role in 2:91–92
'gift economy model' 2:91–92,
2:92*F*
'use-inspired research' 2:91
University-based providers,
definition 2:206
University of Toronto, Eli Lilly
collaboration 1:585
'Unlikelihood' plotting 3:403*F*, 3:405,
3:405*F*
Unnatural amino acids
peptidomimetics 2:606–608
proteolysis resistance 2:584

Unnatural oligomers, solid support synthesis **2**:478, **2**:480*F*

Unnatural peptide backbone, proteolysis resistance **2**:584–585

Unnatural product synthesis, Hirschmann, R F **1**:2

Unoprostone, structure **6**:309*F*

Unoprostone isopropyl ester **1**:**241** introduction of **1**:108

Unsaturated nitrile synthesis, PASP **3**:796–797

Unstirred water layer (UWL) intestinal absorption **5**:261 intestinal membrane **5**:458 permeability **5**:672 permeation theory **5**:464

Unsupervised data clustering, virtual screening, large data sets **4**:476

Untranslated regions (UTRs), messenger RNA (mRNA) **2**:1044–1045

Upjohn, mergers and acquisitions **1**:541

UPOV Convention **1**:813

Uppsala Electron Density Server (EDS), protein 3D structure **3**:547

Upstate Biotechnology **1**:515

Upstream conserved regions (UCRs), phosphodiesterase-4 **2**:924–925

Uptake models, physiologically based pharmacokinetics **5**:881

Uracil gonadotropin-releasing hormone (GnRH) agonist/antagonists **7**:176 structure **1**:256*F*; **5**:371*F*

Uracil-DNA glycosylase, quantum mechanical calculations **4**:404

Urapidil bioavailability **5**:712*T* oral absorption **5**:680*T*

Urea effective permeability (P_{eff}) **5**:673*T* 'functional analogs' **2**:668 solid-phase discrete libraries **3**:709*T*

Urea/carbonic anhydrase mechanism, *Helicobacter pylori* **6**:607, **6**:610*F*

Urea library construction, PASP **3**:798, **3**:799–800, **3**:799*F*, **3**:800*F*

Ureaplasma infections, tetracyclines **7**:612

Ureido acids solid-phase libraries **3**:737, **3**:737*F* solid-phase pool libraries **3**:705*F*

Ureido hydantoins, solid-phase pool libraries **3**:705*F*

Urge urinary incontinence (UUI) definition **8**:124 duloxetine clinical trials **8**:133

Uric acid central nervous system injury **6**:268 structure **2**:67*F*; **6**:268*F*

Uricosuric diuretics, Hirschmann, R F **1**:13

Uridine, structure **2**:1038*F*

Uridine-5'diphospho-α-D-glucuronic acid, UDP-glucuronyltransferases **5**:150, **5**:151*F*

Uridine kinase, drug metabolism **5**:149

Urinalysis, benign prostatic hyperplasia diagnosis **6**:561

Urinary bladder *see* Bladder

Urinary continence **8**:123

Urinary incontinence **6**:543, **6**:**551–573**, **6**:568; **8**:123 anterior fibromuscular stroma **6**:571 definition **8**:124 5-HT receptor antagonists **8**:124 interstitial cystitis **6**:571 markets **8**:127 Agency for Health Care Policy and Research **8**:127 patient advocacy groups **8**:127–128 Simon Foundation **8**:127–128 prostate surgery **6**:569 prostatitis **6**:570 stress **6**:544 treatment **6**:543–544 antimuscarinics **6**:543–544 darifenacin **6**:543–544 estrogen treatment **6**:544 oxybutynin **6**:543–544 tolterodine **6**:543–544 tricyclic antidepressants **6**:544 treatment preconceptions **8**:127 surgery **8**:127 urge incontinence **6**:571

Urinary tract infections balofloxacin **1**:109 gatifloxacin **1**:114 oxytetracycline **7**:610

Urinary urge incontinence **6**:571

Urine amount eliminated in urine (M_U), pharmacokinetics **5**:19 metabonomics **5**:990

Urocortins, obesity treatment **6**:409

Urogenital diseases/disorders **6**:**541–550** *see also specific diseases/disorders* future work **6**:549 markets **6**:541 surgery **6**:541

Urokinase **6**:767, **6**:769 animal models **6**:766 clinical indication **6**:769 crystallization **3**:883–884 mechanism of action **6**:769 pharmacokinetics **6**:769

Urokinase inhibitors targeted libraries **4**:792

x-ray crystallography **4**:788*F*, **4**:789, **4**:789*F*

Urokinase-type plasminogen activator (uPA) cancer, invasive growth **7**:17–18 thrombolysis **6**:765–766

Urotensin II receptors activation **2**:785*T* in human disease **2**:775*T*

Urticaria, epinastine hydrochloride **1**:204

Urushiol, immunotoxicity **5**:221*T*

USA dominance **2**:94 drug discovery expenditure **1**:726, **1**:738, **1**:740*F* drug introductions **1**:105 governments, in drug discovery **1**:735, **1**:735*F* industry/academia collaborations *see* Collaborations, industry/academia IPA law **1**:560 markets **2**:7 medicine marketing system **1**:632, **1**:633 orphan drugs legislation *see* Orphan drugs Presidential Emergency Plan for AIDS Relief (PEPFAR) **1**:645–646

US Adopted Name (USAN) **2**:161–162

'Use-inspired research,' universities **2**:91

User-defined constraints, product-based library design **4**:352

User fees developing countries **1**:640–641 orphan drugs push and pull incentives **1**:661

User requirement specification (USR), customized automation **5**:347

U-shaped value curve, business development **2**:25–26, **2**:25*F*, **2**:26*F*

US Patent and Trademark Office (USPTO) **1**:782, **1**:795

USPTO **3**:276*T*

Ussing chamber oral absorption **5**:674 VolSurf descriptors *vs.* **5**:540

Uterine bleeding, vascular endothelial growth factor **3**:202

Uterine cancer, nedaplatin **1**:147

Uveitis, treatment **6**:317

V

V-2006 **1**:506–507 Parkinson's disease treatment **6**:217

Vaccine Adverse Event Reporting System (VAERS) **1**:772

Vaccines
see also specific vaccines
buccal administration **5**:284
development **1**:530
influenza viruses **7**:376
introductions **1**:227
malaria *see* Malaria
multiple sclerosis treatment **7**:929
pharmaceutical companies **2**:100
R&D
AIDS **1**:666
orphan incentives **1**:666
respiratory syncytial virus
(RSV) **7**:406–407
safety **1**:772
SMEs **1**:517
Vaccinia virus
antivirals **7**:261
gene therapy **1**:304
Vaccinium myrtillus (bilberry) **1**:424
Vacuum cavities, steric
properties **4**:196
VADAR **3**:547*T*
Vagelos, P Roy **1**:15
Vaginal administration **5**:289
formulations **5**:290
hepatic circumvention **5**:289
hormone therapies **5**:289
in vitro studies **5**:289–290
continuous flow-trough diffusion
cells **5**:289–290
epithelial cell cultures **5**:290
EpiVaginal **5**:290
Reconstituted Human Vaginal
Epithelium **5**:290
in vivo evaluation **5**:290
animal models **5**:290
lipophilic hormones **5**:289–290
local treatment **5**:290
products **5**:281*T*
Vaginismus, female sexual
dysfunction **6**:547
Valaciclovir **1**:**130**; **7**:299
alpha-herpesvirus infections
7:257–258
BCS **5**:973*T*
development **7**:299
herpes simplex virus type1 **7**:257–258
herpes simplex virus type2 **7**:257–258
intestinal absorption **5**:266
medicinal chemistry **7**:299
target information **4**:750*T*
varicella zoster virus **7**:257–258
Valance angle descriptors **4**:172–173
Valance state electronegativity,
electrotopological indices
(E-state index) **5**:557
Valcyte *see* Valganciclovir
Valdecoxib **1**:**216**
arthritis treatment **7**:853
glutathione conjugation **5**:161,
5:161*F*
introduction of **1**:108

safety **1**:684
structure **2**:658*F*, **2**:681*F*
target information **4**:750*T*
withdrawal **1**:757*T*
Valence angle values, CSD
applications **3**:400
Valence electron parameter (NVE),
comparative QSAR **4**:55
Valence electron structure,
electrotopological indices
(E-state index) **5**:559
Valence state electronegativity *see*
Electrotopological indices
(E-state index)
Valganciclovir **1**:**130**
cytomegalovirus **7**:260
cytomegalovirus retinitis **1**:130
Validated prediction, electrotopological
indices (E-state index)
5:561
VALIDATE program **4**:22
regression-based **4**:484
Validation
animal experimentation alternatives
see Animal experimentation
alternatives
cell-based assays **3**:620–621
criteria testing, x-ray structures,
problems **3**:535
CSD **3**:391
electrotopological indices (E-state
index) **5**:564, **5**:565*F*
homology-based protein structure
prediction **3**:326
QSAR *see* Quantitative structure–
activity relationship (QSAR)
solubility models **5**:630
Validation Server, PDB **3**:376
Valinomycin, SARS **7**:273–274
Valium, invention of, Sternbach,
Leo **2**:256–257
Valproic acid **6**:280–281, **6**:288
alcohol addiction **6**:184*T*
amino acid-coenzyme A
conjugation **5**:153*F*, **5**:157,
5:158*F*
β-oxidation **5**:158, **5**:158*F*
bioavailability **5**:712*T*
bipolar affective disorder
treatment **6**:71
blood to plasma ratios **5**:325*T*
blood volume of distribution
($V_{SS,b}$) **5**:325*T*
cholesterol esters/hybrid
triacylglyceride
conjugation **5**:153*F*,
5:156*F*, **5**:157
glucuronic acid conjugation **5**:151,
5:153*F*
oral absorption **5**:680*T*
oxidation **5**:107, **5**:108*F*
plasma clearance (CL_p)
5:325*T*

second generation **6**:291
structure **6**:292*F*
structure **6**:181*F*, **6**:281*F*
Valrocemide **6**:291
structure **6**:292*F*
Valrubicin **1**:**142**
bladder cancer **1**:143
Valsartan **1**:**162**
bioavailability **6**:722*T*
drug action **1**:532*T*
half-life **6**:722*T*
hypertension treatment **6**:721–723
mechanism of action, GPCR
targeting **2**:773*T*
oral absorption **5**:468*T*
structure **6**:587*F*, **6**:722*T*
Valspodar
P-glycoprotein (P-gp) inhibition **5**:58
structure **5**:770*F*
Valstar *see* Valrubicin
Valtorcitabine
hepatitis B **7**:319
hepatitis C **7**:269
Valtrex *see* Valaciclovir
Value chain migration,
outsourcing **2**:212
Value drivers, pharmaceutical industry
organization **2**:106–108
Value per drug **2**:124
Vancomycin **1**:369
analogs, solid-phase discrete
libraries **3**:707*F*
clinical indication **7**:654, **7**:678
methicillin-resistant *Staphylococcus
aureus* (MRSA) **7**:678
discovery/development **1**:369
gentamicin combination **7**:637
in vitro anti-aerobic Gram-positive
activity **7**:661*T*
in vitro anti-enterococcal
activity **7**:659*T*
in vitro anti-streptococci
activity **7**:659*T*
pharmacokinetics **7**:662*T*
SAR **7**:657–658
structure **2**:46*F*
N-(decylaminoethyl)-
Vancomycin **7**:657–658
Vancomycin-resistant enterococci
(VRE) **7**:481, **7**:654, **7**:676,
7:677
development **8**:159
Vancomycin-resistant *Enterococcus faecalis*
dalfopristin **1**:112
quinupristin **1**:112
Vancomycin-resistant *Staphylococcus
aureus* (VRSA) **7**:677
daptomycin **7**:665
development **8**:159
Van der Waals' interactions
force-field molecular modeling *see*
Force-field molecular modeling
steric properties **4**:195

Van der Waals surface, surfaces **5**:581, **5**:582*F*
Vanillin **4**:70–71
Vanillinbananin, SARS **7**:272
Vanilloid receptor modulators, pain **6**:340
Vanilloids, α$_1$-acid glycoprotein (AGP) binding **5**:849
Vanoxerine
  cocaine abuse **6**:184*T*
  structure **6**:181*F*
Vardenafil **1**:245; **2**:927, **2**:938–939
  as phosphodiesterase inhibitor **2**:947*T*
  quinazoline derivatives **2**:939
  target information **4**:750*T*
Varenicline
  nicotine addiction **6**:184*T*
  nicotinic acetylcholine receptor binding **2**:888
  structure **2**:887*F*; **6**:181*F*
Variable data quality, bioinformatics **2**:317–318
Variable surface glycoprotein (VSG), *Trypanosoma brucei* **7**:816–818
Variation of responses, cell-based assays **3**:620–621, **3**:621*F*
Varicella zoster virus (VZV) **7**:295
  adefovir dipivoxil **7**:314
  antiviral drugs **7**:257
    acyclovir **7**:257–258, **7**:297, **7**:299
    brivudine **7**:257–258
    Cf 1742 **7**:260
    Cf 1743 **7**:260
    famciclovir **7**:257–258
    idoxuridine **7**:257–258
    penciclovir **7**:257–258
    trifluridine **7**:257–258
    valaciclovir **7**:257–258
    BCNA **7**:307
Variola virus, antiviral drugs **7**:261
Variolin B
  discovery/development **1**:391
  structure **1**:390*F*
Vascular cell adhesion molecule-1 (VCAM-1), NF-κB pathway **3**:197
Vascular dementia **6**:10
Vascular endothelial growth factor (VEGF) **3**:198
  see also Placental growth factor (PlGF)
  actions **3**:200; **7**:19–20
    angiogenesis **3**:198*F*, **3**:200
    embryogenesis **3**:200–202
    NF-κB pathway **3**:197
  cancer **3**:202; **7**:19
    antisense treatment **3**:203
    expression levels **7**:20
    oncogene activation **3**:202
  cell-signaling pathways **3**:197
    NF-κB pathway **3**:197
  deletion **7**:20

female reproductive tract
  disorders **3**:202
  dysfunctional uterine bleeding **3**:202
  endometrial hyperplasia **3**:202
  endometriosis **3**:202
  polycystic syndrome **3**:202
gene expression regulation **3**:200
  cytokines **3**:200
  growth factors **3**:200
  hypoxia **3**:200
  hypoxia-inducible factor-1 (HIF-1) **3**:200
gene structure **3**:198
inflammatory disorders **3**:202
  psoriasis **3**:202
intraocular neovascular syndromes **3**:202
  age-related macular degeneration **3**:202
  diabetic retinopathy **3**:202
  dry age-related macular degeneration **6**:301–302
  retinopathy of prematurity **3**:202
mutations, amyotrophic lateral sclerosis **6**:222
overexpression **3**:196*F*, **3**:200–202
  psoriasis animal models **7**:971
receptors **7**:20
structure **3**:198
subtypes **3**:198
therapeutic uses **3**:203
  amyotrophic lateral syndrome **3**:203
  cardiovascular disease **3**:203
  neonatal respiratory distress syndrome **3**:203
underexpression **3**:196*F*, **3**:200–202
Vascular endothelial growth factor receptor-1 (VEGF-R1) **3**:198
Vascular endothelial growth factor receptor-2 (VEGF-R2) **3**:198, **3**:200
  signaling pathways **3**:201*F*
  PI3 K activation **3**:200
  Rac activation **3**:200
  Src **3**:200
  VEGFR-associated protein (VRAP) **3**:200
Vascular endothelial growth factor receptor-associated protein (VRAP), vascular endothelial growth factor receptor-2 (VEGF-R2) **3**:200
Vascular endothelial growth factor receptor family kinases **2**:969
  see also specific kinases
  angiogenesis **2**:969
  in disease **2**:963*T*
  inhibitors **2**:969, **2**:970*F*
Vascular endothelial growth factor receptor inhibitors **7**:21, **7**:200
  see also specific drugs

in clinic **7**:201
future work **7**:201
small molecules **7**:200
structure **1**:392*F*; **7**:201*F*, **7**:202*F*
wet age-related macular degeneration treatment **6**:310
Vascular endothelial growth factor receptors (VEGFR) **3**:198; **7**:200
  in cancer **7**:200
  receptor dimer formation **3**:198–199
  structure **3**:198, **3**:199*F*
Vascular injury, thrombus formation **6**:764
Vascular smooth muscle, hypertension **6**:706
Vascular tone regulator, phosphodiesterase-5 **2**:928
Vasoactive intestinal peptide (VIP)
  as drug
    N-terminus modification **2**:583
    structural motif replacement **2**:590
  sleep **6**:147
Vasodilators/vasodilation
  angiotensin II **6**:584
  carvedilol **8**:137
  endothelial dysfunction **6**:580–581
  hypertension treatment **1**:532*T*
  nitric oxide **6**:580–581
Vasopeptidase inhibitors
  see also specific drugs
  hypertension treatment **6**:724
  mechanism of action **6**:724
Vasopressin
  analogs **2**:580
  as peptide/protein drugs **2**:579*T*
  intranasal administration **2**:593–594
Vasopressin 1B receptors, in human disease **2**:775*T*
Vasopressin V$_{1A}$ receptors
  homology models **4**:687–688
  structure-based library design **4**:322–323
Vasopressin V1 receptor antagonists **2**:618–619, **2**:620*F*
Vasopressin V2 receptors, in human disease **2**:775*T*
Vasopressin VP1b receptor antagonists
  see also specific drugs
  affective disorders **6**:78
VAST **3**:557
Vatalanib *see* PTK-787
Vaughan Williams–Singh classification, antiarrhythmic drugs **6**:742
V$_{DSS}$ *see* volume of distribution
Veblen, Thorstein, universities in commerce **2**:93
Vectavir *see* Penciclovir
Vector alignment components, structure-based library design **4**:321

Vector control, malaria **7**:770
Vector inadequacy, gene therapy **1**:300–301
Vector representations, two-dimensional descriptors, similarity searching **4**:171–172
Vector-specific markers, genetically engineered animal characterization **3**:153–154
Vectrine *see* Erdosteine
Velcade *see* Bortezomib
Vendor characteristics, customized automation **5**:347
Venlafaxine **1**:178; **6**:101
  ADHD treatment **6**:133–134
  bioavailability **5**:712*T*
  bladder activity inhibition **8**:127
  cataplexy treatment **6**:154
  introduction of **1**:107
  irritable bowel syndrome treatment **6**:657
  oral absorption **5**:680*T*
  side effects **6**:60*T*
  structure **2**:62*F*; **6**:101*F*
  transporter inhibition **6**:66*T*
Venous equilibrium (well-stirred) model, hepatic metabolic clearance models **5**:871
Ventricular arrhythmias *see* Arrhythmias; *see specific diseases/disorders*
Ventricular fibrillation **6**:732
  ECG **6**:731, **6**:732*F*
Ventricular tachycardia **6**:731
  ECG **6**:731, **6**:732*F*
  mortality **6**:741
Venture capital **1**:553–570
  biotechnology company funding **2**:8
  business plans **1**:560
    content **1**:560
    credibility **1**:561
    development vehicle **1**:561
  definitions **1**:556
  early stage finance **1**:560, **1**:563
    business angels **1**:556, **1**:564
    business plans *see above*
    categories **1**:563
    corporate partnerships **1**:565
    deal structuring **1**:563, **1**:566, **1**:567
    economic development funds **1**:564
    equity *vs.* debt **1**:561
    exit strategies **1**:561
    intermediaries **1**:556
    investment agreement structure **1**:563
    investors **1**:562, **1**:564
    IPR **1**:560, **1**:664–665
    preincorporation finance **1**:563
    project finance **1**:561
    selection **1**:566
    strategic partners/alliances **1**:563

funding history **1**:554, **1**:556
  disaggregation of industry **1**:554
governments, in drug discovery **1**:729, **1**:730*F*
investment trends **1**:557
organizations **1**:565
partnership deals **1**:567
resource providers **1**:555
  early stage finance *see above*
  non-financial **1**:555
risk factors/problems **1**:557
  development time delays **1**:559
  ethical issues **1**:557, **1**:559
  expectation failures **1**:559
  GM foods **1**:558
  stem cell research **1**:558
  technical **1**:557
SME funding **1**:494
valuation assessment **1**:566
Verapamil **6**:747
  BCS **5**:973*T*
  bioavailability **5**:712*T*; **6**:715*T*
  characteristics **2**:829*T*
  derivatives, P-glycoprotein models **5**:779
  dose **6**:715*T*
  effective permeability (P_{eff}) **5**:673*T*
  half-life **6**:715*T*
  hypertension treatment **6**:715
  mechanism of action **6**:715–718
  oral absorption **5**:468*T*; **5**:680*T*
  P-glycoprotein (P-gp) inhibition **5**:58, **5**:781*T*
  structure **2**:75*F*; **2**:829*T*; **5**:778*F*; **6**:587*F*; **6**:715*F*
Verify3D **3**:547*T*
Verlet algorithms, molecular dynamics simulations **4**:652
Verlukast **7**:908
Vernalis **1**:506
Verrucae, atopic dermatitis **7**:949
Verteporfin **1**:237
  structure **6**:307*F*
  wet age-related macular degeneration treatment **6**:306*T*
Vertex Pharmaceuticals **1**:499, **1**:499*T*, **1**:502; **2**:226*T*, **2**:761*T*
Vertical orientation, transmembrane helices **4**:674
Vertical structure, R&D management **2**:249
Very-high density lipoproteins (VHDLs), plasma protein binding **5**:322
Very late antigen-4 (VLA-4), IBD **6**:630
Very-low density lipoproteins (VLDLs) **6**:462, **6**:463
  characteristics **6**:463*T*
  composition **6**:464*T*
  drug binding **5**:849–850
  plasma protein binding **5**:322

Vesicular monoamine transporters (VMATs) **2**:59, **2**:60*T*
Vessel wall remodeling, atherosclerosis **6**:460
Veterinary products, pharmaceutical companies **2**:100
Vexol *see* Rimexolone
VFA (Voronoi Field Analysis), 3D-QSAR alignment **4**:578
Vfend *see* Voriconazole
VH1, three-dimensional structures **2**:636*T*
Viability
  [^{18}F]FDG **3**:668
  small animal imaging **3**:668
Viagra *see* Sildenafil; Sildenafil citrate
Vibramycin *see* Doxycycline
Vibrational circular dichroism, chiral drug analysis **2**:720
Vibrational entropy, ligand–receptor interactions **4**:202
Vibrio cholerae MsbA, P-glycoprotein homology modeling **5**:786
Vidarabine **7**:306
Vidarabine *see* Arabinosyladenine
Videx *see* Didanosine
Vietnam, healthcare systems reform **1**:641
Vigabatrin **6**:289
  cocaine abuse **6**:184*T*
  mechanism of action **6**:282*F*, **6**:284
  structure **6**:282*F*
Vildagliptin, structure **6**:440*F*
Villafranca, Ernest, protein crystallization **3**:887
Villi structure, intestinal absorption **5**:262, **5**:263*F*
Viloxazine
  cataplexy treatment **6**:154
  discovery/development **3**:927–929, **3**:929*F*
  oral absorption **5**:680*T*
Vinblastine **7**:82
  see also AVBD regimen
  structure **5**:778*F*; **7**:83*F*
  toxicity **7**:46*T*
  transporters **7**:45*T*
Vinca alkaloids **7**:82, **7**:83*F*
  see also specific drugs
  discovery/development **1**:380–381
  microtubule-targeting **7**:82
  SAR **7**:82
  structure **1**:381*F*; **7**:83*F*
Vincristine **7**:82
  structure **7**:83*F*
  toxicity **7**:46*T*
  transporters **7**:45*T*
  xenograft models **7**:39*T*
Vindesine, structure **7**:83*F*
Vinflunine, structure **7**:83*F*
Vinograd–McBain treatment, dissolution **5**:413

Vinorelbine 7:82
  structure 7:83F
  toxicity 7:46T
Vinpocetine
  clinical trials 2:947
  phosphodiesterase-1
    inhibition 2:931–932
Vinyl fluorides, carboxamide
  bioisosterism 2:668, 2:669F
Vinyl polymers see Drug–polymer
  conjugates
Viomycin
  mycobacterium treatment 7:705T,
    7:713, 7:714–715
    mechanism of action 7:714
    resistance 7:714
    structure 7:714F
Vioxx see Rofecoxib
Vioxx Gastrointestinal Outcomes
  Research (VIGOR) 1:684
VIPER (Virus Particle Explorer), virus
  structure 3:544
Viracept, structure-based drug
  design 4:77T
Viramidine
  hepatitis C 7:267, 7:390
  structure 7:389F
Viramune see Nevirapine
Viread see Tenofovir; Tenofovir
  disoproxil fumarate
Viread see Tenofovir disoproxil fumarate
Virginiamycin 7:664
Viridicatumtoxin
  biosynthesis 7:602
  structure 7:601F
Virtual cells see Molecular networks,
  bioinformatics
Virtual companies see Start-up
  companies
Virtual dynamic combinatorial
  libraries 3:961
Virtual libraries, product-based library
  design 4:347–348
Virtual screening
  aqueous solubility models 5:642–643
  chemoinformatics see
    Chemoinformatics, virtual
    screening
  de novo drug design 4:297
  descriptor selection 4:476, 4:478F
    DAYLIGHT fingerprints 4:476–
      477
    substructure information 4:477
  drug discovery 1:331
  explicit molecular alignment 4:472
    feature-based alignment 4:474
    Gaussian function
      evaluations 4:473, 4:474F
    shape-based alignment 4:474,
      4:475F
    SPERM software 4:472–473
  factor Xa inhibitors 4:10
  future work 1:82

library design see Library design,
  structure-based
ligand-based see Ligand-based virtual
  screening
nuclear receptors, drug design 4:742
structure-based see Structure-based
  virtual screening
structure-based fragment
  screening 4:789F, 4:790
Virtual screening, large data sets 4:475
  classification techniques 4:476
    CLASSPHARMER program 4:476
    recursive partitioning (RP) 4:476,
      4:477F
    unsupervised data
      clustering 4:476
  descriptor weighing paradigms 4:475
    data fusion 4:475–476
    rank fusion 4:476
    Tanimoto coefficient 4:475
Virtue ethics 1:710
Virulence factors
  aspergillosis 7:427
  suppression, oxazolidinones 7:688
Virulence-related targets, antibacterial
  target identification 7:738,
    7:743
Viruses
  coat proteins, x-ray structures 3:540
  entry inhibitors 1:79
  microRNA precursors 3:175
  replication
    NF-κB pathway 3:192
    Saccharomyces cerevisiae model
      system 3:6
  structure
    problems with 3:544
    VIPER (Virus Particle
      Explorer) 3:544
  transcription/translation
    mechanisms 3:118
  vectors see Gene therapy
Virus infections
  cancer risk factors 7:4
  RNAi (RNA interference) 3:173
Visceral leishmaniasis 7:824
Vista data analysis see Cambridge
  Structural Database (CSD)
Vistide see Cidofovir
Visual analog scale, pain 6:335
Visual diseases see Ophthalmic diseases
Visual impairments, calcium
  channels 2:853
Visual inspection, microarrays 3:97–98,
  3:99F
Visudyne see Verteporfin
Vitamin B₁₂ transporter, P-glycoprotein
  (P-gp) vs. 5:769
Vitamin C see Ascorbic acid (vitamin C)
Vitamin D
  adverse effects 4:735
  bone disease treatment 6:513
  conformation 4:734–735

deficiency, bone disease 6:500
derivatives
  see also specific drugs
  introductions 1:221
  psoriasis treatment 7:944–945
  osteomalacia treatment 6:514
  osteoporosis treatment 6:513
  primary hyperparathyroidism 6:504
Vitamin D₃ analogs
  psoriasis treatment see Psoriasis
    treatment
  skin inflammation treatment 7:976
  solid-phase pool libraries 3:705F
Vitamin D receptor (VDR) 2:1007
  analogs, benign prostatic hyperplasia
    (BPH) treatment 6:568
  benign prostatic hyperplasia
    treatment 6:564
  biological functions 4:734
  bone mineralization 2:1007
  calcium homeostasis 2:1007
  homology modeling 4:740
  inhibitors, focused library
    design 4:328
  ligands 2:1018, 2:1018F
  see also specific drugs
  nomenclature 2:996T
  orthogonal ligand–receptor
    pairs 3:228–229, 3:228F
  polymorphisms
    bone disease 6:506
    pharmacodynamics 1:287T
  retinoid X receptor (RXR)
    heterodimerization 2:1007
  structure-based drug design 4:734
    ligand design/docking 4:742
Vitamin E bifunctional drugs
  coronary heart disease
    treatment 6:478
  sodium channel blockers 6:751
Vitex agnus-castus (chaste berry) 1:424
Vitravene see Formivirsen sodium
Vitravene, cytomegalovirus retinitis
  treatment 1:601
VLOGP 5:661T
Vogel, Friedrich 1:281
Vogel conflict test (VCT) 6:93
Voglibose 1:230
VolSurf 5:537, 5:538, 5:581T, 5:679T
  absorption model
    comparisons 5:538–540, 5:541F
    Caco-2 models vs. 5:540
    first principal component 5:540
    second principal
      component 5:540–541
    Ussing chamber methods vs. 5:540
  computational oral absorption
    prediction 5:687
  diffusivity (DIFF) 5:540
  elongation (Elong) 5:540
  environmental interaction 5:540
  hydrogen bond acceptor (O)
    probes 5:538, 5:538F, 5:539

hydrogen bonding descriptors (HB1-8) **5**:539
polar regions (WO1-8) **5**:539
hydrophobic (DRY) probes **5**:538, **5**:538*F*, **5**:539
common with water probes **5**:539
hydrophobic integy moments (ID1-ID8) **5**:539
hydrophobic regions (D1-D8) **5**:539
library design **4**:363
passive transcellular transport **5**:540
permeation modeling **5**:595
plasma protein binding, prediction **5**:856*T*
polarizability (POL) **5**:540
solubility **5**:540
3D QSAR descriptors **4**:155
transporter interaction modeling **5**:775–776
water probes **5**:538, **5**:538*F*
amphiphilic moment (*A*) **5**:539
best volumes **5**:539
capacity factors **5**:539
critical packing diameter (CP) **5**:539
hydrophilic–lipophilic balances (HL1/2) **5**:539
hydrophilic regions (W1-W8) **5**:538, **5**:539*F*
integy moments **5**:539
local interaction energy minima (Emin1-3) **5**:539
local interaction energy minima distances (d12/13/23) **5**:539
molecular globularity (*G*) **5**:538
molecular surface (*S*) **5**:538
molecular volume (*V*) **5**:538
rugosity (*R*) **5**:538
Voltage clamp, cell-based electrophysiology assays **3**:624–625
Voltage-dependent potassium channels *see* K⁺ ion channels, voltage-dependent
Voltage-gated calcium channels
anesthetics **6**:357, **6**:358
in pain **6**:323–324, **6**:342
Voltage-gated currents, cardiac ion channels **6**:735
Voltage-gated ion channels *see* Ion channels, voltage-gated
Voltage-gated potassium channels *see* K⁺ ion channels, voltage-dependent
Voltage sensor
Kv1.2 **4**:708
KvAP **4**:708
flexibility **4**:717, **4**:718*F*
Volume of distribution (*V*ₛₛ) **5**:335
allometry **5**:335
interspecies variation **5**:335
definition **5**:335, **5**:848, **5**:869*T*

half-life *vs.* **5**:848
in vitro screens **5**:959
Ole–Tozer equation **5**:335, **5**:863
species dependence **5**:335
prediction models **5**:862
Lobell and Sivarajah model **5**:863
Lombardo model **5**:863–864
Oie–Tozer equation **5**:863
physiologically based pharmacokinetics (PBPK) **5**:877
Poulin and Theil model **5**:862–864
tissue binding **5**:330
tissue plasma partition coefficients **5**:336
Volume of distribution at a steady state (*V*d-ss), distribution studies **5**:42
Volume of output, substructure similarity searching **4**:168
Vomiting *see* Emesis/nausea
Von Frey monofilaments, acute pain models **6**:332
Von Hippel–Lindau tumor suppresser **7**:20
von Laue, Max **3**:449–450
Voriconazole **1**:**133**; **7**:430*T*, **7**:432–433
aspergillosis **1**:133–**1**:134; **7**:432–433
candidosis **1**:133–**1**:134; **7**:426, **7**:432–433
clinical profiles **7**:453*T*, **7**:455
cryptococcosis **1**:133–134
dimorphic systemic mycoses **7**:423
pharmacokinetics **7**:454
Voronoi techniques, 3D QSAR descriptors **4**:155
VP-013 **6**:75–76
VP-14637
paramyxoviruses **7**:277
respiratory syncytial virus (RSV), fusion inhibitors **7**:409*F*, **7**:410
vRNP complex, influenza viruses **7**:374
v-Src, orthogonal ligand–receptor pairs **3**:218, **3**:220*F*
*V*ₛₛ *see* volume of distribution (*V*ₛₛ)
VX-497 **7**:207
transplant rejection **7**:920
VX-680, structure **7**:207*F*
VX-745
arthritis treatment **7**:861
IBD treatment **6**:627–628
VX-765, IBD treatment **6**:628
VX-950, hepatitis C **7**:268

W

264W94 **6**:487
Waddington, C H **1**:255–256
Wadsworth–Horner–Emmons alkenylation, PASP **3**:796–797

WAG/Rij rat, epilepsy models **6**:287
Wake After Sleep Onset (WASO), somnolence measures/sleep testing **6**:142
Walker, John E **3**:451
Wall effect elimination, microwave-assisted chemistry **3**:842
Wang–Flanagan model, spherical powder dissolution **5**:416*F*, **5**:420
Warburg effect, apoptosis **7**:13
Ward's clustering
cluster-based compound selection **3**:254–255
diversity analysis **4**:185
Warfarin
bioavailability **5**:712*T*
metabolism
CYP 2C9 **5**:545*T*
CYP 2C19 **5**:546*T*
oral absorption **5**:468*T*, **5**:680*T*
receptor binding, pH dependency **5**:382
reduction **5**:127
structure **5**:370*F*; **6**:776*F*
thrombolytic drugs **6**:775
S-Warfarin, structure **4**:787*F*
Warner Lambert, mergers and acquisitions **1**:541
Water
microwave heating, loss tangent **3**:840*T*
pH *see* pH
production, mitochondria complex IV **2**:1077
Water Aid **1**:642
Water–chloroform system, partition coefficient systems **4**:49
Water–cyclohexane system, partition coefficient systems **4**:49
Water-filled cavity, KcsA **4**:707
WaterLOGSY, NMR, drug discovery **3**:908*T*, **3**:913, **3**:914*F*
Water molecules
displacement, ligand–receptor interactions **4**:202
ligand–receptor interaction models *see* Ligand–receptor interactions, models
protein 3D structure validation **3**:512
x-ray structures, problems **3**:539
Water probes *see* VolSurf
Water removal
chemical stability **5**:490
protein crystallization **3**:438
Water resorption, renal dysfunction **6**:584
Water solubility
buparvaquone **5**:1013*F*, **5**:1014*T*
dapsone **5**:1012, **5**:1013*T*
dermal/transdermal administration **5**:292

Water solubility (*continued*)
  hydrophilic depoting **2**:587
  paclitaxel **5**:1012
Water (solvent) suppression,
    NMR **2**:514
Watson, Herman, protein
    crystallization **3**:884
Watson, James D, protein
    crystallography **3**:450
Wave bioreactors, protein
    fermentation **3**:419, **3**:419*F*
WAY-466 **6**:76
WAY-163426 **6**:75–76
WAY-163909 **6**:76
  structure **6**:414*F*
WAY-185229, structure **7**:512*F*
WB-PK **5**:679*T*
Weak acids
  definition **5**:358, **5**:361
  disk intrinsic dissolution rate (DIDR)
    see Disk intrinsic dissolution rate
      (DIDR)
  ionization constants (pK_a) **5**:361
Weak bases
  definition **5**:358, **5**:361
  disk intrinsic dissolution rate (DIDR)
    see Disk intrinsic dissolution rate
      (DIDR)
  ionization constants (pK_a) **5**:361
Weak binding detection, NMR, drug
    discovery **3**:901–902
Weak hits, fragment-based drug
    discovery **3**:941
Weak hydrogen bonds, CSD **3**:404
WEALD, product-based library
    design **4**:352–353
Wear particle foreign-body reactions
    (protheses), bone diseases/
    disorders **6**:506
Web-based programs, raw MS/MS data
    searching **2**:346
WebCite, CSD **3**:398
Web-enabled databases **3**:294
Weber effect, postmarketing
    surveillance **1**:763
Weighted holistic invariant molecular
    descriptors (WHIM)
  molecular descriptors **5**:548
  3D-QSAR **5**:591
Weighting, library design **4**:356
Weight-loss drugs, oxidative
    phosphorylation
    uncoupling **2**:1079
Weight management, diabetes mellitus
    management **6**:438
Welchol *see* Colesevelam hydrochloride
Well types, HTS data
    acquisition **2**:451
Western blotting
  genetically engineered animal
    characterization **3**:153–154
  protein analysis **3**:425
  RNAi detection **3**:178

West Nile virus, antiviral drugs **7**:267
West of Scotland Coronary Prevention
    Study (WOSCOPS) **6**:474*T*,
      6:475
Wet age-related macular degeneration
  clinical trials **6**:305, **6**:306*T*
    drug-delivery systems **6**:306–307
    gene therapy **6**:306–307
    nepafenac **6**:306–307
    treatment goals **6**:306
  disease state **6**:299
  models **6**:304
    laser-induced CNV model **6**:304
    oxygen-induced retinopathy
      (OIR) **6**:304
  treatment **6**:307*F*, **6**:309
    pegaptanib sodium **6**:310
    photodynamic treatment
      6:309–310
    vascular endothelial growth factor
      blockade **6**:310
  unmet medical needs **6**:311
    drug delivery **6**:312
WHAT_CHECK
  protein 3D structure **3**:546
    validation **3**:510, **3**:516–517, **3**:518
  x-ray structures, problems **3**:537
Wheat germ, cell-free expression
    systems **3**:122
WHIM (weighted holistic invariant
    molecule) **5**:581*T*
  3D-QSAR alignment **4**:584
White, Tom, management
    strategies **2**:261
White piedra **7**:421*T*
Whole blood fraction,
    pharmacokinetics **5**:14
Whole-body models, physiologically
    based pharmacokinetics **5**:870,
      5:873, **5**:875
Whole cell currents, arrhythmias **6**:739
Whole cells, dynamic combinatorial
    libraries **3**:973
Whole-genome sequence data, protein
    function prediction,
    bioinformatics **3**:334
Whole-molecule properties, similarity
    searching *see* Similarity searching,
    representations
Wiener index, similarity searching,
    representations **4**:170
Wikström, Håkan, collaborators,
    Lindberg, P **8**:220
Wild-grown, traditional
    medicine **1**:418
Wilkins, Maurice, protein
    crystallography **3**:450
Wilms' tumors
  epigenetics **1**:270*T*
  faulty genomic imprinting **1**:270–271
WIN-55212, as antiemetic agent **6**:684
WIN-57273 **7**:583
  structure **7**:583*T*

WIN-57294
  development **7**:578–579
  structure **7**:583*T*
Wincef *see* Cefoselis
WinNonlin, metabolic drug–drug
    interaction software **5**:841
Winter, Jill, peptoid-based antibiotics,
    discovery of **2**:259
Wisconsin Package (GCG), molecular
    sequence analysis,
    bioinformatics **3**:318
Withdrawal syndrome,
    definition **6**:170*T*
Wittig reaction
  PASP **3**:793, **3**:793*F*
  solid support synthesis **2**:468–469,
      2:469*F*
WIZARD program, 3D structure
    generation **4**:460–461
Wnt antagonists, solid-phase
    libraries **3**:751–752
Wnt signaling
  bone resorption/remodelling **6**:502,
      6:516
  cDNA functional screening
      3:15–16
  Drosophila melanogaster model
    system **3**:9
Wolbachia infections
  oxytetracycline **7**:612
  tetracyclines **7**:610, **7**:612
'Wolf-and-lamb' approach, PASP
    3:796
WOMBAT-PK database **3**:309
Women's Health Initiative
  hormone replacement therapy
      (HRT) **8**:104
  Memory Study **2**:53
Worcester Foundation for Experimental
    Biology **8**:88
Working climate
  effective leaders **2**:242
  ineffective leaders **2**:242
  Lindberg, P (personal
    opinion) **8**:221
Workstations, HTS automation
      3:681–682
World Bank, developing countries,
    economic reform **1**:640
World Health Organization (WHO)
  Commission on Social Determinants
    of Health **1**:650
  essential drugs approach **1**:643
  natural products, use of **1**:357
  Three by Five program **1**:645–646
World Trade Organization (WTO)
  intellectual property
    protection **2**:131
  TRIPS agreement *see* Trade-Related
    Aspects of Intellectual Property
      (TRIPS)
Wormwood, history of medicinal
    use **1**:530

Wortmannin
  modification **1**:366
  structure **1**:364*F*
Wound healing, drug
    introductions **1**:236
WPI (World Patent Index) **3**:276*T*
WR243251, malaria treatment **7**:800*T*
Written project reports, project
    management teams **2**:148
WSKOWWIN, aqueous solubility
    models **5**:644*T*
Wuchereria bancrofti see Filariasis
Wyeth
  R&D spending **2**:101*T*
  sales/R&D expenditure **1**:542*T*
  SME acquisition **1**:496*T*

X

Xalatan *see* Latanoprost
Xaliproden
  amyotrophic lateral sclerosis
      treatment **6**:224
Xamoterol
  absorption studies **5**:42
  bioavailability **5**:712*T*
  clinical trials, early
      termination **8**:141
Xanomeline, schizophrenia
    treatment **6**:36
Xanthine oxidase, famciclovir
    oxidation **7**:301, **7**:301*F*
Xanthine oxidoreductase
    (XOR) **5**:117, **5**:176
  biological role **5**:117–118, **5**:117*F*
  mechanism **5**:118, **5**:118*F*
  reductions **5**:125
  specificity **5**:176
  structure **5**:117, **5**:177
  tissue distribution **5**:177
Xanthines, solid-phase discrete
    libraries **3**:709*T*
X chromosome inactivation **1**:258
  diseases/disorders **1**:258
  epigenetics **1**:256
Xefo *see* Lornoxicam
Xeloda *see* Capecitabine
Xenical *see* Orlistat
Xenografts, antimetabolites **7**:60
Xenon, as anesthetic **6**:360
Xenopus **3**:649
  GPCR studies **3**:649
  history **3**:647
  ion channel studies **3**:649
  oocytes, cell-based assays **3**:619
Xention **1**:515
Xerosis, atopic dermatitis **7**:948
Xigris *see* Drotrecogin alfa
Ximelagatran, reduction **5**:124, **5**:125*F*
Xing L., method used by ionization
    prediction, in silico tools **5**:614,
    5:614*F*

XL-119 **7**:120
XL665, homogeneous time-resolved
    fluorescence **2**:413, **2**:414*F*
X-linked severe combined
    immunodeficiency, gene
    therapy **1**:300
XLOGP program, log P
    calculation **5**:660, **5**:661*T*
XML, database integration **3**:303
XMP.629, acne treatment **7**:966
Xolair *see* Omalizumab
XOMA **1**:515
Xopenex *see* Levalbuterol hydrochloride
XR-11576 **7**:122
X-ray, osteoarthritis (OA) **7**:847–848
X-ray absorption spectroscopy (XAS),
    metalloproteomics **3**:556
X-ray crystallography **3**:**449–472**
  see also Protein crystallization
  atomic model derivation **4**:763
    electron density map **4**:763
    'phase problem' **4**:763
    R-factor **4**:763
    x-ray scattering **4**:763
  atomic order **3**:441
    anisotropic diffraction
        patterns **3**:441
    birefringent crystals **3**:441
    crystal agglomerates **3**:441
    crystalline defects **3**:441
    focusing effects **3**:440
    protein quality effects **3**:440–441
  CADD with **4**:6–8
  cryocrystallography **3**:466*F*, **3**:468
  crystal handling **3**:465
    cryocrystallography **3**:466*F*, **3**:468
    equipment **3**:467
    harvesting buffer **3**:465–467
    room temperature **3**:467
  crystals **3**:451
    asymmetric unit **3**:451
    *International Tables for Crystallography
        Volume A* **3**:452
    symmetry elements **3**:451,
        3:453*F*
    unit cell **3**:451, **3**:452*F*
    unit cell axis restrictions
        3:451–452
  crystal shape **3**:439*F*, **3**:440
  crystal size **3**:440
    accuracy **3**:440
    intensity effects **3**:440
    neutron diffraction studies **3**:440
  cytochrome P450s **5**:811
  data collection **3**:468
    crystal characterization **3**:468–469
    diffraction pattern indexing **3**:469
    integration **3**:468–469
    scaling **3**:468–470
    space group assignment **3**:469
    strategy **3**:468–469
    twinning **3**:470
  diffraction theory **3**:451

dynamic combinatorial libraries *see*
    Dynamic combinatorial libraries
    (DCLs)
  enzymes **4**:752, **4**:761
  fragment-based drug discovery *see*
    Fragment-based drug discovery
  'fragnomics' **3**:892
  future work **3**:470
  history **3**:433–434, **3**:449
    Arndt, Ulie **4**:66–67
    Astbury, W T **3**:450
    Bernal, J D **3**:450
    Bragg, Lawrence **4**:66
    Bragg, Sir William Henry **3**:449–
        450
    Bragg, William Lawrence **3**:449–
        450
    Crick, Francis **3**:450
    Deisenhofer, Johann **3**:451
    Franklin, Rosalind **3**:450
    Hodgkin, Dorothy Crowfoot **3**:450
    Huber, Robert **3**:451
    Kendrew, John C **3**:450; **4**:66
    MacKinnon, Roderick **3**:451
    Michel, Hartmut **3**:451
    Pauling, Linus **3**:450
    PDB **3**:450
    Perutz, Max F **3**:450; **4**:66
    Röntgen, Wilhelm Conrad **3**:449–
        450
    Sanger, Fred **4**:66
    von Laue, Max **3**:449–450
    Walker, John E **3**:451
    Watson James D **3**:450
    Wilkins, Maurice **3**:450
  model generation **3**:462
    automation **3**:462–463
    boundary location **3**:462, **3**:463*F*
    density modification **3**:462,
        3:463*F*
    density modification
        procedures **3**:462
    electron map scaling **3**:464,
        3:464*F*
    free R factor monitoring **3**:464
    histogram matching **3**:462
    map skeletonization **3**:462
    model bias **3**:463–464
    molecular dynamics **3**:463–464
    noise **3**:462
    pseudoenergy **3**:463–464
    refinement **3**:463
    resolution **3**:462–463, **3**:464*F*
    simulated annealing **3**:463
  NMR *vs.* **3**:540
  packing arrangements **3**:443
    contacts *vs.* interfaces **3**:442
    noncrystallographic
        symmetry **3**:443
  packing densities **3**:442
  peptidomimetics **2**:622–623
  protein kinases **2**:964, **2**:965*T*,
      2:966*F*

X-ray crystallography (*continued*)
 pyruvate dehydrogenase
  kinase **4:**749–752, **4:**755
 quality control **3:**470
  NMR structures *vs.* **3:**542
  phase quality **3:**470
  reflection intensity strength **3:**470
 signal transduction **2:**964
 space groups **3:**441*F*, **3:**442
  diffraction symmetry **3:**441–442
 Src homology-2 domains **2:**634,
  2:637*F*, **2:**638*F*
 Src kinases **2:**964, **2:**966*F*
 structural genomics *see* Structural
  genomics
 structure-based drug design
  (SBDD) **4:**66
 structure-based fragment
  screening **4:**789, **4:**789*F*
 structure-based library design **4:**323
 structure determination **3:**948
 terminology **4:**763
  Cartesian coordinates **4:**763
  resolution **4:**763
  temperature (B-) factor **4:**763–
   764
  thermal vibration **4:**763–764
 x-ray–crystal interaction **3:**452,
  3:454*F*
  charge distribution **3:**452–453
  diffraction data resolution **3:**456
  diffraction pattern **3:**452
  electron oscillation **3:**452, **3:**455*F*
  image formation **3:**454
  phase shifts **3:**452
  reciprocal lattice **3:**456–457
  reflection measurement **3:**454
  scattering factor **3:**452–453,
   3:455*F*
 x-ray detection **3:**468
  charge coupled devices
   (CCDs) **3:**468
  film **3:**468
  image plates **3:**466*F*, **3:**468
  integrating detectors **3:**468
  photon counters **3:**468
 x-ray generation **3:**465, **3:**466*F*
  heat dissipation **3:**465, **3:**466*F*
  monochromatic radiation **3:**465
  target **3:**465
X-ray crystallography, phase problem
  solutions **3:**457
 anomalous dispersion **3:**459
  form factor (*f*) **3:**455*F*, **3:**459
  multiple wavelength anomalous
   dispersion (MAD) **3:**459–460
  phase shift **3:**459
  single wavelength anomalous
   dispersion (SAD) **3:**460
  synchrotron radiation **3:**460
 isomorphous replacement **3:**457
  electron density map **3:**457,
   3:458*F*

 electron density map
  autocorrelation **3:**457
  figure of merit (FoM) **3:**459
  Friedel's law **3:**459
  Harker construction **3:**457–459,
   3:458*F*
  multiple isomorphous replacement
   (MIR) **3:**459
  Patterson map **3:**457
  phase ambiguity **3:**459
 molecular replacement **3:**460
  Laue symmetry **3:**461–462
  Matthews coefficient **3:**460
  molecular transform
   generation **3:**460–461
  molecule numbers **3:**460
  Patterson map **3:**460–461, **3:**461*F*
  Patterson Search model **3:**460
  rotation component **3:**462
  structural *vs.* sequence
   similarity **3:**462
  translation component **3:**462
  translation function **3:**461–462,
   3:461*F*
X-ray crystallography, problems **3:**532
 atomic coordinate reliability **3:**538
  isotropic *B*-factors **3:**538
  protein dynamics **3:**538
  RMSDs **3:**538
 current structures **3:**536
 data overinterpretation **3:**535, **3:**536*F*
 deliberate **3:**535
  R-factor testing **3:**535
  validation criteria testing **3:**535
 improvement reliability **3:**537
 incomplete quaternary
  structure **3:**540, **3:**541*F*
  asymmetric unit **3:**540
  dimers **3:**540
  Protein InTerfaces and assemblies
   (PITA) program **3:**540
  Protein Quaternary Structure
   (PQS) server **3:**540
  tetramers **3:**540
  trimers **3:**540
  virus coat proteins **3:**540
 ligand errors **3:**538
  nomenclature errors **3:**539
  structural errors **3:**538
  target value errors **3:**538
 obsolete structures **3:**535
 remaining errors **3:**537
  bond angles **3:**537
  bond lengths **3:**537
  disallowed Ramachandran plot
   regions **3:**537
  electron density **3:**537
  His/Asn/Gln side chain flips **3:**537
  WHAT_CHECK **3:**537
 sequence–register shift errors **3:**534,
  3:537
  oligosaccharide-binding fold **3:**537
  ProsaII **3:**537

 severe errors **3:**532
  see also specific errors
 space group errors **3:**533
  phasing information **3:**533–534
  topology **3:**532, **3:**534*F*
  totally incorrect structures **3:**532,
   3:533*F*
 water molecules **3:**539
X-ray crystallography, rational drug
  design **3:875–900**, **3:**879*F*
 cost **3:**875
 gene construct design **3:**877, **3:**878*F*
  affinity tag **3:**880
  domain mapping **3:**879, **3:**879*F*
  expression system **3:**880
  multiple construct
   screening **3:**877–878
 history **3:**876
  ACE inhibitor design **3:**876,
   3:877*F*
  Cushman, David **3:**876
  Lipscomb, William **3:**876
  Ondetti, Miguel **3:**876
 lead identification/
  characterization **3:**890, **3:**891*F*
  co-crystallization **3:**891
  complexes **3:**891
  fragment evolution **3:**893, **3:**893*F*
  fragment linking **3:**893, **3:**893*F*
  fragment optimization **3:**893,
   3:893*F*
  fragment self-assembly **3:**893,
   3:893*F*
  'fragnomics' **3:**892
  HTS **3:**890
  lead optimization **3:**890
  ligand soaking **3:**891
  'scaffold-based' discovery **3:**892
 phasing methods **3:**889
  anomalous scattering **3:**890
  multiple isomorphous replacement
   (MIR) **3:**889–890
  multiple-isomorphous replacement
   with anomalous scattering
   (MIRAS) **3:**889–890
  multiple-wavelength anomalous
   dispersion (MAD) **3:**889–890
  single-isomorphous replacement
   with anomalous scattering
   (SIRAS) **3:**889–890
 protein characterization **3:**881
  purity measurement **3:**881
X-ray diffraction
 carvedilol structure **8:**139
 chiral drug analysis **2:**720
 solid-state physicochemistry **5:**514
X-ray LIGA *see* Lithography/
  photolithography
X-ray scattering, atomic model
  derivation **4:**763
X-ray structure-based assembly, protein
  tyrosine phosphatase-1B
  inhibitor design **4:**643

X-SITE program, intermolecular
    interactions 3:406
Xue C.X., model used by albumin
    binding prediction
    models 5:858
Xusal *see* Levocetirizine

Y

^{86}Y, PET 3:664*T*
Y-688 7:590
Y-27632 7:203
    structure 7:203*F*
Yama *see* Caspase-3
Yasmin *see* Drospirenone
Yasuda–Shedlovsky plots, ionization
    constants (pK_a)
    measurement 5:388
Y-BOCS (Yale–Brown obsessive
    compulsive scale), obsessive
    compulsive disorder (OCD) 6:96
Yeast, protein expression *see* Protein
    expression
Yeast expression systems 3:415, 3:416
    cell lysis 3:420
Yeast Structural Genomics
    (YSG) 3:552*T*
Yeast-two-hybrid method, protein
    interaction networks,
    bioinformatics 3:335
Yellow fever vaccine,
    development 1:531
Yellow fever virus, antiviral
    drugs 7:267
Yentreve *see* Duloxetine
Yersinia pestis, streptomycin 7:637
Yes kinase, in disease 2:963*T*
YH 1885, structure 6:610*F*
Yields, microwave-assisted
    chemistry 3:857
YM-14673, structure 6:274*F*
YM-53389, as prokinetic agent 6:678
Yohimbine
    mechanism of action 6:145
    P-glycoprotein interaction 5:781*T*
Y-randomization, QSAR 4:157
'Y'-shaped branched PEG, polyethylene
    glycol conjugates 5:1052,
    5:1053*F*

Z

Z-338, as prokinetic agent 6:680
Zacopride, as prokinetic agent 6:677–
    678
Zactima *see* ZD-6474
Zadaxine *see* Thymosin α1
Zafirlukast 1:**209**; 7:907, 7:908
    arthritis treatment 7:854
    atopic dermatitis treatment 7:979,
    7:980*T*

    discovery 7:908
    introduction of 1:108
    structure 7:980*T*
Zalcitabine 7:279–282
    bioavailability 5:712*T*
    development 1:69
    drug regimens 7:342*T*
    hepatitis B 7:319
    medicinal chemistry 7:338–340
    structure 1:70*F*; 7:338*F*
Zaleplon 1:**195**
    introduction of 1:107
    primary insomnia treatment 6:158,
    6:159*T*
    scaffold hopping 2:682*F*
    structure 2:682*F*
Zambia
    healthcare system history 1:639
    health personnel 1:648
    life expectancy 1:638
    user fees impact 1:640–641
Zanamivir 1:**130**; 7:379
    approval 7:378
    clinical studies 7:384–385
    DMPK/efficacy 7:384
    influenza 1:131
    in vitro activity 7:383–384
    Molecular Interaction Fields 4:245
    neuraminidase binding
        structure 7:379, 7:380*F*
    oral absorption 5:1014
    orthomyxoviruses 7:275
    oseltamivir *vs.* 7:381
    resistance 7:276, 7:385
        mechanism 4:623
        transmission efficacy 7:385
    structure 5:1015*F*;
        7:377*F*
    structure-based design 4:617,
        4:618*F*
    target information 4:750*T*
Zantac *see* Ranitidine
Zantipres *see* Zofenopril calcium
Zap70 SH2, three-dimensional
    structures 2:636*T*
Zaprinast
    derivatives, phosphodiesterase-1
        inhibitors 2:932
    phosphodiesterase-11
        inhibitors 2:931
Zarnestra *see* Tipifarnib
Zatebradine 6:747
Zavesca *see* Miglustat
ZD-1839 *see* Gefitinib (ZD-1839)
ZD-2171 7:201
ZD-6474 2:969–970
    characteristics 2:984*T*
    clinical trials 7:201
    KIT inhibitor 2:972
    structure 2:970*F*; 7:201*F*
    VEGFR family kinase
        inhibitors 2:969
ZD-7288 6:747

Zealand Pharma 1:508
Zebrafish (*Danio rerio*) 3:10,
    3:137–138, 3:652
    advantages 3:10
    affective disorder models 6:54*T*
    antisense phosphorodiamidate
        morpholino
        oligonucleotides 3:10
    blindness 3:10
    BRAF studies 3:652–653
    cDNA 3:10
    deafness 3:10
    developmental biology 3:652
    disease biology 3:652
    forward genetics 3:10
        anemia 3:10
        ferroportin 1 identification 3:10
        target identification 3:10
    gastrointestinal motility
        models 6:675
    gene knockdown 3:10
    genome sequencing 3:10,
        3:647–649, 3:648*T*, 3:652–653
    heart disease 3:10
        drug identification 3:10
        titin mutation 3:10
    history 3:647
    HTS 3:10
    leukemia 3:10
    melanoma 3:10
    neurodegeneration 3:10
    physical maps 3:10
    polycystic kidney disease 3:10
    positional cloning 3:10
    small-molecule screening 3:653
    TILLING 3:10
    transcription profiling 3:10
Zefnart *see* Liranaftate
Zeldox *see* Ziprasidone
Zemplar *see* Paricalcitol
Zemuron *see* Rocuronium bromide
Zenapax *see* Diclizumab
Zeolites, PASP 3:793
Zerit *see* Stavudine
Zerit *see* Stavudine
Zero-valued coefficients, similarity
    searching 4:175
Zevalin *see* Ibritumomab tiuxetan
Zhang R.S., model used by tissue
    binding, prediction
    models 5:861–862
Ziagen *see* Abacavir
Ziconotide 6:324, 6:342
    discovery/development 1:379
    structure 1:379*F*
Zidovudine (azidothymidine:
    AZT) 7:279–282
    clinical use 1:606–607
    discovery/development 1:69, 1:70*F*,
        1:372, 1:528–529
    DNA combination structure
        7:338–341, 7:339*F*, 7:341*F*
    drug compatibility 7:342

Zidovudine (azidothymidine: AZT)
    (*continued*)
    drug regimens   7:342*T*
    exclusivity period   1:532*T*
    glucuronic acid conjugation   5:150*F*,
        5:151
    hepatitis B   7:319
    mechanism of action   7:338–341
    mitochondria, effects on   2:1082
    molecular chemistry   7:338–340
    mother-to-child transmission
        treatment   7:340
    oral absorption   5:468*T*
    phosphate conjugation   5:149, 5:154*F*
    pricing policy   1:719
    prodrugs of, carbonate/carbamate
        esters   5:1024
    resistance   7:336, 7:340–341
    structure   1:70*F*
Zifrosilone, structure   2:695*F*
Zileuton   1:209
    arthritis treatment   7:854
    asthma models   7:880
    asthma treatment   7:882, 7:883
    clinical trials   7:882
    COPD treatment   7:882
    introduction of   1:108
Zimbabwe
    health insurance coverage   1:641
    user fees impact   1:640–641
Zinc
    protein crystallization   3:885–887,
        3:886*F*
    tetracyclines   7:618
ZINC database   3:307
    structure-based virtual
        screening   3:252
Zinc finger flexible loops, scaffold-based
    design   4:25
Zinc finger proteins
    computer-assisted drug design   4:26,
        4:27*F*
    expression profiling   2:312
    SMEs   1:516

Zinc-gated ion channel   2:896
    characteristics   2:881*T*
Zinc ion mediated ligand binding,
        quantum mechanics   4:414–415
Zinc metalloproteinases
    as drug targets   3:3
Zinostatin stimalamer   1:149
    hepatoma   1:149
Zinsser Sophas robotic synthesizer,
        PASP   3:805
ZipA-Fts protein–protein interaction,
        pharmacophores   4:533*F*
ZipDock algorithm, QXP (Quick
        Explore) program   4:259
Ziprasidone   1:186; 6:32
    oral absorption   5:680*T*
    side effects   1:696; 6:33
Zitozonium *see* Tamoxifen
ZK1350 corrosivity test, non-animal
        skin corrosivity testing   1:467*T*,
        1:472
ZK-159222
    structure   2:1018*F*
    as vitamin D receptor ligand   2:1019
ZK-216348   2:1016–1017
    structure   2:1017*F*
ZM44739   7:207–208
    structure   7:207*F*
Zocor *see* Simvastatin
Zofenoprilat, bioavailability   5:712*T*
Zofenopril calcium   1:162
Zoladex
    controlled release
        formulations   2:592
    FDA approval   7:175–176
Zoledronate   1:153; 6:512
    structure   6:510*F*
    target information   4:750*T*
Zolmitriptan   1:184
    migraine treatment   6:373*T*
Zolpidem
    primary insomnia treatment   6:158,
        6:159*T*
    scaffold hopping   2:682*F*

shift work sleep disorder
        treatment   6:162
    structure   2:652*F*, 2:682*F*
Zometa *see* Zoledronate
Zomig *see* Zolmitriptan
Zonisamide   2:850; 6:290
    characteristics   2:829*T*
    structure   2:829*T*; 6:282*F*
Zoom gels, protein expression
        mapping   2:385
Zopiclone
    metabolism, CYP 2C8   5:547*T*
    primary insomnia treatment   6:158,
        6:159*T*
    scaffold hopping   2:682*F*
    shift work sleep disorder
        treatment   6:162
    structure   2:652*F*, 2:682*F*
Zorac *see* Tazoretene
Zosuquidar, structure   5:770*F*
ZP123   6:755
Z-scores, PDB errors   3:523, 3:524*F*
ZSM-I-62
    cytomegalovirus   7:259
    herpesvirus infections   7:259
Zucker Diabetic Fatty (ZDF) rat,
        diabetes mellitus type 2 animal
        models   6:435
Zuclomiphene, structure   8:84*F*
z-VAD-fmk, Huntington's disease
        treatment   6:222*T*
Zwitterions   5:366, 5:366*T*,
        5:367*T*
    hERG blockers   5:952, 5:952*F*
    ionization constants (pK_a) *see*
        Ionization constants (pK_a)
    lipophilicity   5:438
Zyflo *see* Zileuton
Zygomycosis, subcutaneous   7:422*T*
    opportunistic infections   7:425*T*
Zygote microinjection, transgenic
        animal generation   3:152
Zyprexa *see* Olanzapine
Zyvox *see* Linezolid